W. Greiner · B. Müller

QUANTUM MECHANICS

Symmetries

Greiner
Quantum Mechanics
An Introduction 3rd Edition

Greiner
Quantum Theory
Special Chapters
(in preparation)

Greiner · Müller
Quantum Mechanics
Symmetries 2nd Edition

Greiner
Relativistic Quantum Mechanics
Wave Equations

Greiner · Reinhardt
Field Quantization
(in preparation)

Greiner · Reinhardt
Quantum Electrodynamics
2nd Edition

Greiner · Schäfer
Quantum Chromodynamics

Greiner · Maruhn
Nuclear Models
(in preparation)

Greiner · Müller
Gauge Theory of Weak Interactions

Greiner
Mechanics I
(in preparation)

Greiner
Mechanics II
(in preparation)

Greiner
Electrodynamics
(in preparation)

Greiner · Neise · Stöcker
Thermodynamics and Statistical Mechanics

Walter Greiner · Berndt Müller

QUANTUM MECHANICS

Symmetries

With a Foreword by
D. A. Bromley

Second Revised Edition
With 81 Figures,
and 127 Worked Examples and Problems

Springer

Professor Dr. Walter Greiner

Institut für Theoretische Physik der
Johann Wolfgang Goethe-Universität Frankfurt
Postfach 11 19 32
D-60054 Frankfurt am Main
Germany

Street address:

Robert-Mayer-Strasse 8–10
D-60325 Frankfurt am Main
Germany

Professor Dr. Berndt Müller

Physics Department
Duke University
P. O. Box 90305
Durham, NC 27708-0305
USA

Title of the original German edition: *Theoretische Physik,* Band 5: *Quantenmechanik II, Symmetrien 3. Aufl.* © Verlag Harri Deutsch, Thun 1984, 1992

ISBN 3-540-58080-8 2. Auflage Springer-Verlag Berlin Heidelberg New York
ISBN 0-387-58080-8 2nd edition Springer-Verlag New York Berlin Heidelberg

ISBN 3-540-19201-8 1. Auflage Springer-Verlag Berlin Heidelberg New York
ISBN 0-387-19201-8 1st edition Springer-Verlag New York Berlin Heidelberg

This volume originally appeared in the series *"Theoretical Physics* – Text and Exercise Books Vol. 2"

CIP data applied for

Printed and bound by Hamilton Printing Co., Rensselaer, NY., U.S.A.

9 8 7 6 5 4 3 2

Typesetting: Macmillan, India
SPIN 10123169 56/3140 - 5 4 3 2 1 - Printed on acid-free paper

Foreword to Earlier Series Editions

More than a generation of German-speaking students around the world have worked their way to an understanding and appreciation of the power and beauty of modern theoretical physics – with mathematics, the most fundamental of sciences – using Walter Greiner's textbooks as their guide.

The idea of developing a coherent, complete presentation of an entire field of science in a series of closely related textbooks is not a new one. Many older physicists remember with real pleasure their sense of adventure and discovery as they worked their ways through the classic series by Sommerfeld, by Planck and by Landau and Lifshitz. From the students' viewpoint, there are a great many obvious advantages to be gained through use of consistent notation, logical ordering of topics and coherence of presentation: beyond this, the complete coverage of the science provides a unique opportunity for the author to convey his personal enthusiasm and love for his subject.

The present five volume set, *Theoretical Physics*, is in fact only that part of the complete set of textbooks developed by Greiner and his students that presents the quantum theory. I have long urged him to make the remaining volumes on classical mechanics and dynamics, on electromagnetism, on nuclear and particle physics, and on special topics available to an English-speaking audience as well, and we can hope for these companion volumes covering all of theoretical physics some time in the future.

What makes Greiner's volumes of particular value to the student and professor alike is their completeness. Greiner avoids the all too common "it follows that ... " which conceals several pages of mathematical manipulation and confounds the student. He does not hesitate to include experimental data to illuminate or illustrate a theoretical point and these data, like the theoretical content, have been kept up to date and topical through frequent revision and expansion of the lecture notes upon which these volumes are based.

Moreover, Greiner greatly increases the value of his presentation by including something like one hundred completely worked examples in each volume. Nothing is of greater importance to the student than seeing, in detail, how the theoretical concepts and tools under study are applied to actual problems of interest to a working physicist. And, finally, Greiner adds brief biographical sketches to each chapter covering the people responsible for the development of the theoretical ideas and/or the experimental data presented. It was Auguste Comte (1798–1857) in his *Positive Philosophy* who noted, "To understand a science it is necessary to know its history". This is all too often forgotten in modern physics teaching and the bridges that Greiner builds to the pioneering figures of our science upon whose work we build are welcome ones.

Greiner's lectures, which underlie these volumes, are internationally noted for their clarity, their completeness and for the effort that he has devoted to making physics an integral whole; his enthusiasm for his science is contagious and shines through almost every page.

These volumes represent only a part of a unique and Herculean effort to make all of theoretical physics accessible to the interested student. Beyond that, they are of enormous value to the professional physicist and to all others working with quantum phenomena. Again and again the reader will find that, after dipping into a particular volume to review a specific topic, he will end up browsing, caught up by often fascinating new insights and developments with which he had not previously been familiar.

Having used a number of Greiner's volumes in their original German in my teaching and research at Yale, I welcome these new and revised English translations and would recommend them enthusiastically to anyone searching for a coherent overview of physics.

Yale University
New Haven, CT, USA
1989

D. Allan Bromley
Henry Ford II Professor of Physics

Preface to the Second Edition

We are pleased to note that our text *Quantum Mechanics – Symmetries* has found many friends among physics students and researchers so that the need for a second edition has arisen. We have taken this opportunity to make several amendments and improvements to the text. We have corrected a number of misprints and minor errors and have added explanatory remarks at various places. In addition to many other smaller changes the sections 8.6, 8.11, and 11.4 and the exercises 3.9, 7.8, and 9.5 have been expanded. Two new exercises on the Wigner–Eckart theorem (Ex. 5.8) and on the completeness relation for the $SU(N)$ generators (Ex. 11.3) have been added. Finally, the Mathematical Supplement on Lie groups (Chap. 12) has been carefully checked and received a new introductory section.

We thank several colleagues for helpful comments, especially Prof. L. Wilets (Seattle) for providing a list of errors and misprints. We are greatly indebted to Prof. P. O. Hess (University of Mexico) for making available corrections and valuable material for Chap. 12. We also thank Dr. R. Mattiello who has supervised the preparation of the second edition of the book. Finally we acknowledge the agreeable collaboration with Dr. H. J. Kölsch and his team at Springer-Verlag, Heidelberg.

Frankfurt am Main and Durham, NC, USA
July 1994

Walter Greiner
Berndt Müller

Preface to the First Edition

Theoretical physics has become a many-faceted science. For the young student it is difficult enough to cope with the overwhelming amount of new scientific material that has to be learned, let alone obtain an overview of the entire field, which ranges from mechanics through electrodynamics, quantum mechanics, field theory, nuclear and heavy-ion science, statistical mechanics, thermodynamics, and solid-state theory to elementary-particle physics. And this knowledge should be acquired in just 8–10 semesters, during which, in addition, a Diploma or Master's thesis has to be worked on or examinations prepared for. All this can be achieved only if the university teachers help to introduce the student to the new disciplines as early on as possible, in order to create interest and excitement that in turn set free essential, new energy. Naturally, all inessential material must simply be eliminated.

At the Johann Wolfgang Goethe University in Frankfurt we therefore confront the student with theoretical physics immediately, in the first semester. Theoretical Mechanics I and II, Electrodynamics, and Quantum Mechanics I – An Introduction are the basic courses during the first two years. These lectures are supplemented with many mathematical explanations and much support material. After the fourth semester of studies, graduate work begins, and Quantum Mechanics II – Symmetries, Statistical Mechanics and Thermodynamics, Relativistic Quantum Mechanics, Quantum Electrodynamics, the Gauge Theory of Weak Interactions, and Quantum Chromodynamics are obligatory. Apart from these, a number of supplementary courses on special topics are offered, such as Hydrodynamics, Classical Field Theory, Special and General Relativity, Many-Body Theories, Nuclear Models, Models and Elementary Particles, and Solid-State Theory. Some of them, for example the two-semester courses Theoretical Nuclear Physics and Theoretical Solid-State Physics, are also obligatory.

The form of the lectures that comprise *Quantum Mechanics – Symmetries* follows that of all the others: together with a broad presentation of the necessary mathematical tools, many examples and exercises are worked through. We try to offer science in a way as interesting as possible. With symmetries in quantum mechanics we are dealing with a particularly beautiful theme. The selected material is perhaps unconventional, but corresponds, in our opinion, to the importance of this field in modern physics.

After a short reminder of some symmetries in classical mechanics, the great importance of symmetries in quantum mechanics is outlined. In particular, the consequences of rotational symmetry are described in detail, and we are soon led to the general theory of Lie groups. The isospin group, hypercharge, and SU(3) symmetry and its application in modern elementary-particle physics are broadly outlined. Essential mathematical theorems are first quoted without proof and heuristically illustrated to show their importance and meaning. The proof can then be found in detailed examples and worked-out exercises.

A mathematical supplement on root vectors and classical Lie algebras deepens the material, the Young-tableaux technique is broadly outlined, and, by way of a chapter on group characters and another on charm, we lead up to very modern questions of physics. Chapters on special discrete symmetries and dynamical symmetries round off these lectures. These are all themes which fascinate young physicists, because they show them that as early as the fifth semester they can properly address and discuss questions of frontier research.

Many students and collaborators have helped during the years to work out examples and exercises. For this first English edition we enjoyed the help of Maria Berenguer, Snježana Butorac, Christian Derreth, Dr. Klaus Geiger, Dr. Matthias Grabiak, Carsten Greiner, Christoph Hartnack, Dr. Richard Herrmann, Raffaele Mattiello, Dieter Neubauer, Jochen Rau, Wolfgang Renner, Dirk Rischke, Thomas Schönfeld, and Dr. Stefan Schramm. Miss. Astrid Steidl drew the graphs and prepared the figures. To all of them we express our sincere thanks. We are also grateful to Dr. K. Langanke and Mr. R. Könning of the Physics Department of the University in Münster for their valuable comments on the German edition.

We would especially like to thank Mr. Béla Waldhauser, Dipl.-Phys., for his overall assistance. His organizational talent and his advice in technical matters are very much appreciated.

Finally, we wish to thank Springer-Verlag; in particular, Dr. H.-U. Daniel, for his encouragement and patience, and Mr. Michael Edmeades, for expertly copy-editing the English edition.

Frankfurt am Main
July 1989

Walter Greiner
Berndt Müller

Contents

Contents of Examples and Exercises

1. Symmetries in Quantum Mechanics

1.1 Symmetries in Classical Physics

Symmetries play a fundamental role in physics, and knowledge of their presence in certain problems often simplifies the solution considerably. We illustrate this with the help of three important examples.

a) Homogeneity of Space. We assume space to be *homogeneous*, i.e. of equal structure at all positions $\boldsymbol{r}$. This is synonymous with the assumption that the solution of a given physical problem is invariant under translations, because in this case the area surrounding any point can be mapped exactly by a translation from a similar area surrounding an arbitrary point (Fig. 1.1). This "translation-invariance" implies the conservation of momentum for an isolated system. Here we define homogeneity of space to mean that the Lagrange function $L(\boldsymbol{r}_i, \dot{\boldsymbol{r}}_i, t)$ of a system of particles remains invariant if the particle coordinates $\boldsymbol{r}_i$ are replaced by $\boldsymbol{r}_i + \boldsymbol{a}$, where $\boldsymbol{a}$ is an arbitrary constant vector. (A more general concept of "homogeneity of space" would require only the invariance of the equations of motion under spatial translation. In this case a conserved quantity can also be shown to exist, but it is not necessarily the canonical momentum. See Exercises 1.3 and 1.5 for a detailed discussion of this aspect.) Thus

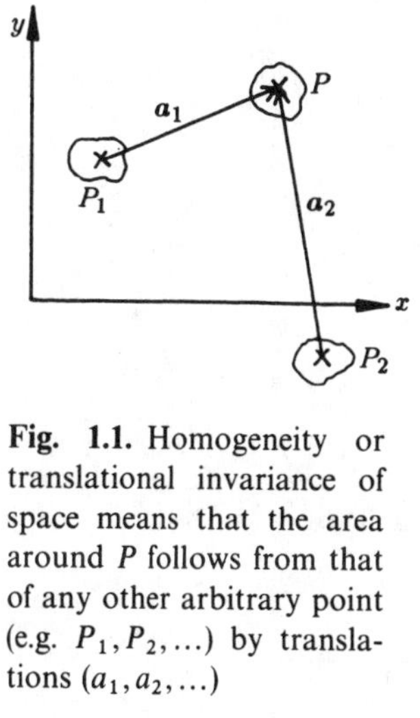

Fig. 1.1. Homogeneity or translational invariance of space means that the area around P follows from that of any other arbitrary point (e.g. $P_1, P_2, \ldots$) by translations $(a_1, a_2, \ldots)$

$$\delta L = \sum_i \frac{\partial L}{\partial \boldsymbol{r}_i} \cdot \delta \boldsymbol{r}_i = \boldsymbol{a} \cdot \sum_i \frac{\partial L}{\partial \boldsymbol{r}_i} = 0 \tag{1.1}$$

must be valid. Since $\boldsymbol{a}$ is arbitrary this implies

$$\sum_i \frac{\partial L}{\partial \boldsymbol{r}_i} = 0 = \left\{ \sum_i \frac{\partial L}{\partial x_i}, \sum_i \frac{\partial L}{\partial y_i}, \sum_i \frac{\partial L}{\partial z_i} \right\} \quad . \tag{1.2}$$

Here we have abbreviated

$$\frac{\partial L}{\partial \boldsymbol{r}_i} = \left\{ \frac{\partial L}{\partial x_i}, \frac{\partial L}{\partial y_i}, \frac{\partial L}{\partial z_i} \right\} \quad ,$$

the gradient of L with respect to $\boldsymbol{r}_i$. From the Euler-Lagrange equations

$$\frac{d}{dt} \frac{\partial L}{\partial \dot{x}_i} - \frac{\partial L}{\partial x_i} = 0 \quad , \quad \text{etc.}$$

it follows immediately with (1.2) that

$$\frac{d}{dt}\sum_i \frac{\partial L}{\partial \dot{x}_i} = \frac{d}{dt} P_x = 0 \quad , \quad \text{thus} \quad P_x = \text{const.} \quad .$$

Here we have used the relation $\partial L/\partial \dot{x}_i = P_{x_i}$ for the canonical momentum and $\sum_i P_{x_i} = P_x$. P_x is the x component of the total momentum

$$\boldsymbol{P} = \left\{ \sum_i P_{x_i}, \sum_i P_{y_i}, \sum_i P_{z_i} \right\} = \sum_i \boldsymbol{P}_i \quad . \tag{1.3}$$

This is the *law of momentum conservation in classical mechanics.* In nonrelativistic physics, it allows for the definition of a centre of mass. This is due to the fact that this law holds in all inertial systems, because in all of these space is homogeneous. Let $\boldsymbol{P} = \sum_i m_i \boldsymbol{v}_i$ denote the total momentum in the system K. Then in the system, K', which moves with the velocity $\boldsymbol{v}$ with respect to K, it is given by

$$\boldsymbol{P}' = \sum_i m_i \boldsymbol{v}_i' = \sum_i m_i (\boldsymbol{v}_i - \boldsymbol{v}) = \boldsymbol{P} - \boldsymbol{v} \sum_i m_i \quad ,$$

because nonrelativistically $\boldsymbol{v}_i' = \boldsymbol{v}_i - \boldsymbol{v}$. The centre-of-mass system is defined by the condition that the total momentum $\boldsymbol{P}'$ vanishes. In K it moves with the velocity

$$\begin{aligned} \boldsymbol{v}_S &= \frac{\boldsymbol{P}}{\sum_i m_i} = \left(\sum_i m_i \boldsymbol{v}_i\right) \Big/ \left(\sum_i m_i\right) \\ &= \left(\sum_i m_i \frac{d\boldsymbol{r}_i}{dt}\right) \Big/ \left(\sum_i m_i\right) = \frac{d}{dt}\left[\left(\sum_i m_i \boldsymbol{r}_i'\right) \Big/ \left(\sum_i m_i\right)\right] \equiv \frac{d\boldsymbol{R}}{dt} \quad , \end{aligned} \tag{1.4}$$

where

$$\boldsymbol{R} = \left(\sum_i m_i \boldsymbol{r}_i'\right) \Big/ \left(\sum_i m_i\right) \tag{1.5}$$

is the coordinate of the nonrelativistic centre of mass.

b) Homogeneity of Time. The homogeneity of time has no less importance than the homogeneity of space. It stands for the invariance of the laws of nature in isolated systems with respect to translations in time, i.e. at time $t + t_0$ they have the same form as at time t. This is expressed mathematically by the fact that the Lagrange function does not depend on time explicitly, i.e.

$$L = L(q_i, \dot{q}_i) \quad . \tag{1.6}$$

Then it follows that

$$\frac{dL}{dt} = \sum_i \frac{\partial L}{\partial q_i} \dot{q}_i + \sum_i \frac{\partial L}{\partial \dot{q}_i} \ddot{q}_i \quad . \tag{1.7}$$

[Note that if L depends explicitly on time, the term $\partial L/\partial t$ has to be added on the right-hand side (rhs) of (1.7).]

Making use of the Euler–Lagrange equations

$$\frac{d}{dt}\frac{\partial L}{\partial \dot{q}_i} - \frac{\partial L}{\partial q_i} = 0$$

one finds

$$\frac{dL}{dt} = \sum_i \dot{q}_i \frac{d}{dt}\frac{\partial L}{\partial \dot{q}_i} + \sum_i \frac{\partial L}{\partial \dot{q}_i}\ddot{q}_i = \sum_i \frac{d}{dt}\left(\dot{q}_i \frac{\partial L}{\partial \dot{q}_i}\right) \quad ,$$

or

$$\frac{d}{dt}\left(\sum_i \dot{q}_i \frac{\partial L}{\partial \dot{q}_i} - L\right) = 0 \quad . \tag{1.8}$$

This expresses the conservation of the quantity

$$E \equiv \sum_i \dot{q}_i \frac{\partial L}{\partial \dot{q}_i} - L = \sum_i \dot{q}_i \pi_i - L = H \quad , \tag{1.9}$$

which represents the total energy (Hamilton function H). The quantities $\pi_i = \partial L/\partial \dot{q}_i$ are the *canonical momenta*. Since the energy (1.9) is linear in L it is additive, so that for two systems which are described by L_1 and L_2, respectively, the energy is $E = E_1 + E_2$. This is valid as long as there exists no interaction L_{12} between the two systems, i.e. if L_1 and L_2 depend on different dynamical variables q_{i1} and q_{i2}. The law of energy conservation is valid not only for isolated systems, but also *in any time-independent external field*, because then L is still independent [the only requirement of L was time-independence, which led to the conservation of energy 1.9)]. Systems in which the total energy is conserved are called *conservative systems.*

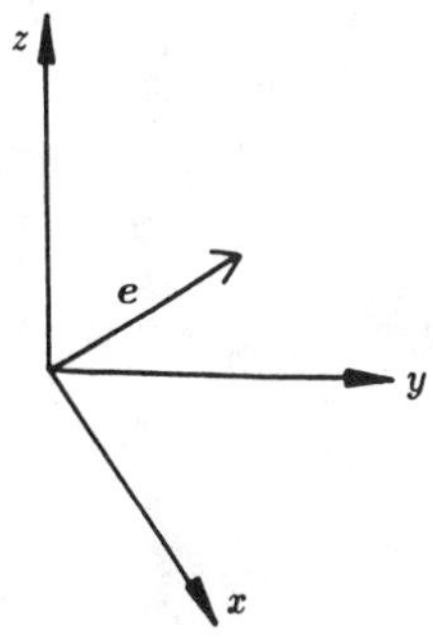

Fig. 1.2. Isotropic space is equally structured in every direction of $\boldsymbol{e}$

c) Isotropy of Space. Isotropy of space means that space has the same structure in all directions (Fig. 1.2). In other words: The mechanical properties of an isolated system remain unchanged if the whole system is arbitrarily rotated in space, i.e. *the Lagrangian is invariant under rotations.* Let us now consider infinitesimal rotations (Fig. 1.3)

$$\delta\boldsymbol{\phi} = \{\delta\phi_x, \delta\phi_y, \delta\phi_z\} \quad . \tag{1.10}$$

The modulus $\delta\phi$ characterizes the size of the rotation angle, and the direction $\delta\boldsymbol{\phi}/\delta\phi$ defines the axis of rotation. The radius vector $\boldsymbol{r}$ changes under the rotation $\delta\boldsymbol{\phi}$ by $\delta\boldsymbol{r}$. We have

$$\delta r = |\delta\boldsymbol{r}| = r \sin\theta\, \delta\phi \quad ,$$

and the direction of $\delta\boldsymbol{r}$ is perpendicular to the plane spanned by $\delta\phi$ and $\boldsymbol{r}$. Hence

$$\delta\boldsymbol{r} = \delta\boldsymbol{\phi} \times \boldsymbol{r} \quad . \tag{1.11}$$

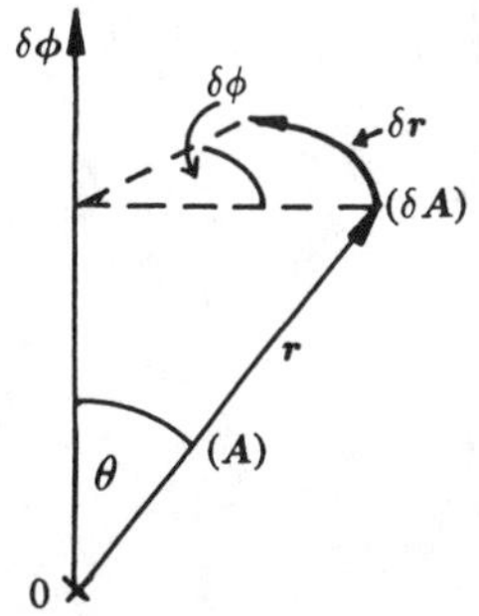

Fig. 1.3. Illustration of an infinitesimal rotation of the position vector $\boldsymbol{r}$ and an arbitrary vector $\boldsymbol{A}$

In addition to the position vectors $\boldsymbol{r}_i$, the particle velocities $\boldsymbol{v}_i$ are also altered by the rotation; they change their direction. In fact all vectors are changed in the same manner by a rotation. Thus the velocity change $\delta\boldsymbol{v}_i$ is given by

$$\delta\boldsymbol{v}_i = \delta\boldsymbol{\phi} \times \boldsymbol{v}_i \quad . \tag{1.12}$$

Since the infinitesimal rotation is assumed not to change the lagrangian, we have

$$\delta L = \sum_i \left(\frac{\partial L}{\partial \boldsymbol{r}_i} \cdot \delta \boldsymbol{r}_i + \frac{\partial L}{\partial \boldsymbol{v}_i} \cdot \delta \boldsymbol{v}_i \right) = 0 \quad . \tag{1.13}$$

The canonical momenta are

$$\boldsymbol{\pi}_i = \frac{\partial L}{\partial \boldsymbol{v}_i} = \left\{ \frac{\partial L}{\partial v_{ix}}, \frac{\partial L}{\partial v_{iy}}, \frac{\partial L}{\partial v_{iz}} \right\}$$

and according to the Lagrange equations we obtain

$$\dot{\boldsymbol{\pi}}_i = \frac{d}{dt} \frac{\partial L}{\partial \boldsymbol{v}_i} = \frac{\partial L}{\partial \boldsymbol{r}_i} \quad .$$

After substituting these quantities and also using (1.11) and (1.12), Eq. (1.13) becomes

$$\sum_i [\dot{\boldsymbol{\pi}}_i \cdot (\delta \boldsymbol{\phi} \times \boldsymbol{r}_i) + \boldsymbol{\pi}_i \cdot (\delta \boldsymbol{\phi} \times \boldsymbol{v}_i)]$$

$$= \delta \boldsymbol{\phi} \cdot \sum_i (\boldsymbol{r}_i \times \dot{\boldsymbol{\pi}}_i + \boldsymbol{v}_i + \boldsymbol{\pi}_i) = \delta \boldsymbol{\phi} \cdot \frac{d}{dt} \left(\sum_i \boldsymbol{r}_i \times \boldsymbol{\pi}_i \right) = 0 \quad . \tag{1.14}$$

Since

$$\boldsymbol{L} = \sum_i \boldsymbol{r}_i \times \boldsymbol{\pi}_i \tag{1.15}$$

is the *classical angular momentum*, and the infinitesimal rotation vector $\delta \phi$ is arbitrary, it follows that

$$\frac{d\boldsymbol{L}}{dt} = 0 \quad , \tag{1.16}$$

thus

$$\boldsymbol{L} = \text{const.}$$

Because the sum in (1.15) extends over all particles, the angular momentum is additive – like the momentum, noted earlier in (1.3) –, i.e. if further particles are added their contribution to the total angular momentum is simply summed according to (1.15). This is valid, no matter whether the additional particles interact with the old particles or not.

We may get further insight into these conservation laws by solving the following two problems.

EXERCISE

1.1 Angular Momenta in Different Reference Frames

Problem. a) What is the connection between the angular momenta in two reference systems which are at rest relative to each other and whose origins are separated by the distance $\boldsymbol{a}$?

b) What is the relation between the angular momenta in two inertial systems K and K' which move with velocity $\boldsymbol{V}$ relative to each other?

Solution. a) Let us consider a system of particles with the position vectors $\boldsymbol{r}_i$ in one coordinate system, and with the position vectors $\boldsymbol{r}_i'$ in another system. Since the origins of the coordinate systems are separated by $\boldsymbol{a}$, we have

$$\boldsymbol{r}_i = \boldsymbol{r}_i' + \boldsymbol{a} \quad . \tag{1}$$

The total angular momentum of the system is given by

$$\boldsymbol{L} = \sum_i \boldsymbol{r}_i \times \boldsymbol{p}_i \quad . \tag{2}$$

Inserting (1) into (2) it follows that

$$\boldsymbol{L} = \sum_i \boldsymbol{r}_i \times \boldsymbol{p}_i = \sum_i \boldsymbol{r}_i' \times \boldsymbol{p}_i + \boldsymbol{a} \times \sum_i \boldsymbol{p}_i \quad . \tag{3}$$

Now $\sum_i \boldsymbol{r}_i' \times \boldsymbol{p}_i = \boldsymbol{L}'$, because the momentum of a particle does not change if we transform between systems which are at rest relative to each other, and $\sum_i \boldsymbol{p}_i = \boldsymbol{P}$ is the total momentum of the system; thus,

$$\boldsymbol{L} = \boldsymbol{L}' + \boldsymbol{a} \times \boldsymbol{P} \quad . \tag{4}$$

The total angular momentum is composed of the internal total angular momentum and of the angular momentum of the entire system with respect to the origin, at a distance $|\boldsymbol{a}|$. $\boldsymbol{L} = \boldsymbol{L}'$ only if $\boldsymbol{P}$ is parallel to $\boldsymbol{a}$ (or $\boldsymbol{P} = 0$), i.e. if the whole system moves in the direction of the translation. However, the angular momentum of the system is a conserved quantity, because the linear momentum $\boldsymbol{P}$ is conserved also!

b) Consider K and K' at the time when the system origins coincide, i.e. $\boldsymbol{r}_i = \boldsymbol{r}_i'$. The velocities are $\boldsymbol{v}_i = \boldsymbol{v}_i' + \boldsymbol{V}$; hence

$$\boldsymbol{L} = \sum_i m_i \boldsymbol{r}_i \times \boldsymbol{v}_i = \sum_i m_i \boldsymbol{r}_i \times \boldsymbol{v}_i' + \sum_i m_i \boldsymbol{r}_i \times \boldsymbol{V} \quad . \tag{5}$$

Since $\boldsymbol{r}_i = \boldsymbol{r}_i'$, we have $\boldsymbol{L}' = \sum_i m_i \boldsymbol{r}_i \times \boldsymbol{v}_i$, and the position vector of the centre of mass reads

$$\boldsymbol{R} = \left(\sum_i m_i \boldsymbol{r}_i\right) \Big/ \left(\sum_i m_i\right) = \frac{1}{M} \sum_i m_i \boldsymbol{r}_i \quad ,$$

Exercise 1.1

where M is the total mass of the system. Therefore, (5) yields

$$L = L' + M(R \times V) \quad . \tag{6}$$

If the particle system is at rest in the system K', then V is the velocity of the centre of mass and $P = MV$ is the total momentum of the system with respect to K, i.e. $L = L' + R \times P = L' + L_S$. This means that angular momentum is composed of the angular momentum L' in the rest frame and of the angular momentum of the centre of mass L_S.

EXERCISE

1.2 Conserved Quantities of Specified Fields

Problem. What components of the momentum P and the angular momentum L are conserved when moving in the following fields?

a) field of an infinite homogeneous plane,
b) field of an infinite homogeneous cylinder,
c) field of an infinite homogeneous prism,
d) field of two points,
e) field of an infinite homogeneous semiplane,
f) field of a homogeneous cone,
g) field of a homogeneous circular ring,
h) field of an infinite homogeneous helix.

Solution. The projection of momentum and of angular momentum onto a symmetry axis of the given field remain conserved, because the mechanical properties (Lagrangian and equation of motion) are not changed by a translation along this axis, or by a rotation around it. For the component of angular momentum, this is only valid if the angular momentum is defined with respect to the centre of the field, and not with respect to an arbitrary spatial point. The momentum or the component of momentum, respectively, remains conserved in the sense of Lagrangian mechanics, if, and only if, the potential of the field does not depend on the corresponding generalized coordinate.

a) *field of an infinite homogeneous plane.* We choose the xy plane. Because of the translational invariance of the plane, the potential does not depend on x and y so that p_x and p_y are conserved. In addition, the Lagrange function does not change when a rotation around the z axis is performed, i.e. L_z is conserved.

b) *field of an infinite homogeneous cylinder.* Due to the infinite extent of the cylinder, the potential does not change under a translation along its axis (z axis); thus, p_z is conserved. In addition, we have rotational symmetry around the z axis, i.e. L_z is conserved.

Exercise 1.2

c) *field of an infinite homogeneous prism* (edges parallel to the z axis). As in b), p_z is conserved, but there is no rotational symmetry around the z axis, i.e. L_z is not conserved.

d) *field of two points* (*points on the z axis*). Here we have only rotational symmetry around the z axis. The sole conserved quantity is L_z.

e) *infinite homogeneous semiplane.* Again we choose the xy plane, which is now bordered by the y axis, giving translational invariance only along the y axis, i.e. p_y is conserved.

f) *homogeneous cone* (z axis along the cone axis). This time rotational symmetry lies around the z axis; L_z is conserved.

g) *homogeneous circular ring* (z axis along the axis of the ring). Again rotational symmetry lies around the z axis; L_z is conserved.

h) *infinite homogeneous helix* (z axis along the axis of the helix). The potential (Lagrange function) does not change under a rotation through $\delta\phi$ about the z axis, as long as one simultaneously shifts along the z axis by δz. If the pitch of the helix is h (for a rotation through 2π on the helix, the change in the z direction is h), then a translation of $\delta z = (h/2\pi)\delta\phi$ and a simultaneous rotation through $\delta\phi$ just conserves the symmetry of the potential; consequently, the variation of the Lagrange function vanishes, giving

$$\delta L = 0 = \frac{\partial L}{\partial z}\delta z + \frac{\partial L}{\partial \phi}\delta\phi \quad . \tag{1}$$

Now we have

$$\frac{d}{dt}p_z = \frac{\partial L}{\partial z} \quad \text{and} \quad \frac{d}{dt}L_z = \frac{\partial L}{\partial \phi} \quad ;$$

hence,

$$\frac{d}{dt}\left(p_z\frac{h}{2\pi} + L_z\right)\delta\phi = 0 \quad .$$

For arbitrary $\delta\phi$ it follows that

$$\frac{d}{dt}\left(p_z\frac{h}{2\pi} + L_z\right) = 0 \quad ,$$

and, therefore

$$\left(p_z\frac{h}{2\pi} + L_z\right) = \text{const.} \quad ,$$

i.e. for the helix a certain linear combination of p_z and L_z remains conserved.

As in classical mechanics, the homogeneity of space and time, and the isotropy of space also play an important role in quantum mechanics. However, in quantum mechanical systems there also exist other symmetries. For that reason we want to develop a uniform approach to symmetry properties. In addition we distinguish *geometric symmetries* – which correspond to the invariance of the system under translation and reflections in space and time, and under rotation, – from *dynamic symmetries*, which are often the reason for unexpected degeneracies of energy states, e.g. of the hydrogen atom or of the isotropic harmonic oscillator. Beyond that we have to mention other symmetries in different branches of physics (e.g. special relativity[1]). Usually we consider only the single-particle problem (or the nonrelativistic two-particle problem in the centre-of-mass system, which is equivalent to the single-particle problem). However, most results can be transformed without difficulty into the problem of several interacting particles, as long as the fundamental symmetries are valid for all particles.

EXAMPLE

1.3 Noether's Theorem (for Improved Insight)

Noether's theorem, which we will now prove, says the following:

If the Euler-Lagrange equations of motion are invariant under a coordinate transformation t, $\boldsymbol{q} \to t'(t)$, $\boldsymbol{q}'(\boldsymbol{q}, t)$, then there exists an integral of motion, i.e. a conserved quantity.

Given a Lagrange function $L(\boldsymbol{q}, \dot{\boldsymbol{q}}, t)$ of the coordinates q_i $(i = 1, \dots, l)$ and time t, we introduce new coordinates t', $\boldsymbol{q}'$ by defining

$$t' := t'(t) \quad , \quad q_i' := q_i'(\boldsymbol{q}, t) \quad . \tag{1}$$

This transformation shall be uniquely invertible. We can then write

$$t' := t + \delta t(t) \quad , \quad q_i' := q_i + \delta q_i(\boldsymbol{q}, t) \quad . \tag{2}$$

Initially the functions δt and δq_i are arbitrary. The velocities $\dot{\boldsymbol{q}}$, $\dot{\boldsymbol{q}}'$ are given by

$$\dot{q}_i := \frac{d}{dt} q_i \quad , \quad \dot{q}_i' := \frac{d}{dt'} q_i' \quad .$$

The connection between these two quantities reads as

$$\begin{aligned}\dot{q}_i' &= \frac{d}{dt'} q_i' = \frac{d}{dt} q_i' \frac{dt}{dt'} = \frac{d}{dt}(q_i + \delta q_i) \frac{dt}{dt'} \\ &= \left(\dot{q}_i + \frac{d}{dt} \delta q_i\right) \frac{1}{1 + (d/dt)\,\delta t} \quad ,\end{aligned} \tag{3}$$

[1] For a full treatment see Vol. 3 of this series, *Relativistic Quantum Mechanics* (Springer, Berlin, Heidelberg, to be published).

Example 1.3

where we have used

$$\frac{dt}{dt'} = \frac{1}{dt'/dt} = \frac{1}{1 + (d/dt)\delta t} \quad . \tag{3'}$$

For infinitesimal transformations this becomes

$$\delta \dot{q}_i := \dot{q}'_i - \dot{q}_i = \frac{d}{dt}\delta q_i - \dot{q}_i \frac{d}{dt}\delta t \quad . \tag{4}$$

Since physics may not change under this coordinate transformation, the action has to remain invariant:

$$S(t_1, t_2) := \int_{t_1}^{t_2} L(\boldsymbol{q}(t), \dot{\boldsymbol{q}}(t), t)\, dt = S'(t_1, t_2) := \int_{t'(t_1)}^{t'(t_2)} L'(\boldsymbol{q}'(t'), \dot{\boldsymbol{q}}'(t'), t')\, dt' \quad .$$

To achieve this aim the following must hold:

$$L(\boldsymbol{q}', \dot{\boldsymbol{q}}', t) := L[\boldsymbol{q}(\boldsymbol{q}', t'), \dot{\boldsymbol{q}}(\boldsymbol{q}', \dot{\boldsymbol{q}}', t'), t(t')] \frac{dt}{dt'} \quad . \tag{5}$$

If the equations of motion are invariant in form under such coordinate transformation, we call this transformation a symmetry transformation. In the simplest case the Lagrange function itself remains invariant, i.e.

$$L'(\boldsymbol{q}', \dot{\boldsymbol{q}}', t') = L(\boldsymbol{q}', \dot{\boldsymbol{q}}', t') \quad ,$$

but this is not a necessity. It is sufficient that

$$L'(\boldsymbol{q}', \dot{\boldsymbol{q}}', t') = L(\boldsymbol{q}', \dot{\boldsymbol{q}}', t') + \frac{d}{dt'}\Omega(\boldsymbol{q}', t') \quad , \tag{6}$$

or in other words, that both Lagrange functions differ by a total derivative with respect to time. It is easily proved that for $\bar{L} = d[\Omega(\boldsymbol{q}, t)]/dt$, the equations of motion

$$\begin{aligned}
\frac{d}{dt}\frac{\partial \bar{L}}{\partial \dot{q}_i} - \frac{\partial \bar{L}}{\partial q_i} &= \frac{d}{dt}\frac{\partial}{\partial \dot{q}_i}\left(\sum_j \frac{\partial \Omega}{\partial q_j}\dot{q}_j + \frac{\partial \Omega}{\partial t}\right) - \frac{\partial}{\partial q_i}\left(\sum_j \frac{\partial \Omega}{\partial q_j}\dot{q}_j + \frac{\partial \Omega}{\partial t}\right) \\
&= \frac{d}{dt}\frac{\partial \Omega}{\partial q_i} - \left(\sum_j \frac{\partial^2 \Omega}{\partial q_i \partial q_j}\dot{q}_j + \frac{\partial^2 \Omega}{\partial t \partial q_i}\right) \\
&= \sum_j \frac{\partial^2 \Omega}{\partial q_i \partial q_j}\dot{q}_j + \frac{\partial^2 \Omega}{\partial t_i \partial q_i} - \left(\sum_j \frac{\partial^2 \Omega}{\partial q_i \partial q_j}\dot{q}_j + \frac{\partial^2 \Omega}{\partial t \partial q_i}\right) \\
&= 0 \quad ,
\end{aligned}$$

are fulfilled. Inserting (6) in (5), we get

$$L(\boldsymbol{q}(\boldsymbol{q}', t'), \ldots, t(t')) \frac{dt}{dt'} = L(\boldsymbol{q}', \dot{\boldsymbol{q}}', t') + \frac{d}{dt'}\Omega(\boldsymbol{q}', t') \quad ,$$

and, reverting to the old coordinates,

$$L(\boldsymbol{q}, \dot{\boldsymbol{q}}, t) = L(\boldsymbol{q}'(\boldsymbol{q}, t), \dot{\boldsymbol{q}}'(\boldsymbol{q}, \dot{\boldsymbol{q}}, t), t'(t)) \frac{dt'}{dt} + \frac{d}{dt}\Omega(\boldsymbol{q}'(\boldsymbol{q}, t), t'(t)) \quad ,$$

Example 1.3

which together with (2) or (3′), respectively, yields the equation

$$L(\boldsymbol{q},\dot{\boldsymbol{q}},t) - L(\boldsymbol{q}'(\boldsymbol{q},t),\dots,t'(t)) = L(\boldsymbol{q}'(\boldsymbol{q},t),\dots)\frac{d}{dt}\delta t + \frac{d}{dt}\Omega(\boldsymbol{q}'(\boldsymbol{q},t),t'(t)) \quad . \tag{7}$$

If the transformation is continuous, it is sufficient to consider infinitesimal transformations in (2). Then (7) may be written (to a first order approximation),

$$\begin{aligned} -\delta L &:= L(\boldsymbol{q},\dot{\boldsymbol{q}},t) - L(\boldsymbol{q}+\delta\boldsymbol{q},\dot{\boldsymbol{q}}+\delta\dot{\boldsymbol{q}},t+\delta t) \\ &= L(\boldsymbol{q},\dot{\boldsymbol{q}},t)\frac{d}{dt}\delta t + \frac{d}{dt}\Omega(\boldsymbol{q}+\delta\boldsymbol{q},t+\delta t) \quad . \end{aligned}$$

In particular, if we choose $\delta\boldsymbol{q}$, $\delta t = 0$, then $\boldsymbol{q} = \boldsymbol{q}'$, $t = t'$ and [from (6)] it follows that $d[\Omega(\boldsymbol{q},t)]/dt = 0$. We may use this to rewrite $(-\delta L)$ as

$$\begin{aligned} -\delta L &= L\frac{d}{dt}\delta t + \frac{d}{dt}[\Omega(\boldsymbol{q}+\delta\boldsymbol{q},t+\delta t) - \Omega(\boldsymbol{q},t)] \\ &= L\frac{d}{dt}\delta t + \frac{d}{dt}\delta\Omega(\boldsymbol{q},t) \quad . \end{aligned} \tag{8}$$

Now if

$$-\delta L = -\sum_i\left[\frac{\partial L}{\partial q_i}\delta q_i + \frac{\partial L}{\partial \dot{q}_i}\delta\dot{q}_i\right] - \frac{\partial L}{\partial t}\delta t \tag{9}$$

is inserted in (8) then, in view of (4),

$$\sum_i\left(\frac{\partial L}{\partial q_i} + \frac{\partial L}{\partial \dot{q}_i}\frac{d}{dt}\right)\delta q_i + \frac{\partial L}{\partial t}\delta t + \left(L - \sum_i\frac{\partial L}{\partial \dot{q}_i}\dot{q}_i\right)\frac{d}{dt}\delta t = -\frac{d}{dt}\delta\Omega(\boldsymbol{q},t) \tag{10}$$

is, for arbitrary $\boldsymbol{q}, t$, *the condition that a mechanical system described by* $\boldsymbol{L}$ *remains invariant under the infinitesimal symmetry transformation (2).* In particular, if $\delta\Omega = 0$, $d(\delta t)/dt = 0$, then $\delta L = 0$ and the Lagrange function itself is invariant under the transformation. If (10) is fulfilled, then, using the equations of motion $\partial L/\partial q_i = d(\partial L/\partial\dot{q}_i)/dt$, it follows that

$$\begin{aligned} &\frac{d}{dt}\left[\sum_i\frac{\partial L}{\partial \dot{q}_i}\delta q_i + \left(L - \sum_i\frac{\partial L}{\partial \dot{q}_i}\dot{q}_i\right)\delta t + \delta\Omega\right] \\ &\quad = \sum_i\left[\frac{\partial L}{\partial q_i}\delta q_i + \frac{\partial L}{\partial \dot{q}_i}\frac{d}{dt}\delta q_i + \left(\frac{\partial L}{\partial q_i}\dot{q}_i + \frac{\partial L}{\partial \dot{q}_i}\ddot{q}_i - \frac{\partial L}{\partial q_i}\dot{q}_i - \frac{\partial L}{\partial \dot{q}_i}\ddot{q}_i\right)\delta t\right] \\ &\qquad + \frac{\partial L}{\partial t}\delta t + \left(L - \sum_i\frac{\partial L}{\partial \dot{q}_i}\dot{q}_i\right)\frac{d}{dt}\delta t + \frac{d}{dt}\delta\Omega = 0 \quad , \end{aligned}$$

i.e. the quantity

$$\sum_i\frac{\partial L}{\partial \dot{q}_i}\delta q_i + \left(L - \sum_i\frac{\partial L}{\partial \dot{q}_i}\dot{q}_i\right)\delta t + \delta\Omega = \text{const.} \tag{11}$$

is an integral of the motion (conserved quantity).

EXERCISE

1.4 Time-Invariant Equations of Motion: The Lagrange Function and Conserved Quantities

Problem. Which condition must be satisfied by the Lagrange function $L(\boldsymbol{q}, \dot{\boldsymbol{q}}, t)$, and what conserved quantities can be found, if the equations of motion are invariant under translations in time?

Solution. A translation in time is parametrised by the coordinate transformation $\delta q = \delta \dot{q} = 0$, $\delta t(t) = \delta\tau =$ const. and the condition [Example 1.3, Eq. (10)] reads

$$\frac{\partial L}{\partial t}\,\delta\tau = -\frac{d}{dt}\,\delta\Omega \quad . \tag{1}$$

If L is not explicitly time-dependent then $\delta\Omega = 0$, and the Lagrange function itself has translational invariance in time. The corresponding conserved quantity [Example 1.3, Eq. (11)] is the total energy,

$$E = L - \sum_i \frac{\partial L}{\partial \dot{q}_i}\,\dot{q}_i \quad . \tag{2}$$

If the kinetic energy T in $L = T - V$ is explicitly time-independent, and a time-dependent potential V is present, then $\delta\Omega$ ought to be found in such a way that

$$\frac{\partial V}{\partial t} = \frac{1}{\delta\tau}\frac{d}{dt}\,\delta\Omega \tag{3}$$

is valid. In general this is not possible, because $\delta V/\partial t$ need not be a total derivative with respect to time.

EXERCISE

1.5 Conditions for Translational, Rotational and Galilean Invariance

Problem. Given the Lagrange function (in cartesian coordinates)

$$L = \tfrac{1}{2}\, m\dot{\boldsymbol{r}}^2 - V(\boldsymbol{r}) \tag{1}$$

and the following transformations:

a) spatial translations

$$\delta x_1 = \delta x_2 = 0 \quad , \quad \delta x_3 = \text{const.} \quad , \quad \delta t = 0 \quad ; \tag{2a}$$

Exercise 1.5

b) spatial rotations,

$$\delta x_1 = -\delta\phi x_2 \quad , \quad \delta x_2 = +\delta\phi x_1 \quad , \quad \delta x_3 = 0 \quad ,$$

$$\delta t = 0 \quad , \quad (\delta\phi = \text{const.}) \quad ; \tag{2b}$$

c) Galilei transformations,

$$\delta x_1 = \delta x_2 = 0 \quad , \quad \delta x_3 = \delta v_3 t \quad , \quad \delta t = 0 \quad , \quad (\delta v_3 = \text{const.}) \quad . \tag{2c}$$

Which conditions must hold, so that these are symmetry transformations? What are the conserved quantities?

Solution. For $\delta t = 0$ the conditional equation for a symmetry transformation (see Example 1.3, Eq. (11)) reads

$$\sum_i \left(\frac{\partial L}{\partial x_i} + \frac{\partial L}{\partial \dot{x}_i} \frac{d}{dt} \right) \delta x_i = -\frac{d}{dt} \delta\Omega(\boldsymbol{r}, t) \quad . \tag{3}$$

The left-hand side (lhs) has to be a total derivative with respect to time. If such a $\delta\Omega$ can be found then (from Noether's theorem – see Example 1.3, Eq. (12)) the quantity

$$\sum_i \frac{\partial L}{\partial \dot{x}_i} \delta x_i + \delta\Omega = \text{const.} \tag{4}$$

is conserved. In our case we have

$$\frac{\partial L}{\partial x_i} = -\frac{\partial V}{\partial x_i} \quad , \quad \frac{\partial L}{\partial \dot{x}_i} = m\dot{x}_i \quad . \tag{5}$$

a) The condition for invariance in the form of the equations of motion reads here as

$$\begin{aligned} -\frac{\partial V}{\partial x_3} \delta x_3 &= -\frac{d}{dt} \delta\Omega(\boldsymbol{r}, t) \\ &= -\sum_i \frac{\partial}{\partial x_i} \delta\Omega(\boldsymbol{r}, t) \dot{x}_i - \frac{\partial}{\partial t} \delta\Omega(\boldsymbol{r}, t) \quad . \end{aligned} \tag{6}$$

The lhs does not contain $\dot{x}_i$; thus,

$$\frac{\partial}{\partial x_i} \delta\Omega(\boldsymbol{r}, t) = 0 \quad \text{or} \quad \delta\Omega(\boldsymbol{r}, t) = \delta\Omega(t) \quad . \tag{7}$$

Hence our condition is

$$\frac{\partial V}{\partial x_3} \delta x_3 = \frac{\partial}{\partial t} \delta\Omega(t) \quad . \tag{8}$$

It follows immediately that $\partial V / \partial x_3$ must be constant (independent of $\boldsymbol{x}, t$), and with

$$\delta\Omega = \frac{\partial V}{\partial x_3} \delta x_3 t \quad , \tag{9}$$

Exercise 1.5

the equation of motion is form-invariant. In this case the spatial translation is a symmetry transformation, and the conserved quantity is (see (4), after reduction by δx_3)

$$m\dot{x}_3 + \frac{\partial V}{\partial x_3} t = \text{const} \quad . \tag{10}$$

In particular, if the constant force in the x_3 direction vanishes, then

$$\frac{\partial V}{\partial x_3} = 0 \quad , \tag{11}$$

and the conserved quantity is the momentum

$$p_3 = \frac{\partial L}{\partial \dot{x}_3} = m\dot{x}_3 \quad . \tag{12}$$

Since now $\delta\Omega = 0$, the Lagrange function itself is invariant.

We thus have shown that the conservation of momentum follows from the spatial translational invariance of the Lagrange function, but not from the invariance of the equations of motion against spatial translations. In this (general) case the conserved quantity is

$$\tilde{\boldsymbol{P}} = \boldsymbol{p} - \boldsymbol{F}t = m\dot{\boldsymbol{r}} - \boldsymbol{F}t \quad , \tag{10'}$$

implying that the momentum is a *linear function* of time. Now $\boldsymbol{F}$ is a constant and homogeneous force field.

From a broader perspective, the constant force field illustrates the difference between *local* and *global* homogeneity of space. The field-filled space is *locally* homogeneous, because no point in space can be distinguished from any other by *local measurements.* However, the force field must be generated by some source, e.g. a distant mass for a gravity field, or distant capacitor plates for a constant electric field. This source configuration destroys the *global* homogeneity of space (see figure below).

b) Here we have

$$\frac{\partial L}{\partial x_1}\delta x_1 + \frac{\partial L}{\partial x_2}\delta x_2 = -\frac{d}{dt}\delta\Omega(\boldsymbol{x}, t) \quad \text{or} \tag{13}$$

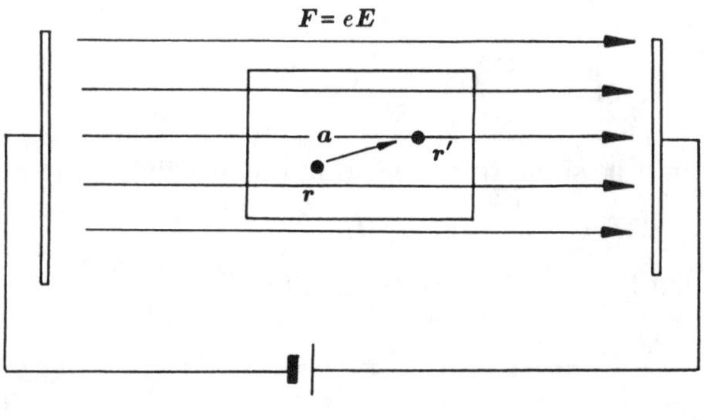

Difference between *local* and *global* homogeneity of space. Global homogeneity implies conservation of momentum, while in a locally homogeneous field the momentum changes as a linear function of time

Exercise 1.5

$$\left(\frac{\partial V}{\partial x_1}x_2 - \frac{\partial V}{\partial x_2}x_1\right)\delta\phi = -\frac{d}{dt}\delta\Omega(\boldsymbol{x},t) \quad . \tag{14}$$

The lhs is essentially the torque around the x_3 axis,

$$(\boldsymbol{r}\times\nabla V)_3\delta\phi = \frac{d}{dt}\delta\Omega(\boldsymbol{r},t) \quad . \tag{15}$$

With the same argument as in case (a) we conclude that only for

$$(\boldsymbol{r}\times\nabla V)_3 = \text{const.} \tag{16}$$

are the Euler-Lagrange equations invariant in form. The student should check that if

$$(\boldsymbol{r}\times\nabla V)_3 = 0 \tag{17}$$

is valid, then we have $\frac{d}{dt}\delta\Omega = 0$ and

$$\delta\Omega = 0 \quad . \tag{18}$$

The corresponding conserved quantity is

$$-m\dot{x}_1x_2 + m\dot{x}_2x_1 = \text{const.} \tag{19}$$

which is just the x_3 component of angular momentum, i.e.

$$L_3 = (\boldsymbol{r}\times\boldsymbol{p})_3 = \text{const.} \quad . \tag{20}$$

Thus the conservation of angular momentum follows from rotational invariance of the lagrangian.

c) For Galilei transformations (1) becomes

$$\left(\frac{\partial L}{\partial x_3} + \frac{\partial L}{\partial \dot{x}_3}\frac{d}{dt}\right)\delta v_3 t = -\frac{d}{dt}\delta\Omega \quad \text{or} \tag{21}$$

$$\left(-\frac{\partial V}{\partial x_3}t + m\dot{x}_3\right)\delta v_3 = -\frac{d}{dt}\delta\Omega = -\sum_i\left(\frac{\partial\delta\Omega}{\partial x_i}\right)\dot{x}_i - \frac{\partial}{\partial t}\delta\Omega \quad . \tag{22}$$

Thus,

$$\frac{\partial}{\partial x_1}\delta\Omega = \frac{\partial}{\partial x_2}\delta\Omega = 0 \quad , \quad \frac{\partial}{\partial x_3}\delta\Omega = -m\delta v_3 \quad ,$$

and

$$\frac{\partial}{\partial t}\delta\Omega = \frac{\partial V}{\partial x_3}\delta v_3 t \tag{23}$$

must be valid. It follows from the first three of these equations that

$$\delta\Omega = -mx_3\delta v_3 + f(t) \tag{24}$$

and from the fourth equation that

$$\frac{df(t)}{dt} = \frac{\partial V}{\partial x_3}\delta v_3 t \quad . \tag{25}$$

Thus $\partial V/\partial x_3 = \text{const.}$ must be valid, implying that

$$\delta\Omega = \left(- m x_3 + \frac{1}{2} \frac{\partial V}{\partial x_3} t^2 \right) \delta v_3 \quad . \tag{26}$$

If L also has translational invariance, ($\partial V/\partial x_3 = 0$), then the corresponding conserved quantity is

$$m\dot{x}_3 t - m x_3 = \text{const.} \tag{27}$$

or

$$x_3 - \dot{x}_3 t = x_3 - \frac{P_3}{m} t = x_3(0) = \text{const.} \tag{28}$$

and the particle moves with constant velocity.

EXERCISE

1.6 Conservation Laws in Homogeneous Electromagnetic Fields

Problem. Derive the conservation laws corresponding to translational symmetry for a charged particle in
(a) a homogeneous electrical field $\boldsymbol{E}$
(b) a homogeneous magnetic field $\boldsymbol{B}$.

Solution. The Lagrangian of a (nonrelativistic) point particle with mass m and charge q in an electromagnetic potential described by the scalar potential ϕ and the vector potential $\boldsymbol{A}$ is given by[2]:

$$L = \frac{m}{2} \dot{\boldsymbol{x}}^2 + \frac{q}{c} \boldsymbol{A} \cdot \dot{\boldsymbol{x}} - q\phi \quad . \tag{1}$$

The canonical momentum is

$$\boldsymbol{p} = \frac{\partial L}{\partial \dot{\boldsymbol{x}}} = m\dot{\boldsymbol{x}} + \frac{q}{c} \boldsymbol{A} \quad . \tag{2}$$

Furthermore one finds

$$\frac{\partial L}{\partial x_i} = \frac{q}{c} \frac{\partial \boldsymbol{A}}{\partial x_i} \cdot \dot{\boldsymbol{x}} - q \frac{\partial \phi}{\partial x_i} \quad . \tag{3}$$

---

[2] See J.D. Jackson: *Classical Electrodynamics*, 2nd ed. (Wiley, New York 1985).

Exercise 1.6

From this one gets the equations of motion

$$0 = \frac{d}{dt}\left(\frac{\partial L}{\partial \dot{x}_i}\right) - \frac{\partial L}{\partial x_i}$$

$$= m\ddot{x}_i + \frac{q}{c}\left(\frac{\partial A_i}{\partial t} + \sum_k \dot{x}_k \frac{\partial A_i}{\partial x_k}\right) - \frac{q}{c}\sum_k \frac{\partial A_k}{\partial x_i}\dot{x}_k + q\frac{\partial \phi}{\partial x_i}$$

$$= m\ddot{x}_i + q\left(\frac{1}{c}\frac{\partial A_i}{\partial t} - \frac{\partial \phi}{\partial x_i}\right) + \frac{q}{c}\sum_k \dot{x}_k\left(\frac{\partial A_i}{\partial x_k} - \frac{\partial A_k}{\partial x_i}\right) \quad , \tag{4}$$

or, written in vectorial notation

$$m\ddot{\boldsymbol{x}} = q\left(-\frac{1}{c}\dot{\boldsymbol{A}} - \vec{\nabla}\phi\right) + \frac{q}{c}\dot{\boldsymbol{x}} \times (\vec{\nabla} \times \boldsymbol{A})$$

$$= q\boldsymbol{E} + \frac{q}{c}\dot{\boldsymbol{x}} \times \boldsymbol{B} \quad . \tag{5}$$

Whenever it is possible that the generalized force $\partial L/\partial x_i$ can be written as the total time derivate $\partial L/\partial x_i = (d/dt)\,G_i$, it follows due to the Euler-Lagrange equation that there is a conservation law:

$$0 = \frac{d}{dt}\frac{\partial L}{\partial \dot{x}_i} - \frac{\partial L}{\partial x_i} = \frac{d}{dt}\left[\frac{\partial L}{\partial \dot{x}_i} - G_i\right] = \frac{d}{dt}[P_i - G_i] \quad . \tag{6}$$

Here P_i is the canonical momentum.

(a) A homogeneous electric field $\boldsymbol{E}$ can either be described by the potential

$$\phi = -\boldsymbol{E}\cdot\boldsymbol{x} \quad , \quad \boldsymbol{A} = 0 \tag{7a}$$

or by

$$\phi' = 0 \quad , \quad \boldsymbol{A}' = -\boldsymbol{E}t \quad . \tag{7b}$$

Both representations correspond to different gauges. In the first case we can write

$$\frac{\partial L}{\partial x_i} = -q\frac{\partial \phi}{\partial x_i} = +qE_i = \frac{d}{dt}(+qE_i t) = \frac{d}{dt}G_i \quad , \tag{8a}$$

from which we get the conservation law

$$\frac{d}{dt}(\boldsymbol{p} - q\boldsymbol{E}t) = 0 \quad . \tag{9a}$$

In this gauge the conserved quantity is *not* equal to the canonical momentum. However, in the second case we have

$$\frac{\partial L'}{\partial x_i} = 0 \;\rightarrow\; \boldsymbol{G}' = 0 \quad , \tag{8b}$$

so that the canonical momentum is conserved:

$$\frac{d}{dt}\boldsymbol{p} = 0 \quad . \tag{9b}$$

Exercise 1.6

The apparent discrepancy between (9a) and (9b) can be explained by the fact that (2) yields two different expressions for the canonical momentum. In the first case we have

$$p = m\dot{x} \quad , \tag{10a}$$

whereas in the second case we get

$$p = m\dot{x} - qEt \quad . \tag{10b}$$

In both cases one finds that the quantity

$$m\dot{x} - qEt \tag{11}$$

is conserved. From that we learn that the physical meaning of a law containing the canonical momentum in the presence of external electromagnetic fields may depend on the gauge!

(b) A homogeneous magnetic field $\boldsymbol{B}$ can, e.g., be described by the vector potential

$$\boldsymbol{A} = \tfrac{1}{2}\boldsymbol{B}\times\boldsymbol{x} \quad , \quad \phi = 0 \quad . \tag{12}$$

Here we get

$$\begin{aligned}\frac{\partial L}{\partial x_i} &= \frac{q}{c}\sum_k \dot{x}_k \frac{\partial A_k}{\partial x_i} \\ &= \frac{q}{2c}\sum_k \dot{x}_k \frac{\partial}{\partial x_i}\left(\sum_{lm} \varepsilon_{klm} B_l x_m\right) \\ &= \frac{q}{2c}\sum_k \dot{x}_k \sum_l \varepsilon_{kli} B_l \quad ,\end{aligned} \tag{13}$$

or, written vectorially:

$$\frac{\partial L}{\partial \boldsymbol{x}} = \frac{q}{2c}\dot{\boldsymbol{x}}\times\boldsymbol{B} = \frac{d}{dt}\left(\frac{q}{2x}\boldsymbol{x}\times\boldsymbol{B}\right) = \frac{d}{dt}\boldsymbol{G} \quad . \tag{14}$$

Thus the conservation law corresponding to translational symmetry reads

$$\frac{d}{dt}\left(\boldsymbol{p} - \frac{q}{2c}\boldsymbol{x}\times\boldsymbol{B}\right) = 0 \quad . \tag{15}$$

Note that in this case the conserved quantities are *not* equal to the components of the canonical momentum. Due to (2) the canonical momentum is given by

$$\boldsymbol{p} = m\dot{\boldsymbol{x}} + \frac{q}{2c}\boldsymbol{B}\times\boldsymbol{x} = m\dot{\boldsymbol{x}} - \frac{q}{2c}\boldsymbol{x}\times\boldsymbol{B} \quad . \tag{16}$$

Thus we can express the conserved quantity by the velocity:

$$m\dot{\boldsymbol{x}} - \frac{q}{c}\boldsymbol{x}\times\boldsymbol{B} \quad . \tag{17}$$

1.2 Spatial Translations in Quantum Mechanics

Consider a state which is given by the ket $|\alpha\rangle$ or by the wave function $\psi_\alpha(\boldsymbol{r}, t)$. If this state is spatially shifted by the vector $\boldsymbol{\varrho}$, then a new state is created. We call it $|\alpha'\rangle$ or $\psi_{\alpha'}(\boldsymbol{r}, t)$, respectively. More precisely, we have

$$\psi_{\alpha'}(\boldsymbol{r} + \boldsymbol{\varrho}, t) = \psi_\alpha(\boldsymbol{r}, t) \quad . \tag{1.17}$$

This is made evident by Fig. 1.4, for the wave function $\psi_\alpha(\boldsymbol{r}, t)$ which has its maximum at $\boldsymbol{r} = \boldsymbol{r}_0$.

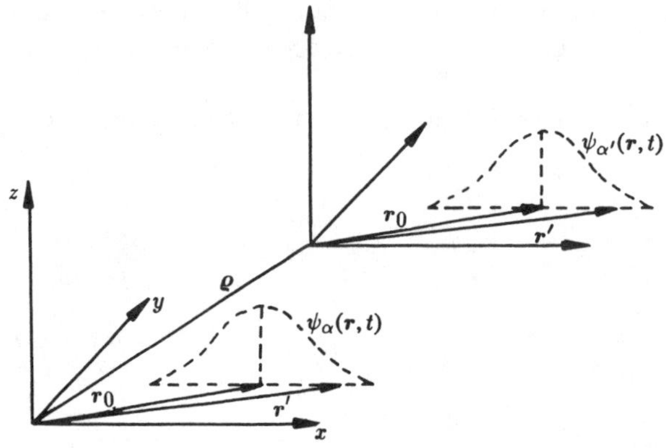

Fig. 1.4. Illustration of the translation of a wave packet by $\boldsymbol{\varrho}$. $\psi_{\alpha'}(\boldsymbol{r})$ has the same value at $\boldsymbol{r} = \boldsymbol{r}' + \boldsymbol{\varrho}$ as $\psi_\alpha(\boldsymbol{r})$ at $\boldsymbol{r} = \boldsymbol{r}'$

Then the shifted wave function $\psi_{\alpha'}(\boldsymbol{r})$ has a maximum at $\boldsymbol{r} = \boldsymbol{r}_0 + \boldsymbol{\varrho}$. In this operation we have shifted the entire wave function ψ_α, although the coordinate system is x, y, z remains the same.

This is a so-called *active execution* of the symmetry operation (in our case a shift). The *passive execution* of the symmetry operation, where the wave function ψ_α is imagined to be unchanged (fixed), but the coordinate system is shifted by the vector $-\boldsymbol{\varrho}$ with respect to the original coordinate system, is

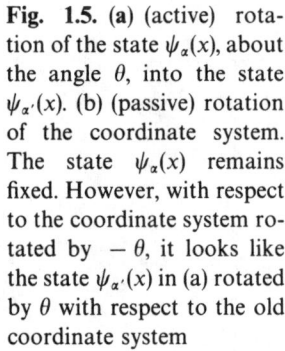

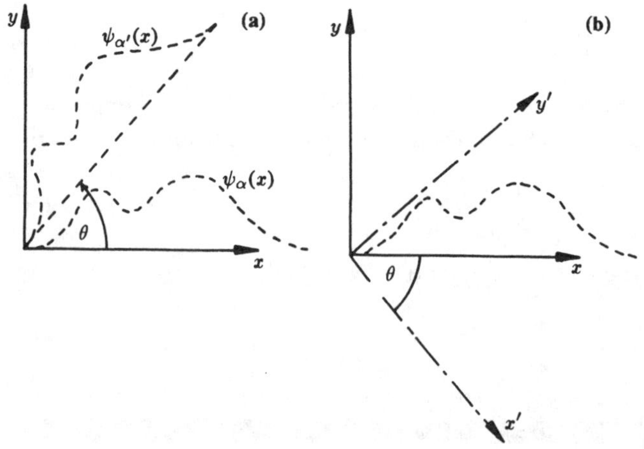

Fig. 1.5. (**a**) (active) rotation of the state $\psi_\alpha(x)$, about the angle θ, into the state $\psi_{\alpha'}(x)$. (**b**) (passive) rotation of the coordinate system. The state $\psi_\alpha(x)$ remains fixed. However, with respect to the coordinate system rotated by $-\theta$, it looks like the state $\psi_{\alpha'}(x)$ in (a) rotated by θ with respect to the old coordinate system

equivalent to the active execution. (See Fig. 1.5 above, where the same concept is illustrated for rotations).

Usually we will deal with the active execution, except later[3] where the passive execution of, say, Lorentz transformations seems to be more suited: There the same physical system is observed from different inertial systems.

1.3 The Unitary Translation Operator

Earlier[4] we became acquainted with two kinds of transformations in Hilbert space, namely

$$\phi_\mu(\boldsymbol{r},t) = \sum_n S_{n\mu}\psi_n(\boldsymbol{r},t) \tag{1.18}$$

and

$$\psi(\boldsymbol{r},t) = \exp\left(-\frac{\mathrm{i}}{\hbar}\hat{H}t\right)\psi(\boldsymbol{r},0) \quad . \tag{1.19}$$

The first equation expresses a transformation of the base vectors $\psi_n(\boldsymbol{r},t)$ of Hilbert space to new base vectors $\phi_\mu(\boldsymbol{r},t)$. Thus it describes a coordinate transformation (rotation of the axes) in Hilbert space. Contrary to this the second transformation (1.19) retains the axes and changes the vector $\psi(\boldsymbol{r},0)$ into the vector $\psi(\boldsymbol{r},t)$ of the same Hilbert space. Therefore (1.19) describes a generalized *rotation of the state vector* ψ in Hilbert space *without changing the axes.*

Here we shall consider transformations of the second type, namely *active translations of the state vector* in coordinate space, which correspond to rotations of the state vector in Hilbert space. Since, during the translation of the state vector in position space, its length remains unchanged, we expect that the corresponding spatial translation operator $\hat{U}_r(\boldsymbol{\varrho})$ is unitary (conservation of probability). The subscript r indicates spatial translations. Subsequently, we will become acquainted with the time-translation operator $\hat{U}_\mathrm{t}(\tau)$ (which shifts by the time interval τ), where t is the usual subscript for time. For the shifted state we write

$$\psi_{\alpha'}(\boldsymbol{r}) = \hat{U}_r(\boldsymbol{\varrho})\psi_\alpha(\boldsymbol{r}) \tag{1.20}$$

and because of (1.17) we conclude that

$$\psi_{\alpha'}(\boldsymbol{r}) = \psi_\alpha(\boldsymbol{r}-\boldsymbol{\varrho}) = \hat{U}_r(\boldsymbol{\varrho})\psi_\alpha(\boldsymbol{r}) \quad . \tag{1.21}$$

To determine the operator $\hat{U}_r(\boldsymbol{\varrho})$ explicitly, we orientate the translation vector

---

[3] See Vol. 3 of this series, *Relativistic Quantum Mechanics* (Springer, Berlin, Heidelberg 1989).

[4] Vol. 1 of this series, *Quantum Mechanics I – An Introduction* (Springer, Berlin, Heidelberg 1989).

$\boldsymbol{\varrho}$ parallel to the x axis; thus $\boldsymbol{\varrho} = \varrho \boldsymbol{e}_1$, and by a Taylor expansion we get

$$\begin{aligned}\psi_\alpha(\boldsymbol{r} - \boldsymbol{\varrho}) &= \psi_\alpha(x - \varrho, y, z)\\ &= \psi_\alpha(x, y, z) - \varrho \frac{\partial}{\partial x} \psi_\alpha(x, y, z) + \frac{\varrho^2}{2!} \frac{\partial^2}{\partial x^2} \psi_\alpha(x, y, z) - \dots \quad .\end{aligned} \tag{1.22}$$

Clearly the rhs of (1.22) may be written in the form

$$\begin{aligned}&\psi_\alpha(x, y, z) - \varrho \frac{\partial}{\partial x} \psi_\alpha(x, y, z) + \frac{1}{2!} \varrho^2 \frac{\partial^2}{\partial x^2} \psi_\alpha(x, y, z) - \dots\\ &= \mathrm{e}^{-\varrho(\partial/\partial x)} \psi_\alpha(x, y, z) \quad .\end{aligned}$$

If the translation vector $\boldsymbol{\varrho}$ has an arbitrary direction, i.e. $\boldsymbol{\varrho} = \varrho_1 \boldsymbol{e}_1 + \varrho_2 \boldsymbol{e}_2 + \varrho_3 \boldsymbol{e}_3$, we may generalize the method just applied [$\varrho(\partial/\partial x)$ must be replaced by $\boldsymbol{\varrho} \cdot \nabla$] and we obtain by a three-dimensional Taylor expansion,

$$\begin{aligned}\psi_\alpha(\boldsymbol{r} - \boldsymbol{\varrho}) &= \psi_\alpha(x - \varrho_1, y - \varrho_2, z - \varrho_3)\\ &= \exp(-\boldsymbol{\varrho} \cdot \nabla) \psi_\alpha(x, y, z) = \exp\left(-\frac{\mathrm{i} \boldsymbol{\varrho} \cdot \hat{\boldsymbol{p}}}{\hbar}\right) \psi_\alpha(\boldsymbol{r}) \quad .\end{aligned} \tag{1.23}$$

Here we have introduced the momentum operator $\hat{\boldsymbol{p}} = -\mathrm{i}\hbar\nabla$. Comparing (1.21) with (1.23) yields the *translation operator*

$$\hat{U}_r(\boldsymbol{\varrho}) = \exp\left(-\frac{\mathrm{i} \boldsymbol{\varrho} \cdot \hat{\boldsymbol{p}}}{\hbar}\right) \quad . \tag{1.24}$$

Equation (1.24) is valid for arbitrary $\psi_\alpha(\boldsymbol{r})$, i.e. for all state vectors. Because the ∇ operator (which is defined only in coordinate representation) was replaced by the momentum operator, equation (1.24) is valid in every representation. Using the hermiticity of $\hat{\boldsymbol{p}}$, we shall now verify that $\hat{U}_r(\boldsymbol{\varrho})$ is unitary.

$$\begin{aligned}\hat{U}_r^{-1}(\boldsymbol{\varrho}) &= \exp\left(+\frac{\mathrm{i} \boldsymbol{\varrho} \cdot \hat{\boldsymbol{p}}}{\hbar}\right) = \left[\exp\left(-\frac{\mathrm{i} \boldsymbol{\varrho} \cdot \hat{\boldsymbol{p}}^\dagger}{\hbar}\right)\right]^\dagger\\ &= \left[\exp\left(-\frac{\mathrm{i} \boldsymbol{\varrho} \cdot \hat{\boldsymbol{p}}}{\hbar}\right)\right]^\dagger = \hat{U}_r^+(\boldsymbol{\varrho}) \quad .\end{aligned} \tag{1.25}$$

It will turn out that for all Lie-groups the group operators can be written in the form $\hat{U}(\alpha_1, \alpha_2, \dots) = \mathrm{e}^{\mathrm{i}(\alpha_1 \hat{L}_1 + \alpha_2 \hat{L}_2 + \dots)}$ with certain operators $\hat{L}_1, \hat{L}_2, \dots$. These operators will be called *generators* of the group. In this sense the momentum operators $\hat{p}_i$ in (24) are the generators of the translation group.

1.4 The Equation of Motion for States Shifted in Space

The state $\psi_\alpha(\boldsymbol{r}, t)$ is transformed to $\psi_{\alpha'}(\boldsymbol{r}, t)$ by a spatial displacement at a fixed time t. Also $\psi_\alpha(\boldsymbol{r}, t)$ satisfies the time dependent Schrödinger equation,

$$\mathrm{i}\hbar \frac{\partial \psi_\alpha(\boldsymbol{r}, t)}{\partial t} = \hat{H} \psi_\alpha(\boldsymbol{r}, t) \quad . \tag{1.26}$$

Equation (1.26) describes the evolution in time $\psi_\alpha(\boldsymbol{r},t)$, and gives rise to the question: Under what conditions do we have the same temporal evolution of the shifted state $\psi_{\alpha'}(\boldsymbol{r},t)$ and the initial state $\psi_\alpha(\boldsymbol{r},t)$? This is precisely the meaning of the homogeneity of space: All spatially displaced states satisfy the same laws of nature (here the same Schrödinger equation). No difference between shifted and unshifted states is observed. Hence, it follows that

$$\begin{aligned} \mathrm{i}\hbar\frac{\partial}{\partial t}\psi_{\alpha'}(\boldsymbol{r},t) &= \mathrm{i}\hbar\frac{\partial}{\partial t}\hat{U}_r(\boldsymbol{\varrho})\psi_\alpha(\boldsymbol{r},t) = \hat{U}_r(\boldsymbol{\varrho})\mathrm{i}\hbar\frac{\partial\psi_\alpha(\boldsymbol{r},t)}{\partial t} \\ &= \hat{U}_r(\boldsymbol{\varrho})\hat{H}\psi_\alpha(\boldsymbol{r},t) = \hat{U}_r(\boldsymbol{\varrho})\hat{H}\hat{U}_r^{-1}(\boldsymbol{\varrho})\psi_{\alpha'}(\boldsymbol{r},t) \quad , \end{aligned} \tag{1.27}$$

and using (1.25) we get

$$= \hat{U}_r(\boldsymbol{\varrho})\hat{H}\hat{U}_r^\dagger(\boldsymbol{\varrho})\psi_{\alpha'}(\boldsymbol{r},t) \quad .$$

Clearly $\psi_\alpha(\boldsymbol{r},t)'$ only satisfies the same Schrödinger equation as $\psi_\alpha(\boldsymbol{r},t)$, if

$$\hat{U}_r(\boldsymbol{\varrho})\hat{H}\hat{U}_r^\dagger(\boldsymbol{\varrho}) = \hat{H}$$

or using (1.25)

$$\hat{U}_r(\boldsymbol{\varrho})\hat{H} = \hat{H}\hat{U}_r(\boldsymbol{\varrho}) \quad . \tag{1.28}$$

Consequently the Hamiltonian $\hat{H}$ and the displacement operator $\hat{U}_r(\boldsymbol{\varrho})$ must commute, i.e.

$$[\hat{H},\hat{U}_r(\boldsymbol{\varrho})]_- = 0 \quad . \tag{1.29}$$

According to (1.24) we can write this as

$$\left[\hat{H},\exp\left(-\frac{\mathrm{i}}{\hbar}\boldsymbol{\varrho}\cdot\hat{\boldsymbol{p}}\right)\right]_- = 0 \quad . \tag{1.30}$$

Since $\boldsymbol{\varrho}$ is an arbitrary vector, this equation leads to the condition that

$$[\hat{H},\hat{\boldsymbol{p}}]_- = 0 \quad , \tag{1.31}$$

i.e. the momentum operator $\hat{\boldsymbol{p}}$ must commute with the hamiltonian $\hat{H}$. Hence[5] the momentum $\boldsymbol{p}$ is a constant of motion; this equation is analogous to the classical equation (1.3). Furthermore, it follows from (1.31) that $\hat{H}$ and $\hat{\boldsymbol{p}}$ can be diagonalized simultaneously. Therefore, states can be constructed that have fixed eigenvalues of the energy as well as of the momentum operator.

According to these considerations, if the displaced state $\psi_{\alpha'}(\boldsymbol{r},t)$ satisfies the same Schrödinger equation as the initial state $\psi_\alpha(\boldsymbol{r},t)$, the momentum is a constant of motion. Hence the requirement for the homogeneity of space is equivalent to stating that *each spatially displaced wave function has to satisfy the Schrödinger equation* (i.e. the laws of nature) as well. This symmetry of the laws of nature (here the Schrödinger equation) with respect to displacements in space implies the law of momentum conversation. States are characterized by constant energy and constant momentum, and such systems are called *space-displacement invariant* or *space-displacement symmetric.*

---

[5] See Vol. 1 of this series, *Quantum Mechanics – An Introduction* (Springer, Berlin, Heidelberg 1989).

Free particles are examples of such systems. However, this symmetry will be lost if a force is introduced (e.g. a particle in a localized potential), when the shifted wave function will be outside the potential: therefore, it decays because it is no longer an eigenstate. The potential disturbs the homogeneity of space. Thus the eigenfunctions of such a potential are not translation invariant (see Fig. 1.6).

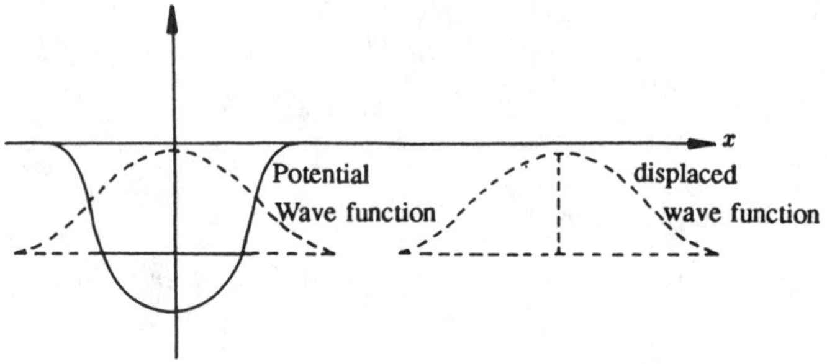

Fig. 1.6. The displaced wave function is not an eigenfunction of the potential shown. Such a state would decay immediately

1.5 Symmetry and Degeneracy of States

The initial state $\psi_\alpha(\boldsymbol{r},t)$ is assumed to satisfy the stationary Schrödinger equation:

$$\hat{H}\psi_\alpha(\boldsymbol{r},t)f = E_\alpha\psi_\alpha(\boldsymbol{r},t) \quad . \tag{1.32}$$

Furthermore, the operator $\hat{S}$ should commute with $\hat{H}$ so that

$$[\hat{H},\hat{S}]_- = 0 \quad . \tag{1.33}$$

For (1.32) also, $\hat{S}\hat{\psi}_\alpha$ is an eigenstate of $\hat{H}$ with the same eigenvalue E_α, i.e.

$$\hat{H}(\hat{S}\psi_\alpha) = \hat{S}\hat{H}\psi_\alpha = \hat{S}E_\alpha\psi_\alpha = E_\alpha(\hat{S}\psi_\alpha) \quad . \tag{1.34}$$

If the state $\chi_\alpha = \hat{S}\hat{\psi}_\alpha$ is linearly independent of $\hat{\psi}_\alpha$, the eigenvalue E_α is degenerate. In the case of symmetry with regard to spatial displacement, the operator $\hat{\boldsymbol{p}}$ commutes with the Hamiltonian according to (1.31). Hence, for all states $\chi_{\boldsymbol{p}_0}$,

$$\begin{aligned}\chi\boldsymbol{p}_0 &= \boldsymbol{a}\cdot\hat{\boldsymbol{p}}\psi_{\boldsymbol{p}_0}(\boldsymbol{r},t) = \boldsymbol{a}\cdot\hat{\boldsymbol{p}}\frac{1}{\sqrt{2\pi\hbar}^3}\exp\left[+\frac{\mathrm{i}}{\hbar}(\boldsymbol{p}_0\cdot\boldsymbol{r} - Et)\right] \\ &= \frac{\boldsymbol{a}\cdot\boldsymbol{p}_0}{\sqrt{2\pi\hbar}^3}\exp\left[\frac{\mathrm{i}}{\hbar}(\boldsymbol{p}_0\cdot\boldsymbol{r} - Et)\right] \quad ,\end{aligned}$$

with the arbitrary constant vector $\boldsymbol{a}$ will have the same energy eigenvalue $E = \boldsymbol{p}_0^2/2m$. In this case, however, the wave functions χ differ from Ψ only by a normalization factor, and we cannot draw a conclusion with regard to degeneracy. Degeneracy becomes apparent if we investigate rotations of the wave functions as later in (1.72). There we will utilize the rotation operator $\hat{U}_{\mathrm{R}}(\boldsymbol{\phi})$ similarly as we introduced the translation operator in Sect. 1.3. Accordingly, the plane wave rotated by the infinitesimal angle $\delta\boldsymbol{\phi}$ is written

$$\begin{aligned}
\chi_p(\boldsymbol{r},t) &= \hat{U}_{\mathrm{R}}(\delta\boldsymbol{\phi})\psi_{p_0}(\boldsymbol{r},t) \\
&= \exp\left(-\frac{\mathrm{i}}{\hbar}\delta\boldsymbol{\phi}\cdot\boldsymbol{L}\right)\frac{1}{\sqrt{2\pi\hbar}^3}\exp\left[\frac{\mathrm{i}}{\hbar}(\boldsymbol{p}_0\cdot\boldsymbol{r}-Et)\right] \\
&\simeq \left[\mathbb{1}-\frac{\mathrm{i}}{\hbar}\delta\boldsymbol{\phi}\cdot(\boldsymbol{r}\times\hat{\boldsymbol{p}})\right]\frac{1}{\sqrt{2\pi\hbar}^3}\exp\left[\frac{\mathrm{i}}{\hbar}(\boldsymbol{p}_0\cdot\boldsymbol{r}-Et)\right] \\
&= \left[\mathbb{1}-\frac{\mathrm{i}}{\hbar}\delta\boldsymbol{\phi}\cdot(\boldsymbol{r}\times\boldsymbol{p}_0)\right]\frac{1}{\sqrt{2\pi\hbar}^3}\exp\left[\frac{\mathrm{i}}{\hbar}(\boldsymbol{p}_0\cdot\boldsymbol{r}-Et)\right] \\
&= \left[\mathbb{1}+\frac{\mathrm{i}}{\hbar}(\delta\boldsymbol{\phi}\times\boldsymbol{p}_0)\cdot\boldsymbol{r}\right]\frac{1}{\sqrt{2\pi\hbar}^3}\exp\left[\frac{\mathrm{i}}{\hbar}(\boldsymbol{p}_0\cdot\boldsymbol{r}-Et)\right] \\
&= \exp\left[\frac{\mathrm{i}}{\hbar}(\delta\boldsymbol{\phi}\times\boldsymbol{p}_0)\cdot\boldsymbol{r}\right]\frac{1}{\sqrt{2\pi\hbar}^3}\exp\left[\frac{\mathrm{i}}{\hbar}(\boldsymbol{p}_0\cdot\boldsymbol{r}-Et)\right] \\
&= \frac{1}{\sqrt{2\pi\hbar}^3}\exp\left\{\frac{\mathrm{i}}{\hbar}[(\boldsymbol{p}_0+\delta\boldsymbol{\phi}\times\boldsymbol{p}_0)\cdot\boldsymbol{r}-Et]\right\} \\
&= \frac{1}{\sqrt{2\pi\hbar}^3}\exp\left[\frac{\mathrm{i}}{\hbar}(\boldsymbol{p}\cdot\boldsymbol{r}-Et)\right] \quad ,
\end{aligned} \tag{1.35}$$

where $\boldsymbol{p}=\boldsymbol{p}_0+\delta\boldsymbol{\phi}\times\boldsymbol{p}_0$ is the rotated momentum, cf. (1.11). Furthermore we have $|\boldsymbol{p}|=|\boldsymbol{p}_0|$, as can be seen in (1.35). The rotated wave $\chi_p(\boldsymbol{r},t)$ is like Ψ_{p_0} part of the set of linearly independent plane waves and can as such not be constructed from these by linear combination of other plane waves. Thus the waves χ_p are degenerate if $\hat{U}_{\mathrm{R}}(\delta\boldsymbol{\phi})$ commutes with the hamiltonian, and $\psi_{p_0}(\boldsymbol{r},t)$ is an eigenstate of the Hamiltonian. This is evident, because the energy of a particle depends on the square of the momentum, but does not depend on its direction.

Resumé: The considerations presented above will be of interest if *two operators $\hat{S}$ and $\hat{A}$ exist, which commute with $\hat{H}$, but do not commute with each other, i.e.*

$$[\hat{H},\hat{S}]_- = [\hat{H},\hat{A}]_- = 0 \quad , \quad [\hat{A},\hat{S}]_- \neq 0 \quad .$$

If ψ_α is an eigenvector of $\hat{A}$ and $\hat{H}$, ψ_α *is not* eigenvector of $\hat{S}$, i.e. $\psi_\alpha(\boldsymbol{r},t)$ and $\hat{S}\psi_\alpha$ are linearly independent. This is utilized in the example above, where $\hat{A}$ is the momentum operator, and $\hat{S}$ the angular momentum operator.

EXAMPLE

1.7 Matrix Elements of Spatially Displaced States

Let us investigate the matrix element of an operator $\hat{A}$ between translated states,

$$\langle\psi_{\alpha'}(\boldsymbol{r},t)|\hat{A}|\psi_{\beta'}(\boldsymbol{r},t)\rangle = \langle\psi_\alpha(\boldsymbol{r},t)|\hat{U}_r^\dagger(\boldsymbol{\varrho})\hat{A}\hat{U}_r(\boldsymbol{\varrho})|\psi_\beta(\boldsymbol{r},t)\rangle \quad . \tag{1}$$

Example 1.7

If $\hat{A} = \hat{A}(\hat{\boldsymbol{p}})$, i.e. $\hat{A}$ is a function of the momentum operator, the translated operator can be written as

$$\hat{U}_r^\dagger(\boldsymbol{\varrho})\hat{A}(\hat{\boldsymbol{p}})\hat{U}_r = \hat{A}(\hat{\boldsymbol{p}})\hat{U}_r^\dagger\hat{U}_r = \hat{A}(\hat{\boldsymbol{p}}) \quad ,$$

i.e. the matrix elements of $\hat{A}$ between shifted and unshifted states are equal. If $\hat{A} = \hat{A}(\boldsymbol{r})$ is a function of position it follows by (1) that

$$\begin{aligned}\hat{U}_r^\dagger(\boldsymbol{\varrho})\hat{A}(\boldsymbol{r})\hat{U}_r(\boldsymbol{\varrho}) &= \exp\left(+\frac{\mathrm{i}\boldsymbol{\varrho}\cdot\hat{\boldsymbol{p}}}{\hbar}\right)\hat{A}(\boldsymbol{r})\exp\left(-\frac{\mathrm{i}\boldsymbol{\varrho}\cdot\hat{\boldsymbol{p}}}{\hbar}\right)\\ &= \hat{A}(\boldsymbol{r}+\boldsymbol{\varrho}) \quad .\end{aligned}$$

This will be proved in Exercises 1.8 and 1.9. Hence, the matrix elements of $\hat{A}(\boldsymbol{r})$ between displaced states are equal to the matrix elements between the original states, but in this case, applying the displaced operator $\hat{A}'(\boldsymbol{r}) = \hat{A}(\boldsymbol{r}+\boldsymbol{\varrho})$.

EXERCISE

1.8 The Relation $(\mathrm{i}\hat{p}/\hbar)^n\hat{B}(x)$ and Transformation Operators

Problem. Prove that the relation

$$\left(\frac{\mathrm{i}}{\hbar}\hat{p}\right)^n\hat{B}(x) = \sum_{\nu=0}^{n}\binom{n}{\nu}\frac{\partial^\nu\hat{B}}{\partial x^\nu}\left(\frac{\mathrm{i}}{\hbar}\hat{p}\right)^{n-\nu} \quad , \tag{1}$$

with $\hat{p} = (\hbar/\mathrm{i})(\partial/\partial x)$, is valid for every differentiable operator $\hat{B}(x)$. Use this relation to calculate

$$\hat{U}^\dagger\hat{A}(x)\hat{U} \quad \text{where} \tag{2}$$

$$\hat{U} = \exp\left(-\frac{\mathrm{i}}{\hbar}\boldsymbol{\varrho}\cdot\hat{\boldsymbol{r}}\right) \quad .$$

Solution. a) We prove (1) by mathematical induction. For $n = 0$ the relation is valid, because it yields the identity

$$\hat{B}(x) = \hat{B}(x) \quad .$$

Now we assume (1) to be valid for $n-1$, which means that

$$\left(\frac{\mathrm{i}}{\hbar}\hat{p}\right)^{n-1}\hat{B}(x) = \sum_{\nu=0}^{n-1}\binom{n-\nu}{\nu}\frac{\partial^\nu\hat{B}}{\partial x^\nu}\left(\frac{\mathrm{i}}{\hbar}\hat{p}\right)^{n-1-\nu} \quad , \tag{3}$$

should be correct. We show that we can deduce (1) for all n by one further application of $(\mathrm{i}\hat{p}/\hbar)$, *Exercise 1.8*

$$\begin{aligned}
\left(\frac{\mathrm{i}}{\hbar}\hat{p}\right)^n \hat{B}(x) &= \left(\frac{\mathrm{i}}{\hbar}\hat{p}\right)\left\{\left(\frac{\mathrm{i}}{\hbar}\hat{p}\right)^{n-1}\hat{B}(x)\right\} \\
&= \left(\frac{\mathrm{i}}{\hbar}\hat{p}\right)\sum_{\nu=0}^{n-1}\binom{n-\nu}{\nu}\frac{\partial^\nu \hat{B}}{\partial x^\nu}\left(\frac{\mathrm{i}}{\hbar}\hat{p}\right)^{n-1-\nu} \\
&= \sum_{\nu=0}^{n-1}\binom{n-\nu}{\nu}\left\{\frac{\mathrm{i}}{\hbar}\left[\hat{p},\frac{\partial^\nu \hat{B}}{\partial x^\nu}\right]\left(\frac{\mathrm{i}}{\hbar}\hat{p}\right)^{n-1-\nu}+\frac{\partial^\nu \hat{B}}{\partial x^\nu}\left(\frac{\mathrm{i}}{\hbar}\hat{p}\right)^{n-\nu}\right\} \\
&= \sum_{\nu=0}^{n-1}\binom{n-1}{\nu}\frac{\partial^{\nu+1}\hat{B}}{\partial x^{\nu+1}}\left(\frac{\mathrm{i}}{\hbar}\hat{p}\right)^{n-(\nu+1)}+\sum_{\nu=0}^{n-1}\binom{n-1}{\nu}\frac{\partial^\nu \hat{B}}{\partial x^\nu}\left(\frac{\mathrm{i}}{\hbar}\hat{p}\right)^{n-\nu} \\
&= \sum_{\nu=0}^{n-2}\binom{n-1}{\nu}\frac{\partial^{\nu+1}\hat{B}}{\partial x^{\nu+1}}\left(\frac{\mathrm{i}}{\hbar}\hat{p}\right)^{n-(\nu+1)}+\binom{n-1}{n-1}\frac{\partial^n \hat{B}}{\partial x^n} \\
&\quad+\binom{n-1}{0}\hat{B}\left(\frac{\mathrm{i}}{\hbar}\hat{p}\right)^n+\sum_{\nu=1}^{n-1}\binom{n-1}{\nu}\frac{\partial^\nu \hat{B}}{\partial x^\nu}\left(\frac{\mathrm{i}}{\hbar}\hat{p}\right)^{n-\nu} \quad .
\end{aligned} \tag{4}$$

We have separated the summations above. Now we use the identity

$$\binom{n-1}{0}=\binom{n-1}{n-1}=\binom{n}{0}=\binom{n}{n}=1 \quad ,$$

and introduce $\mu=\nu+1$ into the first summation; hence, we obtain

$$\begin{aligned}
\left(\frac{\mathrm{i}}{\hbar}\hat{p}\right)^n \hat{B}(x) &= \binom{n}{n}\frac{\partial^n \hat{B}}{\partial x^n}+\sum_{\mu=0}^{n-1}\binom{n-1}{\mu-1}\frac{\partial^\mu \hat{B}}{\partial x^\mu}\left(\frac{\mathrm{i}}{\hbar}\hat{p}\right)^{n-\mu} \\
&\quad+\sum_{\nu=1}^{n-1}\binom{n-1}{\nu}\frac{\partial^\nu \hat{B}}{\partial x^\nu}\left(\frac{\mathrm{i}}{\hbar}\hat{p}\right)^{n-\nu}+\binom{n}{0}\hat{B}\left(\frac{\mathrm{i}}{\hbar}\hat{p}\right)^n \\
&= \binom{n}{n}\frac{\partial^n \hat{B}}{\partial x^n}+\sum_{\nu=0}^{n-1}\left\{\binom{n-1}{\nu-1}+\binom{n-1}{\nu}\right\} \\
&\quad\times\frac{\partial^\nu \hat{B}}{\partial x^\nu}\left(\frac{\mathrm{i}}{\hbar}\hat{p}\right)^{n-\nu}+\binom{n}{0}\hat{B}\left(\frac{\mathrm{i}}{\hbar}\hat{p}\right)^n \quad .
\end{aligned} \tag{5}$$

In the last equation we relabelled μ as ν. Using

$$\binom{n-1}{\nu-1}+\binom{n-1}{\nu}=\binom{n}{\nu} \quad , \tag{6}$$

we can write (5) in the following way,

$$\left(\frac{\mathrm{i}}{\hbar}\hat{p}\right)^n \hat{B}=\sum_{\nu=0}^{n}\binom{n}{\nu}\frac{\partial^\nu \hat{B}}{\partial x^\nu}\left(\frac{\mathrm{i}}{\hbar}\hat{p}\right)^{n-\nu} \quad ,$$

which corresponds to (1).

To prove (2) we calculate

$$\hat{U}^\dagger \hat{A}(x)=\exp\left(\frac{\mathrm{i}}{\hbar}\varrho\hat{p}\right)\hat{A}(x) \quad . \tag{7}$$

Exercise 1.8

With the series expansion of the exponential function, it follows that

$$\begin{aligned}\hat{U}^{\dagger}\hat{A}(x) &= \sum_{n=0}^{\infty}\frac{1}{n!}\left(\frac{\mathrm{i}}{\hbar}\varrho\hat{p}\right)^{n}\hat{A}(x)\\ &= \sum_{n=0}^{\infty}\frac{\varrho^{n}}{n!}\sum_{\nu=0}^{n}\binom{n}{\nu}\frac{\partial^{\nu}\hat{A}}{\partial x^{\nu}}\left(\frac{\mathrm{i}}{\hbar}\hat{p}\right)^{n-\nu}\\ &= \sum_{n=0}^{\infty}\sum_{n=0}^{n}\frac{\varrho^{\nu}}{\nu!}\frac{\partial^{\nu}\hat{A}}{\partial x^{\nu}}\frac{1}{(n-\nu)!}\left(\frac{\mathrm{i}}{\hbar}\varrho\hat{p}\right)^{n-\nu}\\ &= \sum_{n=0}^{\infty}\frac{\varrho^{\nu}}{\nu!}\frac{\partial^{\nu}\hat{A}}{\partial x^{\nu}}\sum_{n=0}^{\infty}\frac{1}{m!}\left(\frac{\mathrm{i}}{\hbar}\varrho\hat{p}\right)^{m} \quad .\end{aligned} \tag{8}$$

Here we have used (1) and then separated ϱ^{n} into two parts. Finally we have written the result as a product of two infinite series. The first one represents the Taylor expansion of $\hat{A}(x+\varrho)$ and the second is $\hat{U}^{\dagger}$. Thus we obtain

$$\hat{U}^{\dagger}\hat{A}(x) = \hat{A}(x+\varrho)\hat{U}^{\dagger}$$

and therefore

$$\hat{U}^{\dagger}\hat{A}(x)\hat{U} = \hat{A}(x+\varrho) \quad . \tag{9}$$

In an analogous way, we find in three dimensions that for an operator $\hat{A}(\boldsymbol{r})$, under the transformation

$$\hat{U} = \exp[-(\mathrm{i}/\hbar)\boldsymbol{\varrho}\cdot\hat{\boldsymbol{p}}]$$

with the vector $\boldsymbol{\varrho}$,

$$U^{\dagger}\hat{A}(\boldsymbol{r})\hat{U} = \hat{A}(\boldsymbol{r}+\boldsymbol{\varrho}) \quad . \tag{10}$$

EXERCISE

1.9 Translation of an Operator $\hat{A}(x)$

Problem. Prove the validity of

$$\hat{U}^{\dagger}\hat{A}(\boldsymbol{r})\hat{U} = \hat{A}(\boldsymbol{r}+\boldsymbol{\varrho}) \quad \text{for} \tag{1}$$

$$\hat{U} = \exp\left(-\frac{\mathrm{i}}{\hbar}\boldsymbol{\varrho}\cdot\boldsymbol{p}\right) \quad ,$$

using the relations

$$\hat{U}b(\boldsymbol{r}) = b(\boldsymbol{r}-\boldsymbol{\varrho}) \quad \text{and} \tag{2}$$

$$\hat{U}^{\dagger}b(\boldsymbol{r}) = b(\boldsymbol{r}+\boldsymbol{\varrho}) \tag{3}$$

which hold for all functions $b(\boldsymbol{r})$ that depend only on $\boldsymbol{r}$.

Exercise 1.9

Solution. We choose an arbitrary function $\psi(\boldsymbol{r})$ and calculate $\hat{U}^{\dagger}\hat{A}(\boldsymbol{r})\hat{U}\psi(\boldsymbol{r})$. Using (2) this yields

$$\hat{U}^{\dagger}\hat{A}(\boldsymbol{r})\hat{U}\psi(\boldsymbol{r}) = \hat{U}^{\dagger}\hat{A}(\boldsymbol{r})\psi(\boldsymbol{r}-\boldsymbol{\varrho}) \quad . \tag{4}$$

Because the function $\hat{A}(\boldsymbol{r})\psi(\boldsymbol{r}-\boldsymbol{\varrho})$ depends only on the variable $\boldsymbol{r}$, we can apply (3) and arrive at the result that

$$\begin{aligned}\hat{U}^{\dagger}\hat{A}(\boldsymbol{r})\hat{U}\psi(\boldsymbol{r}) &= \hat{A}(\boldsymbol{r}+\boldsymbol{\varrho})\psi([\boldsymbol{r}-\boldsymbol{\varrho}]+\boldsymbol{\varrho}) \\ &= \hat{A}(\boldsymbol{r}+\boldsymbol{\varrho})\psi(\boldsymbol{r}) \quad . \end{aligned}\tag{5}$$

Since $\psi(\boldsymbol{r})$ is an arbitrary function, (5) will hold only if

$$\hat{U}^{\dagger}\hat{A}(\boldsymbol{r})\hat{U} = \hat{A}(\boldsymbol{r}+\boldsymbol{\varrho}) \quad .$$

EXERCISE

1.10 Generators for Translations in a Homogeneous Field

Problem. Derive the quantum mechanical operators for the translational symmetry of a charged particle in a homogeneous electric and magnetic field. Suppose that the generator of an infinitesimal transformation has an explicit time dependence. Which relation must be fulfilled if this transformation is a symmetry transformation? Discuss this for a constant electric field [cf. Exercise 1.6, case (a)].

Solution. (a) We consider the infinitesimal transformation

$$\psi \to \psi' = \psi - \mathrm{i}\hat{F}\delta a\psi \quad \text{or} \tag{1a}$$

$$\psi = \psi' + \mathrm{i}\hat{F}\delta a\psi' \tag{1b}$$

up to terms of order δa^2, with the time-independent operator $\hat{F}(t)$. The original wave function ψ obeys the Schrödinger equation

$$\mathrm{i}\hbar\frac{\partial}{\partial t}\psi = \hat{H}\psi \quad . \tag{2}$$

The transformed wave function then fulfills

$$\begin{aligned}\mathrm{i}\hbar\frac{\partial}{\partial t}\psi' &= \mathrm{i}\hbar\frac{\partial}{\partial t}\psi + \hbar\frac{\partial}{\partial t}(\hat{F}\delta a\psi) \\ &= \hat{H}\psi + \hbar\frac{\partial\hat{F}}{\partial t}\delta a\psi + \hbar\hat{F}\delta a\frac{\partial\psi}{\partial t} \\ &= \hat{H}\psi + \hbar\frac{\partial\hat{F}}{\partial t}\delta a\psi - \mathrm{i}\hat{F}\delta a\hat{H}\psi \\ &= \hat{H}\psi' + \left(\hbar\frac{\partial\hat{F}}{\partial t} + \mathrm{i}\hat{H}\hat{F} - \mathrm{i}\hat{F}\hat{H}\right)\delta a\psi' \end{aligned}\tag{3}$$

Exercise 1.10

neglecting terms of the order $(\delta a)^2$. This is just the original Schrödinger equation if the operator $\hat{F}$ satisfies the condition

$$\frac{d\hat{F}}{dt} \equiv \frac{\partial \hat{F}}{\partial t} + \frac{\mathrm{i}}{\hbar}[\hat{H}, \hat{F}] = 0 \quad , \tag{4}$$

i.e. if the total time derivative of the operator $\hat{F}(t)$ vanishes. Otherwise the Hamiltonian is transformed according to

$$\hat{H} \to \hat{H}' = \hat{H} + \hbar \frac{d\hat{F}}{dt} \delta a \quad . \tag{5}$$

If $\hat{F}$ has no explicit time dependence, then (4) just means that $\hat{F}$ and $\hat{H}$ have to commute with each other.

(b) In the gauge

$$\phi = -\boldsymbol{E}\cdot\boldsymbol{x} \quad , \quad \boldsymbol{A} = 0 \tag{6}$$

the classical conservation law

$$\frac{d}{dt}(\boldsymbol{p} - q\boldsymbol{E}t) = 0 \tag{7}$$

that we found in Exercise 1.6, Eq. (9a) suggests that we regard the quantity

$$\hat{\boldsymbol{F}}(t) = \hat{\boldsymbol{p}} - q\boldsymbol{E}t \tag{8}$$

as the operator which generates a symmetry transformation. In this case the Hamiltonian is given by

$$\hat{H} = \frac{1}{2m}\left(\hat{\boldsymbol{p}} - \frac{q}{c}\hat{\boldsymbol{A}}\right)^2 + q\phi = \frac{1}{2m}\hat{\boldsymbol{p}}^2 - q\boldsymbol{E}\cdot\boldsymbol{x} \quad . \tag{9}$$

One easily verifies that the total derivative of $\hat{F}$ vanishes:

$$\begin{aligned}
\frac{d\hat{\boldsymbol{F}}}{dt} &= \frac{\partial \hat{\boldsymbol{F}}}{\partial t} + \frac{\mathrm{i}}{\hbar}[\hat{H}, \hat{\boldsymbol{F}}] \\
&= -q\boldsymbol{E} + \frac{\mathrm{i}}{\hbar}[-q\boldsymbol{E}\cdot\hat{\boldsymbol{x}}, \hat{\boldsymbol{p}}] \\
&= -q\boldsymbol{E} + \frac{\mathrm{i}}{\hbar}(-q\boldsymbol{E}\mathrm{i}\hbar) = 0 \quad .
\end{aligned} \tag{10}$$

Note that the commutation relations of $\hat{\boldsymbol{F}}$ and $\hat{\boldsymbol{x}}$ are the same as those of $\hat{\boldsymbol{p}}$ and $\hat{\boldsymbol{x}}$:

$$[\hat{F}_i, \hat{x}_k] = [\hat{p}_i - qE_i t, \hat{x}_k] = [\hat{p}_i, \hat{x}_k] = -\mathrm{i}\hbar\delta_{ik} \tag{11}$$

so that the operators $\hat{\boldsymbol{x}}$ change in the same way for the transformations generated by $\hat{F}_i$ and $\hat{p}_i$, respectively.

Exercise 1.10

(c) For the case of a constant magnetic field we have found (cf. Exercise 1.6):

$$A = \tfrac{1}{2}\boldsymbol{B}\times\boldsymbol{x} \quad , \quad \phi = 0 \text{ and} \tag{12}$$

$$\frac{d}{dt}\left(\boldsymbol{p} - \frac{q}{2c}\boldsymbol{x}\times\boldsymbol{B}\right) = 0 \quad . \tag{13}$$

Therefore we consider the operator

$$\hat{\boldsymbol{F}} = \hat{\boldsymbol{p}} - \frac{q}{2c}\hat{\boldsymbol{x}}\times\boldsymbol{B} \tag{14}$$

which does not depend explicitly on the time. The Hamiltonian is given by

$$\begin{aligned}\hat{H} &= \frac{1}{2m}\left(\hat{\boldsymbol{p}} - \frac{q}{c}\hat{\boldsymbol{A}}\right)^2 = \frac{1}{2m}\left(\hat{\boldsymbol{p}} - \frac{q}{2c}\boldsymbol{B}\times\hat{\boldsymbol{x}}\right)^2 \\ &= \frac{1}{2m}\left(\hat{\boldsymbol{p}} + \frac{q}{2c}\hat{\boldsymbol{x}}\times\boldsymbol{B}\right)^2 \quad ,\end{aligned} \tag{15}$$

and we calculate the commutator

$$\begin{aligned}\left[\hat{p}_k + \frac{q}{2c}(\hat{\boldsymbol{x}}\times\boldsymbol{B})_k, F_i\right] &= \left[\hat{p}_k + \frac{q}{2c}(\hat{\boldsymbol{x}}\times\boldsymbol{B})_k, \hat{p}_i - \frac{q}{2c}(\hat{\boldsymbol{x}}\times\boldsymbol{B})_i\right] \\ &= -\frac{q}{2c}([\hat{p}_k, (\hat{\boldsymbol{x}}\times\boldsymbol{B})_i] + [\hat{p}_i, (\hat{\boldsymbol{x}}\times\boldsymbol{B})_k]) \quad .\end{aligned} \tag{16}$$

Obviously it suffices to calculate only the first commutator explicitly:

$$\begin{aligned}[\hat{p}_k, (\hat{\boldsymbol{x}}\times\boldsymbol{B})_i] &= \sum_{lm}[\hat{p}_k, \varepsilon_{ilm}\hat{x}_l B_m] \\ &= \sum_{lm}\varepsilon_{ilm}(-\mathrm{i}\hbar\delta_{kl})B_m \\ &= -\mathrm{i}\hbar\sum_m \varepsilon_{ikm}B_m \quad .\end{aligned} \tag{17}$$

Thus we get

$$\left[\hat{p}_k + \frac{q}{2c}(\hat{\boldsymbol{x}}\times\boldsymbol{B})_k, F_i\right] = -\frac{q}{2c}(-\mathrm{i}\hbar)\sum_m(\varepsilon_{ikm} + \varepsilon_{kim})B_m = 0 \tag{18}$$

because of the antisymmetry of ε_{ikm}. Therefore we find

$$[\hat{H}, \hat{\boldsymbol{F}}] = 0 \quad , \tag{19}$$

so that $\hat{\boldsymbol{F}}$ is the operator which generates a symmetry transformation as we have conjectured. Obviously, again we find

$$[\hat{F}_i, \hat{x}_k] = [\hat{p}_i, \hat{x}_k] = -\mathrm{i}\hbar\delta_{ik} \quad . \tag{20}$$

Exercise 1.10

However, in contrast with $\hat{\boldsymbol{p}}$ the individual components of $\hat{\boldsymbol{F}}$ do not commute with each other:

$$
\begin{aligned}
[\hat{F}_i, \hat{F}_k] &= \left[\hat{p}_i - \frac{q}{2c}(\hat{\boldsymbol{x}} \times \boldsymbol{B})_i, \hat{p}_k - \frac{q}{2c}(\hat{\boldsymbol{x}} \times \boldsymbol{B})_k\right] \\
&= -\frac{q}{2c}([\hat{p}_i, (\hat{\boldsymbol{x}} \times \boldsymbol{B})_k] - [\hat{p}_k, (\hat{\boldsymbol{x}} \times \boldsymbol{B})_i]) \\
&= -\frac{q}{2c}(-\mathrm{i}\hbar)(\varepsilon_{ikm} - \varepsilon_{kim}) B_m \\
&= \mathrm{i}\hbar \frac{q}{c} \varepsilon_{ikm} B_m \quad . \qquad (21)
\end{aligned}
$$

The two generators perpendicular to the magnetic field do not commute with each other, so that they cannot be diagonalized simultaneously.

1.6 Time Displacements in Quantum Mechanics

We shall now investigate the temporal displacement of a state $\psi_\alpha(\boldsymbol{r}, t)$ by a time interval τ (Fig. 1.7). We call the time shifted state $\psi_{\alpha'}(\boldsymbol{r}, t)$. With analogy to (1.7) it follows that

$$\psi_{\alpha'}(\boldsymbol{r}, t + \tau) = \psi_\alpha(\boldsymbol{r}, t) \quad . \qquad (1.36)$$

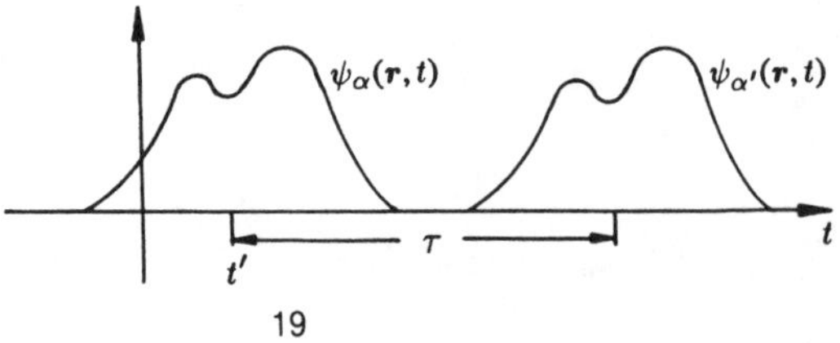

Fig. 1.7. The time-displaced wave function $\psi_{\alpha'}(\boldsymbol{r}, t)$ possesses the same value for $t = t' + \tau$, as the initial wave function $\psi_\alpha(\boldsymbol{r}, t)$ for $t = t'$

Let us better understand this: If the initial wave function has its maximum at $t = t_0$, the time-shifted wave function $\psi_{\alpha'}(\boldsymbol{r}, t)$ will have its maximum at $t = t_0 + \tau$, etc. We express the connection between the time-displaced and initial state by the time evolution operator $\hat{U}_t(\tau)$, with analogy to (1.21),

$$\psi_{\alpha'}(\boldsymbol{r}, t) = \hat{U}_t(\tau)\psi_\alpha(\boldsymbol{r}, t) = \psi_\alpha(\boldsymbol{r}, t - \tau) \quad . \qquad (1.37)$$

In comparison to our former investigations (1.22) to (1.24), the Taylor expansion of $\psi_\alpha(\boldsymbol{r}, t-\tau)$ yields

$$\begin{aligned}\hat{U}_t(\tau)\psi_\alpha(\boldsymbol{r},t) &= \psi_\alpha(\boldsymbol{r},t) + \frac{(-\tau)}{1!}\frac{d}{dt}\psi_\alpha(\boldsymbol{r},t) + \frac{(-\tau)^2}{2!}\frac{d}{dt^2}\psi_\alpha(\boldsymbol{r},t) + \dots \\ &= \left[\mathbb{1} + \frac{(-\tau)}{1!}\frac{d}{dt} + \frac{(-\tau)^2}{2!}\frac{d^2}{dt^2} + \dots\right]\psi_\alpha(\boldsymbol{r},t) \\ &= \mathrm{e}^{-\tau(\partial/\partial t)}\psi_\alpha(\boldsymbol{r},t) \quad . \end{aligned} \tag{1.38}$$

Hence it follows that

$$\hat{U}_t(\tau) = \mathrm{e}^{-\tau(\partial/\partial t)} = \exp\left(\frac{\mathrm{i}}{\hbar}\tau\hat{E}\right) = \exp\left(\frac{\mathrm{i}}{\hbar}\tau\hat{H}\right) \quad , \tag{1.39}$$

where we have introduced the energy operator $\hat{E} = \mathrm{i}\hbar(\partial/\partial t)$ $(=\hat{H})$. $\hat{U}_t(\tau)$ is unitary, because

$$\hat{U}_t^\dagger(\tau) = \exp\left(\frac{\mathrm{i}}{\hbar}\tau\hat{E}^\dagger\right) = \exp\left(\frac{\mathrm{i}}{\hbar}\tau\hat{E}\right) = \hat{U}_t^{-1}(\tau) \quad .$$

There we used the fact that $\hat{E}$ is hermitian so that $\hat{E} = \hat{E}^\dagger$. We can use each one of the forms of (1.39), but the last expression will only be valid, if the Hamiltonian $\hat{H}$ is time independent. This can be seen in the following way: From the Schrödinger equation we have

$$\frac{\partial}{\partial t}\psi = \frac{1}{\mathrm{i}\hbar}\hat{H}\psi \quad , \quad \text{but}$$

$$\frac{\partial^2}{\partial t^2}\psi = \frac{\partial}{\partial t}\left(\frac{1}{\mathrm{i}\hbar}\hat{H}\psi\right) = \frac{1}{\mathrm{i}\hbar}\hat{H}\frac{\partial\psi}{\partial t} + \frac{1}{\mathrm{i}\hbar}\frac{\partial\hat{H}}{\partial t}\psi = \frac{1}{(\mathrm{i}\hbar)^2}\hat{H}^2\psi + \frac{1}{\mathrm{i}\hbar}\frac{\partial\hat{H}}{\partial t}\psi \quad .$$

We notice that the higher derivatives with respect to time can only be replaced by powers of $\hat{H}$ if $\partial\hat{H}/\partial t = 0$. Then $\hat{H}$ is time independent and so (1.39) holds, implying that

$$[\hat{H}, \hat{U}_t(\tau)]_- = \left[\hat{H}, \exp\left(\frac{\mathrm{i}}{\hbar}\tau\hat{H}\right)\right]_- = 0 \quad . \tag{1.40}$$

This relation is analogous to (1.30). It implies that the temporally displaced state will also satisfy the Schrödinger equation. Indeed, from

$$\mathrm{i}\hbar\frac{\partial\psi_\alpha(\boldsymbol{r},t)}{\partial t} = \hat{H}\psi_\alpha(\boldsymbol{r},t) \quad ,$$

it follows that

$$\begin{aligned}\mathrm{i}\hbar\frac{\partial\psi_{\alpha'}(\boldsymbol{r},t)}{\partial t} &= \mathrm{i}\hbar\frac{\partial}{\partial t}\hat{U}_t\psi_\alpha(\boldsymbol{r},t) = \mathrm{i}\hbar\hat{U}_t(\tau)\frac{\partial\psi_\alpha}{\partial t} \\ &= \hat{U}_t(\tau)\hat{H}\psi_\alpha = \hat{H}\hat{U}_t(\tau)\psi_\alpha = \hat{H}\psi_{\alpha'}(\boldsymbol{r},t) \quad . \end{aligned} \tag{1.41}$$

Alternatively, we may say that the assumption that the time-displaced state satisfies the Schrödinger equation implies (1.40), and vice versa. The assumption that the Hamiltonian $\hat{H}$ commutes with the time displacement operator $\hat{U}_t(\tau)$ ensures that the temporally shifted state $\psi_{\alpha'}(\boldsymbol{r}, t)$ also satisfies the Schrödinger equation. *Then the system exhibits symmetry* (also called *invariance) against time displacements.* Because this is equivalent to the time independence of the Hamiltonian, *energy conservation* is valid, as we know from classical physics, due to the homogeneity of time [cf. (1.8)]. Otherwise, if $\hat{H}$ is explicitly time dependent, the conclusion (1.40) does not hold and the time-shifted state does not satisfy the same Schrödinger equation as the original state.

Let us draw attention to an apparent contradiction; we know from (1.19) that the time development of the state $\psi_\alpha(\boldsymbol{r}, 0)$, known at the time $t = 0$, may be written

$$\psi_\alpha(\boldsymbol{r}, t) = \exp\left(-\frac{\mathrm{i}}{\hbar}\hat{H}t\right)\psi_\alpha(\boldsymbol{r}, 0) \quad . \tag{1.42}$$

On the other hand we find from (1.37) and (1.39) *with temporally constant* $\hat{H}$

$$\psi_{\alpha'}(\boldsymbol{r}, t) = \exp\left(\frac{\mathrm{i}}{\hbar}\hat{H}\tau\right)\psi_\alpha(\boldsymbol{r}, t) \quad . \tag{1.43}$$

At first glance both statements may appear inconsistent. This is clarified when we consider that the time shift τ in (1.42) has been denoted by t and the time-shifted state has the index α while it carries the index α' in (1.37)! Hence we replace $t = -\tau$ in (1.42) and obtain

$$\psi_\alpha(\boldsymbol{r}, -\tau) = \exp\left(\frac{\mathrm{i}}{\hbar}\hat{H}\tau\right)\psi_\alpha(\boldsymbol{r}, 0) \quad . \tag{1.44}$$

Now, in (1.43) we also fix the initial time at $t = 0$. This yields

$$\psi_{\alpha'}(\boldsymbol{r}, 0) = \exp\left(\frac{\mathrm{i}}{\hbar}\hat{H}\tau\right)\psi_\alpha(\boldsymbol{r}, 0) \quad . \tag{1.45}$$

and finally we express $\psi_{\alpha'}(\boldsymbol{r}, 0)$ through $\psi_\alpha(\boldsymbol{r}, -\tau)$, using (1.32),

$$\psi_\alpha(\boldsymbol{r}, -\tau) = \exp\left(\frac{\mathrm{i}}{\hbar}\hat{H}\tau\right)\psi_\alpha(\boldsymbol{r}, 0) \quad .$$

which is clearly identical with (1.44). Thus we can state that $\psi_{\alpha'}(\boldsymbol{r}, t)$ has the same structure at $t = \tau$, as $\psi_\alpha(\boldsymbol{r}, t)$ at $t = 0$. Thereby one can obtain $\psi_{\alpha'}(\boldsymbol{r}, \tau)$ from $\psi_\alpha(\boldsymbol{r}, \tau)$ by reverse evolution from time $t = \tau$ to time $t = 0$.

1.7 Mathematical Supplement: Definition of a Group

Symmetries are conveniently described by group theory, and in order to get acquainted with group theoretical methods we need to discuss some definitions. The elements of the set $\{a, b, c, \dots\}$ form a group, if a combination $a \circ b$ of these elements can be found, which is called *multiplication* and satisfies the following conditions*:

1. The product (ab) is also an element of the group $G = \{a, b, c, \dots\}$ for all a and b. In other words, the group is *closed* with respect to multiplication.
2. The set $G = \{a, b, c \dots\}$ contains a *unit element* e, which satisfies
   $$ae = ea = a \quad ,$$
   where a is an arbitrary element of the group.
3. To each element of the group there exists an *inverse element* a^{-1}, which fulfills the condition that
   $$a^{-1}a = aa^{-1} = e \quad .$$
4. The multiplication is *associative*, i.e.
   $$(ab)c = a(bc) \quad .$$

Examples. The set of the integers $N = \{0, \pm 1, \pm 2, \dots\}$ form a group with respect to addition $+$. The group multiplication is normal addition, and the unit element is obviously $e = 0$. The inverse element of a is $-a$ and the associativity of addition is well known.

Some further terms associated with groups are:

1) We will call a group *abelian*, if we can write

$$ab = ba \quad ,$$

for every element of the group, i.e. if group multiplication is commutative.

Examples. The addition of integers yields an abelian group. However, the group of square matrices with dimension N under multiplication is not abelian.

2) We will call a group *continuous*, if its elements are functions of one or more continuous variables, e.g. $G = \{a(t),\ b(t),\ c(t), \dots\}$, where t is a continuous parameter.

Examples. The group of spatial displacements within a plane (vector addition) is continuous along with the group of temporal displacements.

3) We will call a group *continuously connected*, if a continuous variation of the group parameters leads from any arbitrary element of the group to any other.

Examples. The translation group of the elements $\{\boldsymbol{a} = a_x\boldsymbol{e}_1 + a_y\boldsymbol{e}_2 + a_z\boldsymbol{e}_3\}$ possesses three continuous parameters a_x, a_y, a_z. We can generate each displacement vector in space by continuous variation of these parameters. However, rotations combined with reflection in space [called O(3)], form a continuous,

---

* The combination-sign $\circ$ is often omitted and instead of $a \circ b$ we write ab. We will use the combination-sign only if misunderstandings might be expected.

not connected group. We obtain the same result for the group of Lorentz transformations, which contain reflections in time and in space as discontinuous parts.

4) In each sequence within a *compact* group there exists an infinite partial sequence $\{a_n\}$ of group elements, which converges to an element of the group, i.e.

$$\lim_{n\to\infty} a_n = a \quad , \quad a \in G$$

Examples. (a) The group of the translation vectors on a lattice

$$\bar{T} = \{\boldsymbol{a}_n = n_1\boldsymbol{e}_1 + n_2\boldsymbol{e}_2 + n_3\boldsymbol{e}_3\} \quad , \quad \text{with } n_1, n_2, n_3 \in N$$

is discontinuous and not compact, because $\lim\limits_{n\to\infty} a_n$ does not belong to $\bar{T}$ and (in usual metrics) does not even exist.

(b) The group of rotations in three-dimensional space $\mathrm{SO}(3) = \{\boldsymbol{n}, \varphi$ with $|\boldsymbol{n}| = 1, 0 \leq \varphi \leq \pi\}$ is compact. Here $\boldsymbol{n}$ denotes the axis of rotation, φ the angle. Any vector $\boldsymbol{n}' = \lim\limits_{i\to\infty} \boldsymbol{n}_i$ is again a unit vector, and the range of angles is closed, hence containing the limit of any series of angles.

5) Two groups $\{a, b, c, \ldots\}$ and $\{a', b', c', \ldots\}$ are called isomorphic, if a bijective transformation between the elements of both groups exists (e.g. $a \leftrightarrow a'$, $b \leftrightarrow b'$, etc.), such that

$$ab \leftrightarrow a'b' \quad , \quad \text{etc.}$$

Products of the first group are thereby uniquely associated with products of the second group and vice versa.

Example. The group of spatial displacements is isomorphic to the group of three-dimensional vectors with vector addition as the group operation. Also the group of the translation operators $U = \{\hat{U}_r(\boldsymbol{\varrho})\}$ is isomorphic to the group of three-dimensional vectors and their addition as group operation.

6) If a group G_1 is isomorphic to another group G_2, whose elements are matrices, G_2 is said to be the *matrix representation* of G_1.

Example. If $G_1 = \{\hat{U}_r(\boldsymbol{\varrho})\}$ is the group of the displacement operators with three parameters and $\psi_n(\boldsymbol{r})$ is a complete set of wave functions, then the matrices

$$\langle m|\hat{U}_r(\boldsymbol{\varrho})|n\rangle = \left\langle m\left|\exp\left(-\frac{\mathrm{i}}{\hbar}\boldsymbol{\varrho}\cdot\hat{\boldsymbol{p}}\right)\right|n\right\rangle = \int \psi_m^*(\boldsymbol{r})\exp\left(-\frac{\mathrm{i}}{\hbar}\boldsymbol{\varrho}\cdot\hat{\boldsymbol{p}}\right)\psi_n(\boldsymbol{r})\,d^3r$$

will be a matrix representation of the displacement operators.

More examples and explanations appear in the following section.

1.8 Mathematical Supplement: Rotations and their Group Theoretical Properties

As in classical mechanics, isotropy of space means invariance of the laws of nature – in our case the Schrödinger equation – against any rotation of the wave function. Let us call the initial wave function $\psi_\alpha(\boldsymbol{r}, t)$. It is transformed to $\psi_{\alpha'}(\boldsymbol{r}, t)$ by an (active) rotation, in which the position vector is transformed from $\boldsymbol{r}$ into $\boldsymbol{r}'$ (see Fig. 1.8). This section covers the mathematical knowledge required to discuss the physical consequences of the isotropy of space.

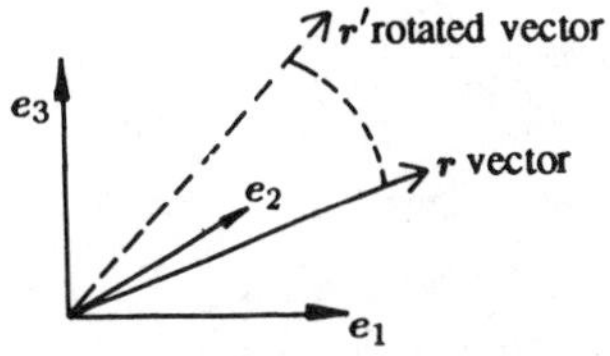

Fig. 1.8. The rotation of a vector $\boldsymbol{r}$ yields $\boldsymbol{r}'$. The rotated vector $\boldsymbol{r}'$ is described within the initial frame of the unrotated basis $\{\boldsymbol{e}_i\}$

To describe the additional rotation of a vector $\boldsymbol{r}$ towards $\boldsymbol{r}'$ we write

$$\boldsymbol{r}' = \hat{\tilde{R}}\boldsymbol{r} \quad . \tag{1.46}$$

We elucidate this equation with the following considerations. The three orthonormal basis vectors $\boldsymbol{e}_i$ are rotated into $\boldsymbol{e}_i'$ (Fig. 1.9). We know[6] that

$$\boldsymbol{e}_i' = R_{ij}\boldsymbol{e}_j \quad , \tag{1.47}$$

where we use Einstein's sum convention and always sum over all indices occurring twice. Thus (1.47) signifies that

$$\boldsymbol{e}_i' = R_{i_1}\boldsymbol{e}_1 + R_{i_2}\boldsymbol{e}_2 + R_{i_3}\boldsymbol{e}_3 \quad .$$

The 3×3 matrix R_{ij} must be real, because the basis vectors are real. We will denote the inverse transformation of (1.47) as

$$\boldsymbol{e}_i = U_{ij}\boldsymbol{e}_j' \quad . \tag{1.48}$$

From (1.47) it follows by scalar multiplication with $\boldsymbol{e}_k$ that

$$\boldsymbol{e}_i' \cdot \boldsymbol{e}_k = R_{ij}\boldsymbol{e}_j \cdot \boldsymbol{e}_k = R_{ij}\delta_{jk} = R_{ik} \tag{1.49}$$

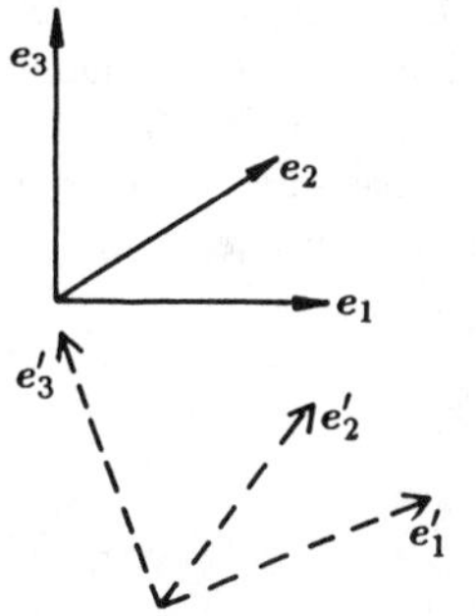

Fig. 1.9. Rotation of the coordinate system

and simply from (1.48) by scalar multiplication with $\boldsymbol{e}_k'$

$$\boldsymbol{e}_i \cdot \boldsymbol{e}_k' = U_{ij}\boldsymbol{e}_j' \cdot \boldsymbol{e}_k' = U_{ij}\delta_{jk} = U_{ik} \quad . \tag{1.50}$$

Hence the matrix elements R_{ik} and U_{ik}, respectively, are the *directional cosines*. Making a comparison between the final parts of (4) and (5) yields the result that

$$U_{ik} = R_{ki} \quad ,$$

or since $\hat{U}$ is the inverse matrix of $\hat{R}$ by definition (1.48),

$$(\hat{R}^{-1})_{ik} = R_{ki} \quad . \tag{1.51a}$$

Written in matrix representation,

$$\hat{R}^{-1} = \hat{\tilde{R}} \quad , \tag{1.51b}$$

[6] See H. Goldstein: *Classical Mechanics*, 2nd ed. (Addison-Wesley, Reading 1980); W. Greiner: *Theoretische Physik, Mechanik I* (Harri Deutsch, Frankfurt 1989) Chap. 30.

$\hat{\tilde{R}}$ is the transpose (i.e. reflected along the main diagonal) of the matrix $\hat{R}$. From the orthonormality of both basis sets we conclude that

$$\delta_{jk} = e'_i \cdot e'_k = R_{im} R_{kn} e_m \cdot e_n = R_{im} R_{kn} \delta_{mn} = R_{im} R_{km} \quad . \tag{1.52}$$

This implies the *orthogonality of the rows* of the matrix $\hat{R}$. Equally from (1.48) and (1.51) it follows that

$$\delta_{ik} = e_i \cdot e_k = U_{im} U_{kn} e'_m \cdot e'_n = U_{im} U_{kn} \delta_{mn} = U_{im} U_{km} = R_{mi} R_{mk} \quad , \tag{1.53}$$

implying the *orthogonality of the columns* of $\hat{R}$. In order to obtain an expression for the transformation of the vector components of $\boldsymbol{r}$ we note that $\boldsymbol{r}$ rotates along with its frame. Accordingly $\boldsymbol{r}'$ has the same components in $\{e'_i\}$ as $\boldsymbol{r}$ in $\{e_i\}$ (Fig. 1.10). These considerations lead us to the following equations:

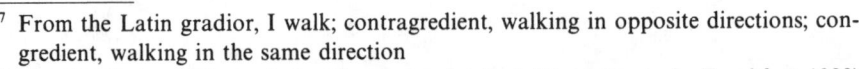

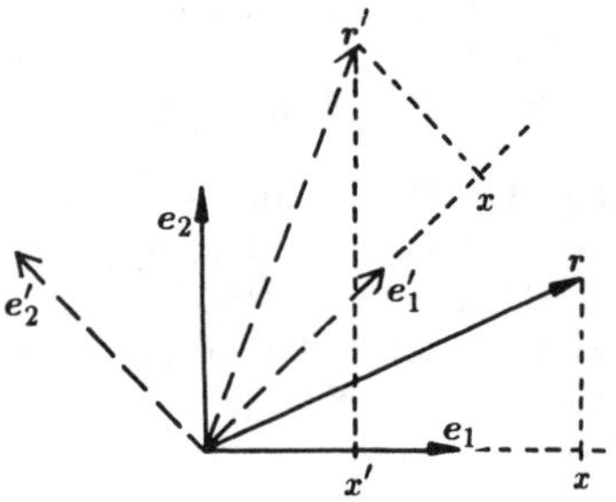

Fig. 1.10. The rotated vector $\boldsymbol{r}'$ has the same components in the rotated base $\{e'_i\}$ as the initial vector $\boldsymbol{r}$ in the initial base $\{e_i\}$

$$\boldsymbol{r} = x_i e_i \quad , \quad \boldsymbol{r}' = x_i e'_i = x'_i e_i \quad \text{or} \quad x_i e'_i = x'_i (\hat{R}^{-1})_{ij} e'_j \quad .$$

Relabelling the summation indices, we find

$$x_i e'_i = x'_j (\hat{R}^{-1})_{ji} e'_j \quad , \tag{1.54}$$

which yields due to (1.51b)

$$x_i = (\hat{R}^{-1})_{ji} x'_j = (\hat{\tilde{R}})_{ji} x'_j = R_{ij} x'_j \quad . \tag{1.55}$$

Thus the components are transformed *contragrediently*[7] with respect to the basis set, if an *active* rotation is performed. Elsewhere[8] we have only investigated *passive* rotations of the coordinate frame, where the vector $\boldsymbol{r}$ remained fixed in space ($\boldsymbol{r} = \boldsymbol{r}'$), whilst the coordinate system rotated. Accordingly components and basis set were transformed in the same way, i.e. *congrediently*.

The inverse transformation of (1.55) can be calculated immediately with respect to the orthonormality of the matrix R_{ij} or by the substitution of e'_i in (1.54), i.e.

$$x'_i = R_{ji} x_j = (\hat{\tilde{R}})_{ij} x_j \quad , \tag{1.56}$$

which is identical to (1.46). Matrices with row and column orthogonality, such as

$$R_{ij} R_{ij'} = \delta_{jj'} \quad , \quad R_{ji} R_{j'i} = \delta_{jj'} \quad , \tag{1.57}$$

are called *orthogonal matrices*. Because the determinant of a product of square matrices is equal to the product of the determinants of the individual square matrices, it follows that

$$\det(R_{ij} R'_{ij}) = \det(R_{ij}) \det(R'_{ij}) = \det \delta_{jj'} \quad ,$$

[7] From the Latin gradior, I walk; contragredient, walking in opposite directions; congredient, walking in the same direction

[8] See W. Greiner: *Theoretische Physik, Mechanik I* (Harri Deutsch, Frankfurt 1989) Chap. 30 or H. Goldstein: *Classical Mechanics*, 2nd ed. (Addison-Wesley, Reading 1980)

which implies $(\det(R_{ij}))^2 = 1$, and hence

$$\det(R_{ij}) = \pm 1 \quad . \tag{1.58}$$

Thus the rotations can be separated into two disconnected sets; namely, one with $\det(R_{ij}) = +1$, which forms a group, and the other with $\det(R_{ij}) = -1$, which does not form a group by itself (it is not closed and contains no unit element). The first are called *proper*, the second *improper* rotations. The rotation matrix R_{ij} contains $3 \times 3 = 9$ real arguments. Only three of them are independent due to the six requirements (1.57). Accordingly the rotations can be described by three independent parameters (three components of the rotation vector). They form a *continuously connected, three-parameter group*[9].

This can easily be seen, because matrix multiplication of the matrices R_{ij} is associative, the R_{ij} contain the unity matrix δ_{ij} as identity element and each R_{ij} has an inverse element $(R^{-1})_{ij} = R_{ji}$. The rotation group is not Abelian, because in general two rotations in different directions do not commute. This is illustrated in Fig. 1.11: The cases (a) and (b) show the same two rotations carried out in different order. The resulting vector $\hat{a}'$ is different in both case.

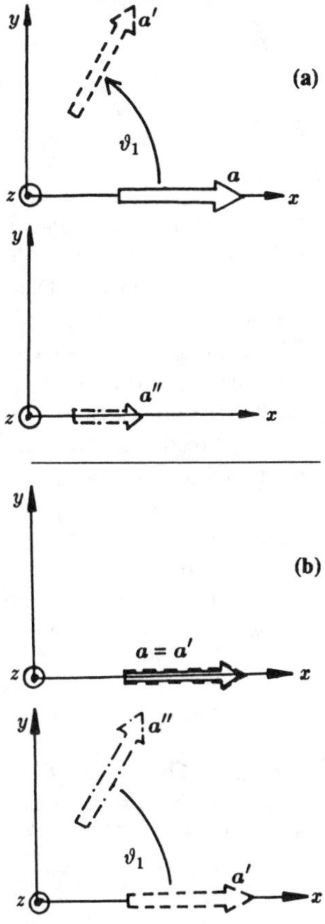

Fig. 1.11. (a). Rotation about the z axis through θ_1, $\boldsymbol{a} \to \boldsymbol{a}'$. 2. Rotation about the x axis through $\pi/2$: $\boldsymbol{a}' \to \boldsymbol{a}''$. (The vector $\boldsymbol{a}''$ lies in the xz plane). **(b)**. 1. Rotation about the $\boldsymbol{x}$ axis through $\pi/2$ does not change $\boldsymbol{a}$, $\boldsymbol{a} = \boldsymbol{a}'$. 2. Rotation about the z axis through θ_1, $\boldsymbol{a}' \to \boldsymbol{a}''$. (The vector $\boldsymbol{a}''$ lies in the xy plane)

The mathematical reason for a non-abelian character of the rotation group is given by the fact that matrix multiplication is not commutative in general, so that

$$R^{(1)}_{ij} R^{(2)}_{jk} \neq R^{(2)}_{ij} R^{(1)}_{jk} \quad .$$

It will be shown later that the rotation matrices $\hat{R}^{(1)}$ and $\hat{R}^{(2)}$ really do not commute. The (special) rotation group is denoted as SO(3). This indicates an *orthogonal group in three dimensions*, which includes all real 3×3 matrices with determinant equal to $+1$. Here S stands for *special*, denoting orthonormal rotations with $\det R_{ij} = +1$. SO(3) is a typical example of a ***Lie group***, which is defined as a continuous group for which the elements are differentiable functions of their parameters. For example the group of space and time displacements is a non-compact Lie group, while rotations form a compact Lie group. The Lorentz group[10] is a six-parameter Lie group (3 parameters for rotations, 3 for velocities $\boldsymbol{v} = \{v_x, v_y, v_z\}$ of the inertial systems). It is non-compact, because there exists no Lorentz transformation for $v = c$, whereas a sequence with $v_n \to c$ can be found.

1.9 An Isomorphism of the Rotation Group

Each orthonormal 3×3 matrix $\hat{R}$ corresponds to a rotation, which can be represented by a rotation vector $\boldsymbol{f} = \{\phi_x, \phi_y, \phi_z\}$, giving the direction of the

[9] Only proper rotations represent a continuously connected group. Improper rotations contain a reflection in space, which is a discrete transformation. Therefore all rotations including improper ones form a disconnected group.

[10] See H. Goldstein: *Classical Mechanics*, 2nd ed. (Addison-Wesley, Reading 1980) or W. Greiner: *Theoretische Physik, Mechanik I* (Harri Deutsch, Frankfurt 1989)

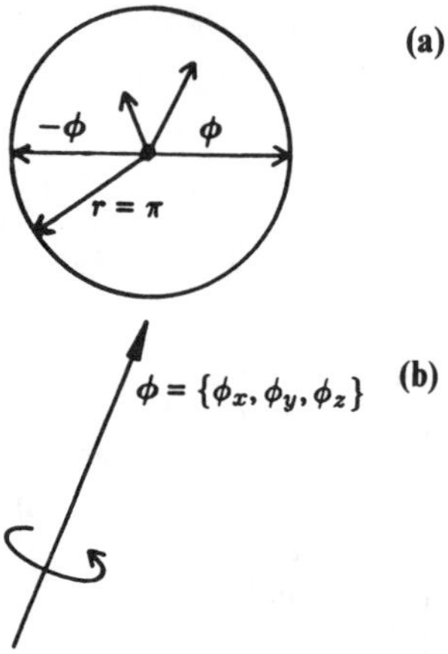

Fig. 1.12. **(a)** The end points of the rotation vectors ϕ fill a sphere of radius π. Diametrically opposed points on the surface of the sphere describe the same rotation. **(b)** The rotation vector describes direction and magnitude of the rotation angle $\phi = [\phi_x^2 + \phi_y^2 + \phi_z^2]^{1/2}$. Rotation and rotation vector are connected in the sense of a right handed screw

rotation axis and the magnitude of the rotation. Arranging all possible rotation vectors ϕ around the origin we find the end points of all vectors ϕ lying within a sphere of radius π, because the maximal rotation is given by $\phi = |\boldsymbol{\phi}| = \pi$: Each rotation about an axis can be expressed through a rotation angle $-\pi \leqslant \phi \leqslant \pi$. The rotations $\boldsymbol{\phi}$ and $-\boldsymbol{\phi}$ with $|\boldsymbol{\phi}| = \pi$ describe the same rotation. The end points lie diametrically opposed on the surface of the sphere (see Figs. 1.12,13).

Note that the result $\boldsymbol{\phi}$ of two successive rotations $\boldsymbol{\phi}_1$ and $\boldsymbol{\phi}_2$ is not obtained by vector addition. $\boldsymbol{\phi} \neq \boldsymbol{\phi}_1 + \boldsymbol{\phi}_2$, except for rotations about the same axis, i.e. collinearity of $\boldsymbol{\phi}_1$ and $\boldsymbol{\phi}_2$. In general the resulting rotation vector can only be determined by matrix multiplication of the matrices $R_{ij}^{(1)}(\boldsymbol{\phi}_1)$ and $R_{ij}^{(2)}(\boldsymbol{\phi}_2)$ and subsequent calculation of the rotation vector from the complete rotation matrix

$$R_{ij}(\boldsymbol{\phi}) = R_{ik}^{(1)}(\boldsymbol{\phi}_1)\, R_{kj}^{(2)}(\boldsymbol{\phi}_2) \quad . \tag{1.59}$$

Clearly, however, there exists a one-to-one correspondence between $\hat{R}_{(1)}$, $\hat{R}_{(2)}$ and $\hat{R}$ on the one hand and $\boldsymbol{\phi}_1, \boldsymbol{\phi}_2$ and $\boldsymbol{\phi}$ within the "rotation sphere" on the other. This obvious isomorphism between rotation group ($\hat{R}$) and rotation vectors ($\boldsymbol{\phi}$) can be used to show that *the rotation group is not simply connected.* Although SO(3) is continuously connected, it is not *simply connected.* This means that one can reach every element $[R_{ik}(\boldsymbol{\phi}_2)]$ of the group, starting from another element $[R_{ik}(\boldsymbol{\phi}_1)]$, by variation of the group parameters (the rotational angles ϕ_x, ϕ_y, ϕ_z), but that different paths from $\boldsymbol{\phi}_1$ to $\boldsymbol{\phi}_2$ cannot always be transformed into each other by continuous deformation.

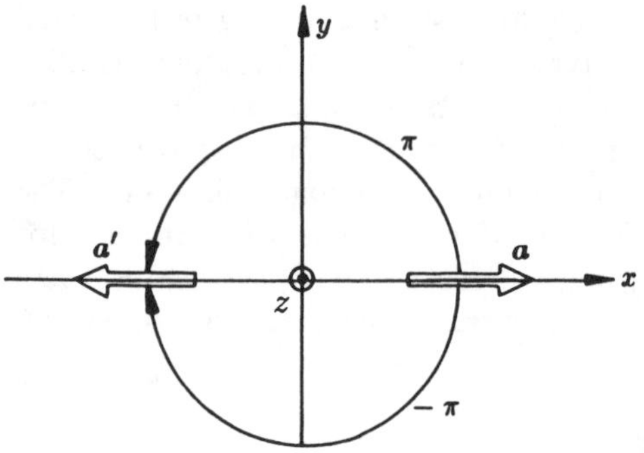

Fig. 1.13. The vector $\boldsymbol{a}' = -\boldsymbol{a}$ can be obtained by a rotation $\boldsymbol{\phi} = (0, 0, \pi)$ or $\boldsymbol{\phi}' = -\boldsymbol{\phi} = (0, 0, -\pi)$, respectively

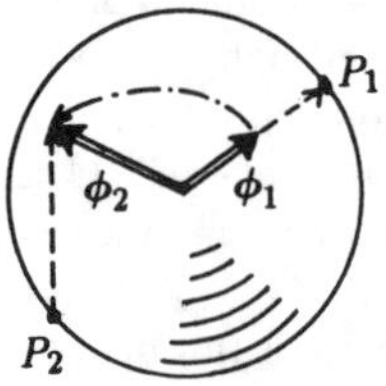

Fig. 1.14. Illustration of two paths from $\boldsymbol{\phi}_1$ to $\boldsymbol{\phi}_2$ that are not continuously deformable into each other

For the rotational group there exist two paths from $\hat{R}(\boldsymbol{\phi}_1)$ to $\hat{R}(\boldsymbol{\phi}_2)$ which cannot be continuously transformed into each other: The first path leads from $\boldsymbol{\phi}_1$ to $\boldsymbol{\phi}_2$ while completely remaining within the rotation sphere. The second one, first leads from $\boldsymbol{\phi}_1$ to the point P_1 on the surface of the sphere, then appears at the diametrically opposed point P_2, and from there finally leads to $\boldsymbol{\phi}_2$ (Fig. 1.14). If a possible third path makes a second jump from P_2 back to P_1, it will be equivalent to the first path. This can be seen from the fact that paths of the type (1) and (2), cannot be produced from one other and are the *only* paths, which are not continuously deformable into each other. Hence, the rotation group is *doubly continuously connected.*

1.9.1 Infinitesimal and Finite Rotations

From the example of the rotation group we can explicitly study many general characteristics of Lie groups. The properties of the group elements (rotations) under infinitesimal variations of the group parameters (infinitesimal rotations) are of particular interest. In this case much is simplified. Let $\delta\boldsymbol{\phi} = \{\delta\phi_x, \delta\phi_y, \delta\phi_z\}$ be an infinitesimal rotation vector, then according to (1.11) the rotated vector becomes

$$\boldsymbol{r}' = \boldsymbol{r} + \delta\boldsymbol{\phi} \times \boldsymbol{r} \tag{1.60}$$

or expressed in components

$$x_i' = x_i + \varepsilon_{ijk}\delta\phi_j x_k = (\delta_{ik} + \varepsilon_{ijk}\delta\phi_j)x_k = R_{ki}x_k = \hat{\hat{R}}_{ik}x_k \quad . \tag{1.61}$$

This can be written symbolically or explicitly as components in matrix representation respectively, i.e.

$$\boldsymbol{r}' = \hat{\hat{R}}\boldsymbol{r}\ ; \quad \begin{pmatrix} x_1' \\ x_2' \\ x_3' \end{pmatrix} = \widetilde{\begin{pmatrix} R_{11} & R_{12} & R_{13} \\ R_{21} & R_{22} & R_{23} \\ R_{31} & R_{32} & R_{33} \end{pmatrix}} \begin{pmatrix} x_1 \\ x_2 \\ x_3 \end{pmatrix} = \begin{pmatrix} R_{11} & R_{21} & R_{31} \\ R_{12} & R_{22} & R_{32} \\ R_{13} & R_{23} & R_{33} \end{pmatrix} \begin{pmatrix} x_1 \\ x_2 \\ x_3 \end{pmatrix} \ .$$

R_{ik} is the rotation matrix according to (1.56) and is given explicitly by

$$\hat{\hat{R}} = \begin{pmatrix} 1 & -\delta\phi_z & \delta\phi_y \\ \delta\phi_z & 1 & -\delta\phi_x \\ -\delta\phi_y & \delta\phi_x & 1 \end{pmatrix} , \quad \hat{R} = \begin{pmatrix} 1 & \delta\phi_z & -\delta\phi_y \\ -\delta\phi_z & 1 & \delta\phi_x \\ \delta\phi_y & -\delta\phi_x & 1 \end{pmatrix} . \tag{1.62}$$

Again we define the rotation operator $\hat{U}_{\mathrm{R}}(\delta\phi)$ in such a way that the rotated scalar state $\psi_{\alpha'}(\boldsymbol{r},t)$ follows from the initial state $\psi_\alpha(\boldsymbol{r},t)$ by

$$\psi_{\alpha'}(\boldsymbol{r},t) = \hat{U}_{\mathrm{R}}(\delta\boldsymbol{\phi})\psi_\alpha(\boldsymbol{r},t) \quad . \tag{1.63}$$

The rotated scalar wave function has to satisfy the requirement analogous to (1.17) and (1.36), namely that

$$\psi_{\alpha'}(\hat{\hat{R}}\boldsymbol{r},t) = \psi_\alpha(\boldsymbol{r},t) \tag{1.64}$$

or equivalently

$$\psi_{\alpha'}(\boldsymbol{r},t) = \psi_\alpha(\hat{\hat{R}}^{-1}\boldsymbol{r},t) = \psi_\alpha(\boldsymbol{r},t) \quad . \tag{1.65}$$

Equation (1.64) shows the value of the rotated wave function for the rotated position vector, $\boldsymbol{r}' = \hat{\hat{R}}\boldsymbol{r}$, to be equal to the value of the unrotated wave function at the unrotated position $\boldsymbol{r}$. From (1.65) we obtain

$$\begin{aligned}\psi_{\alpha'}(\boldsymbol{r},t) &= \psi_\alpha(\hat{\hat{R}}^{-1}\boldsymbol{r},t) = \psi_\alpha(\boldsymbol{r}-\delta\boldsymbol{r},t) = \psi_\alpha(\boldsymbol{r}-\delta\boldsymbol{\phi}\times\boldsymbol{r},t)\\ &\simeq \psi_\alpha(\boldsymbol{r},t) \;-(\delta\boldsymbol{\phi}\times\boldsymbol{r})\cdot\nabla\psi_\alpha(\boldsymbol{r},t)\\ &= \psi_\alpha(\boldsymbol{r},t) \;-\frac{\mathrm{i}}{\hbar}(\delta\boldsymbol{\phi}\times\boldsymbol{r})\cdot\hat{\boldsymbol{p}}\psi_\alpha(\boldsymbol{r},t)\\ &= \left(\mathbb{1}-\frac{\mathrm{i}}{\hbar}\delta\boldsymbol{\phi}\cdot\hat{\boldsymbol{L}}\right)\psi_\alpha(\boldsymbol{r},t) \quad .\end{aligned} \tag{1.66}$$

Comparison with (1.63) yields

$$\hat{U}(\delta\phi) = \mathbb{1} - \frac{\mathrm{i}}{\hbar}\delta\boldsymbol{\phi}\cdot\hat{\boldsymbol{L}} \quad , \tag{1.67}$$

$\hat{\boldsymbol{L}} = \boldsymbol{r}\times\hat{\boldsymbol{p}}$ being the angular momentum operator. This holds for infinitesimal rotations. For finite rotations $\boldsymbol{\phi}$ the rotation operator can easily be found according to the following consideration: We choose the coordinates in such a way that, say, the x axis lies along the rotational axis. Carrying out first the rotation $\hat{U}_{\mathrm{R}}(\phi_x)$ and then an infinitesimal rotation $\hat{U}_{\mathrm{R}}(\Delta\phi_x) = 1-(\mathrm{i}/\hbar)\Delta\phi_x\hat{L}_x$ yields the complete rotation

$$\begin{aligned}\hat{U}_{\mathrm{R}}(\phi_x+\Delta\phi_x,\phi_y,\phi_z) &= \hat{U}_{\mathrm{R}}(\Delta\phi_x)\hat{U}_{\mathrm{R}}(\phi_x,\phi_y,\phi_z)\\ &= \left(1-\frac{\mathrm{i}}{\hbar}\Delta\phi_x\hat{L}_x\right)\hat{U}_{\mathrm{R}}(\phi_x,\phi_y,\phi_z) \quad .\end{aligned}$$

Therefore,

$$\begin{aligned}\frac{\Delta\hat{U}_{\mathrm{R}}(\phi_x,\phi_y,\phi_z)}{\Delta\phi_x} &= \frac{\hat{U}_{\mathrm{R}}(\phi_x+\Delta\phi_x,\phi_y,\phi_z)-\hat{U}_{\mathrm{R}}(\phi_x,\phi_y,\phi_z)}{\Delta\phi_x}\\ &= -\frac{\mathrm{i}}{\hbar}\hat{L}_x\hat{U}_{\mathrm{R}}(\phi_x,\phi_y,\phi_z),\end{aligned}$$

i.e. in the limit $\Delta\phi_x\to 0$ the differential equation

$$\frac{\partial\hat{U}_{\mathrm{R}}(\phi_x,\phi_y,\phi_z)}{\partial\phi_x} = -\frac{\mathrm{i}}{\hbar}\hat{L}_x\hat{U}_{\mathrm{R}}(\phi_x,\phi_y,\phi_z) \quad , \tag{1.68}$$

is obtained. Analogous relations hold with respect to the other coordinate directions

$$\frac{\partial\hat{U}_{\mathrm{R}}(\boldsymbol{\phi})}{\partial\phi_y} = -\frac{\mathrm{i}}{\hbar}\hat{L}_y\hat{U}_{\mathrm{R}}(\boldsymbol{\phi}) \quad , \tag{1.69}$$

$$\frac{\partial\hat{U}_{\mathrm{R}}(\boldsymbol{\phi})}{\partial\phi_y} = -\frac{\mathrm{i}}{\hbar}\hat{L}_y\hat{U}_{\mathrm{R}}(\boldsymbol{\phi}) \quad . \tag{1.70}$$

Integration of (1.69, 70) with respect to the boundary condition $\hat{U}_{\mathrm{R}}(0) = 1$ yields

$$\hat{U}_{\mathrm{R}}(\boldsymbol{\phi}) = \exp\left(-\frac{\mathrm{i}}{\hbar}\boldsymbol{\phi}\cdot\hat{\boldsymbol{L}}\right) \tag{1.71}$$

for finite rotations. This result is completely analogous to our former results (1.24) and (1.39). It shows explicitly that the rotation group possesses three parameters (ϕ_x, ϕ_y, ϕ_z). The *rotation operator* (1.71) is unitary since

$$\hat{U}_{\mathrm{R}}^{-1}(\boldsymbol{\phi}) = \hat{U}_{\mathrm{R}}(\boldsymbol{\phi}) = \exp\left(\frac{\mathrm{i}}{\hbar}\boldsymbol{\phi}\cdot\hat{\boldsymbol{L}}\right)$$

and, because $\hat{\boldsymbol{L}}$ is hermitian

$$\hat{U}_{\mathrm{R}}^{\dagger}(\boldsymbol{\phi}) = \exp\left(\frac{\mathrm{i}}{\hbar}\boldsymbol{\phi}\cdot\hat{\boldsymbol{L}}^{\dagger}\right) = \exp\left(\frac{\mathrm{i}}{\hbar}\boldsymbol{\phi}\cdot\hat{\boldsymbol{L}}\right) \quad .$$

Hence,

$$\hat{U}_{\mathrm{R}}^{\dagger}(\boldsymbol{\phi}) = \hat{U}_{\mathrm{R}}^{-1}(\boldsymbol{\phi}) \quad . \tag{1.72}$$

The three operators $\hat{L}_x, \hat{L}_y, \hat{L}_z$ are called *generators* of rotations about the three coordinate axes. This is analogous to our former results (1.25) and (1.39a) for the operators $\hat{\boldsymbol{p}}$ and $\hat{E}$ for space and time displacements, respectively. From the fact that three generators $\hat{p}_x$, $\hat{p}_y, \hat{p}_z$ commute for translations in space, we can conclude that the translation group is abelian (the order of two translations can be interchanged). The rotation group, however, is non-abelian, which is expressed formally by the three non-commuting generators of this group $\hat{L}_x, \hat{L}_y, \hat{L}_z$,

1.9.2 Isotropy of Space

Isotropy of space means symmetry of the laws of nature (here of the Schrödinger equation) with respect to rotations in space. Therefore, the following has to be valid

$$\begin{aligned} \mathrm{i}\hbar\frac{\partial\psi_{\alpha'}(\boldsymbol{r},t)}{\partial t} &= \mathrm{i}\hbar\frac{\partial\hat{U}_{\mathrm{R}}(\boldsymbol{\phi})\hat{U}_r^{\dagger}}{\partial t} = \hat{U}_{\mathrm{R}}(\boldsymbol{\phi})\mathrm{i}\hbar\frac{\delta\psi_{\alpha}(\boldsymbol{r},t)}{\partial t} = \hat{U}_{\mathrm{R}}(\boldsymbol{\phi})\hat{H}\psi_{\alpha}(\boldsymbol{r},t) \\ &= \hat{U}_{\mathrm{R}}(\boldsymbol{\phi})\hat{H}\hat{U}_{\mathrm{R}}^{-1}(\boldsymbol{\phi})\psi_{\alpha'}(\boldsymbol{r},t) \stackrel{!}{=} \hat{H}\psi_{\alpha'}(\boldsymbol{r},t) \quad . \end{aligned} \tag{1.73}$$

Here we have used the fact that the initial state $\psi_\alpha(\boldsymbol{r},t)$ satisfies the Schrödinger equation

$$\mathrm{i}\hbar\frac{\partial\psi_{\alpha}(\boldsymbol{r},t)}{\partial t} = \hat{H}\psi_{\alpha}(\boldsymbol{r},t) \quad . \tag{1.74}$$

The last step made in (1.73) expresses the requirement of invariance of the Schrödinger equation concerning rotations or, equivalently, isotropy of configuration space. Hence, for arbitrary rotation vectors $\boldsymbol{\phi}$,

$$\hat{U}_{\mathrm{R}}(\boldsymbol{\phi})\hat{H}\hat{U}_{\mathrm{R}}^{-1}(\boldsymbol{\phi}) = \hat{H} \quad \text{or} \tag{1.75}$$

$$[\hat{U}_{\mathrm{R}}(\boldsymbol{\phi}), \hat{H}]_{-} = 0 \tag{1.76}$$

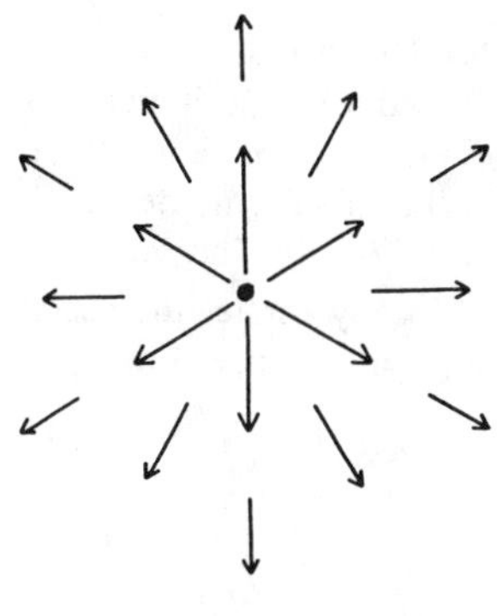

All directions of an isotropic force field are equivalent

and, since ϕ is arbitrary,

$$[\hat{L}, \hat{H}]_{-} = 0 \quad . \tag{1.77}$$

The equations (1.75–1.77) are equivalent and express conservation of angular momentum. As in classical mechanics [cf. (1.16)], we can infer the conservation of angular momentum from spatial isotropy in quantum mechanics.

Isotropy of space means that all directions are equivalent. Of course, a field of force may be present, but only one that does not exhibit a preferred direction, i.e. the field must be spherically symmetric. Otherwise isotropy would be destroyed. Spherically symmetric fields always allow for conserved angular momentum with respect to the centre of the force.

EXAMPLE

1.11 Transformation of Vector Fields Under Rotations

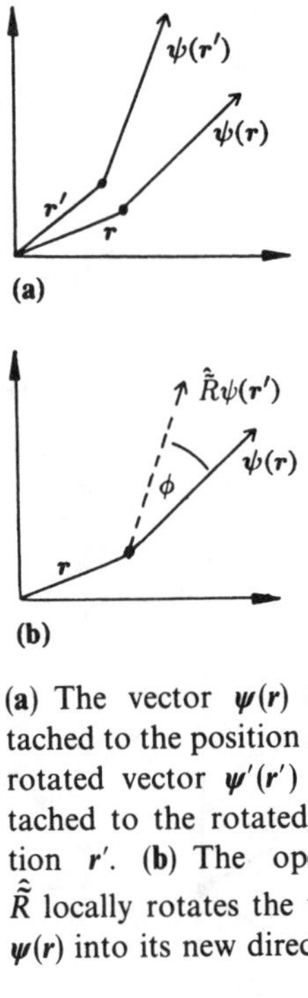

(a) The vector $\boldsymbol{\psi}(\boldsymbol{r})$ is attached to the position $\boldsymbol{r}$. The rotated vector $\boldsymbol{\psi}'(\boldsymbol{r}')$ is attached to the rotated position $\boldsymbol{r}'$. (b) The operator $\hat{\hat{R}}$ locally rotates the vector $\boldsymbol{\psi}(\boldsymbol{r})$ into its new direction

We examine a vector field $\boldsymbol{\psi} = \{\psi_1, \psi_2, \psi_3\}$ as another example for the behaviour of wave fields under rotations. An example of a particle that is described by a vector field is the photon. Its wave field is described by the vector potential $\boldsymbol{A}(\boldsymbol{r}, t)$, which obeys Maxwell's equations. Vector mesons are other examples, being subnuclear particles with a non-vanishing rest mass, which are also described by a vector field.

The equation for the rotation operator $\hat{U}(\phi)$ of vector fields is more complicated than that for scalar fields (1.64). It is important to recognize that the rotation operator not only shifts the vector $\boldsymbol{\psi}$ from $\boldsymbol{r}$ to $\boldsymbol{r}' = \hat{\hat{R}}\boldsymbol{r}$, but also rotates the direction of $\boldsymbol{\psi}$ into a new direction $\boldsymbol{\psi}'$ as in (a) of the above figure. The operator $\hat{U}$ transforms the vector $\boldsymbol{\psi}$ into the new vector $\boldsymbol{\psi}'$. We call this transformation an *active* rotation of the wave field. The operation $\hat{\hat{R}}$ rotates the vector $\boldsymbol{\psi}$ at $\boldsymbol{r}$ (locally) into the same direction as $\boldsymbol{\psi}'(\boldsymbol{r}')$ (Fig. b). If the wave field is scalar function we do not need this additional operation, since the local rotation at $\boldsymbol{r}$ transforms the scalar to itself. On the other hand, the vector $\boldsymbol{\psi}(\boldsymbol{r})$ is rotated. Hence,

$$\boldsymbol{\psi}'(\boldsymbol{r}', t) = \hat{\hat{R}}\boldsymbol{\psi}(\boldsymbol{r}, t) \quad \text{or} \tag{1}$$

$$\boldsymbol{\psi}'(\hat{\hat{R}}\boldsymbol{r}, t) = \hat{\hat{R}}\boldsymbol{\psi}(\boldsymbol{r}, t) \quad \text{or} \tag{2}$$

$$\boldsymbol{\psi}'(\boldsymbol{r}, t) = \hat{\hat{R}}\boldsymbol{\psi}(\hat{\hat{R}}^{-1}\boldsymbol{r}, t) \quad . \tag{3}$$

On the other hand, the defining equation for the rotation operator is

$$\boldsymbol{\psi}'(\boldsymbol{r}, t) = \hat{U}_{\mathrm{R}}(\boldsymbol{\phi})\boldsymbol{\psi}(\boldsymbol{r}, t) \quad , \tag{4}$$

so that by comparing with (3) we obtain

$$\hat{U}_{\mathrm{R}}(\boldsymbol{\phi})\boldsymbol{\psi}(\boldsymbol{r}, t) = \hat{\hat{R}}\boldsymbol{\psi}(\hat{\hat{R}}^{-1}\boldsymbol{r}, t) \quad . \tag{5}$$

Example 1.11

From this relation, we derive the operator for infinitesimal rotations $\hat{U}_{\mathbf{R}}(\delta\boldsymbol{\phi})$

$$\begin{aligned}\hat{U}_{\mathbf{R}}(\delta\boldsymbol{\phi})\boldsymbol{\psi}(\boldsymbol{r},t) &= \hat{\hat{R}}\boldsymbol{\psi}(\hat{\hat{R}}^{-1}\boldsymbol{r},t)\\ &= \boldsymbol{\psi}(\hat{\hat{R}}^{-1}\boldsymbol{r},t) + \delta\phi\times\boldsymbol{\psi}(\hat{\hat{R}}^{-1}\boldsymbol{r},t)\\ &= \boldsymbol{\psi}(\boldsymbol{r}-\delta\boldsymbol{\phi}\times\boldsymbol{r},t) + \delta\boldsymbol{\phi}\times\boldsymbol{\psi}(\boldsymbol{r}-\delta\boldsymbol{\phi}\times\boldsymbol{r},t)\\ &\simeq \boldsymbol{\psi}(\boldsymbol{r},t) - \frac{\mathrm{i}}{\hbar}(\delta\boldsymbol{\phi}\cdot\hat{\boldsymbol{L}})\boldsymbol{\psi}(\boldsymbol{r},t) + \delta\boldsymbol{\phi}\times\boldsymbol{\psi}(\boldsymbol{r},t) \quad .\end{aligned} \tag{6}$$

In the second line of (6) we have neglected quantities of second order in $\delta\boldsymbol{\phi}$. We can write the last term more explicitly as

$$\begin{aligned}(\delta\boldsymbol{\phi}\times\boldsymbol{\psi}(\boldsymbol{r},t))_i = \varepsilon_{ijk}\delta\phi_j\psi_k &= -\frac{\mathrm{i}}{\hbar}\delta\phi_j(\hat{S}_j)_{ik}\psi_k\\ &= -\frac{\mathrm{i}}{\hbar}\delta\boldsymbol{\phi}\cdot(\hat{\boldsymbol{S}})_{ik}\psi_k \quad ,\end{aligned} \tag{7}$$

where we have introduced the matrices:

$$(\hat{S}_j)_{ik} = \mathrm{i}\hbar\varepsilon_{ijk} \tag{8}$$

$$\begin{aligned}(\hat{S}_1)_{ik} &= \mathrm{i}\hbar\varepsilon_{i1k} = \mathrm{i}\hbar\begin{pmatrix}0 & 0 & 0\\ 0 & 0 & -1\\ 0 & 1 & 0\end{pmatrix} \quad ,\\ (\hat{S}_2)_{ik} &= \mathrm{i}\hbar\varepsilon_{i2k} = \mathrm{i}\hbar\begin{pmatrix}0 & 0 & 1\\ 0 & 0 & 0\\ -1 & 0 & 0\end{pmatrix} \quad ,\\ (\hat{S}_3)_{ik} &= \mathrm{i}\hbar\varepsilon_{i3k} = \mathrm{i}\hbar\begin{pmatrix}0 & -1 & 0\\ 1 & 0 & 0\\ 0 & 0 & 0\end{pmatrix} \quad .\end{aligned} \tag{9}$$

We can regard these three matrices as components of a vector $\hat{\boldsymbol{S}} = \{\hat{S}_1, \hat{S}_2, \hat{S}_3\}$. Returning to vector notation, we can now write (6) in the form

$$\hat{U}_{\mathrm{R}}(\delta\boldsymbol{\phi})\boldsymbol{\psi}(\boldsymbol{r},t) \simeq \left[\mathbb{1} - \frac{\mathrm{i}}{\hbar}\delta\boldsymbol{\phi}\cdot(\hat{\boldsymbol{L}}+\hat{\boldsymbol{S}})\right]\boldsymbol{\psi}(\boldsymbol{r},t) \quad . \tag{10}$$

We thus obtain an expression for the infinitesimal rotation operator

$$\hat{U}_{\mathrm{R}}(\delta\boldsymbol{\phi}) = \mathbb{1} - \frac{\mathrm{i}}{\hbar}\delta\boldsymbol{\phi}\cdot(\hat{\boldsymbol{L}}+\hat{\boldsymbol{S}}) \quad . \tag{11}$$

Since $\hat{\boldsymbol{L}} = \hat{\boldsymbol{r}}\times\hat{\boldsymbol{p}}$ is a differential operator, while, according to (9), $\hat{\boldsymbol{S}}$ is a matrix operator with position-independent components, the two operators obviously commute so that

$$[\hat{L}_i, \hat{S}_k]_- = 0 \quad . \tag{12}$$

Example 1.11

In order to obtain $\hat{U}_{\mathrm{R}}(\boldsymbol{\phi})$ for finite rotations we could make use of the method expressed in (1.67–70), but here we choose to follow a different way. By dividing the finite angle $\boldsymbol{\phi}$ into N small rotations $\delta\boldsymbol{\phi} = \boldsymbol{\phi}/N$, we can obtain the operator $\hat{U}_{\mathrm{R}}$ by N successive infinitesimal rotations. Hence,

$$\begin{aligned}\hat{U}_{\mathrm{R}}(\boldsymbol{\phi}) &= \lim_{N\to\infty}\left[\mathbb{1} - \frac{\mathrm{i}}{\hbar}\left(\frac{\boldsymbol{\phi}}{N}\cdot(\hat{\boldsymbol{L}} + \hat{\boldsymbol{S}})\right)\right]^N \\ &= \exp\left[-\frac{\mathrm{i}}{\hbar}\boldsymbol{\phi}\cdot(\hat{\boldsymbol{L}} + \hat{\boldsymbol{S}})\right] = \exp\left(-\frac{\mathrm{i}}{\hbar}\boldsymbol{\phi}\cdot\hat{\boldsymbol{J}}\right) \quad . \end{aligned} \tag{13}$$

Here we have introduced the operator $\hat{\boldsymbol{J}} = \hat{\boldsymbol{L}} + \hat{\boldsymbol{S}}$, whose components $\hat{J}_i$ are the generators of the rotations for vector fields. As before, isotropy of space implies that

$$[\hat{\boldsymbol{J}}, \hat{H}]_- = [\hat{\boldsymbol{L}} + \hat{\boldsymbol{S}}, \hat{H}]_- = 0 \quad . \tag{14}$$

Thus, for vector fields the quantity

$$\hat{\boldsymbol{J}} = \hat{\boldsymbol{L}} + \hat{\boldsymbol{S}}$$

is conserved. Since $\hat{\boldsymbol{L}}$ is the operator of orbital angular momentum, it seems to be natural to interpret $\hat{\boldsymbol{S}}$ as an *intrinsic angular momentum* or *spin of the vector field*. The commutation relation (14) expresses the *conservation of the total angular momentum* $\hat{\boldsymbol{J}}$.

We know that the components of orbital angular momentum satisfy the commutation relations

$$[\hat{L}_i, \hat{L}_j]_- = \mathrm{i}\hbar\varepsilon_{ijk}\hat{L}_k \quad . \tag{16}$$

From (9) we can easily infer, by explicit calculation, that the components of $\hat{\boldsymbol{S}}$ obey analogous commutation relations, i.e.

$$[\hat{S}_i, \hat{S}_j]_- = \mathrm{i}\hbar\varepsilon_{ijk}\hat{S}_k \quad . \tag{17}$$

For instance, in the case of $i = 1, j = 2, k = 3$ we have

$$\begin{aligned}\hat{S}_1\hat{S}_2 - \hat{S}_2\hat{S}_1 &= (\mathrm{i}\hbar)^2\left[\begin{pmatrix}0 & 0 & 0\\ 0 & 0 & -1\\ 0 & 1 & 0\end{pmatrix}\begin{pmatrix}0 & 0 & 1\\ 0 & 0 & 0\\ -1 & 0 & 0\end{pmatrix}\right. \\ &\quad \left. - \begin{pmatrix}0 & 0 & 1\\ 0 & 0 & 0\\ -1 & 0 & 0\end{pmatrix}\begin{pmatrix}0 & 0 & 0\\ 0 & 0 & -1\\ 0 & 1 & 0\end{pmatrix}\right] \\ &= (\mathrm{i}\hbar)^2\left[\begin{pmatrix}0 & 0 & 0\\ 1 & 0 & 0\\ 0 & 0 & 0\end{pmatrix} - \begin{pmatrix}0 & 1 & 0\\ 0 & 0 & 0\\ 0 & 0 & 0\end{pmatrix}\right] \\ &= (\mathrm{i}\hbar)(\mathrm{i}\hbar)\begin{pmatrix}0 & -1 & 0\\ 1 & 0 & 0\\ 0 & 0 & 0\end{pmatrix} \\ &= \mathrm{i}\hbar\hat{S}_3 \quad . \end{aligned}$$

Hence, the components $\hat{S}_i$ of the spin vector satisfy the commutation relations for angular momentum operators too. This also holds for $\hat{J}_i$, as from $[\hat{L}_i, \hat{S}_j]_- = 0$, it follows that

$$[\hat{J}_i, \hat{J}_j]_- = [\hat{L}_i + \hat{S}_i, \hat{L}_j + \hat{S}_j]_- = [\hat{L}_i, \hat{L}_j]_- + [\hat{S}_i, \hat{S}_j]_-$$
$$= \mathrm{i}\hbar\varepsilon_{ijk}\hat{L}_k + \mathrm{i}\hbar\varepsilon_{ijk}\hat{S}_k = \mathrm{i}\hbar\varepsilon_{ijk}\hat{J}_k \quad . \tag{18}$$

This result provides further justification for the interpretation of $\hat{\boldsymbol{J}}$ as the operator of total angular momentum and $\hat{\boldsymbol{S}}$ as the spin operator. Remember that we accepted the commutation relations (16)–(18) as general definitions for angular momentum operators, when we introduced the electron spin. Now we will calculate the absolute value of the spin $\hat{\boldsymbol{S}}$ (intrinsic angular momentum) of vector fields. From (9) we get

$$\hat{\boldsymbol{S}}^2 = \hat{\boldsymbol{S}}\cdot\hat{\boldsymbol{S}} = \hat{S}_1^2 + \hat{S}_2^2 + \hat{S}_3^2$$
$$= (\mathrm{i}\hbar)^2\left[\begin{pmatrix} 0 & 0 & 0 \\ 0 & -1 & 0 \\ 0 & 0 & -1 \end{pmatrix} + \begin{pmatrix} -1 & 0 & 0 \\ 0 & 0 & 0 \\ 0 & 0 & -1 \end{pmatrix}\right] + \left[\begin{pmatrix} -1 & 0 & 0 \\ 0 & -1 & 0 \\ 0 & 0 & 0 \end{pmatrix}\right]$$
$$= 2\hbar^2\begin{pmatrix} 1 & 0 & 0 \\ 0 & 1 & 0 \\ 0 & 0 & 1 \end{pmatrix} = 1(1+1)\hbar^2\mathbb{1} \quad . \tag{19}$$

Here we have written the value 2 in the form $S(S+1)$, known to hold for the orbital angular momentum. Obviously, we have to attribute to S the value 1. This result is very important: *vector fields have a spin (intrinsic angular momentum) of* 1, *i.e. they describe spin*-1 *particles.* Hence photons and vector mesons have spin $S = 1$.

Comment. In the previous example we learnt that the rotation operator

$$\hat{U}_{\mathrm{R}}(\boldsymbol{\phi}) = \exp\left(-\frac{\mathrm{i}}{\hbar}\boldsymbol{\phi}\cdot\hat{\boldsymbol{J}}\right) \tag{20}$$

contains the sum of the orbital angular momentum and the spin (intrinsic angular momentum) of the considered field, because $\hat{U}_{\mathrm{R}}(\boldsymbol{\phi})$ contains the operator $\hat{\boldsymbol{J}}$. If we examine a general wave field (scalar, vector or tensor), in order to determine the spin of the particles described by this field we must study the transformation properties of the field under rotations. This means that we have to construct the operator $\hat{U}_{\mathrm{R}}(\boldsymbol{\phi})$. The generators $\hat{J}_i$ we obtain indicate the angular momentum properties, in particular the spin of the particles described by the field. This was exactly the method applied before. In the case of a scalar field we obtained $\hat{\boldsymbol{J}} = \hat{\boldsymbol{L}} = \hat{\boldsymbol{r}}\times\hat{\boldsymbol{p}}$, i.e. it describes particles which have only orbitan angular momentum, or particles without spin ($S = 0$). In the case of a vector field we obtained $\hat{\boldsymbol{J}} = \hat{\boldsymbol{L}} + \hat{\boldsymbol{S}}$, where $\hat{\boldsymbol{S}}^2 = S(S+1)\hbar^2 = 1(1+1)\hbar^2$. Hence, vector fields describe particles with spin $S = 1$ (cf. Example 1.10).

EXAMPLE

1.12 Transformation of Two-Component Spinors Under Rotations

The rotation of scalar and spinor wave fields was evident and easy to comprehend, as easy to imagine as a rotated scalar or vector. In the case of two-component spinors it is not so easy. We have to modify the direct method of determining the rotation operator, which we developed in previous sections. We have to study not only a single wave function, but also the behaviour of an equation for two-component spinors under rotation. For simplicity we choose the Pauli equation,

$$\begin{aligned} \mathrm{i}\hbar\frac{\partial\psi}{\partial t} &= \hat{H}\psi \\ &= \left\{\frac{1}{2m}\left(\hat{\boldsymbol{p}}-\frac{e}{c}\boldsymbol{A}\right)^2 + V(\boldsymbol{r}) - \mu_{\mathrm{B}}\hat{\boldsymbol{\sigma}}\cdot\boldsymbol{B}\right\}\psi \quad . \end{aligned} \tag{1}$$

We denote the two-component spinor by $\psi = \binom{\psi_1}{\psi_2}$. The rotation operator is again denoted by $\hat{U}_{\mathrm{R}}(\boldsymbol{\phi})$ and the rotated spinor by ψ'. Both are related by

$$\psi'(\boldsymbol{r},t) = \hat{U}_{\mathrm{R}}(\boldsymbol{\phi})\psi(\boldsymbol{r},t) \quad . \tag{2}$$

We do not know enough to construct $\hat{U}_{\mathrm{R}}(\phi)$ directly, but we can follow an alternative route: $\int\psi^\dagger\psi d^3x$ must be invariant. Hence, $\hat{U}_{\mathrm{R}}(\boldsymbol{\phi})$ is a unitary operator,

$$\begin{aligned} \int\psi^\dagger\psi d^3x &= \int\psi'^\dagger\psi' d^3x \\ &= \int\psi^\dagger\hat{U}_{\mathrm{R}}^\dagger(\boldsymbol{\phi})\hat{U}_{\mathrm{R}}(\boldsymbol{\phi})\psi d^3x \quad , \end{aligned} \tag{3}$$

and

$$\hat{U}_{\mathrm{R}}^\dagger(\boldsymbol{\phi})\hat{U}_{\mathrm{R}}(\boldsymbol{\phi}) = \mathbb{1} \quad . \tag{4}$$

From (1.65), we can infer that the value of rotated scalar function $f(\boldsymbol{r})$ at the rotated location equals that of the original function at the original location. If $f(\boldsymbol{r})$ is a product, e.g. $\psi^\dagger\psi$, this statement holds too, and we must take the related locations of the individual factors [cf. (1.65)]. This means that the scalar density $\psi^\dagger\psi(\boldsymbol{r},t)$ behaves under active rotations like a scalar wave function, and

$$\psi'^\dagger(\boldsymbol{r})\psi'(\boldsymbol{r}) = \psi^\dagger(\hat{R}^{-1}\boldsymbol{r})\psi(\hat{R}^{-1}\boldsymbol{r}) \quad . \tag{5}$$

We can write the rhs of (5) as a sum over the two spinor components. The terms of the sum are scalar functions which are transformed by $\exp(-\mathrm{i}/\hbar)\boldsymbol{\phi}\cdot\hat{\boldsymbol{L}})$, as in (1.71), i.e.

$$\begin{aligned} \psi^\dagger(\hat{R}^{-1}\boldsymbol{r})\psi(\hat{R}^{-1}\boldsymbol{r}) &= \sum_{m=1}^{2}\psi_m^*(\hat{R}^{-1}\boldsymbol{r})\psi_m(\hat{R}^{-1}\boldsymbol{r}) \\ &= \sum_{m=1}^{2}\left[\exp\left(-\frac{\mathrm{i}}{\hbar}\boldsymbol{\phi}\cdot\hat{\boldsymbol{L}}\right)\psi_m(\boldsymbol{r})\right]^*\left[\exp\left(-\frac{\mathrm{i}}{\hbar}\boldsymbol{\phi}\cdot\hat{\boldsymbol{L}}\right)\psi_m(\boldsymbol{r})\right] \\ &= [\hat{U}_L(\boldsymbol{\phi})\psi(\boldsymbol{r})]^\dagger[\hat{U}_L(\boldsymbol{\phi})\psi(\boldsymbol{r})] \quad . \end{aligned}$$

In the last step we have introduced the operator

Example 1.12

$$\hat{U}_L(\boldsymbol{\phi}) = \exp\left(-\frac{\mathrm{i}}{\hbar}\boldsymbol{\phi}\cdot\hat{\boldsymbol{L}}\right) \tag{6}$$

known from (1.70). $\hat{U}_L$ takes into account the orbital angular momentum of the components. We denoted this part by $\hat{U}_L$, as there will be another part originating from the spin. Clearly $\hat{U}_L(\boldsymbol{\phi})$ determines the behaviour under transformations dependent on the coordinates. Hence, we need another term that depends on the spinorial properties, as in the case of the vector field (cf. Example 1.9). This term $\hat{U}_S$ must not be coordinate-dependent and, we can now write for the total rotation operator

$$\hat{U}_R(\boldsymbol{\phi}) = \hat{U}_S(\boldsymbol{\phi})\hat{U}_L(\boldsymbol{\phi}) \quad . \tag{7}$$

Relations (3) and (4) only hold if $\hat{U}_S(\boldsymbol{\phi})$ is a *unitary* operator, because $\hat{U}_R$ and $\hat{U}_L$ are unitary. So we make the following ansatz for $\hat{U}_S(\boldsymbol{\phi})$:

$$\hat{U}_S(\boldsymbol{\phi}) = \exp\left(-\frac{\mathrm{i}}{\hbar}\boldsymbol{\phi}\cdot\hat{\boldsymbol{a}}\right) \quad , \tag{8}$$

where the hermitian vector operator $\hat{\boldsymbol{a}}$ has still to be determined.

In order to find $\hat{\boldsymbol{a}}$, we have to study the transformation of the Pauli equation (1). We require that the spin vector $\boldsymbol{s}$ or – being identical up to a constant factor – the Pauli vector operator $\boldsymbol{\sigma} = [\hat{\sigma}_1, \hat{\sigma}_2, \hat{\sigma}_3]$ transforms under space rotations as a vector. This leads us to the relation

$$\hat{\sigma}_i' = \hat{U}_S^\dagger(\boldsymbol{\phi})\hat{\sigma}_i\hat{U}_S(\boldsymbol{\phi}) = R_{ji}\hat{\sigma}_j \quad , \tag{9}$$

which simply expresses that the matrices $\hat{\sigma}_i$ transform in the same manner as the components of a vector under space rotations. If (9) holds, we can be sure that $\boldsymbol{\sigma}$ really is a vector. Restricting ourselves to infinitesimal rotations $\delta\boldsymbol{\phi}$ by writing [cf. (1.61)]

$$R_{ji} = \delta_{ji} + \varepsilon_{imj}\delta\phi_m \tag{10}$$

and

$$\hat{U}_S = \left\{\mathbb{1} - \frac{\mathrm{i}}{\hbar}\delta\phi_n\hat{a}_n\right\} \quad . \tag{11}$$

Writing (9) in the form

$$R_{ji}\hat{U}_S^\dagger(\boldsymbol{\phi})\hat{\sigma}_i\hat{U}_S(\boldsymbol{\phi}) = \hat{\sigma}_j \tag{9a}$$

and using (10) and (11) yields

$$\begin{aligned}
\hat{\sigma}_j &\simeq \{\delta_{ji} + \varepsilon_{imj}\delta\phi_m\}\left\{\mathbb{1} + \frac{\mathrm{i}}{\hbar}\delta\phi_l\hat{a}_l\right\}\hat{\sigma}_i\left\{\mathbb{1} - \frac{\mathrm{i}}{\hbar}\delta\phi_n\hat{a}_n\right\} \\
&\simeq \{\delta_{ji} + \varepsilon_{imj}\delta\phi_m\}\left\{\hat{\sigma}_i + \frac{\mathrm{i}}{\hbar}\delta\phi_n[\hat{a}_n, \hat{\sigma}_i]_-\right\} \\
&\simeq \hat{\sigma}_j - \varepsilon_{imj}\delta\phi_m\hat{\sigma}_i + \frac{\mathrm{i}}{\hbar}\delta\phi_n[\hat{a}_n, \hat{\sigma}_j]_-
\end{aligned} \tag{12}$$

Example 1.12

where we have omitted terms of higher order in $\delta\phi$. Since the $\delta\phi_m$ can be chosen arbitrarily, (12) implies the relation

$$[\hat{a}_n, \hat{\sigma}_j]_- = \mathrm{i}\hbar\varepsilon_{knj}\hat{\sigma}_k \quad . \tag{13}$$

The operators $\hat{a}_n$ must be represented by 2×2 matrices. Since the three Pauli matrices combined with the two-dimensional unit matrix form a complete basis for all hermitian 2×2 matrices, the $\hat{a}_n$ can be written as a linear combination, i.e.

$$\hat{a}_n = \alpha_{nm}\hat{\sigma}_m + \beta_n \mathbb{1}, \quad , \tag{14}$$

where the coefficients α_{nm} and β_n are real (*c*-numbers). Without loss of generality, we may set $\beta_n = 0$, as the unit matrix only generates a (trivial) phase transformation in (8). Furthermore, it commutes with $\hat{\sigma}_j$ in (13) which now becomes an equation for the coefficients α_{nm}:

$$\alpha_{nm}[\hat{\sigma}_m, \hat{\sigma}_j]_- = \mathrm{i}\varepsilon_{njk}\hat{\sigma}_k \quad . \tag{15}$$

Taking the commutation relations of the Pauli matrices

$$[\hat{\sigma}_m, \hat{\sigma}_j]_- = 2\mathrm{i}\hbar\varepsilon_{mjk}\hat{\sigma}_k \tag{16}$$

into account, (15) reads

$$2\alpha_{nm}\varepsilon_{mjk}\hat{\sigma}_k = \hbar\varepsilon_{njk}\hat{\sigma}_k \quad . \tag{17}$$

This can only be fulfilled by

$$\alpha_{nm} = \tfrac{1}{2}\hbar\delta_{nm} \quad . \tag{18}$$

Thus, we get

$$\hat{a}_n = \tfrac{1}{2}\hbar\hat{\sigma}_n \tag{19}$$

and, after inserting this into (8)

$$\hat{U}_S(\boldsymbol{\phi}) = \exp(-\tfrac{1}{2}\mathrm{i}\hbar\boldsymbol{\phi}\cdot\hat{\boldsymbol{\sigma}}) \quad . \tag{20}$$

Introducing the spin operator

$$\hat{\boldsymbol{s}} = \tfrac{1}{2}\hbar\hat{\boldsymbol{\sigma}} \quad , \tag{21}$$

we get the *complete rotation operator for two-component spinors* according to (7) and (8), i.e.

$$\begin{aligned}\hat{U}_R(\boldsymbol{\phi}) &= \hat{U}_L(\boldsymbol{\phi})\hat{U}_S(\boldsymbol{\phi}) = \exp\left[-\frac{\mathrm{i}}{\hbar}\boldsymbol{\phi}\cdot\left(\hat{\boldsymbol{L}} + \frac{\hbar}{2}\hat{\boldsymbol{\sigma}}\right)\right] \\ &= \exp\left[-\frac{\mathrm{i}}{\hbar}\boldsymbol{\phi}\cdot(\hat{\boldsymbol{L}} + \hat{\boldsymbol{s}})\right] \quad .\end{aligned} \tag{22}$$

We conclude therefore, that the two-component spinors transform according to

$$\psi'(\boldsymbol{r}) = \exp\left(-\frac{\mathrm{i}}{\hbar}\boldsymbol{\phi}\cdot\hat{\boldsymbol{J}}\right)\psi(\boldsymbol{r}) \quad , \tag{23}$$

where the total angular momentum is given by

$$\hat{\boldsymbol{J}} = \hat{\boldsymbol{L}} + \hat{\boldsymbol{s}} \quad . \tag{24}$$

Such an expression is known to us from scalar (total angular momentum = orbital angular momentum) and vector (total angular momentum = orbital angular momentum + spin $1\hbar$) wave functions, but now the spin eigenvalue is $|s| = 1/2\,\hbar$.

Example 1.12

EXERCISE

1.13 Measuring the Direction of Electron Spins

Consider a device that allows the measurement of the component of spin of an electron in an arbitrary direction. Assume that such a measurement has been performed on a certain electron with respect to a certain direction (e_2), and the result is spin up ($\uparrow$) in the direction of e_2.

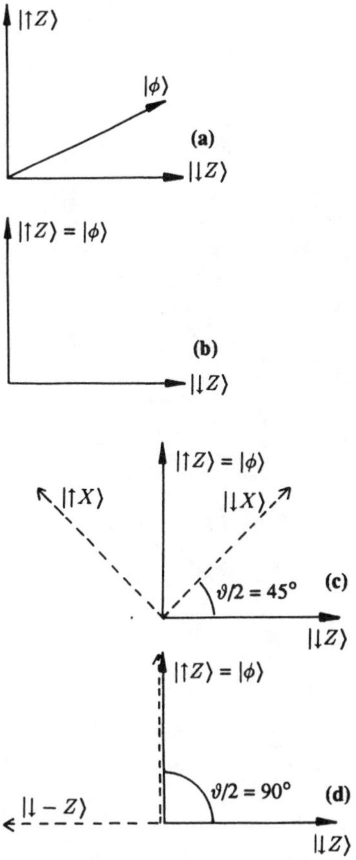

Spin state vector before (**a**) and after (**b**) the measurement. (**c**) and (**d**) show the appearance of the state vector from coordinate systems that are rotated by 90 and 180 degrees, respectively

Problem. In which state is the electron after this experiment? Draw the state vector in two-dimensional (complex) spin space.

Let the spin component of the same electron now be measured in another direction ($e_{z'}$), which is inclined to the original direction by an angle of $\theta = 90°$, $\theta = 180°$. What is the probability of the result being spin $\uparrow$ or spin $\downarrow$? What are the positions of the vectors spin $\uparrow$ and spin $\downarrow$ in the z' direction in spin space? By what angle are they rotated compared with spin $\uparrow$ and spin $\downarrow$ in the z direction?

Solution. The spin of the electron is represented by a vector (spinor) in two dimensional complex vector space with two orthogonal basis vectors, which choose as $|\uparrow z\rangle$ (spin up in z direction) and $|\downarrow z\rangle$ (spin down in z direction). To simplify thus we restrict ourselves to the graphical representation of real linear combinations $a_+|\uparrow z\rangle + a_-|\downarrow z\rangle$ ($a_+, a_- \in \mathbb{R}$). The state vector $|\phi\rangle$ is an arbitrary linear combination with length 1 (see Fig. a). With the first measurement it is projected onto one of the basis vectors, in our case to $|\uparrow z\rangle$ (Fig. b). Therefore, after the first measurement, the state vector is

$$|\phi'\rangle = \frac{\langle\uparrow z|\phi\rangle}{|\langle\uparrow z|\phi\rangle|}|\uparrow z\rangle = \frac{a_+}{|a_+|}|\uparrow z\rangle = \pm\,|z\rangle \quad . \tag{1}$$

For a measurement of the spin component in the x direction ($\theta = 90°$) we assume 50% spin $\uparrow$ and 50% spin $\downarrow$.

The vectors $|\uparrow z\rangle$ and $|\downarrow z\rangle$ can be written as linear combinations of $|\uparrow x\rangle$and $|\downarrow x\rangle$,

$$|\uparrow z\rangle = a|\uparrow x\rangle + b|\downarrow x\rangle \quad , \quad |\downarrow z\rangle = c|\uparrow x\rangle + d|\downarrow x\rangle \tag{2}$$

with

$$|a|^2 = |b|^2 = |c|^2 = |d|^2 = \tfrac{1}{2} \quad . \tag{3}$$

Exercise 1.13

Therefore,

$$|\langle\phi|\uparrow x\rangle|^2 = |\langle\uparrow z\rangle|\uparrow x\rangle|^2 = |a|^2 \quad \text{etc.}$$

is the probability of finding the result "spin ↑" in the second measurement. The vectors $|\uparrow x\rangle$, $|\downarrow x\rangle$ are inclined by $\theta/2 = 45°$ compared with $|\uparrow z\rangle$, $|\downarrow z\rangle$ (Fig. c). We can infer directly from the figures that by a projection of $|\phi\rangle = |\uparrow z\rangle$ onto $|\uparrow x\rangle$ or $|\downarrow x\rangle$, respectively, we obtain the probability 1/2. If we rotate the measuring device by $\theta = 180°$, we get the result "spin ↓" with the same probability as for spin ↑ before. Clearly, the vectors $|\uparrow -z\rangle$ and $|\downarrow -z\rangle$ are rotated by $\theta/2 = 90°$ compared with $|\uparrow z\rangle$ and $|\downarrow z\rangle$ (Fig. d). Instead of rotating the apparatus we could have rotated the electron in the opposite direction by the angle θ (for instance, with the help of a magnetic field), and then we could have measured the z component.

From the same considerations we can infer that the rotation of the electron in configuration space about an angle θ is related to a rotation of the state vector in spin space about the angle $\theta/2$. Configuration space and spin space are not directly related, their dimensions are different. It is important to keep this fact in mind, especially in the case of spin 1, when both spaces have three dimensions.

1.10 The Rotation Operator for Many-Particle States

We consider a scalar (Schrödinger), many-particle wave function

$$\psi_\alpha = \psi_\alpha(\boldsymbol{r}_1\boldsymbol{r}_2,\dots,t) \equiv \psi_\alpha(\boldsymbol{r}_i,t) \quad . \tag{1.78}$$

It is dependent on the various particle coordinates $\boldsymbol{r}_i$ and on time, and obeys the many-particle Schrödinger equation

$$\mathrm{i}\hbar\frac{\partial}{\partial t}\psi_\alpha(\boldsymbol{r}_i,t) = \hat{H}(\hat{\boldsymbol{p}}_i,\boldsymbol{r}_i)\psi_\alpha(\boldsymbol{r}_i,t) \quad . \tag{1.79}$$

The rotation operator $\hat{U}_{\mathrm{R}}(\boldsymbol{\phi})$ may be defined in precisely the same way as for the single-particle case by the equation

$$\psi'_\alpha(\boldsymbol{r}_i,t) = \hat{U}_{\mathrm{R}}(\boldsymbol{\phi})\psi_\alpha(\boldsymbol{r}_i,t) \quad , \tag{1.80}$$

and can be determined in analogy with the method used in (1.63–1.71). For instance, the determining equation (1.66) for infinitesimal rotations is

$$\begin{aligned}\psi'_\alpha(\boldsymbol{r}_i,t) &= \psi_\alpha(\hat{R}^{-1}\boldsymbol{r}_i,t)\\ &= \psi_\alpha(\boldsymbol{r}_i - \delta\boldsymbol{\phi}\times\boldsymbol{r}_i,t) \simeq \psi_\alpha(\boldsymbol{r}_i,t) - \sum_i(\delta\boldsymbol{\phi}\times\boldsymbol{r}_i)\cdot\nabla_i\psi_\alpha(\boldsymbol{r}_i,t)\\ &= \psi_\alpha(\boldsymbol{r}_i,t) - \delta\boldsymbol{\phi}\cdot\frac{\mathrm{i}}{\hbar}\sum_i\hat{\boldsymbol{L}}_i\psi_\alpha(\boldsymbol{r}_i,t) = \left(\mathbb{1} - \frac{\mathrm{i}}{\hbar}\delta\boldsymbol{\phi}\cdot\hat{\boldsymbol{L}}\right)\psi_\alpha(\boldsymbol{r}_i,t) \quad ,\end{aligned} \tag{1.81}$$

where

$$\hat{\boldsymbol{L}} = \sum_i \hat{\boldsymbol{L}}_i = \sum_i \boldsymbol{r}_i \times \hat{\boldsymbol{p}}_i \qquad (1.82)$$

is the total angular momentum of the system. Hence, the rotation operator is now

$$U_{\mathbf{R}}(\boldsymbol{\phi}) = \exp\left(-\frac{\mathrm{i}}{\hbar} \delta\boldsymbol{\phi} \cdot \hat{\boldsymbol{L}} \right) \quad . \qquad (1.83)$$

Formally, it is identical with the operator for single-particle states (1.71), but now $\hat{\boldsymbol{L}}$ stands for the *total angular momentum* of all particles. Again the symmetry of spatial isotropy, as in the single particle case, leads to $[\hat{\boldsymbol{L}}, \hat{H}(\hat{\boldsymbol{p}}_i, \boldsymbol{r}_i)]_- = 0$. This means that the total angular momentum is conserved, a result which is not surprising.

From a mathematical point of view, the operators of angular momentum $\hat{J}_i$ in (1.71) and $\hat{L}_i$ in (1.83) are the *generators of the Lie group* SO(3). This will be explained in detail in the third chapter, but from now on we will occasionally use the language of group theory. They satisfy a *Lie algebra*, which is defined by the commutators

$$[\hat{J}_i, \hat{J}_j]_- = \mathrm{i}\hbar\varepsilon_{ijk}\hat{J}_k \quad . \qquad (1.84)$$

The Lie algebra is *closed*, i.e. the commutator of an arbitrary pair of generators $\hat{J}_i$ can be expressed as a combination of the generators (in our case, due to the ε_{ijk}, by just one generator). We will show in the next chapter that the Lie algebra largely determines the possible matrix representations of the generators $\hat{J}_i$.

1.11 Biographical Notes

NOETHER, Emmy, * Erlangen 23.3.1882, † Bryn Mawr 14.4.1935. N. studied in Göttingen and Erlangen, becoming associate professor in Göttingen in 1922. After her emigration to the United States in 1933, she obtained a guest professorship at the small college of Bryn Mawr. N. has deeply influenced various branches of algebra by her work. It may be ascribed to her influence that structural theoretical thinking became a dominant principle of modern mathematics.

LIE, Sophus, *Nordfjordeid 17.2.1842, † Kristiana (Oslo) 18.2.1899. L. passed his examination as a teacher in Kristiana in 1965. In 1869/1870 he visited Berlin and in 1870, together with Klein, Paris. From 1872 L. was professor in Kristiana, and between 1886/1898 in Leipzig, until he returned to Kristiana as professor in the theory of transformation groups. In Paris L. had perceived the fundamental importance of group theory for mathematical investigations, had discovered tangential transformations and thereby shown that dynamics can be interpreted as a part of group theory. Nearly all his further investigations dealt with the theory *continuous transformation groups* and their applications.

2. Angular Momentum Algebra Representation of Angular Momentum Operators — Generators of SO(3)

2.1 Irreducible Representations of the Rotation Group

In many applications of atomic and nuclear physics, states of particles in central potentials are examined. Angular momentum is a conserved quantity in central potentials, i.e. its eigenvalues can be used to classify the states. Due to this significance of angular momentum in the applications of quantum mechanics, in this chapter we study once again the angular momentum operators and their eigenfunctions. Furthermore, the commutation relations of the angular momentum operators represent the Lie algebra of SO(3).

Angular momentum is classically defined by the relation

$$\boldsymbol{L} = \boldsymbol{r} \times \boldsymbol{p} \quad . \tag{2.1}$$

If we replace the momentum variable by the operator

$$\hat{\boldsymbol{p}} = -\mathrm{i}\hbar\nabla \tag{2.2}$$

in (2.1), we obtain the commutation relations

$$[\hat{L}_x, \hat{L}_y] = \mathrm{i}\hbar\hat{L}_z \tag{2.3}$$

for the components of the angular momentum operator. If the total angular momentum $\boldsymbol{L}$ is the sum of angular momenta $\boldsymbol{L}^{(n)}$ of single systems (particles, spin, orbital angular momentum, etc.). the commutation relation (2.3) holds for the sum as well. With

$$\boldsymbol{L} = \sum_n \boldsymbol{L}^{(n)} \text{ and } [\hat{L}_i^{(m)}, \hat{L}_j^{(n)}]_- = \mathrm{i}\hbar\varepsilon_{ijk}\hat{L}_k^{(n)}\delta_{nm} \quad ,$$

we obtain

$$[\hat{L}_x, \hat{L}_y]_- = \left[\sum_n \hat{L}_x^{(n)}, \sum_m \hat{L}_y^{(m)}\right]_- = \sum_n [\hat{L}_x^{(n)}, \hat{L}_y^{(n)}]_- = \sum_n \mathrm{i}\hbar\hat{L}_z^{(n)} = \mathrm{i}\hbar\hat{L}_z \quad .$$

We emphasize once more that the angular momentum operators belonging to different systems commute, as they act in different spaces, i.e. they act on different coordinates. As the commutation relation (2.3) holds generally, we may define:

Any vector operator $\hat{\boldsymbol{J}}$, whose components are observables and satisfy the commutation relation (2.3), is called an angular momentum operator.

Based only upon the commutation relation [the Lie algebra of SO(3)] we can infer several fundamental properties of the angular momentum operator and its eigenfunctions.

The square of the angular momentum commutes with all components, that is

$$[\hat{\boldsymbol{J}}, \hat{\boldsymbol{J}}^2]_- = 0, \quad \text{with} \quad \hat{\boldsymbol{J}}^2 = \hat{J}_x^2 + \hat{J}_y^2 + \hat{J}_z^2 \quad . \tag{2.4}$$

We know this to be valid for the orbital angular momentum operator from elementary quantum mechanics and we can easily check (2.4) in the generalized case, making use of the commutation relation (2.3).

From (2.4) we may deduce that one component of the angular momentum and its total square can be measured simultaneously, i.e. they have a common eigenfunction.

To obtain the spectrum of these operators it is appropriate to use the following Hermitian conjugate operators:

$$\hat{J}_+ = \hat{J}_x + \mathrm{i}\hat{J}_y \quad , \quad \hat{J}_- = \hat{J}_x - \mathrm{i}\hat{J}_y \quad . \tag{2.5}$$

These operators, which are not Hermitian, are called *step operators* or *shift operators*.

From relation (2.4) it follows that

$$[\hat{J}_+, \hat{\boldsymbol{J}}^2]_- = [\hat{J}_-, \hat{\boldsymbol{J}}^2]_- = 0 \quad . \tag{2.6}$$

The three operators $\hat{J}_+, \hat{J}_-, \hat{J}_z$ determine the vector operator $\hat{\boldsymbol{J}}$ entirely and are more convenient for algebraic transformations than the operators $\hat{J}_x, \hat{J}_y, \hat{J}_z$. From (2.3) we can easily infer the commutation relations of the operators (2.5),

$$[\hat{J}_z, \hat{J}_+]_- = \hbar\hat{J}_+ \quad , \quad [\hat{J}_z, \hat{J}_-]_- = -\hbar\hat{J}_- \quad , \quad [\hat{J}_+, \hat{J}_-]_- = 2\hbar\hat{J}_z \quad . \tag{2.7a}$$

To remove the constant $\hbar$ from the right-hand side (rhs) we use the transformation $\hat{J}_i = \hbar\hat{J}'_i$. The operators $\hat{J}'_i$ clearly satisfy the dimensionless commutation relations.

$$[\hat{J}'_z, \hat{J}'_+]_- = \hat{J}'_+ \quad , \quad [\hat{J}'_z, \hat{J}'_-]_- = -\hat{J}'_- \quad , \quad [\hat{J}'_+, \hat{J}'_-]_- = 2\hat{J}'_z \quad . \tag{2.7b}$$

They are identical to the relations in (2.7a) with $\hbar = 1$. From now on we will implicitly use the operators $\hat{J}'_i$, but will denote them by $\hat{J}_i$ for simplicity. All further work in this section will be based on the relations (2.7b).

Expressing the square of the angular momentum operator by the step operators we get

$$\hat{\boldsymbol{J}}^2 = \tfrac{1}{2}(\hat{J}_+\hat{J}_- + \hat{J}_-\hat{J}_+) + \hat{J}_z^2 \tag{2.8}$$

as well as the relations

$$\hat{J}_+\hat{J}_- = \hat{\boldsymbol{J}}^2 - \hat{J}_z(\hat{J}_z - 1) \quad , \quad \hat{J}_-\hat{J}_+ = \hat{\boldsymbol{J}}^2 - \hat{J}_z(\hat{J}_z + 1) \quad . \tag{2.9}$$

As $\hat{\boldsymbol{J}}^2$ commutes with all of its Cartesian coordinates (2.4), we can find a set of common eigenfunctions for one component (we take $\hat{J}_z$) and for $\hat{\boldsymbol{J}}^2$ with the corresponding eigenvalues $j(j+1)$ and m. We write, therefore,

$$\hat{\boldsymbol{J}}^2\psi_{jm} = j(j+1)\psi_{jm} \quad , \quad \hat{J}_z\psi_{jm} = m\psi_{jm} \quad . \tag{2.10}$$

As the eigenvalue of $\hat{\boldsymbol{J}}^2$, which we denoted by $j(j+1)$, must be positive, j can be chosen as positive without loss of generality since all positive numbers can be represented in the form $j(j+1)$ for positive j.

We want now to obtain the spectrum of the operators by algebraic methods; hence, using (2.9) we get

$$\begin{aligned} \hat{J}_+\hat{J}_-\psi_{jm} &= [j(j+1) - m(m-1)]\psi_{jm} \quad , \\ \hat{J}_-\hat{J}_+\psi_{jm} &= [j(j+1) - m(m+1)]\psi_{jm} \quad . \end{aligned} \tag{2.11}$$

The step operators are Hermitian conjugate to each other, $\hat{J}_-^+ = \hat{J}_+$; and thus,

$$0 \leq \int |\hat{J}_+\psi_{jm}|^2 dV = \int \psi_{jm}^* \hat{J}_-\hat{J}_+\psi_{jm} dV \tag{2.12}$$

holds. Using (2.11) we obtaine

$$\int \psi_{jm}^* \hat{J}_-\hat{J}_+\psi_{jm} dV = [j(j+1) - m(m+1)] \int |\psi_{jm}|^2 dV \geq 0 \quad , \tag{2.13}$$

and, analogously,

$$\int |\hat{J}_-\psi_{jm}|^2 dV = \int \psi_{jm}^* \hat{J}_+\hat{J}_-\psi_{jm} dV = [j(j+1) - m(m-1)] \int |\psi_{jm}|^2 dV \geq 0 \quad .$$

Hence, we can infer the following relations for the quantum numbers j and m:

$$\begin{aligned} j(j+1) - m(m+1) &= (j-m)(j+m+1) \geq 0 \quad , \\ j(j+1) - m(m-1) &= (j+m)(j-m+1) \geq 0 \quad . \end{aligned} \tag{2.14}$$

For the quantum number m they set the bounds

$$-j \leq m \leq j \quad . \tag{2.15}$$

For $m = \pm j$ we can infer from (2.12), using (2.13), that

$$\hat{J}_+\psi_{jm} = 0 \text{ for } m = j \quad \text{and} \quad \hat{J}_-\psi_{jm} = 0 \text{ for } m = -j \quad .$$

In order to examine the effect of the step operators acting on the eigenfunctions ψ_{jm}, we examine the relation

$$\hat{\boldsymbol{J}}^2\hat{J}_+\psi_{jm} = \hat{J}_+\hat{\boldsymbol{J}}^2\psi_{jm} = j(j+1)\hat{J}_+\psi_{jm} \quad .$$

This signifies that $\hat{J}_+\psi_{jm}$ is an eigenfunction to the eigenvalue $j(j+1)$, as indeed is $\hat{J}_-$. The step operators do not change the eigenvalue of the square of the angular momentum.

From (2.7b) we see that

$$\hat{J}_z\hat{J}_+ = \hat{J}_+(\hat{J}_z + 1)$$

or, written as an eigenvalue equation for $\hat{J}_z$,

$$\hat{J}_z(\hat{J}_+\psi_{jm}) = (m+1)(\hat{J}_+\psi_{jm}) \quad . \tag{2.16}$$

The step operators $\hat{J}_+$ and $\hat{J}_-$ raise and lower, respectively, the eigenvalue m by one unit. With step by step operations of $\hat{J}_+$ on the wave function ψ_{jm}, we obtain the eigenfunctions of $\hat{\boldsymbol{J}}^2$ and $\hat{J}_z$ which correspond to an eigenvalue increased by one in each case. We can arrange these functions with respect to increasing m:

$$\psi_{jm}, \hat{J}_+\psi_{jm}, \hat{J}_+^2\psi_{jm}, \ldots, \hat{J}_+^p\psi_{jm} \quad ,$$

and the corresponding eigenvalues of $\hat{J}_z$ are

$$m, m+1, m+2, \ldots, m+p=j \quad .$$

The number p is a positive integer and the maximal value of $m+p$ must equal j. We can understand this in the the following way: consider ψ_{jm*} to be the non-trivial state corresponding to the maximum value of m. Due to (2.16) and (2.13) we then have $\hat{J}_+\psi_{jm*} \sim \psi_{j,m*+1}$ and $\psi_{j,m*+1} \sim \hat{J}_+\psi_{jm*} \equiv 0$. Therefore it also holds that $0 = \hat{J}_-\hat{J}_+\psi_{jm*} = (\hat{\boldsymbol{J}}^2 - \hat{J}_z(\hat{J}_z+1))\psi_{jm*} = [j(j+1) - m^*(m^*+1)]\psi_{jm*}$. Because ψ_{jm*} is a non-trivial state, this equation can be fulfilled only if $m^* = j$. Correspondingly, we get for the operator $\hat{J}_-$ the sequence

$$\psi_{jm}, \hat{J}_-\psi_{jm}, \ldots, \hat{J}_-^q\psi_{jm}$$

with the eigenvalues

$$m, m-1, \ldots, m-q = -j \quad .$$

From the construction used, q is also a positive integer number. We can use the same arguments for the minimal value $m-q$ and get $-q = -j$. From these conditions we can infer that

$$p + q = 2j \quad .$$

As p and q are positive integers we get the following values for the quantum number j;

$$j = 0, \tfrac{1}{2}, 1, \tfrac{3}{2}, 2, \tfrac{5}{2}, 3, \ldots \quad .$$

This is a very important result since it means that from the commutation relation (2.7) we can infer directly that in quantum mechanics there exist only systems with integer ($j = 0, 1, 2, \ldots$) or half-integer ($j = \frac{1}{2}, \frac{3}{2}, \frac{5}{2}, \ldots$) angular momentum. Other values, for instance $\hbar/3$ or $\hbar/5$ are not allowed by the commutation relations.

The quantum numbers are

$$m = 0, \pm\tfrac{1}{2}, \pm 1, \pm\tfrac{3}{2}, \ldots \quad .$$

The eigenvalues of the angular momentum operator $\hat{\boldsymbol{J}}^2$ are $j(j+1)$, and for each fixed value j there are $2j+1$ different eigenvalues m of $\hat{J}_z$. They are

$$-j, -j+1, \ldots, 0, 1, \ldots, +j \quad .$$

We have shown that we can generate from a single eigenfunction ψ_{jm}, all $2j+1$ eigenfunctions of a fixed eigenvalue j by the step operators. We now show that they are uniquely defined in this way except for a phase factor.

Since the eigenfunctions are to be normalized to unity,

$$\hat{J}_+\psi_{jm} = a_m\psi_{j,m+1}$$

holds and from (2.13) for the constant factor we obtain

$$|a_m|^2 = [j(j+1) - m(m+1)] \quad .$$

We choose the phase so that a_m is positive and real (*Condon-Shortley Phase*)[1]. With the same choice of the phase we apply $\hat{J}_+$ and $\hat{J}_-$ and obtain $2j+1$ orthonormal eigenfunctions

$$\psi_{j,-j}, \psi_{j,-j+1}, \ldots, \psi_{jm}, \ldots, \psi_{jj}$$

satisfying the eigenvalue equations

$$\hat{\boldsymbol{J}}^2\psi_{jm} = j(j+1)\psi_{jm} \quad , \quad \hat{J}_z\psi_{jm} = m\psi_{jm} \quad . \tag{2.17}$$

According to the choice of the phase they also satisfy

$$\hat{J}_+\psi_{jm} = \sqrt{j(j+1)-m(m+1)}\,\psi_{j,m+1} \quad \text{and} \tag{2.18a}$$

$$\hat{J}_-\psi_{jm} = \sqrt{j(j+1)-m(m-1)}\,\psi_{j,m-1} \quad \text{with} \tag{2.18b}$$

$$\hat{J}_+\psi_{jj} = \hat{J}_-\psi_{j,-j} = 0 \quad .$$

The $2j+1$ eigenvectors are transformed into one another by the operators $\hat{J}_+, \hat{J}_-, \hat{\boldsymbol{J}}^2$ and $\hat{J}_z$. Applying these operators, i.e. in the context of the rotation operator $\hat{U}_R(\boldsymbol{\phi})$, we always get these eigenvectors and their linear combinations as a result. In other words, the $(2j+1)$-dimensional vector space spanned by the eigenfunctions ψ_{jm} is invariant under the application of the angular momentum operators. We therefore call it an *invariant subspace* of Hilbert space. It cannot be separated into smaller invariant subspaces and every one of the $(2j+1)$ vectors can be transformed into any other vector by applying the step operators. This leads us to the expression of the *irreducible representation* of the rotation group.

Note. In general there are several states ψ_{jm} for one pair of values j and m, which we denote by ψ_{njm}. Here n is a new quantum number which distinguishes between these states (e.g. the principal quantum number of the hydrogenic atom eigenfunctions). The operators $\hat{J}_\pm, \hat{J}_z$ transform the wave functions ψ_{njm} with the same n and j into one another, but wave functions $\psi_{n'j'm'}$ with $n' \neq n$ and/or $j' \neq j$ are never obtained.

2.2 Matrix Representations of Angular Momentum Operators

Utilizing the relations obtained up to now, we can easily calculate the matrix representations of the angular momentum operators $\hat{J}_x, \hat{J}_y, \hat{J}_z$. We choose the basis of eigenvectors ψ_{jm} which are diagonal in $\hat{J}_z$. To calculate the matrices for $\hat{J}_x$ and $\hat{J}_y$ we apply the expressions

$$\hat{J}_x = \tfrac{1}{2}(\hat{J}_+ + \hat{J}_-) \quad , \quad \hat{J}_y = \frac{1}{2\mathrm{i}}(\hat{J}_+ - \hat{J}_-) \quad . \tag{2.19}$$

[1] The original definition was first given in the book by E.U. Condon, G.H. Shortley: *Theory of Atomic Spectra* (Cambridge University Press, Cambridge 1935)

The matrix elements are the integrals

$$\int \psi^*_{j'm'} \hat{J}_i \psi_{jm} dV = \delta_{jj'} (\hat{J}_i)_{m'm} \quad . \tag{2.20}$$

In the case of angular momentum $j = \frac{1}{2}$, we obtain the Pauli matrices

$$(\hat{J}_x)_{mm'} = \hat{S}_x = \frac{1}{2}\begin{pmatrix} 0 & 1 \\ 1 & 0 \end{pmatrix} = \frac{1}{2}\hat{\sigma}_x \quad ,$$

$$(\hat{J}_y)_{mm'} = \hat{S}_y = \frac{1}{2}\begin{pmatrix} 0 & -\mathrm{i} \\ +\mathrm{i} & 0 \end{pmatrix} = \frac{1}{2}\hat{\sigma}_y \quad ,$$

$$(\hat{J}_z)_{mm'} = \hat{S}_z = \frac{1}{2}\begin{pmatrix} 1 & 0 \\ 0 & -1 \end{pmatrix} = \frac{1}{2}\hat{\sigma}_z \quad . \tag{2.22}$$

and

$$(\hat{\boldsymbol{J}}^2)_{mm'} = \frac{3}{4}\begin{pmatrix} 1 & 0 \\ 0 & 1 \end{pmatrix} = \frac{3}{4}\mathbb{1} = \frac{1}{2}\left(\frac{1}{2}+1\right)\begin{pmatrix} 1 & 0 \\ 0 & 1 \end{pmatrix} \quad .$$

In the case of angular momentum $j = 1$ (absolute value $\sqrt{2}\hbar$) we obtain three-dimensional matrices with $m = -1, 0, 1$. Using the same notation as above, we infer from (2.19), (2.20) and (2.18) that

$$\hat{S}_x = \frac{1}{\sqrt{2}}\begin{pmatrix} 0 & 1 & 0 \\ 1 & 0 & 1 \\ 0 & 1 & 0 \end{pmatrix} \quad , \quad \hat{S}_y = \frac{1}{\sqrt{2}}\begin{pmatrix} 0 & -\mathrm{i} & 0 \\ -\mathrm{i} & 0 & -\mathrm{i} \\ 0 & -\mathrm{i} & 0 \end{pmatrix} \quad ,$$

$$\hat{S}_z = \begin{pmatrix} 1 & 0 & 0 \\ 0 & 0 & 0 \\ 0 & 0 & -1 \end{pmatrix} \quad , \quad \hat{\boldsymbol{S}}^2 = 1(1+1)\begin{pmatrix} 1 & 0 & 0 \\ 0 & 1 & 0 \\ 0 & 0 & 1 \end{pmatrix} \quad . \tag{2.22a}$$

In the same way as we used the spinors $\chi_{\frac{1}{2},m}$ (χ_+ and χ_-) to describe the states with spin $\frac{1}{2}$, we now may use the vectors χ_{1m}; i.e.

$$\chi_{11} = \begin{pmatrix} 1 \\ 0 \\ 0 \end{pmatrix} \quad , \quad \chi_{10} = \begin{pmatrix} 0 \\ 1 \\ 0 \end{pmatrix} \quad , \quad \chi_{1-1} = \begin{pmatrix} 0 \\ 0 \\ 1 \end{pmatrix} \quad ,$$

which represent all possible states for spin 1.

The vectors χ_{1m} are eigenvectors of the matrix $\hat{S}_z$. Hence,

$$\hat{S}_z \chi_{1m} = m \chi_{1m}$$

holds. An angular momentum vector with $j = 1$ may take three distinct states, corresponding to the vectors χ_{1m}. Formally, there is a complete analogy between the spin vectors and the unit vectors of the three-dimensional space, implying that

$$e_x = \begin{pmatrix} 1 \\ 0 \\ 0 \end{pmatrix} \quad , \quad e_y = \begin{pmatrix} 0 \\ 1 \\ 0 \end{pmatrix} \quad , \quad e_z = \begin{pmatrix} 0 \\ 0 \\ 1 \end{pmatrix} \quad .$$

This will be discussed in detail in the next exercise. In spherical representation the vectors of three-dimensional space transform under space rotations just like the spin functions χ_{1m}. We have show (cf. Example 1.11) that it follows from the behaviour of vector fields under rotation that they possess spin 1, e.g. the photon as a 'particle' of the electromagnetic field (a vector field) has spin 1. The relation between the $\hat{S}_i$-operators [Example 1.11, Eq. (9)] and the representation in (2.21) will be clarified in Exercise 2.1.

For spin $\frac{3}{2}$ we get the $2j+1$-dimensional matrices

$$\hat{S}_x = \frac{1}{2}\begin{pmatrix} 0 & \sqrt{3} & 0 & 0 \\ \sqrt{3} & 0 & 2 & 0 \\ 0 & 2 & 0 & \sqrt{3} \\ 0 & 0 & \sqrt{3} & 0 \end{pmatrix} \quad , \quad \hat{S}_y = \frac{1}{2}\begin{pmatrix} 0 & -\mathrm{i}\sqrt{3} & 0 & 0 \\ -\mathrm{i}\sqrt{3} & 0 & -2\mathrm{i} & 0 \\ 0 & 2\mathrm{i} & 0 & -\mathrm{i}\sqrt{3} \\ 0 & 0 & \mathrm{i}\sqrt{3} & 0 \end{pmatrix} \quad ,$$

$$\hat{S}_z = \frac{1}{2}\begin{pmatrix} 3 & 0 & 0 & 0 \\ 0 & 1 & 0 & 0 \\ 0 & 0 & -1 & 0 \\ 0 & 0 & 0 & 3 \end{pmatrix} \quad , \quad \hat{S}^2 = \frac{3}{2}\left(\frac{3}{2}+1\right)\begin{pmatrix} 1 & 0 & 0 & 0 \\ 0 & 1 & 0 & 0 \\ 0 & 0 & 1 & 0 \\ 0 & 0 & 0 & 1 \end{pmatrix} \qquad (2.22b)$$

which will not be discussed further here. We can get the matrices for higher angular momenta in a similar manner. The corresponding eigenfunctions are column vectors (spinors) with $(2j+1)$ components.

We have constructed the matrix representations as matrix elements of the angular momentum operators, with the states of the invariant subspaces which are not further separable. As we have mentioned above, such representations are called "irreducible". They play an important role in quantum mechanics, as in general every matrix representation is separable into a product of irreducible representations.

EXERCISE

2.1 Special Representation of the Spin-1 Operators

Problem. Find the basis in which the spin-1 operators $\hat{S}^2$ and $\hat{S}_3$ are diagonal. Derive $\hat{S}_1$, $\hat{S}_2$, and $\hat{S}_3$ in this representation. Show that the generators of SO(3)

Exercise 2.1

for vector fields can be transformed by a unitary transformation to the form (2.22).

Solution. From Example 1.11, Eqs. (9) and (19) we know (setting $\hbar = 1$) that

$$\hat{\boldsymbol{S}}^2 = 2\begin{pmatrix} 1 & 0 & 0 \\ 0 & 1 & 0 \\ 0 & 0 & 1 \end{pmatrix} \quad \text{and}$$

$$\hat{S}_3 = \mathrm{i}\begin{pmatrix} 0 & -1 & 0 \\ 1 & 0 & 0 \\ 0 & 0 & 0 \end{pmatrix} \quad . \tag{1}$$

Obviously $\hat{\boldsymbol{S}}^2$ is already diagonal with eigenvalues 2. Since (1) [respectively (9), (19) from Example 1.11] was based on the vector $\boldsymbol{\psi} = \{\psi_1, \psi_2, \psi_3\}$ in *Cartesian representation*, $\{\psi_1, \psi_2, \psi_3\}$ are the Cartesian components of $\boldsymbol{\psi}$. Thus,

$$\hat{\boldsymbol{S}}^2\boldsymbol{\psi} = 2\begin{pmatrix} 1 & 0 & 0 \\ 0 & 1 & 0 \\ 0 & 0 & 1 \end{pmatrix}\begin{pmatrix} \psi_1 \\ \psi_2 \\ \psi_3 \end{pmatrix} = 2\begin{pmatrix} \psi_1 \\ \psi_2 \\ \psi_3 \end{pmatrix} = 2\boldsymbol{\psi} \quad ,$$

$$\hat{S}_3\boldsymbol{\psi} = \mathrm{i}\begin{pmatrix} 0 & -1 & 0 \\ 1 & 0 & 0 \\ 0 & 0 & 0 \end{pmatrix}\begin{pmatrix} \psi_1 \\ \psi_2 \\ \psi_3 \end{pmatrix} = \mathrm{i}\begin{pmatrix} -\psi_2 \\ \psi_1 \\ 0 \end{pmatrix} \quad . \tag{2}$$

We are looking for a unitary transformation of the spinors

$$\begin{pmatrix} \psi'_1 \\ \psi'_2 \\ \psi'_3 \end{pmatrix} = \begin{pmatrix} U_{11} & U_{12} & U_{13} \\ U_{21} & U_{22} & U_{23} \\ U_{31} & U_{32} & U_{33} \end{pmatrix}\begin{pmatrix} \psi_1 \\ \psi_2 \\ \psi_3 \end{pmatrix}$$

or, in shorthand notation,

$$\boldsymbol{\psi}' = \hat{U}\boldsymbol{\psi} \tag{3}$$

such that

$$\hat{\boldsymbol{S}}^2\boldsymbol{\psi}' = 2\boldsymbol{\psi}' \quad , \tag{4a}$$

$$\hat{S}_3\boldsymbol{\psi}' = \mu\boldsymbol{\psi}' \quad . \tag{4b}$$

Equation (4a) is always fulfilled, since by inserting (3) in (4a) it follows

$$\hat{\boldsymbol{S}}^2\hat{U}\boldsymbol{\psi} = 2\hat{U}\boldsymbol{\psi} \quad , \quad \text{or} \quad \hat{U}^{-1}\hat{\boldsymbol{S}}^2\hat{U}\boldsymbol{\psi} = 2\boldsymbol{\psi} \quad ,$$

$$\hat{U}^{-1}(2\cdot\mathbb{1})\hat{U}\boldsymbol{\psi} = 2\boldsymbol{\psi} \quad , \quad 2\boldsymbol{\psi} = 2\boldsymbol{\psi} \quad .$$

Independent of the unitary transformation $\hat{U}$, (4a) is always fulfilled because of (1). Equation (4b) is an eigenvalue equation with μ as eigenvalue. Explicitly it reads,

Exercise 2.1

$$\mathrm{i}\begin{pmatrix}0 & -1 & 0\\ 1 & 0 & 0\\ 0 & 0 & 0\end{pmatrix}\begin{pmatrix}\psi'_1\\ \psi'_2\\ \psi'_3\end{pmatrix}_\mu = \mu\begin{pmatrix}\psi'_1\\ \psi'_2\\ \psi'_3\end{pmatrix}_\mu \quad \text{or}$$

$$\left[\mathrm{i}\begin{pmatrix}0 & -1 & 0\\ 1 & 0 & 0\\ 0 & 0 & 0\end{pmatrix} - \mu\begin{pmatrix}1 & 0 & 0\\ 0 & 1 & 0\\ 0 & 0 & 1\end{pmatrix}\right]\begin{pmatrix}\psi'_1\\ \psi'_2\\ \psi'_3\end{pmatrix}_\mu = 0 \quad ,$$

thus

$$\begin{pmatrix}-\mu & -\mathrm{i} & 0\\ \mathrm{i} & -\mu & 0\\ 0 & 0 & -\mu\end{pmatrix}\begin{pmatrix}\psi'_1\\ \psi'_2\\ \psi'_3\end{pmatrix}_\mu = 0 \quad .$$

This is a linear, homogeneous system of equations, which has a non-trivial solution only if the determinant of its coefficients is zero, i.e.

$$\begin{vmatrix}-\mu & -\mathrm{i} & 0\\ \mathrm{i} & -\mu & 0\\ 0 & 0 & -\mu\end{vmatrix} = -\mu^3 - (\mathrm{i})^2\mu = 0 \quad .$$

Hence, the three eigenvalues μ are

$$\mu_0 = 0 \quad , \quad \mu_+ = +1 \quad , \quad \mu_- = -1 \quad . \tag{5}$$

The corresponding eigenvectors can be easily determined. Let us first consider the eigenvalue $\mu = +1$:

$$\begin{pmatrix}-1 & -\mathrm{i} & 0\\ \mathrm{i} & -1 & 0\\ 0 & 0 & -1\end{pmatrix}\begin{pmatrix}\psi'_1\\ \psi'_2\\ \psi'_3\end{pmatrix}_+ = 0 \quad ,$$

$$\left.\begin{aligned}-\psi'_1 - \mathrm{i}\psi'_2 &= 0\\ \mathrm{i}\psi'_1 - \psi'_2 &= 0\\ -\psi'_3 &= 0\end{aligned}\right\} \Rightarrow \left\{\begin{aligned}\psi'_1 &= A'\\ \psi'_2 &= \mathrm{i}A'\\ \psi'_3 &= 0\end{aligned}\right. \quad .$$

where A' is arbitrary. Thus,

$$\begin{pmatrix}\psi'_1\\ \psi'_2\\ \psi'_3\end{pmatrix}_+ = \begin{pmatrix}A'\\ \mathrm{i}A'\\ 0\end{pmatrix} \quad .$$

Exercise 2.1

After normalization we find, up to a phase factor, that

$$\begin{pmatrix} \psi'_1 \\ \psi'_2 \\ \psi'_3 \end{pmatrix}_+ = -\frac{1}{\sqrt{2}}\begin{pmatrix} 1 \\ \mathrm{i} \\ 0 \end{pmatrix} \quad . \tag{6a}$$

Analogously, one obtains

$$\begin{pmatrix} \psi'_1 \\ \psi'_2 \\ \psi'_3 \end{pmatrix}_0 = \begin{pmatrix} 0 \\ 0 \\ 1 \end{pmatrix} \quad \text{and} \quad \begin{pmatrix} \psi'_1 \\ \psi'_2 \\ \psi'_3 \end{pmatrix}_- = \frac{1}{\sqrt{2}}\begin{pmatrix} 1 \\ -\mathrm{i} \\ 0 \end{pmatrix} \quad . \tag{6b}$$

The phases here have been determined in complete analogy to the eigenfunctions of the orbital angular momentum $1\hbar$, i.e. $Y_{1m}(\theta,\phi)$. In order to make this clear, we will introduce *spherical basis vectors* $\xi_\mu(\mu = \pm 1, 0)$ and compare them with the vectors (6a) and (6b). The spherical basis vectors ξ_μ are constructed from the Cartesian basis vectors in the following way:

$$\begin{aligned} \xi_1 &= -\frac{1}{\sqrt{2}}(\boldsymbol{e}_1 + \mathrm{i}\boldsymbol{e}_2) \\ &= -\frac{1}{\sqrt{2}}\left\{\begin{pmatrix} 1 \\ 0 \\ 0 \end{pmatrix} + \mathrm{i}\begin{pmatrix} 0 \\ 1 \\ 0 \end{pmatrix}\right\} = -\frac{1}{\sqrt{2}}\begin{pmatrix} 1 \\ +\mathrm{i} \\ 0 \end{pmatrix} \quad , \\ \xi_0 &= \boldsymbol{e}_3 = \begin{pmatrix} 0 \\ 0 \\ 1 \end{pmatrix} \quad , \\ \xi_{-1} &= \frac{1}{\sqrt{2}}(\boldsymbol{e}_1 - \mathrm{i}\boldsymbol{e}_2) \\ &= \frac{1}{\sqrt{2}}\left\{\begin{pmatrix} 1 \\ 0 \\ 0 \end{pmatrix} - \mathrm{i}\begin{pmatrix} 0 \\ 1 \\ 0 \end{pmatrix}\right\} = \frac{1}{\sqrt{2}}\begin{pmatrix} 1 \\ -\mathrm{i} \\ 0 \end{pmatrix} \quad . \end{aligned} \tag{7}$$

These spherical unitary vectors ξ_μ, according to (4) and (6) represent the eigenvectors of $\hat{\boldsymbol{S}}^2$ and $\hat{S}_3$. Comparing them with the spherical harmonics $Y_{1m}(\theta,\phi)$[2] for $m = 1, 0, -1$,

$$Y_{1m}(\theta,\phi) = \sqrt{3/(4\pi)}\,\frac{1}{r}\begin{cases} -(x+\mathrm{i}y)/\sqrt{2}, & m = 1 \\ z, & m = 0 \\ (x-\mathrm{i}y)/\sqrt{2}, & m = -1 \end{cases} \quad ,$$

[2] See J.D.Jackson: *Classical Electrodynamics*, 2nd ed. (Wiley, New York 1975) or W. Greiner: *Theoretische Physik, Elektrodynamik* (Harri Deutsch, Frankfurt 1986), Chap. 3.

the analogy becomes evident. The spherical harmonics $Y_{1m}(\theta, \phi)$ correspond – up to the factor $\sqrt{3/(4\pi)}/r$ – to the spherical representation of the position vector $\boldsymbol{r} = \{x, y, z\}$. The transformation (7) is achieved by the matrix $\hat{U}$

$$\begin{pmatrix} \xi_1 \\ \xi_0 \\ \xi_{-1} \end{pmatrix} = \hat{U} \begin{pmatrix} e_1 \\ e_2 \\ e_3 \end{pmatrix}$$

$$= \begin{pmatrix} -1/\sqrt{2} & -\mathrm{i}/\sqrt{2} & 0 \\ 0 & 0 & 1 \\ 1/\sqrt{2} & -\mathrm{i}/\sqrt{2} & 0 \end{pmatrix} \begin{pmatrix} e_1 \\ e_2 \\ e_3 \end{pmatrix} \quad . \tag{8}$$

It is the same matrix as in (3) leading to (6). The transformation (3) describes the transition from Cartesian components $\boldsymbol{\psi} = \{\psi_1, \psi_2, \psi_3\}$ to spherical coordinates $\{\psi'_1, \psi'_0, \psi'_{-1}\}$:

$$\begin{aligned} \psi'_1 &= -(\psi_1 + \mathrm{i}\psi_2)/\sqrt{2} \quad , \\ \psi'_0 &= \psi_3 \quad , \\ \psi'_{-1} &= (\psi_1 - \mathrm{i}\psi_2)/\sqrt{2} \quad \text{or} \end{aligned} \tag{9}$$

$$\begin{pmatrix} \psi'_1 \\ \psi'_0 \\ \psi'_{-1} \end{pmatrix} = \begin{pmatrix} -1/\sqrt{2} & -\mathrm{i}/\sqrt{2} & 0 \\ 0 & 0 & 1 \\ 1/\sqrt{2} & -\mathrm{i}/\sqrt{2} & 0 \end{pmatrix} \begin{pmatrix} \psi_1 \\ \psi_2 \\ \psi_3 \end{pmatrix}$$

An arbitrary vector can be represented in the spherical basis with the help of (7) so that

$$\boldsymbol{A} = \sum_{\mu=1}^{-1} A'_\mu \xi^*_\mu = \sum_\mu A'_\mu (-)^\mu \xi_{-\mu} \quad . \tag{10}$$

Since

$$\xi^*_\mu \cdot \xi_{\mu'} = \delta_{\mu\mu'} \quad ,$$

which can easily be proven from (7), it follows that

$$A'_\mu = \boldsymbol{A} \cdot \xi_\mu \quad .$$

Thus,

$$\begin{aligned} A'_1 &= \boldsymbol{A} \cdot \xi_1 = -(A_1 + \mathrm{i}A_2)/\sqrt{2} \quad , \\ A'_0 &= \boldsymbol{A} \cdot \xi_0 = A_3 \quad , \\ A'_{-1} &= \boldsymbol{A} \cdot \xi_{-1} = (A_1 - \mathrm{i}A_2)/\sqrt{2} \quad , \end{aligned}$$

in accordance with (9). Due to the relations

$$\xi^*_\mu \cdot \xi_{\mu'} = \delta_{\mu\mu'} \quad \text{and} \quad \xi^*_\mu = (-)^\mu \xi_{-\mu} \quad ,$$

Exercise 2.1

the scalar product of two vectors $\boldsymbol{A}$ and $\boldsymbol{B}$ in spherical representation simply reads,

$$\begin{aligned}\boldsymbol{A}\cdot\boldsymbol{B} &= \sum_{\mu} A'_\mu \xi^*_\mu \cdot \sum_{\nu} B'_\nu \xi^*_\nu = \sum_{\mu,\nu} A'_\mu B'_\nu \xi^*_\mu (-)^\nu \xi_{-\nu} \\ &= \sum_{\mu,\nu} A'_\mu B'_\nu (-1)^\nu \delta_{\mu,-\nu} = \sum_{\mu} (-1)^\mu A'_\mu B'_{-\mu} \\ &= \sum_{\mu} A'_\mu B'^*_\mu \quad .\end{aligned}$$

The spin vectors $\hat{S}_i$ in the eigenrepresentation of $\hat{\boldsymbol{S}}^2$ and $\hat{S}_3$ can be calculated in two different ways:

a) $(\hat{S}'_i)_{\mu\nu} = \xi^\dagger_\mu \hat{S}_i \xi_\nu \quad .$ (12)

This gives

$$(\hat{S}'_3)_{\mu\nu} = \xi^\dagger_\mu \hat{S}_3 \xi_\nu = \begin{pmatrix} 1 & 0 & 0 \\ 0 & 0 & 0 \\ 0 & 0 & -1 \end{pmatrix} \quad . \tag{13}$$

As an example we check the element $(\hat{S}'_3)_{11}$, giving

$$\begin{aligned}(\hat{S}'_3)_{11} &= \xi^\dagger_1 \hat{S}_3 \xi_1 \\ &= \frac{1}{\sqrt{2}}(1 \quad -\mathrm{i} \quad 0)\mathrm{i}\begin{pmatrix} 0 & -1 & 0 \\ 1 & 0 & 0 \\ 0 & 0 & 0 \end{pmatrix}\left(-\frac{1}{\sqrt{2}}\right)\begin{pmatrix} 1 \\ \mathrm{i} \\ 0 \end{pmatrix} \\ &= \frac{\mathrm{i}}{2}(-\mathrm{i} \quad -1 \quad 0)\begin{pmatrix} 1 \\ \mathrm{i} \\ 0 \end{pmatrix} = \frac{\mathrm{i}}{2}(-2\mathrm{i}) = 1 \quad .\end{aligned}$$

The diagonal representation of $\hat{S}_3$ [in (13)] clearly exhibits the possible eigenvalues (spin orientations along the z axis), namely $1, 0, -1$. For the other components of $\hat{\boldsymbol{S}}$ we find analogously

$$(\hat{S}'_1)_{\mu\nu} = \frac{1}{\sqrt{2}}\begin{pmatrix} 0 & 1 & 0 \\ 1 & 0 & 1 \\ 0 & 1 & 0 \end{pmatrix} \quad ,$$

$$(\hat{S}'_2)_{\mu\nu} = \frac{1}{\sqrt{2}}\begin{pmatrix} 0 & -\mathrm{i} & 0 \\ \mathrm{i} & 0 & -\mathrm{i} \\ 0 & \mathrm{i} & 0 \end{pmatrix} \quad ,$$

$$(\hat{S}'_3)_{\mu\nu} = \begin{pmatrix} 1 & 0 & 0 \\ 0 & 0 & 0 \\ 0 & 0 & -1 \end{pmatrix} \quad .$$

We have already seen this representation in (2.21).

Exercise 2.1

b) The diagonal representation of $\hat{S}_3$ can also be obtained in the following way: Let

$$\hat{S}'_3 = \mathrm{i}\begin{pmatrix} 0 & -1 & 0 \\ 1 & 0 & 0 \\ 0 & 0 & 0 \end{pmatrix} \tag{15}$$

be the Cartesian representation of the $\hat{S}_3$ operator. We next evaluate the matrix elements

$$(\hat{S}'_3)_{\mu\nu} = \langle \xi_\mu | \hat{S}'_3 | \xi_\nu \rangle = \xi^\dagger_\mu \hat{S}'_3 \xi_\nu \quad (\mu, \nu = 1, 0, -1) \quad .$$

According to (8) we have

$$\xi_\mu = \sum_{n=1}^{3} U_{\mu n} e_n \quad , \quad \text{where}$$

$$\hat{U} = \begin{pmatrix} -1/\sqrt{2} & -\mathrm{i}/\sqrt{2} & 0 \\ 0 & 0 & 1 \\ 1/\sqrt{2} & -\mathrm{i}/\sqrt{2} & 0 \end{pmatrix} \quad , \quad \text{i.e.}$$

$$\xi^\dagger_\mu = \sum_{m=1}^{3} U^*_{\mu n} e^\dagger_n \quad \text{with} \quad e^\dagger_n = (\delta_{1n}, \delta_{2n}, \delta_{3n}) \quad .$$

Using $\xi_\nu = \sum_{m=1}^{3} U_{\nu m} e_m$, the matrix elements follow as

$$(\hat{S}'_3)_{\mu\nu} = \sum_{n,m=1}^{3} U^*_{\mu n} e^\dagger_n \hat{S}'_3 e_m U_{\nu m} \quad .$$

$\hat{S}'^{(3)}_{nm} = e^\dagger_n \hat{S}'_3 e_m$, however, is just the Cartesian representation of the $\hat{S}'_3$ operator, given by (15) above, i.e. we obtain

$$(\hat{S}'_3)_{\mu\nu} = \sum_{n,m=1}^{3} U^*_{\mu n} S'^{(3)}_{nm} U_{\nu m} \quad .$$

Introducing a matrix $\hat{V} = \hat{U}^*$, we can write in a more compact form

$$(\hat{S}'_3)_{\mu\nu} = \sum_{n,m=1}^{3} V_{\mu n} \hat{S}'^{(3)}_{nm} V^*_{\nu m}$$

$$= \sum_{n,m=1}^{3} V_{\mu n} \hat{S}'^{(3)}_{nm} V^\dagger_{m\nu} \quad ,$$

or in operator notation

$$\hat{S}'_{3,\,\mathrm{spher.}} = \hat{V} \hat{S}'_{3,\,\mathrm{cart.}} \hat{V}^\dagger \quad . \tag{16}$$

Exercise 2.1

We verify it by writing out explicitly

$$\hat{S}'_{3,\,\mathrm{cart.}}\hat{V}^{\dagger} = \mathrm{i}\begin{pmatrix} 0 & -1 & 0 \\ 1 & 0 & 0 \\ 0 & 0 & 0 \end{pmatrix}\begin{pmatrix} -1/\sqrt{2} & 0 & 1/\sqrt{2} \\ -\mathrm{i}/\sqrt{2} & 0 & -\mathrm{i}/\sqrt{2} \\ 0 & 1 & 0 \end{pmatrix}$$

$$= \mathrm{i}\begin{pmatrix} \mathrm{i}/\sqrt{2} & 0 & \mathrm{i}/\sqrt{2} \\ -1/\sqrt{2} & 0 & -1/\sqrt{2} \\ 0 & 0 & 0 \end{pmatrix} \quad ,$$

and, therefore,

$$\hat{V}\hat{S}'_{3,\,\mathrm{cart.}}\hat{V}^{\dagger} = \mathrm{i}\begin{pmatrix} -1/\sqrt{2} & \mathrm{i}/\sqrt{2} & 0 \\ 0 & 0 & 1 \\ 1/\sqrt{2} & -\mathrm{i}/\sqrt{2} & 0 \end{pmatrix}\begin{pmatrix} \mathrm{i}/\sqrt{2} & 0 & \mathrm{i}/\sqrt{2} \\ -1/\sqrt{2} & 0 & 1/\sqrt{2} \\ 0 & 0 & 0 \end{pmatrix}$$

$$= \mathrm{i}\begin{pmatrix} +(\mathrm{i}+\mathrm{i})/2 & 0 & 0 \\ 0 & 0 & 0 \\ 0 & 0 & -(\mathrm{i}+\mathrm{i})/2 \end{pmatrix} \quad ,$$

i.e.

$$\hat{S}'_{3,\,\mathrm{spher.}} = \hat{V}\hat{S}'_{3,\,\mathrm{cart.}}\hat{V}^{\dagger} = \begin{pmatrix} 1 & 0 & 0 \\ 0 & 0 & 0 \\ 0 & 0 & -1 \end{pmatrix} \quad .$$

Thus we have explicitly shown that the unitary transformation $\hat{V}\hat{S}'_{3,\,\mathrm{cart.}}\hat{V}^{\dagger}$ diagonalizes the z component of the spin operator.

2.3 Addition of Two Angular Momenta

Now we want to consider the case of two angular momenta $\hat{\boldsymbol{J}}_1$ and $\hat{\boldsymbol{J}}_2$ which are combined to form a resulting total angular momentum $\hat{\boldsymbol{J}}$,

$$\hat{\boldsymbol{J}} = \hat{\boldsymbol{J}}_1 + \hat{\boldsymbol{J}}_2 \quad .$$

At the beginning of this chapter we showed that the sum of angular momentum operators obeys the same commutation relations as the individual angular momentum operators. Let $\psi_{j_1m_1}$ and $\psi_{j_2m_2}$ be the orthonormal set of eigenfunctions of the operators $\hat{\boldsymbol{J}}_1^2$, $\hat{J}_{1z}$ and $\hat{\boldsymbol{J}}_2^2$, $\hat{J}_{2z}$, respectively. We then have that

$$\hat{\boldsymbol{J}}_1^2\psi^{(1)}_{j_1m_1} = j_1(j_1+1)\psi^{(1)}_{j_1m_1} \quad \text{and} \quad \hat{\boldsymbol{J}}_2^2\psi^{(2)}_{j_2m_2} = j_2(j_2+2)\psi^{(2)}_{j_2m_2} \quad ,$$

$$\hat{J}_{1z}\psi^{(1)}_{j_1m_1} = m_1\psi^{(1)}_{j_1m_1} \quad \text{and} \quad \hat{J}_{2z}\psi^{(2)}_{j_2m_2} = m_2\psi^{(2)}_{j_2m_2} \quad .$$

The arguments $\boldsymbol{r}_1, t_1$ and $\boldsymbol{r}_2, t_2$ are here abbreviated by "1" and "2", respectively. However, these will generally be omitted in the following unless necessary for clearer understanding.

This way of coupling the angular momentum also arises in the theory of many-body problems, e.g. the two-electron problem, in which the single electrons can be described by the wave function $\psi_{j_1m_1}(1)$ and $\psi_{j_2m_2}(2)$. The total wave function of the two-electron system is then given by $\psi_{jm}(1,2)$, with a total angular momentum of j and its 2-component m. But also in the two-electron problem there is a coupling between spin and orbital angular momentum to a total angular momentum and we have to use the technique which will be introduced now.

The eigenfunctions of the total angular momentum operators $\hat{\boldsymbol{J}}^2$ and $\hat{J}_z$ are denoted by $\psi_{jm}(1,2)$. As long as there is no coupling between the two systems, the angular momenta $\hat{J}_1$, $\hat{J}_2$, and $\hat{J}$ are fixed. The wave function ψ separates into $\psi(1)$ and $\psi(2)$, and we can write ψ_{jm} as a product of $\psi_{j_1m_1} \times \psi_{j_2m_2}$. If there is any coupling, ψ_{jm} can always be described as a linear combination of products $\psi_{j_1m_1} \times \psi_{j_2m_2}$. We write the coefficients in the form $(j_1 j_2 j \,|\, m_1 m_2 m)$ in order to show the dependence on the various quantum numbers. These coefficients are called ***Clebsch-Gordan*** *coefficients.*

Thus, we write for the total wave function

$$\psi_{jm}(1,2) = \sum_{m_1 m_2} (j_1 j_2 j \,|\, m_1 m_2 m)\, \psi^{(1)}_{j_1m_1} \psi^{(2)}_{j_2m_2} \quad . \tag{2.23}$$

If there is coupling between the angular momenta, then $\psi_{j_1m_1}$ and $\psi_{j_2m_2}$ are no longer good eigenfunctions (which means that m_1 and m_2 are not conserved quantum numbers), since the constituent momenta are precessing around the total angular momentum. This is already expressed in the sum over m_1 and m_2 in (2.23). The relation (2.23) gives a transformation of the Hilbert space, spanned by the orthonormal vectors $\psi_{j_1m_1}$ and $\psi_{j_2m_2}$, to the new orthonormal basis set ψ_{jm} of the same subspace. The total product space is invariant, but can be further decomposed, whereas the invariant subspaces, spanned by ψ_{jm} for a fixed j, cannot be decomposed. We decompose, mathematically speaking, the representation of the rotation group generated by the product space into its irreducible parts.

If ψ_{jm} denotes an eigenfunction of the operators $\hat{\boldsymbol{J}}^2$ and $\hat{J}_z$, then conditions for the evaluation of the coefficients $(j_1 j_2 j \,|\, m_1 m_2 m)$ can be found, and we write

$$\begin{aligned} \hat{J}_z \psi_{jm} &= (\hat{J}_{1z} + \hat{J}_{2z}) \sum_{m_1, m_2} (j_1 j_2 j \,|\, m_1 m_2 m)\, \psi_{j_1m_1} \psi_{j_2m_2} \\ &= \sum_{m_1, m_2} (m_1 + m_2)(j_1 j_2 j \,|\, m_1 m_2 m)\, \psi_{j_1m_1} \psi_{j_2m_2} \end{aligned}$$

and in the same way

$$\hat{J}_z \psi_{jm} = m \psi_{jm} \sum_{m_1, m_2} m (j_1 j_2 j \,|\, m_1 m_2 m)\, \psi_{j_1m_1} \psi_{j_2m_2} \quad .$$

The m_1 sum is performed over all values $-j_1 \leq m_1 \leq j_1$, and analogously in the m_2 case over the interval $-j_2 \leq m_2 \leq j_2$. By identifying the above relations we obtain, because of linear independence, the condition

$$(m - m_1 - m_2)(j_1 j_2 j | m_1 m_2 m) = 0 \quad . \tag{2.24}$$

Thus it follows that the Clebsch-Gordan coefficient vanishes if $m \neq m_1 + m_2$. This means that the double sum in (2.23) reduces to a single sum, since either the coefficient vanishes, or m_2 can be determined by $m_2 = m - m_1$. Equation (2.23) thus becomes

$$\psi_{jm} = \sum_{m_1} (j_1 j_2 j | m_1 m_2 m) \psi_{j_1 m_1} \psi_{j_2, m - m_1} \quad . \tag{2.25}$$

The conservation of angular momentum (more precisely: of the projection of the angular momentum on the quantization direction) is expressed by the relation $m_1 + m_2 = m$. In the next step we want to calculate the possible values of the quantum number j defined by

$$\hat{\boldsymbol{J}}^2 \psi_{jm} = j(j+1)\psi_{jm} \quad .$$

Since $\hat{\boldsymbol{J}}$ is an angular momentum and obeys the commutation relations (2.23) we have

$$-j \leq m \leq j \quad .$$

In the following we assume that there exist several values of j (at least two). The states ψ_{jm} are assumed to be orthonormal. Since the wave function contains the coordinates of two particles (angular momenta), we have

$$\begin{aligned}\delta_{jj'}\delta_{mm'} &= \int \psi^*_{jm} \psi_{j'm'} dV_1\, dV_2 \\ &= \int dV_1\, dV_2 \left\{ \sum_{m_1} (j_1 j_2 j | m_1 m - m_1 m) \psi_{j_1 m_1} \psi_{j_2, m - m_1} \right\}^* \\ &\quad \times \left\{ \sum_{m'_1} (j_1 j_2 j' | m'_1 m' - m'_1 m') \psi_{j_1 m'_1} \psi_{j_2, m' - m'_1} \right\} \quad .\end{aligned}$$

From this equation it follows that

$$\delta_{jj'}\delta_{mm'} = \sum_{m_1, m'_1} (j_1 j_2 j | m_1 - m_1 m) * (j_1 j_2 j' | m'_1 m' - m'_1 m') \delta_{m_1 m'_1} \delta_{m - m_1, m' - m'_1}$$

or

$$\delta_{jj'} = \sum_{m_1} (j_1 j_2 j | m_1 - m_1 m) * (j_1 j_2 j' | m_1 - m_1 m) \quad . \tag{2.26}$$

Equation (2.26) expresses the orthogonality of rows of the Clebsch-Gordan coefficients. Since they are real – as we will see – we may omit the complex conjugation sign "*" in (2.26).

From the relation $m = m_1 + m_2$, derived above, it follows that the greatest value of m, which we call $m_{\max}$ is

$$m_{\max} = j_1 + j_2 \quad . \tag{2.27}$$

This value appears only once in (2.25), namely if $m_1 = j_1$ and $m_2 = j_2$. This shows that the largest eigenvalue j, which we call $j_{\max}$, must be

$$j_{\max} = j_1 + j_2 \quad . \tag{2.28}$$

The second largest m value is $m_{\max} - 1$. It occurs twice, for

$$m_1 = j_1 \quad , \quad m_2 = j_2 - 1 \quad \text{and} \quad m_1 = j_1 - 1 \quad , \quad m_2 = j_2 \quad . \tag{2.29}$$

One of the two possible linearly independent combinations of the two states (2.29),

$$\psi_{j_1 j_1}\psi_{j_2, j_2 - 1} \quad \text{and} \quad \psi_{j_1, j_1 - 1}\psi_{j_2 j_2} \quad , \tag{2.30}$$

must belong to $\psi_{j=j_1+j_2,m}(1,2)$, since m assumes all values $j_1 + j_2 \geq m \geq -j_1 - j_2$ in integer steps. The other possible combination necessarily belongs to the state

$$\psi_{jm} \quad \text{with} \quad j = j_1 + j_2 - 1 \tag{2.31}$$

because there are no states with $j > j_{\max} = j_1 + j_2$. There can be only one state of the kind (2.31), i.e. with $j = j_1 + j_2 - 1$, since for a second state of this kind the corresponding basis combination (2.30) with $m = j_1 + j_2 - 1$ is missing.

Continuing with this argument we notice that for j, all values

$$j = \begin{cases} j_1 + j_2 \\ j_1 + j_2 - 1 \\ \vdots \\ |j_1 - j_2| \end{cases} \tag{2.32}$$

appear precisely once. This is the so called *triangle rule*,

$$\Delta(j_1 j_2 j) \quad ,$$

which tells us that two angular momenta j_1, j_2 can only be combined to form such a resulting total angular momentum j as is compatible with *vector addition* (*triangle*). This model, illustrated in Fig. 2.1, is also called the *vector model* of angular momentum coupling.

We can now count the number of coupled states $\psi_{jm}(1,2)$, namely,

$$\sum_{j=|j_1-j_2|}^{j_1+j_2} (2j+1) = (2j_1+1)(2j_2+1) \quad . \tag{2.33}$$

As expected, it equals the number of basis states $\psi_{j_1 m_1} \times \psi_{j_2 m_2}$.

It is difficult to derive algebraically the lower limit for the resulting angular momentum j in (2.32). A simpler approach is given by the following consideration on dimensions: Since the dimensions of Hilbert spaces spanned by ψ_{jm} and $\psi_{j_1 m_1} \times \psi_{j_2 m_2}$ must be equal, and $j_{\max} = j_1 + j_2$, one may write (2.33) in the form,

$$\sum_{j=j_{\min}}^{j_{\max}} (2j+1) = (2j_1+1)(2j_2+1) \quad .$$

This equation determines $j_{\min}$, for which we find $j_{\min} = j_1 - j_2$ (if $j_1 > j_2$), or $j_{\min} = j_2 - j_1$ (for $j_1 < j_2$).

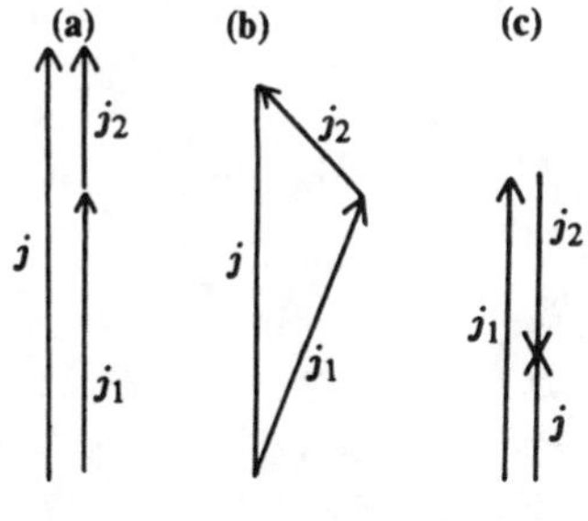

Fig. 2.1a–c. Illustration of the triangle rule: **(a)** maximal coupling by equally oriented j_1 and j_2; **(b)** the angular momentum is between $j_{\max}$ and $j_{\min}$; **(c)** minimal coupling by opposite j_1 and j_2

2.4 Evaluation of Clebsch-Gordan Coefficients

Equation (2.23) defines a unitary transformation from the orthonormal basis set $\psi_{j_1m_1}(1)\,\psi_{j_2m_2}(2)$, consisting of the direct product of *single-particle wave functions* $\psi_{j_1m_1}(1)$ and $\psi_{j_2m_2}(2)$, to the orthonormal basis set $\psi_{jm}(1,2)$. The $\psi_{jm}(1,2)$ are *two-particle wave functions.* In order to simplify the notation, we will now employ Dirac's bra $\langle|$and ket$|\rangle$ vectors. Equation (2.23) then reads[3]

$$|jm\rangle = |m_1m_2\rangle\langle m_1m_2|jm\rangle \quad . \tag{2.34}$$

Here $|m_1m_2\rangle$ stands for the product $|m_1m_2\rangle = j_1m_1|j_2m_2\rangle$. The quantum numbers j_1 and j_2 are omitted, and must be considered as fixed numbers wherever m_1 and m_2 appear. The matrix element

$$\langle m_1m_2|jm\rangle = \int \psi^*_{j_1m_1}(1)\,\psi^*_{j_2m_2}(2)\,\psi_{jm}(1,2)\,dV_1\,dV_2 \tag{2.35}$$

corresponds, according to our previous notation (2.23), to the Clebsch-Gordan coefficient

$$\langle m_1m_2|jm\rangle \equiv (j_1j_2j|m_1m_2m) \quad . \tag{2.36}$$

From the orthonormality of both basis sets we deduce that

$$\delta_{jj'}\delta_{mm'} = \langle jm|j'm'\rangle = \langle jm|m_1m_2\rangle\langle m_1m_2|j'm'\rangle \quad , \tag{2.37}$$

which, together with

$$\langle jm|m_1m_2\rangle = \langle m_1m_2|jm\rangle^* \quad , \tag{2.38}$$

forms our earlier relation (2.26). For the inverse transformation of (2.34), describing the transition from the two-particle basis $|jm\rangle$ to the product basis $|j_1m_1\rangle|j_2m_2\rangle = |m_1m_2\rangle$, we have

$$|m_1m_2\rangle = |jm\rangle\langle jm|m_1m_2\rangle \quad , \tag{2.39}$$

where we now sum over the double indices j and m. From this relation follows

$$\delta_{m_1m_1'}\delta_{m_2m_2'} = \langle m_1'm_2'|m_1m_2\rangle = \langle m_1'm_2'|jm\rangle\langle jm|m_1m_2\rangle \quad , \tag{2.40}$$

i.e. the complementary relation of (2.37), namely the orthogonality of columns of the unitary transformation matrix $\langle jm|m_1m_2\rangle$. The sum over j and m is performed, according to (2.32), over all possible quantum numbers

$$|j_1 - j_2| \le j \le j_1 + j_2 \quad , \quad -j \le m \le j \quad . \tag{2.41}$$

In our original notation (2.40) reads

$$\sum_{j=|j_1-j_2|}^{j_1+j_2}\ \sum_{m=-j}^{j} (j_1j_2j|m_1m_2m)^*(j_1j_2j|m_1'm_2'm) = \delta_{m_1m_1'}\delta_{m_2m_2'} \quad . \tag{2.42}$$

As we have already mentioned, one may construct the Clebsch-Gordan coefficients, i.e. the matrix elements $\langle m_1m_2|jm\rangle$, as real numbers. This will be shown

---

[3] We are also making use of Einstein's sum convention, i.e. we sum over all indices occurring twice on the same side of an equation.

in the following section, and for this reason the asterisk in (2.42) and all other formulae can from now on be omitted.

2.5 Recursion Relations for Clebsch-Gordan Coefficients

It is possible to deduce explicit, albeit quite complicated, relations for Clebsch-Gordan coefficients[4]. There also exist extensive tables that are especially useful for practical work[5]. Here we intend only to show how to calculate them with the help of recursion relations, and to give some examples. The starting point is the operator of total angular momentum,

$$\hat{J} = \hat{J}_1 + \hat{J}_2 \quad ,$$

with its spherical components

$$\hat{J}_\pm = \hat{J}_{1\pm} + \hat{J}_{2\pm} = \hat{J}_x \pm \mathrm{i}\hat{J}_y = (\hat{J}_{1x} + \hat{J}_{2x}) \pm \mathrm{i}(\hat{J}_{1y} + \hat{J}_{2y})$$
$$\hat{J}_0 = \hat{J}_{10} + \hat{J}_{20} = \hat{J}_z = (\hat{J}_{1z} + \hat{J}_{2z}) \quad .$$

Applying it to (2.34) we obtain

$$\begin{aligned}\hat{J}_+|jm\rangle &= (\hat{J}_{1+} + \hat{J}_{2+})|jm\rangle = (\hat{J}_{1+} + \hat{J}_{2+})|m_1m_2\rangle\langle m_1m_2|jm\rangle \\ &= [(\hat{J}_{1+}|m_1m_2\rangle + (\hat{J}_{2+}|m_1m_2\rangle]\langle m_1m_2|jm\rangle \quad ,\end{aligned}$$

and by threefold application of (2.18) we find that

$$\begin{aligned}&[j(j+1) - m(m+1)]^{1/2}|j,m+1\rangle \\ &\quad = \{[j_1(j_1+1) - m_1(m_1+1)]^{1/2}|m_1+1,m_2\rangle \\ &\qquad + [j_2(j_2+1) - m_2(m_2+1)]^{1/2}|m_1,m_2+1\rangle\} \times \langle m_1m_2|jm\rangle \quad . \end{aligned} \tag{2.43}$$

Substitution of (2.34) on the lhs yields

$$\begin{aligned}&[j(j+1) - m(m+1)]^{1/2}|m_1m_2\rangle\langle m_1m_2|jm+1\rangle \\ &\quad = [j_1(j_1+1) - m_1'(m_2'-1)]^{1/2}|m_1'm_2\rangle\langle m_1'-1,m_2|jm\rangle \\ &\qquad + [j_2(j_2+1) - m_2'(m_2'-1)]^{1/2}|m_1m_2'\rangle|m_1m_2'\rangle\langle m_1,m_2'-1|jm\rangle \quad . \end{aligned} \tag{2.44}$$

[4] M.E. Rose: *Elementary Theory of Angular Momentum* (Wiley, New York 1957); A.R. Edmonds: *Angular Momentum in Quantum Mechanics* (Princeton University Press, Princeton, N.J. 1957); D.M. Brink and G.R. Satchler: *Angular Momentum* (Clarendon, Oxford 1962).

[5] M. Rottenberg, R. Bivins, N. Metropolis, and J.K. Wooten, Jr.: *The 3j- and 6j-symbols* (Technology, Cambridge, Mass. 1959).

Here we have introduced $m_1' = m_1 + 1$ in the first term on the rhs, and in the second term $m_2' = m_2 + 1$. The sum goes, just as before, over m_1' from $-j_1$ to j_1, and over m_2' from $-j_2$ to j_2, the reason being the vanishing factor in front of the terms with $m_1 = j_1$ and $m_2 = j_{12}$ in (2.43). Thus, in (2.44), the terms with $m_1' = j_1 + 1$ and $m_2' = j_2 + 1$ do not contribute.

Similarly, we find that the terms with $m_1' = -j_1$ and $m_2' = -j_2$ in (2.44) belong to the null vectors $|-j_1 - 1, m_2\rangle$ and $|m_1, -j_2 - 1\rangle$ in (2.43) and, therefore, do not contribute to the result. Since m_1' and m_2' are summation indices, they can be renamed m_1 for m_1' and m_2 for m_2'.

This yields

$$\begin{aligned}[]&[j(j+1) - m(m+1)]^{1/2} |m_1 m_2\rangle\langle m_1 m_2 | jm+1\rangle \\ &\quad = [j_1(j_1+1) - m_1(m_1-1)]^{1/2} |m_1 m_2\rangle\langle m_1 - 1, m_2 | jm\rangle \\ &\qquad + [j_2(j_2+1) - m_2(m_2-1)]^{1/2} \langle m_1, m_2 - 1 | jm\rangle \end{aligned} \tag{2.45a}$$

is obtained, Here m_1 and m_2 are fixed numbers and not summation indices. Repeating this procedure with $\hat{J}_-$ gives the analogous result

$$\begin{aligned}[]&[j(j+1) - m(m-1)]^{1/2} \langle m_1 m_2 | jm-1\rangle \\ &\quad = [j_1(j_1+1) - m_1(m_1+1)]^{1/2} \langle m_1 + 1, m_2 | jm\rangle \\ &\qquad + [j_2(j_2+1) - m_2(m_2+1)]^{1/2} \langle m_1, m_2 + 1 | jm\rangle \quad . \end{aligned} \tag{2.45b}$$

This recursion relation allows us to derive the Clebsch-Gordan coefficients for the same total angular momenta j, having same j_1 and j_2 but different m. This is shown in the following section.

2.6 Explicit Calculation of Clebsch-Gordan Coefficients

With the help of the recursion relations (2.45a, b) one can now evaluate step by step the quadratic Clebsch-Gordan matrix $\langle m_1 m_2 | jm\rangle$ of dimension $(2j_1 + 1)(2j_2 + 1)$. In deriving (2.33) we proved the dimensionality. However, the matrix now breaks up into *disconnected quadratic submatrices* in accordance with the value of $m = m_1 + m_2$, which can be seen in the following way:

For the different m values we have:

$m = j_1 + j_2$	only a single value each for $j = j_1 + j_2$ and for $(m_1, m_2) = (j_1, j_2)$, i.e. a 1×1 submatrix.
$m = j_1 + j_2 - 1, j = j_1 + j_2$, $j = j_1 + j_2 - 1$	and two possibilities $\begin{cases}(m_1, m_2) = (j_1, j_2 - 1), \\ (m_1, m_2) = (j_1 - 1, j_2),\end{cases}$ i.e. a 2×2 submatrix;

Table 2.1. Structure of the Clebsch-Gordan matrix. The Clebsch-Gordan coefficients $(j_1 j_2 j | m_1 m_2 m) \equiv \langle m_1 m_2 | jm \rangle$ are given for fixed j_1, j_2. The rows are characterized by the pair m_1, m_2, the columns by m. For a given m only a finite number of m_1, m_2 pairs, obeying the condition $m = m_1 + m_2$ are possible; moreover, only j values with $j_1 + j_2 \geq j \geq m$ are allowed. The first $(N-1)$ columns of the $N \times N$ submatrix can be calculated via the recursion relation (2.45b) applied to the known $(N-1)\times(N-1)$ submatrix. The N'th column results from the normalization and orthogonality relations of these rows.

	$m = j_1 + j_2$	$m = j_1 + j_2 - 1$	$m = j_1 + j_2 - 2$	$\cdots$
$(m_1 m_2) = (j_1 j_2)$	$\langle j_1 j_2 \| j_1 + j_2, j_1 + j_2 \rangle$	0	0	$\cdots$
$(m_1 m_2) = \begin{cases} (j_1, j_2 - 1) \\ (j_1 - 1, j_2) \end{cases}$	0	(2×2)*)	0	$\cdots$
$(m_1 m_2) = \begin{cases} (j_1, j_2 - 2) \\ (j_1 - 1, j_2 - 1) \\ (j_1 - 2, j_2) \end{cases}$	0	0	(3×3)**)	$0\cdots$
$\vdots$	$\vdots$	$\vdots$	0 $\vdots$	$(\ldots)\cdots$ $\vdots\ \ddots$

*) $(2\times 2) = \begin{pmatrix} \langle j_1, j_2 - 1 | j_1 + j_2, j_1 + j_2 - 1 \rangle, & \langle j_1, j_2 - 1 | j_1 + j_2 - 1, j_1 + j_2 - 1 \rangle \\ \langle j_1 - 1, j_2 | j_1 + j_2, j_1 + j_2 - 1 \rangle, & \langle j_1 - 1, j_2 | j_1 + j_2 - 1, j_1 + j_2 - 1 \rangle \end{pmatrix}$

**) (3×3)

$$= \begin{pmatrix} \langle j_1, j_2 - 2 | j_1 + j_2, j_1 + j_2 - 2 \rangle, & \langle j_1, j_2 - 2 | j_1 + j_2 - 1, j_1 + j_2 - 2 \rangle, & \langle j_1, j_2 - 2 | j_1 + j_2 - 2, j_1 + j_2 - 2 \rangle \\ \langle j_1 - 1, j_2 - 1 | j_1 + j_2, j_1 + j_2 - 2 \rangle, & \langle j_1 - 1, j_2 - 1 | j_1 + j_2 - 1, j_1 + j_2 - 2 \rangle, & \langle j_1 - 1, j_2 - 1 | j_1 + j_2 - 2, j_1 + j_2 - 2 \rangle \\ \langle j_1 - 2, j_2 | j_1 + j_2, j_1 + j_2 - 2 \rangle, & \langle j_1 - 2, j_2 | j_1 + j_2 - 1, j_1 + j_2 - 2 \rangle, & \langle j_1 - 2, j_2 | j_1 + j_2 - 2, j_1 + j_2 - 2 \rangle \end{pmatrix}$$

$$\begin{aligned} &m = j_1 + j_2 - 2, j = j_1 + j_2, \\ &j = j_1 + j_2 - 1, j = j_1 + j_2 - 2, \end{aligned} \quad \text{and} \begin{cases} (m_1, m_2) = (j_1, j_2 - 2), \\ (m_1, m_2) = (j_1 - 1, j_2 - 1), \\ (m_1, m_2) = (j_1 - 2, j_2), \end{cases}$$

i.e. a 3×3 submatrix,
etc. (2.46)

The structure of the matrix is shown in Table 2.1. The rank of the diagonalized submatrices at first increases in steps of one until a maximum rank is reached for a certain value of m (if $j_1 = j_2$ at $m = 0$), thereafter it decreases in steps of one. The last 1×1 submatrix has $m = -j_1 - j_2$ and $j = j_1 + j_2$. Since the whole matrix is unitary, each of these *submatrices* on the main diagonal must be *unitary*. From this observation it follows that the 1×1 submatrices must be numbers of modulus 1. *We choose the value* $+1$ *by convention*[6]. This is the

[6] See Footnote 1, Section 2.1.

reason for

$$\langle j_1 j_2 | j_1 + j_2, j_1 + j_2 \rangle = 1 \quad . \tag{2.47}$$

This is at once evident, since coupling $\psi_{j_1 m_1}$ and $\psi_{j_2 m_2}$ to maximum angular momentum

$$\psi_{j=j_1+j_2, m=j_1+j_2} = \psi_{j_1 j_1} \psi_{j_2 j_2}$$

can be done only with the maximally oriented single wave functions $\psi_{j_1 j_2}$ and $\psi_{j_2 j_2}$.

The relation (2.47) may be used as the basis of a calculational algorithm if the recursion relation (2.45) is taken into account. We start with (2.45b) and set $m_1 = j_1$, $m_2 = j_2 - 1$, $j = j_1 + j_2$, and $m = j_1 + j_2$. This gives

$$\begin{aligned}
&[(j_1 + j_2)(j_1 + j_2 + 1) - (j_1 + j_2)(j_1 + j_2 - 1)]^{1/2} \langle j_1, j_2 - 1 | j_1 + j_2, j_1 + j_2 - 1 \rangle \\
&\quad = [j_2(j_2 + 1) - (j_2 - 1) j_2]^{1/2} \langle j_1 j_2 | j_1 + j_2, j_1 + j_2 \rangle \\
&\quad = [j_2(j_2 + 1) - (j_2 - 1) j_2]^{1/2} \quad .
\end{aligned}$$

Thus

$$\langle j_1, j_2 - 1 | j_1 + j_2, j_1 + j_2 - 1 \rangle = \left(\frac{j_2}{j_1 + j_2} \right)^{1/2} \quad . \tag{2.48}$$

In similar fashion, by setting $m_1 = j_1 - 1$, $m_2 = j_2$, $j = j_1 + j_2$, $m = j_1 + j_2$, we obtain from (2.45b)

$$\langle j_1 - 1, j_2 | j_1 + j_2, j_1 + j_2 - 1 \rangle = \left(\frac{j_1}{j_1 + j_2} \right)^{1/2} \quad . \tag{2.49}$$

Thus, we have determined the first column of the 2×2 submatrix of the Clebsch-Gordan matrix given in Table 2.1. The second column of this submatrix, belonging to $m = j_1 + j_2 - 1$, $j = j_1 + j_2 - 1$, may be calculated by use of the orthonormality relation (2.37). We must then have:

$$\begin{aligned}
&\sqrt{\frac{j_2}{j_1 + j_2}} \langle j_1, j_2 - 1 | j_1 + j_2 - 1, j_1 + j_2 - 1 \rangle \\
&\quad + \sqrt{\frac{j_1}{j_1 + j_2}} \langle j_1 - 1, j_2 | j_1 + j_2 - 1, j_1 + j_2 - 1 \rangle = 0
\end{aligned}$$

and

$$\begin{aligned}
&|\langle j_1, j_2 - 1 | j_1 + j_2 - 1, j_1 + j_2 - 1 \rangle|^2 \\
&\quad + |\langle j_1 - 1, j_2 | j_1 + j_2 - 1, j_1 + j_2 - 1 \rangle|^2 = 1 \quad .
\end{aligned}$$

The solution of both equations is unique up to an arbitrary phase factor of modulus one. We choose it in such a way that the *first matrix element of the form* $\langle j_1, j - j_1 | jj \rangle$ *is real and positive.* Obviously, our considerations deal with the matrix elements which can be found in the upper right corner of the various submatrices. This yields

$$\begin{aligned}
&\langle j_1, j_2 - 1 | j_1 + j_2 - 1, j_1 + j_2 - 1 \rangle = \left(\frac{j_1}{j_1 + j_2} \right)^{1/2} \quad , \\
&\langle j_1 - 1, j_2 | j_1 + j_2 - 1, j_1 + j_2 - 1 \rangle = - \left(\frac{j_2}{j_1 + j_2} \right)^{1/2} \quad .
\end{aligned} \tag{2.50}$$

We continue our discussion with the 3×3 submatrix of the Clebsch-Gordan matrix (Table 2.1). By inserting (2.48) and (2.49) into the rhs of (2.45b), we obtain, after a brief calculation,

$$\begin{aligned}
&\langle j_1, j_2 - 1 | j_1 + j_2, j_1 + j_2 - 2 \rangle = \left(\frac{j_2(2j_2 - 1)}{(j_1 + j_2)(2j_1 + 2j_2 - 1)} \right)^{1/2} \quad , \\
&\langle j_1 - 1, j_2 - 1 | j_1 + j_2, j_1 + j_2 - 2 \rangle = \left(\frac{4j_1 j_2}{(j_1 + j_2)(2j_1 + 2j_2 - 1)} \right)^{1/2} \quad , \\
&\langle j_1 - 2, j_2 | j_1 + j_2, j_1 + j_2 - 2 \rangle = \left(\frac{j_1(2j_1 - 1)}{(j_1 + j_2)(2j_1 + 2j_2 - 1)} \right)^{1/2} \quad .
\end{aligned} \tag{2.51}$$

This corresponds to the first column of the 3×3 submatrix. Similarly, we obtain the second column by inserting (2.50) into (2.45b), giving

$$\begin{aligned}
&\langle j_1, j_2 - 2 | j_1 + j_2 - 1, j_1 + j_2 - 2 \rangle = \left(\frac{j_1(2j_2 - 1)}{(j_1 + j_2)(j_1 + j_2 - 1)} \right)^{1/2} \quad , \\
&\langle j_1 - 1, j_2 - 1 | j_1 + j_2 - 1, j_1 + j_2 - 2 \rangle = \left(\frac{j_1 - j_2}{[(j_1 + j_2)(j_1 + j_2 - 1)]} \right)^{1/2} \quad , \\
&\langle j_1 - 2, j_2 | j_1 + j_2 - 1, j_1 + j_2 - 2 \rangle = - \left(\frac{j_2(2j_1 - 1)}{(j_1 + j_2)(j_1 + j_2 - 1)} \right)^{1/2} \quad .
\end{aligned} \tag{2.52}$$

We continue in the same spirit: the third column of the 3×3 submatrix follows from the orthonormality relation (2.37). The column vectors (2.51) and (2.52) must be orthonormal to the third one, and the latter must be normalized. In order to fix the *arbitrary phase factor*, we choose, as before, *the first component of the column vector to be real and positive.* The result of this longer calculation is that

$$\begin{aligned}
&\langle j_1, j_2 - 2 | j_1 + j_2 - 2, j_1 + j_2 - 2 \rangle = \left(\frac{j_1(2j_1 - 1)}{(j_1 + j_2 - 1)(2j_1 + 2j_2 - 1)} \right)^{1/2} \quad , \\
&\langle j_1 - 1, j_2 - 1 | j_1 + j_2 - 2, j_1 + j_2 - 2 \rangle = \left(\frac{(2j_1 - 1)(2j_2 - 1)}{(j_1 + j_2 - 1)(2j_1 + 2j_2 - 1)} \right)^{1/2} \quad , \\
&\langle j_1 - 2, j_2 | j_1 + j_2 - 2, j_1 + j_2 - 2 \rangle = \left(\frac{j_2(2j_2 - 1)}{(j_1 + j_2 - 1)(2j_1 + 2j_2 - 1)} \right)^{1/2} \quad .
\end{aligned}$$

The procedure continues in this way. At the next step the 4×4 submatrix of Table 2.1 is evaluated. The calculation of the last column becomes, due to the use of the orthonormality relation (2.38), more and more complicated as the rank of the submatrix increases. One solves this problem in practice by calculating the coefficients numerically for given j_1 and j_2, abandoning closed expressions[7]. As we have already mentioned, there also exist tables of Clebsch-Gordan coefficients. By means of the first recursion relation (2.45a) the Clebsch-Gordan matrix can also be calculated by starting at the other end: For the 1×1 submatrix at the lower and the assignment is $m = -j_1 - j_2$ and $j = j_1 + j_2$. In the following example we apply the above method to a simple problem.

EXAMPLE

2.2 Calculation of the Clebsch-Gordan Coefficients for Spin-Orbit Coupling

The calculation of the coefficients for vector coupling is performed by the application of step operators on the wave functions. For the general case one gets long and cumbersome expressions; we will demonstrate the working principle with a simple example.

We consider an electron with spin $\hbar/2$ and orbital angular momentum $1\hbar$. We thus have

$$l = 1 \quad , \quad s = \tfrac{1}{2} \quad ,$$

and the magnetic quantum numbers can take the values

$$m_l = 0, \pm 1 \quad , \quad m_s = \pm \tfrac{1}{2} \quad .$$

The total angular momentum $\boldsymbol{j} = \boldsymbol{l} + \boldsymbol{s}$, according to (2.32), must lie in the range

$$\tfrac{1}{2} \leq j \leq \tfrac{3}{2} \quad .$$

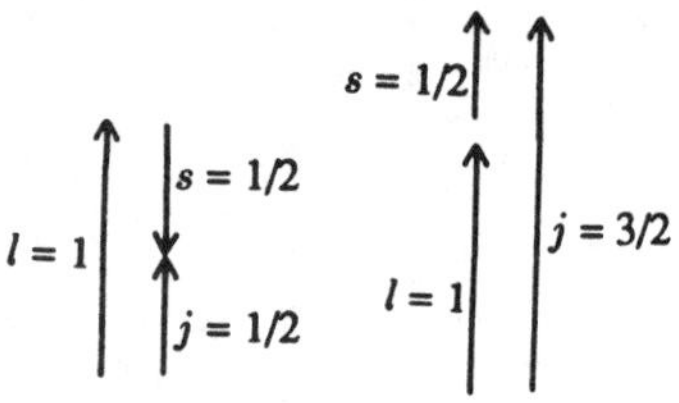

Possibilities for j if $l = 1$ and $s = \frac{1}{2}$

Since the angular momentum quantum numbers always differ by one unit, there are two different states $j = \frac{1}{2}$ ($\boldsymbol{s}$ and $\boldsymbol{l}$ parallel) and $j = \frac{3}{2}$ ($\boldsymbol{s}$ and $\boldsymbol{l}$ antiparallel) (see above figure). Here we restrict ourselves to the state $j = \frac{1}{2}$. The wave functions of the first excited state, with the orbital angular momentum l, of an electron in a hydrogen atom are

$$\psi_{n_l l m_l m_s} = \psi_{n_1 l m_1 m_s} = Y_{l m_l} \chi_{\frac{1}{2} m_s} r^{-1} U_{n_1}(r) \quad .$$

Since we are only interested in the angle-dependent part, we neglect the radial part and write for the total wave function, according to (2.25),

$$\psi_{lm} = \sum_{m_l} (1 \tfrac{1}{2} \tfrac{1}{2} | m_l m - m_l m) \, Y_{1 m_l} \chi_{\frac{1}{2} m - m_l} \quad . \tag{1}$$

For the spin function we write here $\chi_{\frac{1}{2}, m_s}$ instead of $\chi_{\pm}$, in order to unify the notation. Since, for now, we only want to construct a state with the total angular

[7] E.P. Wigner gave the first closed analytical expression for the Clebsch-Gordan coefficients in 1931 [E.P. Wigner: *Gruppentheorie* (Vieweg, Wiesbaden 1931)].

Example 2.2

momentum $j = \frac{1}{2}$ and $l = 1$, m can assume only the two values $+\frac{1}{2}$ or $-\frac{1}{2}$. For the wave functions (1) this yields:

$$\psi_{l=1,\frac{1}{2},\frac{1}{2}} = \sum_{m_l=0,\pm 1} (1\tfrac{1}{2}\tfrac{1}{2}|m_l\tfrac{1}{2}-m_l\tfrac{1}{2})\, Y_{1m_l}\chi_{\frac{1}{2},\frac{1}{2}-m_l} \quad , \tag{2}$$

$$\psi_{l=1,\frac{1}{2},-\frac{1}{2}} = \sum_{m_l=0,\pm 1} (1\tfrac{1}{2}\tfrac{1}{2}|m_l-\tfrac{1}{2}-m_l-\tfrac{1}{2})\, Y_{1m_l}\chi_{\frac{1}{2},-\frac{1}{2}-m_l} \quad . \tag{3}$$

Since m has only two values, the sums contain only two terms ($\chi_{\frac{1}{2}\frac{3}{2}}$ and $\chi_{\frac{1}{2}-\frac{3}{2}}$ do not exist) which are

$$\psi_{\frac{1}{2}\frac{1}{2}} = (1\tfrac{1}{2}\tfrac{1}{2}|0\tfrac{1}{2}\tfrac{1}{2})\, Y_{10}\chi_{\frac{1}{2}\frac{1}{2}} + (1\tfrac{1}{2}\tfrac{1}{2}|1-\tfrac{1}{2}\tfrac{1}{2})\, Y_{11}\chi_{\frac{1}{2}-\frac{1}{2}} \quad , \tag{4}$$

and

$$\psi_{\frac{1}{2}-\frac{1}{2}} = (1\tfrac{1}{2}\tfrac{1}{2}|-1\tfrac{1}{2}-\tfrac{1}{2})\, Y_{1-1}\chi_{\frac{1}{2}\frac{1}{2}} + (1\tfrac{1}{2}\tfrac{1}{2}|0-\tfrac{1}{2}-\tfrac{1}{2})\, Y_{10}\chi_{\frac{1}{2}-\frac{1}{2}} \quad . \tag{5}$$

On the left hand side the index $l = 1$ is suppressed because it is redundant for the following considerations. From the normalization condition follow the two equations

$$(1\tfrac{1}{2}\tfrac{1}{2}|0\tfrac{1}{2}\tfrac{1}{2})^2 + (1\tfrac{1}{2}\tfrac{1}{2}|1-\tfrac{1}{2}\tfrac{1}{2})^2 = 1 \quad , \tag{6}$$

$$(1\tfrac{1}{2}\tfrac{1}{2}|-1\tfrac{1}{2}-\tfrac{1}{2})^2 + (1\tfrac{1}{2}\tfrac{1}{2}|0-\tfrac{1}{2}-\tfrac{1}{2})^2 = 1 \quad . \tag{7}$$

By application of $\hat{J}_\pm = \hat{L}_\pm + \hat{S}_\pm$, (4) and (5) are transformed into each other. The step operators yield the result that

$$\hat{J}_+\psi_{jm} = \sqrt{j(j+1) - m(m+1)}\,\psi_{jm+1} \quad , \tag{8}$$

$$\hat{J}_-\psi_{jm} = \sqrt{j(j+1) - m(m-1)}\,\psi_{jm-1} \quad . \tag{9}$$

Similar results hold for $\hat{L}_\pm$ and $\hat{S}_\pm$. In our case the expressions are as follows

$$\hat{J}_+\psi_{\frac{1}{2}-\frac{1}{2}} = \psi_{\frac{1}{2}\frac{1}{2}} \quad ,$$

$$\hat{L}_+Y_{10} = \sqrt{2}Y_{11} \quad , \quad \hat{L}_+Y_{1-1} = \sqrt{2}Y_{10} \quad ,$$

$$\hat{S}_+\chi_{\frac{1}{2}\frac{1}{2}} = 0 \quad , \quad \hat{S}_+\chi_{\frac{1}{2}-\frac{1}{2}} = \chi_{\frac{1}{2}\frac{1}{2}} \quad .$$

Applying $\hat{J}_+$ on $\psi_{\frac{1}{2}-\frac{1}{2}}$ and inserting $\psi_{\frac{1}{2}\frac{1}{2}}$ we obtain

$$\begin{aligned}
&(1\tfrac{1}{2}\tfrac{1}{2}|0\tfrac{1}{2}\tfrac{1}{2})\, Y_{10}\chi_{\frac{1}{2}\frac{1}{2}} + (1\tfrac{1}{2}\tfrac{1}{2}|1-\tfrac{1}{2}\tfrac{1}{2})\, Y_{11}\chi_{\frac{1}{2}-\frac{1}{2}} \\
&\quad = \sqrt{2}(1\tfrac{1}{2}\tfrac{1}{2}|-1\tfrac{1}{2}-\tfrac{1}{2})\, Y_{10}\chi_{\frac{1}{2}\frac{1}{2}} + \sqrt{2}(1\tfrac{1}{2}\tfrac{1}{2}|0-\tfrac{1}{2}-\tfrac{1}{2})\, Y_{11}\chi_{\frac{1}{2}-\frac{1}{2}} \\
&\qquad + (1\tfrac{1}{2}\tfrac{1}{2}|0-\tfrac{1}{2}-\tfrac{1}{2})\, Y_{10}\chi_{\frac{1}{2}\frac{1}{2}} \quad .
\end{aligned}$$

The wave functions are linearly independent, thus we can compare the coefficients, obtaining

$$(1\tfrac{1}{2}\tfrac{1}{2}|0\tfrac{1}{2}\tfrac{1}{2}) = \sqrt{2}(1\tfrac{1}{2}\tfrac{1}{2}|-1\tfrac{1}{2}-\tfrac{1}{2}) + (1\tfrac{1}{2}\tfrac{1}{2}|0-\tfrac{1}{2}-\tfrac{1}{2}) \quad ,$$

$$(1\tfrac{1}{2}\tfrac{1}{2}|1-\tfrac{1}{2}\tfrac{1}{2}) = \sqrt{2}(1\tfrac{1}{2}\tfrac{1}{2}|0-\tfrac{1}{2}-\tfrac{1}{2}) \quad .$$

For clarity we write these equations in the abbreviated form, i.e.

$$a = \sqrt{2}c + d \quad , \quad b = \sqrt{2}d \quad . \tag{10}$$

Example 2.2

The same abbreviation yields, if we apply $\hat{J}_- = \hat{L}_- + \hat{S}_-$ on $\psi_{\frac{1}{2}\frac{1}{2}}$ and insert $\psi_{\frac{1}{2}-\frac{1}{2}}$,

$$c = \sqrt{2}a \quad , \quad d = a + \sqrt{2}b \quad . \tag{11}$$

The numerical values of the Clebsch-Gordan coefficients [with the phase factors given by (2.50)] follow easily from (10)/(11) and (6)/(7):

$$a = -d = -\sqrt{\tfrac{1}{3}} \quad , \quad b = -c = +\sqrt{\tfrac{2}{3}} \quad .$$

Substituting these values into (2) and (3), the eigenfunctions of the total angular momentum read

$$\psi_{\frac{1}{2}\frac{1}{2}} = -\sqrt{\tfrac{1}{3}}\, Y_{10}\chi_{\frac{1}{2}\frac{1}{2}} + \sqrt{\tfrac{2}{3}}\, Y_{11}\chi_{\frac{1}{2}-\frac{1}{2}}$$

and

$$\psi_{\frac{1}{2}-\frac{1}{2}} = -\sqrt{\tfrac{2}{3}}\, Y_{1-1}\chi_{\frac{1}{2}\frac{1}{2}} + \sqrt{\tfrac{1}{3}}\, Y_{10}\chi_{\frac{1}{2}-\frac{1}{2}} \quad .$$

As a further exercise one may calculate the Clebsch-Gordan coefficients for the case $j = \frac{3}{2}$. The following wave functions are obtained:

$$\psi_{\frac{3}{2}\frac{3}{2}} = Y_{11}\chi_{\frac{1}{2}\frac{1}{2}} \quad ,$$
$$\psi_{\frac{3}{2}\frac{1}{2}} = \sqrt{\tfrac{2}{3}}\, Y_{10}\chi_{\frac{1}{2}\frac{1}{2}} + \sqrt{\tfrac{1}{3}}\, Y_{11}\chi_{\frac{1}{2}-\frac{1}{2}} \quad ,$$
$$\psi_{\frac{3}{2}-\frac{1}{2}} = \sqrt{\tfrac{1}{3}}\, Y_{1-1}\chi_{\frac{1}{2}\frac{1}{2}} + \sqrt{\tfrac{2}{3}}\, Y_{10}\chi_{\frac{1}{2}-\frac{1}{2}} \quad ,$$
$$\psi_{\frac{3}{2}-\frac{3}{2}} = Y_{1-1}\chi_{\frac{1}{2}-\frac{1}{2}} \quad .$$

The complete wave functions for these ($p_{\frac{3}{2}}$ and $p_{\frac{1}{2}}$) states of the hydrogen atom follow by multiplication with the radial part $R_{n_1}(r) = r^{-1}U_{n_1}(r)$. These states are energetically degenerate in the pure Coulomb potential, but they split up into two states if there is spin-orbit coupling. Furthermore, the state $p_{\frac{3}{2}}$ is four times, and the state $p_{\frac{1}{2}}$ twice, degenerate. If one populates these states by irradiation of the atom with light of a broad frequency spectrum, then one obtains a value of 2:1 for the ratio of intensities of the emitted fine-structure doublet.

In general, the wave function of total angular momentum j in a spherically symmetric potential can be constructed by coupling the orbital [$Y_{lm}(\theta, \phi)$] and spin ($\chi_{\frac{1}{2},m_s}$) angular momentum eigenfunctions, giving

$$\psi_{ljM} = \sum_{m,m_s} (l\tfrac{1}{2}j|mm_sM)\, Y_{lm}(\theta,\phi)\chi_{\frac{1}{2}m_s} \quad .$$

The allowed values are $j = l \pm \frac{1}{2}$, so that in the case $l = 2$, j can take the values $\frac{5}{2}$ and $\frac{3}{2}$; for $l = 3$ it is $j = \frac{7}{2}$ and $\frac{5}{2}$. In the first case the states are called "$d_{\frac{5}{2}}$" and "$d_{\frac{3}{2}}$", respectively ("f", "d", "p" and "s" represent the $l = 3, 2, 1, 0$ states, respectively). In the second case we refer to $f_{\frac{7}{2}}$ and $f_{\frac{5}{2}}$ states. In the Coulomb potential all states are degenerate, and only after introducing the spin-orbit coupling potential $V(r)\boldsymbol{L}\cdot\boldsymbol{s}$ is the degeneracy lifted. This leads to the *fine structure of atomic spectra.*

2.7 Biographical Notes

CLEBSCH, Rudolf Friedrich Alfred, mathematician, *Königsberg 19. 1.1833, †Göttingen 7.11.1872, professor in Karlsruhe, Giessen and Göttingen. He worked on mathematical physics, variational calculus, partial differential equations, the theory of curves and surfaces, applications of abelian functions in geometry, surface mappings and the theory of invariants. Together with C. Neumann, C. founded the journal *Mathematische Annalen* in 1868.

3. Mathematical Supplement: Fundamental Properties of Lie Groups

3.1 General Structure of Lie Groups

The rotation group is composed of the infinite number of operators (setting $\hbar = 1$)

$$\hat{U}_{\mathrm{R}}(\boldsymbol{\phi}) = \exp(-\mathrm{i}\phi_\mu \hat{J}_\mu) = \exp(-\mathrm{i}\boldsymbol{\phi}\cdot\hat{\boldsymbol{J}}) \quad . \tag{3.1}$$

The fact that the operators are functions of only three fundamental operators $\{\hat{J}_\nu\} = \{\hat{J}_1, \hat{J}_2, \hat{J}_3\}$, allows us to represent them in a simple way. Every operator $\hat{U}_{\mathrm{R}}(\boldsymbol{\phi})$ is characterized by three real numbers, the parameters ϕ_1, ϕ_2, ϕ_3. We immediately recognize that the fundamental operators $\hat{J}_\nu$ can be obtained from the continuous group elements $\hat{U}_{\mathrm{R}}(\boldsymbol{\phi})$ by differentiation, so that

$$-\mathrm{i}\hat{J}_\mu \hat{U}_{\mathrm{R}}(\boldsymbol{\phi})|_{\phi=0} = -\mathrm{i}\hat{J}_\mu = \partial \hat{U}_{\mathrm{R}}(\boldsymbol{\phi})/\partial\phi_\mu|_{\boldsymbol{\phi}=0} \quad . \tag{3.2}$$

Obviously the differentiability (analyticity) of $\hat{U}_{\mathrm{R}}(\boldsymbol{\phi})$ is required with respect to the parameters ϕ_μ in the "neighbourhood" of the identity operator $\hat{U}_{\mathrm{R}}(\boldsymbol{\phi}) = \mathbb{1}$.

We now generalize this concept. Continuous groups,whose elements are given by operators $\hat{U}(\alpha_1, \alpha_2, \dots, \alpha_n; \boldsymbol{r})$, which depend on n parameters (coordinates), are called *Lie groups*, after the Norwegian mathematician Sophus Lie. Their elements depend analytically on the n parameters α_i and the argument $\boldsymbol{r}$ stands symbolically for a possible coordinate dependence. For example, in (3.1) the operators $\hat{J}_k = -\mathrm{i}\varepsilon_{ijk}x_j\partial/\partial x_j$ occur which depend on the x_i and the corresponding derivatives. In the following, we drop the argument $\boldsymbol{r}$ as a reference to this dependence, although we always continue to bear it in mind.

It is advantageous to choose the parameters such that

$$\hat{U}(0) = \mathbb{1} \tag{3.3}$$

holds. As we shall now show, analogy to (3.1) one can represent the *operators of the group* in the form

$$\hat{U}(\alpha_1 \dots \alpha_n; \boldsymbol{r}) = \exp\left(-\mathrm{i}\sum_{\mu=1}^{n} \alpha_\mu \hat{L}_\mu\right) \quad , \tag{3.4}$$

where the operator functions $\hat{L}_\mu$ are still unknown. Referring to (3.2), we have

$$-\mathrm{i}\hat{L}_\mu = \partial\hat{U}(\alpha_i)/\partial\alpha_\mu|_{\alpha=0} \quad . \tag{3.5}$$

The $\hat{L}_\mu$ are called *generators of the group*. Indeed, one obtains for the infinitesimal transformation in the neighbourhood of the identity

$$\hat{U}(\delta\alpha_\mu) = \hat{U}(0) + \partial\hat{U}(\alpha_i)/\partial\alpha_\mu|_{\alpha=0}\partial\alpha_\mu = 1\!\!1 - \mathrm{i}\hat{L}_\mu\delta\alpha_\mu = 1\!\!1 + d\hat{A} ,$$

where we put $d\hat{A} = -\mathrm{i}\hat{L}_\mu\delta\alpha_\mu$. We proceed as previously [cf. Example 1.11, Eq. (13)], and set

$$d\hat{A} = \hat{A}/N = -\mathrm{i}\hat{L}_\mu\alpha_\mu/N ,$$

where N is an integer. Performing N successive infinitesimal transformations, in order to obtain the finite transformation $\hat{U}(\alpha_\mu)$, yields

$$\hat{U}(\alpha_\mu) = \lim_{N\to\infty} [1\!\!1 + \hat{A}/N]^N = \mathrm{e}^{\hat{A}} = \exp(-\mathrm{i}\hat{L}_\mu\alpha_\mu) . \tag{3.6}$$

Here we have made substantial use of a group property, by constructing a finite operator (a finite group element) from the product of infinitesimal elements. Because of the group property, the operators of the Lie group must always be representable in the form (3.4). We shall illustrate this from another point of view in the following Example 3.3.

From the assumed analyticity, it follows for small $\delta\alpha_i$ that

$$\hat{U}(\delta\alpha_\mu) = 1\!\!1 - \mathrm{i}\sum_{\mu=1}^{n} \delta\alpha_\mu\hat{L}_\mu - \tfrac{1}{2}\sum_{\mu,\nu=1}^{n} \delta\alpha_\mu\delta\alpha_\nu\hat{L}_\mu\hat{L}_\nu + \dots . \tag{3.6a}$$

One recognizes that the $\hat{L}_i$ must be linearly independent ($\sum_i \delta\alpha_i\hat{L}_i = 0$, only if all $\delta\alpha_i = 0$), since $\hat{U}$ for $\delta\boldsymbol{\alpha} = \{\delta\alpha_i\} = 0$ has to be identical with the identity operators $1\!\!1$ in a *unique way*. The existence of a $\delta\boldsymbol{\alpha} \neq 0$ with $\hat{\boldsymbol{L}}\cdot\delta\boldsymbol{\alpha} = 0$ would imply that at least two operators $\hat{U}(\delta\boldsymbol{\alpha}) = 1\!\!1$ exist.

If $\hat{U}(\alpha_\mu)$ is *unitary*, i.e. $\hat{U}^\dagger(\alpha_\mu) = \hat{U}^{-1}(\alpha_\mu)$, then from (3.6) we obtain

$$\hat{U}^\dagger(\alpha_\mu) = 1\!\!1 + \mathrm{i}\sum_{\mu=1}^{n} \delta\alpha_\mu\hat{L}_\mu^\dagger = \hat{U}^{-1}(\delta\alpha_\mu) = 1\!\!1 + \mathrm{i}\sum_{\mu=1}^{n} \delta\alpha_\mu\hat{L}_\mu . \tag{3.7}$$

The parameters α_μ are chosen to be real in (3.4), i.e. $\alpha_\mu^* = \alpha_\mu$. From this follows the *hermiticity of the generators*

$$\hat{L}_\mu^\dagger = \hat{L}_\mu . \tag{3.8}$$

Next we calculate the inverse operator up to the second order in α_μ according to (3.6a),

$$\hat{U}^{-1}(\delta\alpha_\mu) = 1\!\!1 + \mathrm{i}\sum_{\mu=1}^{n} \delta\alpha_\mu\hat{L}_\mu - \frac{1}{2}\sum_{\mu,\nu=1}^{n} \delta\alpha_\mu\delta\alpha_\nu\hat{L}_\mu\hat{L}_\nu , \tag{3.9}$$

and with that, exploiting of Einstein's summation convention,

$$\begin{aligned}
&\hat{U}^{-1}(\delta\beta_\mu)\hat{U}^{-1}(\delta\alpha_\mu)\hat{U}(\delta\beta_\mu)\hat{U}(\delta\alpha_\mu)\\
&= (1\!\!1 + \mathrm{i}\delta\beta_i\hat{L}_i - \tfrac{1}{2}\delta\beta_i\delta\beta_j\hat{L}_i\hat{L}_j)(1\!\!1 + \mathrm{i}\delta\alpha_k\hat{L}_k - \tfrac{1}{2}\delta\alpha_k\delta\alpha_l\hat{L}_k\hat{L}_l)\\
&\quad\times(1\!\!1 - \mathrm{i}\delta\beta_m\hat{L}_m - \tfrac{1}{2}\delta\beta_m\delta\beta_n\hat{L}_m\hat{L}_n)(1\!\!1 - \mathrm{i}\delta\alpha_\mu\hat{L}_\mu - \tfrac{1}{2}\delta\alpha_\mu\delta\alpha_\nu\hat{L}_\mu\hat{L}_\nu)\\
&= (1\!\!1 - \delta\beta_m\delta\alpha_\mu\hat{L}_m\hat{L}_\mu + \delta\beta_m\delta\alpha_k\hat{L}_k\hat{L}_m - \delta\beta_i\delta\alpha_k\hat{L}_i\hat{L}_k + \delta\beta_i\delta\alpha_\mu\hat{L}_i\hat{L}_\mu)\\
&= 1\!\!1 + \delta\alpha_k\delta\beta_m(\hat{L}_k\hat{L}_m - \hat{L}_m\hat{L}_k) .
\end{aligned} \tag{3.10}$$

Since all four factors on the lhs of (3.10) are elements of the group, the whole product (3.10) is a group element located in the vicinity of $\alpha = 0$. We denote it by $\hat{U}(\delta\gamma_\nu)$, where

$$\hat{U}(\delta\gamma_\nu) = \mathbb{1} - \mathrm{i}\delta\gamma_j\hat{L}_j + \dots \quad . \tag{3.11}$$

Therefore, we have the relation

$$\hat{U}^{-1}(\delta\beta_\mu)\hat{U}^{-1}(\delta\alpha_\mu)\hat{U}(\delta\beta_\mu)\hat{U}(\delta\alpha_\mu) = \hat{U}(\delta\gamma_\nu) \quad , \tag{3.12}$$

which is illustrated in Fig. 3.1, and hence by comparing (3.10 and 3.11) we write

$$\delta\alpha_k\delta\beta_m(\hat{L}_k\hat{L}_m - \hat{L}_m\hat{L}_k) = -\mathrm{i}\delta\gamma_j\hat{L}_j \quad .$$

We now put

$$-\mathrm{i}\delta\gamma_j = C_{kmj}\delta\alpha_k\delta\beta_m \quad ,$$

as $\delta\gamma_j$ must tend to zero with $\delta\alpha_m$ or with $\delta\beta_m$, and consequently has to be proportional to both. We obtain, therefore,

$$[\hat{L}_k, \hat{L}_m]_- = C_{kmj}\hat{L}_j \quad , \tag{3.13}$$

i.e. the generators have to satisfy commutation relations of the form (3.13). They form a *closed commutator algebra*, as once again only the generators $\hat{L}_j$ appear on the rhs of (3.13). Since

$$[\hat{L}_i, \hat{L}_j]_- = -[\hat{L}_j, \hat{L}_i]_- \quad ,$$

the *structure constants* C_{ijk} are antisymmetric in the first two indices, i.e.

$$C_{ijk} = -C_{jik} \quad . \tag{3.14}$$

Furthermore, the *Jacobi identity*

$$[[\hat{L}_i, \hat{L}_j]_-, \hat{L}_k]_- + [[\hat{L}_j, \hat{L}_k]_-, \hat{L}_i]_- + [[\hat{L}_k, \hat{L}_i]_-, \hat{L}_j]_- = 0 \tag{3.15}$$

holds, as can easily be verified by inserting the definition of the commutator $[\hat{L}_i, \hat{L}_j]_- = \hat{L}_i\hat{L}_j - \hat{L}_j\hat{L}_i$. Substituting (3.13) into (3.15) yields an additional condition for the structure constants,

$$[C_{ijm}\hat{L}_m, \hat{L}_k]_- + [C_{jkm}\hat{L}_m, \hat{L}_i]_- + [C_{kim}\hat{L}_m, \hat{L}_j]_- = 0$$

and furthermore,

$$C_{ijm}C_{mkn}\hat{L}_n + C_{jkm}C_{min}\hat{L}_n + C_{kim}C_{mjn}\hat{L}_n$$
$$= [C_{ijm}C_{mkn} + C_{jkm}C_{min} + C_{kim}C_{mjn}]\hat{L}_n = 0 \quad .$$

Therefore, due to the linear independence of the $\hat{L}_n$, we have

$$C_{ijm}C_{mkn} + C_{jkm}C_{min} + C_{kim}C_{mjn} = 0 \quad . \tag{3.16}$$

The relations (3.13) to (3.16) form the fundamental relations of the *Lie algebra*, which is characteristic of the group. The structure constants contain all of the information concerning the group, since they stipulate the exchangeability of the infinitesimal group operations performed in different order. All finite operators can be successively constructed from the infinitesimal operators.

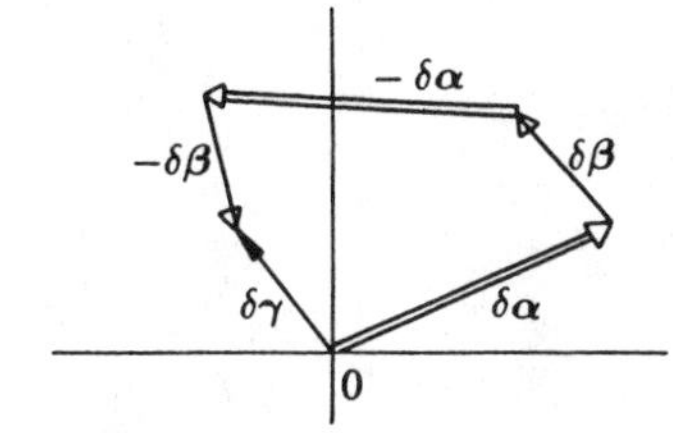

Fig. 3.1. Illustration of the relation (3.12). The four rotations $\delta\boldsymbol{\alpha}$, $\delta\boldsymbol{\beta}$, $-\delta\boldsymbol{\alpha}$, $-\delta\boldsymbol{\beta}$ from (3.10) are equivalent to a single rotation δ_γ, which again belongs to the neighbourhood of the identity (0). Notice that a rotation refers to the respectively valid coordinate system. Therefore $\delta\boldsymbol{\alpha}$ and $-\delta\boldsymbol{\alpha}$ are not antiparallel to each other, for example. δ_γ might be zero, but needs not to be

EXAMPLE

3.1 Lie Algebra of SO(3)

We remind the reader of the Lie algebra of SO(3),

$$[\hat{J}_i, \hat{J}_j]_- = \mathrm{i}\varepsilon_{ijk}\hat{J}_k \quad , \tag{1}$$

which is just the relation (3.13) applied to the group SO(3). The structure constants are the components of the ε_{ijk} tensor. In general they change if we form linear combinations of the generators (or the parameters α_μ). For example, if we pass from $\hat{J}_1, \hat{J}_2, \hat{J}_3$, to $\hat{J}_\pm = \hat{J}_1 \pm \mathrm{i}\hat{J}_2, \hat{J}_3$, then the commutation relations are modified to

$$[\hat{J}_3, \hat{J}_\pm]_- = \pm\hat{J}_\pm \quad , \quad [\hat{J}_+, \hat{J}_-]_- = 2\hat{J}_3 \quad , \tag{2}$$

and obviously the structure constants differ from the above ε_{ijk}. Thus, we can state that the structure constants depend on the representation of the Lie group, i.e. on the choice of parameters. The ε_{ijk} apparently fulfill the antisymmetry relation (3.14). Likewise the Jacobi identity relation (3.16) can be explicitly verified in this case [according to (1) we have $C_{ijk} = \mathrm{i}\varepsilon_{ijk}$],

$$\begin{aligned} &C_{ijm}C_{mkn} + C_{jkm}C_{min} + C_{kim}C_{mjn} \\ &\quad = -(\varepsilon_{ijm}\varepsilon_{mkn} + \varepsilon_{jkm}\varepsilon_{min} + \varepsilon_{kim}\varepsilon_{mjn}) \\ &\quad = -(\delta_{ik}\delta_{jn} - \delta_{in}\delta_{jk} + \delta_{ji}\delta_{kn} - \delta_{jn}\delta_{ki} + \delta_{kj}\delta_{in} - \delta_{kn}\delta_{ij}) = 0 \quad , \end{aligned}$$

since $\varepsilon_{ijk} = 1$ if the triple (ijk) can be transformed into the triple (123) by an even permutation, and $\varepsilon_{ijk} = -1$ if an odd permutation leads to (123). Otherwise $\varepsilon_{ijk} = 0$ holds.

EXERCISE

3.2 Calculation with complex $n \times n$ Matrices

Problem. Let $\hat{A}$, $\hat{B}$ be two complex $n \times n$ matrices. Show that:

(a) if $[\hat{A}, \hat{B}]_- = 0$, then

$$\mathrm{e}^{\hat{A}}\mathrm{e}^{\hat{B}} = \mathrm{e}^{\hat{A}+\hat{B}};$$

(b) if $\hat{B}$ is regular, then

$$\hat{B}\mathrm{e}^{\hat{A}}\hat{B}^{-1} = \mathrm{e}^{\hat{B}\hat{A}\hat{B}^{-1}};$$

(c) if $\lambda_1, \ldots, \lambda_n$ are the eigenvalues of $\hat{A}$, then $\mathrm{e}^{\lambda_1}, \ldots, \mathrm{e}^{\lambda_n}$ are the eigenvalues of $\mathrm{e}^{\hat{A}}$;

(d) $[e^{\hat{A}}]^* = e^{\hat{A}^*}$, $[e^{\hat{A}}]^T = e^{\hat{A}^T}$, *Exercise 3.2*

$[e^{\hat{A}}]^\dagger = e^{\hat{A}^\dagger}$, $[e^{\hat{A}}]^{-1} = e^{-\hat{A}}$;

(e) $\det e^{\hat{A}} = e^{\mathrm{Tr}\hat{A}}$.

Solution. To begin with, we remind the reader of the definition of an exponential function

$$e^A = \sum_{k=0}^{\infty} \frac{1}{k!} A^k = \mathbb{1} + A + \frac{1}{2!} A^2 + \frac{1}{3!} A^3 + \dots \tag{1}$$

With this, the solution offers no further difficulties.

(a) In the first instance we expand the product $e^{\hat{A}} e^{\hat{B}}$ to obtain

$$e^{\hat{A}} e^{\hat{B}} = \sum_{k=0}^{\infty} \frac{1}{k!} \hat{A}^k \times \sum_{j=0}^{\infty} \frac{1}{j!} \hat{B}^j$$

and introduce a new summation index $m = k + j$, such that

$$\begin{aligned} e^{\hat{A}} e^{\hat{B}} &= \sum_{m=0}^{\infty} \sum_{k=0}^{\infty} \frac{1}{k!} \frac{1}{(m-k)!} \hat{A}^k \hat{B}^{m-k} \\ &= \sum_{m=0}^{\infty} \frac{1}{m!} \sum_{k=0}^{\infty} \frac{m!}{[k!(m-k)!]} \hat{A}^k \hat{B}^{m-k} \\ &= \sum_{m=0}^{\infty} \frac{1}{m!} (\hat{A} + \hat{B})^m = e^{\hat{A}+\hat{B}} \quad . \end{aligned} \tag{2}$$

In the last but one step we exploited the assumption that $\hat{A}$ and $\hat{B}$ commute.

(b) We start with an expansion of the product on the lhs

$$\begin{aligned} \hat{B} e^{\hat{A}} \hat{B}^{-1} &= \hat{B} \sum_{k=0}^{\infty} \frac{1}{k!} \hat{A}^k \hat{B}^{-1} \\ &= \sum_{k=0}^{\infty} \frac{1}{k!} \underbrace{\hat{B}(A \times \hat{A} \times \dots \times \hat{A})\, \hat{B}^{-1}}_{k\text{-times}} \\ &= \sum_{k=0}^{\infty} \frac{1}{k!} \underbrace{(\hat{B}\hat{A}\hat{B}^{-1})(\hat{B}\hat{A}\hat{B}^{-1}) \dots (\hat{B}\hat{A}\hat{B}^{-1})}_{k\text{-times}} \\ &= \sum_{k=0}^{\infty} \frac{1}{k!} (\hat{B}\hat{A}\hat{B}^{-1})^k = e^{\hat{B}\hat{A}\hat{B}^{-1}} \quad . \end{aligned} \tag{3}$$

(c) If $\boldsymbol{a}_i$ $(i = 1, \dots, n)$ are the eigenvectors belonging to λ_i, that is

$$\hat{A}\boldsymbol{a}_i = \lambda_i \boldsymbol{a}_i \quad , \tag{4}$$

then obviously

$$e^{\hat{A}} \boldsymbol{a}_i = \sum_{k=0}^{\infty} \frac{1}{k!} \hat{A}^k \boldsymbol{a}_i = \sum_{k=0}^{\infty} \frac{1}{k!} \lambda_i^k \boldsymbol{a}_i = e^{\lambda_i} \boldsymbol{a}_i \tag{5}$$

holds, i.e. $\boldsymbol{a}_i$ is an eigenvector of $e^{\hat{A}}$ with eigenvalue e^{λ_i}.

Exercise 3.2

(d) We solely want to consider the case $[e^{\hat{A}}]^* = e^{\hat{A}^*}$, therefore we write

$$[e^{\hat{A}}]^* = \left[\sum_{k=0}^{\infty} \frac{1}{k!} \hat{A}^k\right]^* = \sum_{k=0}^{\infty} \frac{1}{k!} (\hat{A}^k)^* = \sum_{k=0}^{\infty} \frac{1}{k!} (\hat{A}^*)^k = e^{\hat{A}^*} \quad . \tag{6}$$

The cases "†" and "T" are proven similarly.

According to (a) $e^{\hat{A}} e^{-\hat{A}} = e^0 = 1$, i.e.

$$[e^{\hat{A}}]^{-1} = e^{-\hat{A}} \quad . \tag{7}$$

(e) The determinant is the product of the eigenvalues, the trace is the sum of the eigenvalues, hence by (c),

$$\det e^{\hat{A}} = e^{\lambda_1} \ldots e^{\lambda_n} = e^{\lambda_1 + \ldots \lambda_n} = e^{\mathrm{Tr} \hat{A}} \quad . \tag{8}$$

EXERCISE

3.3 Proof of a Commutation Relation

Problem. For $\hat{L}, \hat{M} \in GL(n)$, i.e. elements out of the general linear group of all $n \times n$-matrices, prove the relation

$$e^{\hat{L}} \hat{M} e^{-\hat{L}} = \sum_{n=0}^{\infty} \frac{1}{n!} [\hat{L}, \hat{M}]_{(n)} \quad ,$$

where

$$[\hat{L}, \hat{M}]_{(0)} = \hat{M}$$
$$[\hat{L}, \hat{M}]_{(1)} = [\hat{L}, \hat{M}]_-$$
$$[\hat{L}, \hat{M}]_{(2)} = [\hat{L}, [\hat{L}, \hat{M}]]_-$$
$$[\hat{L}, \hat{M}]_{(n)} = [\hat{L}, [\hat{L}, \hat{M}]_{(n-1)}]_- \quad .$$

Solution. We define a matrix-valued function of the real parameter α,

$$\hat{F}(\alpha) = e^{\alpha \hat{L}} \hat{M} e^{-\alpha \hat{L}} \quad . \tag{1}$$

$\hat{F}(\alpha)$ can be expanded into a Taylor series at $\alpha = 0$

$$\hat{F}(\alpha) = \sum_{n=0}^{\infty} \frac{1}{n!} \left(\frac{d^n \hat{F}}{d\alpha^n}\right)_{\alpha=0} \alpha^n \quad . \tag{2}$$

Now we obtain (observe the ordering of factors!),

$$
\begin{aligned}
\frac{d}{d\alpha}\{\hat{A}(\alpha)\hat{B}(\alpha)\} &= \lim_{\Delta\alpha\to 0}\left\{\frac{1}{\Delta\alpha}[\hat{A}(\alpha+\Delta\alpha)\hat{B}(\alpha+\Delta\alpha)-\hat{A}(\alpha)\hat{B}(\alpha)]\right\} \\
&= \lim_{\Delta\alpha\to 0}\left\{\frac{1}{\Delta\alpha}\hat{A}(\alpha+\Delta\alpha)[\hat{B}(\alpha+\Delta\alpha)-\hat{B}(\alpha)]\right. \\
&\qquad \left. +[\hat{A}(\alpha+\Delta\alpha)-\hat{A}(\alpha)]\hat{B}(\alpha)\right\} \\
&= \hat{A}(\alpha)\frac{d}{d\alpha}\{\hat{B}(\alpha)\}+\frac{d}{d\alpha}\{\hat{A}(\alpha)\}\hat{B}(\alpha)
\end{aligned} \tag{3}
$$

and

$$
\frac{d}{d\alpha}\mathrm{e}^{\alpha\hat{L}} = \hat{L}\mathrm{e}^{\alpha\hat{L}} = \mathrm{e}^{\alpha\hat{L}}\hat{L} \quad . \tag{4}
$$

Therefore

$$
\begin{aligned}
\frac{d}{d\alpha}\hat{F}(\alpha) &= \mathrm{e}^{\alpha\hat{L}}\hat{L}\hat{M}\mathrm{e}^{-\alpha\hat{L}} - \mathrm{e}^{\alpha\hat{L}}\hat{M}\hat{L}\mathrm{e}^{-\alpha\hat{L}} \\
&= \mathrm{e}^{\alpha\hat{L}}[\hat{L},\hat{M}]\mathrm{e}^{-\alpha\hat{L}} \quad ,
\end{aligned} \tag{5}
$$

$$
\begin{aligned}
\frac{d^2}{d\alpha^2}\hat{F}(\alpha) &= \frac{d}{d\alpha}\{\mathrm{e}^{\alpha\hat{L}}[\hat{L},\hat{M}]\mathrm{e}^{-\alpha\hat{L}}\} \\
&= \mathrm{e}^{\alpha\hat{L}}\hat{L}[\hat{L},\hat{M}]\mathrm{e}^{-\alpha L} - \mathrm{e}^{\alpha\hat{L}}[\hat{L},\hat{M}]\hat{L}\mathrm{e}^{-\alpha\hat{L}} \\
&= \mathrm{e}^{\alpha\hat{L}}[\hat{L},[\hat{L},\hat{M}]]\mathrm{e}^{-\alpha\hat{L}} \\
&= \mathrm{e}^{\alpha\hat{L}}[\hat{L}\hat{M}]_{(2)}\mathrm{e}^{-\alpha\hat{L}} \quad .
\end{aligned} \tag{6}
$$

Now assume

$$
\frac{d^{n-1}}{d\alpha^{n-1}}\hat{F}(\alpha) = \mathrm{e}^{\alpha\hat{L}}[\hat{L},\hat{M}]_{(n-1)}\mathrm{e}^{-\alpha\hat{L}} \quad . \tag{7}
$$

To complete the proof by induction, we show that

$$
\begin{aligned}
\frac{d^n}{d\alpha^n}\hat{F}(\alpha) &= \frac{d}{d\alpha}\left\{\frac{d^{n-1}}{d\alpha^{n-1}}\hat{F}(\alpha)\right\} \\
&= \frac{d}{d\alpha}\{\mathrm{e}^{\alpha\hat{L}}[\hat{L},\hat{M}]_{(n-1)}\mathrm{e}^{-\alpha\hat{L}}\} \\
&= \mathrm{e}^{\alpha\hat{L}}\hat{L}[\hat{L},\hat{M}]_{(n-1)}\mathrm{e}^{-\alpha\hat{L}} - \mathrm{e}^{\alpha\hat{L}}[\hat{L},\hat{M}]_{(n-1)}\hat{L}\mathrm{e}^{-\alpha\hat{L}} \\
&= \mathrm{e}^{\alpha\hat{L}}[\hat{L},\hat{M}]_{(n)}\mathrm{e}^{-\alpha\hat{L}} \quad ,
\end{aligned} \tag{8}
$$

Exercise 3.3

whence follows

$$\hat{F}(\alpha) = \sum_{n=0}^{\infty} \frac{1}{n!} e^{\alpha\hat{L}}[\hat{L}, \hat{M}]_{(n)} e^{-\alpha\hat{L}}|_{\alpha=0}\alpha^n$$

$$= \sum_{n=0}^{\infty} \frac{1}{n!}[\hat{L}, \hat{M}]_{(n)}\alpha^n \quad . \tag{9}$$

EXERCISE

3.4 Generators and Structure Constants of Proper Lorentz Transformations

The (proper) Lorentz group L is composed of all 4×4 matrices $\hat{a} = (a^\mu_\nu)$ which describe a (proper) Lorentz transformation, i.e. for which the following holds:

i) Invariance of the scalar product;

$$(a^\mu{}_\alpha x^\alpha) g_{\mu\nu} (a^\nu{}_\beta x^\beta) = x^\mu g_{\mu\nu} x^\nu \quad , \tag{1}$$

where $g_{\mu\nu}$ is the Lorentz-Minkowski metric.

ii) $\hat{a}$ is contained in the same set of Lorentz transformations as $\mathbb{1}$ ($\hat{a}$ arises continuously from $\mathbb{1}$).

The lorentz transformations form a six-parameter group. One can take as parameters, e.g. the rotation angle ω and the rapidity ξ.

Hint: Compare with Sect. 3.13 and Exercises 3.16 and 3.17!

Problem. a) Determine the corresponding generators; b) Find the structure constants!

Solution. As usual we make the ansatz

$$\hat{\alpha}(\boldsymbol{\omega}, \boldsymbol{\xi}) = \exp(-\mathrm{i}\boldsymbol{\omega}\cdot\hat{\boldsymbol{S}} - \mathrm{i}\boldsymbol{\xi}\cdot\hat{\boldsymbol{K}}) \quad , \tag{2}$$

where $\boldsymbol{\omega}$, $\boldsymbol{\xi}$ represent the parameters and $\hat{\boldsymbol{S}}$, $\hat{\boldsymbol{K}}$ are the corresponding generators. For infinitesimal transformations this becomes

$$\hat{a}(\delta\boldsymbol{\omega}, \delta\boldsymbol{\xi}) = \mathbb{1} - \mathrm{i}\delta\boldsymbol{\omega}\cdot\hat{\boldsymbol{S}} - \mathrm{i}\delta\boldsymbol{\xi}\cdot\hat{\boldsymbol{K}} \quad . \tag{3}$$

Here $\hat{\boldsymbol{S}}$ and $\hat{\boldsymbol{K}}$ are three-vectors of 4×4 matrices. Let us start by investigating the spatial rotations, i.e. setting $\delta\boldsymbol{\xi} = 0$. The transformation matrix for a rotation of the coordinate system around the x axis ($\delta\omega_1 \neq 0$, $\delta\omega_2 = \delta\omega_3 = 0$), in the positive sense of rotation is given by

$$\hat{a} = \left[\begin{array}{c|ccc} 1 & 0 & 0 & 0 \\ \hline 0 & 1 & 0 & 0 \\ 0 & 0 & 1 & +\delta\omega_1 \\ 0 & 0 & -\delta\omega_1 & 1 \end{array}\right] \quad ,$$

Exercise 3.4

hence according to (3) $\delta\boldsymbol{\omega}\cdot\hat{\boldsymbol{S}} = \mathrm{i}(\hat{a} - \mathbb{1}) = \delta\omega_1\hat{S}_1$ and therefore

$$\hat{S}_1 = -\mathrm{i}\left[\begin{array}{c|ccc} 0 & 0 & 0 & 0 \\ \hline 0 & 0 & 0 & 0 \\ 0 & 0 & 0 & -1 \\ 0 & 0 & +1 & 0 \end{array}\right] . \tag{4}$$

Similarly, we obtain

$$\hat{S}_2 = -\mathrm{i}\left[\begin{array}{c|ccc} 0 & 0 & 0 & 0 \\ \hline 0 & 0 & 0 & +1 \\ 0 & 0 & 0 & 0 \\ 0 & -1 & 0 & 0 \end{array}\right] , \quad \hat{S}_3 = -\mathrm{i}\left[\begin{array}{c|ccc} 0 & 0 & 0 & 0 \\ \hline 0 & 0 & -1 & 0 \\ 0 & 1 & 0 & 0 \\ 0 & 0 & 0 & 0 \end{array}\right] . \tag{5}$$

We see that the spin-1 matrices appear as submatrices, but with negative sign [see Example 1.11, Eq. (9)]. Concerning the Lorentz "boosts", we first recall the definition of the rapidity variable $\boldsymbol{\xi}$, for which

$$\boldsymbol{\xi} = \frac{\boldsymbol{\beta}}{|\boldsymbol{\beta}|}\operatorname{artanh}|\boldsymbol{\beta}| \quad , \quad \boldsymbol{\beta} = \frac{\boldsymbol{v}}{c} . \tag{6}$$

With that we obtain, after a straight forward calculation,

$$\beta = \tanh\xi \quad , \quad \gamma = \cosh\xi \quad , \quad \gamma\beta = \sinh\xi . \tag{7}$$

Then the matrix, describing a boost in the x direction, is simply

$$\hat{a}(\xi) = \left[\begin{array}{c|ccc} \cosh\xi & -\sinh\xi & 0 & 0 \\ \hline -\sinh\xi & \cosh\xi & 0 & 0 \\ 0 & 0 & 1 & 0 \\ 0 & 0 & 0 & 1 \end{array}\right] . \tag{8}$$

Observe that there is a certain similarity between the boost and the rotation matrices, the differences originating in the asymmetry of temporal and spatial directions in Minkowski space.

A benefit of the "rapidity representation" consists in the property

$$\hat{a}(\xi_1)\hat{a}(\xi_1') = \hat{a}(\xi_1 + \xi_1') . \tag{9}$$

that is, rapidities in the same spatial direction are additive just as rotation angles. For an infinitesimal boost ($\xi \to \delta\xi_1$, $\delta\xi_2 = \delta\xi_3 = 0$) we obtain

$$\hat{a} = \left[\begin{array}{c|ccc} 1 & -\delta\xi_1 & 0 & 0 \\ \hline -\delta\xi_1 & 1 & 0 & 0 \\ 0 & 0 & 1 & 0 \\ 0 & 0 & 0 & 1 \end{array}\right] ,$$

Exercise 3.4

hence according to (3) $\delta\boldsymbol{\xi}\cdot\hat{\boldsymbol{K}} = \mathrm{i}(\hat{a} - \mathbb{1}) = \delta\xi_1\hat{K}_1$ and therefore

$$\hat{K}_1 = -\mathrm{i}\left[\begin{array}{c|ccc} 0 & 1 & 0 & 0 \\ \hline 1 & & & \\ 0 & & & \\ 0 & & 0 & \\ 0 & & & \end{array}\right] \tag{10}$$

and by analogy

$$\hat{K}_2 = -\mathrm{i}\left[\begin{array}{c|ccc} 0 & 0 & 1 & 0 \\ \hline 0 & & & \\ 1 & & 0 & \\ 0 & & & \end{array}\right] , \quad \hat{K}_3 = -\mathrm{i}\left[\begin{array}{c|ccc} 0 & 0 & 0 & 1 \\ \hline 0 & & & \\ 0 & & 0 & \\ 1 & & & \end{array}\right] . \tag{11}$$

We have hereby determined the generators of the Lorentz group. Of course one should bear in mind that these matrices depend to a high degree on the definition of the parameters.

b) If we enumerate the generators $\hat{S}_i$, $\hat{K}_i$ $(i = 1, 2, 3)$ in some way, combining them to form a six-component vector $\hat{L}_i(i = 1, \dots, 6)$, then the structure constants C_{ijk} are defined by

$$[\hat{L}_i, \hat{L}_j] = C_{ijk}\hat{L}_k \quad . \tag{12}$$

Hence we have to calculate all commutators between generators, and as we already know,

$$[\hat{S}_i, \hat{S}_j] = \mathrm{i}\varepsilon_{ijk}\hat{S}_k \tag{13}$$

holds for the negative spin matrices. The reader should note the sign here. If we had chosen $-\omega$ as a parameter instead of the rotation angle ω, a " + "-sign should have resulted. A commutator between boost generators is, for example

$$\begin{aligned} [\hat{K}_1, \hat{K}_2]_- = -\Bigg\{ & \left[\begin{array}{c|ccc} 0 & 1 & 0 & 0 \\ \hline 1 & & & \\ 0 & & 0 & \\ 0 & & & \end{array}\right] \left[\begin{array}{c|ccc} 0 & 0 & 1 & 0 \\ \hline 0 & & & \\ 1 & & 0 & \\ 0 & & & \end{array}\right] \\ & - \left[\begin{array}{c|ccc} 0 & 0 & 1 & 0 \\ \hline 0 & & & \\ 1 & & 0 & \\ 0 & & & \end{array}\right] \left[\begin{array}{c|ccc} 0 & 1 & 0 & 0 \\ \hline 1 & & & \\ 0 & & 0 & \\ 0 & & & \end{array}\right] \Bigg\} \\ = & \left[\begin{array}{c|ccc} 0 & 0 & 0 & 0 \\ \hline 0 & 0 & -1 & 0 \\ 0 & 1 & 0 & 0 \\ 0 & 0 & 0 & 0 \end{array}\right] = \mathrm{i}\hat{S}_3 \quad . \end{aligned} \tag{14}$$

Exercise 3.4

Calculation yields

$$[\hat{K}_1, \hat{K}_2] = -\,\mathrm{i}\hat{S}_3 \quad , \tag{15}$$

and altogether we obtain

$$[\hat{K}_i, \hat{K}_j] = -\,\mathrm{i}\varepsilon_{ijk}\hat{S}_k \quad . \tag{16}$$

Eventually $[\hat{S}_i, \hat{K}_j]$ must be calculated, e.g. we get

$$[\hat{S}_1, \hat{K}_2] = -\left\{\left[\begin{array}{c|ccc} 0 & 0 & 0 & 0 \\ \hline 0 & 0 & 0 & 0 \\ 0 & 0 & 0 & -1 \\ 0 & 0 & +1 & 0 \end{array}\right]\left[\begin{array}{c|ccc} 0 & 0 & 1 & 0 \\ \hline 0 & & & \\ 1 & & 0 & \\ 0 & & & \end{array}\right] - \left[\begin{array}{c|ccc} 0 & 0 & 1 & 0 \\ \hline 0 & & & \\ 1 & & 0 & \\ 0 & & & \end{array}\right]\left[\begin{array}{c|ccc} 0 & 0 & 0 & 0 \\ \hline 0 & 0 & 0 & 0 \\ 0 & 0 & 0 & -1 \\ 0 & 0 & +1 & 0 \end{array}\right]\right\}$$

$$= -\left[\begin{array}{c|ccc} 0 & 0 & 0 & 0 \\ \hline 0 & & & \\ 0 & & 0 & \\ +1 & & & \end{array}\right] + \left[\begin{array}{c|ccc} 0 & 0 & 0 & -1 \\ \hline 0 & & & \\ 0 & & 0 & \\ 0 & & & \end{array}\right]$$

$$= \mathrm{i}\hat{K}_3 \tag{17}$$

and more generally,

$$[\hat{S}_i, \hat{K}_j] = +\,\mathrm{i}\varepsilon_{ijk}\hat{K}_k \quad . \tag{18}$$

The structure constants are thus given by the coefficients of the rhs of (13), (16), and (18), i.e. essentially by $\pm\,\varepsilon_{ijk}$. This completes the exercise.

3.2 Interpretation of Commutators as Generalized Vector Products, Lie's Theorem, Rank of Lie Group

The commutation relations (3.13) can be comprehended as a direct *generalization of the vectorial cross product* of two vectors. In fact Eq. (12) in Exercise 3.4

can be written as

$$\hat{L}_i \times \hat{L}_j = C_{ijk} \hat{L}_k \quad , \tag{3.17}$$

meaning that the cross product of two "basis vectors" $\hat{L}_i$ and $\hat{L}_j$ is also a vector of the same space, and thus a linear combination of the basis vectors. This is the generalized relation $\mathbf{e}_i \times \mathbf{e}_j = \mathbf{e}_k$ $(i, j, k = 1, 2, 3,$ cyclic), valid in three-dimensional vector space, where the e_i are the basis vectors. The Lie algebra of rank 2 is particular noteworthy with regard to this viewpoint; it contains two vectors, the cross product of which vanishes, i.e. $[\hat{L}_1, \hat{L}_2]_- = 0$, although $\hat{L}_1$ and $\hat{L}_2$ are linearly independent. Therefore, the comparison with a normal vector product in three-dimensional configuration space must not be taken too far. With the exception of the Lie algebra of the group SO(3) it is not an isomorphism, merely an analogy.

Up to now we have started with the Lie group, determined its generators and then calculated the commutators (3.13). In this way we were led to the Lie algebra. *This procedure can be reversed*: If a set of N hermitian operators $\hat{L}_i$ is given, which is closed under commutation [hence fulfills equations of the form (3.13)], then these operators $\hat{L}_i$ specify a Lie group, whose generators they are. This is called *Lie's theorem.*

This theorem is evident, for we can immediately write down the infinitesimal group operators $\hat{U}(\delta\alpha_\nu)$

$$\hat{U}(\delta\alpha_\nu) = \mathbb{1} - \mathrm{i} \sum_{n=1}^{N} \delta\alpha_n \hat{L}_n = \mathbb{1} - \mathrm{i} \sum_{n=1}^{N} \frac{\alpha_n}{M} \hat{L}_n \quad , \quad \alpha_n = M\delta\alpha_n \quad . \tag{3.18}$$

Successive M-fold application of these operators in the limit $M \to \infty$ then yields the operators of the pertinent Lie group, corresponding to finite rotations,

$$\hat{U}(\alpha_\nu) = \lim_{M \to \infty} \left(\mathbb{1} - \mathrm{i} \sum_{n=1}^{N} \frac{\alpha_n}{M} \hat{L}_n \right)^M = \exp\left(-\mathrm{i} \sum_{n=1}^{N} \alpha_n \hat{L}_n \right) \quad . \tag{3.19}$$

This construction of $\hat{U}(\alpha_r)$ is exactly the same as the one which lead to (3.6).

An essential characteristic of a Lie group is its *rank*, which is defined as the *largest number of generators commuting with each other*. For example, the abelian *translation group* has three generators $\hat{p}_\nu = -\mathrm{i}\partial/\partial x_\nu$, which all commute with each other, and hence has *rank 3*. In this case the rank is to a certain extent trivial since, in view of the commutativity of the generators $\hat{p}_\nu$, one can write

$$\mathrm{e}^{-\mathrm{i}\boldsymbol{\varrho}\cdot\hat{\boldsymbol{p}}} = \mathrm{e}^{-\mathrm{i}\varrho_1\hat{p}_1} \mathrm{e}^{-\mathrm{i}\varrho_2\hat{p}_2} \mathrm{e}^{-\mathrm{i}\varrho_3\hat{p}_3} \quad . \tag{3.20}$$

This means that this group of rank 3 is actually a product of three groups of rank 1. More interesting examples are provided by the *rotation group* SO(3), in which none of the generators $\hat{J}_\nu$ commutes with any other (*rank 1*); and by the group SU(3) (which we shall introduce later), in which the third component of the isospin $\hat{T}_3$ and the hypercharge $\hat{Y}$ (which is connected with strangeness) commute (*rank 2*).

EXAMPLE

3.5 Algebra of $\hat{p}_\nu$ and $\hat{J}_\nu$

a) The three operators $\hat{p}_\nu = -\mathrm{i}\partial/\partial x_\nu$ fulfill the trivial commutation relations

$$[\hat{p}_\nu, \hat{p}_\mu]_- = 0 \quad . \tag{1}$$

If the group parameters are denoted by $\varrho_1, \varrho_2, \varrho_3$, then

$$\hat{U}(\boldsymbol{\varrho}) = \exp(-\mathrm{i}\varrho_\nu \hat{p}_\nu) = \exp(-\boldsymbol{\varrho}\cdot\hat{\nabla}) \tag{2}$$

are the operators of the translation group, as we already know. These operators $\hat{U}(\boldsymbol{\varrho})$ represent an Abelian group. All three generators commute; hence the rank of the translation group is 3.
b) The three angular momentum operators $\hat{J}_\nu$ satisfy

$$[\hat{J}_i, \hat{J}_j]_- = \mathrm{i}\varepsilon_{ijk}\hat{J}_k \quad . \tag{3}$$

Therefore

$$\hat{U}_\mathrm{R}(\boldsymbol{\phi}) = \exp(-\mathrm{i}\phi_i \hat{J}_i) \tag{4}$$

are the known group operators of SO(3), and the angles ϕ_i are the parameters of the group. A generator $\hat{J}_i$ commutes only with itself; hence the rank of the special rotations group is 1.

3.3 Invariant Subgroups, Simple and Semisimple Lie Groups, Ideals

Lie groups which have special importance in physics are *semisimple*. In the following, we shall explain this concept and the reason for its particular relevance.

Let us investigate a group $g = \{\hat{g}_\nu\}$ with an *abelian* subgroup $a = \{\hat{a}_i\}$ (Fig. 3.2). Then, if

$$\hat{g}_\nu \hat{a}_i \hat{g}_\nu^{-1} = \hat{a}_j \quad , \tag{3.21}$$

$$\hat{a}_i \hat{a}_k = \hat{a}_k \hat{a}_i \tag{3.22}$$

hold for every $\hat{a}_i$ and for every element $\hat{g}_\nu$ of the full group, i.e. $\hat{g}_\nu \hat{a}_i \hat{g}_\nu^{-1}$ is *again an element of the abelian subgroup*, the group a is called an *abelian invariant subgroup*. If the subgroup a is invariant but not abelian, i.e. (3.21) holds, but not (3.22), then it is simply named an *invariant subgroup*.

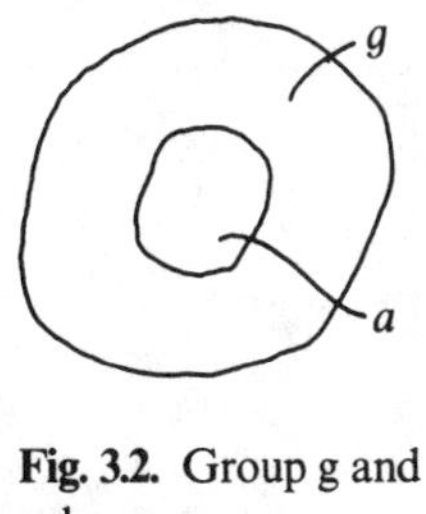

Fig. 3.2. Group g and abelian subgroup a

EXAMPLE

3.6 Translation-Rotation Group

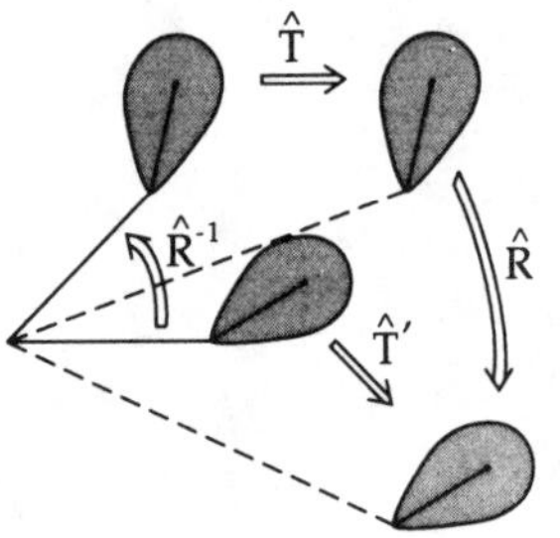

The group of all translations and rotations has six generators $\hat{p}_1, \hat{p}_2, \hat{p}_3, \hat{J}_1, \hat{J}_2, \hat{J}_3$. More concisely, it is called a *translation-rotation group* and has the translations $(\hat{p}_\nu)$ as an abelian subgroup. Intuition shows us that

$$\hat{R}\hat{T}\hat{R}^{-1} = \hat{T}' \quad ,$$

where $\hat{R}$ is a rotation and $\hat{T}$ a translation, is again a pure translation $\hat{T}'$ (see next figure). Consequently the translation group is an invariant abelian subgroup of the translation-rotation group. This property will also be pursued analytically in Exercise 3.18.

We now arrive at the definition of simple and semisimple Lie groups and state that Lie group is called *simple*, if it does not posses a *continuous* (hence Lie type) *invariant subgroup*. A Lie group is called *semisimple*, if it does *not* possess *a continuous abelian invariant subgroup*. One should notice that a semisimple group can well contain a continuous invariant subgroup, although it is *not* allowed to have *a continuous abelian invariant subgroup*.

Henceforth, we want to study *Lie algebras of groups with continuous invariant subgroups*. We argue that with

$$\hat{a}_j = \hat{g}_\nu a_i g_\nu^{-1}$$

and $\hat{a}_i^{-1}$ also

$$\hat{a}_l = \hat{a}_j \hat{a}_i^{-1} = \hat{g}_\nu a_i \hat{g}_\nu^{-1} \hat{a}_i^{-1} \tag{3.23}$$

is an element of the invariant subgroup. Since g and a are Lie groups, we term $\hat{G}_k$ the generators belonging to g, and $\hat{A}_j$ the generators belonging to a (note the operator sign ˆ). It then follows that

$$G_\nu = \mathbb{1} - \mathrm{i}\delta\alpha_i \hat{G}_i - \tfrac{1}{2}\delta\alpha_i \delta\alpha_j \hat{G}_i \hat{G}_j \tag{3.24}$$

and

$$A_i = \mathbb{1} - \mathrm{i}\delta\beta_k \hat{A}_k - \tfrac{1}{2}\delta\beta_k \delta\beta_l \hat{A}_k \hat{A}_l \tag{3.25}$$

for the rhs of (3.23) by complete analogy to our previous calculations (3.10)

$$A_l = \mathbb{1} - \delta\alpha_i \delta\beta_k [\hat{G}_i, \hat{A}_k]_- \quad . \tag{3.26}$$

Since A_l has to be located within the group A in the neighbourhood of the unit element, we can write

$$A_l = \mathbb{1} - \mathrm{i}\delta\gamma_m \hat{A}_m \quad . \tag{3.27}$$

If we put – as previously –

$$\mathrm{i}\delta\gamma_l = a_{ikl}\delta\alpha_i\delta\beta_k \quad ,$$

because $\delta\gamma_l$ has to vanish for $\delta\alpha_i \to 0$ as well as for $\delta\beta_k \to 0$, then from (3.26) and (3.27) follows

$$[\hat{G}_i, \hat{A}_k]_- = a_{ikl}\hat{A}_l \tag{3.28}$$

for all $\hat{G}_i$. Therefore, if one can linearly combine M generators $\hat{A}_l$ ($M < N$) out of the N generators $\hat{G}_i$ of a Lie group, so that (3.28) holds, then the Lie group possesses an invariant subgroup. Obviously the M generators $\hat{A}_l = \{A_1, A_2, \dots, A_M\}$ of the invariant subgroup form a *subalgebra* of the original Lie algebra. Indeed, the set of commutators $[\hat{A}_l, \hat{A}_k]_-$ is evidently closed among itself. Such a subalgebra, for which (3.28) holds in addition, is called *ideal.* One says by analogy to the definitions for groups:

A Lie algebra is named *simple*, if it does *not* possess *an ideal* apart from the null ideal $\{0\}$, and *semisimple*, if it does *not* possess *an abelian ideal.* Thus (3.28) may hold in the latter case, but not all commutators $[\hat{A}_i, \hat{A}_j]_-$ within the ideal are allowed to vanish.

EXAMPLE

3.7 Simple and Semisimple Lie Groups

a) The three generators of the translation group $\hat{p}_\nu$ form a closed subalgebra within the TR group (translation-rotation group), since $[\hat{p}_\nu, \hat{p}_\mu]_- = 0$. Together with the generators $\hat{J}_\nu = (\hat{r} \times \hat{p})_\nu$ of the rotation group, one gets, for example

$$\hat{p}_x\hat{J}_x - \hat{J}_x\hat{p}_x = 0 \quad , \quad \hat{p}_x\hat{J}_y - \hat{J}_y\hat{p}_x = \mathrm{i}\hat{p}_z \quad ,$$
$$\hat{p}_x\hat{J}_z - \hat{J}_z\hat{p}_x = -\mathrm{i}\hat{p}_y \quad , \quad \text{and so forth} \quad . \tag{1}$$

These are relations of the form (3.28). Therefore, the $\hat{p}_\nu$ algebra is an ideal within the TR group (more precisely, an abelian ideal). Consequently the algebra of $\hat{p}_i$ and $\hat{J}_\nu$ is neither simple nor semisimple. The relations (1) explain again very precisely the statement of Example 3.6.

b) If one tries to apply (3.28) to the rotation group, it is clear to see that these equations can only be fulfilled, if the $\hat{A}_k$ span the total Lie algebra. Namely, if one puts $\{\hat{G}_k\} = \{\hat{J}_k\}$, then the commutators

$$[\hat{J}_1, \hat{J}_2]_- = \mathrm{i}\hat{J}_3 \quad , \quad [\hat{J}_3, \hat{J}_1]_- = \mathrm{i}\hat{J}_2 \quad , \quad [\hat{J}_2, \hat{J}_3]_- = \mathrm{i}\hat{J}_1 \tag{2}$$

do not generate a subalgebra analogous to (3.28). Solely the three $\hat{J}_i$ operators together are closed. Therefore, the angular momentum algebra does *not* possess *an ideal, it is simple.*

Example 3.7

c) Typical examples of semisimple groups are the *direct products* (*Kronecker-products*) of simple groups. For example, in many physical discussions of angular momentum couplings, the semisimple group

$$\mathrm{SO}(3) \times \mathrm{SO}(3) \tag{3}$$

plays an important role. The elements of this group have the form

$$\exp(-\mathrm{i}\boldsymbol{\phi}\cdot\hat{\boldsymbol{J}}) \times \exp(-\mathrm{i}\boldsymbol{\phi}'\cdot\hat{\boldsymbol{J}}') \quad . \tag{4}$$

where $\boldsymbol{\phi} = \{\phi_1, \phi_2, \phi_3\}$ and $\boldsymbol{\phi}' = \{\phi'_1, \phi'_2, \phi'_3\}$ are six independent parameters and the generators (separately) fulfill the algebras

$$[\hat{J}_i, \hat{J}_j]_- = \mathrm{i}\varepsilon_{ijk}\hat{J}_k \quad \text{and} \quad [\hat{J}'_i, \hat{J}'_j]_- = \mathrm{i}\varepsilon_{ijk}\hat{J}'_k \quad . \tag{5}$$

In addition they commute, i.e.

$$[\hat{J}_i, \hat{J}'_k]_- = 0 \quad , \quad i, k = \{1, 2, 3\} \quad . \tag{6}$$

The generators $\hat{J}_\nu$ and $\hat{J}'_\nu$ act in different spaces, e.g. in the well known spaces of orbital and spin angular momentum or in the configuration space of particle 1 $(\boldsymbol{r}_1, s_1)$ and particle 2 $(\boldsymbol{r}_2, s_2)$. Here s denotes the spin variable.

Thus, we recognize by comparison with equation (3.28) that the generators $\{\hat{J}_1, \hat{J}_2, \hat{J}_3\}$ as well as the generators $\{\hat{J}'_1, \hat{J}'_2, \hat{J}'_3\}$ constitute – each set for itself – a non-abellian ideal of the Lie algebra, which is spanned by $\{\hat{J}_1, \hat{J}_2, \hat{J}_3, \hat{J}'_1, \hat{J}'_2, \hat{J}'_3\}$.

The algebra and the group is *semisimple*, since indeed there is an ideal (i.e. an invariant subgroup), but the latter is not abelian. As may be seen from this example, it is a typical property of semisimple Lie groups that they can be set up as a direct product of simple Lie groups. This holds generally for *every* semisimple Lie group.

d) Note that the definition of a *simple group* requires that there exists no Lie type (i.e. continuous) invariant subgroup. Quite possibly, however, there may be a *discrete invariant subgroup*, although this subgroup must not form a Lie group, as is of course the case for a continuous invariant subgroups. As an example we mention the SU(2), with which we became acquainted earlier, in connection with the *rotation of spinors* (Example 1.12). Its elements are given by

$$\begin{aligned}\hat{U}_\mathrm{R} &= \exp(-\mathrm{i}\boldsymbol{\phi}\cdot\hat{\boldsymbol{s}}) = \exp(-\tfrac{1}{2}\mathrm{i}\boldsymbol{\phi}\cdot\hat{\boldsymbol{\sigma}}) \\ &= \exp(-\tfrac{1}{2}\mathrm{i}\phi\boldsymbol{n}\cdot\hat{\boldsymbol{\sigma}}) \\ &= \mathbb{1}\cos(\tfrac{1}{2}\phi) - \mathrm{i}\boldsymbol{n}\cdot\hat{\boldsymbol{\sigma}}\sin(\tfrac{1}{2}\phi)\end{aligned} \tag{7}$$

(see also Exercise 3.8). However, the two discrete matrices

$$\begin{pmatrix}1 & 0\\ 0 & 1\end{pmatrix} \quad \text{and} \quad \begin{pmatrix}-1 & 1\\ 0 & -1\end{pmatrix}$$

form a discrete invariant subgroup for these *continuous* 2×2 matrices. Nevertheless, the group is simple, since indeed there is an invariant subgroup, but it is discrete, not continuous. Furthermore, the generators $\hat{s}_\nu$ possess the same Lie algebra as the generators of the rotation group.

Example 3.7

e) We now draw attention to a physically important Lie algebra, which is not semisimple, Let $\hat{P}$ and $\hat{Q}$ be momentum and position operators in a well-defined direction and $\hat{E}$ the identity operator. Then obviously

$$[\hat{E}, \hat{P}]_- = [\hat{E}, \hat{Q}]_- = 0 \quad ,$$
$$[\hat{P}, \hat{Q}]_- = -\mathrm{i}\hat{E} \tag{8}$$

holds. The algebra of these three operators (sometimes called *Heisenberg algebra*) possesses an abelian subalgebra which is spanned by $\hat{E}$ and $\hat{P}$, for example. Simultaneously, it is an ideal, for it fits into the scheme (3.28). The commutators

$$[\hat{E}, \hat{P}]_- = 0 = [\hat{E}, \hat{E}]_- = [\hat{P}, \hat{P}]_- \tag{9}$$

define the abelian subalgebra. At the same time the relation (3.28) is fulfilled with $[\hat{E}, \hat{Q}]_- = 0$ and $[\hat{P}, \hat{Q}]_- = \mathrm{i}\hat{E}$. Therefore, the Heisenberg algebra is a Lie algebra, which is *neither simple nor semisimple.*

EXERCISE

3.8 Reduction of $\exp\{-\frac{1}{2}\mathrm{i}\boldsymbol{n}\cdot\hat{\boldsymbol{\sigma}}\}$

Problem. Show that

$$\exp(-\tfrac{1}{2}\phi\boldsymbol{n}\cdot\hat{\boldsymbol{\sigma}}) = \mathbb{1}\cos(\tfrac{1}{2}\phi) - \mathrm{i}\boldsymbol{n}\cdot\hat{\boldsymbol{\sigma}}\sin(\tfrac{1}{2}\phi)$$

holds [see Example 3.7, (7)] for any angle ϕ and any unit vector $\boldsymbol{n}$.

Solution. According to the assumption that $\boldsymbol{n}$ is a unit vector, i.e.

$$n_i n_i = 1 \quad , \tag{1}$$

then, furthermore for the Pauli matrices,

$$\hat{\sigma}_i\hat{\sigma}_j = \mathrm{i}\varepsilon_{ijk}\hat{\sigma}_k + \delta_{ij}\mathbb{1} \tag{2}$$

holds. Hence, with (1) it follows that (one should bear in mind Einstein's summation convention)

$$(\boldsymbol{n}\cdot\hat{\boldsymbol{\sigma}})^2 = n_i n_j \hat{\sigma}_i\hat{\sigma}_j = \mathbb{1}n_i n_j \delta_{ij} + \mathrm{i}n_i n_j \varepsilon_{ijk}\hat{\sigma}_k$$
$$= \mathbb{1}n_i n_i = \mathbb{1} \quad , \tag{3}$$

because $n_i n_j \varepsilon_{ijk} = 0$ is true for every k, since $n_i n_j$ is symmetric and ε_{ijk} is antisymmetric in i and j. The result (3) can obviously be generalized. Thus

$$(\boldsymbol{n}\cdot\hat{\boldsymbol{\sigma}})^{2n} = \mathbb{1} \quad \text{and} \tag{4}$$
$$(\boldsymbol{n}\cdot\hat{\boldsymbol{\sigma}})^{2n+1} = \mathbb{1}(\boldsymbol{n}\cdot\hat{\boldsymbol{\sigma}}) = (\boldsymbol{n}\cdot\hat{\boldsymbol{\sigma}}) \quad . \tag{5}$$

Exercise 3.8

Now the desired transformation may be carried out:

$$\begin{aligned}\exp(-\tfrac{1}{2}\mathrm{i}\phi\boldsymbol{n}\cdot\hat{\boldsymbol{\sigma}}) &= \sum_{n=0}^{\infty}\frac{1}{n!}(-\tfrac{1}{2}\mathrm{i}\phi)^n(\boldsymbol{n}\cdot\hat{\boldsymbol{\sigma}})^n\\ &= \sum_{n=0}^{\infty}\frac{1}{(2n)!}(-\tfrac{1}{2}\mathrm{i}\phi)^{2n}(\boldsymbol{n}\cdot\hat{\boldsymbol{\sigma}})^{2n}\\ &\quad+\sum_{n=0}^{\infty}\frac{1}{(2n+1)!}(-\tfrac{1}{2}\mathrm{i}\phi)^{2n+1}(\boldsymbol{n}\cdot\hat{\boldsymbol{\sigma}})^{2n+1}\\ &= \mathbb{1}\sum_{n=0}^{\infty}\frac{1}{(2n)!}(-1)^n(\tfrac{1}{2}\phi)^{2n}\\ &\quad-\mathrm{i}(\boldsymbol{n}\cdot\hat{\boldsymbol{\sigma}})\sum_{n=0}^{\infty}\frac{1}{(2n+1)!}(-1)^n(\tfrac{1}{2}\phi)^{2n+1}\\ &= \mathbb{1}\cos(\tfrac{1}{2}\phi)-\mathrm{i}(\boldsymbol{n}\cdot\hat{\boldsymbol{\sigma}})\sin(\tfrac{1}{2}\phi)\quad,\end{aligned}$$

for the two series just represent the trigonometric functions.

EXAMPLE

3.9 Cartan's Criterion for Semisimplicity

There is an elementary criterion for finding out whether a Lie algebra is semisimple, which originates from ***Cartan***. To that end we define the symmetric tensor

$$g_{\sigma\lambda} = g_{\lambda\sigma} = C_{\sigma\varrho\tau}C_{\lambda\tau\varrho}\quad, \tag{1}$$

which is built from structure constants. $g_{\sigma\lambda}$ is called a *metric tensor*, sometimes also *Killing form.* This tensor can be defined for any Lie group and its associated Lie algebra. Sometimes the metric is likewise defined as

$$(\hat{L}_i, \hat{L}_j) = \mathrm{Tr}(\hat{L}_i, \hat{L}_j)\quad,$$

where the $\hat{L}_i$ are generators. Now, we have the so-called *regular representation* of a Lie group [see (5.34) ff. later on] in which the matrix elements of $\hat{L}_i$ are defined by

$$(\hat{L}_i)_{\alpha\beta} = C_{i\alpha\beta}\quad.$$

From this follows

$$\begin{aligned}(\hat{L}_i, \hat{L}_j) = \mathrm{Tr}(\hat{L}_i, \hat{L}_j) &= \sum_{\alpha\beta}(\hat{L}_i)_{\alpha\beta}(\hat{L}_j)_{\beta\alpha}\\ &= C_{i\alpha\beta}C_{j\beta\alpha} = g_{ij}\quad.\end{aligned}$$

Example 3.9

$(\hat{L}_i, \hat{L}_j)$ fulfills all properties of a *metric*, for example

$$(\hat{L}_i, \hat{L}_j) = (\hat{L}_j, \hat{L}_i) \quad ,$$

$$(\hat{L}_i + \hat{L}_j, \hat{L}_k) = (\hat{L}_i, \hat{L}_k) + (\hat{L}_j, \hat{L}_k) \quad ,$$

but it is not necessarily positive definite [i.e. $(\hat{L}_i, \hat{L}_i) > 0$], as is evidently the case for, e.g. the Minkowski metric. Now we show that a *Lie algebra is semisimple if and only if*

$$\det(g_{\sigma\lambda}) \neq 0 \quad . \tag{2}$$

For this we have to demonstrate that $\det(g_{\sigma\lambda}) = 0$, if there is a nontrivial subalgebra (an abelian ideal). Let us therefore suppose that the Lie algebra possesses an abelian ideal. We discern the generators belonging to the ideal from the remaining generators by marked indices. Then for the λ' column of the metric tensor,

$$g_{\sigma\lambda'} = C_{\sigma\varrho\tau} C_{\lambda'\tau\varrho} = C_{\sigma\varrho'\tau} C_{\lambda'\tau\varrho'} \tag{3}$$

holds, since $C_{\lambda'\tau\varrho} = 0$ for those values $\varrho \neq \varrho'$ which do not belong to the ideal [cf. (3.28)]. Furthermore, it follows from (3) that

$$g_{\sigma\lambda'} = C_{\varrho'\sigma\tau} C_{\lambda'\tau\varrho'} = -\, C_{\varrho'\sigma\tau'} C_{\lambda'\tau'\varrho'} \quad , \tag{4}$$

since $C_{\varrho'\sigma\tau} = 0$ for all $\tau \neq \tau'$ for the same reason. But now for an abelian ideal

$$C_{\lambda'\tau'\varrho'} = 0 \tag{5}$$

holds. Thus follows

$$g_{\sigma\lambda'} = 0 \quad . \tag{6}$$

Therefore, the λ' column of the metric tensor vanishes and we obtain

$$\det(g_{\sigma\lambda}) = 0 \quad . \tag{7}$$

We here remark that Cartan's condition (2) has the meaning that the metric tensor must have an inverse tensor $g^{\sigma\lambda} = (g_{\sigma\lambda})^{-1}$, for which then

$$g^{\sigma\lambda} g_{\sigma\tau} = \delta^{\lambda}_{\tau} \tag{8}$$

holds. We now prove the reverse.

Let

$$\det g_{\sigma\lambda} = 0 \tag{9}$$

We show that the algebra defined by

$$[\hat{L}_i, \hat{L}_j] = C_{ijk} \hat{L}_k \quad , \quad \hat{L} \in G \tag{10}$$

possesses an abelian subalgebra

$$[\hat{L}'_i, \hat{L}'_j] = 0 \quad \hat{L}' \in U \subset G \quad . \tag{11}$$

If det $g_{\sigma\lambda} = 0$, the equation

$$a^{\sigma} g_{\sigma\lambda} = 0 \tag{12}$$

Example 3.9

yields nontrivial solutions. The subspace of generators defined in this way

$$\hat{L}'_i = a_i^\sigma \hat{L}_\sigma \quad , \quad a_i^\sigma g_{\sigma\lambda} = 0 \tag{13}$$

again forms an algebra. If $\hat{L}_k$ is an arbitrary generator of $G_i \hat{L}'_i$, $\hat{L}'_j$ are generators of U, we have

$$\mathrm{Tr}[\hat{L}'_i, \hat{L}'_j]\hat{L}_k = \mathrm{Tr}\,\hat{L}_i[\hat{L}'_j, \hat{L}_k] \quad . \tag{14}$$

substituting the expansion (13) for $\hat{L}'_i$, one obtains

$$\begin{aligned}\mathrm{Tr}[\hat{L}'_i, \hat{L}'_j]\hat{L}_k &= a_i^\sigma a_j^\delta \mathrm{Tr}\,\hat{L}_\sigma[\hat{L}_\delta, \hat{L}_k]_- \\ &= a_i^\sigma a_j^\delta C_{\delta kl} \mathrm{Tr}\hat{L}_\sigma \hat{L}_l \\ &= a_i^\sigma a_j^\delta C_{\delta kl}\, g_{\sigma l} \\ &= 0 \quad .\end{aligned} \tag{15}$$

Here, we have used $a_i^\delta g_{\sigma l} = 0$.

Hence,

$$\mathrm{Tr}[L'_i, L'_j]\hat{L}_k = 0 \quad . \tag{16}$$

Since $\hat{L}_k$ is chosen arbitrarily, the trace vanishes if the commutator vanishes

$$[L'_i, L'_j] = 0 \tag{17}$$

which was what we set out to prove.

EXAMPLE

3.10 Semisimplicity of SO(3)

Problem. Show by means of Cartan's criterion that the group SO(3) is semisimple.

Solution. The structure constants of SO(3) are identical with the antisymmetric tensor $\mathrm{i}\varepsilon_{ijk}$ (see Example 3.1). Therefore the metric tensor of the SO(3) is given by

$$g_{\sigma\lambda} = -\,\varepsilon_{\sigma jk}\varepsilon_{\lambda kj} = \varepsilon_{\sigma jk}\varepsilon_{\lambda jk} = 2\delta_{\sigma\lambda} \quad . \tag{1}$$

Then

$$\det(g_{\sigma\lambda}) = 8 \neq 0 \tag{2}$$

follows, which demonstrates that SO(3) is semisimple. Indeed, already by means of our earlier considerations we know that SO(3) is simple (cf. Example 3.7).

3.4 Compact Lie Groups and Lie Algebras

Here we want to introduce one further concept: A Lie group is usually called *compact*, if its parametrisation consists of a finite number of bounded parameter domains. Otherwise the group is called *noncompact*. The related Lie algebra is correspondingly called compact and noncompact, respectively. Any compact Lie algebra is semisimple. Earlier we defined the compactness more precisely (Sect. 1.7). The present definition is less precise, but it is also correct: Each group element is uniquely related by the parameters. For example, for O(2), ϕ describes the group element $e^{i\phi \hat{L}_z}$.

If there are a finite number of parameters (i.e. finite number of generators), and if the parameter domain is compact (if it is bounded, it can be compactified by including in the parameter domain every element, which can be obtained by limiting process), then the definition of Sect. 1.7 holds.

3.5 Invariant Operators (Casimir Operators)[1]

The spherical harmonics $Y_{lm}(\theta, \phi)$ are characterized by the "quantum numbers" l and m. More precisely, one says that the $Y_{lm}(\theta, \phi)$ are simultaneous eigenfunctions of the two operators of orbital angular momentum $\hat{\boldsymbol{L}}^2$ and $\hat{L}_3$, which are connected to the generators $\hat{J}_i$ of the rotation group *for spinless fields* by

$$\hat{\boldsymbol{L}}^2 = \sum_i \hat{J}_i^2 = \hat{\boldsymbol{J}}^2 \quad , \quad \hat{L}_3 = \hat{J}_3 \quad .$$

$\hat{\boldsymbol{L}}^2$ (and also $\hat{\boldsymbol{J}}^2$) *is not a generator* of the group, but it is a *bilinear function of all generators*. $\hat{\boldsymbol{L}}^2$ has the distinctive property of commuting with all generators, i.e.

$$[\hat{\boldsymbol{J}}^2, \hat{J}_i]_- = 0 \quad . \quad \text{Therefore}$$

$$[\hat{\boldsymbol{J}}^2, \hat{U}_{\mathrm{R}}(\boldsymbol{\phi})]_- = 0$$

holds too, i.e. $\hat{\boldsymbol{J}}^2$ also commutes with all group operators $\hat{U}_{\mathrm{R}}(\boldsymbol{\phi})$. Therefore $\hat{\boldsymbol{J}}^2$ is called an *invariant operator* of the group or ***Casimir*** *operator*.

The importance of the $\hat{\boldsymbol{J}}^2$ rests in the fact that its $(2j+1)$-fold degenerated eigenvectors exactly represent the *multiplets of the rotation group*. $j = 0$ is a singlet, $j = \frac{1}{2}$ a doublet, $j = 1$ a triplet and so forth. This property is no peculiarity of the rotation group, but is a general characteristic of semisimple Lie groups in a slightly generalized form. Namely, ***Racah*** proved the following theorem:

[1] To some extent, we follow the article by K.W. McVoy: Reviews of Modern Physics **37** (No. 1), 84 (1965).

3.6 Theorem of Racah[2]

For any semisimple Lie group of *rank* l, there exists a set of l Casimir operators. These are functions $\hat{C}_\lambda(\hat{L}_1, \hat{L}_1, \dots, \hat{L}_n)$ $(\lambda = 1, 2, \dots, l)$ of the generators $\hat{L}_i$, and commute with every operator of the group and therefore also amongst themselves: $[\hat{C}_\lambda, \hat{C}'_\lambda] = 0$. The eigenvalues of the $\hat{C}_\lambda$ uniquely characterize the multiplets of the group.

This theorem provides us with the possibility of a precise formulation of the notion *multiplet*. Let us, therefore, begin to explain this concept.

3.7 Comments on Multiplets

We begin with the concept of an *invariant subspace of the total Hilbert space*, i.e. of all states on which the operators of the symmetry group act. By that, one understands a set of states which reproduce themselves by application of some operator of the group, i.e. yield other states of the same set. *The operators of the group transform the states of the invariant subspace among themselves.* In other words, the matrix elements of group operators (i.e. generators) between states of the invariant subspace and states outside of it vanish. A *multiplet* is an *irreducible invariant subspace* of a group, i.e. a subspace which does not contain a further invariant subspace.

EXAMPLE

3.11 An Invariant Subspace to the Rotation Group

The orthogonal vectors $\{Y_{00}, Y_{11}, Y_{10}, Y_{1-1}\}$ form a four-dimensional invariant subspace with respect to the rotation group, because any of the generators $\hat{J}_k$ (= angular momentum operators $\hat{L}_k$ in this case) of this group can only change – as we know – the quantum number m, but not l. Consequently the $\hat{J}_k$ transform these vectors among themselves, as do the $\hat{U}_\mathrm{R}(\boldsymbol{\phi})$, which are functions of the $\hat{J}_k$(cf. Vol. 4, Chap. 4), so that

$$\hat{J}_\pm Y_{lm} = (\hat{J}_1 \pm \mathrm{i}\hat{J}_2) Y_{lm} = \sqrt{l(l+1) - m(m \pm 1)}\, Y_{lm\pm 1} \quad ,$$

$$\hat{J}_3 Y_{lm} = m Y_{lm} \quad .$$

[2] G. Rach: *Group Theory and Spectroscopy*, Princeton Lectures (CERN Reprint).

EXAMPLE

3.12 Reduction of an Invariant Subspace

The four-dimensional invariant subspace $(Y_{00}, Y_{11}, Y_{10}, Y_{1-1}\}$, which we have already mentioned in the previous example, is a reducible subspace, because it contains smaller subspaces (which are not further reducible, i.e. which are true multiplets), namely the *triplet*

$$\{Y_{11}, Y_{10}, Y_{1-1}\} \tag{1}$$

and the *singlet*

$$\{Y_{00}\} \quad . \tag{2}$$

This is so because the operators $\hat{J}_\nu$ do not change the l quantum number, for example

$$\hat{J}_\pm Y_{lm} = \sqrt{l(l+1) - m(m \pm 1)}\, Y_{lm \pm 1}$$

$$\hat{J}_0 Y_{lm} = m Y_{lm} \quad . \tag{3}$$

The triplet (and also the singlet) cannot be further reduced. From (3) it also follows that

$$\langle Y_{00}|\hat{U}_{\mathrm{R}}(\boldsymbol{\phi})|Y_{lm}\rangle = 0 \quad (l \neq 0) \tag{4}$$

for every rotation operator $\hat{U}_{\mathrm{R}}(\boldsymbol{\phi})$ of the group. There exists no group operator with nonvanishing matrix elements between singlet and triplet (more generally, between one multiplet and another). The complete decomposition of the invariant subspace $\{Y_{00}, Y_{11}, Y_{10}, Y_{1-1}\}$ into a singlet and a triplet is expressed by the notation

$$\{Y_{00}, Y_{11}, Y_{10}, Y_{1-1}\} = \{Y_{00}\} \oplus \{Y_{11}, Y_{10}, Y_{1-1}\} \quad . \tag{5}$$

We recognize from the last example that the states contained in a single multiplet are obviously related to each other. This becomes clearer by considering the following *alternative construction of a multiplet*:

We start from a normalized state ψ_0, which lies completely within a multiplet of the group. Then all vectors

$$\psi_{\boldsymbol{a}}(\boldsymbol{r}) = \hat{U}(\boldsymbol{a})\psi_0 \tag{3.29}$$

are constructed which can be reached from ψ_0 by application of the group operators $\hat{U}(\boldsymbol{a})$. Each successive application of a group operator $\hat{U}(\boldsymbol{\beta})$ has to transform the vectors $\psi_{\boldsymbol{a}}(\boldsymbol{r})$ among themselves because $\hat{U}(\boldsymbol{\beta})\hat{U}(\boldsymbol{a}) = \hat{U}(\boldsymbol{\gamma})$ again must be a group operator. The sphere of vectors $\psi_{\boldsymbol{a}}$ is transformed into itself (cf.

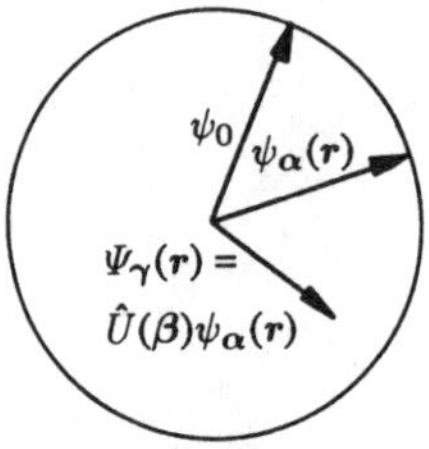

Fig. 3.3. The states $\psi_\alpha(\boldsymbol{r})$ can be interpreted as vectors on a unit sphere (normalized states) [$\psi_\gamma(\boldsymbol{r})$ appears only perspectively shortened]

Fig. 3.3). The $\psi_\alpha(\boldsymbol{r})$ alone do not form a vector space (because each element is normalized to unity), but all linear combinations of $\psi_\alpha(\boldsymbol{r})$ together span a vector space. The later is obviously invariant under the group. *It forms exactly that multiplet*, which contains ψ_0.

The name "multiplet" originated in atomic spectroscopy, where the invariant subspaces are characterized by the total angular momentum j and the orbital momentum l.

For example

$2p_{3/2}$ means	$n = 2$ (principal quantum number)
	$l = 1$ (p state Y_{1m})
	$j = 3/2$ (total angular momentum)
$3d_{5/2}$ means	$n = 3$ (principal quantum number)
	$l = 3$ (d state Y_{2m})
	$j = 5/2$ (total angular momentum)
$3p_{3/2}$ means	$n = 3$ (principal quantum number)
	$l = 1$ (p state Y_{1m})
	$j = 3/2$ (total angular momentum etc.)

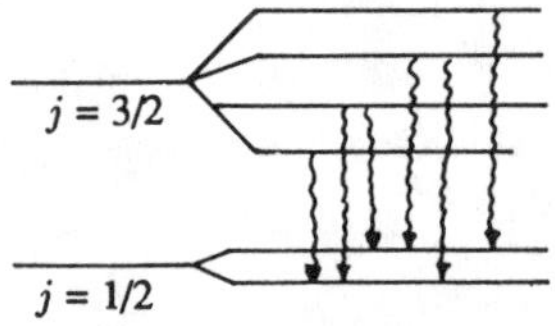

Fig. 3.4. Line multiplet obtained by splitting of the state multiplet. The splitting of the multiplet states indicates that the symmetry underlying these multiplet is not exactly valid. One says: the symmetry is (slightly) broken. Otherwise it would be impossible to recognize the multiplet (here: the multiplet of X-ray lines). See also the following discussion of invariance under a symmetry (Sect. 3.8)

In the absence of an external field, all these multiplets are $(2j + 1)$-times degenerate. In the presence of an external field (for example a magnetic field), these states split up, which can be demonstrated, for example, by the Zeeman effect, and lead to a series of *multiplets of neighbouring lines* in the spectrum of the atom (Fig. 3.4).

This splitting indicates that the symmetry, which is represented by the multiplets, is no longer exact. Usually one speaks of a (weakly) broken symmetry. Otherwise, one would not be able to recognize the multiplicity of states in the multiplet, in our case the multiplet of lines, (compare also to the following discussion on the invariance under a symmetry group).

In terms of group theory, a *set of degenerate states* is called a multiplet. The multiplets depend on the symmetry group. So in atomic physics we deal with angular momentum (spin) multiplets which follow from the rotational invariance. Y_{00} is a singlet with respect to the rotation group; the Λ particle (cf. Table 6.1) is an isospin-singlet; neutron and proton (n, p) form an isospin doublet (cf. Chap. 5), as do the ground states of the nuclei ^{3}H and ^{3}He; the pions $\{\pi^+, \pi^0, \pi^-\}$ form an isospin-triplet, etc.

The important qualitative insight that we gain from this discussion is that each group has a well-defined, unique and partly characteristic set of multiplets. Although these multiplets are determined by the structure of the group, *there exists no general method to find them for arbitrary continuous groups.* Only for semi-simple Lie groups, we have the Racah theorem at hand.

3.8 Invariance Under a Symmetry Group

Let $\hat{U}(\boldsymbol{\alpha})$ be the operators of a *symmetry group* (for example, the rotation operators). Then the invariance of the system under the group $\hat{U}(\boldsymbol{\alpha})$ means that both the initial state ψ which fulfills[3]

$$\mathrm{i}\frac{\partial}{\partial t}\psi = \hat{H}\psi \quad , \tag{3.30}$$

and also the state generated by the *symmetry operation* (rotation)

$$\psi'(\boldsymbol{r}) = \hat{U}(\boldsymbol{\alpha})\psi(\boldsymbol{r}) \tag{3.31}$$

fulfill the same Schrödinger equation (3.30) with the same Hamiltonian $\hat{H}$, hence,

$$\mathrm{i}\frac{\partial}{\partial t}\psi' = \hat{H}\psi' \quad . \tag{3.32}$$

In order to recognize the consequences of this requirement, we multiply (3.30) by $\hat{U}(\alpha)$

$$\mathrm{i}\frac{\partial}{\partial t}\hat{U}(\boldsymbol{\alpha})\psi(\boldsymbol{r}) = \hat{U}(\boldsymbol{\alpha})\hat{H}\hat{U}^{-1}(\boldsymbol{\alpha})\hat{U}(\alpha)\psi(\boldsymbol{r}) \quad . \tag{3.33}$$

Above, we have inserted $\mathbb{1} = \hat{U}^{-1}(\boldsymbol{\alpha})\hat{U}(\boldsymbol{\alpha})$ between $\hat{H}$ and ψ on the rhs. The group parameters $\boldsymbol{\alpha}$ are fixed numbers; they are especially independent of time. Therefore, because of (3.31), (3.33) takes the form

$$\mathrm{i}\frac{\partial}{\partial t}\psi'(\boldsymbol{r},t) = \hat{U}(\boldsymbol{\alpha})\hat{H}\hat{U}^{-1}(\boldsymbol{\alpha})\psi'(\boldsymbol{r},t) \quad , \tag{3.34}$$

so that a comparison with the requirement (3.32) yields

$$\hat{H} = \hat{U}(\boldsymbol{\alpha})\hat{H}\hat{U}^{-1}(\boldsymbol{\alpha}) \quad \text{or} \tag{3.35}$$

$$[\hat{U}(\boldsymbol{\alpha}),\hat{H}]_- = \hat{U}(\boldsymbol{\alpha})\hat{H} - \hat{H}\hat{U}(\boldsymbol{\alpha}) = 0 \quad . \tag{3.36}$$

Hence, the invariance of the system under the group $\hat{U}$ necessarily means that $\hat{H}$ commutes with all group operators $\hat{U}(\boldsymbol{\alpha})$, and, therefore, that it also commutes with all generators $\hat{L}_i$ of the group. One immediately recognizes that the inversion is valid, too. From

$$[\hat{L}_i,\hat{H}]_- = 0 \tag{3.37}$$

and $\hat{U}(\boldsymbol{\alpha}) = \exp(-\mathrm{i}\alpha_i\hat{L}_i)$, it follows that

$$[\hat{U}(\boldsymbol{\alpha}),\hat{H}]_- = 0 \quad . \tag{3.38}$$

Whenever a state, for example ψ_0, is an eigenstate of the Hamiltonian

$$\hat{H}\psi_0 = E_0\psi_0 \quad , \tag{3.39}$$

then, according to (3.36), we obtain

$$\hat{U}(\boldsymbol{\alpha})\hat{H}\psi_0 = \hat{U}(\boldsymbol{\alpha})E_0\psi_0 \quad , \quad \hat{H}\hat{U}(\boldsymbol{\alpha})\psi_0 = E_0\hat{U}(\boldsymbol{\alpha})\psi_0 \quad , \tag{3.40}$$

[3] Note that here we set $\hbar = 1$.

i.e. all other states $\hat{U}(\boldsymbol{a})\psi_0$ of the multiplet are eigenstates of the Hamiltonian with the same eigenvalue E_0. In other words, *the Hamiltonian is degenerate on each multiplet of a symmetry group.* This also holds for the eigenvalues C_λ of the l Casimir operators $\hat{C}_\lambda$ which commute with $\hat{H}$ because the generators $\hat{L}_i$ of the symmetry group $\hat{U}$ commute with $\hat{H}$, and because the $\hat{C}_\lambda(\hat{L}_1, \dots, \hat{L}_l)$ commute with $\hat{L}_i$. This can be understood in the following way: From

$$[\hat{C}_\lambda(\hat{L}_1, \dots, \hat{L}_l), \hat{L}_i]_- = 0 \quad , \tag{3.41}$$

it immediately follows that

$$[\hat{C}_\lambda(\hat{L}_\nu), \hat{C}_{\lambda'}(\hat{L}_\nu)]_- = 0 \quad , \tag{3.42}$$

i.e. all l Casimir operators commute with each other and of course also with $\hat{H}$. Since commuting operators can be simultaneously diagonalized, i.e. they have the same eigenfunctions, we conclude from (3.40) that *the $\hat{C}_\lambda$ are also degenerate on the multiplet.* In other words, for a given multiplet the operators $\hat{C}_\lambda$ possess a common set of eigenvalues

$$C_1, C_2, \dots, C_l \quad . \tag{3.43}$$

Thus the Racah theorem guarantees that each multiplet is related uniquely to a set of eigenvalues $C_1, C_2, \dots, C_l$. We can summarise this as follows: *Each multiplet of semisimple Lie group can be uniquely characterized by the eigenvalues $C_1, C_2, \dots, C_l$ of the l Casimir operators $\hat{C}_1, \hat{C}_2, \dots, \hat{C}_l$.*

EXAMPLE

3.13 Casimir Operator of the Rotation Group

The multiplets of the rotation group (rank 1) are uniquely characterized by the eigenvalues of the Casimir operator $\hat{\boldsymbol{J}}^2$, i.e. by $j(j+1)$ (usually one simply says by j).

In (3.29) we saw that each state ψ of the multiplet can be represented by the application of a group operator $\hat{U}(\boldsymbol{a})$ or, to the arbitrary state ψ_0 of the multiplet more generally, by an appropriate linear combination

$$\psi = \sum_{\boldsymbol{a}} a_{\boldsymbol{a}} \hat{U}(\boldsymbol{a})\psi_0 \tag{1}$$

of such multiplet vectors $\hat{U}(\boldsymbol{a})\psi_0$. Now, with the help of the Racah theorem, we are able to understand this relation from another point of view: As ψ_0 is an eigenstate of $\hat{H}$ and all $\hat{C}_\lambda$, with the corresponding eigenvalues E_0 and $C_\lambda(\lambda = 1, 2, \dots, l)$ respectively, i.e.

$$\begin{aligned} \hat{H}\psi_0 &= E_0\psi_0 \\ \hat{C}_\lambda\psi_0 &= C_\lambda\psi_0 \quad \lambda = 1, 2, \dots, l \end{aligned} \tag{2}$$

and because $[\hat{H}, \hat{C}_\lambda]_- = 0$, $[\hat{H}, \hat{L}_i]_- = 0$ and also $[\hat{C}_\lambda, \hat{L}_i]_- = 0$ [cf. (3.36–42)], it follows that

$$[\hat{H}, \hat{U}(\boldsymbol{a})]_- = 0 \quad , \quad \text{and} \tag{3}$$

$$[\hat{C}_\lambda, \hat{U}(\boldsymbol{a})]_- = 0 \quad , \tag{4}$$

and, therefore, with (1)

$$\hat{H}\psi = \hat{H}\left(\sum_{\boldsymbol{a}} a_{\boldsymbol{a}}\hat{U}(\boldsymbol{a})\psi_0\right) = \sum_{\boldsymbol{a}} a_{\boldsymbol{a}}\hat{U}(\boldsymbol{a})\hat{H}\psi_0 = E_0\psi \quad , \tag{5}$$

$$\hat{C}_\lambda\psi = \hat{C}_\lambda\left(\sum_{\boldsymbol{a}} a_{\boldsymbol{a}}\hat{U}(\boldsymbol{a})\psi_0\right) = \sum_{\boldsymbol{a}} a_{\boldsymbol{a}}\hat{U}(\boldsymbol{a})\hat{C}_\lambda\psi_0 = C_\lambda\psi \quad . \tag{6}$$

Hence, we conclude: Each state ψ of the multiplet is simultaneously an eigenstate of $\hat{H}$ and of all Casimir operators with the same eigenvalues of E_0 and $C_\lambda(\lambda = 1, 2, \ldots, l)$ respectively. This set of eigenvalues comprises all symmetry properties of the multiplet with respect to the symmetry group $\hat{U}(\boldsymbol{a})$. We will soon understand (cf. "Completeness Relation for Casimir Operators", Sect. 3.11) that the multiplet does not possess any further symmetry properties with respect to the group $\hat{U}(\boldsymbol{a})$. This leads us to recognize the *fundamental role of the invariant operators* (*Casimir opertors*) *of the symmetry group*. With their eigenvalues they represent the symmetry properties of the group.

EXAMPLE

3.14 Some Groups with Rank 1 or 2

a) The rotation group is a special one because it is of rank one and therefore possesses only one invariant operator $\hat{\boldsymbol{J}}^2$.

b) The group SU(3), which we will later study in more detail (see Chap.7), is of rank two and consequently has two Casimir operators. One of them corresponds to $\hat{\boldsymbol{J}}^2$. It can have the same eigenvalue on different multiplets of the SU(3) group. In this case we need the second invariant operator, in order to uniquely classify the SU(3) multiplets. Then Racah's theorem tells us that with the two invariant operators, the SU(3) multiplets are completely classified. So, for example, consider the following notations (compare Chap. 7), in particular Exercise 7.9):

the	$\bar{3}$-triplet	by: $D^3(0, 1)$
or the	octet	by: $D^8(1, 1)$
or the	15-dimensional multiplet	by: $D^{15}(2, 1)$

The two numbers in brackets are the eigenvalues of the two Casimir operators. Note that all three multiplets considered here have the same eigenvalue for the

Example 3.14

second Casimir operator (namely 1). All of them are distinguished by the first eigenvalue (0, 1, 2).

3.9 Construction of the Invariant Operators

In general there exists no method to construct the Casimir operators for arbitrary semisimple groups. Each group must be studied separately. Only for the SU(n) *groups*, i.e. for the group of unitary, unimodular $n \times n$ matrices (cf. Chap. 4), was Biedenharn[4] able to show that the Casimir operators have to be simple homogeneous polynomials in the generators.

$$\hat{C}_\lambda = \sum_{ij} a^\lambda_{ij} \ldots \hat{L}_i \hat{L}_j \ldots \quad . \quad (\lambda - \text{factors}) \tag{3.44}$$

where the a^λ_{ij} are well defined functions of the structure constants. The simplest Casimir operator $\hat{C}_1$ is a quadratic function of the generators. In the case of the SU(2), for example, this is the operator $\hat{\boldsymbol{J}}^2 = \frac{1}{2}(\hat{J}_+\hat{J}_- + \hat{J}_-\hat{J}_+) + \hat{J}_3^2 . \hat{C}_2$ is of third order in $\hat{L}_i$ etc. We shall construct $\hat{C}_1$ below [see 3.50)ff.] for the general case. For the other Casimir operators a general construction procedure is not known. The invariant operators of the SU(3) are $\hat{C}_1$ and $\hat{C}_2$, as we will see later, in Chap. 7. In this case we will be successful to construct both Casimir operators.

Note that the *Casimir operators are not unique*. Assume, for example, $\hat{C}$ and $\hat{C}'$ to be invariant operators of a group of rank l, then

$$\hat{\hat{C}} = \hat{C} + \hat{C}' \quad , \quad \hat{\hat{C}}' = \hat{C} - \hat{C}' \tag{3.45}$$

are invariant operators of the same group, too. They are as well suited as the original Casimir operators $\hat{C}$ and $\hat{C}'$ to classify the multiplets. In the same way, any powers of $\hat{C}$ and $\hat{C}'$ or products of both are again Casimir operators. If the group operators $\hat{U}(\boldsymbol{a})$ are unitary and hence the generators $\hat{L}_i$ are hermitian, then one can make use of this freedom in order always to *construct the Casimir operators of a unitary, semi-simple Lie group als hermitian operators.* Namely, if $\hat{C}$ is an invariant operator, then for any group operator $\hat{U}(\boldsymbol{a})$ [cf. Exercise 3.13, Eq. (4)], we have

$$\hat{C}\hat{U}(\boldsymbol{a}) = \hat{U}(\boldsymbol{a})\hat{C} \tag{3.46}$$

and, therefore,

$$\hat{U}^\dagger(\boldsymbol{a})\hat{C}^\dagger = \hat{C}^\dagger\hat{U}^\dagger(\boldsymbol{a}) \quad . \tag{3.47}$$

[4] L.C. Biedenharn: J. Math. Phys. **4**, 436 (1963).

Since $\hat{U}^\dagger = \hat{U}^{-1}$ for unitary group operators, it follows that

$$\hat{U}^{-1}(\boldsymbol{a})\hat{C}^\dagger = \hat{C}^\dagger \hat{U}^{-1}(\boldsymbol{a}) \quad . \tag{3.48}$$

In words: $\hat{C}^\dagger$ commutes with all inverse group operators $\hat{U}^{-1}(\boldsymbol{a})$. This means that $\hat{C}^\dagger$ commutes with all group operators $\hat{U}(\boldsymbol{a})$, because the inverse of every group operator is also a group operator by definition of a group. So $\hat{C}$ *is an invariant operator* too, and it follows from the note to [cf. Example 3.13, Eq. (2)] that

$$\hat{C}'' = \hat{C} + \hat{C}^\dagger \tag{3.49}$$

is a Casimir operator, too. Obviously, it is hermitian. Thus, from now on we assume all invariant operators of unitary groups to be hermitian.

One of the Casimir operators is always given by

$$\hat{C}_1 = g^{\varrho\sigma}\hat{L}_\varrho\hat{L}_\sigma \tag{3.50}$$

where $g^{\varrho\sigma}$ is the inverse metric tensor (see Example 3.9). This inverse metric tensor always exists for semi-simple Lie groups, because we know from Example 3.9, Eq. (2) that $\det(g_{\sigma\lambda}) \neq 0$. The $\hat{L}_\varrho$ are the generators of the Lie algebra. Starting from the following commutator, we now show that $\hat{C}_1$ is a Casimir operator:

$$\begin{aligned}
[\hat{C}_1, \hat{L}_\tau]_- &= g^{\varrho\sigma}[\hat{L}_\varrho\hat{L}_\sigma, \hat{L}_\tau]_- = g^{\varrho\sigma}\hat{L}_\varrho[\hat{L}_\sigma, \hat{L}_\tau]_- + g^{\varrho\sigma}[\hat{L}_\varrho\hat{L}_\tau]_-\hat{L}_\sigma \\
&= g^{\varrho\sigma}C^\lambda_{\sigma\tau}\hat{L}_\varrho\hat{L}_\lambda + g^{\varrho\sigma}C^\lambda_{\varrho\tau}\hat{L}_\lambda\hat{L}_\sigma = g^{\varrho\sigma}C^\lambda_{\sigma\tau}\hat{L}_\varrho\hat{L}_\lambda + g^{\varrho\sigma}C^\lambda_{\sigma\tau}\hat{L}_\lambda\hat{L}_\varrho \\
&= g^{\varrho\sigma}C^\lambda_{\sigma\tau}(\hat{L}_\varrho\hat{L}_\lambda + \hat{L}_\lambda\hat{L}_\varrho) \quad .
\end{aligned} \tag{3.51}$$

Note that we exchanged the summation indices σ and ϱ in the last but one line. We also have denoted structure constants in the form $C_{\sigma\tau\lambda} \equiv C^\lambda_{\sigma\tau}$, i.e. we have raised the third index. This is convenient in the following and also sometimes used in literature. The tensor

$$a^{\varrho\lambda}_\tau = g^{\varrho\sigma}C^\lambda_{\sigma\tau} \tag{3.52}$$

is antisymmetric in ϱ and λ. In order to understand this we introduce the tensor

$$b_{\sigma\mu\nu} = g_{\sigma\lambda}C^\lambda_{\mu\nu} = C^\tau_{\sigma\varrho}C^\varrho_{\lambda\tau}C^\lambda_{\mu\nu} = C^\tau_{\sigma\varrho}C^\lambda_{\mu\nu}C^\varrho_{\lambda\tau} \quad , \tag{3.53}$$

which can, with the help of the Jacobi identity (3.16), be written in the form

$$b_{\sigma\mu\nu} = -\,C^\tau_{\sigma\varrho}(C^\lambda_{\nu\tau}C^\varrho_{\lambda\mu} + C^\lambda_{\tau\mu}C^\varrho_{\lambda\nu}) = C^\tau_{\sigma\varrho}C^\lambda_{\nu\tau}C^\varrho_{\mu\lambda} + C^\tau_{\varrho\sigma}C^\lambda_{\tau\mu}C^\varrho_{\lambda\nu} \quad . \tag{3.54}$$

The summation convention applies to repeated indices. The rhs is invariant under any cyclic permutation of the indices σ, μ and ν. We will show only that this is true for cyclic permutations

$$(\sigma\mu\nu) \rightarrow (\mu\nu\sigma) \quad , \tag{3.55}$$

since all other cases can be easily verified by the reader. Thus, we consider

$$b_{\mu\nu\sigma} = C^\tau_{\mu\varrho}C^\lambda_{\sigma\tau}C^\varrho_{\nu\lambda} + C^\tau_{\varrho\mu}C^\lambda_{\tau\nu}C^\varrho_{\lambda\sigma} = C^\varrho_{\mu\lambda}C^\tau_{\sigma\varrho}C^\lambda_{\nu\tau} + C^\lambda_{\tau\sigma}C^\varrho_{\lambda\mu}C^\tau_{\varrho\nu} = b_{\sigma\mu\nu} \quad . \tag{3.56}$$

Here we renamed the summation indices in the second line, namely $(\varrho\tau\lambda) \rightarrow (\lambda\varrho\tau)$ in the first term and $(\varrho\tau\lambda) \rightarrow (\tau\lambda\varrho)$ in the second term.

As in (3.53), $g_{\sigma\lambda}$ is symmetric in σ and λ; $C^{\lambda}_{\mu\nu}$ is antisymmetric in μ and ν; the tensor $b_{\sigma\mu\nu}$ has to be antisymmetric in μ and ν, i.e. in the last two indices. Because of (3.56), $b_{\sigma\mu\nu}$ has to be antisymmetric in σ and ν, too, and due to the general cyclic invariance of $b_{\sigma\mu\nu}$, it has to be antisymmetric with respect to an exchange of any two indices.

Let us return to (3.51) and (3.52) once more. The tensor $a_{\varrho\tau\lambda}$ can be expressed in another way, using $C^{\lambda}_{\sigma\tau} = g^{\nu\lambda} b_{\nu\sigma\tau}$ to obtain

$$a^{\varrho\lambda}_{\tau} = g^{\varrho\sigma} C^{\lambda}_{\sigma\tau} = g^{\varrho\sigma} g^{\nu\lambda} b_{\nu\sigma\tau} \quad , \tag{3.57}$$

hence, (3.52) reads

$$[\hat{C}_1, \hat{L}_\tau]_- = g^{\varrho\sigma} g^{\nu\lambda} b_{\nu\sigma\tau} (\hat{L}_\varrho \hat{L}_\lambda + \hat{L}_\lambda \hat{L}_\varrho) \quad . \tag{3.58}$$

The sum over σ and ν on the rhs vanishes because of the total antisymmetry property of $b_{\nu\sigma\tau}$. Hence,

$$[\hat{C}_1, \hat{L}_\tau]_- = 0 \tag{3.59}$$

for each $\hat{L}_\tau$, and we have shown that $\hat{C}_1$ is a Casimir operator.

Let us emphasize once more that the Casimir operators are defined in the sense of the Racah Theorem for semi-simple Lie groups. This does not mean that for other Lie groups such invariant operators, which commute with each group operator, cannot be constructed. For example, consider the translation-rotation group (which is sometimes called the *euclidian group*) which is not semi-simple with the algebra.

$$\begin{aligned} &[\hat{J}_i, \hat{J}_j]_- = \varepsilon_{ijk} \hat{J}_k \quad , \quad [\hat{P}_i, \hat{P}_j]_- = 0 \quad , \\ &[\hat{P}_i, \hat{J}_j]_- = \varepsilon_{ijk} \hat{P}_k \quad , \quad [\hat{P}_i, \hat{J}_i]_- = 0 \quad . \quad i, j, k = 1, 2, 3 \end{aligned} \tag{3.60}$$

One easily verifies that in this case there exist three invariant operators, namely $\hat{\boldsymbol{J}}^2 = \hat{J}_1^2 + \hat{J}_2^2 + \hat{J}_3^2$, $\hat{\boldsymbol{p}} \cdot \hat{\boldsymbol{J}} = \hat{p}_1 \hat{J}_1 + \hat{p}_2 \hat{J}_2 + \hat{p}_3 \hat{J}_3$ and $\hat{\boldsymbol{P}}^2 = \hat{P}_1^2 + \hat{P}_2^2 + \hat{P}_3^2$.

We will come back to Casimir operators and to the Racah theorem in a more general context in Chap. 12 (mathematical supplement).

3.10 Remark on Casimir Operatorts of Abelian Lie Groups

If a Lie group is abelian, then its rank is identical with its number of generators $\hat{L}_i$. These are invariant operators themselves, and therefore Casimir operators. The completeness relation, for which a proof will be given in the next section, is valid in this case. So the Racah theorem can be extented (that is to say, trivially) to all abelian Lie groups.

3.11 Completeness Relation for Casimir Operators

Although the l invariant operators are not uniquely determined, they form a complete set. More precisely this means: *Each operator $\hat{A}$ which commutes with* *all operators of a Lie group (hence with all generators $\hat{L}_i$ of this group) is necessarily a function of the Casimir operators $\hat{C}_\lambda$ of the group*

$$\hat{A} = \hat{A}(\hat{C}_\lambda) \quad . \tag{3.61}$$

In other words, the Casimir operators are the largest set of independent operators which commute with the group, i.e. with the group operators $\hat{U}(\boldsymbol{a})$. Specifically, if we consider an operator $\hat{A}$ with $[\hat{A}, \hat{L}_i]_- = 0$, then it has to be a function of the generators $\hat{L}_i$

$$\hat{A} = \hat{A}(\hat{L}_i) \quad . \tag{3.62}$$

because the operation

$$\hat{A}\psi_{\boldsymbol{a}} = \hat{A}\hat{U}(\boldsymbol{a})\psi_0 = \hat{U}(\boldsymbol{a})\hat{A}\psi_0 \tag{3.63}$$

does not lead out of the multiplet of ψ_0. The latter is easily seen noting that, by virtue of their commutativity, $\hat{L}_i$ and $\hat{A}$ have a common eigenstate:

$$\hat{L}_i\psi_0 = l_i\psi_0 \quad , \quad \hat{A}\psi_0 = a\psi_0 \quad . \tag{3.64}$$

Therefore, according to (3.63), $\psi_{\boldsymbol{a}}$ is also an eigenstate of $\hat{A}$ with the same eigenvalue a:

$$A\psi_{\boldsymbol{a}} = \hat{U}(\boldsymbol{a})a\psi_0 = a\psi_{\boldsymbol{a}} \quad . \tag{3.65}$$

So we have $\hat{A}\psi_{\boldsymbol{a}} \propto \psi_{\boldsymbol{a}}$ and thus it lies in the ψ_0-multiplet. Since all vectors in this multiplet can be reached from ψ_0 by appropriate rotations $\hat{U}(\boldsymbol{a})$, $\hat{A}$ has to be a combination of the generators $\hat{L}_i$ of these rotations. Together with (3.65), $\hat{A}$ is diagonal for each state $\psi_{\boldsymbol{a}}$ of the multiplet which contains ψ_0. Hence, $\hat{A}$ fulfills all criteria of an invariant operator; thus, either it must be one itself, or an (eventually nonlinear) combination of the Casimir operators, because according to the Racah theorem there exist exactly l independent invariant operators.

This theorem on the completeness of the Casimir operators is rather useful. If a system is endowed with a certain symmetry, then the corresponding Hamiltonian must commute with the generators and with the Casimir operators of the symmetry group. Together with the point stressed above, however, this means that $\hat{H}$ *itself has to be built up from invariant operators of the symmetry group.*

EXAMPLE

3.15 Construction of the Hamiltonian from the Casimir Operators

a) The spherically symmetric Hamiltonian of a spinless particles in a central field commutes with the operators of the rotation group $\hat{U}_{\mathbf{R}}(\boldsymbol{\phi})$. Therefore, the Hamiltonian must be of the structure

$$\begin{aligned} \hat{H} &= T(\boldsymbol{r}^2, \boldsymbol{p}^2) + f(\boldsymbol{r}^2, \boldsymbol{p}^2)\hat{\boldsymbol{L}}^2 \\ &= T(\boldsymbol{r}^2, \boldsymbol{p}^2)\mathbb{1} + f(\boldsymbol{r}^2, \boldsymbol{p}^2)\hat{\boldsymbol{L}}^2 \quad , \end{aligned} \tag{1}$$

Example 3.15

where $\hat{\boldsymbol{L}}^2$ (the square of the angular momentum operator) is the invariant operator of the rotation group. The term T is proportional to the unit-operator $\mathbb{1}$ of the rotation group (and, therefore, a trivial invariant). Of course, any higher power of $\hat{\boldsymbol{L}}^2$ (for example $\hat{\boldsymbol{L}}^4$) can also appear, we have only considered the most simple structure here.

b) A Hamiltonian which is invariant under translations must contain the Casimir operators of the translation group $\hat{\boldsymbol{p}} = (\hat{p}_1, \hat{p}_2, \hat{p}_3)$; hence, it must be of the form

$$\hat{H} = \alpha\mathbb{1} + \boldsymbol{\beta}\cdot\hat{\boldsymbol{p}} + \gamma\hat{\boldsymbol{p}}^2 + \dots \tag{2}$$

where $\boldsymbol{\alpha}$, $\boldsymbol{\beta}$ and γ are constants. $\mathbb{1}$ is the unit operator of the translation group. As in a), we have only noted the most simple structure. If, additionally, one requires the invariance of $\hat{H}$ under time reversal ($t \to -t$, which implies $\boldsymbol{p} \to -\boldsymbol{p}$), then only even powers of $\hat{p}_i$ are allowed to appear in $\hat{H}$.

c) The only invariant operator of the isospin group of a two-particle system (cf. Chap. 5, section on "Isospin operators for a many-nucleon system") is

$$\hat{\boldsymbol{T}}^2 = \tfrac{1}{4}(\hat{\boldsymbol{\tau}}_1 + \hat{\boldsymbol{\tau}}_2)^2 = \tfrac{1}{4}(6\cdot\mathbb{1} + 2\hat{\boldsymbol{\tau}}_1\cdot\hat{\boldsymbol{\tau}}_2) \quad , \tag{3}$$

where

$$\hat{\boldsymbol{\tau}} = \left\{ \begin{bmatrix} 0 & 1 \\ 1 & 0 \end{bmatrix} , \begin{bmatrix} 0 & -\mathrm{i} \\ \mathrm{i} & 0 \end{bmatrix} , \begin{bmatrix} 1 & 0 \\ 0 & -1 \end{bmatrix} \right\} \tag{4}$$

is the isospin vector which has as components the three Pauli matrices which act in isospin space. As $\hat{\boldsymbol{\tau}}_1^2 = \hat{\boldsymbol{\tau}}_1\cdot\hat{\boldsymbol{\tau}}_1 = 3\cdot\mathbb{1}$ and $\hat{\boldsymbol{\tau}}_2\cdot\hat{\boldsymbol{\tau}}_2 = 3\cdot\mathbb{1}$, we may understand the above result. A hamiltonian which is invariant under an isospin group must contain besides the trivial invariant $\mathbb{1}$ also terms proportional $\hat{\boldsymbol{\tau}}_1\cdot\hat{\boldsymbol{\tau}}_2$. Hence, the most simple structure is

$$\hat{H} = f(\boldsymbol{r})\cdot\mathbb{1} + G(\boldsymbol{r})\hat{\boldsymbol{\tau}}_1\cdot\hat{\boldsymbol{\tau}}_2 \quad . \tag{5}$$

3.12 Review of Some Groups and Their Properties

In Table 3.1 we summarize some groups and their properties. In this context it is remarkable that the *simplest invariant operators of abelian groups are the generators themselves*, because they commute by definition with all group operators and thus also with all generators.

The inversion group, the generator of which is the parity operator $\hat{P}$ which replaces $\boldsymbol{r} \to -\boldsymbol{r}$ (or more precisely $\{x, y, z\} \to \{-x, -y, -z\}$), and the isobaric-spin group will be discussed in detail in Chaps. 5 and 11.

We see that in all cases the rank of the group coincides with the number of invariant operators (as it should according to the Racah theorem) except for the

Table 3.1

Group	Generator	Rank	Invariant Operators	Type
Translation	$\hat{\boldsymbol{p}} = \{\hat{p}_1, \hat{p}_2, \hat{p}_3\}$	3	$\hat{\boldsymbol{p}} = \{\hat{p}_1, \hat{p}_2, \hat{p}_3\}$	abelian
Rotation	$\hat{\boldsymbol{J}} = (\hat{J}_1, \hat{J}_2, \hat{J}_3)$	1	$\hat{\boldsymbol{J}}^2 = \hat{J}_1^2 + \hat{J}_2^2 + \hat{J}_3^2$	simple and hence semi simple too
Rotation-Translation (Euclidean group)	$\hat{\boldsymbol{J}} = (\hat{J}_1, \hat{J}_2, \hat{J}_3)$ $\hat{\boldsymbol{p}} = (p_1, \hat{p}_2, \hat{p}_3)$	3	$\hat{\boldsymbol{p}}^2 = p_1^2 + \hat{p}_2^2 + \hat{p}_3^2$ $\hat{\boldsymbol{J}} \cdot \hat{\boldsymbol{p}} = \hat{J}_1\hat{p}_1 + \hat{J}_1\hat{p}_2 + \hat{J}_3\hat{p}_3$	neither simple nor semi simple[a]
Inversion Invariance	$\hat{P}$	1	$\hat{P}$	discrete (no Lie-Group)
Rotation Inversion	$\hat{\boldsymbol{J}}^2 = (\hat{J}_1, \hat{J}_2, \hat{J}_3)$ $\hat{P}$	2	$\hat{\boldsymbol{J}}^2 = \hat{J}_1^2 + \hat{J}_2^2 + \hat{J}_3^2$ $\hat{P}$	unconnected
Isotopic Spin	$\hat{\boldsymbol{T}} = \{\hat{T}_1, \hat{T}_2, \hat{T}_3\}$	1	$\hat{\boldsymbol{T}}^2 = \hat{T}_1^2 + \hat{T}_2^2 + \hat{T}_3^2$	simple and hence semi simple too

rotation-translation group. However, this group may violate the Racah theorem since it is not semi simple, containing an abelian subgroup. Note once more that the Racah theorem only holds for semi simple groups.

3.13 The Connection Between Coordinate Transformations and Transformations of Functions

Consider a Lie group consisting of transformations which transform the coordinates x_i into x'_i. In compact form these read as

$$\boldsymbol{x}' = f(\boldsymbol{x}, \boldsymbol{a}) \quad , \tag{3.66}$$

where $\boldsymbol{x}$ and $\boldsymbol{x}'$ are the space vectors in an n-dimensional space, and $\boldsymbol{a}$ represents the r group parameters. In detail, the transformation given above reads

$$x'_i = f_i(x_1, x_2, \dots, x_n; a_1, a_2, \dots, a_r) \quad , \quad i = 1, 2, \dots, n \quad . \tag{3.67}$$

The parameters are chosen such that $\boldsymbol{a} = 0$ coincides with the identity

$$\boldsymbol{x} = f(\boldsymbol{x}, 0) \quad . \tag{3.68}$$

Now, if we perform an infinitesimal rotation $d\boldsymbol{a}$ starting from the identity, $\boldsymbol{x}$ is transformed into $\boldsymbol{x}' = \boldsymbol{x} + d\boldsymbol{x}$, i.e.

$$\boldsymbol{x} + d\boldsymbol{x} = f(\boldsymbol{x}, d\boldsymbol{a}) \quad . \tag{3.69}$$

Taking the lowest order of da, this yields

$$d\boldsymbol{x} = f(\boldsymbol{x}, d\boldsymbol{a}) - f(\boldsymbol{x}, 0) = \left[\frac{\partial}{\partial \boldsymbol{a}} f(\boldsymbol{x}, \boldsymbol{a})\right]\bigg|_{\boldsymbol{a}=0} \cdot d\boldsymbol{a} \quad . \tag{3.70}$$

Introducing the abbreviation

$$\boldsymbol{u}(\boldsymbol{x}) = \left[\frac{\partial}{\partial \boldsymbol{a}} f(\boldsymbol{x}, \boldsymbol{a})\right]\bigg|_{\boldsymbol{a}=0} \quad , \quad \text{we can write} \tag{3.71}$$

$$d\boldsymbol{x} = \boldsymbol{u}(\boldsymbol{x}) \cdot d\boldsymbol{a} \quad , \tag{3.72}$$

or in detail

$$dx_i = \left[\frac{\partial}{\partial a_\mu} f_i(\boldsymbol{x}, \boldsymbol{a})\right]\bigg|_{\boldsymbol{a}=0} da_\mu = u_{i\mu} da_\mu \quad , \quad \text{where} \tag{3.73}$$

$$u_{i\mu}(\boldsymbol{x}) = \left[\frac{\partial}{\partial a_\mu} f_i(\boldsymbol{x}, \boldsymbol{a})\right]\bigg|_{\boldsymbol{a}=0} \quad . \tag{3.74}$$

Here we made use of the summation convention in the sense that roman summation indices (such as i) have the range from 1 to n and greek indices (such as μ) from 1 to r.

We now discuss the change of a function $F(\boldsymbol{x})$ under the rotation $d\boldsymbol{a}$ introduced above. We have

$$dF = \frac{\partial F(\boldsymbol{x})}{\partial \boldsymbol{x}} \cdot d\boldsymbol{x} = \sum_i \frac{\partial F(\boldsymbol{x})}{\partial x_i} dx_i \tag{3.75}$$

and with (3.72–75)

$$\begin{aligned} dF &= \frac{\partial F(\boldsymbol{x})}{\partial \boldsymbol{x}} \cdot d\boldsymbol{x} = \frac{\{\partial F(\boldsymbol{x})\}}{\partial \boldsymbol{x}} \cdot \boldsymbol{u}(\boldsymbol{x}) \cdot d\boldsymbol{a} = d\boldsymbol{a} \cdot \left\{\boldsymbol{u}(\boldsymbol{x}) \cdot \frac{\partial F(\boldsymbol{x})}{\partial \boldsymbol{x}}\right\} \\ &= \sum_{\mu,i} da_\mu \left\{u_{i\mu}(\boldsymbol{x}) \frac{\partial}{\partial x_i}\right\} F(\boldsymbol{x}) = -\mathrm{i} \sum_{\mu,i} da_\mu \hat{L}_\mu(\boldsymbol{x}) F(\boldsymbol{x}) \quad . \end{aligned} \tag{3.76}$$

Obviously, the r quantities

$$\hat{L}_\mu = \mathrm{i} \sum_i u_{i\mu}(\boldsymbol{x}) \frac{\partial}{\partial x_i} = \sum_i \left[\frac{\partial}{\partial a_\mu} f_i(\boldsymbol{x}, \boldsymbol{a})\right]\bigg|_{\boldsymbol{a}=0} \frac{\partial}{\partial x_i} \tag{3.77}$$

are the generators of the group [compare (3.4, 3.5)]. This finding can be understood in the following way: Because of the group property, the rotated quantity $F'(\boldsymbol{x}, \boldsymbol{a})$ has to be obtained by successive rotations from $F(\boldsymbol{x})$. Together with (3.76), this implies that

$$da_\mu = a_\mu / N \quad , \quad (\text{with } N \text{ a large integer}) \tag{3.78}$$

$$\begin{aligned} F(\boldsymbol{x}, \boldsymbol{a}) &= \lim_{N\to\infty} (1 - \mathrm{i}\hat{L}_\mu da_\mu)^N F(\boldsymbol{x}, 0) \\ &= \lim_{N\to\infty} (1 - \mathrm{i}\hat{L}_\mu a_\mu / N)^N F(\boldsymbol{x}, 0) \\ &= \mathrm{e}^{-\mathrm{i}\hat{L}_\mu a_\mu} F(\boldsymbol{x}, 0) = \hat{U}(\boldsymbol{x}, \boldsymbol{a}) F(\boldsymbol{x}, 0) \quad . \end{aligned} \tag{3.79}$$

So it is clear that

$$\hat{U}(\boldsymbol{x}, \boldsymbol{a}) = \exp(-\mathrm{i} a_\mu \hat{L}_\mu)$$

are the group operators and $\hat{L}_\mu$ the group generators.

We shall now illustrate all of these points with some examples and problems.

EXERCISE

3.16 Transformations with r Parameters of an n-Dimensional Space

Consider a group of transformations with r parameters of an n-dimensional space ($\mathbb{R}^n$ or $\mathbb{C}^n$), i.e. to each $\boldsymbol{x} = (x^i, i, \dots, n)$ we relate an

$$x_i' = f_i(\boldsymbol{x}; a_1, \dots, a_r) \tag{1}$$

with the group parameters $a_1 \dots . a_r$. As usual, let us assume that

$$\boldsymbol{x} = f(\boldsymbol{x}; 0, 0, \dots, 0) \quad . \tag{2}$$

We know that a function $F(\boldsymbol{x})$ transforms under an infinitesimal transformation $d\boldsymbol{a}$ according to

$$\begin{aligned} F(\boldsymbol{x}') &= F(f(\boldsymbol{x}, d\boldsymbol{a})) = F(\boldsymbol{x} + d\boldsymbol{x}) \\ &= \sum_{\mu=1}^{r} (-\mathrm{i}) da_\mu \hat{L}_\mu(\boldsymbol{x}) F(\boldsymbol{x}) \quad , \end{aligned} \tag{3}$$

with the operators (generators)

$$\begin{aligned} \hat{L}_\mu &= \mathrm{i} \sum_{j=1}^{n} u_{j\mu}(\boldsymbol{x}) \frac{\partial}{\partial x_j} \\ &= \mathrm{i} \sum_{j=1}^{n} \frac{\partial}{\partial a_\mu} f_j(\boldsymbol{x}, \boldsymbol{a})|_{\boldsymbol{a}=0} \frac{\partial}{\partial x_j} \quad , \quad \mu = 1, \dots, r \end{aligned} \tag{4}$$

Problem. a) Calculate the $\hat{L}_\mu$ for the two-parameter group given by

$$x' = ax + b \quad (a, b \in \mathbb{R}) \quad . \tag{5}$$

b) Calculate the commutators $[\hat{L}_\mu, \hat{L}_\nu]_- (\mu, \nu = a, b)$.

Solution. a) The two generators $\hat{L}_\mu$ are denoted $\hat{L}_a$ and $\hat{L}_b$. From the definition we obtain

$$\hat{L}_a = \mathrm{i} \frac{\partial}{\partial a} (ax + b)\bigg|_{a=b=0} \frac{\partial}{\partial x} = \mathrm{i} x \frac{\partial}{\partial x} \quad , \tag{6a}$$

$$\hat{L}_b = \mathrm{i} \frac{\partial}{\partial b} (ax + b)\bigg|_{a=b=0} \frac{\partial}{\partial x} = \mathrm{i} \frac{\partial}{\partial x} \quad . \tag{6b}$$

b) The only nonvanishing commutator is

$$\begin{aligned} [\hat{L}_a, \hat{L}_b]_- &= -\left\{ \left(x \frac{\partial}{\partial x} \right) \left(\frac{\partial}{\partial x} \right) - \left(\frac{\partial}{\partial x} \right) \left(x \frac{\partial}{\partial x} \right) \right\} \\ &= \frac{\partial}{\partial x} = -\mathrm{i} \hat{L}_b \quad . \end{aligned} \tag{7}$$

EXERCISE

3.17 Generators and Infinitesimal Operators of SO(n)

Problem. a) How many parameters has the special orthogonal group in n dimensions, SO(n), i.e. the group of rotation matrices in the $\mathbb{R}^n$? Find a set of generators.

b) Show that the infinitesimal operators $\hat{L}_{pr}$ [see (3.77)] can be written as

$$\hat{L}_{pr} = -\mathrm{i}\left(x_p \frac{\partial}{\partial x_r} - x_r \frac{\partial}{\partial x_p}\right) \quad p, r = 1, \dots, n; \quad r > p \quad . \tag{1}$$

Solution. a) A $n \times n$ matrix in the vicinity of the unity matrix $\mathbb{1}$ is

$$\hat{A} = \begin{bmatrix} 1 + a_{11} & a_{12} & \cdots & a_{1n} \\ a_{21} & \ddots & & \vdots \\ \vdots & & \ddots & \vdots \\ a_{1n} & \cdots & \cdots & 1 + a_{nn} \end{bmatrix}$$

$$= \mathbb{1} + \begin{bmatrix} a_{11} & \cdots & a_{1n} \\ \vdots & \ddots & \vdots \\ a_{n1} & \cdots & a_{nn} \end{bmatrix} = \mathbb{1} + \delta\hat{A} \tag{2}$$

where all of the a_{pq} $(p, q = 1, \dots, n)$ are infinitesimal. Such a matrix is a member of the SO(n), if it leaves the scalar product $\boldsymbol{x} \cdot \boldsymbol{y} = \sum_i x_i y_i$ $(x_i, y_i \in \mathbb{R})$ invariant, i.e. if $(A\boldsymbol{x})^{\mathrm{T}} \cdot (A\boldsymbol{y}) = \boldsymbol{x} \cdot \boldsymbol{y}$ holds. This leads to

$$\begin{aligned} \boldsymbol{x} \cdot \boldsymbol{y} &= [(\mathbb{1} + \delta\hat{A}) \cdot \boldsymbol{x}]^{\mathrm{T}} \cdot [\mathbb{1} + \delta\hat{A}) \cdot \boldsymbol{y}] \\ &= \boldsymbol{x} \cdot \boldsymbol{y} + (\delta\hat{A} \cdot \boldsymbol{x})^{\mathrm{T}} \cdot \boldsymbol{y} + \boldsymbol{x} \cdot (\delta\hat{A} \cdot \boldsymbol{y}) + O(a^2) \end{aligned} \tag{3}$$

or

$$\boldsymbol{x} \cdot \delta\hat{A}^{\mathrm{T}} \cdot \boldsymbol{y} + \boldsymbol{x} \cdot \delta\hat{A} \cdot \boldsymbol{y} = 0 \quad . \tag{3'}$$

Condition (3′) must hold for all $\boldsymbol{x}$ and $\boldsymbol{y}$, hence we require

$$\delta\hat{A}^{\mathrm{T}} = -\delta\hat{A} \quad , \tag{4}$$

i.e. $\delta\hat{A}$ must be an antisymmetric matrix

$$\delta\hat{A} = \begin{bmatrix} 0 & a_{12} & \cdots & a_{1n} \\ -a_{12} & 0 & & \\ \vdots & \vdots & & \vdots \\ \vdots & \vdots & & a_{n-1,n} \\ -a_{1n} & \cdots & -a_{n-1,n} & 0 \end{bmatrix} \tag{5}$$

Exercise 3.17

with $\frac{1}{2}n(n-1)$ free parameters a_{ij}. The generators $\hat{S}_{pr}$ $(p, r = 1, \dots, n; p < r)$ are therefore

$$\hat{S}_{pr} = \mathrm{i}\frac{\partial}{\partial a_{pr}}\delta\hat{A} = \mathrm{i}\begin{bmatrix} 0 & & & \\ & \ddots & 1 & \\ & -1 & \ddots & \\ & & & 0 \end{bmatrix} , \tag{6}$$

where the 1 and -1 stand at the position of a_{pr} resp. $-a_{pr}$ in (5), or in detail

$$(\hat{S}_{pr})_{ij} = \mathrm{i}(\delta_{ip}\delta_{jr} - \delta_{ir}\delta_{jp}) \quad . \tag{6'}$$

b) For small a_{pr}, the coordinate transformation is given by

$$\begin{aligned} \boldsymbol{x}' = A\boldsymbol{x} \;\Leftrightarrow\; x'_k &= A_{kj}x_j \\ &= x_k - \mathrm{i}\sum_{j=1}^{n}\sum_{p,r=1}^{n}(a_{pr}\hat{S}_{pr})_{kj}x_j \quad . \end{aligned} \tag{7}$$

The corresponding infinitesimal operators (generators) are, for the sake of brevity, denoted as $\hat{L}_{pr}$:

$$\begin{aligned} \hat{L}_{pr} &= \mathrm{i}\sum_{j=1}^{n}\frac{\partial}{\partial a_{pr}}x'_j|_{\delta A=0}\frac{\partial}{\partial x_j} \\ &= \mathrm{i}\sum_{j=1}^{n}(-\mathrm{i})\sum_{k=1}^{n}(\hat{S}_{pr})_{jk}x_k\frac{\partial}{\partial x_j} \\ &= \mathrm{i}\left(x_r\frac{\partial}{\partial x_p} - x_p\frac{\partial}{\partial x_r}\right) \quad . \end{aligned} \tag{8}$$

This is a generalization of the angular momentum operator for arbitrary n.

EXERCISE

3.18 Matrix Representation for the Lie Algebra of Spin-1

Problem. Show that the matrices $S_{\alpha\beta}$ from Exercise 3.17 are a representation of the Lie algebra of spin-1 objects.

Solution. We have already shown that the group SO(3) is isomorphic to SU(2), which means that the elements from both groups obey the same Lie algebra. In Exercise 3.17 we derived a matrix representation $S_{\alpha\beta}$ and the infinitesimal generators of rotations in $\mathbb{R}^n$. We interpreted these operators as a generalization of the angular-momentum operator for arbitrary space dimension n. It is important to stress that these objects transform, in general, as tensors of rank 2.

Exercise 3.18

Consequently, only in the special case $n = 3$, there exist three independent elements (L_{12}, L_{13}, L_{23}), which may be interpreted as an axial vector in this 3-dimensional space $\boldsymbol{L}$. Already for $n = 4$ there are six elements which may no longer be interpreted as the components of a vector in $\mathbb{R}^4$. Hence, it is more instructive to interpret the $L_{\alpha\beta}$ as rotations in the $\alpha\beta$-plane (but note for instance the possible decomposition into two three-component vectors, which is used, for example, in special relativity, and is called spatial rotation with Lorentz boosts).

Nevertheless, we use the language developed for SO(3) and try to generalize for arbitrary n. We first calculate the commutation relations for the generalized angular momentum operators $L_{\alpha\beta}$:

$$\begin{aligned}[L_{\alpha\beta}, L_{\mu\nu}] &= \mathrm{i}^2((x_\beta\partial_\alpha - x_\alpha\partial_\beta), (x_\nu\partial_\mu - x_\mu\partial_\nu)) \\ &= \mathrm{i}^2(x_\beta\partial_\alpha x_\nu\partial_\mu - x_\beta\partial_\alpha x_\mu\partial_\nu - x_\alpha\partial_\beta x_\mu\partial_\nu + x_\alpha\partial_\beta x_\nu\partial_\mu \\ &\quad - x_\nu\partial_\mu x_\beta\partial_\alpha + x_\nu\partial_\mu x_\alpha\partial_\beta + x_\mu\partial_\nu x_\beta\partial_\alpha - x_\mu\partial_\nu x_\alpha\partial_\beta) \\ &= \mathrm{i}^2(x_\beta\delta_{\alpha\nu}\partial_\mu + x_\beta x_\nu\partial_\alpha\partial_\mu - x_\beta\delta_{\alpha\mu}\partial_\nu - x_\beta x_\mu\partial_\alpha\partial_\nu \\ &\quad - x_\alpha\delta_{\beta\nu}\partial_\mu - x_\alpha x_\nu\partial_\beta\partial_\mu + x_\alpha\delta_{\beta\mu}\partial_\nu + x_\alpha x_\mu\partial_\beta\partial_\nu \\ &\quad - x_\nu\delta_{\mu\beta}\partial_\alpha - x_\nu x_\beta\partial_\mu\partial_\alpha + x_\nu\delta_{\mu\alpha}\partial_\beta + x_\nu x_\alpha\partial_\mu\partial_\beta \\ &\quad + x_\mu\delta_{\nu\beta}\partial_\alpha + x_\mu x_\beta\partial_\nu\partial_\alpha - x_\mu\delta_{\nu\alpha}\partial_\beta - x_\mu x_\alpha\partial_\nu\partial_\beta) \quad . \end{aligned} \tag{1}$$

The coordinate and derivative components each commute: $[x_\alpha, x_\beta] = 0$, $[\partial_\alpha, \partial_\beta] = 0$ and consequently terms of the form $x_\alpha x_\beta\partial_\alpha\partial_\beta$ cancel pairwise. As a result of the symmmetry property of the Kronecker delta $\delta_{\alpha\beta} = \delta_{\beta\alpha}$ it follows that

$$\begin{aligned}[L_{\alpha\beta}, L_{\mu\nu}] &= \mathrm{i}^2(\delta_{\alpha\nu}(x_\beta\partial_\mu - x_\mu\partial_\beta) - \delta_{\alpha\mu}(x_\beta\partial_\nu - x_\nu\partial_\beta) \\ &\quad + \delta_{\beta\nu}(x_\alpha\partial_\mu - x_\mu\partial_\alpha) - \delta_{\beta\mu}(x_\alpha\partial_\nu - x_\nu\partial_\alpha)) \end{aligned} \tag{2}$$

Using the definition (3.17(8)) we get

$$[L_{\alpha\beta}, L_{\mu\nu}] = \mathrm{i}(+ \delta_{\alpha\nu}L_{\mu\beta} - \delta_{\alpha\mu}L_{\nu\beta} - \delta_{\beta\nu}L_{\mu\alpha} - \delta_{\beta\mu}L_{\nu\alpha}) \tag{3}$$

or, employing the antisymmetry of the generators $L_{\alpha\beta} = -L_{\beta\alpha}$, the standard form

$$[L_{\alpha\beta}, L_{\mu\nu}] = \mathrm{i}(+ \delta_{\alpha\mu}L_{\beta\nu} + \delta_{\beta\nu}L_{\alpha\mu} - \delta_{\alpha\nu}L_{\beta\mu} - \delta_{\beta\mu}L_{\alpha\nu} \quad . \tag{4}$$

This is the Lie algebra of SO(n). The next step is the explicit construction of a Casimir operator. By definition it commutes with all group elements $L_{\alpha\beta}$. We try the ansatz

$$\Lambda^2 = \frac{1}{2}\sum_{\alpha,\beta}^{k=N} (L_{\alpha\beta})^2 = \frac{1}{2}\delta^{\alpha\mu}\delta^{\beta\nu}L_{\alpha\beta}L_{\mu\nu} \tag{5}$$

which is just the sum over all squares of the group elements. Calculation of the commutation relation yields

Exercise 3.18

$$[\Lambda^2, L_{\sigma\tau}] = \frac{1}{2}\delta^{\alpha\mu}\delta^{\beta\nu}[L_{\alpha\beta}L_{\mu\nu}, L_{\sigma\tau}]$$

$$= \frac{1}{2}\delta^{\alpha\mu}\delta^{\beta\nu}(L_{\alpha\beta}L_{\mu\nu}L_{\sigma\tau} - L_{\sigma\tau}L_{\alpha\beta}L_{\mu\nu})$$

$$= \frac{1}{2}\delta^{\alpha\mu}\delta^{\beta\nu}(L_{\alpha\beta}[L_{\mu\nu}L_{\sigma\tau}] + [L_{\alpha\beta}, L_{\sigma\tau}]L_{\mu\nu}) \quad . \tag{6}$$

Inserting the commutation relations (4) gives

$$[\Lambda^2, L_{\sigma\tau}] = \frac{\mathrm{i}}{2}\delta^{\alpha\mu}\delta^{\beta\nu}(L_{\alpha\beta}(\delta_{\mu\sigma}L_{\nu\tau} + (\delta_{\nu\tau}L_{\mu\sigma} - \delta_{\mu\tau}L_{\nu\sigma} - \delta_{\nu\sigma}L_{\mu\tau})$$

$$+ (\delta_{\mu\sigma}L_{\nu\tau} + \delta_{\nu\tau}L_{\mu\sigma} - \delta_{\mu\tau}L_{\nu\sigma} - \delta_{\nu\sigma}L_{\mu\tau})L_{\mu\nu})$$

$$= \frac{\mathrm{i}}{2}(L_{\alpha\beta}(\delta^{\alpha}{}_{\sigma}\delta^{\beta\mu}L_{\nu\tau} + \delta^{\alpha\mu}\delta^{\beta}{}_{\tau}L_{\mu\sigma} - \delta^{\alpha}{}_{\tau}\delta^{\beta\nu}L_{\nu\sigma} - \delta^{\alpha\mu}\delta^{\beta}{}_{\sigma}L_{\mu\tau})$$

$$+ (\delta^{\mu}{}_{\sigma}\delta^{\beta\nu}L_{\beta\tau} + \delta^{\alpha\mu}\delta^{\nu}{}_{\tau}L_{\alpha\sigma} - \delta^{\mu}{}_{\tau}\delta^{\beta\nu}L_{\beta\sigma} - \delta^{\alpha\mu}\delta^{\nu}{}_{\sigma}L_{\alpha\tau})L_{\mu\nu}) \tag{7}$$

We now employ the mixed Kronecker δs:

$$[\Lambda^2, L_{\sigma\tau}] = \frac{\mathrm{i}}{2}(\delta^{\beta\mu}L_{\sigma\beta}L_{\nu\tau} + \delta^{\alpha\mu}L_{\alpha\tau}L_{\mu\sigma} - \delta^{\beta\mu}L_{\tau\beta}L_{\nu\sigma} - \delta^{\alpha\mu}L_{\alpha\sigma}L_{\mu\tau}$$

$$+ \delta^{\beta\nu}L_{\beta\tau}L_{\sigma\nu} + \delta^{\alpha\mu}L_{\alpha\sigma}L_{\mu\tau} - \delta^{\beta\nu}L_{\beta\sigma}L_{\tau\nu} - \delta^{\alpha\mu}L_{\alpha\tau}L_{\mu\sigma})$$

$$= \frac{\mathrm{i}}{2}(\delta^{\beta\nu}(L_{\sigma\beta}L_{\nu\tau} - L_{\tau\beta}L_{\nu\sigma} + L_{\beta\tau}L_{\sigma\nu} - L_{\beta\sigma}L_{\tau\nu})$$

$$+ \delta^{\alpha\mu}(L_{\alpha\tau}L_{\mu\sigma} - L_{\alpha\sigma}L_{\mu\tau} + L_{\alpha\sigma}L_{\mu\tau} - L_{\alpha\tau}L_{\mu\sigma}))$$

$$= 0 \tag{8}$$

Let us construct a Hilbert space and calculate the spectrum of Λ^2 algebraically. The Casimir operator can be written explicitly using the definition of $L_{\alpha\beta}$ as

$$\Lambda^2 = \frac{1}{2}L^{\mu\nu}L_{\mu\nu}$$

$$= -\frac{1}{2}(x^{\nu}\partial^{\mu} - x^{\mu}\partial^{\nu})(x_{\nu}\partial_{\mu} - x_{\mu}\partial_{\nu})$$

$$= -\frac{1}{2}(x^{\nu}\partial^{\mu}x_{\nu}\partial_{\mu} - x^{\nu}\partial^{\mu}x_{\mu}\partial_{\nu} - x^{\mu}\partial^{\nu}x_{\nu}\partial_{\mu} + x^{\mu}\partial^{\nu}x_{\mu}\partial_{\nu})$$

$$= -\frac{1}{2}(x^{\nu}\delta^{\mu}{}_{\nu}\partial_{\mu} + x^{\nu}x_{\nu}\partial^{\mu}\partial_{\mu} - x^{\nu}\delta^{\mu}{}_{\mu}\partial_{\nu} - x^{\nu}x_{\mu}\partial^{\mu}\partial_{\nu}$$

$$- x^{\mu}\delta^{\nu}{}_{\nu}\partial_{\mu} - x^{\mu}x_{\nu}\partial^{\nu}\partial_{\mu} + x^{\mu}\delta^{\nu}{}_{\mu}\partial_{\nu} + x^{\mu}x_{\mu}\partial^{\nu}\partial_{\nu})$$

$$= -(x^{\mu}\partial_{\mu} + x^{\nu}x_{\nu}\partial^{\mu}\partial_{\mu} - Nx^{\nu}\partial_{\mu} - x^{\nu}x^{\mu}\partial_{\mu}\partial_{\nu}) \tag{9}$$

Exercise 3.18

we define the homogeneous Euler operator $J_e = x^\mu \partial_\mu$, with

$$J_e^2 = x^\mu \partial_\mu x^\nu \partial_\nu = x^\mu \delta_\mu{}^\nu \partial_\nu + x^\mu x^\nu \partial_\mu \partial_\nu = J_e + x^\mu x^\nu \partial_\mu \partial_\nu \tag{10}$$

and thus rewrite (9) as

$$\begin{aligned} \Lambda^2 &= -(J_e + x^\nu x_\nu \partial^\mu \partial_\mu - NJ_e - J_e(J_e - 1)) \\ &= -(x^\nu x_\nu \partial^\mu \partial_\mu - J_e(J_e + N - 2)) \end{aligned} \tag{11}$$

Defining a Hilbert space H_p as the space of all homogenous polynomials of degree l which fulfil the Laplacian,

$$H_p = \{f\colon f(\lambda x) = \lambda^l f(x);\ \partial^\mu \partial_\mu f = 0\} \quad , \tag{12}$$

the eigenwertspektrum of Λ^2 results as

$$\Lambda^2 f = l(l + N - 2) f \quad . \tag{13}$$

Let us now examine the properties of the matrix representation (3.17.6). We can calculate the commutation relations for these matrices

$$\begin{aligned} ([S_{\alpha\beta}, S_{\mu\nu}])_{ik} &= \mathrm{i}^2 \sum_j^N \{(\delta_{i\alpha}\delta_{j\beta} - \delta_{i\beta}\delta_{j\alpha})(\delta_{j\mu}\delta_{k\nu} - \delta_{j\nu}\delta_{k\mu}) \\ &\quad - (\delta_{i\mu}\delta_{j\nu} - \delta_{i\nu}\delta_{j\mu})(\delta_{j\alpha}\delta_{k\beta} - \delta_{j\beta}\delta_{k\alpha})\} \\ &= \mathrm{i}^2 \sum_j^N (\delta_{i\alpha}\delta_{j\beta}\delta_{j\mu}\delta_{k\nu} - \delta_{i\alpha}\delta_{j\beta}\delta_{j\nu}\delta_{k\mu} - \delta_{i\beta}\delta_{j\alpha}\delta_{j\mu}\delta_{k\nu} + \delta_{i\beta}\delta_{j\alpha}\delta_{j\nu}\delta_{k\mu} \\ &\quad - \delta_{i\mu}\delta_{j\nu}\delta_{j\alpha}\delta_{k\beta} + \delta_{i\mu}\delta_{j\nu}\delta_{j\beta}\delta_{k\alpha} + \delta_{i\nu}\delta_{j\mu}\delta_{j\alpha}\delta_{k\beta} - \delta_{i\nu}\delta_{j\mu}\delta_{j\beta}\delta_{k\alpha}) \\ &= \mathrm{i}^2 (\delta_{\beta\mu}\delta_{i\alpha}\delta_{k\nu} - \delta_{\beta\nu}\delta_{i\alpha}\delta_{k\mu} - \delta_{\alpha\mu}\delta_{i\beta}\delta_{k\nu} + \delta_{\alpha\nu}\delta_{i\beta}\delta_{k\mu} \\ &\quad - \delta_{\nu\alpha}\delta_{i\mu}\delta_{k\beta} + \delta_{\nu\beta}\delta_{i\mu}\delta_{k\alpha} + \delta_{\mu\alpha}\delta_{i\nu}\delta_{k\beta} - \delta_{\mu\beta}\delta_{i\nu}\delta_{k\alpha}) \\ &= \mathrm{i}^2 (\delta_{\beta\mu}(\delta_{i\alpha}\delta_{k\nu} - \delta_{i\nu}\delta_{k\alpha}) - \delta_{\beta\nu}(\delta_{i\alpha}\delta_{k\mu} - \delta_{i\mu}\delta_{k\alpha}) \\ &\quad - \delta_{\alpha\mu}(\delta_{i\beta}\delta_{k\nu} - \delta_{i\nu}\delta_{k\beta}) + \delta_{\alpha\nu}(\delta_{i\beta}\delta_{k\mu} - \delta_{i\mu}\delta_{k\beta})) \end{aligned} \tag{14}$$

Inserting the definition for $S_{\alpha\beta}$ yields

$$[S_{\alpha\beta}, S_{\mu\nu}] = \mathrm{i}(\delta_{\beta\mu} S_{\alpha\nu} - \delta_{\beta\nu} S_{\alpha\mu} - \delta_{\alpha\mu} S_{\beta\nu} + \delta_{\alpha\nu} S_{\beta\mu}) \tag{15}$$

Comparison with (4) leads to the conclusion that the elements $S_{\alpha\beta}$ obey the same algebra as the $L_{\alpha\beta}$. Consequently, the Casimir operator for classifying a multiplet is given in analogy to (5) by

$$S^2 = \frac{1}{2} S^{\alpha\beta} S_{\alpha\beta} \tag{16}$$

Exercise 3.18

We deduce that

$$\left(\frac{1}{2}\delta^{\alpha\mu}\delta^{\beta\nu}S_{\alpha\beta}S_{\mu\nu}\right)_{ik}$$

$$=\frac{\mathrm{i}^2}{2}\delta^{\alpha\mu}\delta^{\beta\nu}\sum_j^N(\delta_{i\alpha}\delta_{j\beta}-\delta_{i\beta}\delta_{j\alpha})(\delta_{j\mu}\delta_{k\nu}-\delta_{j\nu}\delta_{k\mu})$$

$$=\frac{\mathrm{i}^2}{2}\delta^{\alpha\mu}\delta^{\beta\nu}\sum_j^N(\delta_{i\alpha}\delta_{j\beta}\delta_{j\mu}\delta_{k\nu}-\delta_{i\alpha}\delta_{j\beta}\delta_{j\nu}\delta_{k\mu}-\delta_{i\beta}\delta_{j\alpha}\delta_{j\mu}\delta_{k\nu}+\delta_{i\beta}\delta_{j\alpha}\delta_{j\nu}\delta_{k\mu})$$

$$=\frac{\mathrm{i}^2}{2}\delta^{\alpha\mu}\delta^{\beta\nu}(\delta_{\beta\mu}\delta_{i\alpha}\delta_{k\nu}-\delta_{\beta\nu}\delta_{i\alpha}\delta_{k\mu}-\delta_{\alpha\mu}\delta_{i\beta}\delta_{k\nu}+\delta_{\alpha\nu}\delta_{i\beta}\delta_{k\mu})$$

$$=\frac{\mathrm{i}^2}{2}\delta^{\alpha\mu}(\delta^{\nu}{}_{\mu}\delta_{i\alpha}\delta_{k\nu}-\delta^{\beta}{}_{\beta}\delta_{i\alpha}\delta_{k\mu}-\delta^{\nu}{}_{i}\delta_{\alpha\mu}\delta_{k\nu}-\delta^{\nu}{}_{i}\delta_{\alpha\nu}\delta_{k\mu})$$

$$=\frac{\mathrm{i}^2}{2}\delta^{\alpha\mu}(\delta_{i\alpha}\delta_{k\mu}-N\delta_{i\alpha}\delta_{k\mu}-\delta_{ik}\delta_{\alpha\mu}+\delta_{i\alpha}\delta_{k\mu})$$

$$=\frac{\mathrm{i}^2}{2}(\delta^{\mu}{}_{i}\delta_{k\mu}-N\delta^{\mu}{}_{i}\delta_{k\mu}-N\delta_{ik}+\delta^{\mu}{}_{i}\delta_{k\mu})=(N-1)\delta_{ik} \tag{17}$$

or

$$\frac{1}{2}\delta^{\alpha\mu}\delta^{\beta\nu}S_{\alpha\beta}S_{\mu\nu}=(N-1)\mathbb{1}_N \tag{18}$$

Obviously the unit-matrix $\mathbb{1}$ indeed commutes with all $N\times N$ matrices. In analogy to (13) we define a quantity s with

$$S^2f=s(s+N-2)\mathbb{1}f \tag{19}$$

Comparison with (18) yields

$$s=1 \tag{20}$$

While l is the generalization of the orbital angular momentum we define s to be the generalization of the intrinsic angular momentum (spin). Hence, we have found that the matrices $S_{\alpha\beta}$ obey a generalized angular-momentum algebra with spin equal to 1.

EXERCISE

3.19 Translations in One-dimensional Space; the Euclidean Group E_3 in Three Dimensions

Problem. a) Show that the infinitesimal operator $\hat{P}$ of the one-dimensional translation $x\to x+a$ is of the form $\hat{P}=-\mathrm{i}d/dx$.

Exercise 3.19

b) Show that the inifinitesimal operators (generators)

$$-\mathrm{i}\frac{\partial}{\partial x} \quad , \quad -\mathrm{i}\frac{\partial}{\partial y} \quad , \quad -\mathrm{i}\frac{\partial}{\partial z} \tag{1}$$

of translations in the three-dimensional Euclidean space, together with the infinitesimal operators (generators)

$$-\mathrm{i}\left(y\frac{\partial}{\partial z}-z\frac{\partial}{\partial y}\right) \quad , \quad -\mathrm{i}\left(z\frac{\partial}{\partial x}-x\frac{\partial}{\partial z}\right) \quad , \quad -\mathrm{i}\left(x\frac{\partial}{\partial y}-y\frac{\partial}{\partial x}\right) \tag{2}$$

of rotations in the same space, are closed with respect to the formation of commutators and hence define a Lie group (the Euclidean group E_3 in three dimensions).

Solution. a) The infinitesimal operators (generators) have been defined in Exercise 3.16. The function $f(x;a)$ for a transformation

$$T\colon x \to x+a \quad \text{or} \quad x' = x-a \tag{3}$$

according to

$$F'(x) = F(x') = F(\hat{T}^{-1}x) = F(x-a) = F(f(x;a)) \tag{4}$$

is given by

$$f(x;a) = x-a \quad . \tag{5}$$

Thus, we have

$$\hat{P} = \mathrm{i}\frac{\partial}{\partial a}f(x;a)|_{a=0}\frac{\partial}{\partial x} = -\mathrm{i}\frac{d}{dx} \quad . \tag{6}$$

b) Define

$$\hat{\boldsymbol{P}} = -\mathrm{i}\left(\frac{\partial}{\partial x}, \frac{\partial}{\partial y}, \frac{\partial}{\partial z}\right) \quad , \tag{1'}$$

$$\hat{\boldsymbol{L}} = -\mathrm{i}\left\{\left(y\frac{\partial}{\partial z}-z\frac{\partial}{\partial y}\right), \left(z\frac{\partial}{\partial x}-x\frac{\partial}{\partial z}\right), \left(x\frac{\partial}{\partial y}-y\frac{\partial}{\partial x}\right)\right\} \quad . \tag{2'}$$

Then we know that

$$[\hat{P}_i, \hat{P}_j]_- = 0 \quad , \quad [\hat{L}_i, \hat{L}_j]_- = \mathrm{i}\varepsilon_{ijk}\hat{L}_k \quad , \tag{7}$$

and also

$$\begin{aligned}[\hat{P}_i, \hat{L}_j]_- &= -\left[\frac{\partial}{\partial x_i}, \varepsilon_{jkm}x_k\frac{\partial}{\partial x_m}\right]_- = -\varepsilon_{jkm}\delta_{ik}\frac{\partial}{\partial x_m} \\ &= -\varepsilon_{jim}\frac{\partial}{\partial x_m} = (-\mathrm{i})\mathrm{i}\varepsilon_{ijm}\frac{\partial}{\partial x_m} = \mathrm{i}\varepsilon_{ijm}\hat{P}_m \quad .\end{aligned} \tag{8}$$

Finite translations and rotations are described by the operators

$$\hat{U}_{\mathrm{T}}(\boldsymbol{a}) = \mathrm{e}^{-\mathrm{i}\boldsymbol{a}\cdot\hat{\boldsymbol{P}}} \quad , \quad \hat{U}_{\mathrm{R}}(\boldsymbol{\phi}) = \mathrm{e}^{-\mathrm{i}\boldsymbol{\phi}\cdot\hat{\boldsymbol{L}}} \quad . \tag{9}$$

Exercise 3.19

The group of translations forms an invariant subgroup of the full group, because one can show that the matrix relations, for which a proof was given in Example 3.2 and 3.3, also hold analogously for operators. It is sufficient to show that this is true for infinitesimal rotations

$$\hat{U}_{\mathrm{R}}(\delta\boldsymbol{\phi})\hat{U}_{\mathrm{T}}(\boldsymbol{a})\hat{U}_{\mathrm{R}}^{-1}(\delta\boldsymbol{\phi}) = \mathrm{e}^{-\mathrm{i}\delta\boldsymbol{\phi}\cdot\hat{\boldsymbol{L}}}\mathrm{e}^{-\mathrm{i}\boldsymbol{a}\cdot\hat{\boldsymbol{P}}}\mathrm{e}^{+\mathrm{i}\delta\boldsymbol{\phi}\cdot\hat{\boldsymbol{L}}}$$
$$= \exp\{-\mathrm{i}[\mathrm{e}^{-\mathrm{i}\delta\boldsymbol{\phi}\cdot\hat{\boldsymbol{L}}}(\boldsymbol{a}\cdot\hat{\boldsymbol{P}})\mathrm{e}^{+\mathrm{i}\delta\boldsymbol{\phi}\cdot\hat{\boldsymbol{L}}}]\} \tag{10}$$

Now, according to Exercise 3.3

$$-\mathrm{i}\mathrm{e}^{-\mathrm{i}\delta\boldsymbol{\phi}\cdot\hat{\boldsymbol{L}}}(\boldsymbol{a}\cdot\hat{\boldsymbol{P}})\mathrm{e}^{+\mathrm{i}\delta\boldsymbol{\phi}\cdot\hat{\boldsymbol{L}}} = -\mathrm{i}\boldsymbol{a}\cdot\hat{\boldsymbol{P}} + [-\mathrm{i}\delta\boldsymbol{\phi}\cdot\hat{\boldsymbol{L}}, -\mathrm{i}\boldsymbol{a}\cdot\hat{\boldsymbol{P}}]_-$$
$$= -\mathrm{i}\boldsymbol{a}\cdot\hat{\boldsymbol{P}} - \sum_{i,j}\delta\phi a_j[\hat{L}_i, \hat{P}_j]_-$$
$$= -\mathrm{i}\boldsymbol{a}\cdot\hat{\boldsymbol{P}} + \mathrm{i}\sum_{i}\delta\phi a_j\varepsilon_{ijk}\hat{P}_k$$
$$= -\mathrm{i}(\boldsymbol{a} - \delta\boldsymbol{\phi}\times\boldsymbol{a})\cdot\hat{\boldsymbol{P}} \quad , \tag{10'}$$

i.e.

$$\hat{U}_{\mathrm{R}}(\delta\boldsymbol{\phi})\hat{U}_{\mathrm{T}}(\boldsymbol{a})\hat{U}_{\mathrm{R}}^{-1}(\delta\boldsymbol{\phi}) = \hat{U}_{\mathrm{T}}(\boldsymbol{a}') = \hat{U}_{\mathrm{T}}(\boldsymbol{a} - \delta\boldsymbol{\phi}\times\boldsymbol{a}) \quad . \tag{11}$$

An arbitrary product of translations and rotations can always be written as

$$\hat{U} = \hat{U}_{\mathrm{T}}\hat{U}_{\mathrm{R}} \quad , \tag{12}$$

because on one hand we have for

$$\hat{U}_{\mathrm{R}}\hat{U}_{\mathrm{T}} = \hat{U}_{\mathrm{R}}\hat{U}_{\mathrm{T}}\hat{U}_{\mathrm{R}}^{-1}\hat{U}_{\mathrm{R}} = \hat{U}'_{\mathrm{T}}\hat{U}_{\mathrm{R}} \quad ,$$

and on the other hand for an n-factor product of the type

$$\hat{U} = \hat{U}_{\mathrm{T}_1}\hat{U}_{\mathrm{R}_1}\hat{U}_{\mathrm{T}_2}\hat{U}_{\mathrm{R}_2}\dots\hat{U}_{\mathrm{T}_n}\hat{U}_{\mathrm{R}_n}$$
$$= \hat{U}_{\mathrm{T}_1}\underbrace{\hat{U}_{\mathrm{R}_1}\hat{U}_{\mathrm{T}_2}\hat{U}_{\mathrm{R}_1}^{-1}}_{\hat{U}'_{\mathrm{T}_2}}\underbrace{\hat{U}_{\mathrm{R}_1}\hat{U}_{\mathrm{R}_2}}_{\hat{U}'_{\mathrm{R}_2}}\hat{U}_{\mathrm{T}_3}\hat{U}_{\mathrm{R}_3}\dots$$
$$= \underbrace{\hat{U}_{\mathrm{T}_1}\hat{U}_{\mathrm{T}_2}}_{\hat{U}''_{\mathrm{T}_2}}\hat{U}'_{\mathrm{R}_2}\hat{U}_{\mathrm{T}_3}\hat{U}_{\mathrm{R}_3}\dots = \hat{U}''_{\mathrm{T}_2}\hat{U}'_{\mathrm{R}_2}\hat{U}_{\mathrm{T}_3}\hat{U}_{\mathrm{R}_3}\dots$$
$$= \dots = \hat{U}''_{\mathrm{T}_n}\hat{U}'_{\mathrm{R}_n} \quad . \tag{13}$$

Here we have utilized the fact that not only do the translations form an invariant subgroup, but also the rotations form an (albeit non-invariant) subgroup. To any such product, however, an element of the group E_3 (rotations and translations in the three-dimensional space) is uniquely related according to

$$\hat{U}_{\mathrm{T}}(\boldsymbol{a})\hat{U}_{\mathrm{R}}(\boldsymbol{\phi})F(\boldsymbol{x}) = \hat{U}_{\mathrm{T}}(\boldsymbol{a})F(\hat{R}^{-1}(\boldsymbol{\phi})\boldsymbol{x}) = F(\hat{R}^{-1}(\boldsymbol{\phi})\boldsymbol{x} - \boldsymbol{a}) \quad . \tag{14}$$

As this relation is uniquely invertible, we are dealing with an isomorphism – the group of our unitary operators is also called a *realization* of the group E_3 (in contrast to a *representation* where the group elements are represented by matrices).

EXERCISE

3.20 Homomorphism and Isomorphism of Groups and Algebra

a) First we discuss the meaning of the terms "homomorphism" and "isomorphism" of groups and algebras.
b) We then show that the Lie algebras of the groups SU(2) and SO(3) are isomorphic.

a) Consider two groups G and G'. A mapping

$$f: G \to G'' \tag{1}$$

is called a

a1) *group homomorphism*, if for any $g_1, g_2 \in G$ the relation

$$f(g_1 \cdot g_2) = f(g_1) \cdot f(g_2) \quad , \tag{2}$$

holds, thus if the structure of the group is conserved under the mapping;

a2) *group isomorphism*, if in addition the mapping f is unique and hence also the inverse mapping f^{-1} exists.

For two algebras A, A', which are vector spaces on the number field K with an inner product $[a_1, a_2]$, $a_1, a_2 \in A$ or A', the mapping

$$f: A \to A'$$

is a

a1′) *homomorphism*, if for all $a_1, a_2 \in A$; $\alpha_1, \alpha_2 \in K$ we have

$$f(\alpha_1 a_1 + \alpha_2 a_2) = \alpha_1 f(a_1) + \alpha_2 f(a_2) \quad , \tag{3}$$

$$f([a_1, a_2]) = [f(a_1), f(a_2)] \quad , \tag{3′}$$

i.e. if the algebraic structure is left unchanged by the mapping;

a2′) *isomorphism*, again if f can be inverted.

Incidentally, one also speaks of a local isomorphism of two Lie groups, if the associated Lie algebras are isomorphic.

b) The generators of SO(3) and SU(2) are the matrices $(\hat{S}_1, \hat{S}_2, \hat{S}_3)$ and $(\hat{\sigma}_1, \hat{\sigma}_2, \hat{\sigma}_3)$ respectively, with the commutation relations

$$[\hat{S}_i, \hat{S}_j]_- = \mathrm{i}\varepsilon_{ijk}\hat{S}_k \quad , \quad [\hat{\sigma}_i, \hat{\sigma}_j]_- = 2\mathrm{i}\varepsilon_{ijk}\hat{\sigma}_k \quad . \tag{4}$$

The mapping

$$f(\alpha \hat{S}_i) = \tfrac{1}{2}\alpha\hat{\sigma}_i \tag{5}$$

is a *homomorphism*, and since

$$f^{-1}(\alpha\hat{\sigma}_i) = 2\alpha\hat{S}_i \tag{6}$$

is its inversion, the algebras of SU(2) and SO(3) are isomorphic, and these two Lie groups are locally isomorphic. *Exercise 3.20*

EXERCISE

3.21 Transformations of the Structure Constants

Problem. a) Show that under a transformation of the basis $\{\hat{X}_i\}$ of an algebra

$$\hat{X}_i \to X'_i = a_{ij}\hat{X}_j \quad , \tag{1}$$

the structure constants transform as

$$C_{ijk} \to C'_{ijk} = \sum_{l,m,n} a_{il}a_{jm}C_{lmn}(a^{-1})_{kn} \quad .$$

b) Construct an a_{ij} in such a way that commutation relation

$$[\hat{X}_i, \hat{X}_j]_- = \varepsilon_{ijk}\hat{X}_k \quad (i,j,k = 1,2,3)$$

is transformed to

$$[\hat{X}'_i, \hat{X}'_j]_- = C_{ijk}\hat{X}'_k \tag{2}$$

with

$$C_{123} = -C_{231} = -C_{312} = 1 \quad . \tag{3}$$

Solution. a) With $X'_i = a_{ij}\hat{X}_j$ we have

$$[\hat{X}'_i, \hat{X}'_j]_- = a_{ik}a_{jl}[\hat{X}_k, \hat{X}_l]_- = a_{ik}a_{jl}C_{klm}\hat{X}_m = C'_{ijn}\hat{X}'_n = C'_{ijn}a_{nm}\hat{X}_m \quad ,$$

from which immediately follows the assertion (the $\hat{X}_m$ are linearly independent).

b) The matrix

$$\hat{a} = \begin{bmatrix} \mathrm{i} & & \\ & -\mathrm{i} & \\ & & 1 \end{bmatrix}$$

satisfies the required relations.

3.14 Biographical Notes

CARTAN, Elie Joseph, French mathematician, *9.4.1869 Dolomien, †6.5.1951 Paris, from 1903 professor at Nancy, from 1909 at the Sorbonne. C. was an eminent representative among those who continued and perfected the theory of continuous Lie groups. He worked on differential geometry, differential forms and groups without parameter representation.

CASIMIR, Hendrik Brught Gerhard, Dutch physicist, *15.7.1909 The Hague, who was the first to work out the quantum mechanics of the rigid rotator. C. in 1942 entered the research laboratory of the Philips B.V. where he became director and in 1957 a member of the board of the company. Besides his early work on the rigid rotator he is known for the Casimir-effect, which is the change of the zero point energy of electromagnetic waves between e.g. two condenser plates.

RACAH, Giulio, *9.2.1909 Florence, †28.8.1965 Jerusalem. R. studied at the universities of Florence and Rome as well as at the Eidgenössische Technische Hochschule in Zurich. Later he taught theoretical physics at Florence and Pisa, unit he emigrated to Jerusalem in 1939. He continued to be active at the Hebrew University; his main fields of activity were atomic and nuclear physics.

4. Symmetry Groups and Their Physical Meaning — General Considerations

In this chapter we again ask the question of what can be learned about a physical system by considering its symmetry properties. The answer to this question was given in 1918 by E. Noether within the framework of variational calculations[1]. We will restrict ourselves, however, to special cases which are of interest in context with the problems discussed in this book[2]. Some facts encountered in this chapter are already well known to us, but this time they will be discussed at a more advanced level.

We take into consideration the case of a semi-simple unitary Lie group with n generators and l invariant operators ($l < n$), which can all be chosen to be hermitian. This case is not as special as one might think, since it includes all symmetry groups with physical relevance, as well as their applications which have been discussed in detail up to now. The answer consists of three parts, which will be discussed one after another. In the following, we will give a proof and then illustrate it with examples:

1) The system possesses $2l$ good quantum numbers. Half of these are generators $\hat{L}_i$ ($i = 1 \dots n$) of the symmetry group, l of which commute with each other. The other half of the quantum numbers are obtained by the l Casimir operators, which classify the multiplets. The states within a single multiplet are characterized by l commuting generators. In order to avoid complications, we first assume only a single symmetry S_α (for example spherical symmetry). Direct products of two or more symmetry groups will be discussed in the next section.

The statement that a system (or an interaction between two systems) has the symmetry S_α is equivalent to the requirement that the hamiltonian describing the system commutes with every operator $\hat{U}(\boldsymbol{\alpha})$ of the symmetry group.

Namely, if the Schrödinger equation

$$i\hbar \frac{\partial \psi(\boldsymbol{r},t)}{\partial t} = \hat{H}\psi(\boldsymbol{r},t) \tag{4.1}$$

holds for the initial state, then it follows immediately by application of the time

---

[1] E. Noether: Nachr. Ges. Wiss. Göttingen, Math.-Phys. Kl. 235 (1918), cf. also Example 1.3.

[2] For a discussion of the general theorem, see Vol. 4 in this series: W. Greiner, J. Reinhardt: Quantum Electrodynamics (Springer, Berlin, Heidelberg) to be published.

independent symmetry operator $\hat{U}(\boldsymbol{\alpha})$ that

$$\mathrm{i}\hbar\frac{\partial\hat{U}(\boldsymbol{\alpha})\psi(\boldsymbol{r},t)}{\partial t}=\hat{U}(\boldsymbol{\alpha})\hat{H}\hat{U}^{-1}(\boldsymbol{\alpha})\hat{U}(\boldsymbol{\alpha})\psi(\boldsymbol{r},t)\quad . \tag{4.2}$$

Hence, the *displaced* wave function $\psi'(\boldsymbol{r},t)=\hat{U}(\boldsymbol{\alpha})\psi(\boldsymbol{r},t)$ obeys exactly the same Schrödinger equation (4.1) as the original wave function $\psi(\boldsymbol{r},t)$, if

$$\hat{U}(\boldsymbol{\alpha})\hat{H}\hat{U}^{-1}(\boldsymbol{\alpha})=\hat{H},\ \text{i.e.}\ [\hat{H},\hat{U}(\boldsymbol{\alpha})]_-=[\hat{H},\mathrm{e}^{-\mathrm{i}\alpha_k\hat{L}_k}]_-=0\quad . \tag{4.3}$$

For a small symmetry displacement $\delta\boldsymbol{\alpha}=\{\delta\alpha_k\}$, we conclude that thus

$$[\hat{H},\delta\alpha_k\hat{L}_k]_-=0\quad , \tag{4.4}$$

and, since $\delta\boldsymbol{\alpha}=\{\delta\alpha_1,\ \dots,\delta\alpha_n\}$ can be chosen arbitrarily, we finally have

$$[\hat{H},\hat{L}_k]_-=0\quad . \tag{4.5}$$

The consequence of the existence of the symmetry group S_α with the operators $\hat{U}(\boldsymbol{\alpha})$ is that all generators $\hat{L}_k$ have to commute with the Hamiltonian of the system. Obviously, according to (4.5), the generators $\hat{L}_k$ describe physical observables which are conserved quantities. This is ensured by the vanishing commutator of the $\hat{L}_k$ with the Hamiltonian $\hat{H}$. Conversely, if (4.5) holds, then (4.4) also and with it (4.3) are valid and, consequently, (4.2) is also true. Thus, if all generators which commute with $\hat{H}$ are known, the symmetry operators $\hat{U}(\boldsymbol{\alpha})=\exp(-\mathrm{i}\alpha_k\hat{L}_k)$ can be constructed without difficulty. In this case we say: *S_α together with the operators $\hat{U}(\boldsymbol{\alpha})$ forms a symmetry group of the Hamiltonian (i.e. the physical system).*

EXAMPLE

4.1 Conservation Laws with Rotation Symmetry and Charge Independent Forces

a) If we have *symmetry under rotations*, then the rotation group SO(3) is the symmetry group and $\hat{J}_1,\hat{J}_2,\hat{J}_3$ are the generators which conserve the eigenvalues, i.e. which are good quantum numbers. We have $[\hat{H},\hat{J}]_-=0$, which is physically plausible, because the invariance of the Schrödinger equation under rotations with respect to any of the three coordinate axes guarantees the invariance of this equation under arbitrary rotations. In this case, we obtain, as already mentioned, conservation laws for $\hat{J}_1,\hat{J}_2$ and $\hat{J}_3$. The fact that these $\hat{J}_i$ do not commute with respect to each other means only that just one of the three operators can be diagonalized, i.e. that its value can be measured precisely. In the general case of a symmetry group of rank l, l generators can be diagonalized together with the energy, i.e. physically measured exactly.

b) If we *disregard charge dependent forces*, then the isotopic-spin group is the symmetry group and one of the isospin components T_1,T_2,T_3 is a good (conserved) quantum number. For instance, nuclei with different values of

$T_3 = \frac{1}{2}(Z - N)$, but equal isospin T and nucleon number A have the same mass (apart from electromagnetic contributions to the energy). *Example 4.1*

One might wonder why the Hamiltonian $\hat{H}$ occupies a special position in its relation to symmetry groups according to (4.1–5). The reason lies in the particular role played by $\hat{H}$ in the fundamental laws of quantum mechanics, e.g. the Schrödinger equation. $\hat{H}$ determines the time evolution of the quantum system. Accordingly, commutativity of the group operators $\{\hat{U}(\boldsymbol{\alpha})\}$ with $\hat{H}$ implies that the time evolution of the original state $\psi(\boldsymbol{r},t)$ and the symmetry shifted one $\psi'(\boldsymbol{r},t) = \hat{U}(\boldsymbol{\alpha})\psi(\boldsymbol{r},t)$ are identical. The simultaneously valid conservation laws due to the n generators are characterized by so called *linear quantum numbers*. This means, e.g. that the total angular momentum in the z direction J_3 is linear with the number of particles of the system. The same holds for the total charge (T_3) and the strangeness (S). This will be discussed in more detail in the next chapter.

2) The system will have another l good quantum numbers defined by the l invariant operators $\hat{C}_\lambda(\lambda = 1, 2, \dots, l)$ of the symmetry group. It is easy to see that these operators commute with each other and with all generators: All n generators $\hat{L}_i$ commute with $\hat{H}$ provided S_α is a symmetry group of the Hamiltonian [see (4.5)] and, therefore, the functions $\hat{C}_\lambda(\hat{L}_i)$ commute with $\hat{H}$, too, so that

$$[\hat{C}_\lambda(\hat{L}_i), \hat{H}]_- = 0 \quad . \tag{4.6}$$

All l quantum numbers $C_1, C_2, \dots, C_l$ can be measured simultaneously, together with the energy (can be "diagonalized") because the $\hat{C}_\lambda$ commute with each other, i.e.

$$[\hat{C}_\lambda, \hat{C}_{\lambda'}]_- = 0 \quad . \tag{4.7}$$

In a sense the l Casimir operators $\hat{C}_\lambda(\lambda = 1, 2, \dots, l)$, therefore, are more important than the generators: In other words they are the observables which characterize the degeneracy of the energy eigenvalues of the system. They uniquely determine the multiplet, particularly its dimension. We understand as a *precise definition of a multiplet* the set of states having the same quantum numbers $C_1, C_2, \dots, C_l$.

It is easy to show that the *transitions of a system from one multiplet to another* are forbidden if the system has the symmetry S_α. Let the two multiplets be denoted by M and M' and the eigenvalues of $\hat{C}_\lambda$ by C_λ and $C_{\lambda'}$ respectively. At least once we must have $C_\lambda \neq C_{\lambda'}$ since the multiplets are different. Due to (4.6) we have

$$0 = \langle C_{\lambda'} | \hat{C}_\lambda \hat{H} - \hat{H}\hat{C}_\lambda | C_\lambda \rangle = (C_{\lambda'} - C_\lambda)\langle C_{\lambda'} | \hat{H} | C_\lambda \rangle \quad .$$

With $C_\lambda \neq C_{\lambda'}$ follows

$$\langle C_{\lambda'} | \hat{H} | C_\lambda \rangle = 0 \quad , \tag{4.8}$$

i.e. the matrix elements of $\hat{H}$ between different multiplets vanish. Therefore, there are no transitions from one multiplet to another one. We can make this clear in a different way: Since the C_λ are good quantum numbers (i.e. conserved quantities), then due to their conservation there are no transitions between states $|C_1', C_2', \ldots, C_l'\rangle$ and $|C_1, C_2, \ldots, C_l\rangle$ with at least one $C_\lambda \neq C_{\lambda'}$. Therefore, all interactions causing transitions from one multiplet into another are forbidden. Interactions only occur within a multiplet. The commutators (4.6) also contain ***Schur***'s *lemma*:

Any operator $\hat{H}$, which commutes with all group operators $\hat{U}(\boldsymbol{\alpha})$, and, therefore, with all generators $\hat{L}_i$ of a group, has every state of a multiplet as an eigenvector and is degenerate on every multiplet. Or equivalently,

$$[\hat{H}, \hat{U}(\boldsymbol{\alpha})]_- = 0 \Leftrightarrow [\hat{H}, \hat{L}_i]_- = 0 \Rightarrow [\hat{H}, C_\lambda]_- = 0 \quad . \tag{4.9}$$

We will prove this in the following way:

Let ψ be an eigenstate of $\hat{H}$, i.e.

$$\hat{H}\psi = E\psi \quad ,$$

then it follows from (4.9) that

$$\hat{H}\hat{U}(\boldsymbol{\alpha})\psi = E\hat{U}(\boldsymbol{\alpha})\psi \quad .$$

This means that $\psi'(\boldsymbol{r}, t) = \hat{U}(\boldsymbol{\alpha})\psi(\boldsymbol{r}, t)$ is also an eigenstate of $\hat{H}$, with the same eigenvalue E. Since $[\hat{H}, \hat{C}_\lambda]_- = 0$, the Casimir operators $\hat{C}_\lambda$ can be diagonalized simultaneously with $\hat{H}$. Then the eigenstate ψ of $\hat{H}$ is an eigenstate of all $\hat{C}_\lambda$, as well, and belongs to the same multiplet. Due to

$$[\hat{C}_\lambda, \hat{U}(\boldsymbol{\alpha})]_- = 0 \quad ,$$

all symmetry shifted states $\psi' = \hat{U}(\boldsymbol{\alpha})\psi$ are also eigenstates of the $\hat{C}_\lambda$, and their eigenvalues, $C_1, C_2, \ldots, C_l$, are the same. Hence, they all belong to the multiplet.

This theorem is important because it makes clear why the eigenstates are degenerate, the reason always being due to a special symmetry of the system. On the other hand, this theorem helps to classify the eigenstates by good quantum numbers $C_1, C_2, \ldots, C_l$.

EXAMPLE

4.2 Energy Degeneracy for Various Symmetries

a) *Rotation Group*: The single Casimir operator is $\hat{J}^2$ with eigenvalues $j(j+1)$, $j = 0, 1, 2, \ldots$. The dimensions of the angular momentum multiplets are $(2j+1)$. In principle there are no transitions of the system between energetically degenerate states $|j'm'\rangle$ and $|jm\rangle$ with $j' \neq j$. Only external disturbances that break the rotational symmetry can cause transitions between the rotational multiplets (e.g. an electromagnetic wave or a passing particle which does not belong to the system). This is illustrated in the following two figures.

Example 4.2

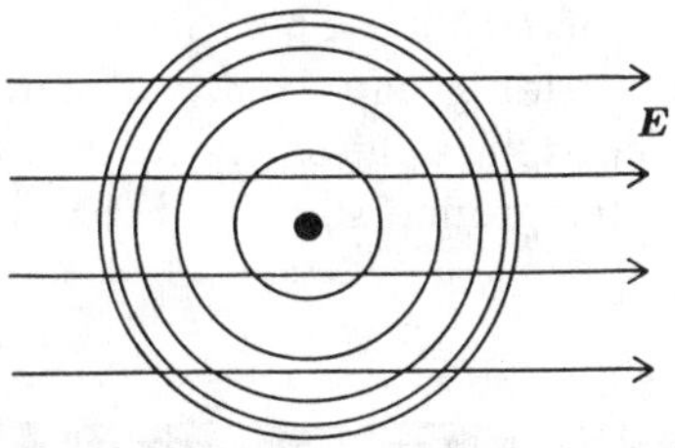

An external field E breaks the rotational symmetry of the closed system. The spherical symmetry of the latter is indicated by circular equipotential lines

b) *Isobaric Spin Group*: The only Casimir operator is $\hat{T}^2$ with the eigenvalues $T(T+1)$. The dimensions of the multiplets are $(2T+1)$.

c) The $T=\frac{1}{2}$ and $T=\frac{3}{2}$ multiplets of the pion–nucleon system scatter independently of one another. The resultant scattering phases depend on T but not on T_3 within the multiplet (see Example 5.7).

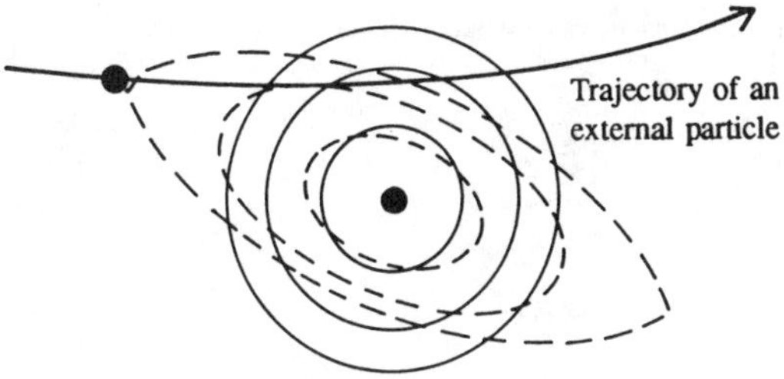

An external particle destroys the rotational symmetry of the closed system and the spherical equipotential lines become deformed. An example is the collision of two heavy ions. There the angular momentum of the electrons in the two-centre Coulomb potential of the two nuclei is no longer a good quantum number, i.e. it is no longer conserved. Due to the presence of two centres, the spherical symmetry of the system is lost for the electrons (*dashed line*)

3) All states of a multiplet of a symmetry group have the same mass (i.e. the same energy eigenvalue), as we have just shown by means of Schur's lemma. If there exist additional symmetry groups like the inversion group or the rotation group, and if they commute with the operators of the original symmetry group, then all states of a multiplet of the original symmetry group have the same quantum numbers with respect to the additional symmetry group.

EXAMPLE

4.3 Degeneracy and Parity of More Symmetries

a) *Rotation Group*. It commutes with the space inversion group (parity operation $\hat{P}: \boldsymbol{r} \to -\boldsymbol{r}$). Hence all rotation multiplets have the same parity. For example, all $8 = (2\times\frac{7}{2}+1)$ states of the $g_{7/2}$-spin multiplet have parity $(-1)^l = (-1)^4 = +1$. These eight states are energetically degenerated.

Example 4.3

b) *Isobaric Spin Group.* It commutes with the rotation group and the inversion group. Hence all states of an isospin multiplet have the same spin as well as the same parity. So the isospin triplet (π^+, π^0, π^-) has the same spin (zero in the ground state) and the same parity (-1).

4.1 Biographical Notes

SCHUR, Issai, *10.1.1875 at Mohilev, †10.1.1941 at Tel Aviv. S. received his Ph.D. in 1901 in Berlin under Frobenius but he did not become a full professor until 1921. In 1935 he had to emigrate. S. predominantly worked on number theory, group theory and on the theory of power series. He developed the representation theory of groups by means of fractional linear substitutions.

5. The Isospin Group (Isobaric Spin)

The isobaric spin (isospin) group is of great importance in nuclear physics as well as in the theory of elementary particles, and we will require it repeatedly in the following discussions. In part, we will follow the historical route, but then quickly come to the modern applications of the isospin group.

Immediately after the discovery of the neutron in 1932, the strong similarity between the proton and neutron led nuclear physicists to interpret *both particles as two states of the same particle*[1]. Of course, this is only meaningful if proton and neutron have the same mass (energy). Table 5.1 summarizes their properties.

Table 5.1. Properties of protons and neutrons

	Mass m_0c^2 [MeV]	Mass difference Δm [MeV]	Spin	Lifetime [s]	Magnetic *moment* $[\mu_N]$[a]
p	983.213		$\frac{1}{2}$	stable	2.793
		1.294			
n	939.507		$\frac{1}{2}$	918 ± 14	-1.913

[a] $\mu_N = e\hbar/(2m_pc)$

Indeed, the masses of the proton and neutron are nearly equal. The small difference in the masses is partly explained by the different electromagnetic interaction of the two particles (see the slightly more detailed discussion in the following Example 5.4: pions). Hence, to a good approximation we can take the masses of proton and neutron to be equal with respect to the strong interaction. The wave function of the nucleons (a collective term for proton and neutron) depends on the space-time coordinates $\boldsymbol{r}, t$ and the spin coordinates s. Over and above that we now have the inner *isospin coordinate* τ, which allows us to distinguish between the two charge states (positive-proton, negative-neutron). This variable has only two possible values, just like the spin coordinate s. We denote them by $\tau = \pm 1$ and define

$$\psi_p = \psi(\boldsymbol{r}, t, s, \tau = +1) = \text{proton state}$$

$$\psi_n = \psi(\boldsymbol{r}, t, s, \tau = -1) = \text{neutron state}$$

[1] Werner Heisenberg: Zeitschrift für Physik **77**, 1 (1932).

Instead of using the inner coordinate τ, the wave function of the nucleon can be represented as a two-component column vector

$$\psi = \begin{pmatrix} u_1(\boldsymbol{r}, t, s) \\ u_2(\boldsymbol{r}, t, s) \end{pmatrix} \quad . \tag{5.1}$$

$|u_1(\boldsymbol{r}, t, s)|^2$ is then the probability density for a proton at position $\boldsymbol{r}$, time t with spin projection s. Analogously, $|u_2(\boldsymbol{r}, t, s)|^2$ denotes the probability density of a neutron. We next introduce the 2×2 matrix operator

$$\hat{\tau}_3 = \begin{pmatrix} 1 & 0 \\ 0 & -1 \end{pmatrix} \quad , \tag{5.2}$$

with the eigenvalue $+1$ corresponding to the proton state and -1 corresponding to the neutron state:

$$\hat{\tau}_3 \psi_{\mathrm{p}} = +1 \psi_{\mathrm{p}} \quad , \quad \hat{\tau}_3 \psi_{\mathrm{n}} = -1 \psi_{\mathrm{n}} \quad . \tag{5.3}$$

This yields, together with (5.1) and (5.2),

$$\psi_{\mathrm{p}} = \begin{pmatrix} u_1(\boldsymbol{r}, t, s) \\ 0 \end{pmatrix} \quad \psi_{\mathrm{n}} = \begin{pmatrix} 0 \\ u_2(\boldsymbol{r}, t, s) \end{pmatrix} \quad . \tag{5.4}$$

We emphasize that $u_1(\boldsymbol{r}, t, s)$ and $u_2(\boldsymbol{r}, t, s)$ are standard two component Pauli spinors, so that the nucleon state ψ of (5.1) has four components in all. (In the relativistic Dirac theory the spinors have also four components[2], describing particles and antiparticles. In this case the nucleon state comprises $2 \times 4 = 8$ components, four each for the proton and neutron states.) We continue by constructing the 2×2 matrix operators which formally transform the proton into a neutron and vice versa. For that reason we first note that

$$\chi_{\mathrm{p}} = \begin{pmatrix} \chi(\boldsymbol{r}, t, s) \\ 0 \end{pmatrix} \tag{5.5}$$

describes a proton in the state $\chi(\boldsymbol{r}, t, s)$ and

$$\chi_{\mathrm{n}} = \begin{pmatrix} 0 \\ \chi(\boldsymbol{r}, t, s) \end{pmatrix} \quad , \tag{5.6}$$

a neutron in the same state $\chi(\boldsymbol{r}, t, s)$. In order to transform the neutron, with wave function χ, into a proton with the same wave function, χ has to be moved from the lower component (5.6) into the upper one (5.5). These requirements can be written in the form

$$\hat{\tau} + \chi_{\mathrm{p}} = 0 \quad , \quad \hat{\tau} + \chi_{\mathrm{n}} = \chi_{\mathrm{p}} \quad , \quad \hat{\tau} - \chi_{\mathrm{p}} = \chi_{\mathrm{n}} \quad , \quad \hat{\tau} - \chi_{\mathrm{n}} = 0 \tag{5.7}$$

By means of (5.5), (5.6) and (5.7), we can easily verify that the matrices

$$\hat{\tau}_+ = \begin{pmatrix} 0 & 1 \\ 0 & 0 \end{pmatrix} \quad \text{and} \quad \hat{\tau}_- = \begin{pmatrix} 0 & 0 \\ 1 & 0 \end{pmatrix} \tag{5.8}$$

satisfy these conditions.

[2] These four-component Dirac spinors are discussed in Vol. 3 of this series, *Relativistic Quantum Mechanics* (Springer, Berlin, Heidelberg 1989)

In place of these singular ($\det \hat{\tau}_+ = \det \hat{\tau}_- = 0$) and non-hermitian operators $\tau_\pm$, it is advantageous to introduce the *non singular, hermitian linear combinations*

$$\hat{\tau}_1 = \hat{\tau}_+ + \hat{\tau}_- = \begin{pmatrix} 0 & 1 \\ 1 & 0 \end{pmatrix} \quad , \quad \hat{\tau}_2 = -\mathrm{i}(\hat{\tau}_+ - \hat{\tau}_-) = \begin{pmatrix} 0 & -\mathrm{i} \\ \mathrm{i} & 0 \end{pmatrix} \quad . \tag{5.9}$$

The three matrices $\hat{\tau}_1$, $\hat{\tau}_2$ and $\hat{\tau}_3$ are the welll known *Pauli matrices.* They satisfy the relations

$$\hat{\tau}_1 \hat{\tau}_2 = \mathrm{i}\hat{\tau}_3$$

and cyclical permutations thereof, and the anticommutation relations

$$\hat{\tau}_i \hat{\tau}_k + \hat{\tau}_k \hat{\tau}_i = 2\delta_{ik} \quad . \tag{5.10}$$

The effect of the $\hat{\tau}_i$ operators on proton and neutron states can easily be calculated by means of (5.5)–(5.7) and (5.9); hence we find that

$$\begin{aligned} &\hat{\tau}_1 \chi_\mathrm{p} = \chi_\mathrm{n} \quad , \quad \hat{\tau}_2 \chi_\mathrm{p} = \mathrm{i}\chi_\mathrm{n} \quad , \quad \hat{\tau}_3 \chi_\mathrm{p} = \chi_\mathrm{p} \quad , \\ &\hat{\tau}_1 \chi_\mathrm{n} = \chi_\mathrm{p} \quad , \quad \hat{\tau}_2 \chi_\mathrm{n} = -\mathrm{i}\chi_\mathrm{p} \quad , \quad \hat{\tau}_3 \chi_\mathrm{n} = -\chi_\mathrm{n} \quad . \end{aligned} \tag{5.11}$$

Earlier, when we discussed the Pauli equation[3], we showed that every hermitian 2×2 matrix can be represented by a linear superposition of the three Pauli matrices and the unit matrix. Hence, it follows that every self-adjoint operator which acts on the two degrees of freedom of the two component nucleon system with the wave function ψ can be represented by a linear combination of the Pauli operators and the unit matrix.

Introducing the operators

$$\hat{T}_k = \tfrac{1}{2}\hat{\tau}_k \quad (k = 1, 2, 3) \tag{5.12}$$

yields the modified commutation relations

$$\hat{T}_i \hat{T}_j - \hat{T}_j \hat{T}_i = \mathrm{i}\varepsilon_{ijk} \hat{T}_k \quad . \tag{5.13}$$

These $\hat{T}_k$ operators are analogous to the spin operators $\hat{S}_k = \frac{1}{2}\hat{\sigma}_k$.

Now we are ready to show that the two components (u_1, u_2) of the nucleon state ψ build up an elementary spinor in an abstract three-dimensional *isospin space.* Acccording to (5.13) the hermitian operators $\hat{T}_i$ form a closed algebra, which we recognize as a Lie algebra. From (3.18), the operators of the associated Lie group can be obtained immediately[4],

$$\begin{aligned} \hat{U}_{\mathrm{Is}}(\varepsilon) &= \hat{U}_{\mathrm{Is}}(\varepsilon_1, \varepsilon_2, \varepsilon_3) = \exp(-\mathrm{i}\varepsilon_\nu \hat{T}_\nu) = \exp[-(\mathrm{i}/2)(\varepsilon_1 \hat{\tau}_1 + \varepsilon_2 \hat{\tau}_2 + \varepsilon_3 \hat{\tau}_3)] \\ &= \exp[-(\mathrm{i}/2)\varepsilon n_\nu \hat{\tau}_\nu] = 1\!\!1 \cos(\varepsilon/2) - \mathrm{i} n_\nu \hat{\tau}_\nu \sin(\varepsilon/2) \quad . \end{aligned} \tag{5.14}$$

The angles

$$\boldsymbol{\varepsilon} = \{\varepsilon_1, \varepsilon_2, \varepsilon_3\} = \varepsilon\{n_1, n_2, n_3\} = \varepsilon \boldsymbol{n}$$

---

[3] See Vol. 1 of this series, *Quantum Mechanics- An Introduction* (Springer, Berlin. Heidelberg 1989)

[4] For the last transformation in (5.14) see also Exercise 3.8

characterize rotations about the three axes of the abstract *iso-space* which can also be interpreted as rotations through ε about this axis in the $\boldsymbol{n}$ direction. The iso-rotation operators $\hat{U}_{\mathrm{Is}}(\varepsilon)$ are clearly unitary, i.e.

$$\hat{U}_{\mathrm{Is}}^{+} = \hat{U}_{\mathrm{Is}}^{-1} = \hat{U}_{\mathrm{Is}}(-\varepsilon) \quad ,$$

and the determinant of this 2×2 matrix [see (5.14)] is

$$\det \hat{U}_{\mathrm{Is}}(\varepsilon) = \det\exp(-\mathrm{i}\varepsilon_\nu \hat{T}_\nu) = \exp[-\mathrm{i}\varepsilon_\nu \mathrm{tr}(\hat{T}_\nu)] = \mathrm{e}^0 = +1 \quad .$$

Here we have used a relation for determinants of unitary matrices which we derived in detail in Exercise 3.2. For a complete understanding of the steps in the calculation, the reader may care to refer to this exercise now once more.

The determinant of $\hat{U}_{\mathrm{Is}}(\varepsilon)$ can be calculated directly by means of (5.14). From

$$\hat{U}_{\mathrm{Is}}(\varepsilon) = \begin{pmatrix} \cos(\varepsilon/2) - \mathrm{i}n_3 \sin(\varepsilon/2) \quad , & -(n_2 + \mathrm{i}n_1)\sin(\varepsilon/2) \\ (n_2 - \mathrm{i}n_1)\sin(\varepsilon/2) \quad , & \cos(\varepsilon/2) + \mathrm{i}n_3 \sin(\varepsilon/2) \end{pmatrix} \quad ,$$

we obtain

$$\begin{aligned} \det \hat{U}_{\mathrm{Is}}(\varepsilon) &= \cos^2(\varepsilon/2) + n_3^2 \sin^2(\varepsilon/2) + (n_2^2 + n_1^2)\sin^2(\varepsilon/2) \\ &= \cos^2(\varepsilon/2) + \sin^2(\varepsilon/2) = 1 \quad . \end{aligned}$$

Hence, the group of the iso-rotation operators (5.14) consists of unitary 2×2 matrices with determinant $+1$. This group is denoted by SU(2), i.e. *s*pecial (determinant $+1$) *u*nitary rotations in 2 dimensions. The *iso-spinor* ψ then transforms according to

$$\begin{aligned} \psi'(\boldsymbol{r}, t, s) = \begin{pmatrix} u_1'(\boldsymbol{r}, t, s) \\ u_2'(\boldsymbol{r}, t, s) \end{pmatrix} &= \exp(-\mathrm{i}\varepsilon_\nu \hat{T}_\nu)\psi(\boldsymbol{r}, t, s) \\ &= \exp(-\mathrm{i}\varepsilon_\nu \hat{T}_\nu) \begin{pmatrix} u_1(\boldsymbol{r}, t, s) \\ u_2(\boldsymbol{r}, t, s) \end{pmatrix} \quad . \end{aligned} \tag{5.15}$$

Obviously, the operator $\hat{U}_{\mathrm{Is}}(\varepsilon)$ is unimodular (determinant $= +1$) and unitary. Due to their formal similarity with the real spin (i.e. angular momentum) the $\hat{T}_i$ (i.e. $\hat{\tau}_i$) operators are also called *operators of the isobaric spin.*

To classify the nucleon states the three-component of the isospin

$$\hat{T}_3 = \begin{pmatrix} \frac{1}{2} & 0 \\ 0 & -\frac{1}{2} \end{pmatrix} \tag{5.16}$$

can be used too. Its eigenstates $+\frac{1}{2}$ and $-\frac{1}{2}$ denote the proton and neutron, respectively. The charge operator of the nucleon then reads

$$\hat{Q} = e(\hat{T}_3 + \tfrac{1}{2}) = \tfrac{1}{2}e(\hat{\tau}_3 + 1) \tag{5.17}$$

with eigenvalues $+e$ and 0 for the proton and neutron.

EXERCISE

5.1 Addition Law for Infinitesimal SU(2) Transformations

Problem. Find the addition law for an infinitesimal SU(2) transformation, i.e. find the connection $\boldsymbol{\Phi}(\boldsymbol{\Theta}, \delta\boldsymbol{\Theta})$ of the rotation angles in the SU(2) space if

$$\exp(\mathrm{i}\boldsymbol{\Phi}\cdot\hat{\boldsymbol{\tau}}/2) = \exp(\mathrm{i}\delta\boldsymbol{\Theta}\cdot\hat{\boldsymbol{\tau}}/2)\exp(\mathrm{i}\boldsymbol{\Theta}\cdot\hat{\boldsymbol{\tau}}/2) \quad .$$

Solution. First of all we define the unit vectors (indicated by a tilde sign over the symbol)

$$\tilde{\boldsymbol{\Theta}} = \frac{\boldsymbol{\Theta}}{\Theta} \quad \delta\tilde{\boldsymbol{\Theta}} = \frac{\delta\boldsymbol{\Theta}}{\delta\Theta} \quad \tilde{\boldsymbol{\Phi}} = \frac{\boldsymbol{\Phi}}{\Phi} \tag{1a}$$

with

$$\Theta = |\boldsymbol{\Theta}| \quad \delta\Theta = |\delta\boldsymbol{\Theta}| \quad \Phi = |\boldsymbol{\Phi}| \quad . \tag{1b}$$

Due to the universal relation

$$(\boldsymbol{a}\cdot\hat{\boldsymbol{\tau}})(\boldsymbol{b}\cdot\hat{\boldsymbol{\tau}}) = \boldsymbol{a}\cdot\boldsymbol{b} + \mathrm{i}(\boldsymbol{a}\times\boldsymbol{b})\cdot\hat{\boldsymbol{\tau}} \quad , \tag{2a}$$

then

$$(\tilde{\boldsymbol{\Theta}}\cdot\hat{\boldsymbol{\tau}})^2 = (\delta\tilde{\boldsymbol{\Theta}}\cdot\hat{\boldsymbol{\tau}})^2 = (\tilde{\boldsymbol{\Phi}}\cdot\hat{\boldsymbol{\tau}})^2 = 1 \quad . \tag{2b}$$

This enables us to convert the exponential representation of the SU(2) transformation into a linear form by expanding the exponential function into its power series:

$$\begin{aligned}\exp(\mathrm{i}\boldsymbol{\Theta}\cdot\hat{\boldsymbol{\tau}}/2) &= \exp[\mathrm{i}\Theta/2(\tilde{\boldsymbol{\Theta}}\cdot\hat{\boldsymbol{\tau}})] \\ &= 1 + \mathrm{i}\frac{\Theta}{2}(\tilde{\boldsymbol{\Theta}}\cdot\hat{\boldsymbol{\tau}}) - \frac{1}{2!}\left(\frac{\Theta}{2}\right)^2 - \frac{\mathrm{i}}{3!}\left(\frac{\Theta}{2}\right)^2(\tilde{\boldsymbol{\Theta}}\cdot\hat{\boldsymbol{\tau}}) + \cdots \\ &= \cos\frac{\Theta}{2} + \mathrm{i}(\tilde{\boldsymbol{\Theta}}\cdot\hat{\boldsymbol{\tau}})\sin\frac{\Theta}{2}\end{aligned} \tag{3a}$$

and, analogously,

$$\begin{aligned}\exp(\mathrm{i}\delta\boldsymbol{\Theta}\cdot\hat{\boldsymbol{\tau}}/2) &= \cos\frac{\delta\Theta}{2} + \mathrm{i}(\delta\tilde{\boldsymbol{\Theta}}\cdot\hat{\boldsymbol{\tau}})\sin\frac{\delta\Theta}{2} \\ &\simeq 1 + \mathrm{i}\frac{\delta\Theta}{2}(\delta\tilde{\boldsymbol{\Theta}}\cdot\hat{\boldsymbol{\tau}})\end{aligned} \tag{3b}$$

$$\exp(\mathrm{i}\boldsymbol{\Theta}\cdot\hat{\boldsymbol{\tau}}/2) = \cos\Phi/2 + \mathrm{i}(\tilde{\boldsymbol{\Phi}}\cdot\hat{\boldsymbol{\tau}})\sin\Phi/2 \quad . \tag{3c}$$

In (3b) we have taken into account that $\delta\Theta$ has to be an infinitesimal angle and therefore neglected all terms of second or higher order.

On the other hand, due to the basic requirement

$$\exp(\mathrm{i}\boldsymbol{\Phi}\cdot\hat{\boldsymbol{\tau}}/2) = \exp(\mathrm{i}\delta\boldsymbol{\Theta}\cdot\hat{\boldsymbol{\tau}}/2)\exp(\mathrm{i}\boldsymbol{\Theta}\cdot\hat{\boldsymbol{\tau}}/2) \quad ,$$

Exercise 5.1

we obtain, in consideration of (3a, 3b) and (2a), that

$$\exp(\mathrm{i}\boldsymbol{\Phi}\cdot\hat{\boldsymbol{\tau}}/2) = \left[1 + \mathrm{i}\frac{\delta\Theta}{2}(\delta\tilde{\boldsymbol{\Theta}}\cdot\hat{\boldsymbol{\tau}})\right]\left[\cos\frac{\Theta}{2} + \mathrm{i}(\tilde{\boldsymbol{\Theta}}\cdot\hat{\boldsymbol{\tau}})\sin\frac{\Theta}{2}\right]$$

$$= \cos\frac{\Theta}{2} - (\delta\tilde{\boldsymbol{\Theta}}\cdot\tilde{\boldsymbol{\Theta}})\frac{\delta\Theta}{2}\sin\frac{\Theta}{2}$$

$$+ \mathrm{i}\left[\delta\tilde{\boldsymbol{\Theta}}\frac{\delta\Theta}{2}\cos\frac{\Theta}{2} + \tilde{\boldsymbol{\Theta}}\sin\frac{\Theta}{2} - (\delta\tilde{\boldsymbol{\Theta}}\times\tilde{\boldsymbol{\Theta}})\frac{\delta\Theta}{2}\sin\frac{\Theta}{2}\right]\hat{\boldsymbol{\tau}} \quad . \tag{4}$$

Since the Pauli matrices are linearly independent, a comparison of (3c) with (4) yields the following relations:

$$\cos\frac{\Phi}{2} = \cos\frac{\Theta}{2} - (\delta\tilde{\boldsymbol{\Theta}}\cdot\tilde{\boldsymbol{\Theta}})\frac{\delta\Theta}{2}\sin\frac{\Theta}{2} \tag{5a}$$

$$\tilde{\boldsymbol{\Phi}}\sin\frac{\Phi}{2} = \tilde{\boldsymbol{\Theta}}\sin\frac{\Theta}{2} + \delta\tilde{\boldsymbol{\Theta}}\frac{\delta\Theta}{2}\cos\frac{\Theta}{2} - (\delta\tilde{\boldsymbol{\Theta}}\times\tilde{\boldsymbol{\Theta}})\frac{\delta\Theta}{2}\sin\frac{\Theta}{2} \quad . \tag{5b}$$

To solve this equation for the vector $\boldsymbol{\Phi}$ we first compare (5a) to the addition theorem for the cosine function,

$$\cos(\alpha + \beta) = \cos\alpha\cos\beta - \sin\alpha\sin\beta$$

$$\simeq \cos\alpha - \beta\sin\alpha \quad ,$$

where β is the infinitesimal angle. Obviously,

$$\Phi = \Theta + \delta\Theta(\delta\tilde{\boldsymbol{\Theta}}\cdot\tilde{\boldsymbol{\Theta}}) \tag{6}$$

must hold. With $\sin(\alpha + \beta) = \sin\alpha\cos\beta + \sin\beta\cos\alpha$ we furher have

$$\sin\frac{\Phi}{2} = \sin\left[\frac{\Theta}{2} + \frac{\delta\Theta}{2}(\delta\tilde{\boldsymbol{\Theta}}\cdot\tilde{\boldsymbol{\Theta}})\right]$$

$$\simeq \sin\frac{\Theta}{2} + \frac{\delta\Theta}{2}(\delta\tilde{\boldsymbol{\Theta}}\cdot\tilde{\boldsymbol{\Theta}})\cos\frac{\Theta}{2} \quad ,$$

or, in linear approximation.

$$\left(\sin\frac{\Phi}{2}\right)^{-1} = \left(\sin\frac{\Theta}{2}\right)^{-1}\left(1 - \frac{\delta\Theta}{2}(\delta\tilde{\boldsymbol{\Theta}}\cdot\tilde{\boldsymbol{\Theta}})\cot\frac{\Theta}{2}\right) \quad . \tag{7}$$

With this and (5b) we can find the direction of Φ, given by

$$\tilde{\boldsymbol{\Phi}} = \left[\tilde{\boldsymbol{\Theta}} + \frac{\delta\Theta}{2}\left(\delta\tilde{\boldsymbol{\Theta}}\cot\frac{\Theta}{2} - \delta\tilde{\boldsymbol{\Theta}}\times\tilde{\boldsymbol{\Theta}}\right)\right]\times\left[1 - \frac{\delta\Theta}{2}(\delta\tilde{\boldsymbol{\Theta}}\cdot\tilde{\boldsymbol{\Theta}})\cot\frac{\Theta}{2}\right]$$

$$= \tilde{\boldsymbol{\Theta}} + \frac{\delta\Theta}{2}\left[\{\delta\tilde{\boldsymbol{\Theta}} - \tilde{\boldsymbol{\Theta}}(\delta\tilde{\boldsymbol{\Theta}}\cdot\tilde{\boldsymbol{\Theta}})\}\cot\frac{\Theta}{2} - \delta\tilde{\boldsymbol{\Theta}}\times\tilde{\boldsymbol{\Theta}}\right] \quad . \tag{8}$$

Multiplication of (8) by the modulus of Φ yields the final result

$$\boldsymbol{\Phi} \simeq \boldsymbol{\Theta} + \delta\boldsymbol{\Theta}\left(\frac{\Theta}{2}\cot\frac{\Theta}{2}\right) - \boldsymbol{\Theta}(\delta\boldsymbol{\Theta}\cdot\boldsymbol{\Theta})\Theta^{-2}\left(1 - \frac{\Theta}{2}\cot\frac{\Theta}{2}\right) - \frac{1}{2}(\delta\boldsymbol{\Theta}\times\boldsymbol{\Theta}) \quad . \tag{9}$$

5.1 Isospin Operators for a Multi-Nucleon System

Let us consider a system of A nucleons (protons and neutrons). The isospin operator for the nth nucleon reads

$$\hat{T}(n) = \tfrac{1}{2}\hat{\tau}(n) \quad , \quad n = 1, 2, \dots, A \quad ,$$

or, in detail,

$$\{\hat{T}_1(n), \hat{T}_2(n), \hat{T}_3(n)\} = \tfrac{1}{2}\{\hat{\tau}_1(n), \hat{\tau}_2(n), \hat{\tau}_3(n)\} \quad . \tag{5.18}$$

$\hat{T}(n)$ only acts on the nth nucleon. For that reason the operators $\hat{T}(n)$ and $\hat{T}(n')$ commute just as in the case of ordinary spin (see Sect. 1.10); therefore,

$$[\hat{\boldsymbol{T}}(n), \hat{\boldsymbol{T}}(n')]_- = 0 \quad , \quad n \neq n' \quad , \tag{5.19}$$

i.e. each component of $\hat{\boldsymbol{T}}$ of the n-space commutes with each other component of $\hat{\boldsymbol{T}}$ of the n' space. Therefore, the isospin of the A-nucleon system can simply be defined as the sum of the isospins of the nucleons.

$$\hat{\boldsymbol{T}} = \sum_{n=1}^{A} \hat{\boldsymbol{T}}(n) = \frac{1}{2} \sum_{n=1}^{A} \hat{\boldsymbol{\tau}}(n) \quad . \tag{5.20a}$$

The components $\hat{T}_i$ of the total isospin fulfill, due to (5.13) and (5.19), the commutation relation

$$[\hat{T}_i, \hat{T}_j] = -\mathrm{i}\varepsilon_{ijk}\hat{T}_k \tag{5.20b}$$

Correspondingly, the charge operator is obtained from (5.17) as a sum,

$$\hat{Q} = \sum_n \hat{Q}(n) = e \sum_{n=1}^{A} \tfrac{1}{2}(\hat{\tau}_3(n) + 1) = e\left(\hat{T}_3 + \tfrac{1}{2}A\right) \quad . \tag{5.21}$$

Since nuclei are characterized by two numbers, the mass number A and charge number (number of protons) $Z = Q/e$, isobars (nuclei with identical mass number A) can only be different in the value of T_3. This, by the way, is the reason for the name *isobaric spin*. The name *isotopic spin*, which is in use too, is less fraught with meaning, because only isobars are classified by T_3, and not isotopes.

We realize that the eigenvalues of $\hat{T}_3$ characterize the states of a given isobaric charge multiplet uniquely, just like the angular momentum operator $\hat{J}_3$ classifies the states of an angular momentum multiplet. We say: *the angular momentum group and the isospin group are isomorphous.* For that reason we can transcribe the results of Chap. 2 directly, obtaining, with the isospin states of a multiplet

$$|TT_3\rangle \quad , \tag{5.22}$$

the relations

$$\begin{aligned} \hat{\boldsymbol{T}}^2|TT_3\rangle &= T(T+1)|TT_3\rangle \quad , \quad T = 0, \tfrac{1}{2}, 1, \tfrac{3}{2}, \dots \quad , \\ \hat{T}_3|TT_3\rangle &= T_3|TT_3\rangle \quad , \quad T \geq T_3 \geq -T \quad . \end{aligned} \tag{5.23}$$

Each isobaric multiplet is $(2T+1)$ times degenerate, just as the angular momentum multiplets have been $(2j+1)$-times degenerate. For $T=\frac{1}{2}$ we have the *fundamental iso-spin doublet*

$$|\tfrac{1}{2}T_3\rangle \quad , \quad \text{where } \tilde{\chi}_{\frac{1}{2}\frac{1}{2}} = |\tfrac{1}{2}, T_3 = \tfrac{1}{2}\rangle \equiv |\mathrm{p}\rangle$$

is the proton state, and

$$\tilde{\chi}_{\frac{1}{2},-\frac{1}{2}} = |\tfrac{1}{2}, T_3 = -\tfrac{1}{2}\rangle \equiv |\mathrm{n}\rangle \tag{5.24}$$

is the neutron state. This is the smallest non-trivial multiplet of SU(2). By this we mean that all higher multiplets can be constructed from this multiplet. Spin 0, as well as all other spins occurring in nature, can be constructed from spin $\frac{1}{2}$. The smallest multiplet of SU(2) with $T=0$ is trivial, since only $T=0$ multiplets can be constructed from it. Accordingly, we speak of an *isospin triplet* in the case of $T=1$, which is realized in nature among the π mesons and Σ baryons.

EXAMPLE

5.2 The Deuteron

The deuteron contains one proton and one neutron. Its wave function consists of a space part $R_{nl}(r)Y_{lm_l}(\vartheta,\varphi)$ (describing the relative motion of the two nucleons), spin part χ_{sm_s} and isospin part $|TT_3\rangle$. Hence,

$$\psi_{\text{deuteron}} = R_{nl}[Y_{lm_l} \times \chi_{sm_s}]^{[j]}|TT_3\rangle \tag{1}$$

The []-bracket indicates the coupling of angular momentum and spin to the total angular momentum,

$$[Y_{lm_l} \times \chi_{sm_s}]^{[j]} = \sum_{m_l,m_s} (lsj|m_l m_s m), \, Y_{lm_l}\chi_{sm_s}(1,2)$$

$$\chi_{sm_s}(1,2) = \sum_{m_1,m_2} (\tfrac{1}{2}\tfrac{1}{2}s|m_1 m_2 m_s)\chi_{\frac{1}{2}m_1}(1)\chi_{\frac{1}{2}m_2}(2) \quad . \tag{2}$$

$|TT_3\rangle$ is constructed from the single iosospin wave fuctions, $\tilde{\chi}_{\frac{1}{2}t_v}$, of the nucleons, i.e.

$$|TT_3\rangle = \sum_{t_1,t_2} (\tfrac{1}{2}\tfrac{1}{2}T|t_1 t_2 T_3)\tilde{\chi}_{\frac{1}{2}t_1}(1)\tilde{\chi}_{\frac{1}{2}t_2}(2) \quad . \tag{3}$$

The Clebsch–Gordan coefficients for coupling the iso-spin functions are identical to the Clebsch–Gordan coefficients for coupling the angular momentum wave functions. This follows immediately from the isomorphism between the Lie algebras SO(3) and SU(2) of the rotation and isospin groups, respectively. Indeed, as remarked earlier the two Lie algebras are identical, for we have

$$[\hat{J}_i, \hat{J}_j]_- = \mathrm{i}\varepsilon_{ijk}\hat{J}_k \tag{4a}$$

Example 5.2

for the rotation group, and

$$[\hat{T}_i, \hat{T}_j]_- = \mathrm{i}\varepsilon_{ijk}\hat{T}_k \tag{4b}$$

for the isospin group.

Since the total charge of the deuteron is $Q = e$, and the nucleon number $A = 2$, there follows, according to (5.21), $T_3 = 0$. However, the total isospin can take two values

$$T = 0 \quad \text{and} \quad T = 1$$

For $T = 0$, i.e. for the iso-singlet, T_3 can only be zero. This is the *ground state of the deuteron.* Only the state $|T = 1, T_3 = 0\rangle$ of the iso-triplet belongs to the deuteron. The two other states of the iso-triplet, $|T = 1, T_3 = 1\rangle$ and $T = 1, T_3 = -1\rangle$, correspond to combinations with the products $\tilde{\chi}_{\frac{1}{2}\frac{1}{2}}\tilde{\chi}_{\frac{1}{2}\frac{1}{2}}$ and $\tilde{\chi}_{\frac{1}{2}-\frac{1}{2}}\tilde{\chi}_{\frac{1}{2}-\frac{1}{2}}$, respectively. Hence, according to (5.24) they are, respectively, states with two protons and two neutrons. They do not belong to the deuteron system (proton-neutron system) but to the di-neutron and di-proton systems which are unstable. In fact, all states of the two nucleon systems with $T = 1$ are unstable, in particular the $T = 1$ state of the deuteron, demonstrating the close connection among the states of the $T = 1$ multiplet.

It may be noted that the isospin wave function (3) for the singlet state

$$\begin{aligned}|T = 0, T_3 = 0\rangle &= \sum_{t_1} (\tfrac{1}{2}\tfrac{1}{2}0|t_1 - t_1\, 0)\tilde{\chi}_{\frac{1}{2}t_1}(1)\tilde{\chi}_{\frac{1}{2}-t_1}(2)\\ &= (\tfrac{1}{2}\tfrac{1}{2}0|\tfrac{1}{2} - \tfrac{1}{2}0)\tilde{\chi}_{\frac{1}{2}\frac{1}{2}}(1)\tilde{\chi}_{\frac{1}{2}-\frac{1}{2}}(2) + (\tfrac{1}{2}\tfrac{1}{2}0| - \tfrac{1}{2}\tfrac{1}{2}0)\tilde{\chi}_{\frac{1}{2}-\frac{1}{2}}(1)\tilde{\chi}_{\frac{1}{2}\frac{1}{2}}(2)\\ &= \frac{1}{\sqrt{2}}\{\tilde{\chi}_{\frac{1}{2}\frac{1}{2}}(1)\tilde{\chi}_{\frac{1}{2}-\frac{1}{2}}(2) - \tilde{\chi}_{\frac{1}{2}-\frac{1}{2}}(1)\tilde{\chi}_{\frac{1}{2}\frac{1}{2}}(2)\}\end{aligned} \tag{5}$$

is obviously antisymmetric with respect to the exchange $\mathrm{p} \leftrightarrow \mathrm{n}$ (or particles $1 \leftrightarrow 2$), whereas the triplet state

$$\begin{aligned}|T = 1, T_3 = 0\rangle &= \sum_{t_1} (\tfrac{1}{2}\tfrac{1}{2}1|t_1 - t_1\, 0)\tilde{\chi}_{\frac{1}{2}t_1}\tilde{\chi}_{\frac{1}{2}-t_1}\\ &= (\tfrac{1}{2}\tfrac{1}{2}1|\tfrac{1}{2} - \tfrac{1}{2}0)\tilde{\chi}_{\frac{1}{2}\frac{1}{2}}(1)\tilde{\chi}_{\frac{1}{2}-\frac{1}{2}}(2) + (\tfrac{1}{2}\tfrac{1}{2}1| - \tfrac{1}{2}\tfrac{1}{2}0)\tilde{\chi}_{\frac{1}{2}-\frac{1}{2}}(1)\tilde{\chi}_{\frac{1}{2}\frac{1}{2}}(2)\\ &= \frac{1}{\sqrt{2}}\{\tilde{\chi}_{\frac{1}{2}\frac{1}{2}}(1)\tilde{\chi}_{\frac{1}{2}-\frac{1}{2}}(2) + \tilde{\chi}_{\frac{1}{2}-\frac{1}{2}}(1)\tilde{\chi}_{\frac{1}{2}\frac{1}{2}}(2)\}\end{aligned} \tag{6}$$

is symmetric with respect to the exchange $\mathrm{p} \leftrightarrow \mathrm{n}$ (or the particles $1 \leftrightarrow 2$). Due to the Pauli principle, the total wave function (1) has to be antisymmetric with respect to particle exchange. Hence, the iso-singlet state must occur in combination with the symmetric spin state $s = 1$ [(2)]. The contrary holds for the symmetric iso-triplet state: it is paired with the spin singlet state ($s = 0$).

For completeness, we still quote the two other states belonging to the iso-triplet,

$$\begin{aligned}|T = 1, T_3 = 1\rangle &= \tilde{\chi}_{\frac{1}{2}\frac{1}{2}}(1)\tilde{\chi}_{\frac{1}{2}\frac{1}{2}}(2)\ \text{(two protons)}\quad ,\\ |T = 1, T_3 = -1\rangle &= \tilde{\chi}_{\frac{1}{2}-\frac{1}{2}}(1)\tilde{\chi}_{\frac{1}{2}-\frac{1}{2}}(2)\ \text{(two neutrons)}\quad ,\end{aligned} \tag{7}$$

although they describe unstable systems as shown above.

Example 5.2

The physical reason for the lack of bound $T = 1$ states of the two nucleon system is closely connected to the Pauli principle. The isospin wave function for $T = 0$ is, as we have seen, antisymmetric with respect to exchanging the two nucleons, whereas the states of the $T = 1$ triplet are symmetric. Since the nucleon-nucleon force in the spin-1 channel is attractive at short range, the wave function in configuration space must have quantum number $l = 0$, i.e. it must be completely symmetric. That is the reason why a bound state can only be constructed with $T = 0$[5]. The best way to make this fact clear is to state that the Hamiltonian of the strong interaction $\hat{H}_{\text{strong}}$ is isospin invariant and thus depends on the Casimir operator $\hat{T}^2$ of the isospin group, i.e. $\hat{H} = \hat{H}(\hat{\boldsymbol{T}}^2)$. For different multiplets, $\hat{H}$ (and, with that, the included potential) can take rather different values. If, for example,

$$\hat{H}(T) = f(r)\mathbb{1} + g(r)\hat{\boldsymbol{T}}^2 \quad ,$$

then the hamiltonians for the iso-singlet and iso-triplet are, respectively,

$$\hat{H}(T=0) = f(r) \quad , \quad \hat{H}(T=1) = f(r) + 2g(r) \quad .$$

The observation that the $T = 1$ state of the deuteron is unbound shows that $f(r)$ has to be an attractive function with a potential pocket, whereas $f(r) + 2g(r)$ is not.

EXERCISE

5.3 The Charge Independence of Nuclear Forces

Problem. Show the charge independence of nuclear forces as a consequence of the isospin invariance of the strong interaction.

Solution. Due to the requirement that the isospin group be a symmetry group of the strong interaction, the following must hold,

$$[\hat{H}_{\text{strong}}, \hat{\boldsymbol{T}}]_- = 0 \quad , \tag{1}$$

or, equivalently,

$$[\hat{H}_{\text{strong}}, \hat{U}_{\text{Is}}(\varepsilon)]_- = 0 \quad .$$

This equation defines precisely the term "iso-invariance". If, as is often the case, the Hamiltonian itself is not known, one has to take recourse to the $\hat{S}$-matrix

[5] For more information about the deuteron and the usage of the isobaric spin formalism see, e.g. J.M. Eisenberg, W. Greiner: *Microscopic Theory of the Nucleus* (2nd ed.), Nuclear Theory, Vol. 3, (North-Holland, Amsterdam 1976).

operator

Exercise 5.3

$$\hat{S} = \exp\left(\frac{\mathrm{i}}{h}\hat{H}t\right) \quad . \tag{2}$$

Accordingly, it then follows from (1) that

$$[\hat{S}_{\text{strong}}, \hat{\boldsymbol{T}}]_- = 0 \quad \text{or} \quad [\hat{S}_{\text{strong}}, \hat{U}_{\text{Is}}(\varepsilon)]_- = 0 \tag{3}$$

and, vice versa, (1) follows from (3) too. We know from general considerations [cf. (3.38–3.40) and Example 3.13, Eq. (2–6)] that then all states $|T, T_3\rangle$ of an iso-multiplet, also called charge multiplet, are energetically degenerate. We use the relations

$$\begin{aligned}
\mathrm{e}^{-\mathrm{i}\pi\hat{T}_2}|\tfrac{1}{2}\tfrac{1}{2}\rangle &= \mathrm{e}^{-\mathrm{i}(\pi/2)\hat{\tau}_2}|\tfrac{1}{2}\tfrac{1}{2}\rangle = (\cos\tfrac{1}{2}\pi - \mathrm{i}\hat{\tau}_2\sin\tfrac{1}{2}\pi)|\tfrac{1}{2}\tfrac{1}{2}\rangle \\
&= -\mathrm{i}\begin{pmatrix} 0 & -\mathrm{i} \\ \mathrm{i} & 0 \end{pmatrix}\begin{pmatrix} 1 \\ 0 \end{pmatrix} = \begin{pmatrix} 0 & -1 \\ 1 & 0 \end{pmatrix}\begin{pmatrix} 1 \\ 0 \end{pmatrix} = \begin{pmatrix} 1 \\ 0 \end{pmatrix} = |\tfrac{1}{2} - \tfrac{1}{2}\rangle
\end{aligned}$$

and

$$\begin{aligned}
\mathrm{e}^{-\mathrm{i}\pi\hat{T}_2}|\tfrac{1}{2} - \tfrac{1}{2}\rangle &= -\mathrm{i}\begin{pmatrix} 0 & -\mathrm{i} \\ \mathrm{i} & 0 \end{pmatrix}\begin{pmatrix} 0 \\ 1 \end{pmatrix} = \begin{pmatrix} 0 & -1 \\ 1 & 0 \end{pmatrix}\begin{pmatrix} 0 \\ 1 \end{pmatrix} \\
&= -\begin{pmatrix} 1 \\ 0 \end{pmatrix} = -|\tfrac{1}{2}\tfrac{1}{2}\rangle \quad ,
\end{aligned}$$

which can be written in the short-hand notation

$$\mathrm{e}^{-\mathrm{i}\pi\hat{T}_2}|\mathrm{p}\rangle = |\mathrm{n}\rangle \quad \text{and} \quad \mathrm{e}^{-\mathrm{i}\pi\hat{T}_2}|\mathrm{n}\rangle = -|\mathrm{p}\rangle \quad . \tag{4}$$

For the two-nucleon states follows

$$\exp[-\mathrm{i}\pi(\hat{T}_2(1) + \hat{T}_2(2))]|\mathrm{p}(1)\mathrm{p}(2)\rangle = |\mathrm{n}(1)\mathrm{n}(2)\rangle$$

and

$$\exp[-\mathrm{i}\pi(\hat{T}_2(1) + \hat{T}_2(2)]|\mathrm{n}(1)\mathrm{n}(2)\rangle = |\mathrm{p}(1)\mathrm{p}(2)\rangle \quad . \tag{5}$$

From this the strong interaction of the two-proton system can be obtained,

$$\begin{aligned}
&\langle \mathrm{p}(1)\mathrm{p}(2)|\hat{H}_{\text{strong}}|\mathrm{p}(1)\mathrm{p}(2)\rangle \\
&\quad = \langle \exp[-\mathrm{i}\pi(\hat{T}_2(1) + \hat{T}_2(2))]\mathrm{n}(1)\mathrm{n}(2)|\hat{H}_{\text{strong}} \\
&\qquad \times \exp[-\mathrm{i}\pi(\hat{T}_2(1) + \hat{T}_2(2))]\mathrm{n}(1)\mathrm{n}(2)\rangle \\
&\quad = \langle \mathrm{nn}|\exp[+\mathrm{i}\pi(\hat{T}_2(1) + \hat{T}_2(2))]\hat{H}_{\text{strong}}\exp[-\mathrm{i}\pi(\hat{T}_2(1) + \hat{T}_2(2))]|\mathrm{nn}\rangle \\
&\quad = \langle \mathrm{nn}|\hat{H}_{\text{strong}}|\mathrm{nn}\rangle \quad .
\end{aligned} \tag{6}$$

Hence, we conclude that the interaction of two protons is equal to the interaction of two neutrons.

EXAMPLE

5.4 The Pion Triplet

Three pions with the following masses (energies) and charges are known from experiment (see table below):

Properties of the pions

Pion	Mass m_0c^2	Mass difference [MeV]	Charge	Lifetime [s]	Spin	Magn. moment
π^+	139.59	4.59	e	$(2.55 \pm 0.03) \times 10^{-8}$	0	0
π^0	135.00	0	0	0.83×10^{-16}	0	0
π^-	139.59	4.59	$-e$	$(2.55 \pm 0.03) \times 10^{-8}$	0	0

Clearly, all pion masses (energies) are nearly equal. In analogy to the mass difference between proton and neutron, this near equality may be interpreted as meaning that the strong interaction (which determines the dominant part of the mass) is invariant in isospin space, and that the small mass differences of a few MeV ($\Delta m = 4.59$ MeV) are caused by the electromagnetic or other interactions. Indeed, the Coulomb energy of a homogeneously charged sphere with radius $r_0 = \hbar/m_\pi c$ (= Compton wave length of the pion) can easily be calculated. One obtains[6] with the finestructure constant $e^2/\hbar c = 1/137$,

$$
\begin{aligned}
|E_c| &= \frac{3}{5}\frac{e^2}{r_0} = \frac{3}{5}\frac{e^2}{\hbar c} m_\pi c^2 \\
&= \frac{3}{5}\frac{1}{137} \times 139 \text{ MeV} \simeq \frac{3}{5} \text{ MeV} \quad .
\end{aligned} \tag{1}
$$

Therefore, we will disregard the small mass difference during the investigation of the strong interaction. We will interpret the three pions as an *iso-triplet* or *charge triplet* (which means the same thing). This suggests the identification

$$
\begin{aligned}
|T = 1, T_3 = 1\rangle &= -|\pi^+\rangle \quad , \\
|T = 1, T_3 = 0\rangle &= |\pi^0\rangle \quad , \\
|T = 1, \hat{T}_3 = -1\rangle &= +|\pi^-\rangle \quad .
\end{aligned} \tag{2}
$$

[6] See J.D. Jackson: *Classical Electrodynamics*, 2nd ed. (Wiley, New York 1975) or W. Greiner: *Theoretische Physik III, Klassische Elektrodynamik* (Harri Deutsch, Frankfurt 1986)

Example 5.4

The choice of phases on the rhs of (2) is arbitrary, but once taken it has to be maintained. The fact that we did not choose the same phase for all pion states but used the factor (-1) in (2) for the positive pions has a profound reason, which we will now explain:

Since the Lie algebra of the isospin group is isomorphic to the angular momentum algebra, it follows, as in (2.18a, b), that

$$\hat{T}_\pm |TT_3\rangle = [T(T+1) - T_3(T_3 \pm 1]^{1/2} |TT_3 \pm 1\rangle \quad ,$$
$$\hat{T}_0 |TT_3\rangle = T_3 |TT_3\rangle \quad , \tag{3}$$

where $\hat{T}_\pm = \hat{T}_1 \pm \mathrm{i}\hat{T}_2$ and $\hat{T}_0 = \hat{T}_3$.

As we know, (3) follows directly from the commutation relations

$$[\hat{T}_3, \hat{T}_\pm]_- = \pm \hat{T}_\pm \quad , \quad [\hat{T}_+, \hat{T}_-]_- = 2\hat{T}_3 \quad , \tag{4}$$

which, in turn, stem from (5.13) with $\hat{T}_\pm = \hat{T}_1 \pm \mathrm{i}\hat{T}_2$ and $\hat{T}_0 = \hat{T}_3$. They are identical to the commutation relations of the angular momentum operators (2.7). From (3) follow the relations

$$\hat{T}_+|11\rangle = 0 \quad , \quad \hat{T}_+|10\rangle = \sqrt{2}|11\rangle \quad , \quad \hat{T}_+|1-1\rangle = \sqrt{2}|10\rangle \quad , \tag{5a}$$

$$\hat{T}_3|11\rangle = 1|11\rangle \quad , \quad \hat{T}_3|10\rangle = 0 \quad , \quad \hat{T}_3|1-1\rangle = -1|1-1\rangle \quad , \tag{5b}$$

$$\hat{T}_-|11\rangle = \sqrt{2}|10\rangle \quad , \quad \hat{T}_-|10\rangle = \sqrt{2}|1-1\rangle \quad , \quad \hat{T}_-|1-1\rangle = 0 \quad , \tag{5c}$$

Using (2), these equations can be written as

$$\begin{aligned} &\hat{T}_+(\sqrt{2}|\pi^+\rangle) = 0 && [\hat{T}_+, \hat{T}_+]_- = 0 \\ &\hat{T}_+(|\pi^0\rangle) = -(\sqrt{2}|\pi^+\rangle) && [\hat{T}_+, \hat{T}_3]_- = -\hat{T}_+ \\ &\hat{T}_+(\sqrt{2}|\pi^-\rangle) = 2(|\pi^0\rangle) && [\hat{T}_+, \hat{T}_-]_- = 2\hat{T}_3 \end{aligned} \tag{6a}$$

$$\begin{aligned} &\hat{T}_3(\sqrt{2}|\pi^+\rangle) = +(\sqrt{2}|\pi^+\rangle) && [\hat{T}_3, \hat{T}_+]_- = +\hat{T}_+ \\ &\hat{T}_3(|\pi^0\rangle) = 0 && [\hat{T}_3, \hat{T}_3]_- = 0 \\ &\hat{T}_3(\sqrt{2}|\pi^-\rangle) = -(\sqrt{2}|\pi^-\rangle) && [\hat{T}_3, \hat{T}_-]_- = -\hat{T}_- \end{aligned} \tag{6b}$$

$$\begin{aligned} &\hat{T}_-(\sqrt{2}|\pi^+\rangle) = -2(|\pi^0\rangle) && [\hat{T}_-, \hat{T}_+]_- = -2\hat{T}_3 \\ &\hat{T}_-(|\pi^0\rangle) = (\sqrt{2}|\pi^-\rangle) && [\hat{T}_-, \hat{T}_3]_- = \hat{T}_- \\ &\hat{T}_-(\sqrt{2}|\pi^-\rangle) = 0 && [\hat{T}_-, \hat{T}_-]_- = 0 \quad . \end{aligned} \tag{6c}$$

In order to better illustrate the object under consideration, we noted on the rhs of (6) analogous formulae for the isospin algebra of (4). We may now understand the choice of the minus sign for positive pions in (2): it ensures the complete agreement of both sets of equations in (6). Clearly,

$$\hat{T}_+ \leftrightarrow \sqrt{2}|\pi^+\rangle \quad , \quad \hat{T}_3 \leftrightarrow |\pi^0\rangle \quad , \quad \hat{T}_- \leftrightarrow \sqrt{2}|\pi^-\rangle \tag{7}$$

Example 5.4

and

$$\hat{T}_\mu|\pi^\nu\rangle \leftrightarrow [\hat{T}_\mu, \hat{T}_\nu]_- \quad , \tag{8}$$

correspond to each other for $\mu, \nu = +, -$ (respectively 3). This correspondence is linear in the first element ($\hat{T}_\mu$) as well as in the second element ($|\pi^\nu\rangle$ and $\hat{T}_\nu$). Therefore, all transformations of both sides of (8) are isomorphic, so that a linear transformation among the $|\pi^\nu\rangle$ corrresponds to the same linear transformations among the $\hat{T}_\nu$ and vice versa. Consequently, the transformation of the generators of the isospin group from spherical representation into cartesian representation, i.e.

$$\begin{pmatrix} \hat{T}_1 \\ \hat{T}_2 \\ \hat{T}_3 \end{pmatrix} = \begin{pmatrix} \frac{1}{2} & (\hat{T}_+ + \hat{T}_-) \\ -\frac{1}{2}\mathrm{i} & (\hat{T}_+ - \hat{T}_-) \\ & \hat{T}_0 \end{pmatrix} \quad , \tag{9}$$

is connected with a transition of the commutation relations (4) into

$$[\hat{T}_i, \hat{T}_j]_- = \mathrm{i}\varepsilon_{ijk}\hat{T}_k \quad . \tag{10}$$

Analogously, we define the *cartesian representation of the pion multiplet*

$$\begin{pmatrix} |\pi_1\rangle \\ |\pi_2\rangle \\ |\pi_3\rangle \end{pmatrix} = \begin{pmatrix} 1/\sqrt{2} & (|\pi^+\rangle + |\pi^-\rangle) \\ -\mathrm{i}/\sqrt{2} & (|\pi^+\rangle - |\pi^-\rangle) \\ & |\pi^0\rangle \end{pmatrix} \quad . \tag{11}$$

From (10) we obtain, by virtue of this isomorphism,

$$\hat{T}_i|\pi_j\rangle = \mathrm{i}\varepsilon_{ijk}|\pi_k\rangle \quad . \tag{12}$$

This relation can be directly obtained from (3), using (9) and (11) The factor $1/\sqrt{2}$ in (11) is not identical to the analogous factor $\frac{1}{2}$ in (9), because the states $|\pi_i\rangle$ are normalized to unity,

$$\langle\pi_i|\pi_j\rangle = \delta_{ij} \quad . \tag{13}$$

5.2 General Properties of Representations of a Lie Algebra

By Eq. (6) of Example 5.4 we became acquainted with a construction that holds for every Lie algebra and is called the *regular* (*or adjoint*) *representation of a Lie algebra*. To understand it, we will first explain the term *representation of a Lie algebra*: In complete correspondence to the representation of a group, the representation of a Lie algebra is defined as a mapping of the algebra onto the linear operators of a vector space, i.e. operators (matrices) $\hat{D}(\hat{L}_i)$ are assigned to the elements of the Lie algebra L_i (generators of the Lie group)

$$\hat{L}_i \to \hat{D}(\hat{L}_i) \quad . \tag{5.25}$$

These operators have to satisfy the requirements of *linearity*,

$$\hat{D}(\alpha\hat{L}_i + \beta\hat{L}_j) = \alpha\hat{D}(\hat{L}_i) + \beta\hat{D}(\hat{L}_j) \quad , \tag{5.26}$$

and must be homomorphic to the Lie algebra

$$\hat{D}([\hat{L}_i, \hat{L}_j]_-) = [\hat{D}(\hat{L}_i), \hat{D}(\hat{L}_j)]_- \quad . \tag{5.27}$$

The last equation states that the operator $\hat{D}([\hat{L}_i, \hat{L}_j]_-)$, assigned to the commutator $[\hat{L}_i, \hat{L}_j]_-$, has to be equal to the commutator of the $\hat{D}$-operators assigned to $\hat{L}_i$ and $\hat{L}_j$. In general, a representation in the vector space with the basis $\{|\phi_k\rangle\}$ is obtained by assigning to every operator $\hat{L}_i$, by means of

$$\hat{L}_i|\phi_j\rangle = D(\hat{L}_i)_{kj}|\phi_k\rangle \tag{5.28}$$

(note the summation convention), a matrix

$$D(\hat{L}_i)_{kj} = \langle\phi_k|\hat{L}_i|\phi_j\rangle = (\hat{L}_i)_{kj} \quad . \tag{5.29}$$

Hence, the $D(\hat{L}_i)_{kj}$ are the matrix elements of the operator $\hat{L}_i$ in the representation $\{|\phi_k\rangle\}$. They satisfy the requirement

$$\hat{D}(\hat{L}_i)\cdot\hat{D}(\hat{L}_j) = \hat{D}(\hat{L}_i\hat{L}_j) \quad , \tag{5.30}$$

which, in detail, reads (with summation over repeated indices m)

$$\hat{D}(\hat{L}_i)_{nm}\hat{D}(\hat{L}_j)_{mk} = D(\hat{L}_i\hat{L}_j)_{nk} \quad , \tag{5.31}$$

or, with (5.29),

$$\langle n|\hat{L}_i|m\rangle\langle m|\hat{L}_j|k\rangle = \langle n|\hat{L}_i\hat{L}_j|k\rangle \quad .$$

Hence, the matrix obtained by simple matrix multiplication of $D(\hat{L}_i)$ and $D(\hat{L}_j)$ is equal to the matrix $D(\hat{L}_i\hat{L}_j)$, assigned to the operator $\hat{L}_i\hat{L}_j$. Equations (5.26) and (5.27) must be satisfied too. But that is automatically the case with (5.28) and (5.30)

$$\begin{aligned} D(\alpha\hat{L}_i + \beta\hat{L}_j)_{nm} &= \langle n|\alpha\hat{L}_i + \beta\hat{L}_j|\mathrm{m}\rangle \\ &= \alpha\langle n|\hat{L}_i|m\rangle + \beta\langle n|\hat{L}_j|m\rangle \\ &= \alpha D(\hat{L}_i)_{nm} + \beta D(\hat{L}_j)_{nm} \quad , \end{aligned} \tag{5.32}$$

where, from (5.32), follows

$$\hat{D}([\hat{L}_i, \hat{L}_j]_-) = \hat{D}(\hat{L}_i\hat{L}_j - \hat{L}_i\hat{L}_j) = \hat{D}(\hat{L}_i\hat{L}_j) - \hat{D}(\hat{L}_i\hat{L}_j) \quad ,$$

and, taking into account (5.30),

$$\hat{D}([\hat{L}_i, \hat{L}_j]_-) = \hat{D}(\hat{L}_i)\hat{D}(\hat{L}_j) - \hat{D}(\hat{L}_j)\hat{D}(\hat{L}_i) = [\hat{D}(\hat{L}_i), \hat{D}(\hat{L}_j)]_- \quad . \tag{5.33}$$

5.3 Regular (or Adjoint) Representation of a Lie Algebra

One possible vector space $\{|\phi_k\rangle\}$, on which the generators $\hat{L}_i$ can be applied, is that with the vectors

$$|L_j\rangle \quad . \tag{5.34}$$

There have to be the same number of such vectors as there are generators and with the requiement that, by analogy to the commutation relations of Lie algebra

$$[\hat{L}_i, \hat{L}_j]_- = C_{ijk}\hat{L}_k \quad , \tag{5.35}$$

the relations

$$\hat{L}_i|L_j\rangle \overset{\text{def}}{\equiv} C_{ijk}|L_k\rangle \quad \text{and} \tag{5.36}$$

$$\langle L_i|L_j\rangle \overset{\text{def}}{\equiv} \delta_{ij} \tag{5.37}$$

shall hold, we define the action of the operators in this vector space. Equation (5.36) defines the effect of the operator $\hat{L}_i$ on the vector $|\hat{L}_j\rangle$ and (5.37) defines the inner product. From the correspondence of (5.36) and (5.28), it follows that

$$D(\hat{L}_i)_{kj} = C_{ijk} \qquad \text{and} \tag{5.38}$$

$$D(\alpha\hat{L}_i)_{kj} = \alpha C_{ijk} \quad . \tag{5.39}$$

Therefore, the structure constants themselves are a matrix representation of the generators, and the matrix of the generator $(\hat{L}_i)_{kj}$ is, according to (5.38), given by C_{ijk}. Now we want to ascertain in detail that the relations (5.30, 31) and (5.32) really do hold, i.e. that there are no contradictions to (5.36) and (5.38).

Equation (5.31) can easily be proven:

$$\begin{aligned} D(\hat{L}_i)_{nm}D(\hat{L}_j)_{mk} &= D(\hat{L}_i\hat{L}_j)_{nk} \quad . \\ C_{imn}C_{jkm} &= \langle L_n|\hat{L}_i\hat{L}_j|L_k\rangle = \langle L_n|\hat{L}_i C_{jkm}|L_m\rangle \\ &= C_{jkm}\langle L_n|C_{iml}|L_l\rangle = C_{jkm}C_{iml}\delta_{nl} \\ &= C_{jkm}C_{imn} \quad . \end{aligned} \tag{5.40}$$

Next we prove (5.33):

$$D(\hat{L}_i\hat{L}_j - \hat{L}_j\hat{L}_i)_{nk} = (\hat{D}(\hat{L}_i)\hat{D}(\hat{L}_j) - \hat{D}(\hat{L}_j)\hat{D}(\hat{L}_i))_{nk} \quad , \tag{5.41}$$

which, with (5.35, 38–40), can be brought into the form

$$D(C_{ijm}\hat{L}_m)_{nk} = \hat{D}(\hat{L}_i)_{nm}\hat{D}(\hat{L}_j)_{mk} - \hat{D}(\hat{L}_j)_{mn}\hat{D}(\hat{L}_i)_{mk} \quad \text{or}$$

$$C_{ijm}C_{mkn} = C_{imn}C_{jkm} - C_{jmn}C_{ikm} \quad ; \quad \text{hence} \quad , \tag{5.42}$$

$$C_{ijm}C_{mkn} + C_{jkm}C_{\min} + C_{kim}C_{mjn} = 0 \quad . \tag{5.43}$$

In the penultimate step, (5.42), we used the evident linearity of (5.35) and in the final step we used the antisymmetry in the first two indices of $C_{ijk} = -C_{jik}$ [see

(3.14)] several times. The relation (5.43) is identically satisfied, as we deduced from the Jacobi identity [see (3.16)]. Hence, (5.41) holds for the special representation (5.35) and (5.36). It is called the *regular representation or adjoint representation.*

The analogy of both sides of Eq. (6) of Example 5.4 is, therefore, of a profound nature: It holds for all Lie algebras. Therefore, this regular (or adjoint) representation can be constructed for every Lie algebra. We have already illustrated this theorem, in (6) of the last example, for the special case of the isospin algebra.

EXERCISE

5.5 Normalization of the Group Generators

Problem. Show that the generators $\hat{L}_i$ of a unitary matrix group, i.e. $U^{-1} = U^\dagger$ for each group element U, can be chosen in such a way that the relation $\mathrm{Tr}(\hat{L}_i\hat{L}_j) = \delta_{ij}/2$ is valid. Show, further, that the resulting strucure constants are purely imaginary and totally antisymmetric in this case.

Solution. According to (3.8) the generators of a unitary group are hermitian, i.e. $\hat{L}_i^\dagger = \hat{L}_i$. With the exception of this property they have been chosen arbitrarily up to now. We investigate the expression

$$\gamma_{ik} = \mathrm{Tr}(\hat{L}_i\hat{L}_k) \tag{1}$$

which, due to the invariance of the trace with respect to cyclic permutation of the matrices, can be written

$$\gamma_{ik} = \mathrm{Tr}(\hat{L}_i\hat{L}_k) = \mathrm{Tr}(\hat{L}_k\hat{L}_i) = \gamma_{ki} \quad . \tag{2}$$

Since the trace of the matrix $\hat{A}$ equals the trace of the transposed matrix $\hat{A}^\mathrm{T}$, we find that

$$\begin{aligned}\gamma_{ik}^* &= \mathrm{Tr}(\hat{L}_i^*\hat{L}_k^*) = \mathrm{Tr}[(\hat{L}_i^*\hat{L}_k^*)^\mathrm{T}] = \mathrm{Tr}[(\hat{L}_i^*)^\mathrm{T}(\hat{L}_k^*)^\mathrm{T}] \\ &= \mathrm{Tr}[\hat{L}_i^\dagger\hat{L}_k^\dagger] = \mathrm{Tr}(\hat{L}_k\hat{L}_i) = \gamma_{ik} \quad . \end{aligned} \tag{3}$$

Hence, the γ_{ik} can be written as the components of a real symmetric matrix. Such a matrix can be diagonalized by an orthogonal transformation, i.e.

$$\sum_{j,k} R_{ij}\gamma_{jk}R_{kl}^{-1} = \lambda_i\delta_{il} \quad , \tag{4}$$

where λ_i are the eigenvalues of the matrix, and

$$\sum_j R_{ij}R_{kj} = \delta_{ik} \Rightarrow R_{kl}^{-1} = R_{lk} \tag{5}$$

Exercise 5.5

from which

$$\sum_{j,k} R_{ij}\gamma_{jk}R_{lk} = \lambda_i\delta_{il} \tag{6}$$

follows. We can also choose any linear combination (with real coefficients) of the operators $\hat{L}_i$ for our group. Hence, we can define a new set of generators by

$$\hat{F}_j = \sum_i R_{ij}\hat{L}_i \quad , \quad \hat{L}_i = \sum_j R_{ij}\hat{F}_j \quad , \tag{7}$$

where the second relation follows from (5). Since the R_{ij} are real, then the generators $\hat{F}_i$ are also hermitian. Now we find that, according to (6),

$$\begin{aligned} \mathrm{Tr}(\hat{F}_i\hat{F}_l) &= \sum_{jk} R_{ij}R_{lk}\,\mathrm{Tr}(\hat{L}_j\hat{L}_k) \\ &= \sum_{jk} R_{ij}R_{lk}\gamma_{jk} = \lambda_i\delta_{il} \quad . \end{aligned} \tag{8}$$

Additionally, we can show that the eigenvalues λ_i are greater than zero. Denoting the matrix indices of $\hat{F}_i$ by greek letters we obtain from (8) for $i = l$,

$$\begin{aligned} \lambda_i = \mathrm{Tr}(\hat{F}_i\hat{F}_i) &= \mathrm{Tr}(\hat{F}_i\hat{F}_i^\dagger) = \sum_{\alpha\beta}(F_i)_{\alpha\beta}(F_i^\dagger)_{\beta\alpha} \\ &= \sum_{\alpha\beta}(F_i)_{\alpha\beta}(F_i)^*_{\alpha\beta} = \sum_{\alpha\beta}|(F_i)_{\alpha\beta}|^2 > 0 \quad , \end{aligned} \tag{9}$$

because at least one component of $\hat{F}_i$ is not zero. Hence, we normalize the new generators by

$$\hat{F}_i \to \hat{T}_i = \frac{1}{\sqrt{2\lambda_i}}\hat{F}_i \tag{10}$$

and, thereby, we obtain

$$\mathrm{Tr}(\hat{T}_i\hat{T}_j) = \tfrac{1}{2}\delta_{ij} \quad , \tag{11}$$

which is the required result. The structure constants are totally antisymmetric, as can be shown by

$$\begin{aligned} [\hat{T}_i, \hat{T}_j]_- &= \sum_k C_{ijk}\hat{T}_k \\ &\Rightarrow \mathrm{Tr}\{[\hat{T}_i, \hat{T}_j]\hat{T}_l\} = \sum_k C_{ijk}\,\mathrm{Tr}(\hat{T}_k\hat{T}_l) \\ &= \tfrac{1}{2}\sum_k C_{ijk}\delta_{kl} = \tfrac{1}{2}C_{ijl} \quad . \end{aligned} \tag{12}$$

Due to the invariance of the trace with respect to permutation, we write

$$\begin{aligned} \mathrm{Tr}\{[\hat{T}_i, \hat{T}_j]\hat{T}_l\} &= \mathrm{Tr}\{\hat{T}_i\hat{T}_j\hat{T}_l - \hat{T}_j\hat{T}_i\hat{T}_l\} \\ &= \mathrm{Tr}\{\hat{T}_j\hat{T}_l\hat{T}_i - \hat{T}_l\hat{T}_j\hat{T}_i\} \\ &= \mathrm{Tr}\{[\hat{T}_j, \hat{T}_l]\hat{T}_i\} = \tfrac{1}{2}C_{jli} \quad . \end{aligned} \tag{13}$$

Exercise 5.5

Comparison with (12) yields

$$C_{ijl} = C_{jli} \quad . \tag{14}$$

Because C_{ijl} is antisymmetric with respect to i and j, this is also valid for C_{jli} and $C_{lji} = -C_{jli}$. Thus, the structure constants are antisymmetric with respect to the exchange of two arbitrary indices. Finally, we show that the C_{ijk} are purely imaginary. By hermitian conjugation of the equation

$$\hat{T}_i\hat{T}_j - \hat{T}_j\hat{T}_i = C_{ijk}\hat{T}_k \quad , \tag{15}$$

we find, due to the hermiticity of the $\hat{T}_i$, that

$$\begin{aligned}(C_{ijk})^*\hat{T}_k &= (\hat{T}_i\hat{T}_j - \hat{T}_j\hat{T}_i)^\dagger = \hat{T}_j^\dagger\hat{T}_i^\dagger - \hat{T}_i^\dagger\hat{T}_j^\dagger \\ &= \hat{T}_j\hat{T}_i - \hat{T}_i\hat{T}_j = C_{jik}\hat{T}_k = -C_{ijk}\hat{T}_k \quad ;\end{aligned} \tag{16}$$

hence,

$$(C_{ijk})^* = -C_{ijk} \tag{17}$$

follows, which means that the C_{ijk} are purely imaginary, i.e. $C_{ijk} = \mathrm{i}f_{ijk}$, with real f_{ijk}. For the metric tensor we obtain

$$g_{ij} = \sum_{kl} C_{ikl}C_{jlk} = -\sum_{kl} f_{ikl}f_{jlk} = \sum_{kl} f_{ikl}f_{jkl} \quad . \tag{18}$$

Let us assume that $\det(g_{ij}) = 0$. Thus, a real eigenvector μ_j exists, which fulfills

$$\sum_j g_{ij}\mu_j = 0 \tag{19}$$

and, thereby, also

$$0 = \sum_{ij} \mu_i g_{ij}\mu_j = \sum_{ijkl} \mu_i f_{ikl} f_{jkl}\mu_j = \sum_{kl}\left(\sum_i \mu_i f_{ikl}\right)^2 \geq 0 \quad . \tag{20}$$

The expression only becomes zero if

$$\sum_i \mu_i f_{ikl} = 0 \tag{21}$$

is valid. This implies that

$$\left[\sum_j \mu_j\hat{T}_j, \hat{T}_k\right] = \mathrm{i}\sum_j u_j f_{jkl}\hat{T}_l = 0 \quad . \tag{22}$$

Accordingly, the linear combination $\sum_j \mu_j\hat{T}_j$ commutes with all other matrices, and thereby, it is a generator for an independent U(1)-subgroup. Hence, it follows Cartan's criterion, i.e. that a unitary matrix group is semi-simple only in the case that no such subgroup exists.

5.4 Transformation Law for Isospin Vectors

The canonical form of the regular representation of isospin algebra is [see Example 5.4, Eq. (12)]

$$\hat{T}_i|\pi_j\rangle = \mathrm{i}\varepsilon_{ijk}|\pi_k\rangle \quad . \tag{5.44}$$

It shall now be used in another illustration of an isospin transformation, in the so called *iso-space* (isospin space). If we consider an infinitesimal isospin transformation, with parameters $\delta\varepsilon_i$, then, from (5.14) and (5.44) it follows that

$$|\pi'_j\rangle = \mathrm{e}^{-\mathrm{i}\delta\varepsilon_i\hat{T}_i}|\pi_j\rangle \simeq (\mathbb{1} - \mathrm{i}\delta\varepsilon_i\hat{T}_i)|\pi_j\rangle = |\pi_j\rangle + \delta\varepsilon_i\varepsilon_{ijk}|\pi_k\rangle$$

or

$$|\pi'_j\rangle = |\pi_j\rangle - \varepsilon_{jik}\delta\varepsilon_i|\pi_k\rangle + \mathrm{O}(\delta\varepsilon^2) \quad . \tag{5.45}$$

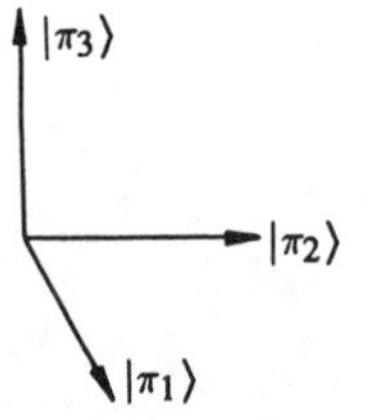

Fig. 5.1. The vectors $|\pi^k\rangle$ span the isospin space

These equations describe infinitesimal, real and orthogonal transformations in a three-dimensional real space, i.e. we are dealing with rotations, as one could already have gathered from the isomorphism between the angular momentum algebra [Lie algebra of O(3)] and the isospin algebra and, hence, from the isomorphism between the rotation group SO(3) and the isospin group SU(2). This space, the *iso-space* $\mathbb{R}^3$, is spanned by the three orthogonal vectors (see Fig. 5.1)

$$\{|\pi_1\rangle, |\pi_2\rangle, |\pi_3\rangle\} = \mathbb{R}^3 \quad , \quad \langle\pi_i|\pi_j\rangle = \delta_{ij} \quad . \tag{5.46}$$

It is *not identical* with the three-dimensional configuration space of daily experience, $\mathbb{R}^3$, *but* it is *isomorphic* to it. It is just this isomorphism that enables us to work out many of the properties of the isospin space and even to understand them. So it is natural to define the iso-vector v, in analogy to vectors in $\mathbb{R}^3$, as some linear combination of the three basis vectors (5.46), i.e.

$$\boldsymbol{v} = v_k|\pi_k\rangle \quad , \tag{5.47}$$

where the components v_k are real. From (5.45) the *transformation properties of the components v_k of the isospin vectors* can be obtained,

$$v'_j|\pi'_j\rangle = v_\alpha|\pi_\alpha\rangle = v_\alpha(|\pi'_\alpha\rangle + \varepsilon_{\alpha ik}\delta\varepsilon_i|\pi'_k\rangle) \quad , \tag{5.48}$$

and thus after projection with $\langle\pi'_j|$

$$v'_j = v_j - \varepsilon_{ji\alpha}\delta\varepsilon_i v_\alpha \quad . \tag{5.49}$$

This transformation (5.49) corresponds to a *passive rotation*, since we did not change the iso-vector in (5.48), but rotated the basis according to (5.45). To obtain an *active rotation* in iso-space we only have to reverse the rotation vector, i.e. $\delta\varepsilon_i \to -\delta\varepsilon_i$, changing (5.49) into

$$v'_i = v_i + \varepsilon_{ijk}\delta\varepsilon_j v_k \quad . \tag{5.50}$$

Compare the active rotation (1.61) with this! An additional factor -1 occurs in the second part on the rhs compared to (5.49). With the rotation vector

$\delta\boldsymbol{\varepsilon} = \{\delta\varepsilon_i\}$, equation (5.50) may be written concisely as

$$\boldsymbol{v}' = \boldsymbol{v} + \delta\boldsymbol{\varepsilon} \times \boldsymbol{v} = \boldsymbol{v} + \delta\boldsymbol{v} \quad , \quad \text{with} \tag{5.51}$$

$$\delta\boldsymbol{v} = \boldsymbol{v}' - \boldsymbol{v} = \delta\boldsymbol{\varepsilon} \times \boldsymbol{v} \quad . \tag{5.52}$$

This emphasizes the analogy to normal rotations in $\mathbb{R}^3$ (see Fig. 5.2), if we remember (1.60). Furthermore, it is easy to prove that

$$\boldsymbol{v}'^2 = \boldsymbol{v}^2 + 2\boldsymbol{v}\cdot\delta\boldsymbol{v} + (\delta\boldsymbol{v})^2$$

and, neglecting second order terms,

$$\delta(\boldsymbol{v}^2) = \boldsymbol{v}'^2 - \boldsymbol{v}^2 = 2\delta\boldsymbol{v}\cdot\boldsymbol{v} = 2\varepsilon_{ijk}\delta\varepsilon_j v_k v_i = 0 \tag{5.53}$$

due to the antisymmetry of ε_{ijk}. The lengths of the iso-vectors remain unchanged (invariant) by rotations in isospin space, a property that is again illuminated by the isomorphism with usual rotations in $\mathbb{R}^3$ (see Fig. 5.2).

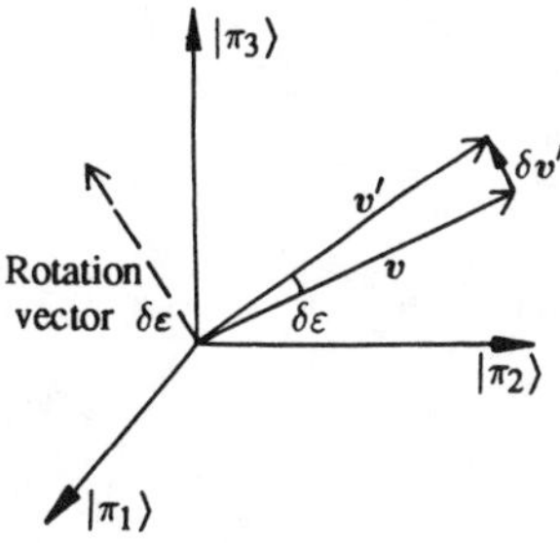

Fig. 5.2. Rotations in isospin space

In isospin space one often deals with iso-vectors, the components of which are not real numbers, but operators in Hilbert space (i.e. they act on states $|\varphi_v\rangle$ of a Hilbert space). The isospin operators $\hat{\boldsymbol{T}} = \{\hat{T}_1, \hat{T}_2, \hat{T}_3\}$ themselves are examples. By comparison the *vector operator* [see e.g. (1.60)] of the $\mathbb{R}^3$ we are led to the *iso-vector operator* $\boldsymbol{V}$. The transformation property is defined with a view to (5.50) as

$$\delta\boldsymbol{V} = \hat{\boldsymbol{V}}' - \hat{\boldsymbol{V}} = \delta\hat{\boldsymbol{\varepsilon}} \times \hat{\boldsymbol{V}} \tag{5.54}$$

or, in components,

$$\delta\hat{V}_i = \hat{V}_i' - \hat{V}_i = \varepsilon_{ijk}\delta\varepsilon_j\hat{V}_k \quad . \tag{5.55}$$

On the other hand an *operator* is transformed according to the general laws of quantum mechanics by

$$\begin{aligned}\hat{V}_i' &= \hat{U}_{\text{Is}}(\delta\varepsilon)\hat{V}_i\hat{U}_{\text{Is}}^{-1}(\delta\varepsilon) = \mathrm{e}^{-\mathrm{i}\delta\varepsilon_k\hat{T}_k}\hat{V}_i\mathrm{e}^{\mathrm{i}\delta\varepsilon_k\hat{T}_k}\\ &= (1 - \mathrm{i}\delta\varepsilon_k\hat{T}_k)\hat{V}_i(1 + \mathrm{i}\delta\varepsilon_k\hat{T}_k) = \hat{V}_i - \mathrm{i}\delta\varepsilon_k[\hat{T}_k, \hat{V}_i]_- \quad .\end{aligned} \tag{5.56}$$

This yields

$$\delta\hat{V}_i = \hat{V}_i' - \hat{V}_i = -\mathrm{i}\delta\varepsilon_j[\hat{T}_j, \hat{V}_i]_- \quad , \tag{5.57}$$

where the summation index has been renamed j. Comparison with (5.55) yields

$$[\hat{T}_j, \hat{V}_i]_- = \mathrm{i}\varepsilon_{ijk}\hat{V}_k \quad . \tag{5.58}$$

This equation *defines* an *iso-vector operator*. It states that for vector operators in iso-space the quantum mechanical transformation law (5.57) must be in accordance with the usual vector transformation law in iso-space (5.55).

To summarize: SU(2) transformations of the isospin group are orthogonal transformations in isospin space wich can be interpreted as rotations in the space of the regular $T = 1$ representation with, e.g., the basis $|\pi_1\rangle, |\pi_2\rangle, |\pi_3\rangle$. We have proven this only for infinitesimal transformations, but it can be generalized for finite $\boldsymbol{\varepsilon} = \{\varepsilon_i\}$, as we showed in (3.18). It should be noted that, because the spin is half-integral, there can occur ambiguities [see (5.14) and associated remarks, in connection with the usual spin].

We have constructed the isospin space on the basis of empirical experience: the charge doublet of the nucleons and the charge triplet of the pions were the starting points which led us to the abstract concept of iso-space. The isomorphism between this and configuration space $\mathbb{R}^3$ have been of great illustrative and practical value. Nevertheless, the difference between both is clear: while the $\mathbb{R}^3$ is physically realized (it led us to the rotation group and the angular momentum multiplets), the isospin space is a formal space which is realized in the symmetries of the elementary particles (charge multiplets). This formal iso-space is spanned by three vectors $|\pi_i\rangle$ ($i = 1, 2, 3$) [Example 5.4, Eq. (11)]. The state vectors of the charged pions

$$|\pi^{\pm}\rangle = \frac{1}{\sqrt{2}}\{|\pi_1\rangle \pm \mathrm{i}|\pi_2\rangle\} \quad , \tag{5.59a}$$

which have complex components, belong to the spherical representation of isospin space, and it is worth noting that solely such iso-vectors in spherical representation are physically realized, hence the states $|\pi^{\pm}\rangle$ and $|\pi^0\rangle$. Linear combinations like, for example,

$$|\pi_1\rangle = \frac{1}{\sqrt{2}}\{|\pi^+\rangle + |\pi^-\rangle\} \quad , \quad |\pi_2\rangle = \frac{-\mathrm{i}}{\sqrt{2}}\{|\pi^+\rangle - |\pi^-\rangle\} \quad , \tag{5.59b}$$

cannot be physically realized by any equipment, because these are not eigenstates of the charge operator $\hat{Q}$ [see (5.21)].

EXAMPLE

5.6 The *G*-Parity

We have just seen that the isospin symmetry may be treated in complete analogy to the symmetry of angular momentum. Rotations in the three-dimensional configuration space correspond to rotations in the three-dimensional isospin space. Therefore, we might ask whether there also exists a quantity in isospin space that is equivalent to parity in configuration space. Indeed, there is such an additional internal quantum number, as we will prove in the following. The property of *isospin parity* will enable us to explain the decay of the ω and ϱ mesons into three and two pions, respectively (see the discussion of the ω and ϱ mesons in Example 5.10). As we have seen in the previous section, the pion triplet $|\pi_j\rangle_{j=1,2,3}$ transforms like a vector in isospin space. Thus, it is straightforward to define the so-called *G*-parity as

$$\hat{G}|\pi_j\rangle = -|\pi_j\rangle \quad . \tag{1}$$

The property of this operator $\hat{G}$ in isospin space corresponds to the space reflection that is performed by the parity operator in configuration space, which inverts the directions of the unit vectors. i.e. $\hat{P}|e_j\rangle = -|e_j\rangle$. We choose the representation

Example 5.6

$$\hat{G} = e^{-i\pi\hat{T}_2}\hat{C} \quad , \tag{2}$$

where $\hat{C}$ denotes the *charge conjugation operator*,

$$\hat{C}|\pi^+\rangle = |\pi^-\rangle \quad , \quad \hat{C}|\pi^-\rangle = |\pi^+\rangle \quad , \quad \hat{C}|\pi^0\rangle = |\pi^0\rangle \quad . \tag{3}$$

The representation of G in (2) is consistent with the definition in (1), as we will now prove. From (3) we can infer the cartesian pion components (5.59),

$$\hat{C}|\pi_1\rangle = |\pi_1\rangle \quad , \quad \hat{C}|\pi_2\rangle = -|\pi_2\rangle \quad , \quad \hat{C}|\pi_3\rangle = |\pi_3\rangle \quad , \tag{4}$$

or, in matrix representation,

$$\hat{C} = \begin{pmatrix} 1 & 0 & 0 \\ 0 & -1 & 0 \\ 0 & 0 & 1 \end{pmatrix} \quad . \tag{5}$$

On the other hand, the rotation matrix in isospin space is given by

$$\hat{T}_2 = -i \begin{pmatrix} 0 & 0 & -1 \\ 0 & 0 & 0 \\ 1 & 0 & 0 \end{pmatrix}$$

$$e^{-i\pi\hat{T}_2} = \begin{pmatrix} 1 & 0 & 0 \\ 0 & 0 & 0 \\ 0 & 0 & 1 \end{pmatrix} \cos\pi + \begin{pmatrix} 0 & 0 & 0 \\ 0 & 1 & 0 \\ 0 & 0 & 0 \end{pmatrix}$$

$$= \begin{pmatrix} -1 & 0 & 0 \\ 0 & 1 & 0 \\ 0 & 0 & -1 \end{pmatrix} \quad . \tag{6}$$

The properties are illustrated in the following figure. Hence, it follows that

$$\hat{G}|\pi_j\rangle = \begin{pmatrix} -1 & 0 & 0 \\ 0 & 1 & 0 \\ 0 & 0 & -1 \end{pmatrix} \begin{pmatrix} 1 & 0 & 0 \\ 0 & -1 & 0 \\ 0 & 0 & 1 \end{pmatrix} |\pi_j\rangle = -|\pi_j\rangle \tag{7}$$

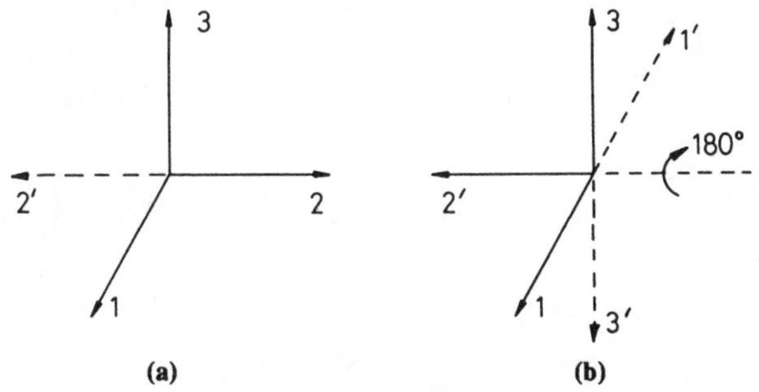

Action of C and $\exp(-i\pi T_2)$

Example 5.6

Reflection in Isospin Space: a) Action of the operator $\hat{C}$: it transforms the cartesian x_2 axis of the isospin space into the x'_2 axis.

b) Succeeded by rotation relative to the original x_2 axis by 180°: transformation of $x_1 \to x'_1$ and $x_3 \to x'_3$.

What is the physical meaning of the G-parity? We first postulate that the G-parity expresses an internal symmetry that represents a conserved quantity in particle reactions. Under this assumption (the possible reactions are restricted to the strong interactions) each meson has either positive or negative G-parity. Consequently, they can only decay into such combinations of other mesons that together also have positive or negative G-parity, respectively. Since the 2-pion system has positive and the 3-pion system negative G-parity, we can conclude from

$$\omega \to 3\pi$$

$\omega \nrightarrow 2\pi$ (through strong interaction)

$$\text{that } \hat{G}|\omega\rangle = -|\omega\rangle \tag{8}$$

and from

$$\varrho \to 2\pi \quad \varrho \nrightarrow 3\pi \quad \text{that } \hat{G}|\varrho\rangle = |\varrho\rangle \quad . \tag{9}$$

A certain value of the G-parity can be assigned to every meson, which enables one to understand many experimental observations.

Applying the G-operation (2) to a baryon state is not a symmetry operation, since charge conjugation leads out of a given multiplet (baryons and antibaryons, in fact, belong to different, conjugate multiplets – see the following Examples 6.3, 6.4). However, a definite G-parity may be assigned to the nucleon-antinucleon system in analogy to the positronium system. The neutral $N\bar{N}$ system is either in an isospin triplet state $|T=1, T_3=0\rangle$ or in a singlet state $|T=0, T_3=0\rangle$. Rotation around the T_2-axis, which is part of the $\hat{G}$ operation, yields a different sign for these two cases:

$$\begin{aligned}
e^{-i\pi\hat{T}_2}|T=1, T_3=0\rangle &= \exp\left[-i\frac{\pi}{2}(\hat{\tau}_2(1)+\hat{\tau}_2(2))\right]\frac{1}{\sqrt{2}}(|p\rangle_1|\bar{p}\rangle_2 + |n\rangle_1|\bar{n}\rangle_2) \\
&= [-i\hat{\tau}_2(1)][-i\hat{\tau}_2(2)]\frac{1}{\sqrt{2}}(|p\rangle_1|\bar{p}\rangle_2 + |n\rangle_1|\bar{n}\rangle_2) \\
&= \frac{1}{\sqrt{2}}[|n\rangle_1(-|\bar{n}\rangle_2) + (-|p\rangle_2)|\bar{p}\rangle_2] \\
&= -|T=1, T_3=0\rangle
\end{aligned} \tag{10}$$

$$\begin{aligned}
e^{-i\pi\hat{T}_2}|T=0, T_3=0\rangle &= [-i\hat{\tau}_2(1)][-i\hat{\tau}_2(2)]\frac{1}{\sqrt{2}}(|p\rangle_1|\bar{p}\rangle_2 - |n\rangle_1|\bar{n}\rangle_2) \\
&= \frac{1}{\sqrt{2}}[|n\rangle_1(-|\bar{n}\rangle_2) - (-|p\rangle_1)|\bar{p}\rangle_2] \\
&= \frac{1}{\sqrt{2}}[|p\rangle_1|\bar{p}\rangle_2 - |n\rangle_1|\bar{n}\rangle_2] \\
&= +|T=0, T_3=0\rangle \quad .
\end{aligned} \tag{11}$$

Example 5.6

Here we use the fact that the generators for group operations in multiplets of antiparticles are equal to the generators for group operations in multiplets of particles with opposite sign.

We will demonstrate this strange property later in Exercise 8.2 for the generators of the SU(3)-group.

For positronium one finds that charge parity

$$\hat{C}|e^+e^-\rangle = (-1)^J|e^+e^-\rangle$$

for a state with total angular momentum J. If we transfer this result to the $N\bar{N}$ system, we have

$$\hat{C}|N\bar{N}\rangle = (-)^J|N\bar{N}\rangle \quad . \tag{12}$$

Together with (10) and (4), we obtain

$$\hat{G}|N\bar{N}\rangle = (-)^{J+T}|N\bar{N}\rangle \quad . \tag{13}$$

The G-parity of the $N\bar{N}$ system has been tested experimentally at CERN, where low-energy antiprotons are captured by protons. Analyzing the resulting decays with respect to the assignment of (13), yields the followng selection rules:

$$|N\bar{N}(J=0, T=0)\rangle \not\to 3\pi$$

$$|N\bar{N}(J=0, T=1)\rangle \not\to 2\pi \quad , \tag{14}$$

indicating that the $T=0$ state has positive G-parity, whereas that of the $T=1$ state must be negative. The excellent agreement with the obtained experimental results confirms the concept of the G-parity.

The fact that the G-parity operation transforms a nucleon into an antinucleon enables one to connect the interaction between a nucleon and a nucleus with the corresponding antinucleon-nucleus interaction. The effective antinucleon potential $V_{\rm eff}(r)$ turns out to be the G-conjugate nucleon-nucleus potential. $V_{\rm eff}(r)$ is a composition of the contributions of the mesons π, ϱ, ω, ..., i.e.

$$V^{(N)}_{\rm eff}(r) = V^\pi(r) + V^\varrho(r) + V^\omega(r) + \cdots \tag{15}$$

The G-parity of the mesons ϱ and ω can be determined from their pionic decays, utilizing the fact that the pion has negative G-parity. Since the ω-meson decays into 3 pions, its G-parity is negative, while that of the ϱ-meson, which decays into 2 pions, is positive. Therefore

$$\begin{aligned} V^{(\bar{N})}_{\rm eff}(r) &= \hat{G}V^{(N)}_{\rm eff}(r)\hat{G} \\ &= G_\pi V^\pi(r) + G_\varrho V^\varrho(r) + G_\omega V^\omega(r) + \cdots \\ &= -V^\pi(r) + V^\varrho(r) - V^\omega(r) + \cdots \quad , \end{aligned} \tag{16}$$

which has also been confirmed experimentally. In particular, the potential V^ω due to the ω meson exchange, which is strongly repulsive at short range in the NN system, becomes strongly attractive in the $N\bar{N}$ system. It has been conjectured that this may allow for strongly bound, highly localized $N\bar{N}$ states (so called baryonium resonances). Up to now such states have not been conclusively identified however.

EXERCISE

5.7 Representation of a Lie Algebra, Regular Representation of the Algebra of Orbital Angular Momentum Operators

Problem. a) Explain the concepts "representation of a Lie algebra" and "regular representation".

b) Find a regulr representation for the algebra of the orbital angular momentum operators.

Solution. (a) The expression "representation of a Lie algebra" $\mathscr{A}$ means that one can assign an $n \times n$ matrix $\hat{a}$ to all $\hat{A} \in \mathscr{A}$, such that the matrices fulfill the same algebraic relations (i.e. homomorphism to a matrix algebra). Hence, the requirement is for

$$\alpha\hat{A} + \beta\hat{B} \rightarrow \alpha\hat{a} + \beta\hat{b} \quad , \quad [\hat{A}, \hat{B}]_- \rightarrow [\hat{a}, \hat{b}]_- \quad .$$

The first relation is the scalar multiplication in $\mathscr{A}$, the second is the ordinary commutator of matrices. Obviously, there always exists a (trivial) representation, namely, $\hat{A} \rightarrow 0$ for all $\hat{A}$. If the algebra is defined by

$$[\hat{A}_i, \hat{A}_j]_- = c_{ijk}\hat{A}_k \quad ,$$

we can assign each $\hat{A}_i$ a matrix $\hat{a}_i$ with

$$(\hat{a}_i)_{kj} = c_{ijk}$$

and, similarly, for the sum $\sum \alpha_i \hat{A}_i$. This representation is called the *regular* or *adjoint representation.* Indeed, we find

$$\begin{aligned}([\hat{a}_i, \hat{a}_j]_-)_{lm} &= (\hat{a}_i\hat{a}_j - \hat{a}_j\,\hat{a}_i)_{lm} \\ &= (\hat{a}_i)_{lk}(\hat{a}_j)_{km} - (\hat{a}_j)_{lk}(\hat{a}_i)_{km}\end{aligned}$$

$$c_{ikl}\,c_{jmk} - c_{jkl}\,c_{imk} = c_{mjk}\,c_{kil} + c_{imk}\,c_{kjl} \quad ,$$

where we have used $c_{ijk} = -c_{jki}$. Using the Jacobi identity, it follows that

$$([\hat{a}_i, \hat{a}_j]_-)_{lm} = -c_{jik}\,c_{kml} = c_{ijk}\,c_{kml} = c_{ijk}(\hat{a}_k)_{lm} \quad .$$

Thus, we have proved that, in fact, the matrices fulfill the algebra

$$[\hat{a}_i, \hat{a}_j]_- = c_{ijk}\,\hat{a}_k \quad .$$

b) The orbital angular momentum operators satisfy the commutation relations

$$[\hat{L}_i, \hat{L}_j]_- = \mathrm{i}\varepsilon_{ijk}\,\hat{L}_k \quad .$$

The regular representation, therefore, is

$$(\hat{L}_i)_{jk} = i\varepsilon_{ikj} = -i\varepsilon_{ijk}$$

Exercise 5.7

or, explicitly,

$$\hat{L}_1 \to -\mathrm{i}\begin{pmatrix} 0 & 0 & 0 \\ 0 & 0 & 1 \\ 0 & -1 & 0 \end{pmatrix} \quad , \quad \hat{L}_2 \to -\mathrm{i}\begin{pmatrix} 0 & 0 & -1 \\ 0 & 0 & 0 \\ 1 & 0 & 0 \end{pmatrix} \quad ,$$

$$\hat{L}_3 \to -\mathrm{i}\begin{pmatrix} 0 & 1 & 0 \\ -1 & 0 & 0 \\ 0 & 0 & 0 \end{pmatrix} \quad .$$

These matrices are spin-1 matrices.

5.5 Experimental Test of Isospin Invariance

Up to now we know of two arguments for the isospin invariance of the strong interaction: first the very small differences between the masses of particles that belong to a certain charge multiplet (see tables for nucleons and pions) and secondly the charge independence of the nuclear forces (see Exercise 5.3). Both facts can be inferred from the hypothesis that the isospin group is a symmetry group of the strong interaction, i.e.

$$[\hat{H}_{\text{strong}}, \hat{T}]_- = 0 \quad ,$$

and from properties of the corresponding isospin Lie algebra. In the following we will discuss further experimental facts that support this hypothesis (isospin group = symmetry group of the strong interaction).

The use of the isomorphism between the isospin group and SO(3) (we have already used it in connection with the deuteron) enables us to infer the composition of a total isospin of two (or more) particles according to the rules of the angular momentum algebra, i.e.

$$|TT_3\rangle = \sum_{T_3(1)+T_3(2)=T_3} (T(1)T(2)T|T_3(1)T_3(2)T_3)|T(1)T_3(1)\rangle|T(2)T_3(2)\rangle \quad , \tag{5.60}$$

where the total isospin T is restricted to the values

$$T = T(1) + T(2),\ T(1) + T(2) - 1, \dots, |T(1) - T(2)| \quad .$$

$T(1)$ and $T(2)$ denote the isospins of particle 1 and particle 2, respectively. Correspondingly $T_3(1)$, $T_3(2)$ are the three-components of isospin.

The Clebsch-Gordan coefficients $(T(1)T(2)T|T_3(1)T_3(2)T_3)$ follow from the Lie algebra of the SO(3) (see Chap. 2). Since the isospin algebra is isomorphic to the angular momentum algebra, we can carry over all results of Chap. 2 to the

Table 5.2. Clebsch-Gordan coefficients for the coupling of 1) $T(1) = \frac{1}{2}$ and $T(2) = 1$, 2) $T(1) = \frac{1}{2}$ and $T(2) = \frac{1}{2}$

	$T = \frac{1}{2}$		$T = \frac{3}{2}$			
$\frac{1}{2} \otimes 1$	$T_3 = \frac{1}{2}$	$T_3 = -\frac{1}{2}$	$T_3 = \frac{3}{2}$	$T_3 = \frac{1}{2}$	$T_3 = -\frac{1}{2}$	$T_3 = -\frac{3}{2}$
$T_3(1) = \frac{1}{2}$	$\sqrt{\frac{1}{3}}$	$\sqrt{\frac{2}{3}}$	1	$\sqrt{\frac{2}{3}}$	$\sqrt{\frac{1}{3}}$	0
$T_3(1) = -\frac{1}{2}$	$-\sqrt{\frac{2}{3}}$	$-\sqrt{\frac{1}{3}}$	0	$\sqrt{\frac{1}{3}}$	$\sqrt{\frac{2}{3}}$	1

	$T = 0$	$T = 1$		
$\frac{1}{2} \otimes \frac{1}{2}$	$T_3 = 0$	$T_3 = 1$	$T_3 = 0$	$T_3 = -1$
$T_3(1) = \frac{1}{2}$	$\sqrt{\frac{1}{2}}$	1	$\sqrt{\frac{1}{2}}$	0
$T_3(1) = -\frac{1}{2}$	$-\sqrt{\frac{1}{2}}$	0	$\sqrt{\frac{1}{2}}$	1

case of isospin; the Clebsch-Gordan coefficients are identical for both algebras, and in Table 5.2 we have listed some of the most important of them[7].

Now let us consider the *decay of a particle* with isospin T into two other particles with isospins $T(1)$ and $T(2)$, respectively. We start from the matrix element of the $\hat{S}$ operator that connects the initial and final states[8],

$$\langle T(1)T_3(1);\, T(2)T_3(2)|\hat{S}|TT_3\rangle \quad . \tag{5.61}$$

Using the orthogonality relation for Clebsch-Gordan coefficients, together with (5.60), the inverse equation follows [see also Chap. 2, Eq. (2.39)] as

$$|T(1)T_3(1)\rangle|T(2)T_3(2)\rangle = \sum_{T'=|T(1)-T(2)|}^{T(1)+T(2)} (T(1)\,T(2)\,T'|T_3(1)T_3(2)T'_3)|T'\,T'_3\rangle \quad , \tag{5.62}$$

where $T'_3 = T_3(1) + T_3(2)$. Hence, the S-matrix element in (5.61) becomes

$$\langle T(1)T_3(1);\, T(2)T_3(2)|\hat{S}|TT_3\rangle = \sum_{T'} (T(1)T(2)T'|T_3(1)T_3(2)T'_3)\langle T'T'_3|\hat{S}|TT_3\rangle \quad . \tag{5.63}$$

The isospin invariance of the Hamiltonian implies the invariance of the operator $\hat{S}$, i.e.

$$[\hat{T}_j, \hat{H}]_- = 0 \quad , [\hat{T}_j, \hat{S}]_- = [\hat{T}_j, \exp(-\mathrm{i}\hat{H}t/\hbar)]_- = 0 \quad . \tag{5.64}$$

[7] See, for example, M. Rotenberg, R. Bivins, N. Metropolis and J.K. Wooten, Jr.: *The 3j- and 6j-symbols* (Technology press, Cambridge, Mass. 1959).

[8] Refer to Vol. 1 in this series, *Quantum Mechanics I-An Introduction* (Springer, Berlin, Heidelberg 1989) Chap. 10.

As we learned in Chap. 3, the isospin-space part of the hamiltonian consists of the $\hat{1}$ operator and the Casimir operator $\hat{\boldsymbol{T}}^2$ of the isospin group. In other words the Hamiltonian is a scalar in iso-space, i.e. an iso-scalar. Indeed we have learned earlier, that it must be built up from the Casimir operators, i.e. in our case $\hat{H} = \hat{H}(\hat{\boldsymbol{T}}^2)$ and hence $\hat{S} = \hat{S}(\hat{\boldsymbol{T}}^2)$. Therefore, it must hold that

$$\langle T'T'_3|\hat{S}|TT_3\rangle = \delta_{TT'}\delta_{T_3T'_3}\underbrace{\langle T\|\hat{S}\|T\rangle}_{f(T(1),\,T(2),\,T)} \quad . \tag{5.65}$$

The matrix element on the rhs of (5.65) $\langle T\|\hat{S}\|T\rangle$, is called the *reduced matrix element*. It is a function of the total isospins T, $T(1)$, $T(2)$ only and independent of the projection T_3. Hence, it is equal for all states with different T'_3 of a certain isospin multiplet. In other words: The reduced matrix element is characteristic of a multiplet with total isospin T, since it is independent of the quantum number T_3 which distinguishes between the states that belong to this multiplet. Accordingly, we are led to the equation

$$\begin{aligned}&\langle T(1)T_3(1);\,T(2)T_3(2)|\hat{S}|TT_3\rangle\\ &\quad = (T(1)T(2)T|\,T_3(1)T_3(2)T_3)\langle T(1)T(2)\|\hat{S}\|T\rangle \quad .\end{aligned} \tag{5.66}$$

Equations (5.65) and (5.66) are a special case of the so-called ***Wigner-Eckart*** *theorem.* [9] The implication of (5.66) is the following: Since the transition probabilities are proportional to the squared matrix element on the lhs of (5.66), i.e. proportional to

$$\begin{aligned}&|\langle T(1)T_3(1);\,T(2)T_3(2)|\hat{S}|TT_3\rangle|^2\\ &\quad = |(T(1)T(2)T|\,T_3(1)T_3(2)T_3)|^2|\langle T(1)T(2)\|\hat{S}\|T\rangle|^2 \quad ,\end{aligned} \tag{5.67}$$

then the ratio of intensities for the various possible charge combinations in the final state $|T(1)T_3(2);\,T(2)T_3(2)\rangle$ are determined by the squares of the Clebsch-Gordan coefficients. In other words, in taking the ratio of two possible decay modes, the reduced matrix element, that is generally unknown, drops out. This fact enables one to make quantitative predictions – in the absence of a complete dynamical theory of the strong interaction – that can be tested by measurement.

EXAMPLE

5.8 The Wigner–Eckart Theorem

Equation (5.65) is a special case of the Wigner-Eckart theorem. Here we want to present and prove the general theorem. It is valuable for a general class of tensor

[9] See textbooks on algebra of angular momentum, for example, M.E. Rose: *Elementary Theory of Angular Momentum* (John Wiley, New York 1957).

Example 5.8

operators which are defined, by means of commutator relations, in terms of the angular momentum or isospin operator.

We define: The $2k+1$ operators $\hat{T}_q^{(k)}, (q = -k, -k+1, \ldots, k)$ form the components of an irreducible tensor operator of rank k, if they fulfill the following commutator relations:

$$[\hat{J}_\pm, \hat{T}_q^{(k)}] = \sqrt{k(k+1) - q(q \pm 1)}\, T_{q\pm 1}^{(k)} \tag{1}$$

$$[\hat{J}_0, \hat{T}_q^{(k)}] = q\, T_q^{(k)} \tag{2}$$

The attribute irreducible indicates that the operators J_q only combine operators of the same rank. If J_q is connected with the ordinary angular momentum then the irreducible tensor operators are called spherical tensors; their transformation properties can be explained by spatial rotations. The definition given above, which uses commutator relations, is equivalent to this, and it has the advantage that the operators $\hat{J}_\pm$, $\hat{J}_0$ have – like $\hat{T}_\pm$, $\hat{T}_0$ within the isospinor space – a physical meaning.

The Wigner-Eckart theorem says that in a representation according to the operators $\hat{J}^2$, $\hat{J}_2$, where basis vectors are given by $|\tau jm\rangle$, the matrix element $\langle \tau' j' m' | \hat{T}_q^{(k)} | \tau jm\rangle$ of an irreducible tensor operator is given by the product of a so-called reduced matrix element $\langle \tau' j' m' \| \hat{T}^{(k)} \| \tau jm\rangle$, which does not depend on m, m' and q, and a Clebsch-Gordan coefficient

$$\langle \tau' j' m' | \hat{T}_q^{(k)} | \tau jm\rangle = (jkj' | mqm') \langle \tau' j' \| T^{(k)} \| \tau j\rangle \tag{3}$$

Here τ means quantum numbers which appertain to operators that do not commute with all $\hat{J}_q$.

In order to prove the Wigner-Eckart theorem we have a look at the $(2b+1) \times (2j+1)$ vectors

$$\hat{T}_q^{(k)} | \tau jm\rangle \tag{4}$$

and linear combinations of them

$$|\tau JM\rangle = \sum_{m,q} (jkJ | mq\, M)\, \hat{T}_q^{(k)} | \tau jm\rangle \quad . \tag{5}$$

Applying the operator $\hat{J}_\pm$ to (4), we get, using the commutator relations which define $\hat{T}_q^{(k)}$,

$$\begin{aligned} \hat{J}_\pm \hat{T}_q^{(k)} | \tau jm\rangle &= [\hat{J}_\pm, \hat{T}_q^{(k)}] | \tau jm\rangle + \hat{T}_q^{(k)} \hat{J}_\pm | \tau jm\rangle \\ &= \sqrt{k(k+1) - q(q \pm 1)}\, \hat{T}_{q\pm 1}^{(k)} | \tau jm\rangle \\ &\quad + \sqrt{j(j+1) - m(m \pm 1)}\, \hat{T}_q^{(k)} | \tau jm \pm 1\rangle \quad . \end{aligned} \tag{6}$$

Now we apply $J_\pm$ to the state $|\tau JM\rangle$ of Eq. (5)

$$\begin{aligned} J_\pm |\tau JM\rangle &= \sum_{m,q} \sqrt{k(k+1) - q(q \pm 1)}\, (jk\, J | mq\, M)\, \hat{T}_{q\pm 1}^{(k)} | \tau\, jm\rangle \\ &\quad + \sum_{mq} \sqrt{j(j+1) - m(m \pm 1)}\, (jk\, J | mq\, M)\, \hat{T}_q^{(k)} | \tau\, j\, m \pm 1\rangle \; . \end{aligned} \tag{7}$$

Changing $q \to q \mp 1$ and $m \to m \mp 1$ yields *Example 5.8*

$$\begin{aligned}\hat{J}_\pm |\tau JM\rangle &= \sum_{m,q} \hat{T}_q^{(k)}|\tau jm\rangle \big(\sqrt{k(k+1)-q(q\mp 1)}\,(jkJ|mq\mp 1\,M) \\ &\quad + \sqrt{j(j+1)-m(m\mp 1)}\,(jkJ|m\mp 1qM)\big) \quad . \end{aligned} \tag{8}$$

The expressions in the bracket are the recursion formula of Clebsch-Gordan coefficients (chapter 2, Eq. 2.45a,b)

$$\begin{aligned}&\sqrt{k(k+1)-q(q\mp 1)}\,(jkJ|mq\mp 1M) \\ &\quad + \sqrt{j(j+1)-m(m\mp 1)}\,(jkJ|m\mp 1qM) \\ &\quad = \sqrt{J(J+1)-M(M\pm 1)}\,(jkJ|mq\,M\pm 1)\,. \end{aligned} \tag{9}$$

Inserting this in (8) it follows

$$\begin{aligned}\hat{J}_\pm |\tau JM\rangle &= \sqrt{J(J+1)-M(M\pm 1)} \sum_{m,q} (jkJ|mq\,M\pm 1)\,\hat{T}_q^{(k)}\,|\tau jm\rangle \\ &= \sqrt{J(J+1)-M(M\pm 1)}\,|\tau JM\pm 1\rangle \quad . \end{aligned} \tag{10}$$

Now we apply $\hat{J}_0$ to (4) and (5) analogously to $J_\pm$

$$\begin{aligned}J_0\,\hat{T}_q^{(k)}|\tau jm\rangle &= [\hat{J}_0, \hat{T}_q^{(k)}]|\tau jm\rangle + \hat{T}_q^{(k)}\hat{J}_0|\tau jm\rangle \\ &= (q+m)\,\hat{T}_q^{(k)}\,|\tau jm\rangle \quad . \end{aligned} \tag{11}$$

Together with (5) this yields

$$\hat{J}_0|\tau JM\rangle = \sum_{m,q} (jkJ|mq\,M)(q+m)\,\hat{T}_q^{(k)}|\tau jm\rangle = M|\tau JM\rangle \quad . \tag{12}$$

Here we have used the fact that all Clebsch-Gordan coefficients $(jkJ|mq\,M)$ vanish if $q+m \neq M$. Eqs. (10) and (12) show that the states $|\tau JM\rangle$ fulfill the angular momentum algebra. Therefore they are unnormalized eigenfunctions of the operators $\hat{J}^2$ and $\hat{J}_2$. This means that the scalar products $\langle\tau' J' M'|\tau JM\rangle$ obey orthogonality:

$$\langle\tau' J' M'|\tau JM\rangle = \delta_{JJ'}\,\delta_{MM'}\langle\tau' JM|\tau JM\rangle \quad . \tag{13}$$

The reduced matrix element $\langle\tau' JM|\tau JM\rangle$ does not depend on M. This can be seen by inserting the ladder operator $\hat{J}_\pm$:

$$\begin{aligned}\langle\tau' JM|\tau JM\rangle &= (J(J+1)-M(M\mp 1))^{-1/2}\,\langle\tau' JM|\hat{J}_\pm|\tau' JM\mp 1\rangle \\ &= \langle\tau' JM\mp 1|\tau JM\mp 1\rangle \quad . \end{aligned} \tag{14}$$

Here we have applied $\hat{J}_\pm = \hat{J}_\pm^\dagger$ to the right and left side. Taking this into consideration we get for (13)

$$\langle\tau' J' M'|\tau JM\rangle = \delta_{JJ'}\,\delta_{MM'}\,\langle\tau' J|\tau J\rangle \quad . \tag{15}$$

With the above equation, the Wigner-Eckart theorem can be proved easily. We transform (5) by means of the orthogonality of the Clebsch-Gordan coefficients

Example 5.8

$$\hat{T}_q^{(k)}\,|\tau\, jm\rangle = \sum_{jm} (jk\,J|mq\,M)|\tau\, JM\rangle \quad . \tag{16}$$

Multiplication with $\langle\tau' j' m'|$ yields

$$\begin{aligned}\langle\tau' j' m'|\hat{T}_q^{(k)}|\tau jm\rangle &= \sum_{JM} |jk\,J|mq\,M)\langle\tau' j' m'|\tau\, JM\rangle \\ &= (jkj'|mq\,m')\langle\tau' j'|\tau\, J\rangle\end{aligned} \tag{17}$$

j' is the angular momentum that arises from j and k. Therefore, we can write

$$\langle\tau' j' m'|\hat{T}_q^{(k)}|\tau\, jm\rangle = (jkj'|mqm')\langle\tau' j'||T^{(k)}||\tau j\rangle \quad . \tag{18}$$

This proves the Wigner-Eckart theorem and gives the following selection rules:

The matrix element $\langle\tau' j' m'|\hat{T}_q^{(k)}|\tau jm\rangle$ is only non-zero if $q + m = M$ and if j, k, j' fulfill the triangular inequality. This is a direct consequence of properties of the Clebsch-Gordan coefficients.

The description of physical processes like radiative transitions in atomic physics, classical electrodynamics, nuclear transitions between excited nuclear states can be divided into two aspects.

1) The symmetry of the problem, which is contained in the selection rules, is given by the Clebsch-Gordan coefficients.
2) Other details of the problem are contained in the reduced matrix element.

Often one is only interested in ratios of two transition matrix elements where it is sufficient to regard only the Clebsch-Gordan coefficients.

As one of the simplest applications we consider the Wigner-Eckart theorem for the Hamiltonian of strong interaction. It is (cf. Example 3.15) a scalar in the iso-space and therefore in the basis $|tt_3\rangle$ it is an irreducible tensor operator of rank 0. The Wigner-Eckart theorem gives in this case:

$$\begin{aligned}\langle\tau' t' t_3'|\hat{H}|\tau\, tt_3\rangle &= \langle\tau' t' t_3'|\hat{H}|\tau tt_3\rangle \\ &= (t0\,t|t_3 0 t_3')\,\langle\tau' t'||\hat{H}||\tau t\rangle \\ &= \delta_{tt'}\,\delta_{t_3 t_3'}\langle\tau' t'||\hat{H}||\tau t\rangle\end{aligned} \tag{19}$$

This indeed is Eq. (5.65) .

EXAMPLE

5.9 Pion Production in Proton-Deuteron Scattering

In proton-deuteron scattering the following exit channels, i.e. reactions with well-defined quantum numbers, are possible (among many others):

$$p + d \begin{cases} \nearrow \pi^0 + {}^3\mathrm{He} \\ \searrow \pi^+ + {}^3\mathrm{H} \end{cases} \tag{1}$$

From the discussion in Example 5.2 we know that in the ground state the deuteron has isospin $T = 0$, while the proton has $T = \frac{1}{2}$. This implies that the initial state of the considered reaction bears isospin $T = 0 + \frac{1}{2} = \frac{1}{2}$. The final state consists of a pion with $T = 1$ and a ^{3}He with $T = \frac{1}{2}$, or of a pion and a ^{3}H that also carries isospin $T = \frac{1}{2}$. In fact, the mirror nuclei ^{3}He and ^{3}H form an iso-doublet. Thus, we may write

$$\begin{aligned} |\text{initial state}\rangle &= |\mathrm{p} + \mathrm{d}\rangle = |\tfrac{1}{2}\tfrac{1}{2}\rangle|0\,0\rangle \\ |\text{final state}\rangle &= |\pi^+ + {}^3\mathrm{H}\rangle = |1\,1\rangle|\tfrac{1}{2} - \tfrac{1}{2}\rangle \\ |\text{final state}\rangle &= |\pi^0 + {}^3\mathrm{He}\rangle = |10\rangle|\tfrac{1}{2}\tfrac{1}{2}\rangle \quad . \end{aligned} \tag{2}$$

The relation (5.67) now says that the ratio of the cross-sections of these two particular final states (1) is

$$R = \frac{\sigma(\mathrm{p} + \mathrm{d} \to \pi^+ + \mathrm{H}^3)}{\sigma(\mathrm{p} + \mathrm{d} \to \pi^0 + \mathrm{He}^3)} = \frac{|(1\,\tfrac{1}{2}\tfrac{1}{2}|1 - \tfrac{1}{2}\tfrac{1}{2})|^2}{|(1\,\tfrac{1}{2}\tfrac{1}{2}|0\,\tfrac{1}{2}\tfrac{1}{2})|^2} = \frac{2/3}{1/3} = 2 \quad .$$

The corresponding experimental results are

$$R = \begin{cases} 1.91 \pm 0.25 \\ 2.26 \pm 0.11 \end{cases} .$$

These two results show that isospin invariance is not exactly valid, but at least to within 10%.

EXAMPLE

5.10 Production of Neutral Pions in Deuteron-Deuteron Scattering

Another experimental verification of the isospin invariance is the reaction

$$\mathrm{d} + \mathrm{d} \to {}^4\mathrm{He} + \pi^0 \tag{1}$$

Since an ^{4}He and a deuteron are both iso-singlets, i.e.

$$|{}^4\mathrm{He}\rangle = |T = 0, T_3 = 0\rangle \quad , \quad |\mathrm{d}\rangle = |T = 0, T_3 = 0\rangle \quad ,$$

then the reaction $\mathrm{d} + \mathrm{d} \to {}^4\mathrm{He} + \pi^0$ should be not allowed, because the states

$$|00\rangle|00\rangle \quad |00\rangle|1\,0\rangle$$

are orthogonal. Experiments show that

$$\sigma(\mathrm{d} + \mathrm{d} \to {}^4\mathrm{He}) < 1.6 \times 10^{-32}\ \mathrm{cm}^2 \quad ,$$

which is a very small cross-section on a nuclear scale (typically 10^{-26} cm^2). On the other hand one finds a noticeable cross-section in the reaction

$$\mathrm{d} + \mathrm{d} \to {}^4\mathrm{He} + \gamma \quad .$$

Example 5.10

This is not a forbidden reaction, because the photon contains an iso-singlet component. (In fact, electromagnetic interactions do not conserve isospin invariance. The photon contains both isospin singlet and triplet components of equal magnitude!)

EXAMPLE

5.11 Pion-Nucleon Scattering

Pion-nucleon scattering involves reactions like

$$\pi + \mathrm{N} \rightarrow \pi' + \mathrm{N}' \quad , \tag{1}$$

where π denotes one of the three pions π^-, π^+, π^0, and N′ a proton, neutron or excited state of the nucleon. The isospin eigenfunctions in the initial and final state are

$$\underbrace{\underset{\text{pion}}{|T=1\rangle} \otimes \underset{\text{nucleon}}{|T=\tfrac{1}{2}\rangle}}_{\text{initial state}} \rightarrow \underbrace{\underset{\text{pion}}{|T=1\rangle} \otimes \underset{\text{nucleon}}{|T=\tfrac{1}{2}\rangle}}_{\text{final state}} \quad . \tag{2}$$

For both initial and final state, the individual isospins of the particles can be coupled to the total isospins $T=\frac{1}{2}$ and $T=\frac{3}{2}$. This is expressed by the notation

$$[1] \otimes [\tfrac{1}{2}] = [\tfrac{1}{2}] \otimes [\tfrac{3}{2}] \quad . \tag{3}$$

Consequently, there are two reduced matrix elements, for $T=\frac{1}{2}$ and $T=\frac{3}{2}$, which will now be discussed in detail.

If $|1,\mu\rangle$ and $|\frac{1}{2},\nu\rangle$ denote the isospin states $T=1$ and $T=\frac{1}{2}$, respectively, the initial state is given by

$$|1\mu\rangle|\tfrac{1}{2}\nu\rangle = (1\tfrac{1}{2}\tfrac{1}{2}|\mu\,\nu\,\mu+\nu)|\tfrac{1}{2},\mu+\nu\rangle + (1\tfrac{1}{2}\tfrac{3}{2}|\mu\,\nu\,\mu+\nu)|\tfrac{3}{2},\mu+\nu\rangle \quad . \tag{4}$$

Similarly, the final state takes the form

$$\begin{aligned}|1\mu'\rangle|\tfrac{1}{2}\nu'\rangle &= (1\tfrac{1}{2}\tfrac{1}{2}|\mu'\,\nu'\,\mu'+\nu')|\tfrac{1}{2},\mu'+\nu'\rangle \\ &\quad + (1\tfrac{1}{2}\tfrac{3}{2}|\mu'\,\nu'\,\mu'+\nu'+\nu')|\tfrac{3}{2},\mu'+\nu'\rangle \quad .\end{aligned} \tag{5}$$

The possible quantum numbers $\mu = \pm 1, 0$ of the initial state characterize the pion charge states $|\pi^\pm\rangle$ and $|\pi^0\rangle$, the quantum numbers $\nu = \pm\frac{1}{2}$ represent $|\mathrm{p}\rangle$ or $|\mathrm{n}\rangle$, respectively. The final state is similarly described. The transition matrix element reads

$$\begin{aligned}\langle 1\mu\tfrac{1}{2}\nu|\hat{S}|1\mu'\tfrac{1}{2}\nu'\rangle &= (1\tfrac{1}{2}\tfrac{1}{2}|\mu\nu\mu+\nu)(1\tfrac{1}{2}\tfrac{1}{2}|\mu'\nu'\mu'+\nu')\langle\tfrac{1}{2}\mu+\nu|\hat{S}|\tfrac{1}{2}\mu'+\nu'\rangle \\ &\quad + (1\tfrac{1}{2}\tfrac{3}{2}|\mu\nu\mu+\nu)(1\tfrac{1}{2}\tfrac{3}{2}|\mu'\nu'\mu'+\nu')\langle\tfrac{3}{2}\mu+\nu|\hat{S}|\tfrac{3}{2}\mu'+\nu'\rangle \ ,\end{aligned} \tag{6}$$

where we have used (5.65), i.e. the fact that only matrix elements between equal total isospin $T=T'$ contribute. Also, from (5.65), we infer that $T_3' = T_3$ which, for (6), means

Example 5.11

$$\mu' + v' = \mu + v \quad . \tag{7}$$

Moreover, the matrix elements $\langle TT_3|S|TT_3 \rangle$ are independent of T_3 due to isospin invariance, only depending on the total isospin T, i.e.

$$\langle TT_3|\hat{S}|TT_3 \rangle = \langle T\|\hat{S}\|T \rangle \quad . \tag{8}$$

Hence, (6) becomes

$$\langle 1\mu\tfrac{1}{2}v|\hat{S}|1\mu'\tfrac{1}{2}v' \rangle = [(1\tfrac{1}{2}\tfrac{1}{2}|\mu v\mu + v)(1\tfrac{1}{2}\tfrac{1}{2}|\mu'v'\mu + v)\langle \tfrac{1}{2}\|\hat{S}\|\tfrac{1}{2} \rangle$$
$$+(1\tfrac{1}{2}\tfrac{3}{2}|\mu v\mu + v)(1\tfrac{1}{2}\tfrac{3}{2}|\mu'v'\mu + v)\langle \tfrac{3}{2}\|\hat{S}\|\tfrac{3}{2} \rangle]\,\delta_{\mu+v,\,\mu'+v'} \quad . \tag{9}$$

The ten possible reactions expressed by (1) are described by only two reduced matrix elements $\langle \tfrac{1}{2}\|S\|\tfrac{1}{2} \rangle$ and $\langle \tfrac{3}{2}\|S\|\tfrac{3}{2} \rangle$, which are generally complex. This implies three real parameters for the above reactions, since a common real phase cancels when taking the square $|\langle 1\mu\tfrac{1}{2}v|S|1\,\mu'\tfrac{1}{2}v' \rangle|^2$.

Analyzing the matrix element in (9), we find ten possible reactions:

$\mu = 1$	$v = \tfrac{1}{2} \rightarrow \quad \mu' = 1 \quad v' = \tfrac{1}{2}$	$\pi^+ + p \rightarrow \pi^+ + p$
	$v = -\tfrac{1}{2} \begin{array}{c} \nearrow \mu' = 1 \quad v' = -\tfrac{1}{2} \\ \searrow \mu' = 0 \quad v' = \tfrac{1}{2} \end{array}$	$\pi^+ + n \begin{array}{c} \nearrow \pi^+ + n \\ \searrow \pi^0 + p \end{array}$
$\mu = 0$	$v = \tfrac{1}{2} \begin{array}{c} \nearrow \mu' = 0 \quad v' = \tfrac{1}{2} \\ \searrow \mu' = 1 \quad v' = -\tfrac{1}{2} \end{array}$	$\pi^0 + p \begin{array}{c} \nearrow \pi^0 + p \\ \searrow \pi^+ + n \end{array}$
	$v = -\tfrac{1}{2} \begin{array}{c} \nearrow \mu' = 0 \quad v' = -\tfrac{1}{2} \\ \searrow \mu' = -1 \quad v' = \tfrac{1}{2} \end{array}$	$\pi^0 + n \begin{array}{c} \nearrow \pi^0 + n \\ \searrow \pi^- + p \end{array}$
$\mu = -1$	$v = \tfrac{1}{2} \begin{array}{c} \nearrow \mu' = -1 \quad v' = \tfrac{1}{2} \\ \searrow \mu' = 0 \quad v' = -\tfrac{1}{2} \end{array}$	$\pi^- + p \begin{array}{c} \nearrow \pi^- + p \\ \searrow \pi^0 + n \end{array}$
	$v = -\tfrac{1}{2} \rightarrow \mu' = -1 \quad v' = -\tfrac{1}{2}$	$\pi^- + n \rightarrow \pi^- + n$

On the rhs of this listing the isospin quantum numbers are reexpressed in terms of the physical particles they represent. Since (9) is quite complicated, the possible general reactions of (1) and (10) are somewhat confusing. The situation becomes clearer in the region of the first resonance, the so-called Δ resonance, located at 1232 MeV. It is also called the $\tfrac{3}{2} - \tfrac{3}{2}$ resonance, because its isospin and spin quantum numbers both take the value $\tfrac{3}{2}$. The first Figure illustrates the

Example 5.11

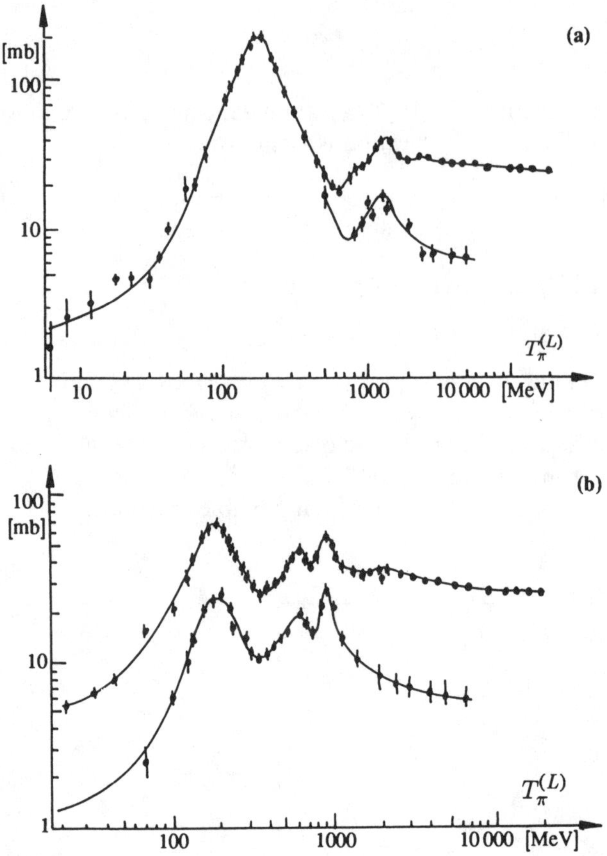

The total scattering cross-sections of (**a**) π^+p and (**b**) π^-p scattering: the various maxima (resonances) have been interpreted very successfully by formations of intermediate particles N* (isobars) that subsequently decay again. In the case of π^-p scattering these are two decay products π^0n and π^-p. In the case of π^+p scattering one finds only π^+p in the outgoing channel. The second curve in part (**a**) implies other excited states in the outgoing channel which arise at a threshold energy of about 800 MeV.

cross-section σ as a function of the pion energy for the reactions

$$\pi^+ + \mathrm{p} \to \pi^+ + \mathrm{p} \qquad \pi^- + \mathrm{p} \to \pi^- + \mathrm{p} \,, \tag{11}$$

where the energy is the kinetic energy of the pion in the laboratory $T_\pi^{(L)}$. Let us reflect shortly on the various energies which are important in high energy reactions and which are specifically of interest in the context here. There is first the total energy W of the pion-nucleon system in the centre of mass. If $p = (\boldsymbol{p}, E_N/c) = (\boldsymbol{p}, p_0)$ and $q = (\boldsymbol{q}, E_\pi/c) = (\boldsymbol{q}, q_0)$ denote the momentum four-vectors for the nucleon and pion in the initial state, then we have, for the Lorentz invariant quantity W^2/c^2,

$$W^2/c^2 = -(p+q)^2 = (p^0 + q^0)^2 - (\boldsymbol{p} + \boldsymbol{q})^2 \quad . \tag{12}$$

In the centre-of-mass frame, by definition, it holds that

$$\boldsymbol{p} + \boldsymbol{q} = 0 \quad , \quad \text{so that} \tag{13}$$

$$W^2 = (p^0 + q^0)^2 c^2 \quad . \tag{14}$$

Therefore, W is indeed the total energy of the π-N system in this frame. In the laboratory system S the situation is different. In this frame the nucleon is at rest

Example 5.11

($p_{\text{lab.}} = 0$) so that the Lorentz invariant of (12) becomes

$$W^2/c^2 = -(p^2 + q^2 + 2p\cdot q) = M^2c^2 + m_\pi^2 c^2 + 2p^0 q^0$$
$$= M^2c^2 + m_\pi^2 c^2 + 2ME_\pi^{(L)} \quad , \tag{15}$$

where we have used[10] $M^2c^2 = -p^2$ and $m_\pi^2 c^2 = -q^2$. M and m_π are the masses of nucleon and pion, $E_\pi^{(L)}$ the total energy of the pion in the laboratory frame, given by

$$(E_\pi^L)^2 = q^2c^2 + m_\pi^2 c^4 \ . \tag{16}$$

Considering the kinetic energy of the pion in the laboratory frame,

$$T_\pi^{(L)} = E_\pi^{(L)} - m_\pi c^2 \quad , \tag{17}$$

we obtain, for (15),

$$W^2 = (Mc^2 + m_\pi c^2)^2 + 2Mc^2 T_\pi^{(L)} \quad . \tag{18}$$

Such resonances can be interpreted as short-lived intermediate states N* that are composed of the pion and the nucleon interacting with each other. In writing this is expressed by

$$\pi + \mathrm{N} \to \mathrm{N}^* \to \pi + \mathrm{N} \quad , \tag{19}$$

i.e. in the pion-nucleon scattering the intermediate state N* is formed. The nature of the resonances is similar to that of compound nuclei in nuclear physics. There the collision of two nuclei yields an intermediate composite nucleus (compound nucleus) which, afterwards, may decay into various possible fragments (see first figure on page 170).

The masses (energies) of these intermediate particles are given by the total energies $W_{\max}$ at the resonance maxima. For example, in the first figure, (b) shows a distinct maximum at $W = 1232$ MeV, corresponding to $T_\pi^{(L)} = 190$ MeV, and, from the experimental data shown, one can conclude that these intermediate particles exhibit double positive charge in the reaction $\pi^+\mathrm{p}$ and are neutral in the reaction $\pi^-\mathrm{p}$.

Other experiments even indicate resonances that carry a single positive or negative charge. An example for such a reaction is the photo-production process

$$\gamma + \mathrm{p} \to \mathrm{N}^* \to \pi^0 + \mathrm{p} \quad . \tag{20}$$

The corresponding cross-section is shown in the below figure.

The striking resonance at $E_\gamma = 330$ MeV corresponds to an intermediate particle with positive charge and mass $M = 1232$ MeV. The negatively charged N* is encountered in the reaction

$$\pi^-\mathrm{p} \to \pi^+\pi^+\pi^-\pi^-\mathrm{n} \tag{21}$$

---

[10] See M. Goldstein: *Classical Mechanics* 2nd ed. (Addison-Wesley, Reading 1980) or W. Greiner: *Theoretische Physik I, Mechanik I* (Harri Deutsch, Frankfurt 1989) Chap. 34.

Example 5.11

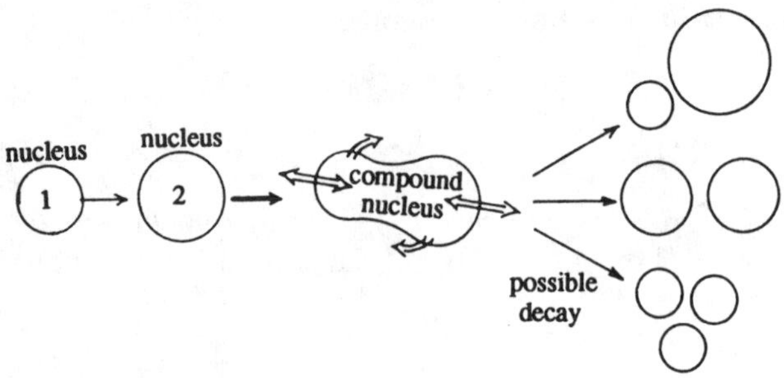

Illustration of a compound nucleus reaction with possible rotations and vibrations. The intermediate compound nucleus subsequently decays into various possible final states.

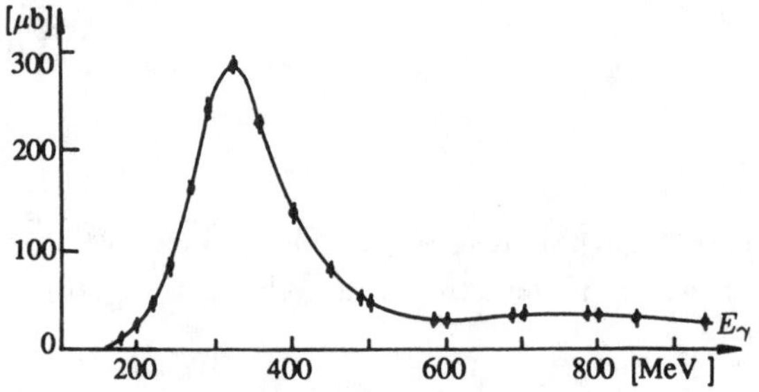

Photo-production of neutral pions; E_γ denotes the energy of the photon with respect to the laboratory frame .

by considering the invariant mass $M_{n\pi^-}$ (i.e. the total mass of neutron and pion) as a function of energy (see figure below).

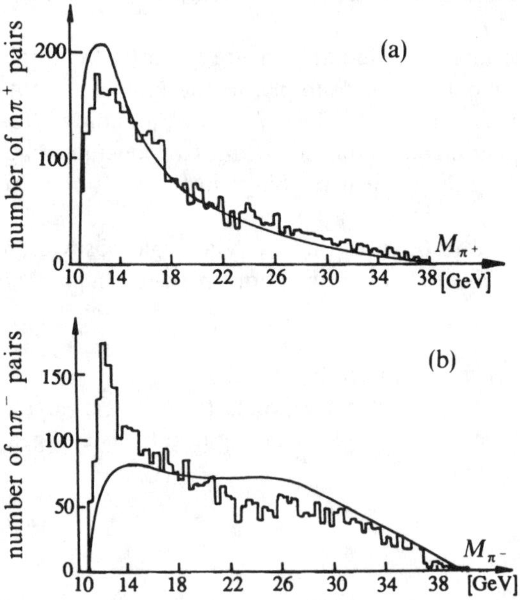

Mass spectrum of $(n\pi)$ systems as a function of the invariant mass $M_{n\pi}$ measured in the reaction $\pi^- p \to \pi\pi^+\pi^+\pi^-\pi^-$ (**a**) $(n\pi^+)$ system (**b**) $(n\pi^-)$ system

In the case of $M_{n\pi^-}$ [part (b) of the figure] there occurs a sharp resonance that stands out significantly against the broad background which can be seen in part (a) of the figure, for $M_{n\pi^+}$. The width of such a resonance contains

Example 5.11

information about the lifetime τ since, due to the uncertainty relation, it holds that $\Gamma\tau \approx \hbar$.

Summarizing these facts, we see that the intermediate state N*, with $M = 1232$ MeV, is actually a charge quartet, denoted by Δ^{charge} (total energy), i.e.

$$\Delta^{++}(1232),\ \Delta^{+}(1232),\ \Delta^{0}(1232),\ \Delta^{-}(1232) \quad . \tag{22}$$

Clearly, the Δ resonance is a $T = \frac{3}{2}$ state since this is the lowest assignment that naturally leads to an isospin multiplet with at least four states, $T_3 = \frac{3}{2}, \frac{1}{2}, -\frac{1}{2}, -\frac{3}{2}$ (see also the following section). From the analysis of the $\pi - \mathrm{N}$ scattering alone, it is not possible to decide whether there are still other charged states of this composite particle, because the states accessible in π-scattering are limited by the total charge of the pion and the nucleon. Besides, other particle systems, e.g. $\pi + \pi + \mathrm{N}$ (nucleon + two pions), have masses close to or larger than 1232 MeV. We will, however, give theoretical arguments that prove the completeness of the Δ quartet (see Exercise 6.5, isospin and hypercharge of baryon resonances and the completeness of the SU(3) decuplet).

Before moving on, some relevant remarks are in order concerning the first figure of this Example, which shows the cross-sections of the $\pi^{\pm}$-p scattering: the figure exhibits additional resonances occurring in the reaction π^{-}p, but not in π^{+}p, namely at

$$M = 1515\ \text{MeV}\ (T_\pi^{(L)} = 605\ \text{MeV}),\ M = 1688\ \text{MeV}\ (T_\pi^{(L)} = 890\ \text{MeV})\ . \tag{23}$$

Obviously, such intermediate particles must be neutral. On the other hand, there also occur charged resonant states. Therefore, these particles seem to belong to charge doublets, termed N(M). In detail, one finds

$$\mathrm{N}^{+}(1518),\ \mathrm{N}^{0}(1515)\ \text{and}\ \mathrm{N}^{+}(1688),\ \mathrm{N}^{0}(1688) \quad . \tag{24}$$

There are still other baryons, which we will study later.

Let us now return to (9). In connection with (10) we have already mentioned that the proportions of the reactions are easily determined in the region of the $\Delta(1232)$ resonance that has been identified as an iso-quartet, i.e. $T = \frac{3}{2}$ in (22). Since the resonance is dominant, one can neglect the $T = \frac{1}{2}$ contribution in (9) and obtain, in this approximation,

$$\langle 1\mu\tfrac{1}{2}\nu|\hat{S}|1\,\mu'\tfrac{1}{2}\nu'\rangle \simeq (1\,\tfrac{1}{2}\tfrac{3}{2}|\mu\,\nu\,\mu+\nu)(1\,\tfrac{1}{2}\tfrac{3}{2}|\mu'\,\nu'\,\mu+\nu)\langle\tfrac{3}{2}||\hat{S}||\tfrac{3}{2}\rangle\,\delta_{\mu+\nu,\,\mu'+\nu'} \quad . \tag{25}$$

Using this relation we evaluate the ratio of the cross-sections to be

$$\begin{aligned}
&\sigma(\pi^{+}\mathrm{p}\to\pi^{+}\mathrm{p}):\sigma(\pi^{-}\mathrm{p}\to\pi^{-}\mathrm{p}):\sigma(\pi^{-}\mathrm{p}\to\pi^{0}\mathrm{n})\\
&=|(1\,\tfrac{1}{2}\tfrac{3}{2}|1\,\tfrac{1}{2}\tfrac{3}{2})|1\,\tfrac{1}{2}\tfrac{3}{2})|^2\\
&:|(1\,\tfrac{1}{2}\tfrac{3}{2}|-1\,\tfrac{1}{2}-\tfrac{1}{2})(1\,\tfrac{1}{2}\tfrac{3}{2}|-1\,\tfrac{1}{2}-\tfrac{1}{2})|^2:|(1\,\tfrac{1}{2}\tfrac{3}{2}|-1\,\tfrac{1}{2}-\tfrac{1}{2})(1\,\tfrac{1}{2}\tfrac{3}{2}|0-\tfrac{1}{2}-\tfrac{1}{2})|^2\\
&=|1\times 1|^2:\left|\frac{1}{\sqrt{3}}\times\frac{1}{\sqrt{3}}\right|^2:\left|\frac{1}{\sqrt{3}}\times\sqrt{\frac{2}{3}}\right|^2=9:1:2 \quad ,
\end{aligned} \tag{26}$$

which is in good agreement with experimental data. A further test of the transition amplitude in (9) and, thus, for the predicted isospin invariance is

Example 5.11

obtained by checking whether the so-called triangle inequality is satisfied. Using the Clebsch-Gordan coefficients, listed in Table 5.2, and by means of the transition amplitude (9), we can prove the identity

$$\sqrt{2}\langle \mathrm{n}\pi^0|\hat{S}|\mathrm{p}\pi^-\rangle + \langle \mathrm{p}\pi^-|\hat{S}|\mathrm{p}\pi^-\rangle = \langle \mathrm{p}\pi^+|\hat{S}|\mathrm{p}\pi^+\rangle \tag{27}$$

since

$$\begin{aligned}
&\sqrt{2}\langle \mathrm{n}\pi^0|\hat{S}|\mathrm{p}\pi^-\rangle + \langle \mathrm{p}\pi^-|\hat{S}|\mathrm{p}\pi^-\rangle - \langle \mathrm{p}\pi^+|\hat{S}|\mathrm{p}\pi^+\rangle \\
&= \sqrt{2}(1\tfrac{1}{2}\tfrac{1}{2}|0-\tfrac{1}{2}-\tfrac{1}{2})(1\tfrac{1}{2}\tfrac{1}{2}|-1\tfrac{1}{2}-\tfrac{1}{2})\langle\tfrac{1}{2}\|\hat{S}\|\tfrac{1}{2}\rangle \\
&\quad + (1\tfrac{1}{2}\tfrac{1}{2}|-1\tfrac{1}{2}-\tfrac{1}{2})(1\tfrac{1}{2}\tfrac{1}{2}|-1\tfrac{1}{2}-\tfrac{1}{2})\langle\tfrac{1}{2}\|\hat{S}\|\tfrac{1}{2}\rangle \\
&\quad + \sqrt{2}(1\tfrac{1}{2}\tfrac{3}{2}|0-\tfrac{1}{2}-\tfrac{1}{2})(1\tfrac{1}{2}\tfrac{3}{2}|-1\tfrac{1}{2}-\tfrac{1}{2})\langle\tfrac{3}{2}\|\hat{S}\|\tfrac{3}{2}\rangle \\
&\quad + (1\tfrac{1}{2}\tfrac{1}{3}|-1\tfrac{1}{2}-\tfrac{1}{2})(1\tfrac{1}{2}\tfrac{3}{2}|-1\tfrac{1}{2}-\tfrac{1}{2})\langle\tfrac{3}{2}\|\hat{S}\|\tfrac{3}{2}\rangle \\
&\quad - (1\tfrac{1}{2}\tfrac{3}{2}|1\tfrac{1}{2}\tfrac{3}{2})(1\tfrac{1}{2}\tfrac{3}{2}|1\tfrac{1}{2}\tfrac{3}{2})\langle\tfrac{3}{2}\|\hat{S}\|\tfrac{3}{2}\rangle \\
&= \langle\tfrac{1}{2}\|\hat{S}\|\tfrac{1}{2}\rangle\left(\sqrt{2}\left(-\frac{1}{\sqrt{3}}\right)\sqrt{\frac{2}{3}}+\frac{2}{3}\right) + \langle\tfrac{3}{2}\|\hat{S}\|\tfrac{3}{2}\rangle\left(\sqrt{2}\sqrt{\frac{2}{3}}\frac{1}{\sqrt{3}}+\frac{1}{3}-1\right) = 0\,.
\end{aligned}$$

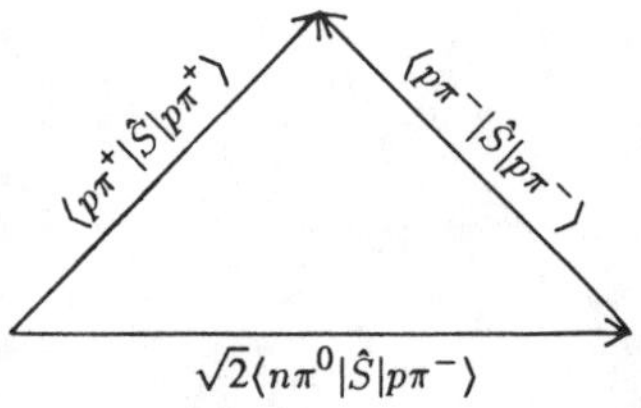

Illustration of the triangle inequality (27) in the complex plane

As each of these three amplitudes contain the two reduced matrix elements, $\langle\frac{1}{2}\|S\|\frac{1}{2}\rangle$ and $\langle\frac{3}{2}\|S\|\frac{3}{2}\rangle$, that are generally complex, (27) is a relation between three complex numbers. Representing these numbers as vectors in the complex plane, the identity (27) stands for a closed triangle of these vectors (vector addition). Each side of the triangle represents the absolute value of one of the amplitudes in (27) (see next figure) and needs to be smaller than the sum of the two others. This property enables us to make further predictions concerning the relations among the scattering cross-sections that must hold for arbitrary (but fixed) energy and angle.

EXAMPLE

5.12 The Decay of the Neutral Rho Meson

The ϱ mesons and the ω meson have been discovered in storage-ring experiments as intermediate states (resonances in the cross-sections), similar to the Δ, N^+, N^0 resonances discussed before, by investigating the reactions

$$\mathrm{e}^- + \mathrm{e}^+ \to \varrho^0 \to \pi^- + \pi^+ \tag{1}$$

$$\mathrm{e}^- + \mathrm{e}^+ \to \omega \to \pi^- + \pi^+ + \pi^0 \quad . \tag{2}$$

The figure below shows the total experimental cross-section of the reaction $\mathrm{e}^- + \mathrm{e}^+ \to \pi^+ + \pi^-$. The resonance at $M = 770$ MeV indicates a neutral intermediate particle (since the total charge of the initial state $e^+ + e^-$ is zero) which is named the ϱ^0 meson.

Similarly, the next figure, illustrating the cross-section of the reaction $e^+ + e^- \rightarrow \pi^- + \pi^+ + \pi^0$, exhibits a resonance at 780 MeV. Consequently, this is interpreted as another neutral intermediate particle, the so-called *ω-meson*. It must be different from the ϱ^0 resonance, because the three-pion final state has negative *G*-parity, whereas the *G*-parity of the ϱ^0, which decays into a pair of pions, is positive. The question arises as to whether there exist further charged particles, completing multiplets that contain the ϱ^0 or ω resonance. Indeed, they have been discovered in the reaction

$$\pi^\pm + p \rightarrow \varrho^\pm + p \rightarrow \pi^\pm + \pi^0 + p$$

as resonances with masses around 770 MeV. The experimental data on the ϱ mesons and the ω meson is summarized in the following table. Since there are three ϱ mesons with almost identical masses, one concludes that they form an isotriplet, whereas the neutral ω meson is an isosinglet.

Example 5.12

$$|F_\pi|^2 = \frac{\Gamma_e^2 m_\varrho^2 \Gamma_\varrho^2}{(4E^2 - m_\varrho^2)^2 + m_\varrho^2 \Gamma_\varrho^2}$$

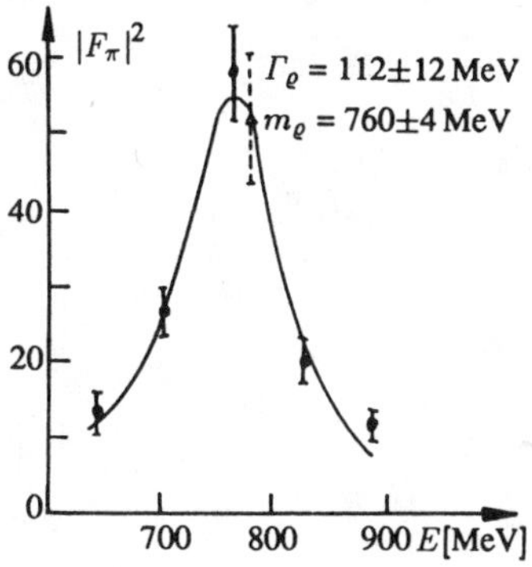

Total cross-section for the reaction $e^+ + e^- \rightarrow \pi^+ + \pi^-$. The resonance at 760 MeV is interpreted as an intermediate ϱ^0 meson

Experimental data of the ϱ and ω mesons

	Spin	Parity	Mass [MeV]	Width Γ [MeV]	Lifetime [s]	Charge
ϱ^+	1	—	770	153	4.3×10^{-24}	e
ϱ^-	1	—	770	153	4.3×10^{-24}	− e
ϱ^0	1	—	770	153	4.3×10^{-24}	0
ω	1	—	783	10	5.5×10^{-23}	0

We next study the decay of the ϱ^0 meson into two pions, i.e. the reactions

$$\varrho^0 \rightarrow \pi^+ + \pi^- \quad , \tag{3a}$$

$$\varrho^0 \rightarrow 2\pi^0 \tag{3b}$$

which are described by the transition matrix elements

$$\langle 1\,\mu\,1 - \mu|\hat{S}|T = 1,\, T_3 = 0\rangle = (1\,1\,1|\mu - \mu\,0)\langle 1\|\hat{S}\|1\rangle \quad . \tag{4}$$

Inserting the μ-values of (3), we find that the decay amplitudes (3a) and (3b) are proportional to the Clebsch-Gordan coefficients $(1\,1\,1|1 - 1\,0)$ and $(1\,1\,1|0\,0\,0)$, respectively. The latter vanishes, i.e. $(1\,1\,1|0\,0\,0) = 0$, due to a general symmetry of the Clebsch-Gordan coefficients, namely[11]

$$(j_1\,j_2\,j_3|m_1\,m_2\,m_3) = (-1)^{j_1+j_2-j_3}(j_2\,j_1\,j_3|m_2\,m_1\,m_3) \quad . \tag{5}$$

The vanishing of the amplitude (3b) implies that the decay of the ϱ^0 meson into two π^0 mesons is a forbidden reaction,

$$\varrho^0 \not\rightarrow 2\pi^0 \quad .$$

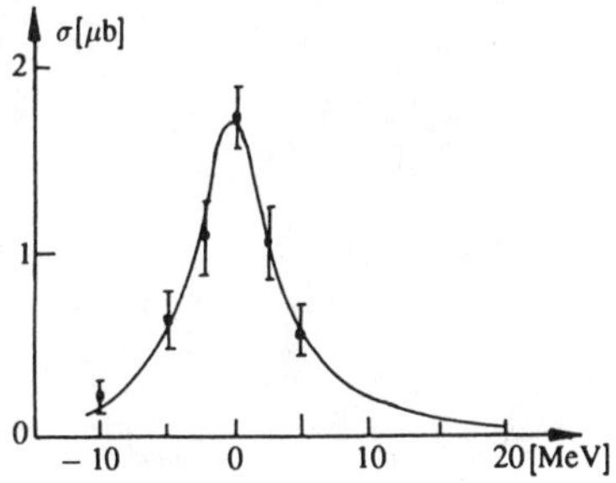

Total cross-section for the reaction $e^+ + e^- \rightarrow \pi^+ + \pi^- + \pi^0$. The resonance is interpreted as an intermediate ω meson. The abscissa shows the energy difference to 780 MeV

[11] See for example M.E. Rose: *Elementary Theory of Angular Momentum* (Wiley, New York 1957).

Example 5.12

This is experimentally confirmed. Similarly, the decay of the ω meson into three pions,

$$\omega \not\rightarrow 3\pi^0 \quad ,$$

turns out to be a forbidden decay mode, due to the vanishing of the relevant Clebsch-Gordan coefficient $(1\,1\,1|0\,0\,0)$.

5.6 Biographical Notes

WIGNER, Eugen Paul, *1902 in Budapest, since 1938 professor in Princeton, received the Nobel Prize in 1963 together with J.H.D. Jensen and Maria Göppert-Mayer for his contributions to the theory of nuclei and elementary particles, especially the discovery and application of fundamental principles of symmetry. After retirement at Princeton he became a distinguished professor at the University of Louisiana in Baton Rouge.

ECKART, Carl Henry, *1902 in St. Louis, †1973 in La Jolla, professor in Chicago from 1928 to 1946 and later in San Diego until 1970. Besides papers on theoretical physics Eckart made numerous contributions to oceanographics.

6. The Hypercharge

In the last chapter we encountered the charge multiplets of the isospin group. Several empirical examples confirmed the validity of this symmetry. The particles in a charge multiplet differ only by charge (and other electromagnetic properties, e.g. the magnetic moment, electric quadrupole moment). In a multiplet all integer multiples of the elementary charge e, from the minimum value $Q_{\min}$ up to the maximum value $Q_{\max}$, are realized. In general, $Q_{\min} + Q_{\max} \neq 0$ and the charge multiplet is not necessarily located symmetrically around the origin of the charge axis. Thus the centre of charge may differ from zero. This is illustrated in Fig. 6.1.

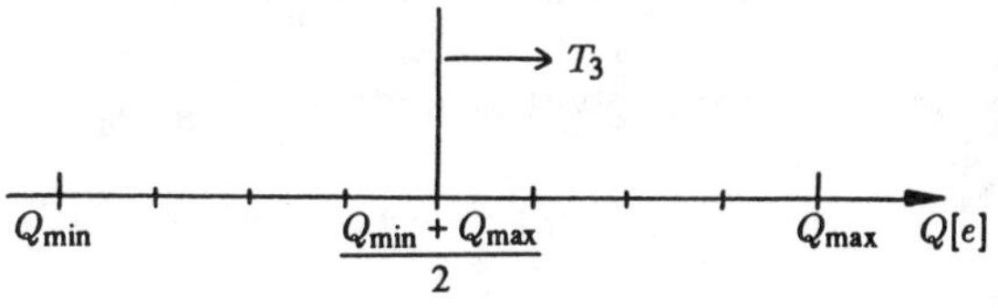

Fig. 6.1. The centre of charge of an isospin multiplet

In order to compensate for such an overall charge shift, the isospin component T_3 is counted from the centre of charge $\frac{1}{2}(Q_{\min} + Q_{\max})$ onwards. We therefore obtain the relation

$$Q = \tfrac{1}{2}(Q_{\min} + Q_{\max}) + T_3 \tag{6.1}$$

between the charge and the 3-component of the isospin of a particle, where T_3 takes the values

$$T_3 = 0,\ \pm 1,\ \pm 2, \ldots, \pm \tfrac{1}{2}(Q_{\max} - Q_{\min}) \quad . \tag{6.2}$$

Clearly one obtains

$$2T = Q_{\max} - Q_{\min} \tag{6.3}$$

or, respectively,

$$2T + 1 = (Q_{\max} - Q_{\min}) + 1 \quad . \tag{6.4}$$

The *centre of charge* of a multiplet, from now on abbreviated by $\frac{1}{2}Y$, where

$$\tfrac{1}{2}Y = \tfrac{1}{2}(Q_{\min} + Q_{\max}) = \text{centre of charge} \quad , \tag{6.5}$$

is a quantity that is not determined in the framework of the isospin symmetry alone. Equation (6.5) defines the so called *hypercharge* Y. The idea of hypercharge goes back to ***Gell-Mann*** and ***Nishijima***[1], who introduced it independently in 1953. Relation (6.1) then takes the form

$$Q = \tfrac{1}{2}Y + T_3 \quad , \qquad T_3 = T, T-1, \dots, -T \quad , \tag{6.6}$$

which is called the *Gell-Mann-Nishijima relation.*

In classifying particles, the role of hypercharge is of the same importance as the isospin. This can be seen from the fact that both quantities, Y and T_3 are part of (6.6), determining the charge Q. The following examples elucidate this point further.

EXERCISE

6.1 Hypercharge of Nuclei

Problem. Determine the hypercharge of nuclei in an isospin multiplet, given that the charge operator $\hat{Q}$ for a nucleus containing Z protons and N neutrons ($A = N + Z$ is the total number of nucleons) is known from (5.21).

Solution. Accordingly the charge of a nucleus in units of the elementary charge e is given by

$$Q = (\tfrac{1}{2}A + T_3) \quad . \tag{1}$$

Comparison with (6.6) yields

$$Y = A \quad . \tag{2}$$

Thus the hypercharge of the nuclei, which belong to a certain isospin multiplet, is equal to their total number of nucleons.

EXAMPLE

6.2 The Hypercharge of the Δ Resonances

In Example (5.8), (22) we became acquainted with the isospin quartet of the Δ resonances.

$$\Delta^{++}, \Delta^{+}, \Delta^{0}, \Delta^{-} \quad . \tag{1}$$

[1] T. Nakuno, K. Nishijima: Prog. Theor. Phys. **10**, 581 (1953); M. Gell-Mann: Phys. Rev. **82**, 833 (1953).

The maximum charge is $Q_{\max} = 2$ and the minimum charge is $Q_{\min} = -1$. According to (6.5), the hypercharge is *Example 6.2*

$$Y = (Q_{\max} + Q_{\min}) = 2 + (-1) = 1 \qquad (2)$$

and, according to (6.3),

$$T = \tfrac{1}{2}(Q_{\max} - Q_{\min}) = \tfrac{1}{2}[2 - (-1)] = \tfrac{3}{2} \quad . \qquad (3)$$

The *Gell-Mann-Nishijima relation*

$$Q = \tfrac{1}{2} + T_3 \qquad (4)$$

with $T_3 = \frac{3}{2}, \frac{1}{2}, -\frac{1}{2}, -\frac{3}{2}$, reproduces all observed charges of the quartet of the Δ resonances.

EXAMPLE

6.3 The Baryons

All of the elementary particles that are compiled in the following table, are called baryons.[2] The internal relationship of baryons (and the baryons resonances too, see Exercise 6.5) is established by the fact that in every decay channel these particles decay into other baryons, i.e. the number of baryons is not changed in any reaction or decay process. The different iso-multiplets (charge multiplets) are evident. The nucleons and the Ξ particles each form an iso-doublet, the Λ^0 and the Ω particle each represent an iso-singlet, and the Σ hyperons an iso-triplet. The hypercharge Y can be deduced in the usual manner and is given in column 6 of the following table.

The figure shows the representation of the quantum numbers Y (hypercharge) and T_3 (third component of isospin) for which the baryons with spin $\frac{1}{2}$ form an octet. All particles carry spin $\frac{1}{2}$ and positive parity. The Ω particle, carrying spin $\frac{3}{2}$ does not fit into this diagram.

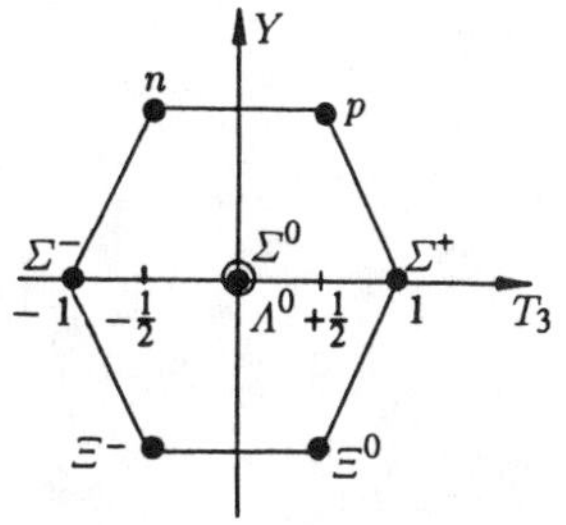

The quantum numbers Y, T_3 of the baryon nonet

Some Facts About Antiparticles: For each particle with half-integer spin there exists a corresponding antiparticle[3]. A consistent relativistic treatment of quantum mechanics predicts these antiparticles. Every antiparticle has the same mass (energy) as its particle, but carries opposite charge; therefore, the antiparticle of the electron (spin $\frac{1}{2}$, e^-) is the positron (spin $\frac{1}{2}$, e^+), the antiparticle of the proton (spin $\frac{1}{2}$, p^+) is the anti proton (spin $\frac{1}{2}$, p^-). Neutral particles have antiparticles too, such as the neutron (spin $\frac{1}{2}$, n) and the anti neutron (spin $\frac{1}{2}$, $\bar{n}$). It is not easy to distinguish between a neutral particle and its antiparticle, because they have

[2] barys (greek = $\beta\alpha\varrho\acute{\upsilon}\zeta$) = heavy.

[3] This is shown in more detail in Vols. 3 and 4 of this series: *Relativistic Quantum Mechanics* and *Quantum Electrodynamics* (Springer, Berlin Heidelberg) to be published.

Properties of low-mass baryons

Name	Symbol	Spin (parity) $J^{(p)}$	T Isospin	T_3 Isospin projection	Hyper charge	Mass [MeV]	τ Lifetime	Main decay	Branching ratio [%]
Nucleon	N p	$1/2^+$	$\frac{1}{2}$	$\frac{1}{2}$	1	938.3	∞	—	
	N n	$1/2^+$	$\frac{1}{2}$	$-\frac{1}{2}$	1	939.6	15 min	$pe^-\bar{\nu}_e$	100
Hyperons									
Lambda	Λ^0	$1/2^+$	0	0	0	1116	2.6×10^{-10} s	$p\pi^-$	64.2
								$n\pi^0$	35.8
Sigma	Σ Σ^+	$1/2^+$	1	1	0	1189	0.8×10^{-10} s	$p\pi^0$, $n\pi^+$	51.6, 48.4
	Σ Σ^0	$1/2^+$	1	0	0	1192	5.8×10^{-20} s	$\Lambda\gamma$	100
	Σ Σ^-	$1/2^+$	1	-1	0	1197	1.5×10^{-10} s	$n\pi^-$	100
Xi	Ξ Ξ^0	$1/2^+$	$\frac{1}{2}$	$\frac{1}{2}$	-1	1315	2.9×10^{-10} s	$\Lambda\pi^0$	100
	Ξ Ξ^-	$1/2^+$	$\frac{1}{2}$	$-\frac{1}{2}$	-1	1321	1.6×10^{-10} s	$\Lambda\pi^-$	100
Omega	Ω^-	$3/2^+$	0	0	-2	1672	0.8×10^{-10} s	ΛK^-, $\Xi^0\pi^-$	68.6, 23.6
								$\Xi^-\pi^0$	8

Example 6.3

very similar properties (there is a difference in the sign of the magnetic moment, however), but they can annihilate each other to form mesons[4], as in the case

$$n + \bar{n} \to \pi^+ + \pi^- \quad . \tag{6.7}$$

Mesons also have antiparticles, but the situation is slightly more complex in this case. For a neutral boson it may happen that particle and antiparticle are identical. This is the case for pions, where we have

$$\overline{\pi^+} = \pi^- \quad , \quad \overline{\pi^-} = \pi^+ \quad , \quad \overline{\pi^0} = \pi^0 \quad . \tag{6.8}$$

On the other hand, the antiparticles of the K mesons (iso-doublet K^+, K^0, spin 1 – see also the discussion on leptons) are different:

$$\overline{K^+} = K^- \quad , \quad \overline{K^0} = \bar{K}^0 \quad . \tag{6.9}$$

Following the standard convention, the antiparticle will be denoted by a bar above the corresponding particle symbol, as we have already done in this section.

[4] From mesos (greek = $\mu\acute{\varepsilon}\sigma o\zeta$) = medium.

EXAMPLE

6.4 Antibaryons

The antibaryons are distinguished from the baryons discussed in the previous example by their charge, whereas mass, spin and total isospin are the same for both baryons and antibaryons.

So besides the iso-triplet of Σ *particles,*

$$\Sigma = \{\Sigma^+, \Sigma^0, \Sigma^-\} \quad , \tag{1}$$

there exists the iso-triplet of *anti-Σ-particles*

$$\overline{\Sigma} = \{\overline{\Sigma^+}, \overline{\Sigma^0}, \overline{\Sigma^-}\} \quad . \tag{2}$$

$\overline{\Sigma^+}$ carries negative charge, but has some mass and spin as the Σ^+. Both annihilate each other, giving

$$\Sigma^+ + \overline{\Sigma^+} \to \begin{cases} 2\gamma \\ \mathrm{p} + \bar{\mathrm{p}} \text{ etc} \end{cases} \quad . \tag{3}$$

It follows from (6.6) that we have to reverse T_3 and Y too, if the charge Q changes sign. So for the antiproton $\bar{p}$ it follows that $T_3 = -\frac{1}{2}$ and $Y = -1$. For antibaryons an octet results in the $Y - T_3$-plane as shown in the figure.

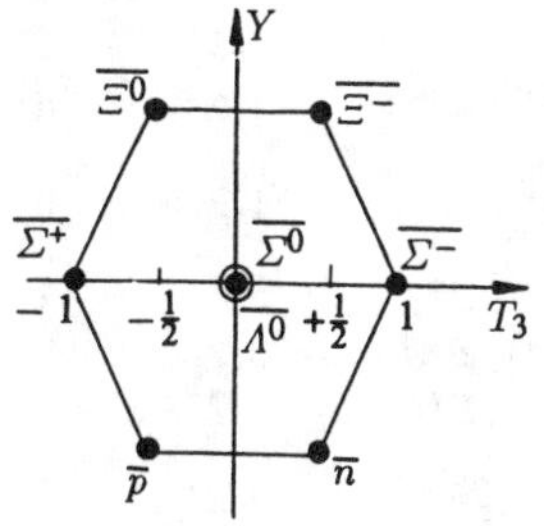

The antibaryon octet. All particles carry spin $\frac{1}{2}$ and positive parity

EXERCISE

6.5 Isospin and Hypercharge of Baryon Resonances

Problem. Based on the facts about baryon resonances given in the following table, deduce their isospin, T, T_3 and hypercharge Y. Draw the corresponding $Y - T_3$ diagram.

Solution. Once again we may recognize the internal relationship of these baryon resonances and the baryons themselves, which becomes obvious as a result of their decay: One of the decay products (9th column) is always a baryon. So, in a way, the baryon resonances are excited baryons. The Δ resonances have already been discussed in Example 6.2. Clearly the N′ and the Ξ resonances form an iso-doublet each, the Λ^* an iso-singlet and the Σ^* an iso-triplet. In particular, if we consider the Σ^* it follows that $Q_{\mathrm{max}} = 1$ and $Q_{\mathrm{min}} = -1$. So, therefore, following (6.3) we obtain

$$T = \tfrac{1}{2}(Q_{\mathrm{max}} - Q_{\mathrm{min}}) = \tfrac{1}{2}[1 - (-1)] = 1 \quad , \tag{1}$$

and according to (6.5)

$$Y = (Q_{\mathrm{max}} + Q_{\mathrm{min}}) = 1 + (-1) = 0 \quad . \tag{2}$$

Properties of the baryon resonances

Symbol		J^P	Q	T	T_3	Y	Mass [MeV]	Lifetime τ [s]	Γ [MeV]	Main decay channels	Resonant partial waves
N*	Δ^{++}	$3/2^+$	2				1232 ± 2	5.49×10^{-24}	120	$N\pi$	$P_{33}\,\pi p$
	Δ^{+}	$3/2^+$	1								
	Δ^{0}	$3/2^+$	0								
	Δ^{-}	$3/2^+$	-1								
N′	N'^{+}	$1/2^+$	1				1440 ± 40	3.13×10^{-24}	210	$N\pi$, $N\pi\pi$	$P_{11}\,\pi p$
	N'^{0}	$1/2^+$	0								
Λ^*		$1/2^-$	0				1405 ± 5	1.65×10^{-23}	40	$\Sigma\pi$	$S_{01}\,K^-p$
Σ^*	Σ^{*1}	$3/2^+$	1				1382.3 ± 0.4	1.78×10^{-25}	37	$\Lambda\pi$, $\Sigma\pi$	$P_{13}\,K^-p$
	Σ^{*0}	$3/2^+$	0				1382.0 ± 2.5				
	Σ^{*-1}	$3/2^+$	-1				1387.4 ± 0.6				
Ξ^*	Ξ^{*0}	$3/2^+$	0				1531.8 ± 0.3	9.4×10^{-23}	7	$\Xi\pi$	P
	Ξ^{*-}	$3/2^+$	-1				1535.0 ± 0.6				

Exercise 6.5

The T_3 component follows from the Gell-Mann-Nishijima relation and the measured charges for

$$\Sigma^{*+} \text{ to } T_3 = 1$$
$$\Sigma^{*0} \text{ to } T_3 = 0$$
$$\Sigma^{*-} \text{ to } T_3 = -1 \quad . \tag{3}$$

Now we fill up the empty columns of the table shown below.

Isospin and hypercharge of the baryon resonances

Particle	T	T_3	Y
Δ^{++}	$\frac{3}{2}$	$+\frac{3}{2}$	1
Δ^{+}	$\frac{3}{2}$	$+\frac{1}{2}$	1
Δ^{0}	$\frac{3}{2}$	$-\frac{1}{2}$	1
Δ^{-}	$\frac{3}{2}$	$-\frac{3}{2}$	1
N'^{+}	$\frac{1}{2}$	$+\frac{1}{2}$	1
N'^{0}	$\frac{1}{2}$	$-\frac{1}{2}$	1
Λ^*	0	0	0
Σ^{*+}	1	1	0
Σ^{*0}	1	0	0
Σ^{*-}	1	-1	0
Ξ^{*0}	$\frac{1}{2}$	$+\frac{1}{2}$	-1
Ξ^{*-}	$\frac{1}{2}$	$-\frac{1}{2}$	-1

The $Y - T_3$ diagram for the $J^P = \frac{3}{2}^+$ baryon-resonances is shown in the following figure.

Exercise 6.5

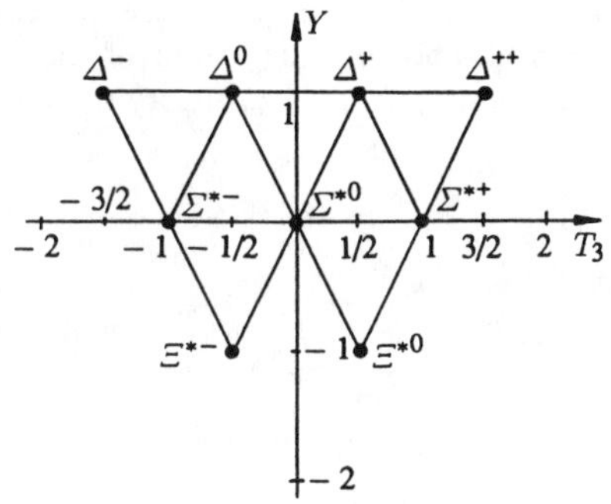

Representation of the $T_3 - Y$ values for the $J^P = 3/2^+$ baryon-resonances

Obviously one particle, with

$$J^P = \tfrac{3}{2}^+ \quad , \qquad Y = -2 \text{ and } T_3 = 0 \quad , \qquad \text{so that } Q = -1 \quad , \tag{4}$$

is missing to establish a figure of higher symmetry. Indeed we allowed for the Ω particle as one of the baryons in Example 6.3. Because of its spin $J = \frac{3}{2}$ it did not fit into that scheme. Now we can insert it at the point $T_3 = 0$ and $Y = -2$ into the multiplet of these baryon resonances. Also its high mass fits well into this diagram now.

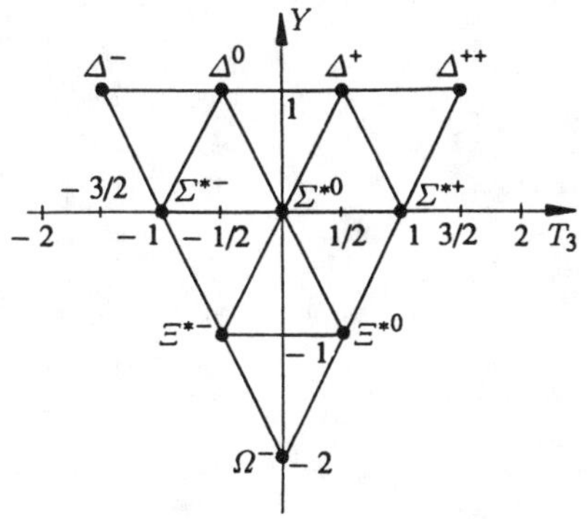

$T_3 - Y$ scheme of the baryon resonances including the Ω particle. This figure is of high symmetry and apparently complete

6.1 Biographical Notes

GELL-MANN, Murray, physicist. *15.9, 1929 in New York, professor at the California Institute of Technology in Pasadena. He wrote articles on the theory of elementary particles, especially on form factors, symmetry groups and Regge poles. At the same time as Y. Ne'eman, G.-M. developed the eight-fold way model of baryons and mesons. For this work he received the Nobel Prize in 1969.

NISHIJIMA, Kazuhiko, physicist, *04.09.1926 in Tsuchiura, Japan. After his education in Tokyo and Osaka, he worked in Göttingen, Princeton and Illinois. In 1966 N. became professor at the University of Tokyo. He is now director of the Yukawa Hall at Kyoto University. N. has made numerous contributions to theoretical particle physics, most notably the suggestion of the strangeness quantum number, the two-neutrino hypothesis and the field theory of bound states.

7. The SU(3) Symmetry

To deepen our understanding of the symmetries in the Y–T_3 plane, which occurred in the last problems investigated in Chap. 6, we turn back to group theory. We suspect that the figures obtained represent the multiplets of a new symmetry. The question arises as to the nature of the symmetry group that is underlying these multiplets. Because isospin multiplets are a part of the larger ones of the obtained multiplets and since isospin is a realization of the SU(2) symmetry, we try the next higher group, the SU(3). Indeed we will find that the multiplets of SU(3) fit exactly the figures obtained in Exercises 6.3–5.

The discovery of the SU(3) symmetry as an ordering principle of elementary particles is a highlight of modern physics, as we will discuss later. First we will give a brief introduction to:

7.1 The Groups U(n) and SU(n)

A *unitary quadratic matrix* $\hat{U}$ with n rows and n columns can be written as

$$\hat{U} = \mathrm{e}^{\mathrm{i}\hat{H}} \quad . \tag{7.1}$$

Here $\hat{H}$ is a *Hermitian quadratic matrix* with n rows and columns. All such matrices form a group under matrix multiplication. This group we call U(n), which stands for "*u*nitary group in n dimensions". Because $\hat{H}$ is Hermitian, the diagonal matrix elements are real,

$$H_{ii}^* = H_{ii} \text{ and} \tag{7.2}$$

$$H_{ij}^* = H_{ji} \quad , \quad i,j = 1,2,\dots,n \quad . \tag{7.3}$$

Thus $\hat{H}$ and therefore $\hat{U}$ allows for n^2 *real independent parameters.* Obviously the group U(n) is a continuously connected group, because the value of every single matrix element can be changed continuously. U(n) represents a *compact Lie group*, because every limit of the complex matrix elements of the form

$$\lim_{\sigma \to \sigma^0} U_{ik}(\sigma) = U_{ik}(\sigma^0) \tag{7.4}$$

yields a complex number $U_{ik}(\sigma^0)$ once again, and the matrix built up with these elements forms a group element once more. The trace of an Hermitian matrix is real because of (7.2). For the unitary matrix $\hat{U}, \hat{U}^\dagger\hat{U} = 1$ holds and it follows that

$$\det \hat{U}^\dagger \det \hat{U} = (\det \hat{U})^* \det \hat{U} = 1 \quad , \quad |\det \hat{U}|^2 = 1 \quad . \tag{7.5}$$

Applying this to (7.1) yields,

$$\begin{aligned} &\mathrm{Tr}\,\hat{H} = \alpha \quad , \quad \alpha \text{ real} \\ &\det \hat{U} = \det(\mathrm{e}^{\mathrm{i}\hat{H}}) = \mathrm{e}^{\mathrm{i}\,\mathrm{Tr}\,\hat{H}} = \mathrm{e}^{\mathrm{i}\alpha} \quad . \end{aligned} \tag{7.6}$$

This is a consequence of

$$\det \hat{U} = \det \hat{U}' = \det \hat{S}\hat{U}\hat{S}^{-1} = \det \begin{pmatrix} U'_{11} & 0 & \cdots & 0 \\ 0 & U'_{22} & \cdots & 0 \\ \vdots & & \ddots & \vdots \\ 0 & \cdots & \cdots & \cdots & U'_{nn} \end{pmatrix} \quad ,$$

assuming that $\hat{S}$ transforms $\hat{U}$ to the diagonal form $\hat{U}'$. So if $\hat{U}'$ is diagonal, $\hat{H}$ has to be diagonal too. Thus,

$$\begin{aligned} \det \hat{U} = \det \hat{U}' = \det \mathrm{e}^{\mathrm{i}\hat{H}'} &= \det \exp \mathrm{i} \begin{pmatrix} H'_{11} & 0 & \cdots & 0 \\ 0 & H'_{22} & \cdots & 0 \\ \vdots & & \ddots & \vdots \\ 0 & \cdots & \cdots & \cdots & H'_{nn} \end{pmatrix} \\ &= \det \begin{pmatrix} \mathrm{e}^{\mathrm{i}H'_{11}} & 0 & \cdots & 0 \\ 0 & \mathrm{e}^{\mathrm{i}H'_{22}} & \cdots & 0 \\ \vdots & & \ddots & \vdots \\ 0 & \cdots & \cdots & \cdots & \mathrm{e}^{\mathrm{i}H'_{nn}} \end{pmatrix} \\ &= \mathrm{e}^{\mathrm{i}(H'_{11}+H'_{22}+\cdots+H'_{nn})} = \mathrm{e}^{\mathrm{i}\,\mathrm{Tr}\,\hat{H}'} = \mathrm{e}^{\mathrm{i}\,\mathrm{Tr}\,\hat{H}} \quad . \end{aligned}$$

The last step is correct because

$$\mathrm{Tr}\,\hat{H}' = \mathrm{Tr}\,\hat{S}\hat{H}\hat{S}^{-1} = \mathrm{Tr}\,\hat{H}\hat{S}\hat{S}^{-1} = \mathrm{Tr}\,\hat{H} \quad ,$$

due to the fact that $\mathrm{Tr}(\hat{A}\hat{B}) = \mathrm{Tr}(\hat{B}\hat{A})$, which reads explicitly as

$$\mathrm{Tr}\,\hat{A}\hat{B} = \sum_{ik} A_{ik}B_{ki} = \sum_{ki} B_{ki}A_{ik} = \mathrm{Tr}\,\hat{B}\hat{A} \quad .$$

If we demand that

$$\det \hat{U} = +1 \tag{7.7}$$

should always hold, this implies a single condition on the n^2 parameters, $\alpha = 0 \bmod 2\pi$. The matrices specified by (7.7) form a continuous compact Lie group. This group is called a *special unitary group in n dimensions.* It depends on $n^2 - 1$ real parameters and is denoted by $\mathrm{SU}(n)$. Obviously $\mathrm{SU}(n)$ is a subgroup of $\mathrm{U}(n)$. If we denote a certain element of $\mathrm{SU}(n)$ by U_0, with

$$\hat{U}_0 = \mathrm{e}^{\mathrm{i}\hat{H}_0} \quad , \quad \mathrm{Tr}\,\hat{H}_0 = 0 \quad , \quad \det \hat{U}_0 = 1 \quad , \tag{7.8}$$

then we always can write an arbitrarily given element U of the group U(n) as [see (7.1)]:

$$\hat{U} = \mathrm{e}^{\mathrm{i}\hat{H}} = \exp\left[\mathrm{i}\left(\hat{H}_0 + \frac{\alpha}{n}\mathbb{1}\right)\right] = \left[\exp\left(\mathrm{i}\frac{\alpha}{n}\right)\mathbb{1}\right]\hat{U}_0 = \hat{U}_0\left[\exp\left(\mathrm{i}\frac{\alpha}{n}\right)\mathbb{1}\right] \quad , \tag{7.9}$$

where $\hat{H} = \hat{H}_0 + (\alpha/n)\mathbb{1}$ and $\mathbb{1}$ is the unit matrix. The factor α/n has been chosen such that

$$\mathrm{Tr}\hat{H} = \mathrm{Tr}\,\hat{H}_0 + \mathrm{Tr}\,\frac{\alpha}{n}\mathbb{1} = \frac{\alpha}{n}\,\mathrm{Tr}\,\mathbb{1} = \frac{\alpha}{n}n = \alpha \quad .$$

In other words the matrix elements U_{ik} of U(n) factorize, forming

$$U_{ik} = \exp\left(\mathrm{i}\frac{\alpha}{n}\right)(\hat{U}_0)_{ik} \quad .$$

Indeed it follows from $\det \hat{U}_0 = 1$ that

$$\det \hat{U} = \mathrm{e}^{\mathrm{i}\alpha}\det \hat{U}_0 = \mathrm{e}^{\mathrm{i}\alpha} \quad .$$

Such a factorized matrix $\hat{U}$ is an element of U(n) and $\hat{U}_0$ is an element of SU(n). The factors $\exp(\mathrm{i}\alpha/n)$ are unitary 1×1 matrices and form the group U(1), *the* $n\times n$ *matrices* $\exp(\mathrm{i}\alpha/n)\mathbb{1}$ *form a possible realization of the group* U(1). We can formulate the result (7.9) in the following way: *An arbitrary element U of* U(n) *can always be written as the matrix product of a suitable* $\exp(\mathrm{i}\alpha/n)\mathbb{1}$, *which is an element of* U(1), *and an element U_0 of* SU(n).

Obviously both U(n) and SU(n) are subgroups of U(m) if $n \le m$. This can be deduced from the fact that an arbitrary element U of U(n) can always be expanded to a $(m\times m)$ matrix via

$$\hat{U}' = \begin{pmatrix} \hat{U} & 0 \\ 0 & \mathbb{1} \end{pmatrix} \quad . \tag{7.10}$$

Here the unit matrix $\mathbb{1}$ has the dimension $(m-n)$ and 0 denotes the rectangular $n\times(m-n)$ zero-matrix. According to (7.1) the corresponding Hermitian matrix $\hat{H}'$ is of the form

$$\hat{H}' = \begin{pmatrix} \hat{H} & 0 \\ 0 & 0 \end{pmatrix} \quad , \tag{7.11}$$

where $\hat{H}$ corresponds to the $(n\times n)$ matrix $\hat{U}$ in (7.1).

7.1.1 The Generators of U(n) and SU(n)

The general statements of Chap. 3 lead to the conclusion [see (3.6)] that for all Lie groups the generators are determined by group elements which are infinitesimally close to the unit element. The group U(n) has n^2 parameters ϕ_j

($j = 1, \ldots, s$) and thus n^2 generators $\hat{\lambda}_j$. The relation then holds that

$$\hat{U}(\delta\phi_j) = e^{i\hat{H}(\delta\phi_j)} = \mathbb{1} + i\hat{H}(\delta\phi_j) = \mathbb{1} + i\sum_{j=1}^{n^2} \delta\phi_j\hat{\lambda}_j \quad . \tag{7.12}$$

Since $\hat{H}$ has to be Hermitian, in the case of U(n) we can use n^2 linearly independent, Hermitian $n \times n$ matrices as generators. In this case it can be deduced directly that

$$[\hat{\lambda}_i, \hat{\lambda}_j]_- = ic_{ijk}\hat{\lambda}_k \tag{7.13}$$

holds, because $i[\hat{\lambda}_i, \hat{\lambda}_j]_-$ is once again a Hermitian $n \times n$ matrix which can be expressed by a linear combination of the linearly independent $\hat{\lambda}_k$. Thus, the Lie algebra of U(n) is closed, a fact which we know already from our general considerations in Chap. 3. For practical reasons we have written the structure constants as ic_{ijk} in (7.13), while in (3.13) the factor i has been absorbed in the c_{ijk}. The generators of SU(n), in analogy to those of U(n), can be chosen as $n^2 - 1$ linearly independent Hermitian ($n \times n$) matrices with trace equal to zero. The tracelessness is necessary to ensure the validity of (7.7) by use of (7.6). Once again we can conclude that $i[\hat{\lambda}_i, \hat{\lambda}_j]_-$ is an Hermitian matrix with vanishing trace and that the Lie algebra of SU(n) is closed, too.

EXAMPLE

7.1 The Lie Algebra of SU(2)

In this particular case we choose $n = 2$. Consequently the SU(2) of two dimensional matrices contains $2^2 - 1 = 3$ parameters. These can be associated with the three real components of a vector $\boldsymbol{\phi}$. As generators we need three linearly independent traceless matrices. From considering the Pauli spin-matrices[1] we know that

$$\hat{\sigma}_1 = \begin{pmatrix} 0 & 1 \\ 1 & 0 \end{pmatrix} \; , \quad \hat{\sigma}_2 = \begin{pmatrix} 0 & -i \\ i & 0 \end{pmatrix} \; ,$$

$$\hat{\sigma}_3 = \begin{pmatrix} 1 & 0 \\ 0 & -1 \end{pmatrix} \; , \quad \mathbb{1} = \begin{pmatrix} 1 & 0 \\ 0 & 1 \end{pmatrix} \tag{1}$$

are linearly independent Hermitian 2×2 matrices which span the 2×2-matrix space

$$\begin{pmatrix} U_{11} & U_{12} \\ U_{21} & U_{22} \end{pmatrix}$$

completely. In addition, the $\hat{\sigma}_1$, $\hat{\sigma}_2$ and $\hat{\sigma}_3$ are traceless. So we can take them as generators of SU(2). The commutation relations of the $\hat{\sigma}_i$ read $[\hat{\sigma}_i, \hat{\sigma}_j] = 2i\varepsilon_{ijk}\sigma_k$.

[1] See Vol. 1 in this series: *Quantum Mechanics I – An Introduction* (Springer, Berlin, Heidelberg 1989).

Example 7.1

Instead of dealing with the $\hat{\sigma}_i$ it is more convenient to use the $\hat{S}_i = \frac{1}{2}\hat{\sigma}_i$ as generators, simplifying the commutation relations to

$$[\hat{S}_i, \hat{S}_j]_- = \mathrm{i}\varepsilon_{ijk}\hat{S}_k \tag{2}$$

without the factor 2. These relations define a Lie algebra which is isomorphic to that of the group SO(3). In view of (7.1), the operators of SU(2) can be expressed as

$$\hat{U} = \exp(-\mathrm{i}\phi_j\hat{S}_j) \quad . \tag{3}$$

Obviously no pair of the generators $\{\hat{S}_1, \hat{S}_2, \hat{S}_3\}$ commutes. So the maximum number of commuting generators is one, which determines the rank of SU(2) to be one. Using Racah's theorem it follows that there exists only one Casimir operator. It reads

$$\hat{S}^2 = \hat{S}_1^2 + \hat{S}_2^2 + \hat{S}_3^2 \quad , \tag{4}$$

which can be proven immediately, employing the homomorphism of the algebra O(3) and SU(2).

7.2 The Generators of SU(3)

The *s*pecial *u*nitary group in *3* dimensions SU(3) has $3^2 - 1 = 8$ generators. We denote them as

$$\hat{\lambda}_1, \hat{\lambda}_2, \ldots, \hat{\lambda}_8 \tag{7.14}$$

and choose them appropriately. Because SU(2) is a subgroup of SU(3), three of the generators of SU(3) can be constructed from those of SU(2) (Pauli matrices) by extending them to three dimensions. Equation (7.10) shows how to proceed, the result being

$$\hat{\lambda}_1 = \begin{pmatrix} 0 & 1 & 0 \\ 1 & 0 & 0 \\ 0 & 0 & 0 \end{pmatrix} \; , \quad \hat{\lambda}_2 = \begin{pmatrix} 0 & -\mathrm{i} & 0 \\ \mathrm{i} & 0 & 0 \\ 0 & 0 & 0 \end{pmatrix} \; , \quad \hat{\lambda}_3 = \begin{pmatrix} 1 & 0 & 0 \\ 0 & -1 & 0 \\ 0 & 0 & 0 \end{pmatrix} \; . \tag{7.15}$$

The trace of the Hermitian $\hat{\lambda}_i$ vanishes as required. The commutation relations of the first 3 generators are similar to those of the 3 Pauli matrices $\hat{\sigma}_j$ from which they are constructed:

$$[\hat{\sigma}_i, \hat{\sigma}_j]_- = 2\mathrm{i}\varepsilon_{ijk}\hat{\sigma}_k \quad , \quad i,j,k = \{1,2,3\} \quad , \tag{7.16}$$

$$[\hat{\lambda}_i, \hat{\lambda}_j]_- = 2\mathrm{i}\varepsilon_{ijk}\hat{\lambda}_k \quad , \quad i,j,k = \{1,2,3\} \quad . \tag{7.17}$$

ε_{ijk} is the well known totally antisymmetric tensor. The remaining five generators can be chosen in different ways. We use the notation from elementary particle physics.[2]

$$\hat{\lambda}_4 = \begin{pmatrix} 0 & 0 & 1 \\ 0 & 0 & 0 \\ 1 & 0 & 0 \end{pmatrix} \quad , \quad \hat{\lambda}_5 = \begin{pmatrix} 0 & 0 & -\mathrm{i} \\ 0 & 0 & 0 \\ \mathrm{i} & 0 & 0 \end{pmatrix} \quad , \quad \hat{\lambda}_6 = \begin{pmatrix} 0 & 0 & 0 \\ 0 & 0 & 1 \\ 0 & 1 & 0 \end{pmatrix} \quad ,$$

$$\hat{\lambda}_7 = \begin{pmatrix} 0 & 0 & 0 \\ 0 & 0 & -\mathrm{i} \\ 0 & \mathrm{i} & 0 \end{pmatrix} \quad , \quad \hat{\lambda}_8 = \frac{1}{\sqrt{3}} \begin{pmatrix} 1 & 0 & 0 \\ 0 & 1 & 0 \\ 0 & 0 & -2 \end{pmatrix} \quad . \tag{7.18}$$

All $\hat{\lambda}$'s are Hermitian and traceless.

$$\hat{\lambda}_i^\dagger = \hat{\lambda}_i \quad , \quad \hat{\lambda}_i = 0 \quad . \tag{7.19}$$

The construction is simple, e.g. $\hat{\lambda}_4$ and $\hat{\lambda}_6$ result from $\hat{\lambda}_1$ by moving the non-zero elements of the Pauli matrix

$$\begin{pmatrix} 0 & 1 \\ 1 & 0 \end{pmatrix}$$

down successively:

$$\hat{\lambda}_1 = \begin{pmatrix} 0 & 1 & 0 \\ 1 & 0 & 0 \\ 0 & 0 & 0 \end{pmatrix} \to \hat{\lambda}_4 = \begin{pmatrix} 0 & 0 & 1 \\ 0 & 0 & 0 \\ 1 & 0 & 0 \end{pmatrix} \to \hat{\lambda}_6 = \begin{pmatrix} 0 & 0 & 0 \\ 0 & 0 & 1 \\ 0 & 1 & 0 \end{pmatrix} \quad . \tag{7.20}$$

Similarly the $\hat{\lambda}_2$, $\hat{\lambda}_5$ and $\hat{\lambda}_7$ are related

$$\hat{\lambda}_2 = \begin{pmatrix} 0 & -\mathrm{i} & 0 \\ \mathrm{i} & 0 & 0 \\ 0 & 0 & 0 \end{pmatrix} \to \hat{\lambda}_5 = \begin{pmatrix} 0 & 0 & -\mathrm{i} \\ 0 & 0 & 0 \\ \mathrm{i} & 0 & 0 \end{pmatrix} \to \hat{\lambda}_7 = \begin{pmatrix} 0 & 0 & 0 \\ 0 & 0 & -\mathrm{i} \\ 0 & \mathrm{i} & 0 \end{pmatrix} \quad . \tag{7.21}$$

Finally the $\hat{\lambda}_3$ and $\hat{\lambda}_8$ are traceless diagonal matrices. In the definition of $\hat{\lambda}_8$ the factor $1/\sqrt{3}$ is necessary, so that the relation

$$\mathrm{Tr}\, \hat{\lambda}_i \hat{\lambda}_j = 2\delta_{ij}$$

holds for all $i,j = 1, 2, \ldots, 8$. We will discuss this relation and its consequences after proving (7.24) below.

EXERCISE

7.2 Linear Independence of the Generators $\hat{\lambda}_i$

Problem. Show that the eight generators $\hat{\lambda}_i$ are linearly independent and span the space of all Hermitian traceless (3×3) matrices.

[2] See for example M. Gell-Mann, Y. Ne'eman: *The Eight-fold Way* (Benjamin, New York 1964).

Exercise 7.2

Solution. The problem is solved if we can show that the most general traceless Hermitian (3 × 3) matrix

$$\hat{H} = \begin{pmatrix} h_{11} & h_{12} & h_{13} \\ h_{12}^* & h_{22} & h_{23} \\ h_{13}^* & h_{23}^* & -(h_{11}+h_{22}) \end{pmatrix}$$

can always be represented as a linear combination of the eight $\hat{\lambda}_i$ matrices (7.15) and (7.18).

$$\sum_{j=1}^{8} \lambda_j \hat{\lambda}_j = \hat{H} \quad . \tag{1}$$

This ansatz leads to linear inhomogeneous equations for the coefficients λ_i

$$\lambda_1 0 + \lambda_2 0 + \lambda_3 1 + \lambda_4 0 + \lambda_5 0 + \lambda_6 0 + \lambda_7 0 + \lambda_8 \frac{1}{\sqrt{3}} = h_{11}$$

$$\lambda_1 1 - \lambda_2 \mathrm{i} + \lambda_3 0 + \lambda_4 0 + \lambda_5 0 + \lambda_6 0 + \lambda_7 0 + \lambda_8 0 = h_{12}$$

$$\lambda_1 0 + \lambda_2 0 + \lambda_3 0 + \lambda_4 1 - \lambda_5 \mathrm{i} + \lambda_6 0 + \lambda_7 0 + \lambda_8 0 = h_{13}$$

$$\lambda_1 1 + \lambda_2 \mathrm{i} + \lambda_3 0 + \lambda_4 0 + \lambda_5 0 + \lambda_6 0 + \lambda_7 0 + \lambda_8 0 = h_{12}^*$$

$$\lambda_1 0 + \lambda_2 0 - \lambda_3 1 + \lambda_4 0 + \lambda_5 0 + \lambda_6 0 + \lambda_7 0 + \lambda_8 \frac{1}{\sqrt{3}} = h_{22}$$

$$\lambda_1 0 + \lambda_2 0 + \lambda_3 0 + \lambda_4 0 + \lambda_5 0 + \lambda_6 1 - \lambda_7 \mathrm{i} + \lambda_8 0 = h_{23}$$

$$\lambda_1 0 + \lambda_2 0 + \lambda_3 0 + \lambda_4 1 + \lambda_5 \mathrm{i} + \lambda_6 0 + \lambda_7 0 + \lambda_8 0 = h_{13}^*$$

$$\lambda_1 0 + \lambda_2 0 + \lambda_3 0 + \lambda_4 0 + \lambda_5 0 + \lambda_6 1 + \lambda_7 \mathrm{i} + \lambda_8 0 = h_{23}^*$$

Since the last equation is not independent, but equals the negative sum of the first and the fifth equation, it can be omitted. So indeed there remain eight equations for the eight coefficients λ_i. The determinant

$$\det \begin{vmatrix} 0 & 0 & 1 & 0 & 0 & 0 & 0 & \frac{1}{\sqrt{3}} \\ 1 & -\mathrm{i} & 0 & 0 & 0 & 0 & 0 & 0 \\ 0 & 0 & 0 & 1 & -\mathrm{i} & 0 & 0 & 0 \\ 1 & +\mathrm{i} & 0 & 0 & 0 & 0 & 0 & 0 \\ 0 & 0 & -1 & 0 & 0 & 0 & 0 & \frac{1}{\sqrt{3}} \\ 0 & 0 & 0 & 0 & 0 & 1 & -\mathrm{i} & 0 \\ 0 & 0 & 0 & 1 & +\mathrm{i} & 0 & 0 & 0 \\ 0 & 0 & 0 & 0 & 0 & 1 & +\mathrm{i} & 0 \end{vmatrix} = \frac{16}{\sqrt{3}} \mathrm{i} \neq 0$$

is not equal to zero. Consequently there always exists a unique solution for this system.

7.3 The Lie Algebra of SU(3)

Table 7.1. Table of the totally antisymmetric structure constants f_{ijk} and the symmetric coefficients d_{ijk} [compare (7.24)]. All nonvanishing structure constants are obtained by permutation of the indices listed above

ijk	f_{ijk}	ijk	d_{ijk}
123	1	118	$\frac{1}{\sqrt{3}}$
147	$\frac{1}{2}$	146	$\frac{1}{2}$
156	$-\frac{1}{2}$	157	$\frac{1}{2}$
246	$\frac{1}{2}$	228	$\frac{1}{\sqrt{3}}$
257	$\frac{1}{2}$	247	$-\frac{1}{2}$
345	$\frac{1}{2}$	256	$\frac{1}{2}$
367	$-\frac{1}{2}$	338	$\frac{1}{\sqrt{3}}$
		344	$\frac{1}{2}$
458	$\frac{\sqrt{3}}{2}$		
678	$\frac{\sqrt{3}}{2}$	355	$\frac{1}{2}$
		366	$-\frac{1}{2}$
		377	$-\frac{1}{2}$
		448	$-\frac{1}{2\sqrt{3}}$
		558	$-\frac{1}{2\sqrt{3}}$
		668	$-\frac{1}{2\sqrt{3}}$
		778	$-\frac{1}{2\sqrt{3}}$
		888	$-\frac{1}{\sqrt{3}}$

With the explicit representation of the generators (7.15) and (7.18), we can easily compute the *commutators* of the $\hat{\lambda}_i$ and the Lie algebra of the SU(3). Using Einstein's summation convention we find that

$$[\hat{\lambda}_i, \hat{\lambda}_j]_- = 2\mathrm{i} f_{ijk}\hat{\lambda}_k \quad , \tag{7.22}$$

where we have here extracted the factor 2i. The structure constants are totally antisymmetric under exchange of any two indices.

$$f_{ijk} = -f_{jik} = -f_{ikj} \quad \text{etc} \quad . \tag{7.23}$$

They are compiled in Table 7.1, where we have also listed the symmetric coefficients d_{ijk} which occur in the *anti-commutation relations*

$$[\hat{\lambda}_i, \hat{\lambda}_j]_+ = \tfrac{4}{3}\delta_{ij}\mathbb{1} + 2d_{ijk}\hat{\lambda}_k \quad ,$$
$$d_{ijk} = d_{jik} = d_{ikj} \quad \text{etc} \quad . \tag{7.24}$$

We have split up the anti-commutator (7.24) into a trace term (first term) and a traceless term. Indeed it can be verified by means of thc explicit representation that

$$\mathrm{Tr}\,\hat{\lambda}_i\hat{\lambda}_j = 2\delta_{ij} \quad ,$$

and accordingly

$$\mathrm{Tr}\,[\hat{\lambda}_i, \hat{\lambda}_j]_+ = 4\delta_{ij} = \mathrm{Tr}\,\tfrac{4}{3}\delta_{ij}\mathbb{1} = \tfrac{4}{3}\delta_{ij}\mathrm{Tr}\,\mathbb{1}$$

because $\mathrm{Tr}(\mathbb{1}) = 3$. We refer to the general considerations contained in Exercises 11.1, 11.2, and also 5.7. The relations (7.22, 24), the antisymmetry relations for the f_{ijk} (7.23) and the symmetry relations for the d_{ijk} are indeed helpful tools in practical calculations. Only the nonvanishing structure constants are listed in Table 7.1.

Equation (7.22) proves explicitly that the Lie algebra of SU(3) is closed, as expected according to our general considerations in Chap. 3. Similarly to SU(2) it is useful to redefine the generators as

$$\hat{F}_i = \tfrac{1}{2}\hat{\lambda}_i \quad . \tag{7.25}$$

From (7.22) we deduce that

$$[\hat{F}_i, \hat{F}_j]_- = \mathrm{i} f_{ijk}\hat{F}_k \quad . \tag{7.26}$$

These are similar commutation relations. However, the factor 2 has now disappeared. The analogy of the relations between $\hat{\lambda}_i$ and $\hat{F}_i$ in SU(3) and $\hat{\sigma}_i$ and $\hat{S}_i$ in SU(2) is obvious. This caused Gell-Mann to call the generators $\hat{F}_i$ collectively *F-spin*, even though there is no direct relation to angular momentum or spin.

As for the considerations concerning the angular momentum algebra [Lie algebra of SU(2) or SO(3), see Chap. 2] we now introduce the spherical representation of the $\hat{F}$-operators

$$\hat{T}_\pm = \hat{F}_1 \pm \mathrm{i}\hat{F}_2 \quad , \quad \hat{T}_3 = \hat{F}_3 \quad ,$$
$$\hat{V}_\pm = \hat{F}_4 \pm \mathrm{i}\hat{F}_5 \quad , \quad \hat{Y} = \frac{2}{\sqrt{3}}\hat{F}_8 \quad , \quad \hat{U}_\pm = \hat{F}_6 \pm \mathrm{i}\hat{F}_7 \quad . \tag{7.27}$$

The definitions for $\hat{T}_+$, $\hat{U}_+$, $\hat{V}_+$ and $\hat{T}_3$ are obviously based on the observation that the $\hat{F}$ operators are constructed from Pauli matrices [cf. our discussion following (7.19)]. The final definition of (7.27), i.e. the $\hat{Y}$-operator will prove useful in the following. The commutation relations for the new operators can be computed as follows:

$$[\hat{T}_3, \hat{T}_\pm]_- = \pm\hat{T}_\pm \quad [\hat{T}_+, \hat{T}_-]_- = 2\hat{T}_3$$
$$[\hat{T}_3, \hat{U}_\pm]_- = \mp\tfrac{1}{2}\hat{U}_\pm \quad [\hat{U}_+, \hat{U}_-]_- = \tfrac{3}{2}\hat{Y} - \hat{T}_3 \stackrel{\text{def}}{=} 2\hat{U}_3$$
$$[\hat{T}_3, \hat{V}_\pm]_- = \pm\tfrac{1}{2}\hat{V}_\pm \quad [\hat{V}_+, \hat{V}_-]_- = \tfrac{3}{2}\hat{Y} + \hat{T}_3 \stackrel{\text{def}}{=} 2\hat{V}_3 \tag{7.28a}$$
$$[\hat{Y}, \hat{T}_\pm]_- = 0 \quad [\hat{Y}, \hat{U}_\pm]_- = \pm\hat{U}_\pm \quad [\hat{Y}, \hat{V}_\pm]_- = \pm\hat{V}_\pm \tag{7.28b}$$
$$[\hat{T}_+, \hat{V}_+]_- = [\hat{T}_+, \hat{U}_-]_- = [\hat{U}_+, \hat{V}_+]_- = 0$$
$$[\hat{T}_+, \hat{V}_-] = -\hat{U}_- \quad [\hat{T}_+, \hat{U}_+]_- = \hat{V}_+$$
$$[\hat{U}_+, \hat{V}_-]_- = \hat{T}_- \quad [\hat{T}_3, \hat{Y}]_- = 0 \quad . \tag{7.28c}$$

The missing commutators can be deduced from (7.27), partly with the help of the relations resulting from the hermiticity of the $\hat{F}_i$, i.e.

$$\hat{T}_+ = (\hat{T}_-)^+ \quad , \quad \hat{U}_+ = (\hat{U}_-)^+ \quad , \quad \hat{V}_+ = (\hat{V}_-)^+ \quad . \tag{7.29}$$

Following (7.28), the maximum number of commuting generators of SU(3) Lie algebra is 2 (e.g. $[\hat{T}_3, \hat{Y}]_-$ or $[\hat{Y}, \hat{T}_\pm]_- = 0$). Thus[3], SU(3) is of rank 2 and as a consequence of Racah's theorem it contains two Casmir operators. These are generally defined as

$$\hat{C}_1(\hat{F}_i) = \sum_{i=1}^{8} \hat{F}_i^2 = -\frac{2i}{3}\sum_{ijk} f_{ijk}\hat{F}_i\hat{F}_j\hat{F}_k \quad \text{and} \tag{7.30}$$
$$\hat{C}_2(\hat{F}_i) = \sum_{ijk} d_{ijk}\hat{F}_i\hat{F}_j\hat{F}_k \quad . \tag{7.31}$$

EXERCISE

7.3 Symmetry of the Coefficient d_{ijk}

Problem. Show that the coefficients d_{ijk} are symmetric by proving the relation:

$$d_{ijk} = \tfrac{1}{4}\mathrm{Tr}([\hat{\lambda}_i, \hat{\lambda}_j]_+ \hat{\lambda}_k) \quad .$$

[3] It can be shown that the group SU(n) in general is of rank ($n-1$).

Exercise 7.3

Solution. We start with the anti-commutation relation

$$[\hat{\lambda}_i, \hat{\lambda}_j]_+ = \tfrac{4}{3}\delta_{ij}\mathbb{1} + 2\sum_{m=1}^{8} d_{ijm}\hat{\lambda}_m \quad , \tag{1}$$

multiply by $\hat{\lambda}_k$

$$[\hat{\lambda}_i, \hat{\lambda}_j]_+\hat{\lambda}_k = \tfrac{4}{3}\delta_{ij}\mathbb{1}\hat{\lambda}_k + 2\sum_m d_{ijm}\hat{\lambda}_m\hat{\lambda}_k \tag{2}$$

and take the trace to give

$$\mathrm{Tr}([\hat{\lambda}_i, \hat{\lambda}_j]_+\hat{\lambda}_k) = \tfrac{4}{3}\delta_{ij}\mathrm{Tr}(\mathbb{1}\hat{\lambda}_k) + 2\mathrm{Tr}\left(\sum_m d_{ijm}\hat{\lambda}_m\hat{\lambda}_k\right) \quad . \tag{3}$$

With $\mathbb{1}\hat{B} = \hat{B}$ it follows that

$$\tfrac{4}{3}\delta_{ij}\mathrm{Tr}(\mathbb{1}\hat{\lambda}_k) = \tfrac{4}{3}\delta_{ij}\mathrm{Tr}(\hat{\lambda}_k) = 0 \tag{4}$$

since the $\hat{\lambda}_k$ are traceless. Furthermore we can see from the explicit form of the $\hat{\lambda}_i$ that

$$\mathrm{Tr}(\hat{\lambda}_i\hat{\lambda}_k) = 2\delta_{ik}$$

$$\mathrm{Tr}([\hat{\lambda}_i, \hat{\lambda}_j]_+\hat{\lambda}_k) = 4\sum_m d_{ijm}\delta_{mk} = 4d_{ijk}$$

$$d_{ijk} = \tfrac{1}{4}\mathrm{Tr}([\hat{\lambda}_i, \hat{\lambda}_j]_+\hat{\lambda}_k) \quad . \tag{5}$$

The symmetry of the d_{ijk}, under exchange of i and j, follows from the symmetry of the anticommutator, because

$$[\hat{\lambda}_i, \hat{\lambda}_j]_+ = [\hat{\lambda}_j, \hat{\lambda}_i]_+ \quad . \tag{6}$$

The symmetry in the remaining indices can be deduced from the fact that two matrices may be exchanged under the trace operation: $\mathrm{Tr}(\hat{A}\hat{B}) = \mathrm{Tr}(\hat{B}\hat{A})$

$$\begin{aligned} 4d_{ijk} &= \mathrm{Tr}(\hat{\lambda}_k\hat{\lambda}_i\hat{\lambda}_j + \hat{\lambda}_k\hat{\lambda}_j\hat{\lambda}_i) = \mathrm{Tr}(\hat{\lambda}_k\hat{\lambda}_i\hat{\lambda}_j + \hat{\lambda}_i\hat{\lambda}_k\hat{\lambda}_j) \\ &= \mathrm{Tr}([\hat{\lambda}_i, \hat{\lambda}_k]_+\hat{\lambda}_j) = 4d_{ikj} \quad . \end{aligned} \tag{7}$$

Thus the d_{ijk} are symmetric in all indices.

EXERCISE

7.4 Antisymmetry of the Structure Constants f_{ijk}

Problem. Show that, analogously to the formula derived in Exercise 7.3 for the d_{ijk}, a similar relation holds for the structure constants f_{ijk}, given by

$$f_{ijk} = \frac{1}{4i}\mathrm{Tr}([\hat{\lambda}_i, \hat{\lambda}_j]_-\hat{\lambda}_k) \quad .$$

Exercise 7.4

Solution. Once again we start with the commutation relation for the λ_i:

$$[\hat{\lambda}_i, \hat{\lambda}_j]_- = 2i \sum_{m=1}^{8} f_{ijm}\hat{\lambda}_m \quad , \tag{1}$$

multiply by λ_k,

$$[\hat{\lambda}_i, \hat{\lambda}_j]_- \hat{\lambda}_k = 2i \sum_{m=1}^{8} f_{ijm}\hat{\lambda}_m\hat{\lambda}_k \tag{2}$$

and take the trace

$$\mathrm{Tr}([\hat{\lambda}_i, \hat{\lambda}_j]_- \hat{\lambda}_k) = 2i\mathrm{Tr}\left(\sum_{m=1}^{8} f_{ijm}\hat{\lambda}_m\hat{\lambda}_k\right) \quad . \tag{3}$$

Because of

$$\mathrm{Tr}(\hat{\lambda}_m\hat{\lambda}_k) = 2\delta_{mk} \tag{4}$$

it follows that

$$\mathrm{Tr}([\hat{\lambda}_i, \hat{\lambda}_j]_- \hat{\lambda}_k) = 4i f_{ijm}\delta_{mk} = 4\mathrm{i} f_{ijk}$$

$$f_{ijk} = \frac{1}{4\mathrm{i}}\mathrm{Tr}([\hat{\lambda}_i, \hat{\lambda}_j]_- \hat{\lambda}_k) \quad . \tag{5}$$

The relations

$$[\hat{\lambda}_i, \hat{\lambda}_j]_- = -[\hat{\lambda}_j, \hat{\lambda}_i]_- \quad \text{and} \quad \mathrm{Tr}(-\hat{A}) = -\mathrm{Tr}(\hat{A}) \quad , \tag{6}$$

yield

$$f_{ijk} = -f_{jik} \quad ,$$

showing that f_{ijk} is antisymmetric under exchange of i and j. The antisymmetry in the other indices follows from our previous considerations, namely,

$$\begin{aligned} 4\mathrm{i}f_{ijk} &= \mathrm{Tr}([\hat{\lambda}_i, \hat{\lambda}_j]_- \hat{\lambda}_k) = \mathrm{Tr}(\hat{\lambda}_i\hat{\lambda}_j\hat{\lambda}_k - \hat{\lambda}_j\hat{\lambda}_i\hat{\lambda}_k) = -\mathrm{Tr}(\hat{\lambda}_j\hat{\lambda}_i\hat{\lambda}_k - \hat{\lambda}_i\hat{\lambda}_j\hat{\lambda}_k) \\ &= -\mathrm{Tr}(\hat{\lambda}_i\hat{\lambda}_k\hat{\lambda}_j - \hat{\lambda}_k\hat{\lambda}_i\hat{\lambda}_j) = -\mathrm{Tr}([\hat{\lambda}_i, \hat{\lambda}_k]_- \hat{\lambda}_j) = -4\mathrm{i}f_{ikj} \quad , \end{aligned} \tag{7}$$

once again using $\mathrm{Tr}(\hat{A}\hat{B}) = \mathrm{Tr}(\hat{B}\hat{A})$. Thus the f_{ijk} are totally antisymmetric in all indices.

EXERCISE

7.5 Calculation of some d_{ijk} Coefficients and Structure Constants

Problem. In order to become more familiar with the SU(3) algebra, calculate the structure constants f_{156}, f_{458} and the coefficients d_{118} and d_{778}, using the explicit representation of the $\hat{\lambda}_i$ and the previously derived trace relations.

Exercise 7.5

Solution. In view of

$$f_{ijk} = \frac{1}{4\mathrm{i}} \mathrm{Tr}([\hat{\lambda}_i, \hat{\lambda}_j]_- \hat{\lambda}_k) \quad f_{156} = \frac{1}{4\mathrm{i}} ([\hat{\lambda}_1, \hat{\lambda}_5]_- \hat{\lambda}_6) \tag{1}$$

we first calculate

$$\begin{aligned} [\hat{\lambda}_1, \hat{\lambda}_5]_- &= \begin{pmatrix} 0 & 1 & 0 \\ 1 & 0 & 0 \\ 0 & 0 & 0 \end{pmatrix} \begin{pmatrix} 0 & 0 & -\mathrm{i} \\ 0 & 0 & 0 \\ \mathrm{i} & 0 & 0 \end{pmatrix} - \begin{pmatrix} 0 & 0 & -\mathrm{i} \\ 0 & 0 & 0 \\ \mathrm{i} & 0 & 0 \end{pmatrix} \begin{pmatrix} 0 & 1 & 0 \\ 1 & 0 & 0 \\ 0 & 0 & 0 \end{pmatrix} \\ &= \begin{pmatrix} 0 & 0 & 0 \\ 0 & 0 & -\mathrm{i} \\ 0 & -\mathrm{i} & 0 \end{pmatrix} \quad . \end{aligned} \tag{2}$$

Matrix multiplication yields

$$[\hat{\lambda}_1, \hat{\lambda}_5]_- \hat{\lambda}_6 = \begin{pmatrix} 0 & 0 & 0 \\ 0 & 0 & -\mathrm{i} \\ 0 & -\mathrm{i} & 0 \end{pmatrix} \begin{pmatrix} 0 & 0 & 0 \\ 0 & 0 & 1 \\ 0 & 1 & 0 \end{pmatrix} = \begin{pmatrix} 0 & 0 & 0 \\ 0 & -\mathrm{i} & 0 \\ 0 & 0 & -\mathrm{i} \end{pmatrix} \quad . \tag{3}$$

Evaluating the trace

$$\mathrm{Tr}([\hat{\lambda}_1, \hat{\lambda}_5]_- \hat{\lambda}_6) = -2\mathrm{i} \quad , \tag{4}$$

the structure constant we are looking for is found to be

$$f_{156} = \frac{1}{4\mathrm{i}}(-2\mathrm{i}) = -\frac{1}{2} \quad . \tag{5}$$

Performing the same calculation for

$$f_{458} = \frac{1}{4\mathrm{i}} \mathrm{Tr}([\hat{\lambda}_4 , \hat{\lambda}_5]_- \hat{\lambda}_8) \tag{6}$$

one obtains

$$\begin{aligned} [\hat{\lambda}_4, \hat{\lambda}_5]_- &= \begin{pmatrix} 0 & 0 & 1 \\ 0 & 0 & 0 \\ 1 & 0 & 0 \end{pmatrix} \begin{pmatrix} 0 & 0 & -\mathrm{i} \\ 0 & 0 & 0 \\ \mathrm{i} & 0 & 0 \end{pmatrix} - \begin{pmatrix} 0 & 0 & -\mathrm{i} \\ 0 & 0 & 0 \\ \mathrm{i} & 0 & 0 \end{pmatrix} \begin{pmatrix} 0 & 0 & 1 \\ 0 & 0 & 0 \\ 1 & 0 & 0 \end{pmatrix} \\ &= 2\mathrm{i} \begin{pmatrix} 1 & 0 & 0 \\ 0 & 0 & 0 \\ 0 & 0 & -1 \end{pmatrix} \end{aligned} \tag{7}$$

Exercise 7.5

and

$$[\hat{\lambda}_4,\hat{\lambda}_5]_-\hat{\lambda}_8 = \frac{2\mathrm{i}}{\sqrt{3}}\begin{pmatrix}1&0&0\\0&0&0\\0&0&-1\end{pmatrix}\begin{pmatrix}1&0&0\\0&1&0\\0&0&-2\end{pmatrix} = \frac{2\mathrm{i}}{\sqrt{3}}\begin{pmatrix}1&0&0\\0&0&0\\0&0&2\end{pmatrix}$$

$$\mathrm{Tr}([\hat{\lambda}_4,\hat{\lambda}_5]_-\hat{\lambda}_8) = \frac{2\mathrm{i}}{\sqrt{3}}3 \quad . \tag{8}$$

Hence, it follows that

$$f_{458} = \frac{1}{4\mathrm{i}}\frac{2\mathrm{i}}{\sqrt{3}}3 = \frac{1}{2}\sqrt{3} \quad . \tag{9}$$

For the d_{ijk},

$$d_{ijk} = \tfrac{1}{4}([\hat{\lambda}_i,\hat{\lambda}_j]_+\hat{\lambda}_k) \quad , \quad d_{118} = \tfrac{1}{4}([\hat{\lambda}_1,\hat{\lambda}_1]_+\hat{\lambda}_8) \quad . \tag{10}$$

Because of

$$[\hat{\lambda}_1,\hat{\lambda}_1]_+ = 2\begin{pmatrix}0&1&0\\1&0&0\\0&0&0\end{pmatrix}\begin{pmatrix}0&1&0\\1&0&0\\0&0&0\end{pmatrix} = 2\begin{pmatrix}1&0&0\\0&1&0\\0&0&0\end{pmatrix} \quad , \tag{11}$$

it follows that

$$[\hat{\lambda}_1,\hat{\lambda}_1]_+\hat{\lambda}_8 = \frac{2}{\sqrt{3}}\begin{pmatrix}1&0&0\\0&1&0\\0&0&0\end{pmatrix}\begin{pmatrix}1&0&0\\0&1&0\\0&0&-2\end{pmatrix} = \frac{2}{\sqrt{3}}\begin{pmatrix}1&0&0\\0&1&0\\0&0&0\end{pmatrix} \quad . \tag{12}$$

The trace gives

$$\mathrm{Tr}([\hat{\lambda}_1,\hat{\lambda}_1]_+\hat{\lambda}_8) = \frac{2}{\sqrt{3}}2 \quad , \tag{13}$$

leading to $d_{118} = 1/\sqrt{3}$.

Analogously one evaluates

$$d_{778} = \tfrac{1}{4}\mathrm{Tr}([\hat{\lambda}_7,\hat{\lambda}_7]_+\hat{\lambda}_8)$$

to find

$$[\hat{\lambda}_7,\hat{\lambda}_7]_+ = 2\begin{pmatrix}0&0&0\\0&0&-\mathrm{i}\\0&\mathrm{i}&0\end{pmatrix}\begin{pmatrix}0&0&0\\0&0&-\mathrm{i}\\0&\mathrm{i}&0\end{pmatrix} = 2\begin{pmatrix}0&0&0\\0&1&0\\0&0&1\end{pmatrix} \tag{14}$$

$$[\hat{\lambda}_7,\hat{\lambda}_7]_+\hat{\lambda}_8 = \frac{2}{\sqrt{3}}\begin{pmatrix}0&0&0\\0&1&0\\0&0&1\end{pmatrix}\begin{pmatrix}1&0&0\\0&1&0\\0&0&2\end{pmatrix} = \frac{2}{\sqrt{3}}\begin{pmatrix}0&0&0\\0&1&0\\0&0&-2\end{pmatrix} \tag{15}$$

$$([\hat{\lambda}_7,\hat{\lambda}_7]_+\hat{\lambda}_8) = -\frac{2}{\sqrt{3}}$$

$$d_{778} = \frac{1}{4}\left(-\frac{2}{\sqrt{3}}\right) = -\frac{1}{2\sqrt{3}} \quad .$$

EXERCISE

7.6 Relations Between the Structure Constants and the Coefficients d_{ijk}

Problem. Show that the identities

$$f_{plm}f_{mkq} + f_{klm}f_{mqp} + f_{pkm}f_{mql} = 0 \quad , \tag{1}$$

and

$$f_{pkm}d_{mlq} + f_{qkm}d_{mlp} + f_{lkm}d_{mpq} = 0 \tag{2}$$

hold.

Solution. In order to prove (1) we employ the Jacobi identity,

$$\begin{aligned}
&[[\hat{\lambda}_p, \hat{\lambda}_l]_-, \hat{\lambda}_k]_- + [[\hat{\lambda}_l, \hat{\lambda}_k]_-, \hat{\lambda}_p]_- + [[\hat{\lambda}_k, \hat{\lambda}_p]_-, \hat{\lambda}_l]_- \\
&\quad = \hat{\lambda}_p\hat{\lambda}_l\hat{\lambda}_k - \hat{\lambda}_l\hat{\lambda}_p\hat{\lambda}_k - \hat{\lambda}_k\hat{\lambda}_p\hat{\lambda}_l + \hat{\lambda}_k\hat{\lambda}_l\hat{\lambda}_p + \hat{\lambda}_l\hat{\lambda}_k\hat{\lambda}_p - \hat{\lambda}_k\hat{\lambda}_l\hat{\lambda}_p - \hat{\lambda}_p\hat{\lambda}_l\hat{\lambda}_k \\
&\qquad + \hat{\lambda}_p\hat{\lambda}_k\hat{\lambda}_l + \hat{\lambda}_k\hat{\lambda}_p\hat{\lambda}_l - \hat{\lambda}_p\hat{\lambda}_k\hat{\lambda}_l - \hat{\lambda}_l\hat{\lambda}_k\hat{\lambda}_p + \hat{\lambda}_l\hat{\lambda}_p\hat{\lambda}_k = 0 \quad .
\end{aligned} \tag{3}$$

Insertion of (7.22) into the l.h.s. of (3) yields

$$\begin{aligned}
0 &= 2if_{plm}[\hat{\lambda}_m, \hat{\lambda}_k]_- + 2if_{lkm}[\hat{\lambda}_m, \hat{\lambda}_p]_- + 2if_{kpm}[\hat{\lambda}_m, \hat{\lambda}_l]_- \\
&= -4(f_{plm}f_{mkn} + f_{lkm}f_{mpn} + f_{kpm}f_{mln})\hat{\lambda}_n \quad .
\end{aligned} \tag{4}$$

We multiply by $\hat{\lambda}_q$ and take the trace. With

$$\mathrm{Tr}(\hat{\lambda}_q\hat{\lambda}_n) = 2\delta_{qn}$$

follows

$$\begin{aligned}
0 &= f_{plm}f_{mkq} + f_{lkm}f_{mpq} + f_{kpm}f_{mlq} \\
&= f_{plm}f_{mkq} + f_{klm}f_{mqp} + f_{pkm}f_{mql} \quad .
\end{aligned} \tag{5}$$

In order to prove (2), we make use of (7.22) and (7.24), obtaining

$$\begin{aligned}
\mathrm{Tr}([\hat{\lambda}_a, \hat{\lambda}_b]_- \{\hat{\lambda}_c, \hat{\lambda}_d\}_+) &= \sum_{e,f} 2\mathrm{i} f_{abe}\, \mathrm{Tr}(\hat{\lambda}_e[\tfrac{4}{3}\delta_{cd} + 2d_{cdf}\hat{\lambda}_f]) \\
&= 4\mathrm{i}\sum_{e,f} f_{abe}d_{cdf}2\delta_{ef} = 8\mathrm{i}\sum_{e} f_{abe}d_{cde} \quad .
\end{aligned} \tag{6}$$

We can now rearrange (2) as follows:

$$\begin{aligned}
&f_{pkm}d_{mlq} + f_{qkm}d_{mlp} + f_{lkm}d_{mpq} \\
&\quad = \frac{1}{8\mathrm{i}}\mathrm{Tr}\{[\hat{\lambda}_p, \hat{\lambda}_k]_-\{\hat{\lambda}_l, \hat{\lambda}_q\}_+ + [\hat{\lambda}_q, \hat{\lambda}_k]_-\{\hat{\lambda}_l, \hat{\lambda}_p\}_+ + [\hat{\lambda}_l, \hat{\lambda}_k]_-\{\hat{\lambda}_p, \hat{\lambda}_q\}_+\} \\
&\quad = \frac{1}{8\mathrm{i}}\mathrm{Tr}\{\hat{\lambda}_p\hat{\lambda}_k\hat{\lambda}_l\hat{\lambda}_q + \underline{\hat{\lambda}_p\hat{\lambda}_k\hat{\lambda}_q\hat{\lambda}_l} - \underline{\hat{\lambda}_k\hat{\lambda}_p\hat{\lambda}_l\hat{\lambda}_q} - \underline{\hat{\lambda}_k\hat{\lambda}_p\hat{\lambda}_q\hat{\lambda}_l} + \underline{\hat{\lambda}_q\hat{\lambda}_k\hat{\lambda}_l\hat{\lambda}_p} \\
&\qquad + \underline{\hat{\lambda}_q\hat{\lambda}_k\hat{\lambda}_p\hat{\lambda}_l} - \underline{\hat{\lambda}_k\hat{\lambda}_q\hat{\lambda}_l\hat{\lambda}_p} - \underline{\hat{\lambda}_k\hat{\lambda}_q\hat{\lambda}_p\hat{\lambda}_l} + \underline{\hat{\lambda}_l\hat{\lambda}_k\hat{\lambda}_p\hat{\lambda}_q} + \underline{\hat{\lambda}_l\hat{\lambda}_k\hat{\lambda}_q\hat{\lambda}_p} \\
&\qquad\qquad - \hat{\lambda}_k\hat{\lambda}_l\hat{\lambda}_p\hat{\lambda}_q - \hat{\lambda}_k\hat{\lambda}_l\hat{\lambda}_q\hat{\lambda}_p\} = 0 \quad .
\end{aligned} \tag{7}$$

Terms equally underlined cancel mutually under the trace.

EXERCISE

7.7 Casimir Operators of SU(3)

Problem. Verify by use of the commutation relations (7.26) and the anticommutation relations (7.24) that $\hat{C}_1$ and $\hat{C}_2$ [(7.30, 31)] are Casimir operators of SU(3).

Solution. 1) We show that

$$\hat{C}_1 = \sum_{i=1}^{8} \hat{F}_i^2$$

is a Casimir operator:

$$\begin{aligned}[\hat{C}_1, \hat{F}_k]_- &= \sum_{i=1}^{8} (\hat{F}_i[\hat{F}_i, \hat{F}_k]_- + [\hat{F}_i, \hat{F}_k]_- \hat{F}_i) \\ &= \mathrm{i} \sum_{i,m=1}^{8} f_{ikm}(\hat{F}_i \hat{F}_m + \hat{F}_m \hat{F}_i) = 0 \quad ,\end{aligned}$$

because the structure constants f_{ikm} are antisymmetric in the indices i and m.

2) We show that

$$\hat{C}_2 = \sum_{ijn} d_{ijn} \hat{F}_i \hat{F}_j \hat{F}_n$$

is a Casimir operator:

$$\begin{aligned}[\hat{C}_2, \hat{F}_k]_- &= \sum_{ijn} d_{ijn} \{\hat{F}_i \hat{F}_j [\hat{F}_n, \hat{F}_k]_- + \hat{F}_i [\hat{F}_j, \hat{F}_k]_- \hat{F}_n + [\hat{F}_i, \hat{F}_k]_- \hat{F}_j \hat{F}_n\} \\ &= \mathrm{i} \sum_{ijn} d_{ijn} \{\hat{F}_i \hat{F}_j f_{nkm} \hat{F}_m + \hat{F}_i \hat{F}_m \hat{F}_n f_{jkm} + \hat{F}_m \hat{F}_j \hat{F}_n f_{ikm}\} \\ &= \mathrm{i} \sum_{ijn} \hat{F}_i \hat{F}_j \hat{F}_m \{d_{ijn} f_{nkm} + f_{nkj} d_{inm} + d_{njm} f_{nki}\} = 0 \quad .\end{aligned}$$

In the last step we have employed (2) from Exercise 7.6. Since $\hat{C}_1$ and $\hat{C}_2$ commute with all operators $\hat{F}_i$, they are indeed Casimir operators.

EXERCISE

7.8 Useful Relations for SU(3) Casimir Operators

Problem. Prove the following relations

$$\text{(a)} \quad \hat{C}_1(\hat{F}_i) \equiv \sum_l \hat{F}_l^2 = -\frac{2\mathrm{i}}{3} \sum_{ijk} f_{ijk} \hat{F}_i \hat{F}_j \hat{F}_k \quad , \tag{1}$$

$$\text{(b)} \quad \hat{C}_2(\hat{F}_i) \equiv \sum_{ijk} d_{ijk} \hat{F}_i \hat{F}_j \hat{F}_k = \hat{C}_1(2\hat{C}_1 - \tfrac{11}{6}) \quad . \tag{2}$$

Exercise 7.8

Solution. (a) For the proof of (1) we use the relations

$$[\hat{F}_i, \hat{F}_j] = \mathrm{i} f_{ijk} \hat{F}_k \quad , \tag{3}$$

$$f_{ijk} f_{ijl} = 3\delta_{kl} \quad . \tag{4}$$

The proof of (4) we will give in Exercise 11.3. We evaluate

$$\begin{aligned} \sum_{ijk} f_{ijk} \hat{F}_i \hat{F}_j \hat{F}_k &= f_{ijk} \hat{F}_j \hat{F}_i \hat{F}_k + \mathrm{i} f_{ijk} f_{ijl} \hat{F}_l \hat{F}_k \\ &= -f_{ijk} \hat{F}_j \hat{F}_i \hat{F}_k + \mathrm{i}^3 \delta_{kl} \hat{F}_l \hat{F}_k \\ &= -f_{ijk} \hat{F}_i \hat{F}_j \hat{F}_k + 3\mathrm{i} \sum_l \hat{F}_l^2 \quad . \end{aligned} \tag{5}$$

Extracting $\sum_l \hat{F}_l^2$ gives

$$\hat{C}_1 = \sum_l \hat{F}_l = -\frac{2\mathrm{i}}{3} \sum_{ijk} f_{ijk} \hat{F}_i \hat{F}_j \hat{F}_k \quad . \tag{6}$$

(b) For a proof of (2) we recall the definition of the d_{ijk}

$$\{\hat{F}_i, \hat{F}_j\}_+ = d_{ijk} \hat{F}_k + \tfrac{1}{3}\delta_{ij} \quad . \tag{7}$$

This yields

$$\begin{aligned} \hat{C}_2 &= \sum_{ijk} d_{ijk} \hat{F}_i \hat{F}_j \hat{F}_k = \sum_{ij} \hat{F}_i \hat{F}_j \{\hat{F}_i, \hat{F}_j\}_+ - \frac{1}{3} \sum_{ij} \hat{F}_i \hat{F}_j \delta_{ij} \\ &= \sum_{ij} \hat{F}_i \hat{F}_j \hat{F}_i \hat{F}_j + \sum_{ij} \hat{F}_i \hat{F}_i \hat{F}_j \hat{F}_j - \frac{1}{3} \sum_i \hat{F}_i^2 \quad . \end{aligned} \tag{8}$$

Now we set $\sum_i \hat{F}_i^2 = \hat{C}_1$ and we apply in the first term again the relation (3). This leads to

$$\hat{C}_2 = 2\hat{C}_1^2 - \tfrac{1}{3}\hat{C}_1 + \mathrm{i} f_{ijk} \hat{F}_i \hat{F}_j \hat{F}_k \quad . \tag{9}$$

For the last term we insert the relation that has been proved in (a) and in this way we get the result

$$\hat{C}_2 = \hat{C}_1 \left(2\hat{C}_1 - \frac{11}{6} \right) \quad . \tag{10}$$

7.4 The Subalgebras of the SU(3)-Lie Algebra and the Shift Operators

In order to become more familiar with the F-spin algebra, we study some of its subalgebras. The first line of (7.28a) shows that the operators $(\hat{T}_1, \hat{T}_2, \hat{T}_3)$ fulfill the Lie algebra of SU(2) [isomorph to SO(3)].

$$[\hat{T}_+,\hat{T}_-]_- = 2\hat{T}_3 \quad , \quad [\hat{T}_3,\hat{T}_\pm]_- = \pm\hat{T}_\pm \quad . \tag{7.32}$$

The operators $\hat{T}_i$ $(i = 1,2,3)$ form a closed subalgebra. The same holds for the operators

$$\{\hat{U}_+,\hat{U}_-,\hat{U}_3\} \quad ,$$

for which we have the relations

$$\{\hat{U}_+,\hat{U}_-]_- = 2\hat{U}_3 \quad , \quad [\hat{U}_3,\hat{U}_\pm]_- = \pm\hat{U}_\pm \quad . \tag{7.33}$$

The last of these identities follows from the commutators, (7.28a, b)

$$[\hat{T}_3,\hat{U}_\pm]_- = \mp\tfrac{1}{2}\hat{U}_\pm \quad | \quad \times(-\tfrac{1}{2})$$

$$[\hat{Y},\hat{U}_\pm]_- = \pm\,\hat{U}_\pm\tfrac{1}{2} \quad | \quad \times\tfrac{3}{4}$$

$$[\hat{U}_3,\hat{U}_\pm] = \pm\tfrac{3}{4}\hat{U}_\pm \pm \tfrac{1}{4}\hat{U}_\pm = \pm\,\hat{U}_\pm \quad .$$

Similarly one obtains

$$[\hat{V}_+,\hat{V}_-]_- = 2\hat{V}_3 \quad , \quad [\hat{V}_3,\hat{V}_\pm]_- = \pm\hat{V}_\pm \quad . \tag{7.34}$$

The *T-spin algebra* (operators $\hat{T}_i$), as well as the *U-spin algebra* (operators $\hat{U}_i$) and the *V-spin algebra* (operators $\hat{V}_i$) are closed. All three of them are subalgebras of SU(3) and each individually matches the algebra of the angular momentum operators [Lie algebra SU(2)]. The operators $\hat{T}_\pm$, $\hat{U}_\pm$, $\hat{V}_\pm$ are also [cf. (2.16)] shift operators, i.e. raising $(\hat{T}_+,\hat{U}_+,\hat{V}_+)$ and lowering $(\hat{T}_-,\hat{U}_-,\hat{V}_-)$ operators.

The question arises as to which quantum number gets raised and lowered, respectively. To answer this we consider the final equation (7.28c),

$$[\hat{Y},\hat{T}_3]_- = 0 \quad . \tag{7.35}$$

Accordingly the operators $\hat{Y}$ and $\hat{T}_3$ may be simultaneously diagnolized. We denote the common eigenstates by

$$|T_3,Y\rangle$$

and have

$$\hat{T}_3|T_3\,Y\rangle = T_3|T_3\,Y\rangle \quad , \tag{7.36}$$

$$\hat{Y}|T_3\,Y\rangle = Y|T_3\,Y\rangle \quad . \tag{7.37}$$

From $[\hat{T}_3,\hat{V}_\pm]_- = \pm\tfrac{1}{2}\hat{V}_\pm$

$$(\hat{T}_3\hat{V}_\pm - \hat{V}_\pm\hat{T}_3)|T_3\,Y\rangle = \pm\tfrac{1}{2}\hat{V}_\pm|T_3\,Y\rangle \quad ,$$

and by making use of (7.36)

$$\hat{T}_3\hat{V}_\pm|T_3\,Y\rangle - T_3\hat{V}_\pm|T_3\,Y\rangle = \pm\tfrac{1}{2}\hat{V}_\pm|T_3\,Y\rangle \quad .$$

follows that

$$\hat{T}_3(\hat{V}_\pm|T_3\,Y\rangle) = (T_3 \pm \tfrac{1}{2})(\hat{V}_\pm|T_3\,Y\rangle) \tag{7.38}$$

implies

$$\hat{V}_\pm |T_3 Y\rangle = \sum_{Y'} N(T_3, Y, Y') |T_3 \pm \tfrac{1}{2}, Y'\rangle \quad . \tag{7.39}$$

The normalization factors $N(T_3, Y, Y')$ occuring in the above equation may depend on the quantum numbers T_3 and Y. Thus, the operators $\hat{V}_\pm$ transform a state with quantum number T_3 into a state with quantum number $T_3 \pm \frac{1}{2}$ and yet unknown hypercharge Y'. $\hat{V}_\pm$ raises and lowers, respectively, the quantum number T_3 by $\frac{1}{2}$. By the same arguments we deduce, from $[\hat{T}_3, \hat{U}_\pm]_- = \mp\frac{1}{2}\hat{U}_\pm$, the relation

$$\hat{T}_3(\hat{U}_\pm |T_3 Y\rangle) = (T_3 \mp \tfrac{1}{2})(\hat{U}_\pm |T_3 Y\rangle) \quad . \tag{7.40}$$

Hence, $\hat{U}_\pm$ lowers and raises, respectively, the quantum number T_3 by $\frac{1}{2}$. From the relation

$$[\hat{Y}, \hat{V}_\pm]_- = \pm \hat{V}_\pm \quad ,$$

follows analogously

$$\hat{Y}(\hat{V}_\pm |T_3, Y\rangle) = (Y \pm 1)(\hat{V}_\pm |T_3 Y\rangle) \quad . \tag{7.41}$$

Hence $\hat{V}_\pm$ raises and lowers the quantum number Y by 1. Finally, the commutator

$$[\hat{Y}, \hat{U}_\pm]_- = \pm \hat{U}_\pm$$

yields the equation for the eigenvalues

$$\hat{Y}(\hat{U}_\pm |T_3, Y\rangle) = (Y \pm 1)(\hat{U}_\pm |T_3 Y\rangle) \quad . \tag{7.42}$$

$\hat{U}_\pm$ raises and lowers, respectively, the quantum number Y by 1. Because of $[\hat{Y}, \hat{T}_\pm]_- = 0$, the operators $\hat{T}_\pm$ do not change the quantum number Y. In view of the algebra of angular momentum (cf. Chap. 2), we already knew that the quantum number T_3 may be integer or half-integer valued. This is evident because $\hat{V}_\pm$ and $\hat{U}_\pm$ shift T_3 by half-integer units and $\hat{T}_\pm$ change it by integer units. We leave open the question of the units in which V has to be measured. We will return to it later, when we consider special examples (cf. the section

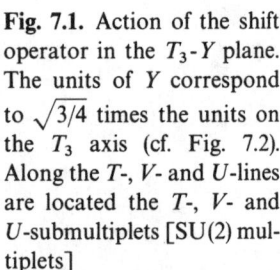

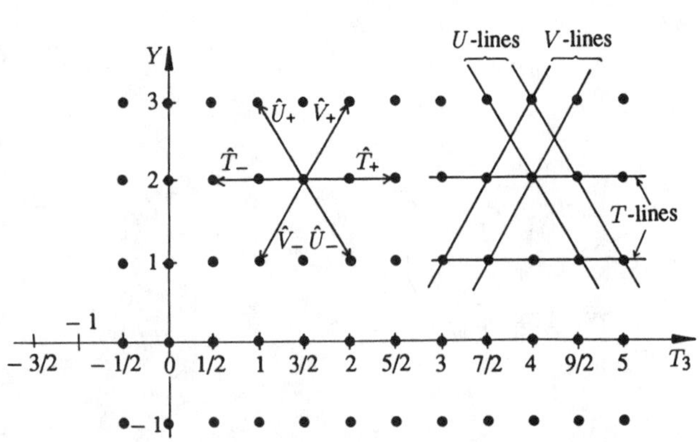

Fig. 7.1. Action of the shift operator in the T_3-Y plane. The units of Y correspond to $\sqrt{3/4}$ times the units on the T_3 axis (cf. Fig. 7.2). Along the T-, V- and U-lines are located the T-, V- and U-submultiplets [SU(2) multiplets]

about quarks). The shift operators either have no effect on the quantum number $Y(\hat{T}_\pm)$, or change it by integer units $(\hat{V}_\pm, \hat{U}_\pm)$. This is illustrated in Fig. 7.1. For example, $\hat{V}_+$ raises Y by 1 and T_3 by $\frac{1}{2}$ whereas $\hat{T}_+$ does not change Y at all, but changes T_3 by 1, etc. We have scaled the integer units on the Y-axis by a factor $(3/4)^{1/2}$ with respect to the units on the T_3-axis. Thus, the shift operators $\hat{T}_-$, $\hat{V}_+$, $\hat{U}_-$ or $\hat{T}_+$, $\hat{V}_-$, $\hat{U}_+$ form equilateral triangles (see Fig. 7.2).

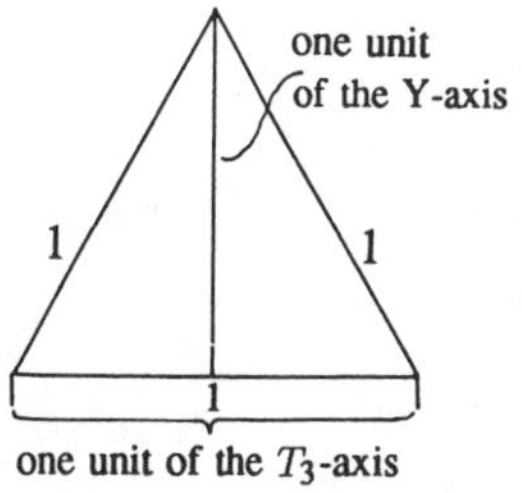

Fig. 7.2. Units in the Y-T_3 diagram

7.5 Coupling of T-, U- and V-Multiplets

Some general features of an SU(3)-multiplet structure may be expressed quite simply:

1) The SU(3) algebra has as subalgebras are T-, U-, and V-algebra. Each of these algebras is isomorphic to the SU(2) algebra, i.e. to the algebra of angular momentum. SU(3) multiplets, therefore, can be constructed by means of coupled T-, U-, and V-multiplets.

2) The operators $\hat{T}_3$, $\hat{Y}$ and, therefore, also $\hat{U}_3 = \frac{1}{2}(\frac{3}{2}\hat{Y} - \hat{T}_3)$ and $\hat{V}_3 = \frac{1}{2}(\frac{3}{2}\hat{Y} + \hat{T}_3)$ may be simultaneously diagonalized. Their eigenvalues are

$$T_3, \quad Y, \quad U_3 = \tfrac{1}{2}(\tfrac{3}{2}Y - T_3) \quad \text{and} \quad V_3 = \tfrac{1}{2}(\tfrac{3}{2}Y + T_3) \quad .$$

3) The shift operators $\hat{T}_\pm$, $\hat{V}_\pm$, $\hat{U}_\pm$ act on the states of a SU(3) multiplet according to Fig. 7.1. The end points of these operators are situated on a regular hexagon.

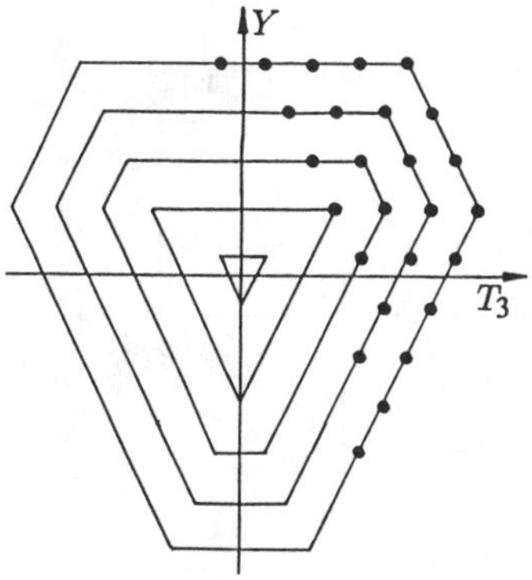

Fig. 7.3. Typical form of a representation of a SU(3) multiplet within the Y-T_3 plane

4) Hence, the SU(3) multiplet is constructed from a T-multiplet (parallel to the T_3 axis), a V-multiplet (along the V-lines of the diagram) and an U-multiplet (along the U-lines of the diagram). These T-, U-, V-submultiplets must be coupled because of the commutation relations (e.g. $[\hat{T}_+, \hat{V}_-]_- = -\hat{U}_-$, $[\hat{T}_+, \hat{U}_+]_- = -\hat{V}_+$).

5) Due to the equivalence of the three subalgebras T, U, V, the (finite) representations of SU(3) multiplets within the Y-T_3 plane have to be regular (but not necessarily equilateral) hexagons or triangles (if three sides of the hexagon have vanishing length) of the form shown in Fig. 7.3.

This is supported by the following considerations: the elements (states) of a T-multiplet are placed along the T-lines and are counted by the quantum numbers T_3. As we know from the angular momentum (and the algebra of isospin) all T_3 values of a given multiplet have to be within the interval $T_{3\,\max} \geq T_3 \geq -T_{3\,\max}$. Therefore, the SU(3) multiplet has to be symmetric with respect to the Y-axis ($T_3 = 0$ axis). Evidently, the T lines make a right angle with the Y-axis.

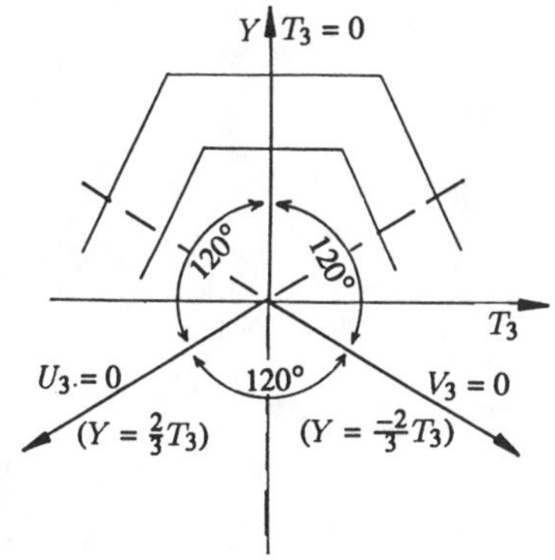

Fig. 7.4. The three axes of an SU(3) multiplet. The angles between the axes amounts to 120°, the Y-axis being scaled by a factor of $\sqrt{3/4}$

6) As the T-, V- and U-algebras are equally symmetric and, hence, equivalent subalgebras of SU(3), a figure representing an SU(3) multiplet has to be symmetric with respect to the axis $U_3 = 0 = \frac{3}{4}(Y) - T_3$ and $V_3 = 0 = \frac{3}{2}(Y) + T_3$. The three symmetric axes intersect, forming an angle of 120°. The V-multiplets are arranged vertically to the ($V_3 = 0$) axis and the U-multiplets are arranged vertically to the ($U_3 = 0$) axis (Fig. 7.4). The three submultiplets are linked to each other (e.g. by the commutation relations of the type $[\hat{T}_+, \hat{V}_-] = -\hat{U}_-$). Thus, a regular hexagonal or triangular structure is formed, representing the SU(3) multiplet.

7) In particular, the origin ($Y = 0$, $T_3 = 0$) is the centre of each SU(3) multiplet. In other words: each SU(3) multiplet is centred around the origin of the Y-T_3 plane and is left unchanged by a rotation through $\pm 120°$ around the origin.

7.6 Quantitative Analysis of Our Reasoning

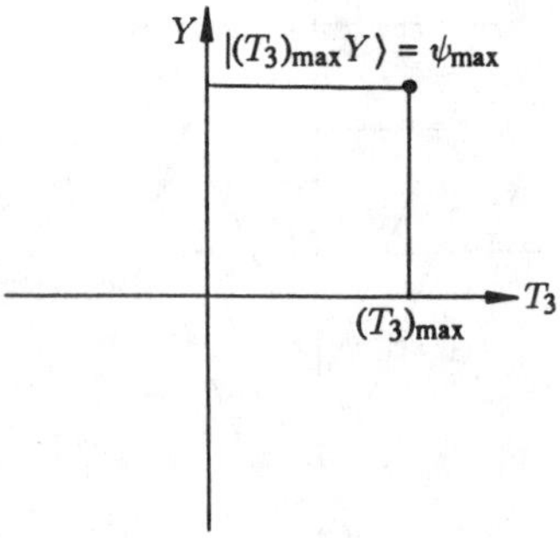

Fig. 7.5. Furthest right of the SU(3)-multiplet states

In the following we want to elaborate on our ideas about the structure of SU(3) multiplets. Our interest is restricted to finite multiplets, i.e. such multiplets containing a finite number of states. Consider first the state which belongs to the largest T_3 value in the SU(3) multiplet. Such a state always exists within a finite multiplet. We denote it by

$$\psi_{max} = |(T_3)_{max} Y\rangle \quad . \tag{7.43}$$

In a T_3-Y diagram (see Fig. 7.5) it is the furthest right of the states, and for this state holds the relation[4]

$$\hat{T}_+ \psi_{max} = \hat{V}_+ \psi_{max} = \hat{U}_- \psi_{max} = 0 \quad , \tag{7.44}$$

since each shift operator would otherwise raise the value of T_3, which is impossible according to the previous assumption.

The boundary of the multiplet may now be constructed by repeated applications of $\hat{V}_-$ on ψ_{max}. Let this process be repeated p times. Assume that the $(p+1)$th action on this state results in

$$(\hat{V}_-)^{p+1} \psi_{max} = 0 \quad . \tag{7.45}$$

This property uniquely defines the integer number p. As soon as the state $(\hat{V}_-)^p \psi_{max}$ is reached, one may follow the boundary of the multiplet by a repeated action of $\hat{T}_-$ on the state. This process may be repeated, say, q times, the $(q+1)$th action of the operator resulting in

$$(\hat{T}_-)^{q+1} (\hat{V}_-)^p \psi_{max} = 0 \quad , \tag{7.46}$$

which determines the integer q. The numbers p and q define a multiplet of the group SU(3).

Until now we have followed the boundary of the multiplet as depicted in Fig. 7.6.

One may wonder why the boundary line turns to the left and not to the right in the lower right corner of our figure. A global answer to this question can already be given based on symmetry groups: The representation of an SU(3)

---

[4] Following S. Gasiorowicz: *Elementary Particle Physics* (Wiley, New York 1967).

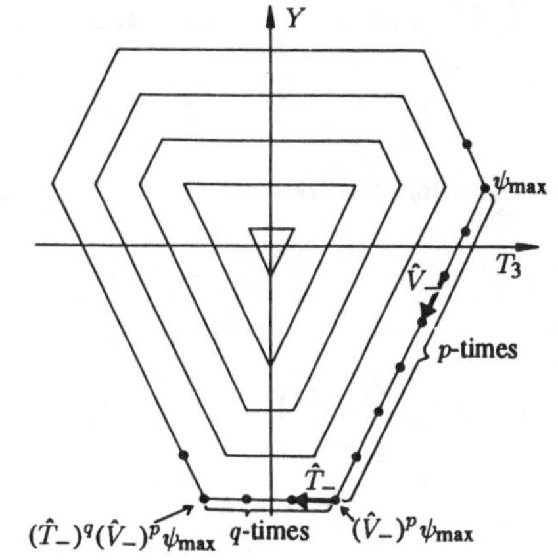

Fig. 7.6. The boundary of the SU(3) multiplet in Y-T_3 space

multiplet has a hexagonal structure with symmetry axes intersecting at 120° and with the origin $Y = 0$, $T_3 = 0$ in the centre. It will now be shown that the boundary of the multiplet always has to be convex. For this purpose, consider Fig. 7.7. The point M, representing the state with the maximal T_3 value, ψ_{max}, is called the state which carries the *maximal weight* T_3. Action of $\hat{V}_-$ on ψ_{max} results in a state N on the mesh.[5]

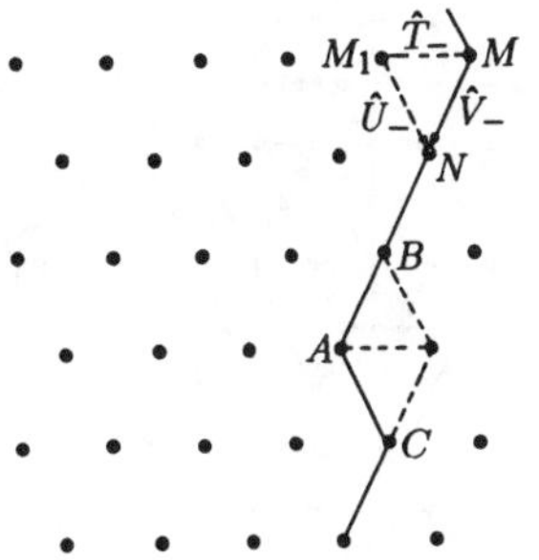

Fig. 7.7. A concave part of the boundary of an SU(3) multiplet. The dashed lines to the right of point A represent forbidden steps

It is important to realize that *there is only one unique state in N, if ψ_{max} in M is unique*. One might be inclined to think that, e.g., the state

$$\hat{U}_-\hat{T}_-\psi_{max}$$

is another, independent state in N, as it has not been reached directly but via the state M_1. However, by means of the commutation relation (7.28c) we show that

$$\begin{aligned}\hat{U}_-\hat{T}_-\psi_{max} &= ([\hat{U}_-\hat{T}_- - \hat{T}_-\hat{U}_-] + \hat{T}_-\hat{U}_-)\psi_{max}\\ &= [\hat{U}_-, \hat{T}_-]_-\psi_{max} \quad (\text{since } \hat{U}_-\Psi_{max} = 0)\\ &= \hat{V}_-\psi_{max} \quad .\end{aligned} \tag{7.47}$$

It may easily be checked that all other detours lead to the same result. Similarly, it follows that

$$\psi_B = \hat{V}_-\psi_N \tag{7.48}$$

is equally unique. The same holds true for the state represented by the point A and denoted by ψ_A, where

$$\psi_A = \hat{V}_-\psi_B \quad . \tag{7.49}$$

If there exists a state at C (which may be denoted by ψ_C), then

$$\hat{U}_+\psi_C = \lambda\psi_A \quad , \tag{7.50}$$

---

[5] If $\hat{V}_-\psi_{max} = 0$, one may argue in a similar way with $\hat{U}_+\psi_{max}$.

since the state in A is uniquely determined to be ψ_A. The relations

$$\lambda\langle\psi_A|\psi_A\rangle = \langle\hat{V}_-\psi_B|\hat{U}_+\psi_C\rangle = \langle\psi_B|\hat{V}_+\hat{U}_+|\psi_C\rangle \quad , \tag{7.51}$$

would then hold and thus

$$\lambda = \frac{\langle\psi_B|[\hat{V}_+,\hat{U}_+]_- + \hat{U}_+\hat{V}_+|\psi_C\rangle}{\langle\psi_A|\psi_A\rangle} = 0 \quad . \tag{7.52}$$

The commutator $[\hat{V}_+, \hat{U}_+]_-$ vanishes identically because of (7.28c), just as the matrix element

$$\begin{aligned}\langle\psi_B|\hat{U}_+\hat{V}_+|\psi_C\rangle &= \langle\hat{U}_-\psi_B|\hat{V}_+\psi_C\rangle \\ &= \langle\hat{U}_-(\hat{V}_-)^2\psi_{\max}|\hat{V}_+\psi_C\rangle\langle(\hat{V}_-)^2\hat{U}_-\psi_{\max}|\hat{V}_+\psi_C\rangle \\ &= 0 \quad , \end{aligned} \tag{7.53}$$

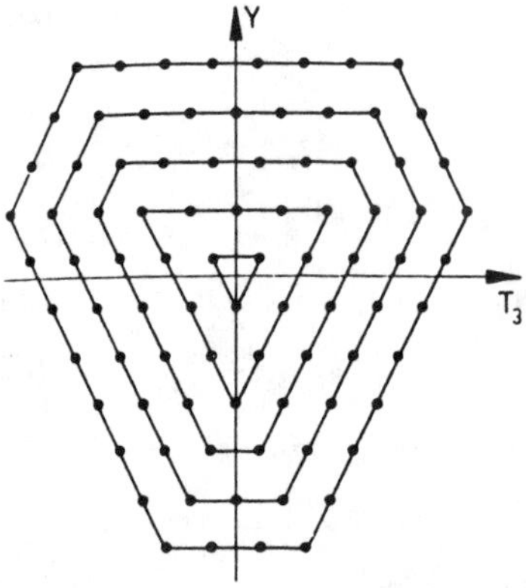

Fig. 7.8. Typical representation of an SU(3) multiplet, characterized by $(p,q) = (7,3)$

because $\hat{V}_-$ and $\hat{U}_-$ commute and $\hat{U}_-\psi_{\max} = 0$ [cf (7.44)]. According to (7.50), a constant $\lambda = 0$ implies the non-existence of a state at C, i.e. ψ_C does not exist. *The outer boundary of the multiplet cannot be concave, it has to be convex.* On the inner side of the boundary all points are occupied. Such an inner point is completely surrounded by occupied points. It should be remarked that the point M carrying the maximal weight must be occupied only by a single state, because it is the outermost point of a T-multiplet. Concerning T-multiplets (which are determined by angular momentum algebra) one knows that all points belonging to a sub-mesh of the quantum number T are occupied only once. This holds especially true for $(T_3)_{\max}$.

However, note that inner mesh-points e.g. M_1, may well be multiply occupied.

7.7 Further Remarks About the Geometric Form of an SU(3) Multiplet

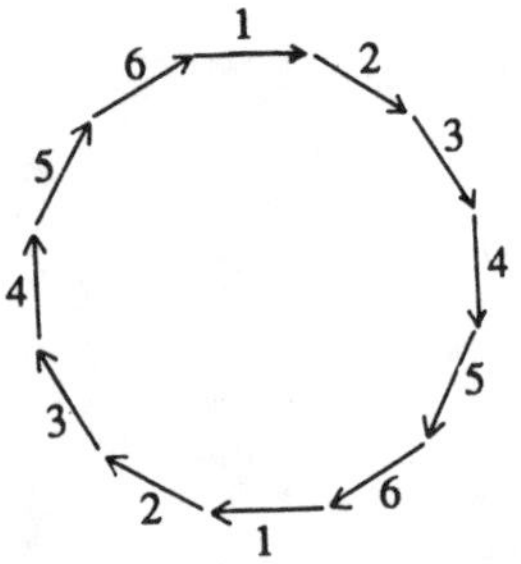

Fig. 7.9. A convex dodecagon contains 12 directions (pairs of which are antiparallel)

The requirements of convexity and 120° symmetry of the boundary of an SU(3) multiplet are met either by a triangle or a hexagon (see Fig. 7.8). Why do not higher polygons appear, e.g., 12 sides or 24 sides? The answer is that the available shift operators only allow for a shift along the V-, T- and U-lines in both directions. Thus, there are only six directions available, pairs of which are antiparallel. The convexity of the polygon allows solely for triangles and hexagons. A convex dodecagon could only be constructed by making use of 12 directions (two at a time antiparallel, see Fig. 7.9). However, as already stated above, there are only six directions available.

7.8 The Number of States on Mesh Points on Inner Shells

The mesh points of the boundary of an SU(3) multiplet are occupied only once, as demonstrated earlier. This means that there is only one state which corresponds to a given mesh point on the boundary. On the next layer or "shell" (by this we refer to the next hexagon on the inner side of the boundary), each mesh point is occupied by two states of the multiplet. Thus, the *next shell* has a *double occupancy, the following shell* has a *triple occupancy* etc. The multiplicity is raised by one each time we pass to the next inner shell, until, after q steps ($q \leq p$), the hexagon has become a triangle. Now every mesh point carries $(q+1)$ states. Each state within the triangle has a multiplicity of $(q+1)$. This is depicted in Fig. 7.10, showing the $(p, q) = (7, 3)$ multiplet.

In this typical SU(3) multiplet, characterized by $(p, q) = (7, 3)$, the multiplicities of states per mesh point are noted on the shells, which are hexagons or triangles, respectively. Every mesh point on a given shell carries the same number of states. This can be proved by first showing that the multiplicity is raised by one when passing from a given hexagonal shell to its next inner neighbour. To this end consider any two neighbouring shells (Fig. 7.11).

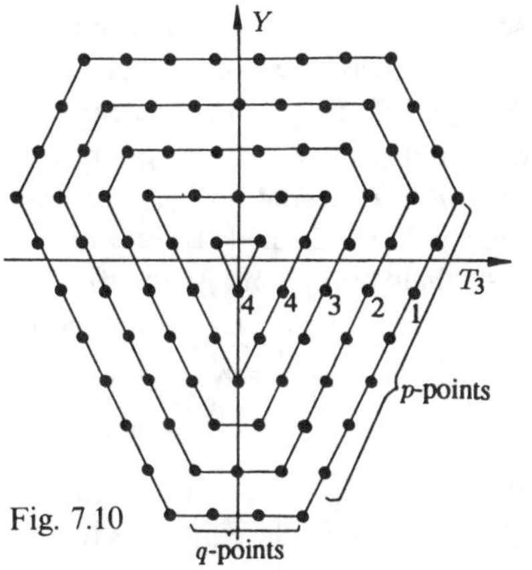

Fig. 7.10

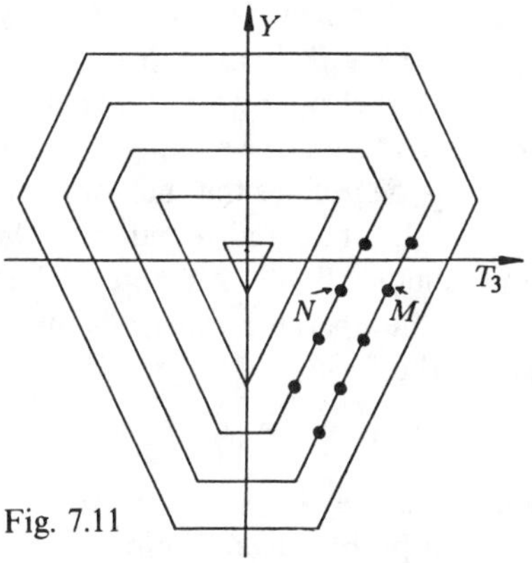

Fig. 7.11

Fig. 7.10. SU(3) multiplet with the multiplicity of the shells. Example of D(7, 3)

Fig. 7.11. The multiplicities of M and N

First it will be demonstrated that constant products of the V-, T- and U-shift operators, which are independent of Y, may be written in a form

$$\hat{P}_{\alpha,\beta} = \hat{T}_-^{\alpha} \hat{U}_+^{\beta} \hat{V}_-^{\beta} \quad (\alpha, \beta \geq 0) \quad . \tag{7.54}$$

Furthermore, for $\mu = \alpha + \beta$ there exist exactly $\mu + 1$ independent products. (These operators transform a state with quantum number T_3 into a state with $T_3 - \mu$). To begin with we consider an example: μ takes on the value $\mu = 1$, only in the cases $\alpha = 1$, $\beta = 0$ or $\alpha = 0$, $\beta = 1$. The products are easy to evaluate

$$\hat{P}_{1,0} = \hat{T}_- \quad \text{and} \quad \hat{P}_{0,1} = \hat{U}_+ \hat{V}_- \quad . \tag{7.55}$$

There is still one combination of operators leading to the same value T_3 and

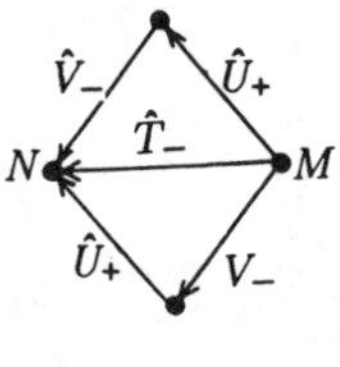

Fig. 7.12. Of the three apparently possible routes from M to N, only two are independent due to the commutation relations

leaving Y unchanged, namely the product $\hat{V}_-\hat{U}_+$ (Fig. 7.12). In view of the identity $\hat{V}_-\hat{U}_+ = -\hat{T}_- + \hat{U}_+\hat{V}_-$, the operator $\hat{V}_-\hat{U}_+$ may be represented in terms of the products (7.55). Thus, the assumption is verified for the case $\mu = 1$.

For the general case consider the products of the shift operators ($\hat{U}_+ \hat{V}_-$) and $\hat{T}_-$ which leave the hypercharge Y unchanged. As $\hat{U}_+$ raises the hypercharge by 1 and $\hat{V}_-$ lowers it by the same amount, such a product must contain an equal number of $\hat{U}_+$ and $\hat{V}_-$ operators. To begin with we show that all of these products may be written as linear combinations of the kind,

$$\hat{P}_{\alpha,\beta} = \hat{T}_-^\alpha \hat{U}_+^\beta \hat{V}_-^\beta \quad . \tag{7.56}$$

By making use of the equations

$$\hat{V}_-\hat{U}_+ = -\hat{T}_- + \hat{U}_+\hat{V}_- \quad , \quad \text{and} \quad \hat{V}_-\hat{T}_- = \hat{T}_-\hat{V}_- \quad , \tag{7.58}$$

any given product may be reordered such that all $\hat{V}_-$ operators appear on the right. Thus, we get a sum of terms in the form

$$\hat{U}_+\hat{T}_- \dots \hat{T}_-\hat{U}_+\hat{V}_-^\beta \quad . \tag{7.59}$$

In a second step all operators $\hat{T}_-$ are moved to the left by making use of

$$\hat{U}_+\hat{T}_- = \hat{T}_-\hat{U}_+ \quad . \tag{7.60}$$

Thus, a decomposition of an arbitrary product into a sum of terms in the form of (7.54) is obtained. The operators $\hat{P}_{\alpha,\beta}$ decrease the isospin quantum number by $\alpha\cdot 1 + \beta\cdot\frac{1}{2} + \beta\cdot\frac{1}{2} = \alpha + \beta \equiv \mu$. On the other hand there are, for any given μ, exactly $\mu + 1$ operators $\hat{P}_{\alpha,\beta}$, namely $\hat{P}_{\alpha,\mu-\alpha}$, $\alpha = 0, 1, 2, \dots, \mu$. Now returning to the general case, if we apply $\hat{P}_{\alpha\beta}$ to a state of the outermost shell then apparently $\mu = \alpha + \beta$ specifies the shell of the multiplet. For $\mu = 0$ we see that $\hat{P}_{00}$ does not change the state and remains on the outermost shell; $\mu = 1$ yields a state of the first inner shell; $\mu = 2$ a state of the second inner shell, and so on. As mentioned above, the operator $P_{\alpha,\beta}$ leads inwards, parallel to the Y axis. All of these paths are of the form

$$(\hat{T}_-)^{n_1}(\hat{V}_-\hat{U}_+)^{n_2}(\hat{U}_+\hat{V}_-)^{n_3}\dots \quad , \tag{7.61}$$

where $n_1 + n_2 + n_3 + \dots = \alpha + \beta$. These different zigzag paths lead from the starting point, towards inner shells parallel to the Y axis. They may be transformed into linear combinations of the shift operators in the form of (7.61) by means of the commutation relations $\hat{V}_-\hat{U}_+ = -\hat{T}_- + \hat{U}_+\hat{V}_-$. All we have to do now is to count the number of paths [i.e. operators in the form of (7.61)] which exist for a given number μ. We elucidate this in the following:

For a displacement of μ units in the T_3 direction, by means of the operator $\hat{P}_{\alpha,\beta}$, various possibilities ($\alpha = \mu$, $\beta = 0$), ($\alpha = \mu - 1$, $\beta = 1$)...($\alpha = 0$, $\beta = \mu$) exist. These are the $\mu + 1$ independent paths. For each state on the outermost shell ($\mu = 0$), we therefore get two states on the next following inner shell, three states on the second inner shell, etc. Since each point on the outermost shell is occupied by a single state [outermost points of the largest T_3 multiplet are singly occupied, because it is an SU(2) multiplet], each point on the next inner shell is occupied by one more state. Why does this reasoning fail to hold when we reach a triangular shell? This either happens after q or p steps, according to $q > p$ or vice versa. In order to obtain a graphical picture, we observe once more the increasing number of paths when going towards inner shells of the multiplets using the example of a $D(7,3)$ multiplet (Fig. 7.13). We start with the state of maximum weight, the outer right (A). There are two independent paths to B, namely $\hat{T}_-$ and $\hat{V}_-\hat{U}_+$. This results in two states at B. From B to C, there is a path for each of the two states in B by applying the operator $\hat{T}_-$, that is $\hat{T}_-\hat{T}_-$ and $\hat{T}_-\hat{V}_-\hat{U}_+$. In addition we have the path $(\hat{V}_-)^2(\hat{U}_+)^2$ to C. This additional path is actually the reason for the increase of the multiplicity by one.

A similar reasoning holds true for D. The paths from C are extended by $\hat{T}_-$. Furthermore the path

$$(\hat{V}_-)^3(\hat{U}_+)^3$$

to D is situated on the triangle. Going from D to E, each existing path to D may be continued by applying the operator $\hat{T}_-$. But the additional path which existed for the previous shells now vanishes. It would be represented by the operator

$$(\hat{V}_-)^4(\hat{U}_+)^4 \quad .$$

This operator applied to the state $|A\rangle$ yields zero, because $(\hat{U}_+)^3|A\rangle$ (i.e. in the general case $(\hat{U}_+)^q|A\rangle$) is the last, nonvanishing state on the boundary of the $D(p,q)$ multiplet in the $\hat{U}_+$ direction. Therefore, the number of independent paths is no longer augmented, once we have reached the triangle. By means of the path $(\hat{U}_+)^p(\hat{V}_+)^p$ we argue in a similar way for the case $p < q$. Example 7.9 demonstrates these facts again using different arguments.

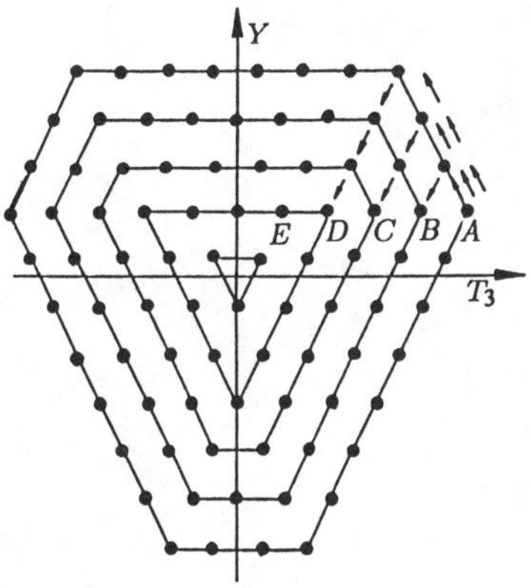

Fig. 7.13. Illustration of the routes from A to B, C, D and E

EXAMPLE

7.9 The Increase of the Multiplicity of States on the Inner Shells of SU(3) Multiplets

In the following we want to determine the multiplicity of states on the inner shells of a multiplet in another way. To this end we start with the state carrying the largest weight in a (p,q) multiplet (see figure). This state has the following values for the V-, U- and T-spin:

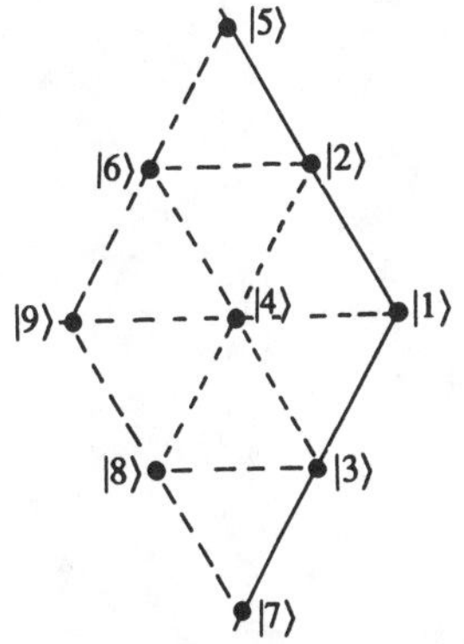

For the derivation of the multiplicities of states on inner shells of a multiplet

$$V = \frac{p}{2} \quad , \qquad V_3 = \frac{p}{2}$$

$$U = \frac{q}{2} \quad , \qquad U_3 = -\frac{q}{2}$$

$$T = \frac{p+q}{2} \quad , T_3 = \frac{p+q}{2} \quad . \tag{1}$$

Example 7.9

The relative phase is set equal to 1, i.e. we define

$$\left|\frac{p}{2},\frac{p}{2}\right\rangle_{V,1}=\left|\frac{q}{2},-\frac{q}{2}\right\rangle_{U,1}=\left|\frac{p+q}{2},\frac{p+q}{2}\right\rangle_{T,1}=|1\rangle\ . \tag{2}$$

In general we choose the phases for all states on the border of the multiplet (outermost shell of the multiplet) equal to 1, e.g.

$$|2\rangle=\left|\frac{p+1}{2},\frac{p+1}{2}\right\rangle_{V,2}=\left|\frac{q}{2},\frac{q-2}{2}\right\rangle_{U,2}=\left|\frac{p+q-1}{2},\frac{p+q-1}{2}\right\rangle_{T,2}$$

$$|3\rangle=\left|\frac{p+1}{2},\frac{p-1}{2}\right\rangle_{V,3}=\left|\frac{q+1}{2},\frac{q+1}{2}\right\rangle_{U,3}$$

$$=\left|\frac{p+q-1}{2},\frac{p+q-1}{2}\right\rangle_{T,3}\ . \tag{3}$$

As we have already seen, all of these states on the outermost shell are unique. How about some arbitrary state $|4\rangle$? To answer this question we apply the operator identity

$$\hat{U}_+\hat{V}_- - \hat{V}_-\hat{U}_+ - \hat{T}_- = 0 \tag{4}$$

to the state $|1\rangle$. This results in

$$\hat{U}_+\hat{V}_-|1\rangle - \hat{V}_-\hat{U}_+|2\rangle - \hat{T}_-|1\rangle$$

$$= \hat{U}+\sqrt{\frac{p}{2}\frac{(p+2)}{2}-\frac{p}{2}\frac{(p-2)}{2}}\,|3\rangle - \hat{V}_-\sqrt{\frac{q}{2}\frac{(q+2)}{2}-\frac{q}{2}\frac{(q-2)}{2}}\,|2\rangle$$

$$-\sqrt{\frac{(p+q)}{2}\frac{(p+q+2)}{2}-\frac{(p+q)}{2}\frac{(p+q-2)}{2}}\left|\frac{p+q}{2},\frac{p+q-2}{2}\right\rangle_{T,4}$$

$$=\sqrt{p}\sqrt{\frac{(q+1)}{2}\frac{(q+3)}{2}-\frac{(q+1)}{2}\frac{(q-1)}{2}}\left|\frac{q+1}{2},-\frac{q-1}{2}\right\rangle_{U,4}$$

$$-\sqrt{q}\sqrt{\frac{(p+1)}{2}\frac{(p+3)}{2}-\frac{(p+1)}{2}\frac{(p-1)}{2}}$$

$$\times\left|\frac{p+1}{2},\frac{p-1}{2}\right\rangle_{V,4}-\sqrt{p+q}\left|\frac{p+q}{2},\frac{p+q-2}{2}\right\rangle$$

$$=\sqrt{p(q+1)}\left|\frac{q+1}{2},-\frac{q-1}{2}\right\rangle_{U,4}-\sqrt{q(p+1)}\left|\frac{p+1}{2},\frac{p-1}{2}\right\rangle_{V,4}$$

$$-\sqrt{p+q}\left|\frac{p+q}{2},\frac{p+q-2}{2}\right\rangle_{T,4}=0\ . \tag{5}$$

For the moment we presume that the state $|4\rangle$ is occupied by one state only. In this case the corresponding U-, V- and T-spin states differ only by factors a, b

$$\left|\frac{p+q}{2},\frac{p+q-2}{2}\right\rangle_{T,4}=\frac{1}{a}\left|\frac{p+1}{2},\frac{p-1}{2}\right\rangle_{Vr,4}$$

$$=\frac{1}{b}\left|\frac{q+1}{2},\frac{q-1}{2}\right\rangle_{U,4}\quad \text{with } a,b=\pm 1\ . \tag{6}$$

Example 7.9

But from (5) it follows that

$$b\sqrt{p(q+1)} - a\sqrt{q(p+1)} = \sqrt{p+q} \quad . \tag{7}$$

We square this equation and obtain

$$ab\sqrt{pq(q+1)(p+1)} = pq \quad . \tag{8}$$

Apparently (8) is valid for the case where p or q equals zero. However, for $p \neq 0$ and $q \neq 0$ (8) has no solution, as squaring and division by pq results in

$$(q+1)(p+1) \neq pq \quad . \tag{9}$$

Thus, we have found that the state $|4\rangle$ is occupied by only one state provided that either p or q equals 0. This means that the corresponding representation has the form of a triangle. Since $|4\rangle$ is positioned on the edge of the multiplet in this case, this fact corresponds to our earlier result.

In the case $p \neq 0$ and $q \neq 0$ one has at least to assume a double occupancy of $|4\rangle$

$$\begin{aligned}
|4\rangle &= t_1 \left|\frac{p+q}{2}, \frac{p+q-2}{2}\right\rangle_{T,4} + t_2 \left|\frac{p+q-2}{2}, \frac{p+q-2}{2}\right\rangle_{T,4} \\
&= u_1 \left|\frac{q+1}{2}, -\frac{q-1}{2}\right\rangle_{U,4} + u_2 \left|\frac{q-1}{2}, -\frac{q-1}{2}\right\rangle_{U,4} \\
&= v_1 \left|\frac{p+1}{2}, \frac{p-1}{2}\right\rangle_{V,4} + v_2 \left|\frac{p-1}{2}, \frac{p-1}{2}\right\rangle_{V,4} \quad .
\end{aligned} \tag{10}$$

The question arises as to whether this ansatz satisfies (5). To prove this we consider the following general ansatz:

$$\begin{aligned}
\left|\frac{p+q}{2}, \frac{p+q-2}{2}\right\rangle_{T,4} &= a_1 \left|\frac{q+1}{2}, -\frac{q-1}{2}\right\rangle_{U,4} + a_2 \left|\frac{q-1}{2}, -\frac{q-1}{2}\right\rangle_{U,4} \\
\left|\frac{p+q-2}{2}, \frac{p+q-2}{2}\right\rangle_{T,4} &= a_2 \left|\frac{q+1}{2}, -\frac{q-1}{2}\right\rangle_{U,4} \\
&\quad - a_1 \left|\frac{q-1}{2}, -\frac{q-1}{2}\right\rangle_{U,A}
\end{aligned} \tag{11}$$

$$a_1^2 + a_2^2 = 1$$

$$\begin{aligned}
\left|\frac{p+q}{2}, \frac{p+q-2}{2}\right\rangle_{T,4} &= b_1 \left|\frac{p+1}{2}, \frac{p-1}{2}\right\rangle_{Vr,4} + b_2 \left|\frac{p-1}{2}, \frac{p-1}{2}\right\rangle_{Vr,4} \\
\left|\frac{p+q-2}{2}, \frac{p+q-2}{2}\right\rangle_{T,4} &= b_2 \left|\frac{p+1}{2}, \frac{p-1}{2}\right\rangle_{Vr,4} - b_1 \left|\frac{p-1}{2}, \frac{p-1}{2}\right\rangle_{V,4}
\end{aligned} \tag{12}$$

$$b_1^2 + b_2^2 = 1 \quad .$$

We consider whether a_1, a_2, b_1 and b_2, may be chosen such that (5) is fulfilled.

Example 7.9

To this end we multiply (5) by

$${}_{T,4}\left\langle \frac{p+q}{2}, \frac{p+q-2}{2} \right| \quad \text{and} \quad {}_{T,4}\left\langle \frac{p+q-2}{2}, \frac{p+q-2}{2} \right| \quad , \quad \text{obtaining:}$$

$$a_1\sqrt{p(q+1)} - b_1\sqrt{q(p+1)} - \sqrt{(p+q)} = 0 \tag{13}$$

$$a_2\sqrt{p(q+1)} - b_2\sqrt{q(p+1)} = 0 \quad . \tag{14}$$

Performing some algebraic manipulations we arrive at

$$b_1^2 q(p+1) = p(q+1)a_1^2 + p + q - 2a_1\sqrt{p(q+1)(p+q)} \tag{15}$$

$$b_1^2 q(p+1) = p(q+1)a_1^2 - p + q \quad . \tag{16}$$

Subtraction of these equations yields

$$a_1 = \sqrt{\frac{p}{(q+1)(p+1)}} \quad . \tag{17}$$

We choose the positive solution for a_2,

$$a_2 = \sqrt{\frac{q(q+p+1)}{(q+1)(p+1)}} \tag{18}$$

and get for b_1 and b_2

$$b_2 = \sqrt{\frac{q(q+p+1)}{(q+1)(p+q)}}, \quad b_1 = -\sqrt{\frac{q}{(q+1)(p+q)}} \quad . \tag{19}$$

Thus, we have shown that in the case $p \neq 0$ and $q \neq 0$, the mesh point must have at least a double occupancy. For the irreducible representations, in which we are interested, this minimal case must be realized.

From the fact that $|4\rangle$ is doubly occupied, it follows immediately, by means of the U, V and T symmetry, that all inner points are occupied by at least two states. In the next step one may derive the decompositions for the states $|6\rangle$ and $|8\rangle$, which are analogous to (11) and (12). As an example we apply the operator identity

$$\hat{T}_-\hat{U}_+ - \hat{U}_+\hat{T}_- = 0 \tag{20}$$

to the state $|1\rangle$. This leads to

$$\begin{aligned} &\sqrt{q(p+q-1)}\left| \frac{p+q-1}{2}, \frac{p+q-3}{2} \right\rangle_{T,6} \\ &- \sqrt{p+q}\left((a_1\sqrt{2q}\left| \frac{q+1}{2}, -\frac{q-3}{2} \right\rangle_{U,6} \right. \\ &\left. + a_2\sqrt{q-1}\left| \frac{q-1}{2}, -\frac{q-3}{2} \right\rangle_{U,6} \right) = 0 \end{aligned} \tag{21}$$

or

Example 7.9

$$\left|\frac{p+q-1}{2}, \frac{p+q-3}{2}\right\rangle_{T,6} = a_1\sqrt{\frac{2(p+q)}{p+q-1}}\left|\frac{q+1}{2}, -\frac{q-3}{2}\right\rangle_{U,6}$$

$$+ a_2\sqrt{\frac{(q-1)(p+q)}{q(p+q-1)}}\left|\frac{q-1}{2}, -\frac{q-3}{2}\right\rangle_{U,6}$$

$$= \sqrt{\frac{2p}{(q+1)(p+q-1)}}\left|\frac{q+1}{2}, -\frac{q-3}{2}\right\rangle_{U,6}$$

$$+ \sqrt{\frac{(q-1)(p+q+1)}{(q+1)(p+q-1)}}\left|\frac{q-1}{2}, -\frac{q-3}{2}\right\rangle_{U,6} \quad .$$

Having derived by this method the decompositions for $|6\rangle$ and $|8\rangle$, the commutator relation (4) is applied to the state $|4\rangle$ which yields the multiplicity and decompositions of $|9\rangle$.

In such a way one is able to derive the multiplicities for simple representations. The general proof for the rules in Sect. 7.8 is carried through by making use of group theoretical methods which cannot be explained in detail here. It should have become clear by now that the multiplicity of states is raised by 1, starting from the state with largest weight, by repeated action of $\hat{T}_-$. This process may be continued as long as the construction is possible, i.e. until a triangle is reached. Within the triangles the multiplicity is not raised.

EXERCISE

7.10 Particle States at the Centre of the Baryon Octet

Problem. With the help of the general relations deduced in the last Example (7.9), specify the combinations of particles which corresponds to the states $|T=0,\ T_3=0\rangle$, $|T=1,\ T_3=0\rangle$, $|V=0,\ V_3=0\rangle$, $|V=1,\ V_3=0\rangle$, $|U=0,\ U_3=0\rangle$ and $|U=1,\ U_3=0\rangle$.

For a physical interpretation of these states consider the octet of baryons from Example 6.3.

Solution. The classification of the particles is based on the isospin eigenstates

$$|\Sigma^0\rangle = |T=1,\ T_3=0\rangle = |1,0\rangle_T$$

$$|\Lambda^0\rangle = |0,0\rangle_T \quad . \tag{1}$$

Exercise 7.10

From (11) and (12), given in Example 7.9, it follows that ($p = q = 1$)

$$|\Sigma^0\rangle = \tfrac{1}{2}|1,0\rangle_U + \frac{\sqrt{3}}{2}|0,0\rangle_U$$

$$|\Lambda^0\rangle = \frac{\sqrt{3}}{2}|1,0\rangle_U - \frac{1}{2}|0,0\rangle_U \quad \text{and} \tag{2}$$

$$|\Sigma^0\rangle = -\frac{1}{2}|1,0\rangle_V + \frac{\sqrt{3}}{2}|0,0\rangle_V$$

$$|\Lambda^0\rangle = \frac{\sqrt{3}}{2}|1,0\rangle_V + \frac{1}{2}|0,0\rangle_V \quad . \tag{3}$$

These relations can be solved for the states on the rhs

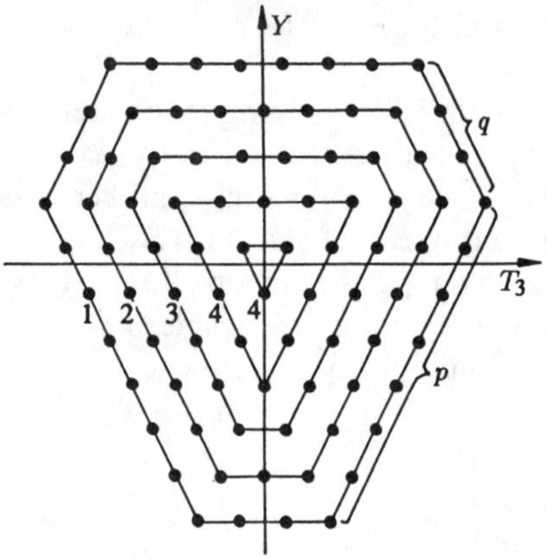

$D(7,3)$ multiplicites
The outermost shell consists of constants with $p = 7$ and $q = 3$ units

$$|0,0\rangle_U = \frac{\sqrt{3}}{2}|\Sigma^0\rangle - \frac{1}{2}|\Lambda^0\rangle$$

$$|1,0\rangle_U = \frac{1}{2}|\Sigma^0\rangle + \frac{\sqrt{3}}{2}|\Lambda^0\rangle$$

$$|0,0\rangle_V = \frac{\sqrt{3}}{2}|\Sigma^0\rangle + \frac{1}{2}|\Lambda^0\rangle$$

$$|1,0\rangle_V = -\frac{1}{2}|\Sigma^0\rangle + \frac{\sqrt{3}}{2}|\Lambda^0\rangle \quad . \tag{4}$$

Note, however, that the signs in (4) depend on the phase convention.

EXERCISE

7.11 Calculation of the Dimension of the Representation $D(p,q)$

Problem. Determine the dimension of the general representation $D(p,q)$ with the help o the known multiplicities of the states of the SU(3) multiplet. The states correspond to the lattice points of the graphical representation of the multiplet within the Y-T_3 plane.

Solution. We remember that p and q are just the number of links at the edge of the weight diagrams [see, for instance, $D(7,3)$ in above figure]. Furthermore we know that each point of the outermost shell has multiplicity 1, the points of the neighbouring inner shell have multiplicity 2, and so on, until we reach a triangular shell ($p=0$ or $q=0$). Then the multiplicity stops increasing. The outermost hexagon contains $3(p+q)$ points (states). The adjacent inner shell has (p and q decreased by one) $3[2(p-1+q-1)] = 3[2(p+q-2)]$ states, the factor of the two originating from the multiplicity of the lattice points. Accordingly the third shell has $3[3(p-2+q-2)] = 3[3(p+q-4)]$ states. In general the number of states of the $(n+1)$th shell is given by

$$3\{(n+1)(p+q-2n)\} \quad .$$

Summing up we obtain

$$3\sum_{n=0}^{q-1}(n+1)(p+q-2n) \quad , \quad (p>q) \quad .$$

The sum does not include the triangle and its interior points. The number of points enclosed by the triangle and lying on the triangle itself is simply (see above figure)

$$\sum_{n=1}^{p-q+1} n = \tfrac{1}{2}(p-q+1)(p-q+2) \quad .$$

The multiplicity of each triangle point is $q+1$. Thus, the contribution of the triangle to the dimension of the representation $D(p,q)$ is

$$\tfrac{1}{2}(q+1)(p-q+1)(p-q+2) \quad .$$

Altogether we have

$$d = \tfrac{1}{2}(q+1)(p-q+1)(p-q+2) + 3\sum_{n=0}^{q-1}(n+1)(p+q-2n) \quad .$$

The summation of the second term can be done by factorizing the expression and ordering with respect to parts proportional to n and n^2. Additionally, using

$$\sum_{n=0}^{q-1} n = \frac{q}{2}(q-1) \quad \text{and} \quad \sum_{n=0}^{q-1} n^2 = \frac{q}{6}(q-1)(2q-1)$$

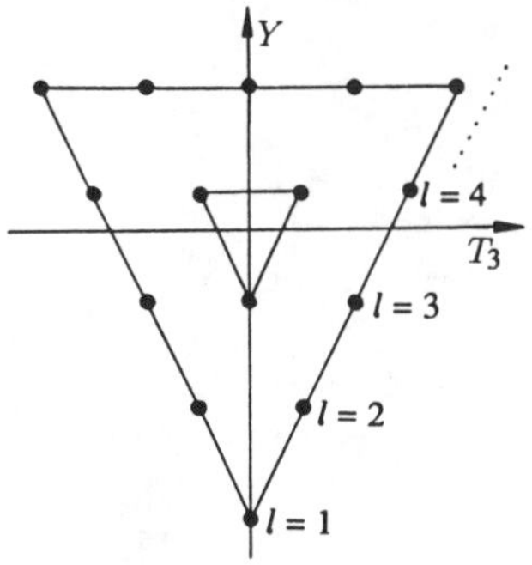

Illustration of the inner triangles. The number of lattice points lying upon the lines parallel to the T_3 axis is $l = 1, 2, 3, \ldots$

Exercise 7.11

the two terms yield the final result

$$d = \tfrac{1}{2}(p+1)(q+1)(p+q+2) \quad .$$

EXERCISE

7.12 Determination of the Dimensions of the Representation D

Problem. Determine the value of the quadratic Casimir operator $\sum_i \hat{F}_i^2$ in the representation $D(p,q)$!

Solution. First we express the $\hat{F}_i$ in terms of the operators defined in (7.27). We use the relation

$$\hat{T}_1^2 + \hat{T}_2^2 = (\hat{T}_+\hat{T}_- + \hat{T}_-\hat{T}_+)/2 \tag{1}$$

which is also valid for U- and V-spin. This is easily verified by explicitly checking for U- and V-spin.

We obtain

$$\sum_i \hat{F}_i^2 = \tfrac{1}{2}(\hat{T}_+\hat{T}_- + \hat{T}_-\hat{T}_+) + \hat{T}_3^2 + \tfrac{1}{2}(\hat{V}_+\hat{V}_- + \hat{V}_-\hat{V}_+)$$
$$+ \tfrac{1}{2}(\hat{U}_+\hat{U}_- + \hat{U}_-\hat{U}_+) + \tfrac{3}{4}\hat{Y}^2 \quad . \tag{2}$$

Since the value of a Casimir operator is fixed for a given multiplet, we can choose an arbitrary state of the multiplet to evaluate it. Again we choose the point where T_3 has its maximum (right outermost point of the multiplet). Acting with $\hat{V}_+$, $\hat{T}_+$ and $\hat{U}_-$ on this state leads out of the multiplet and yields zero as illustrated below.

The idea is now to rearrange the operators in (2) such, that Y_+, T_+ and U_- stand on the right of the operator products. Accordingly such products yield no contribution when acting on a state. For this reason we use (7.28a), which can also be written in the following manner:

Action of the operators $\hat{V}_+, \hat{U}_-$, and $\hat{T}_+$ on the state of maximal weight

$$\hat{T}_+\hat{T}_- = \hat{T}_-\hat{T}_+ + 2\hat{T}_3$$
$$\hat{V}_+\hat{V}_- = \hat{V}_-\hat{V}_+ + \tfrac{3}{2}\hat{Y} + \hat{T}_3$$
$$\hat{U}_-\hat{U}_+ = \hat{U}_+\hat{U}_- - \tfrac{3}{2}\hat{Y} + \hat{T}_3 \quad . \tag{3}$$

Inserting into (2) yields

$$\sum_i \hat{F}_i^2 = \hat{T}_+\hat{T}_- + \hat{T}_3^2 + \hat{V}_-\hat{V}_+ + \tfrac{1}{2}(\tfrac{3}{2}\hat{Y} + \hat{T}_3)$$
$$+ \hat{U}_+\hat{U}_- + \tfrac{1}{2}(-\tfrac{3}{2}\hat{Y} + \hat{T}_3) + \tfrac{3}{4}\hat{Y}^2$$
$$= \hat{T}_3^2 + 2\hat{T}_3 + \tfrac{3}{4}\hat{Y}^2 + \hat{T}_-\hat{T}_+ + \hat{V}_-\hat{V}_+ + \hat{U}_+\hat{U}_- \quad . \tag{4}$$

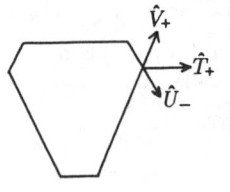

Exercise 7.12

Since the last three terms do not contribute (as mentioned above) it only remains to determine T_3 and Y at the point at the far right of the multiplet by considering the figure below.

Looking at the lhs of the figure, one sees that T_3 is half the length of the side of the dotted triangle, i.e.

$$T_3 = \frac{p+q}{2} \quad . \tag{5}$$

From the rhs of the figure, one infers that Y is one-third of the sidelength of the inner dashed triangle, giving

$$Y = \frac{p-q}{3} \, . \tag{6}$$

We are thus led to the result

$$\begin{aligned} \left(\sum_i \hat{F}_i\right)_{(p,q)} &= \left(\frac{p+q}{2}\right)^2 + 2\,\frac{p+q}{2} + \frac{3}{4}\left(\frac{p-q}{3}\right)^2 \\ &= \frac{p^3 + pq + q^2}{3} + p + q \quad . \end{aligned} \tag{7}$$

As important special cases we obtain the values 4/3 for the representations $D(1,0)$ and $D(0,1)$ and 3 for $D(1,1)$.

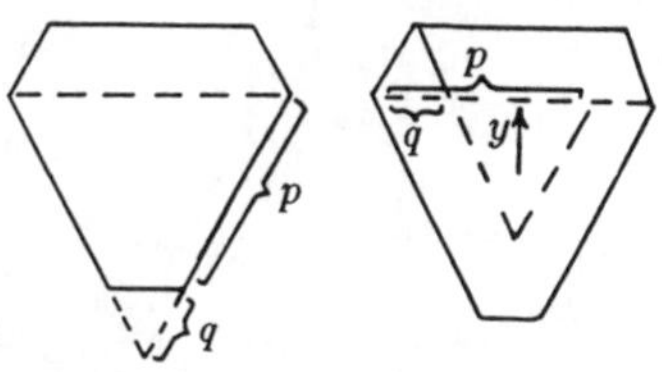

For determining the T_3- and Y-coordinates of the state of maximal weight

8. Quarks and SU(3)

In the previous chapter we have shown that the SU(3) symmetry indeed yields multiplet structures, as we have found before empirically (Examples and Exercises 6.3–5). It therefore appears that the group may represent a new fundamental symmetry for classifying elementary particles. Of course the question remains whether the SU(3) symmetry may predict further multiplets which have not been found until now.

In searching for a physical interpretation of the SU(3) we should understand the implications of the SU(3) representation and its quantum numbers T_3 and Y. The T-, U- and V-spin fulfill the angular momentum algebra [the SU(2) Lie algebra] and turn out to be subalgebras of SU(3). The following considerations will enable us to classify elementary particles within SU(3) multiplets if we interpret Y as hypercharge and T as isospin. Following this hypothesis, we will compare the theoretical implications with experimental facts. Isospin multiplets in a given SU(3) multiplet are represented by parallels to the T_3-axis. As a first step we define the charge operator according to (6.5) as

$$\hat{Q} = \tfrac{1}{2}\hat{Y} + \hat{T}_3 \quad . \tag{8.1}$$

We denote the SU(3) states by

$$|T_3\, Y, \alpha\rangle \quad , \tag{8.2}$$

where the additional quantum numbers abbreviated by α will be specified later (eigenvalues of the two Casimir operators which classify the multiplets uniquely). The eigenvalue equations

$$\hat{Y}|T_3\, Y,\alpha\rangle = Y|T_3\, Y,\alpha\rangle \quad , \quad \hat{T}_3|T_3\, Y,\alpha\rangle = T_3|T_3\, Y,\alpha\rangle \tag{8.3}$$

yield for the charge operator

$$\hat{Q}|T_3\, Y,\alpha\rangle = \left(\frac{Y}{2} + T_3\right)|T_3\, Y,\alpha\rangle \equiv Q|T_3\, Y,\alpha\rangle \quad . \tag{8.4}$$

For SU(3) states the charge Q is a good quantum number. In particular, a singlet state has zero charge (see Fig. 8.1),

$$\hat{Q}|00,\alpha\rangle = 0|00,\alpha\rangle \ , \tag{8.5}$$

because the eigenvalues Y and T_3 vanish in that case.

It becomes natural to interpret the Λ^* hyperon, which has been identified as the baryon resonance (see the list in Example 6.5) at 1405 MeV with spin $\frac{1}{2}$ and

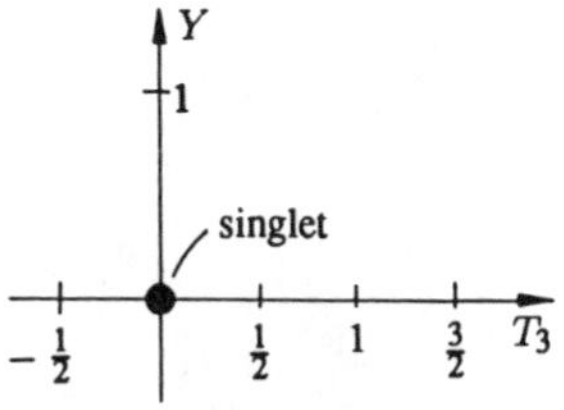

Fig. 8.1. The singlet representation of SU(3): $(p, q) = (0, 0)$. It is the smallest trivial representation

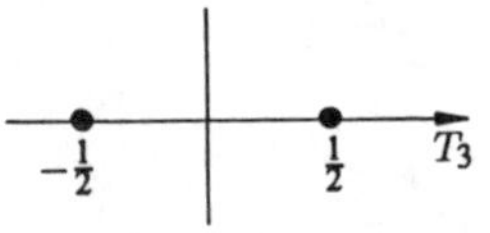

Fig. 8.2. The isospin doublet $T = \frac{1}{2}$ with the states $|TT_3\rangle = |\frac{1}{2}\frac{1}{2}\rangle$ and $|TT_3\rangle = |\frac{1}{2} - \frac{1}{2}\rangle$ is the smallest non-trivial representation of SU(2)

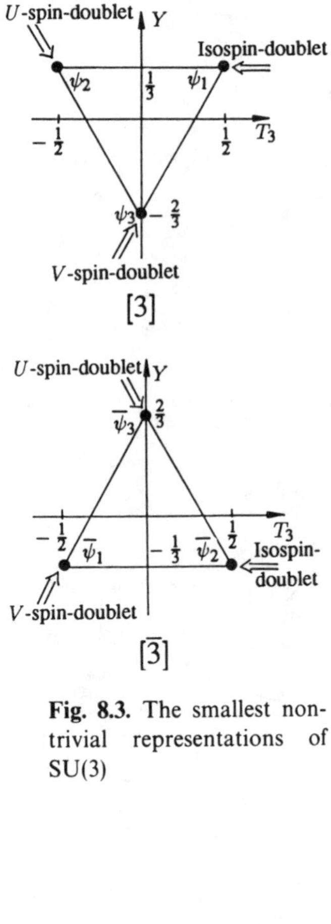

Fig. 8.3. The smallest non-trivial representations of SU(3)

negative parity, as a singlet state.[1] We remember that it did not fit into the highly symmetric scheme of Example 6.5; indeed, it represents, so to speak, the most trivial SU(3) representation. Let us now look at the *smallest non-trivial representation of SU(3)*. We first remind ourselves that the isospin doublet with $T = \frac{1}{2}$ *is the smallest non-trivial* description of the isospin group SU(2) (see Fig. 8.2). This implies that we can construct all higher multiplets from this representation. Technically this is achieved by Clebsch-Gordan coupling of isospins $T = \frac{1}{2}$ to an arbitrary isospin, although this can not be performed with the lowest SU(2)-multiplet $T = 0$. In that sense the $T = 0$ multiplet of the SU(2) is *trivial.*

Because the F-spin algebra [SU(3)] contains the isospin as a subalgebra, the smallest SU(3) representation we are looking for must contain at least one $T = \frac{1}{2}$ charge doublet. As a matter of fact the T-spin, U-spin and V-spin algebra appear fully symmetric in the F-spin algebra. Consequently, the SU(3) multiplet we are interested in *must contain a T-, U- and V-doublet.* By reason of the inherent symmetries of the SU(3) multiplets in the $Y - T_3$ plane (see the conclusions concerning T-U-V multiplet coupling in Chapter 7) we are led to the two equilateral triangles, as shown in Fig. 8.3. As required by general symmetries they are symmetrically centered around the origin ($Y = 0$, $T_3 = 0$). We denote the representations by [3] and [$\bar{3}$], respectively, because each contains 3 states. If [3] stands for particles, then [$\bar{3}$] represents the corresponding antiparticles since the state $\bar{\psi}_\nu$ has opposite hypercharge and opposite T_3-component and thus the opposite charge as compared to ψ_ν:

$$\hat{Q}\psi_\nu = Q_\nu\psi_\nu \quad , \quad \hat{Q}\bar{\psi}_\nu = -Q_\nu\bar{\psi}_\nu \quad . \tag{8.6}$$

Each of the two representations [3] and [$\bar{3}$] contains an isodoublet $T = \frac{1}{2}$ and an isosinglet $T = 0$. For instance the isodoublet for [3] is given by the states

$$\psi_1 \equiv |\tfrac{1}{2}Y\rangle \quad , \quad \psi_2 \equiv |-\tfrac{1}{2}Y\rangle \quad , \tag{8.7}$$

whereas the isosinglet reads

$$\psi_3 \equiv |0\,Y'\rangle \quad . \tag{8.8}$$

Until this point the hypercharges have remained unknown but they may now be determined: The T_3-values of ψ_1, ψ_2, ψ_3 can be directly derived from the eigenvalue equations

$$\hat{T}_3\psi_1 = +\tfrac{1}{2}\psi_1 \quad , \quad \hat{T}_3\psi_2 = -\tfrac{1}{2}\psi_2 \quad , \quad \hat{T}_3\psi_3 = 0\psi_3 \quad , \tag{8.9}$$

and, considering that ψ_1 is a U-spin singlet, the corresponding values of the hypercharge Y can be derived.

$$\hat{U}_3\psi_1 = 0 \quad . \tag{8.10}$$

From $\hat{U}_3 = (3\hat{Y} - 2\hat{T}_3)/4$ [see (7.28)], it follows that

$$\hat{Y}\psi_1 = \tfrac{1}{3}(4\hat{U}_3 + 2\hat{T}_3)\psi_1 = \tfrac{2}{3}\hat{T}_3\psi_1 = \tfrac{1}{3}\psi_1 \quad . \tag{8.11a}$$

[1] See S.L. Glashow, A.H. Rosenfeld: Phys. Rev. Lett. **10**, 192 (1963).

Since ψ_1 and ψ_2 belong to the same isospin doublet (situated perpendicular to the Y axis) and thus have the same hypercharge, it also holds that

$$\hat{Y}\psi_2 = \tfrac{1}{3}(4\hat{U}_3 + 2\hat{T}_3)\psi_2 = \tfrac{1}{3}\psi_2 \quad . \qquad (8.11b)$$

Now one can identify $[4U_3 + 2\times(-\frac{1}{2})]/3 = \frac{1}{3}$ and thus $U_3 = \frac{1}{2}$ for ψ_2. Accordingly, ψ_3 must have the eigenvalue $U_3 = -\frac{1}{2}$. Indeed the $\hat{U}_-$ operator leads from ψ_2 to ψ_3 as seen in Fig. 7.1 and this leads to

$$\hat{Y}\psi_3 = \tfrac{1}{3}(4\hat{U}_3 + 2\hat{T}_3)\psi_3 = \tfrac{1}{3}(4\times(-\tfrac{1}{2}) + 2\times(0))\psi_3 = -\tfrac{2}{3}\psi_3 \quad . \qquad (8.11c)$$

By similar arguments, we obtain for the states of $[\bar{3}]$,

$$\hat{Y}\bar{\psi}_1 = -\tfrac{1}{3}\bar{\psi}_1 \quad , \quad \hat{Y}\bar{\psi}_2 = -\tfrac{1}{3}\bar{\psi}_2 \quad , \quad \hat{Y}\bar{\psi}_3 = +\tfrac{2}{3}\bar{\psi}_3 \quad . \qquad (8.12)$$

We are thus necessarily led to hypercharges which are multiples of one-third. This has far-reaching consequences if one accepts that the charge operator $\hat{Q}$ [see (8.1)] is determined by the Gell-Mann-Nishijima relation. As a consequence the following charge eigenvalues for the states of [3] are found:

$$\begin{aligned}
\hat{Q}\psi_1 &= (\tfrac{1}{2}\hat{Y} + \hat{T}_3)\psi_1 = (\tfrac{1}{2}\times\tfrac{1}{3} + \tfrac{1}{2})\psi_1 = \tfrac{2}{3}\psi_1 \quad , \\
\hat{Q}\psi_2 &= (\tfrac{1}{2}\hat{Y} + \hat{T}_3)\psi_2 = (\tfrac{1}{2}\times\tfrac{1}{3} - \tfrac{1}{2})\psi_2 = -\tfrac{1}{3}\psi_2 \quad , \\
\hat{Q}\psi_3 &= (\tfrac{1}{2}\hat{Y} + \hat{T}_3)\psi_3 = (\tfrac{1}{2}\times(-\tfrac{2}{3}) + 0)\psi_3 = -\tfrac{1}{3}\psi_3 \quad , \qquad (8.13)
\end{aligned}$$

and similarly, for the states $\hat{\psi}_\nu$ of the antitriplet $[\bar{3}]$,

$$\hat{Q}\bar{\psi}_1 = -\tfrac{2}{3}\bar{\psi}_1 \quad , \quad \hat{Q}\bar{\psi}_2 = +\tfrac{1}{3}\bar{\psi}_2 \quad , \quad \hat{Q}\bar{\psi}_3 = +\tfrac{1}{3}\bar{\psi}_3 \quad . \qquad (8.14)$$

The particles which correspond to the states of the [3] multiplet of the SU(3) have fractional charges. Gell-Mann proposed the name "quark" whereas Zweig called these particles "aces".[2] Since ψ_1 and ψ_2 form an isodoublet similar to the proton and neutron, the ψ_1 quark was named "p quark" and the ψ_2 "n quark"; ψ_3 is named "λ quark". A more modern nomenclature is "up quark" (u), "down quark" (d) and "strange quark" (s). From now on the quark states ψ_1, ψ_2, ψ_3 are denoted by q_1, q_2, q_3 to make the association clear. Similarly, we write for the antiquarks:

$$\bar{\psi}_1, \bar{\psi}_2, \bar{\psi}_3 \;\rightarrow\; \bar{q}_1, \bar{q}_2, \bar{q}_3 \quad .$$

8.1 Searching for Quarks

Many physicists have searched for quarks, i.e. particles with one third charges. Jones[3] described attempts to produce quarks in accelerators and to find them in seawater, minerals and in cosmic rays. McCusher and Cairns believed that they

---

[2] G. Zweig: CERN-Preprint 8409/Th. 412 (1964).

[3] L.W. Jones: Int. Conf. on Symmetries and Quark Models, Wayne State University, Detroit (1969).

had found quarks in highly energetic cosmic rays.[4] Nevertheless, up to now no convincing positive result has been achieved. There is also much doubt about the experiment of W. Fairbank[5] et al. who claimed to have found particles with a fractional charge in a modern version of a Milikan experiment. Other authors object to this.[6] Thus, there are strong reasons for assuming that free quarks do not exist in nature. With the great success of the quark model for describing elementary particles on the one hand, one is directly confronted with the fundamental problem of quark confinement on the other hand. This concept tries to explain that quarks can only exist inside particles as bound states (i.e. they are confined) and not as free particles. We will return to this point later.

8.2 The Transformation Properties of Quark States

For the triplet representation [3] the operators $\hat{F}_\alpha$ are given by 3×3 matrices, because we are dealing with three states $|q_i\rangle$ $(i = 1, 2, 3)$, i.e.

$$(\hat{F}_\alpha)_{ij} = \langle q_i|\hat{F}_\alpha|q_j\rangle \quad . \tag{8.15}$$

With the help of the shift operators $\hat{V}_\pm, \hat{U}_\pm$ and $\hat{T}_\pm$ and their properties [see (7.28) and Sect. 7.4] one derives for the quark states:

$$\begin{aligned}
&\hat{T}_-|q_1\rangle = |q_2\rangle \quad , \quad \hat{T}_+|q_2\rangle = |q_1\rangle \quad ,\\
&\hat{U}_-|q_2\rangle = |q_3\rangle \quad , \quad \hat{U}_+|q_3\rangle = |q_2\rangle \quad ,\\
&\hat{V}_-|q_1\rangle = |q_3\rangle \quad , \quad \hat{V}_+|q_3\rangle = |q_1\rangle \quad ,\\
&\hat{T}_3|q_1\rangle = \tfrac{1}{2}|q_1\rangle \quad , \quad \hat{T}_3|q_2\rangle = -\tfrac{1}{2}|q_2\rangle \quad , \quad \hat{T}_3|q_3\rangle = 0|q_3\rangle \quad ,\\
&\hat{Y}|q_1\rangle = \tfrac{1}{3}|q_1\rangle \quad , \quad \hat{Y}|q_2\rangle = \tfrac{1}{3}|q_2\rangle \quad , \quad \hat{Y}|q_3\rangle = -\tfrac{2}{3}|q_3\rangle \quad , \quad \text{etc.}
\end{aligned} \tag{8.16}$$

On this basis we can calculate all matrix elements of the operators $\hat{U}_\pm, \hat{T}_\pm, \hat{V}_\pm$; and from

$$\begin{aligned}
&\hat{F}_1 = \hat{T}_1 = \tfrac{1}{2}(\hat{T}_+ + \hat{T}_-) \quad , && \hat{F}_5 = -\tfrac{1}{2}\mathrm{i}(\hat{V}_+ - \hat{V}_-) \quad ,\\
&\hat{F}_2 = \hat{T}_2 = -\tfrac{1}{2}\mathrm{i}(\hat{T}_+ - \hat{T}_-) \,, && \hat{F}_6 = \tfrac{1}{2}(\hat{U}_+ + \hat{U}_-) \quad ,\\
&\hat{F}_3 = \hat{T}_3 \quad , && \hat{F}_7 = -\tfrac{1}{2}\mathrm{i}(\hat{U}_+ - \hat{U}_-) \quad ,\\
&\hat{F}_4 = \tfrac{1}{2}(\hat{V}_+ + \hat{V}_-) \,, && \hat{F}_8 = \frac{\sqrt{3}}{2}\hat{Y} \quad ,
\end{aligned} \tag{8.17}$$

all matrix elements of the operators $\hat{F}_\alpha$ $(\alpha = 1, \ldots, 8)$ can be evaluated, as well.

---

4 C.B.A. McCusher, I. Cairns: Phys. Rev. Lett. **23**, 658 (1969).

5 G.S. LaRue, W.M. Fairbank, A.F. Hebard: Phys. Rev. Lett. **38**, 1011 (1977).

6 R.G. Milner, B.H. Cooper, K.H. Chang, K. Wilson, J. Labrenz, R.D. McCeown: Phys. Rev. Lett. **54**, 1472 (1985).

After a simple calculation, using

$$\hat{F}_\alpha = \tfrac{1}{2}\hat{\lambda}_\alpha \quad ,$$

one recovers the earlier result for the generators $\hat{\lambda}_\alpha$ of the SU(3) [see Exercise 8.1]. In general one can construct the unitary operators by means of the $\hat{F}_\alpha$ according to

$$\hat{U}(\boldsymbol{\theta}) = \exp\left(-\mathrm{i}\sum_\alpha \theta_\alpha \hat{F}_\alpha\right) \quad . \tag{8.18}$$

These represent the group operators of the SU(3) and transform the states within each SU(3) multiplet (F-spin multiplet). In the case of the triplet representation, the $\hat{F}_\alpha$ and thus the $\hat{U}(\boldsymbol{\theta})$ in (8.18) are unitary 3×3 matrices with the determinant $\det \hat{U}(\boldsymbol{\theta}) = 1$. So we write

$$|q_i\rangle' = \hat{U}(\boldsymbol{\theta})|q_i\rangle = \sum_j |q_j\rangle U_{ji}(\boldsymbol{\theta}) \quad , \quad \text{where}$$

$$U_{ji}(\boldsymbol{\theta}) = \langle q_j|\hat{U}(\boldsymbol{\theta})|q_i\rangle \quad . \tag{8.19}$$

In the following exercises we study the transformation properties of the states $|q_i\rangle$ and $|\bar{q}_i\rangle$ of the representations [3] and $[\bar{3}]$.

EXERCISE

8.1 The Generators of SU(3) in the Representation [3]

Problem. Calculate the generators $\hat{\lambda}_\alpha$ of SU(3) in the representation [3] by applying the shift operators to the quark states.

Solution. We remember the relation $\hat{\lambda}_\alpha = 2\hat{F}_\alpha$ and the fact that the quark states are orthonormal, i.e. $\langle q_i|q_j\rangle = \delta_{ij}$. Together with the relations (8.15–17) we obtain

$$(\hat{\lambda}_\alpha)_{ij} = 2\langle q_i|\hat{F}_\alpha|q_j\rangle \quad . \tag{1}$$

Consequently for $\hat{\lambda}_1$

$$\begin{aligned}(\hat{\lambda}_1)_{ij} &= 2\langle q_i|\hat{F}_1|q_j\rangle = \langle q_i|\hat{T}_+ + \hat{T}_-|q_j\rangle \\ &= \langle q_i|\hat{T}_+|q_j\rangle + \langle q_i|\hat{T}_-|q_j\rangle \quad .\end{aligned} \tag{2}$$

Because of (8.16),

$$\hat{T}_+|q_j\rangle = \delta_{j2}|q_1\rangle \quad \text{and} \quad \hat{T}_-|q_j\rangle = \delta_{j1}|q_2\rangle \quad , \tag{3}$$

$$(\hat{\lambda}_1)_{ij} = \langle q_i|q_1\rangle\delta_{j2} + \langle q_i|q_2\rangle\delta_{j1} = \delta_{i1}\delta_{j2} + \delta_{i2}\delta_{j1} \tag{4}$$

Exercise 8.1

(also using the orthonormality condition); thus, we obtain the only non-vanishing matrix elements, $(\hat{\lambda}_1)_{12} = (\hat{\lambda}_1)_{21} = 1$, which implies that

$$\hat{\lambda}_1 = \begin{pmatrix} 0 & 1 & 0 \\ 1 & 0 & 0 \\ 0 & 0 & 0 \end{pmatrix} \quad . \tag{5}$$

In order to construct the matrix representation of the generators $\hat{\lambda}_2, \dots, \hat{\lambda}_8$ we proceed in the same way:

$$\begin{aligned} (\hat{\lambda}_2)_{ij} &= 2\langle q_i|\hat{F}_2|q_j\rangle = \frac{1}{\mathrm{i}}[\langle q_i|\hat{T}_+|q_j\rangle - \langle q_i|\hat{T}_-|q_j\rangle] \\ &= -\mathrm{i}(\delta_{i1}\delta_{j2} - \delta_{i2}\delta_{j1}) \quad , \end{aligned} \tag{6}$$

$$\text{i.e.} \quad (\hat{\lambda}_2)_{12} = -\mathrm{i} \quad \text{and} \quad (\hat{\lambda}_2)_{21} = \mathrm{i} \quad , \quad \text{i.e.} \quad \hat{\lambda}_2 = \begin{pmatrix} 0 & -\mathrm{i} & 0 \\ \mathrm{i} & 0 & 0 \\ 0 & 0 & 0 \end{pmatrix} \quad ; \tag{7}$$

for $\hat{\lambda}_3$,

$$(\hat{\lambda}_3)_{ij} = 2\langle q_i|\hat{F}_3|q_j\rangle = 2\langle q_i|\hat{T}_3|q_j\rangle \quad , \tag{8}$$

i.e. the matrix elements are

$$\begin{aligned} (\hat{\lambda}_3)_{i1} &= 2\langle q_i|\hat{T}_3|q_1\rangle = \langle q_i|q_1\rangle = \delta_{i1} \quad , \\ (\hat{\lambda}_3)_{i2} &= 2\langle q_i|\hat{T}_3|q_2\rangle = -\langle q_i|q_2\rangle = -\delta_{i2} \quad , \\ (\hat{\lambda}_3)_{i3} &= 2\langle q_i|\hat{T}_3|q_3\rangle = 0 \quad , \end{aligned} \tag{9}$$

and thus

$$\hat{\lambda}_3 = \begin{pmatrix} 1 & 0 & 0 \\ 0 & -1 & 0 \\ 0 & 0 & 0 \end{pmatrix} \quad . \tag{10}$$

Next we have

$$(\hat{\lambda}_4)_{ij} = 2\langle q_i|\hat{F}_4|q_j\rangle = \langle q_i|\hat{V}_+|q_j\rangle + \langle q_i|\hat{V}_-|q_j\rangle \quad . \tag{11}$$

Because

$$\hat{V}_+|q_j\rangle = \delta_{j3}|q_1\rangle \quad \text{and} \quad \hat{V}_-|q_j\rangle = \delta_{j1}|q_3\rangle \tag{12}$$

are valid,

$$(\hat{\lambda}_4)_{ij} = \langle q_i|q_1\rangle\delta_{j3} + \langle q_i|q_3\rangle\delta_{j1} = \delta_{i1}\delta_{j3} + \delta_{i3}\delta_{j1} \quad , \tag{13}$$

i.e.

$$(\hat{\lambda}_4)_{13} = 1 = (\hat{\lambda}_4)_{31} \quad \text{and} \quad \hat{\lambda}_4 = \begin{pmatrix} 0 & 0 & 1 \\ 0 & 0 & 0 \\ 1 & 0 & 0 \end{pmatrix} \quad . \tag{14}$$

The following is valid for $\hat{\lambda}_5$,

$$\begin{aligned}(\hat{\lambda}_5)_{ij} &= \frac{1}{\mathrm{i}}[\langle q_i|\hat{V}_+|q_j\rangle - \langle q_i|\hat{V}_-|q_j\rangle] \\ &= \frac{1}{\mathrm{i}}[\delta_{i1}\delta_{j3} - \delta_{i3}\delta_{j1}] \quad , \end{aligned} \tag{15}$$

$$(\hat{\lambda}_5)_{13} = -\mathrm{i} \quad \text{and} \quad (\hat{\lambda}_5)_{31} = \mathrm{i}$$

(all other matrix elements vanish) and thus

$$\hat{\lambda}_5 = \begin{pmatrix} 0 & 0 & -\mathrm{i} \\ 0 & 0 & 0 \\ \mathrm{i} & 0 & 0 \end{pmatrix} \quad ; \tag{16}$$

Similarly we find for $\hat{\lambda}_6$

$$(\hat{\lambda}_6)_{ij} = 2\langle q_i|\hat{F}_6|q_j\rangle = \langle q_i|\hat{U}_+|q_j\rangle + \langle q_i|\hat{U}_-|q_j\rangle \quad . \tag{17}$$

Because

$$\hat{U}_+|q_j\rangle = \delta_{j3}|q_2\rangle \quad \text{and} \quad \hat{U}_-|q_j\rangle = \delta_{j2}|q_3\rangle \tag{18}$$

we have

$$(\hat{\lambda}_6)_{ij} = \langle q_i|q_2\rangle\delta_{j3} + \langle q_i|q_3\rangle\delta_{j2} = \delta_{i2}\delta_{j3} + \delta_{i3}\delta_{j2} \quad , \tag{19}$$

i.e.

$$(\hat{\lambda}_6)_{23} = (\hat{\lambda}_6)_{32} = 1 \quad \text{and} \quad \hat{\lambda}_6 = \begin{pmatrix} 0 & 0 & 0 \\ 0 & 0 & 1 \\ 0 & 1 & 0 \end{pmatrix} \quad . \tag{20}$$

For $\hat{\lambda}_7$ we obtain

$$\begin{aligned}(\hat{\lambda}_7)_{ij} &= \frac{1}{\mathrm{i}}[\langle q_i|\hat{U}_+|q_j\rangle - \langle q_i|\hat{U}_-|q_j\rangle] \\ &= \frac{1}{\mathrm{i}}[\delta_{i2}\delta_{j3} - \delta_{i3}\delta_{j2}] \quad , \end{aligned} \tag{21}$$

$$(\hat{\lambda}_7)_{23} = -\mathrm{i} \quad \text{and} \quad (\hat{\lambda}_7)_{32} = \mathrm{i} \quad , \quad \text{i.e.} \quad \hat{\lambda}_7 = \begin{pmatrix} 0 & 0 & 0 \\ 0 & 0 & -\mathrm{i} \\ 0 & \mathrm{i} & 0 \end{pmatrix} \quad ; \tag{22}$$

Finally $\hat{\lambda}_8$ is found to be

$$(\hat{\lambda}_8)_{ij} = 2\langle q_i|\hat{F}_8|q_j\rangle = \sqrt{3}\langle q_i|\hat{Y}|q_j\rangle \quad , \tag{23}$$

Exercise 8.1

i.e.

$$(\hat{\lambda}_8)_{i1} = \sqrt{3}\langle q_i|\hat{Y}|q_1\rangle = \frac{1}{3}\sqrt{3}\langle q_i|q_1\rangle = \frac{1}{\sqrt{3}}\delta_{i1} \quad , \tag{24}$$

$$(\hat{\lambda}_8)_{i2} = \sqrt{3}\langle q_i|\hat{Y}|q_2\rangle = \frac{1}{3}\sqrt{3}\langle q_i|q_2\rangle = \frac{1}{\sqrt{3}}\delta_{i2} \quad , \tag{25}$$

$$(\hat{\lambda}_8)_{i3} = \sqrt{3}\langle q_i|\hat{Y}|q_3\rangle = \frac{1}{3}\sqrt{3}(-2)\langle q_i|q_3\rangle = -\frac{2}{\sqrt{3}}\delta_{i3} \quad , \tag{26}$$

$$\hat{\lambda}_8 = \frac{1}{\sqrt{3}}\begin{pmatrix} 1 & 0 & 0 \\ 0 & 1 & 0 \\ 0 & 0 & -2 \end{pmatrix} \quad . \tag{27}$$

The generators $\hat{\lambda}_\alpha$ in the representation [3] are the matrices given by Gell-Mann. Of course this result is not very surprising.

EXERCISE

8.2 Transformation Properties of the States of the Antitriplet $[\bar{3}]$

Problem. Show that the states $|q_i\rangle$ of the antitriplet $[\bar{3}]$ transform in the following way.

$$|\bar{q}_i\rangle' = \hat{U}(\boldsymbol{\theta})|\bar{q}\rangle = \sum_j |\bar{q}_j\rangle \bar{U}_{ji}(\boldsymbol{\theta}) \quad . \tag{1}$$

The unitary transformation operator is given by

$$\hat{\bar{U}}(\boldsymbol{\theta}) = \exp\left(-\mathrm{i}\sum_\alpha \theta_\alpha \hat{\bar{F}}_\alpha\right) \quad , \quad \hat{\bar{F}}_\alpha = -\tfrac{1}{2}\hat{\lambda}^*_\alpha \quad , \tag{2}$$

with the matrix representation

$$\bar{U}_{ji}(\boldsymbol{\theta}) = \langle \bar{q}_j|\hat{\bar{U}}(\boldsymbol{\theta})|\bar{q}_i\rangle = U^*_{ji}(\boldsymbol{\theta}) \quad . \tag{3}$$

Solution. To prove this theorem we start from the transformation of the triplet states:

$$\hat{U}(\boldsymbol{\theta}) = \exp(-\mathrm{i}\boldsymbol{\theta}\cdot\hat{\boldsymbol{F}})$$
$$|q_i\rangle' = \hat{U}(\boldsymbol{\theta})|q_i\rangle \quad . \tag{4}$$

The complex conjugate equation is

$$|q_i\rangle'^* = \hat{U}^*(\boldsymbol{\theta})|q_i^*\rangle \quad , \tag{5}$$

where

$$\hat{U}^*(\boldsymbol{\theta}) = \exp(+\mathrm{i}\boldsymbol{\theta}\cdot(\hat{\boldsymbol{F}})^*) \equiv \exp(-\mathrm{i}\boldsymbol{\theta}\cdot\hat{\bar{\boldsymbol{F}}}) \quad . \tag{6}$$

The operators $-\hat{\boldsymbol{F}}^* = \hat{\bar{\boldsymbol{F}}}$ are the generators of the antitriplet [$\bar{3}$]. They are introduced in such a way that, corresponding to (8.18), the exponent (note the minus sign) reads: $-\mathrm{i}\boldsymbol{\theta}\bar{\boldsymbol{F}}$. Complex conjugation does not change the properties of matrix multiplication, but it changes the sign on the rhs of the commutation relations. This is corrected by the additional minus sign so that the $(-\boldsymbol{F}^*)$ again form a representation. According to (8.1) the generators are given by $\hat{\bar{F}}_i = -(\hat{F}_i)^*$ or $(\hat{\bar{F}}_\alpha = -\hat{\lambda}^*_\alpha/2)$, i.e. $\hat{T}_3$ and $\hat{Y}_3$ (equally their eigenvalues) are multiplied by (-1). Clearly this is exactly the property characterizing the antitriplet. Therefore the states $|q_i\rangle^*$ are called antitriplet states and we denote them by $|\bar{q}_i\rangle$. The change in notation $|q_i\rangle^* \to |\bar{q}_i\rangle$ and $-\hat{F}^*_i - \hat{\bar{F}}_i$ should not be mixed up with hermitian conjugation. In the following exercise we prove that there exists no unitary transformation connecting $\hat{U}$ and $\hat{U}^*$, i.e. [3] and [$\bar{3}$] are independent representations.

EXERCISE

8.3 Non-equivalence of the Two Fundamental Representations of SU(3)

Problem. Show that the representations [3] and [$\bar{3}$] (triplet and antitriplet) are different fundamental representations, which cannot be transformed into each other.

Solution. We have seen in Exercise 8.2 that the generators $\hat{\bar{F}}_\alpha = -\hat{\lambda}^*_\alpha/2$ belong to the generators of [$\bar{3}$], whereas the $\hat{F}_\alpha = \hat{\lambda}_\alpha/2$ are the generators of [3]. This implies that the states $\bar{q}_i$ transform with $\hat{U}^*(\theta_\alpha)$ and not with $\hat{U}(\theta_\alpha)$. If the representations were to be equivalent, then their generators would only differ by a unitary transformation $\hat{S}$, i.e.

$$\hat{S}\hat{F}_\alpha\hat{S}^{-1} \stackrel{!}{=} \hat{\bar{F}}_\alpha \quad \text{or} \quad \hat{S}\hat{\lambda}_\alpha\hat{S}^{-1} = -\hat{\lambda}^*_\alpha \quad . \tag{1}$$

Applying this transformation to the eigenvalue equation of the $\hat{\lambda}_\alpha(\hat{\lambda}_\alpha|q_i\rangle = \lambda|q_i\rangle$, where λ is the eigenvalue),

$$\hat{S}\hat{\lambda}_\alpha|q_i\rangle = \hat{S}\lambda|q_i\rangle = \lambda\hat{S}|q_i\rangle = \hat{S}\hat{\lambda}_\alpha\hat{S}^{-1}\hat{S}|q_i\rangle \quad , \tag{2}$$

with

$$\hat{S}|q_i\rangle = |q_i\rangle' \quad .$$

If (1) is valid one obtains

$$-\hat{\lambda}^*_\alpha|q_i\rangle' = \lambda|q_i\rangle' \quad ,$$

Exercise 8.3

i.e. $\hat{\lambda}_\alpha$ has to have the same eigenvalues as $-\hat{\lambda}_\alpha^*$. Now the $\hat{\lambda}_\alpha$ are hermitian matrices, i.e.

$$\hat{\lambda}_\alpha = \hat{\lambda}_\alpha^\dagger = (\hat{\lambda}_\alpha^*)^{\mathrm{T}} \quad ,$$

where T means the transposition. Since the determinant of a matrix and its transpose are equal, $\hat{\lambda}_\alpha^*$ and $\hat{\lambda}_\alpha$ have the same eigenvalues λ, which are determined by the secular equation

$$\det(\hat{\lambda}_\alpha - \lambda\hat{I}) = \det(\hat{\lambda}_\alpha^* - \lambda\hat{I}) = 0 \quad .$$

Hence the eigenvalues of $-\hat{\lambda}_\alpha^*$ differ from those of $\hat{\lambda}_\alpha$ by a sign. By explicit calculation one finds that the eigenvalues of all $\hat{\lambda}_\alpha$ are $-1, 0$, and $+1$, with the exception of $\hat{\lambda}_8$ which has eigenvalues $1/\sqrt{3}$ twice and $-2/\sqrt{3}$; therefore, $\hat{\lambda}_8$ and $\hat{\lambda}_8^*$ have different eigenvalues. The result is that there exists no transformation which transforms [3] into [$\bar{3}$], i.e. triplet and antitriplet are independent representations. This is of course no surprise, because the operators $\hat{U}$ only transform states of a given multiplet into each other. The states of the antitriplet cannot be transformed into states of the triplet via the generators $\hat{F}_i$ (note that this is different for the group SU(2) where the doublet and the antidoublet are equivalent representations, the reason obviously being that the generators $\hat{\tau}_i$ all have eigenvalues -1 and $+1$!).

8.3 Construction of all SU(3) Multiplets from the Elementary Representations [3] and [$\bar{3}$]

We have just found the smallest non-trivial representations [3] and [$\bar{3}$] of SU(3), which play an important role in constructing higher multiplets (discussed in previous chapters). This should be seen in close analogy to the situation of SU(2), where we were also able to derive the general structure of multiplets of the angular momentum algebra [Lie algebra of the SU(2)]. The result has been that to each value $j = 0, \frac{1}{2}, 1, \frac{3}{2} \ldots$ there corresponds a multiplet of dimension $(2j+1)$ with the states $|jm\rangle$, $m = +j, \ldots, -j$. An alternative way to construct the SU(2) multiplets consists of a successive coupling of the fundamental doublets $j = \frac{1}{2}$, $m = \frac{1}{2}, -\frac{1}{2}$. We already know that each spin j can be represented by coupling spins $\frac{1}{2}$ successively together. Often one uses the following notation to express this procedure in mathematical terms:

$$[\tfrac{1}{2}] \otimes [\tfrac{1}{2}] = [\tfrac{1}{2}]^2 = [1] \oplus [0] \quad \text{or}$$

$$[\tfrac{1}{2}] \otimes [\tfrac{1}{2}] \otimes [\tfrac{1}{2}] = [\tfrac{1}{2}]^3 = [\tfrac{3}{2}] \oplus [\tfrac{1}{2}] \oplus [\tfrac{1}{2}] \quad .$$

By decomposing these Kronecker products (lhs) into direct sums, particular irreducible SU(2) representations, i.e. multiplets of a total angular momentum j, can appear several times, as can be seen above.

The latter is an example where the same total angular momentum $j = \frac{1}{2}$ appears twice. This is a direct consequence of different possibilities of coupling: The first two $j = \frac{1}{2}$ states can couple to intermediate states $j' = 1$ and $j' = 0$. Now these angular momenta j', together with the third $j = \frac{1}{2}$ state, can couple to the total angular momenta $J = \frac{3}{2}, \frac{1}{2}$ (by coupling of $j' = 1$ with $\frac{1}{2}$) and $J = \frac{1}{2}$ (by coupling of $j' = 0$ with $\frac{1}{2}$). This is illustrated in Fig. 8.4, which shows the multiple coupling of the fundamental multiplet $j = \frac{1}{2}$ [SU(2) doublet] to different total angular momenta J. The angular momentum $J = 1$ appears in the Kronecker product $[\frac{1}{2}]^4$ three times. The configurations on the left represent the stretched couplings, i.e. the maximal (parallel) addition of the angular momenta. There is only one unique way to realize this. Therefore the largest angular momentum (stretched scheme) of the product $[\frac{1}{2}]^n$ is $J_{\max} = n/2$. Physically this implies the construction of composite particles with angular momentum J from elementary particles with angular momentum $j = \frac{1}{2}$. Rotational symmetry remains conserved in this process.

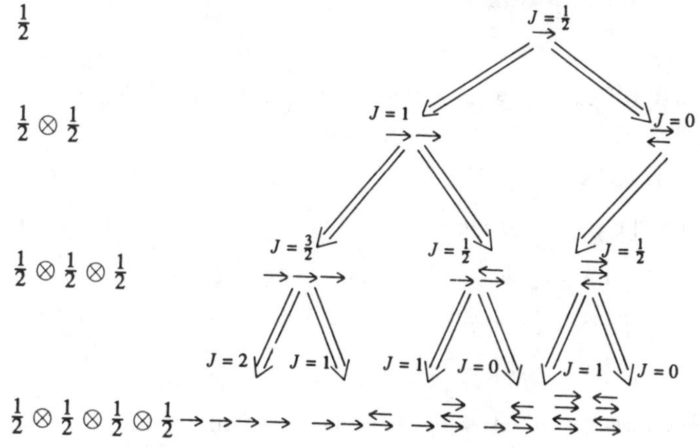

Fig. 8.4. Multiple coupling of spins $\frac{1}{2}$ to total spins J

Similarly, in the case of SU(3) one can construct higher multiplets by means of the fundamental representations [3] or [$\bar{3}$]. The general classifications of SU(3) multiplets and their construction are known from the previous chapters. The method is similar to the general classification of the SU(2) multiplets based on the angular momentum algebra. Now we generate the same SU(3) multiplets using the fundamental multiplets of quarks [3] and antiquarks [$\bar{3}$]. An essential difference to SU(2) becomes clear with the appearance of two fundamental representations.

In principle the construction of all irreducible representations of the SU(3) requires only one of the two fundamental representations 3 and [$\bar{3}$]. The reason is that the states of the representation [$\bar{3}$] can be derived via Kronecker products from the triplet representation [3] and vice versa (we will prove this later, when we discuss the reduction of tensor products of representations):

$$[3] \otimes [3] = [6] \oplus [\bar{3}] \quad , \quad [\bar{3}] \otimes [\bar{3}] = [\bar{6}] \oplus [3] \quad .$$

However, for reasons of physics one needs both fundamental representations, because quarks (represented by [3]) and antiquarks (represented by $[\bar{3}]$) differ by their baryon number ($B = \frac{1}{3}$ for quarks, $B = -\frac{1}{3}$ for antiquarks) and charge.

The general Kronecker product of SU(3) contains p triplets and q antitriplets;

$$\underbrace{[3] \otimes [3] \otimes \dots [3]}_{p \text{ times}} \otimes \underbrace{[\bar{3}] \otimes [\bar{3}] \otimes \dots [\bar{3}]}_{q \text{ times}} \quad . \tag{8.20}$$

In the following we decompose this product by always isolating the largest of the resulting representations, and continue this procedure with the remaining part. In physical terms this means that we construct (complex) particles out of quarks and antiquarks, always conserving the SU(3) symmetry.

8.4 Construction of the Representation $D(p, q)$ from Quarks and Antiquarks

First we explicitly note the basis vectors of the fundamental triplets (quark states)

a) $[3] \quad : \quad |T_3, Y\rangle \quad ,$

with the quantum numbers

$$(T_3, Y) = (\tfrac{1}{2}, \tfrac{1}{3}) \quad , \quad (-\tfrac{1}{2}, \tfrac{1}{3}) \quad , \quad (0, -\tfrac{2}{3}) \quad \text{and} \tag{8.21}$$

b) $[\bar{3}]$: $|\bar{T}_3, \bar{Y}\rangle$ with

$$(\bar{T}_3, \bar{Y}) = (0, \tfrac{2}{3}) \quad , \quad (\tfrac{1}{2}, -\tfrac{1}{3}) \quad , \quad (-\tfrac{1}{2}, -\tfrac{1}{3}) \quad . \tag{8.22}$$

The direct product is given by the set of all product states of the form

$$|T_3(1), Y(1)\rangle |T_3(2), Y(2)\rangle \dots |T_3(p), Y(p)\rangle \otimes |\bar{T}_3(1), \bar{Y}(1)\rangle \dots |\bar{T}_3(q), \bar{Y}(q)\rangle \quad . \tag{8.23}$$

These vectors characterize a *p-quark q-antiquark state*, because of the additivity of $\hat{T}_3$ and the hypercharge $\hat{Y}$,

$$\hat{T}_3 = \sum_i \hat{T}_3(i) \quad , \quad \hat{Y} = \sum_i \hat{Y}(i) \tag{8.24}$$

(the sums run over all particles). These *many-quark states* have eigenvalues of T_3 and Y, where

$$(T_3, Y) = \left(\sum_{i=1}^{p} T_3(i) + \sum_{i=1}^{q} \bar{T}_3(i) \quad , \quad \sum_{i=1}^{p} Y(i) + \sum_{i=1}^{q} \bar{Y}(i) \right) \quad . \tag{8.25}$$

Following the mathematical literature, this pair of eigenvalues is termed the "*weight*" of state (8.23). A weight (T_3, Y) is called "larger" than (T'_3, Y') if

$$T_3 > T'_3 \quad \text{or} \quad T_3 = T'_3 \quad \text{and} \quad Y > Y' \quad .$$

Then one writes

$$(T_3, Y) > (T'_3, Y') \quad . \tag{8.26}$$

We illustrate this in the following examples.

EXAMPLE

8.4 The Weight of a State

According to the definition,

$$(\tfrac{1}{2}, \tfrac{1}{3}) > (-\tfrac{1}{2}, \tfrac{1}{3}) \quad (\tfrac{1}{2}, \tfrac{1}{3}) > (0, -\tfrac{2}{3}) \quad ,$$

as well as

$$(\tfrac{1}{2}, +1) > (\tfrac{1}{2}, -1) \quad .$$

One says the left weights are larger than the right ones.

EXAMPLE

8.5 The Maximum Weight of the Quark Triplet [3] and Antiquark Triplet [$\bar{3}$]

From the $Y - T_3$ diagrams of the representations [$\bar{3}$] and [3] we directly read off their maximum weights [see Fig. (8.3)];

$$[3]: \quad (T_3, Y)_{\max} = (\tfrac{1}{2}, \tfrac{1}{3}) \quad ,$$

$$[\bar{3}]: \quad (T_3, Y)_{\max} = (\tfrac{1}{2}, -\tfrac{1}{3}) \quad .$$

As shown schematically they are the points on the far right.

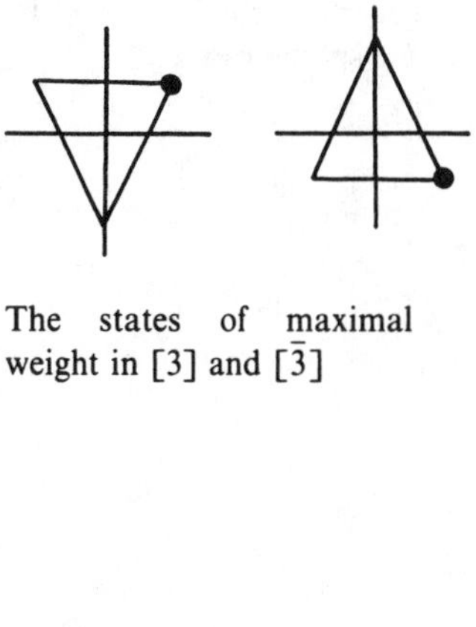

The states of maximal weight in [3] and [$\bar{3}$]

By recalling the results of the previous examples we are led to the conclusion that the (p-quark, q-antiquark) state of maximum weight is that one which consists of p quarks of maximal weight and q antiquarks of maximal weight, i.e. p-quark states $|\tfrac{1}{2}, \tfrac{1}{3}\rangle$ and q-antiquark states $|\tfrac{1}{2}, -\tfrac{1}{3}\rangle$. This state is characterised by

$$(T_3)_{\max} = \frac{p+q}{2} \quad , \quad (Y)_{\max} = \frac{p-q}{3} \quad . \tag{8.27}$$

Accordingly every multiplet contains only one state of maximum weight, as becomes obvious by construction. In contrast to this there exist various possibilities to obtain states with lower weight. For instance states with the weight $[(T_3)_{max} - 1, Y_{max}]$ are obtained by replacing one (and only one) of the factors $|\frac{1}{2}, \frac{1}{3}\rangle$ or $|\frac{1}{2}, \frac{1}{3}\rangle$ in the state of maximum weight,

$$|\tfrac{1}{2}, \tfrac{1}{3}\rangle_1 \cdots |\tfrac{1}{2}, \tfrac{1}{3}\rangle_p |\tfrac{1}{2}, -\tfrac{1}{3}\rangle_1 \cdots |\tfrac{1}{2}, -\tfrac{1}{3}\rangle_q \quad ,$$

by a state $|-\frac{1}{2}, \frac{1}{3}\rangle$ or $|-\frac{1}{2}, -\frac{1}{3}\rangle$.

Based on the state of maximum weight one is now able to generate the whole multiplet by means of the shift operators $\hat{T}_\pm$, $\hat{U}_\pm$, $\hat{V}_\pm$ (in previous chapters we have discussed this procedure in detail). Let us consider an example for the $D(p,q)$ representation, particularly for $p = 2$, $q = 1$ as illustrated in Fig. 8.5. The state of maximal weight is located at the point A. Denoting its weight by $[T_{3\,max} = (p+q)/2,\ Y_{max} = (p-q)/3]$, we know that starting from A the point B is reached along the V line in p' steps. Then, after q' steps along the T line, one ends up at C.

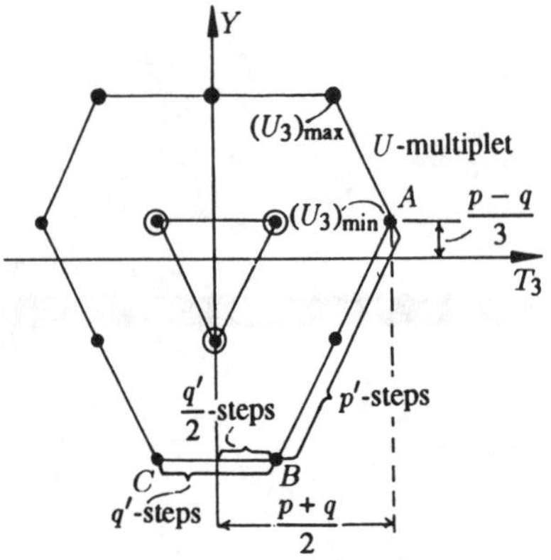

Fig. 8.5. Explanation of the general multiplet $D(p,q)$ in the special case $D(2,1)$

The symbols p' and q' are still unknown, though since $T_3 = 0$ at D, one obtains

$$(T_3)_{max} = \frac{q'}{2} \times 1 + p' \times \frac{1}{2} = \frac{q' + p'}{2} \quad . \tag{8.28}$$

A belongs to a U multiplet

$$(U_3)_{min} = -\tfrac{1}{2}q' \quad . \tag{8.29}$$

On the other hand we know that

$$(T_3)_{max} = \frac{p+q}{2} \quad \text{and}$$

$$(U_3)_{min} = \tfrac{1}{2}(\tfrac{3}{2}Y_{max} - (T_3)_{max}) = \tfrac{1}{2} \times \left(\frac{3}{2}\left(\frac{p-q}{3}\right) - \frac{p+q}{2}\right) = -\tfrac{1}{2}q \quad .$$

Together with (8.28, 29) we can determine p' and q':

$$\frac{p+q}{2} = \frac{p'+q'}{2} \quad , \qquad -\frac{q}{2} = -\frac{q'}{2} \quad ,$$

yielding

$$p' = p \quad , \quad q' = q \quad . \tag{8.30}$$

This proves that $D(p, q)$ represents the largest SU(3) multiplet of the p-quark, q-antiquark configuration (8.21, 23). Thus we have found an important relation between the general multiplet structure and the number of quark and antiquark states, which can form such a maximal multiplet.

8.4.1 The Smallest SU(3) Representations

We show the simplest multiplets of SU(3) in Fig. 8.6 and a higher multiplet in Fig. 8.7.

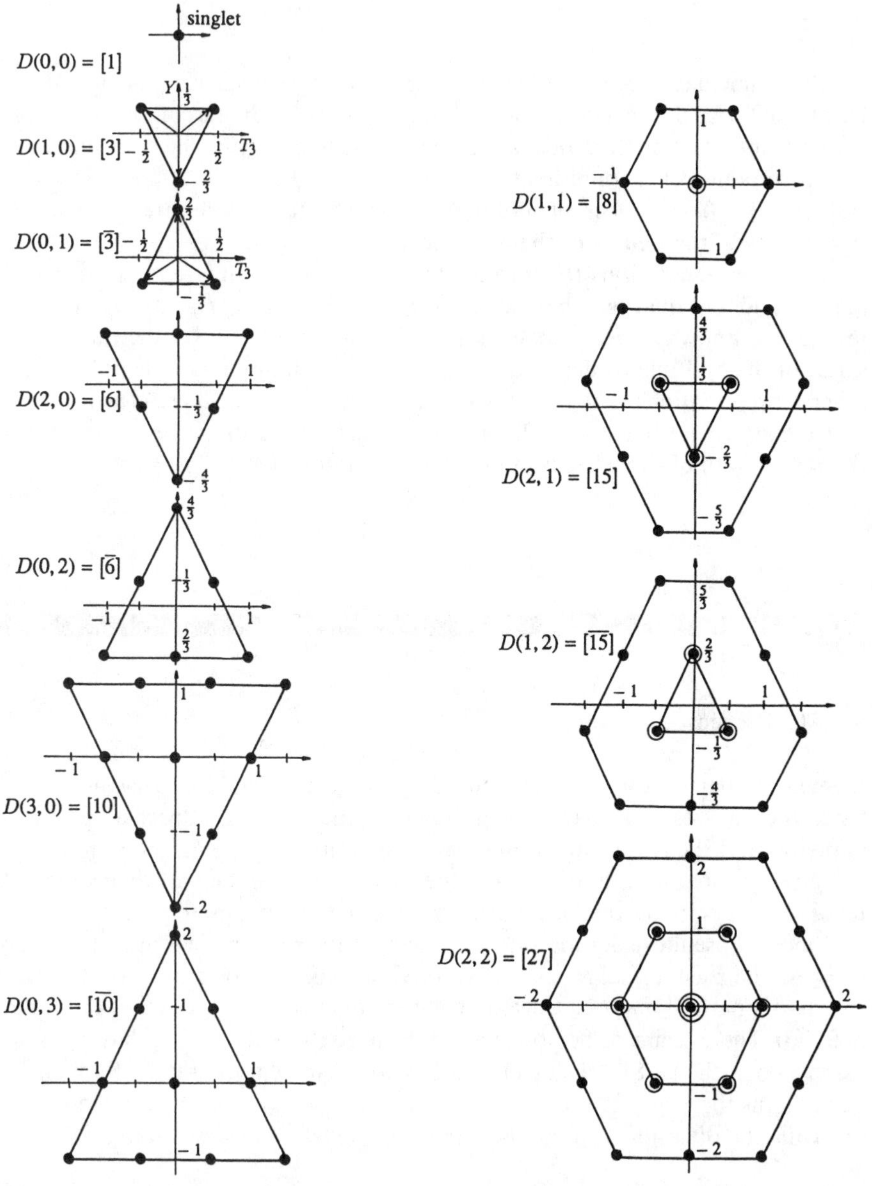

Fig. 8.6. The simplest multiplets of SU(3)

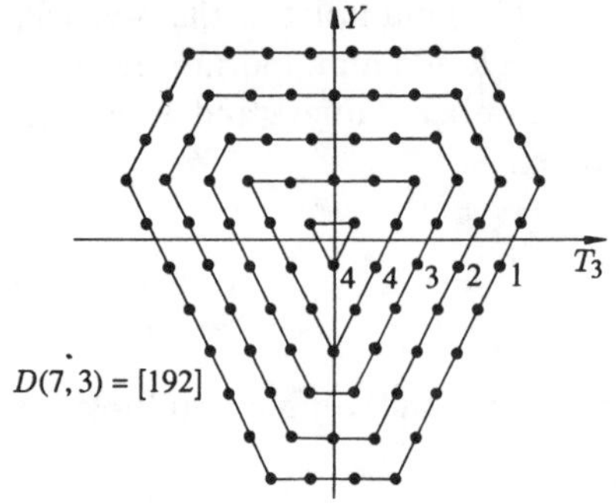

Fig. 8.7. $T_3 - Y$ diagram of a higher SU(3) multiplet. The numbers indicate the multiplicities of states on the various shells

One immediately obtains the values for p and q by considering the boundary. Due to (8.27) we thus get $(T_3)_{\max}$ and $(Y)_{\max}$, i.e. the coordinates of the outermost lattice point. Starting from this point we are able to construct the coordinates of all lattice points of the multiplet in a simple way. The origin of the coordinate system ($T_3 = 0,\ Y = 0$) is of additional help in this task as are the various symmetries of the figures of the multiplets.

In the representations $D(1,0)$ and $D(0,1)$ we again recognize the presence of quarks and antiquarks. The multiplet $D(3,0) = [10]$ represents the baryon resonances known to us from Exercise 6.5. The dimension, i.e. the number of states of the multiplet is denoted by [10]. Corresponding relations hold for all other representations. The baryons known from Example 6.3 are clearly represented by the octet $D(1,1) = [8]$; the same holds for the antibaryons. The centres ($T_3 = 0,\ Y = 0$) of these are occupied by two states each.

EXAMPLE

8.6 The Pseudoscalar Mesons

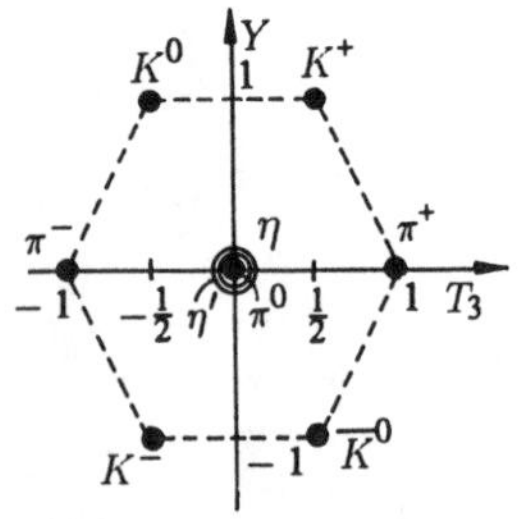

The nonet of the pseudoscalar mesons

Elementary particles with spin 0 and odd parity are called *pseudoscalar mesons*. Since they possess odd parity they are not scalar, but pseudoscalar particles implying that their wave functions change sign under space inversion $\boldsymbol{r}' = -\boldsymbol{r}$. We give a representation of them in the following table. One of their inherent relationships rests on the fact that they have the same spin and parity.

Since all pseudoscalar mesons have no spin their magnetic moments vanish ($\mu = 0$). The pseudoscalar mesons can obviously be arranged into the diagram above in a natural manner; thus, they are members of a $D(1,1) = [8]$ multiplet of SU(3). There seems to be no place for the η' particle, indicating that it can be associated with the SU(3) singlet and has to be shown separately. Sometimes one adjoins the η' particle to the point ($T_3 = 0,\ Y = 0$) in the octet. In so doing, one calls the diagram "a pseudoscalar *meson nonet*", as shown above.

Properties of the Pseudoscalar Mesons (S = 0, Odd Parity)

Name	Symbol	Charge	T spin	T_3 isospin	Y hyper-charge	Mass [MeV]	Lifetime [s]	Main decay mode	Main decay [%]
	π^+	+1	1	1	0	139.57	2.60×10^{-8}	$\pi^+ \to \mu^+ + \nu_\mu$	100
Pions	π^0	0	1	0	0	134.97	0.89×10^{-16}	$\pi^0 \to \gamma + \gamma$	98.8
	π^-	−1	1	−1	0	139.57	2.60×10^{-8}	$\pi^- \to \mu^- + \bar{\nu}_\mu$	100
	K^+	+1		1/2	1	493.82	1.235×10^{-8}	$K^+ \to \mu^+ + \nu_\mu$	63.8
Kaons			1/2					$\to \pi^+ + \pi^0$	20.9
								$\to \pi^+ + \pi^+ + \pi^-$	5.6
	K^0	0		−1/2	1	497.82	50% K_s + 50% K_l	see below	
	K^-	−1		−1/2	−1	493.82	1.235×10^{-8}	$K^- \to \mu^- + \bar{\nu}_\mu$	63.8
			1/2					$\to \pi^- + \pi^0$	20.9
								$\to \pi^- + \pi^+ + \pi^-$	5.6
	$\bar{K}^0$	0		1/2	−1	497.82	50% K_s + 50% K_l	see below	
Short-lived K_0	$K_s \equiv K_1^0$	0	1/2	−1/2	1	497.7	0.88×10^{-10}	$\pi^+ + \pi^-$	70
								$\pi^0\pi^0$	30
Long-lived K_0	$K_l \equiv K_2^0$	0			1	497.7	5.77×10^{-8}	$3\pi^0$	22
								$\pi^+ + \pi^- + \pi^0$	12
								$\pi^\pm + \mu^\mp + \nu_\mu$	27
								$\pi^\pm + e^\mp + \nu_e$	39
								$\eta \to \gamma\gamma$	38
Eta meson	η	0	0	0	0	548.6	ca. 10^{-20}	$\to \pi^0 + \gamma\gamma$	2
								$\to 3\pi^0$	31
								$\to \pi^+ + \pi^- + \pi^0$	23
								$\eta' \to \gamma\gamma$	2
Eta-prime meson	η'	0	0	0	0	958	$> 6.6\times10^{-22}$	$\to \eta + \pi^+\pi^-$	67.6
								$\to \rho^0 + \gamma$	30.4

EXAMPLE

8.7 Example (for Deeper Insight): The K^0- and $\bar{K}^0$-Mesons and Their Decays

The K^0- and $\bar{K}^0$-mesons are peculiar particles, the first one being the particle, the second the corresponding antiparticle:

$$\overline{(K^0)} = \bar{K}^0 \quad .$$

Example 8.7

Looking at the decay modes of the K^0 and $\bar{K}^0$ one finds an amazing feature[7]: There occur *two different lifetimes.* This is a remarkable fact since normally an unstable particle (characterized by its mass M and its lifetime τ) is described by the quantum mechanical state

$$|\psi(t)\rangle = \exp(-\mathrm{i}Mc^2t/\hbar)\exp(-\Gamma t/2\hbar)|\psi(0)\rangle \quad , \tag{1}$$

with $\Gamma = \hbar/\tau$. If, say, the decaying particle is a π^- meson, which decays into $\mu^- + \bar{\nu}_\mu$, then the rate at which $\mu^- - \bar{\nu}_\mu$ pairs are observed at a time t is given by

$$R(\mu^-\bar{\nu}_\mu, t) = R_0 \exp(-\Gamma t/\hbar) \quad , \quad \text{with} \tag{2}$$

$$R_0 = 2\pi|\langle\mu^-\bar{\nu}_\mu|\hat{H}_{\text{int}}|\psi(0)\rangle|^2 \quad ,$$

where $\hat{H}_{\text{int}}$ is the weak interaction Hamiltonian.

Hence, one obtains an *exponential decay law* which is confirmed by other unstable particles (systems). This law also holds if a particle decays – as the pion – in different modes; only the constant N_0 changes. Considering, for instance, the *beta decay* of the π^- particle,

$$\pi^- \to \pi^0 + e^- + \bar{\nu}_e \quad , \tag{3}$$

which occurs very seldom compared to the μ^- decay, one finds that, similarly to (2),

$$R'(\pi^0 e^-\bar{\nu}_e, t) = R'_0 \exp(-\Gamma t/\hbar) \quad , \tag{4}$$

where

$$R'_0 = 2\pi|\langle\pi^0 e^-\bar{\nu}_e|\hat{H}_{\text{int}}|\psi(0)\rangle|^2 \quad .$$

As may be seen, the ratio

$$\frac{R'(\pi^0 e^-\bar{\nu}_e, t)}{R(\mu^-\bar{\nu}_\mu, t)} = \frac{|\langle\pi^0 e^-\bar{\nu}_e|\hat{H}_{\text{int}}|\psi(0)\rangle|^2}{|\langle\mu^-\bar{\nu}_\mu|\hat{H}_{\text{int}}|\psi(0)\rangle|^2} \tag{5}$$

is always *independent of time* and very small (3×10^{-8}). The decay of the charged K mesons is similar: the three possible decay modes of K^+ and K^-, presented in the previous table, are commonly described by the *same* lifetime $\tau_K = 1.235 \times 10^{-8}$ s. The ratios of the various decay modes (branching ratios) are also independent of time.

On the other hand, neutral K mesons ($K^0, \bar{K}^0$) do not have a common lifetime. *They decay faster into two pions (short lifetime) than into three pions (long lifetime)*:

$$\left.\begin{array}{l} K^0 \to \pi^+ + \pi^- \\ \bar{K}^0 \to \pi^+ + \pi^- \end{array}\right\} \quad \tau_s = 0.9 \times 10^{-10}\,\text{s} \tag{6a}$$

---

[7] See W. Greiner, B. Müller: *Theoretical Physics*, Vol. 5 *Gauge Theories of Electroweak Interactions* (Springer Berlin, Heidelberg) to be published; P. Roman: *Theory of Elementary Particles* (North Holland, Amsterdam 1960).

and

Example 8.7

$$\left.\begin{aligned} K^0 &\to \pi^0 + \pi^0 + \pi^0 \quad ,\\ &\to \pi^+ + \pi^- + \pi^0 \quad ,\\ \bar{K}^0 &\to \pi^0 + \pi^0 + \pi^0 \quad ,\\ &\to \pi^+ + \pi^- + \pi^0 \quad . \end{aligned}\right\} \tau_l = 5\times 10^{-8}\,\mathrm{s} \quad . \tag{6b}$$

This observation can be interpreted quantum mechanically such that the states

$$|K^0\rangle \quad \text{and} \quad |\bar{K}^0\rangle \tag{7}$$

may form two linearly independent states[8]

$$\begin{aligned} |K_l^0(t)\rangle &= p|K^0\rangle + q|\bar{K}^0\rangle \quad ,\\ |K_s^0(t)\rangle &= r|K^0\rangle + s|\bar{K}^0\rangle \end{aligned} \tag{8}$$

with different masses (M_l, M_s respectively) and lifetimes (τ_l, τ_s). One speaks of $K_0 - \bar{K}_0$-*mixing*. These new states, given by (8), have a time independence obeying

$$\begin{aligned} |K_l^0(t)\rangle &= \exp\left\{-\left(\mathrm{i}\frac{M_l c^2}{\hbar} + \frac{1}{2}\frac{\Gamma_l}{\hbar}\right)t\right\}|K_l^0(0)\rangle \quad ,\\ |K_s^0(t)\rangle &= \exp\left\{-\left(\mathrm{i}\frac{M_s c^2}{\hbar} + \frac{1}{2}\frac{\Gamma_s}{\hbar}\right)t\right\}|K_s^0(0)\rangle \end{aligned} \tag{9}$$

with the widths $\Gamma_s = \hbar/\tau_s$ and $\Gamma_l = \hbar/\tau_l$. Inversion of (8) leads to

$$\begin{aligned} |K^0(t)\rangle &= a|K_l^0(t)\rangle + b|K_s^0(t)\rangle \quad ,\\ |\bar{K}^0(t)\rangle &= c|K_l^0(t)\rangle + d|K_s^0(t)\rangle \quad , \end{aligned} \tag{10}$$

where

$$\begin{aligned} a &= s/(sp - rq), \quad c = -r/(sp - rq) \quad ,\\ b &= -q/(sp - rq), \quad d = p/(sp - rq) \quad . \end{aligned} \tag{11}$$

Clearly, an arbitrary state in $K^0 - \bar{K}^0$-space can always be written in the form

$$|\psi(t)\rangle = \alpha|K_l^0(t)\rangle + \beta|K_s^0(t)\rangle \quad . \tag{12}$$

Now we can understand in a simple manner the empirical fact that the decay into two pions is characterized by τ_s and the one into three pions by τ_l. Assume the short-lived K_s^0 decays *only into two and not into three pions,*[9]

$$\langle 2\pi|\hat{H}_{\mathrm{int}}|K_s^0\rangle \neq 0 \quad , \quad \langle 3\pi|\hat{H}_{\mathrm{int}}|K_s^0\rangle = 0 \quad , \tag{13}$$

---

[8] The indices l and s are abbreviations for long-lived and short-lived, respectively.

[9] The approximate validity of this relation indicates the CP invariance of the strong interaction – see Vol. 3 of this series, *Relativistic Quantum Mechanics* (Springer, Berlin, Heidelberg 1989).

Example 8.7

and, on the contrary, that K_l^0 decays into three pions[10] only,

$$\langle 2\pi|\hat{H}_{\text{int}}|K_l^0\rangle = 0 \quad , \quad \langle 3\pi|\hat{H}_{\text{int}}|K_l^0\rangle \neq 0 \quad . \tag{14}$$

Then in fact, according to (10), it results that K^0 (and $\bar{K}^0$) has two lifetimes which depend on the decay mode. Additional experiments show that, to quite a good approximation, K^0 as well as $\bar{K}^0$ decay into two pions in 50% of the cases and into three pions in the other half of the events. Thus,

$$|p|^2 = |q|^2 = \tfrac{1}{2} \quad \text{and} \quad |r|^2 = |s|^2 = \tfrac{1}{2} \quad . \tag{15}$$

Since these equations cannot be solved uniquely (phase factor!), one commonly uses the convention (e.g., at time $t = 0$)

$$|K_l^0(0)\rangle \equiv |K_2\rangle = \frac{1}{\sqrt{2}}(|K^0(0)\rangle + |\bar{K}^0(0)\rangle) \quad ,$$

$$|K_s^0(0)\rangle \equiv |K_1\rangle = \frac{1}{\sqrt{2}}(|K^0(0)\rangle - |\bar{K}^0(0)\rangle) \quad . \tag{16}$$

Note that the equations hold only approximately, since the exact solution is

$$|K_{s,l}^0\rangle = (|K_{1,2}\rangle \pm \varepsilon|K_{1,2}\rangle)/\sqrt{1+|\varepsilon|^2} \quad , \tag{17}$$

where $\varepsilon \simeq 10^{-3}$ by experiment.

In the following we shall draw some simple conclusions from (9, 10) and (16) concerning the K^0 mesons. At time $t = 0$ a K^0 meson may be produced, e.g. by the reaction[11]

$$\begin{aligned} \pi^- + p &\to K^0 + \Lambda \\ &\to K^0 + \Sigma^0 \quad . \end{aligned} \tag{18}$$

For the state $|\psi(t)\rangle$ in (12) this yields

$$\begin{aligned} |\psi(t=0)\rangle = |K^0\rangle &= \alpha|K_l^0(t=0)\rangle + \beta|K_s^0(t=0)\rangle \\ &= \frac{1}{\sqrt{2}}(|K_l^0\rangle + |K_s^0\rangle) \quad , \end{aligned} \tag{19}$$

[10] These relations hold only approximately: More precisely,

$$|\langle\pi^+\pi^-|\hat{H}_{\text{int}}|K_l^0\rangle/\langle\pi^+\pi^-|\hat{H}_{\text{int}}|K_s^0\rangle| = 1.95\times 10^{-3} \quad .$$

[11] K mesons and Λ hyperons are always produced in pairs. This is called "associate production". Besides the reactions of (18) there are also others supporting this assumption:

$$\pi^- + p \to K^+ + \Sigma^- \quad , \quad \pi^+ + p \to K^+ + \Sigma^+ \quad .$$

This is due to the fact that on the lhs only up and down quarks and their corresponding antiquarks occur, but on the rhs an s-quark and an anti-s-quark, $\bar{s}$, appear. The anti-$\bar{s}$-quark is contained in K^+, the s-quark in Σ^- and Σ^+. The $s\bar{s}$ pair may be produced in a collision (cf. the tables 11.6 and 11.7 in Sects. 11.4 and 11.5 showing the quark content of the particles).

Example 8.7

leading to (taking (16) into account)

$$\alpha = \beta = \frac{1}{\sqrt{2}} \quad . \tag{20}$$

According to (12), (16) and (9), for arbitrary time t,

$$|\psi(t)\rangle = \frac{1}{\sqrt{2}} \exp\left(-\left(\mathrm{i}\frac{M_l c^2}{\hbar} + \frac{1}{2}\frac{\Gamma_l}{\hbar}\right)t\right)|\mathrm{K}_l^0(0)\rangle + \frac{1}{\sqrt{2}} \exp\left(-\left(\mathrm{i}\frac{M_s c^2}{\hbar} + \frac{1}{2}\frac{\Gamma_s}{\hbar}\right)t\right)|\mathrm{K}_s^0(0)\rangle$$

and, with respect to (16),

$$\begin{aligned}= &\frac{1}{2}\left\{\exp\left[-\left(\mathrm{i}\frac{M_s c^2}{\hbar} + \frac{1}{2}\frac{\Gamma_s}{\hbar}\right)t\right]\right. \\ &\left.+ \exp\left[-\left(\mathrm{i}\frac{M_l c^2}{\hbar} + \frac{1}{2}\frac{\Gamma_l}{\hbar}\right)t\right]\right\}|\mathrm{K}^0\rangle \\ &+ \frac{1}{2}\left\{\exp\left[-\left(\mathrm{i}\frac{M_s c^2}{\hbar} + \frac{1}{2}\frac{\Gamma_s}{\hbar}\right)t\right]\right. \\ &\left.- \exp\left[-\left(\mathrm{i}\frac{M_l c^2}{\hbar} + \frac{1}{2}\frac{\Gamma_l}{\hbar}\right)t\right]\right\}|\bar{\mathrm{K}}^0\rangle \quad .\end{aligned} \tag{21}$$

This state determines the time-evolution of an initially produced K^0 meson. Indeed, one recognizes that at $t = 0$ a pure K^0 meson is present which develops into a superposition of K^0 and $\bar{\mathrm{K}}^0$. If

$$\tau_l > t \gg \tau_s \quad ,$$

then K^0 and $\bar{\mathrm{K}}^0$ are each present at the 50 percent level. Immediately this leads to a paradoxical result: In a beam of K^0 mesons the particles do not only decay, but transform themselves into other particles, namely $\bar{\mathrm{K}}^0$ mesons, without the presence of other matter. The $\bar{\mathrm{K}}^0$ mesons in the beam can be detected experimentally, e.g. by the reaction

$$\bar{\mathrm{K}}^0 + \mathrm{p} \to \pi^+ + \Lambda \quad , \tag{22}$$

if the beam passes through matter. The scheme of the experiment is sketched in the following figure. A π^- beam produces K^0 mesons at A which then transform into a mixture of K_s^0 and K_l^0.

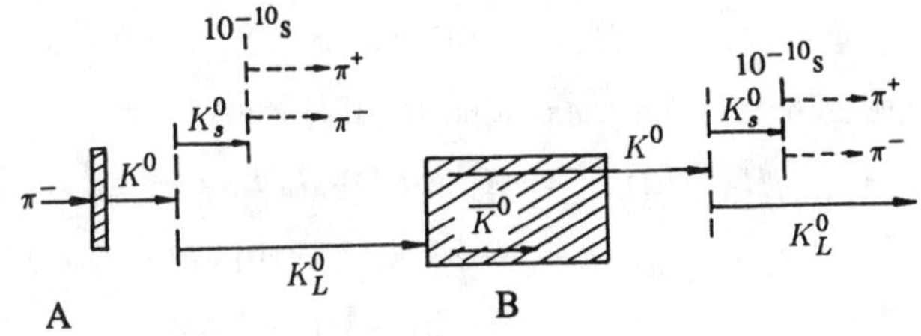

Decay, transformation, and reappearence of K^0 mesons

Example 8.7

Soon all K_s^0 decay into pions and thus only K_l^0 remain. According to (16), this beam corresponds to a linear combination of K^0 and $\bar{K}^0$. The $\bar{K}^0$ are absorbed in the target B due to the reaction of (22), and therefore a pure K^0-beam emerges from behind B and the process repeats itself. Experiments of this kind are called "*regeneration experiments with* K^0 *mesons*" and nowadays they are routine experiments in the laboratories of high-energy physics.

For a better understanding of the regeneration of K^0 mesons one should refer to the analogous experiments with polarized light waves (see next figure). The two directions of polarization of a linear, transversal polarized beam of light correspond to the states K^0 and $\bar{K}^0$. As indicated in the figure, the incoming light may be polarized in the plane of the paper and passes through a medium between A and B which rotates the polarization vector by a certain angle; let us assume the rotation angle at B to be 45°. Next this beam of light is filtered in such a way that at C the polarization vector again coincides with that at A. The experiment starts anew. Thereby, the direction of polarization at A and C obviously corresponds to the K^0 mesons behind A and B in the previous figure, and the direction of polarization at B and D corresponds to the K_l^0.

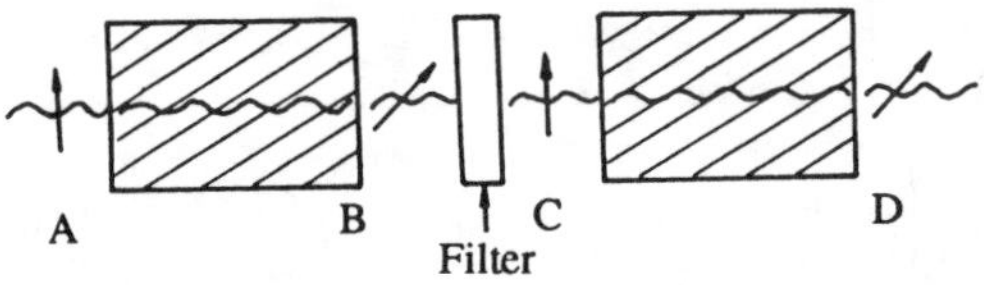

Regeneration experiment with a linear polarized beam of light

This analogy follows from the wave aspect of quantum theory. Both phenomena are essentially based on the superposition principle, which also holds for probability waves (the guiding field of the particles) in quantum mechanics and has been used, e.g., in (8) and (12). As a further consequence, both particles, K_l^0 and K_s^0, can interfere, which is already formally expressed in (12). This interference becomes experimentally observable since the 2π-decay of the K_l^0 does not vanish completely [cf. footnote 4 corresponding to Eq. (14)]. Taking into account (9) and (12), one derives

$$\begin{aligned}|\langle\pi^+\pi^-|\hat{H}_{\mathrm{int}}|\psi(t)\rangle|^2 = |&A\exp[-\mathrm{i}(M_l c^2/\hbar)t]\exp[-\tfrac{1}{2}(\Gamma_l/\hbar)t]\\ &+B\exp[-\mathrm{i}(M_s c^2/\hbar)t]\\ &\times\exp[-\tfrac{1}{2}(\Gamma_s/\hbar)t]|^2 \quad ,\end{aligned} \tag{23}$$

whereby

$$A=\alpha\langle\pi^+\pi^-|\hat{H}_{\mathrm{int}}|K_l^0(0)\rangle \text{ and } B=\beta\langle\pi^+\pi^-|\hat{H}_{\mathrm{int}}|K_s^0(0)\rangle$$

are complex constants. Squaring (23) leads to

$$\begin{aligned}|\langle\pi^+\pi^-|\hat{H}_{\mathrm{int}}|\psi(t)\rangle|^2 &= |A|^2\exp[-(\Gamma_l/\hbar)t]+|B|^2\exp[-(\Gamma_s/\hbar)t]\\ &\quad+2\mathrm{Re}\{AB\}\exp[\mathrm{i}(M_l-M_s)(c^2/\hbar)t]\\ &\quad\times\exp[-\tfrac{1}{2}(\Gamma_l+\Gamma_s)t/\hbar]\end{aligned} \tag{24}$$

Example 8.7

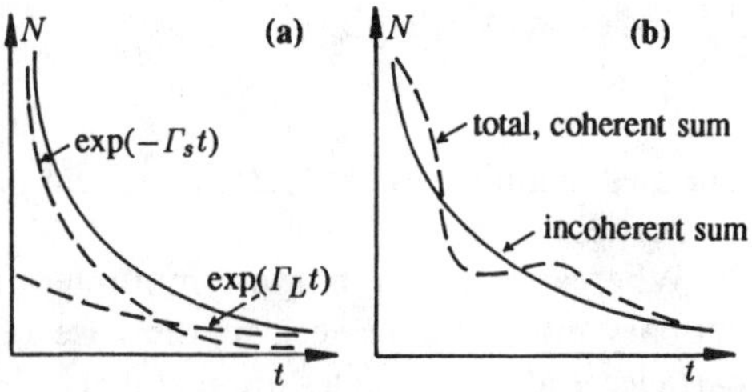

Schematical illustration of the coherent decay function (**a**) of K_s^0 and K_l^0 and the effect of the (coherent) interference term (**b**)

The first two terms on the rhs describe the incoherent superposed decay of two independent particles, shown in the following figure (a). In addition, there still appears an interference term (third term) which modifies the time dependence of the decay and leads to the dashed line of (b). Evidently the maxima and minima depend on the values of the constants A and B. The empirical result (see subsequent figure) clearly confirms the interference of K_l^0 and K_s^0 and allows the determination of the ratio $|\langle\pi^+\pi^-|\hat{H}_{\text{int}}|K_l^0\rangle/\langle\pi^+\pi^-|\hat{H}_{\text{int}}|K_s^0\rangle|$ if α and β are fixed by the regeneration experiment.

Arguments similar to these presented here for the K_s^0 and K_l^0 mesons also lead to oscillations when considering different species of neutrinos. Suppose that ν_e, ν_μ, and ν_τ neutrinos have nonvanishing rest masses and that at least two of these have different masses. If the hypothesis is true that neutrinos may transform into each other, one would expect an expression analogous to (24) for the "neutrino oscillations".[12] At present these effects are being searched at various laboratories, but so far still without success.

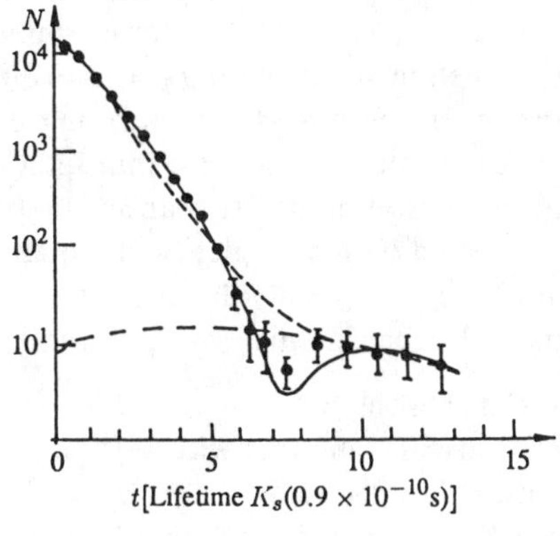

The K_l^0 and K_s^0 interference. Experimentally observed decay as a function of time

[12] See W. Greiner, B. Müller: *Theoretical Physics*, Vol. 5 *Gauge Theories of Electroweak Interactions* (Springer Berlin, Heidelberg) to be published.

8.5 Meson Multiplets

The meson multiples of SU(3) differ from the baryon multiplets in some essential properties:

Whereas to each baryon multiplet there exists a different multiplet of antibaryons (antiparticles), in the case of mesons it has been observed that particles and antiparticles are members of the *same* SU(3) multiplet, i.e. to each state vector with given T_3 and Y there exists a state vector *in the same multiplet* with $-T_3$ and $-Y$. Furthermore, particles and antiparticles possess the same spin and parity. The quantum number which distinguishes particles and antiparticles in the case of baryons, namely the baryon number, vanishes in the case of mesons ($B = 0$). This implies that to each state in a multiplet there exists a state in the same multiplet (either the same or an other vector) which has the quantum number of the antiparticle. Every quark has the baryon number $B = \frac{1}{3}$, every antiquark $B = -\frac{1}{3}$. This follows from the fact that a nucleon, consisting of three quarks, possesses the baryon number $B = +1$ and every anti-nucleon, $B = -1$. Mesons appear to consist of a quark-antiquark pair. In fact, we shall see in Sect. 8.6 that

$$[3] \otimes [\bar{3}] = [8] \oplus [1]$$

which corresponds precisely to the scheme of pseudoscalar mesons provided in Example 8.6.

A further difference to the baryon multiplets is given by a different type of violation of SU(3) symmetry. For explaining this we consider two SU(3) multiplets with equal spin, parity, and baryon number. Furthermore, we assume that to each particle of the first multiplet there exists a particle in the second one with equal, T, T_3, and Y, for example the SU(3) singlet with $T_3 = Y = 0$. Note that a state vector with $T_3 = Y = 0$ is also found in the octet (see the tables of meson properties). By performing experiments one can see that the physical states are mixtures of these multiplet states. This SU(3) *mixing* is more dominant in the meson multiplet than in the baryon multiplets. Each meson multiplet is connected with a singlet (with equal spin and parity). Since the two multiplets are mixed we classify the mesons of both multiplets together in a nonet. Mesons may also be classified by spin and parity:

scalar mesons with	$J^P = 0^+$	vector mesons with	$J^P = 1^-$
pseudoscalar mesons with	$J^P = 0^-$	axial vector mesons with	$J^P = 1^+$
tensor mesons with	$J^P = 2^+$		
pseudotensor mesons with	$J^P = 2^-$	and so on.	

EXAMPLE

8.8 The Scalar Mesons

Informations about masses and main decay modes for these mesons are often uncertain; large widths (several hundred MeV) indicate difficulties in measurements (see table on the next page). The classification into multiplets cannot be made uniquely due to SU(3) mixing. One considers the σ meson as the singlet and combines the other mesons to form an octet. It can be seen right that the octet also contains antiparticles such as $\kappa^0 \to \overline{\kappa^0}$ etc.

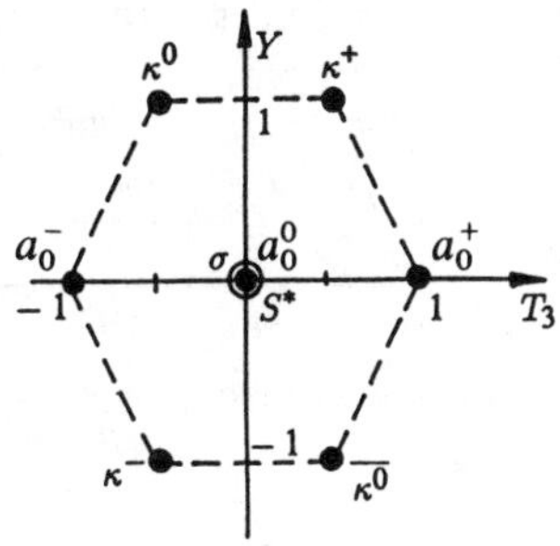

The scalar mesons

Properties of the Scalar mesons

Name	Symbol	Charge [e]	T	T_3	Y	Mass [MeV]	Width $\Gamma = \hbar/\tau$ [MeV]	Main decay modes
a_0 meson	a_0	$\pm 1, 0$	1	± 1 0	0	976	50	$\nu\pi$, $\varrho\pi$(?)
K_π resonance	κ	$+1, 0$	1/2	$\pm 1/2$	± 1	1250	~ 450	$K\pi$
S* resonance	S*	0	0	0	0	~ 993	40	$K\bar{K}$
Sigma resonance	σ	0	0	0	0	(750)	(~ 600)	$\pi\pi$

One should remark that the existence of the σ meson is uncertain; there have been efforts (among them the so-called σ model of M. Gell-Mann and M. Levy)[13] to combine the σ meson and the three pions into a "four-vector" which is invariant under the symmetry group $SU(2)_L \otimes SU(2)_R$ (*chiral group*). The "chiral" model of the pion-nucleon interaction established in this way is remarkably successful. As suggested partially by the names of the particles, all "mesons" have been discovered as resonances in reactions like $\pi + N$, $K + N$, and $\eta + N$ (N = nucleon); therefore, one labels them as pion, kaon, and eta resonances.

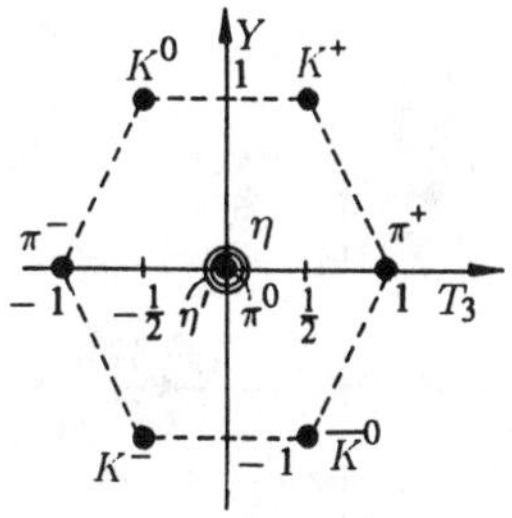

The pseudoscalar mesons

EXAMPLE

8.9 The Vector Mesons

We already know of the rho and omega mesons. In addition to these states there are the kaon resonances $K^{*\pm}$, K^{*0} and the Φ meson. It is shown below that

[13] M. Gell-Hann and M. Levy: Nuovo Cimento **16**, 705 (1960).

Properties of the Vector mesons

Name	Symbol	Charge [e]	T	T_3	Y	Mass [MeV]	Width $\Gamma = \hbar/\tau$ [MeV]	Main decay modes
Rho meson	ϱ^+	+1	1	1	0	773	154	$\pi^0\pi^+$ 100%
	ϱ^-	−1	1	−1	0			$\pi^0\pi^-$
	ϱ^0	0	1	0	0			$\pi^0\pi^0$
Omega	ω	0	0	0	0	782.7	10	$\pi^+\pi^-\pi^0$ 90% $\pi^0\gamma$ 9%
Kaon resonance	$K^{*+}K^{*-}$ $K^{*0}\overline{K^{*0}}$	+1 −1 0 0	1/2	±1/2	+1 −1	892 898	50	$K\pi$ 100%
Phi meson	Φ	0	0	0	0	1019	4.2	K^+K^- 46% $K_1^0K_s^0$ 35% $\pi^+\pi^-\pi^0$ 16.5%

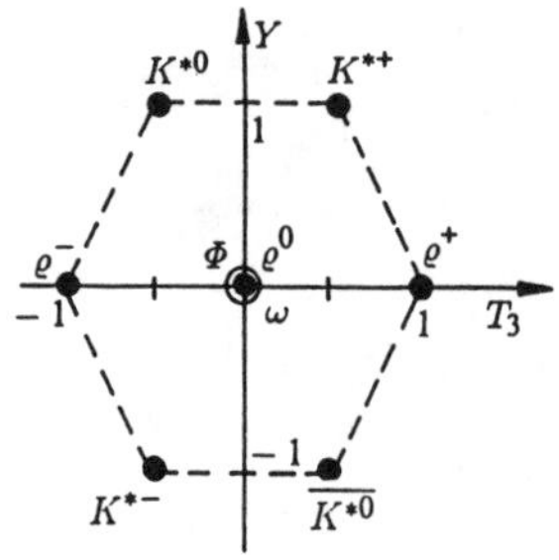

The vector mesons

these resonances again form an octet similar to the corresponding octet of the pseudoscalar mesons (π, η, K).

The η' is a singlet, and thus the Φ is a singlet too. The ω and Φ mesons have to be envisaged as a mixing of the octet ω and the singlet Φ due to SU(3) mixing. This is experimentally verified. The vector mesons, especially the ω and the ϱ mesons, play an important role in the theory of nuclear forces between two nucleons at small distances ($\simeq 0.5 \times 10^{-13}$ cm)[14]. They give rise to a repulsive contribution to the strong interaction between the nucleons if they come sufficiently close together.

EXAMPLE

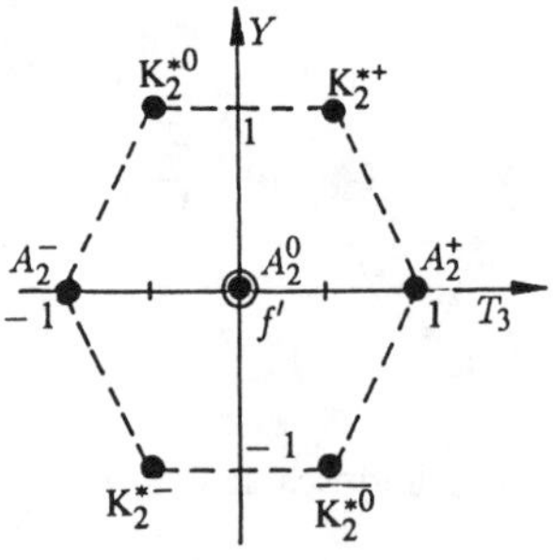

The tensor mesons

8.10 The Tensor Mesons

In this case f is the singlet meson (the $\eta^{*\prime}$ resonance). All other mesons form an octet: A_2 is the pion resonance, K_2^* is the kaon resonance, and f' is the η^* resonance (see table and figure).

The importance of the tensor mesons mainly rests on the fact that they occur as intermediate states in the reactions of the corresponding pseudoscalar particles.

[14] For meson field theory and its applications in nuclear structure physics we refer to J.D. Walecka: Annals of Physics **83**, 491 (1974), and P.G. Reinhard, M. Rufa, J.A. Maruhn, W. Greiner, J. Friedrichs: Z. Phys. A **323**, 13 (1986)

Properties of the tensor mesons

Name	Symbol	Spin/Parity J^P	Charge [e]	T	T_3	Y	Mass [MeV]	Width [MeV]	Main decay modes
A_2 mesons	A_2	2^+	$\pm 1, 0$	1	$\pm 1, 0$	0	1320	110	$\varrho\pi$ 70% $\eta\pi$ 15% $\omega\pi\pi$ 10% $\bar{K}K$ 50%
Kaon resonance	K_2^*	2^+	$\pm 1, 0$	1/2	$\pm 1/2$	± 1	1426	100 ± 3	$K\pi$ 49.1% $K\varrho$ 6.6% $K^*\pi$ 26.9% $K\omega$ 3.7% $K^*\pi\pi$ 11.2% $K\eta$ 2.5%
f′ meson	f_2'	2^+	0	0	0	0	1525	76 ± 10	$K\bar{K}$ 100%
f meson	f_2	2^+	0	0	0	0	1274	185 ± 20	$\pi^+\pi^-$ > 81% $2\pi^+2\pi^-$ 3% $K\bar{K}$ 3%

EXAMPLE

8.11 Other Resonances

There are also other (1^-) resonances belonging to the vector mesons, namely the *J* or ψ particles, discovered in Stanford and Brookhaven as well as at DESY in Hamburg (cf. Chap. 11). The most interesting feature of these particles consists in their surprisingly long lifetime ($\Gamma \simeq 0.067$ MeV, see table below) as in

Properties of some other resonances

Name	Symbol	Spin/Parity J^P	Charge [e]	T	T_3	Y	Mass [MeV]	Width [MeV]	Main decay modes [%]
J/Ψ	J/Ψ	1^-	0	0	0	0	3098	0.067	hadrons 86% e^+e^- 7% $\mu^+\mu^-$ 7%
Ψ (3700)	Ψ (3700)	1^-	0	0	0	0	3684	0.228	hadrons 98 $e^+e^-, \mu^+\mu^-$ 1% each
Ψ (4160)	Ψ (4160)	1^-	0	0	0	0	4160	78 ± 20	hadrons 98% rest $\to J/\Psi + \pi$, e^+e^-
Ψ (4400)	Ψ (4400)	1^-	0	0	0	0	4415	43 ± 20	as Ψ (4100)

Example 8.11

general the lifetime decreases with increasing mass. The long lifetime of the ψ is a hint of a conserved quantum number which prevents their fast decay; this conserved quantity is called "charm". Thus our quark model has to be supplemented by a fourth quark, the "charmed quark" c (u, d, s, c). Within this model the long lifetime of the ψ appears as a consequence of the conservation of charm. This implies that one has to replace the SU(3) by an SU(4) symmetry group which we shall discuss later (cf. Chap. 11).

8.6 Rules for the Reduction of Direct Products of SU(3) Multiplets

This seems a suitable point to show explicitly how to reduce several products of SU(3) multiplets. The rules will be pointed out by considering some examples and then representing them graphically. On the basis of the insight and results provided by the preceeding sections this task will be quite simple.

Let us start with the *product representation* [3] ⊗ [3]. In view of (8.20) and (8.25) we obtain the weight vectors of the corresponding direct product states

$$|T_3(1)\,Y(1)\rangle\,|T_3(2)\,Y(2)\rangle$$

by vector addition of the single vectors in the $Y-T_3$ plane (see Fig. 8.8).

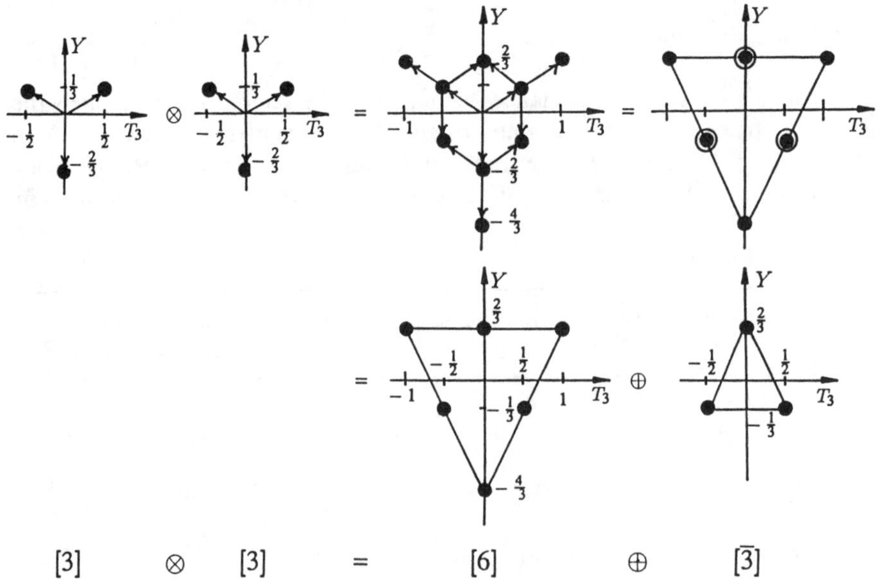

Fig. 8.8. Reduction of the direct product $[3]\otimes[3] = [6]\oplus[\bar{3}]$

To each endpoint of the vectors of the first factor, the multiplet [3], are attached all vectors of the second one. The endpoints of the sum vectors obtained by this construction represent all possible states of the direct product in the $Y-T_3$ lattice. We obtain a scheme of points, partially occupied by two states, which can easily be decomposed into a sextet [6] = $D(2,0)$ and an antitriplet $[\bar{3}]$ = $D(0,1)$. Of course, in this final step one takes into account the properties and the structure of the SU(3) multiplets discussed in the previous paragraphs. Let us investigate $[3] \otimes [\bar{3}] = [8] \oplus [1]$ as the next product (as shown in Fig. 8.9).

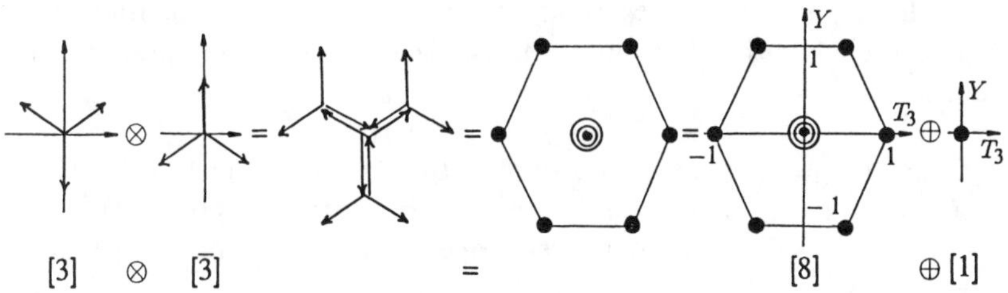

Fig. 8.9. Reduction of the product $[3] \otimes [\bar{3}] = [8] \oplus [1]$

Clearly, the product of a quark triplet and an antiquark triplet can be decomposed into an octet [8] and a singlet [1]. More complicated products like, say, $[3] \otimes [6]$ may be decomposed in a similar way. One starts from the larger multiplet whose "occupied points" may be marked on the lattice. Thereafter the vectors of the lattice are attached to each of these points. The scheme of points obtained by this procedure can be associated with the decuplet [10] and the octet [8] (see Fig. 8.10). Also $[3] \otimes [6] = [10] \oplus [8]$.

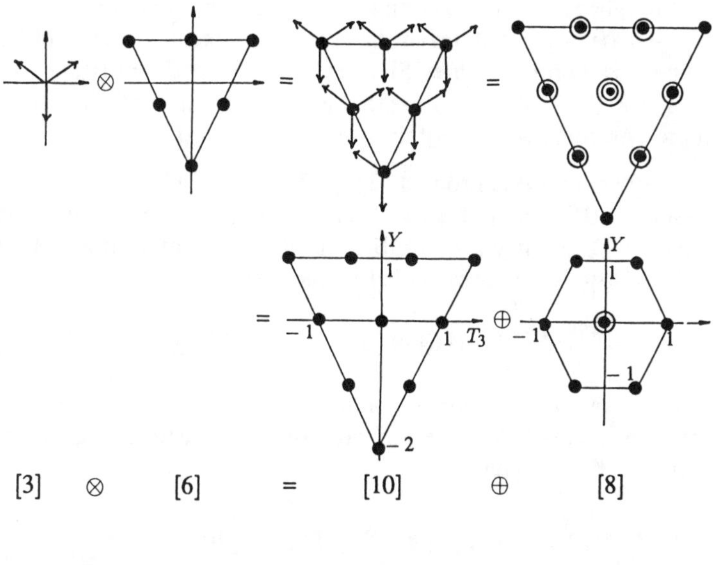

Fig. 8.10. Reduction of the product $[3] \otimes [6] = [10] \oplus [8]$

From the maximum weight of $[3] \otimes [3]$ it becomes evident that $[3] \otimes [3]$ must contain [6] as the largest representation, the remaining states forming the $[\bar{3}]$. One can understand this by the following consideration: In line with (8.27) one has

$$Y_{\max} = (p-q)/3 = (3n+\tau)/3 \quad \begin{cases} \text{with} & n = 0, \pm 1, \pm 2, \ldots \\ \text{and} & \tau = 0, 1, 2 \text{ modulo } 3 \end{cases} \quad . \tag{8.31}$$

Since p and q are integers, $Y_{\max}$ is usually a multiple of one-third. In (8.31) $Y_{\max}$ is expressed as a sum of an integer and the non-integer part (multiple of a third). Accordingly, $Y_{\max}$ is an integer only if the *triality* τ is zero. Equation (8.31) may be interpreted as the division into three classes of all representations of SU(3), namely:

Class 1: $\tau = 0$: Y integer (examples: $D(1,1) = [8]$, $D(3,0) = [10]$),
Class 2: $\tau = 1$: Y is a multiple of one third with the prototype [3],
Class 3: $\tau = 2$: Y is a multiple of one third with the prototype $[\bar{3}]$.

Thus, to define "triality", one can conclude: The triality determines the character of an SU(3) representation, i.e. whether it is a triplet or an antitriplet. *All representations which occur by reduction of a direct product of irreducible representations have equal triality.* For the states of a direct product the equation (8.25),

$$(T_3, Y) = \left(\sum_{i=1}^{p} T_3(i) + \sum_{i=1}^{q} \bar{T}_3(i), \sum_{i=1}^{p} Y(i) + \sum_{i=1}^{q} \bar{Y}(i) \right) \quad , \tag{8.32}$$

is valid. The hypercharge of the various quark states of the product (and thus for the multiplets) differ by $[Y(i) - Y'(i)] = \frac{1}{3} - \frac{1}{3}; \frac{1}{3} - (-\frac{2}{3}); -\frac{2}{3} - \frac{1}{3}$ or a multiple of this [because of (8.21) and (8.22)]. Similar relations hold for the values $\bar{Y}(i) - \bar{Y}'(i)$ of these states in that they are integers. In view of (8.31), τ must be equal for all product states. Since τ is an additive quantity [because of (8.31) and (8.32)], in particular it follows that all representations formed as products of two representations with $\tau = 0$ must also have $\tau = 0$.

Returning to the product $[3] \otimes [3]$; since [3] has the triality $\tau = 1$, then the triality of $D(2,0) = [6]$ is $\tau = 2$. Therefore, the three remaining states must also have $\tau = 2$, i.e. they are members of the representation $[\bar{3}]$. By means of such considerations one is easily led to the result

$$[3] \otimes [3] \otimes [3] = ([6] \oplus [\bar{3}]) \otimes [\bar{3}] = [1] \oplus [8] \oplus [8] \oplus [10] \quad , \tag{8.33a}$$

in which the *conservation of triality* is easily verified. It is a remarkable fact that the octet occurs twice here. Similarly one gets the more complicated result of the product of two octets:

$$[8] \otimes [8] = [1] \oplus [8] \oplus [8] \oplus [10] \oplus [\overline{10}] \oplus [27] \quad . \tag{8.33b}$$

On the left and on the right all multiplets have the trality $\tau = 0$. On the right side $[\overline{10}]$ must occur in addition to [10] because $[8] \otimes [8]$ represents a symmetric set of points around the zero point. All other multiplets on the right are symmetric within themselves.

EXERCISE

8.12 Reduction of SU(2) Multiplets

Problem: Transfer the graphical method for reducing SU(3) multiplets to the simpler case of SU(2) multiplets and split up graphically the direct product of the SU(2) doublets [2]

$$[2] \otimes [2] \otimes [2] \quad . \tag{1}$$

Solution: The SU(2) multiplets can be represented graphically analogously to SU(3) multiplets. The group SU(2) is of rank 1 and has therefore a diagonal operator. In the case of the isospin group this is $\hat{T}_3$. Each irreducible representation, each multiplet, can be represented by a straight line.

$$[2] = \quad \times \!\!-\!\!-\!\!-\!\! \times$$
$$\qquad\qquad -\tfrac{1}{2} \qquad \tfrac{1}{2}$$

$$[3] = \quad \times \!\!-\!\!-\!\!-\!\! \times \!\!-\!\!-\!\!-\!\! \times$$
$$\qquad\qquad -1 \qquad 0 \qquad 1$$

$$[4] = \quad \times \!\!-\!\!-\!\! \times \!\!-\!\!-\!\! \times \!\!-\!\!-\!\! \times$$
$$\qquad\qquad -\tfrac{3}{2} \quad -\tfrac{1}{2} \quad \tfrac{1}{2} \quad \tfrac{3}{2}$$

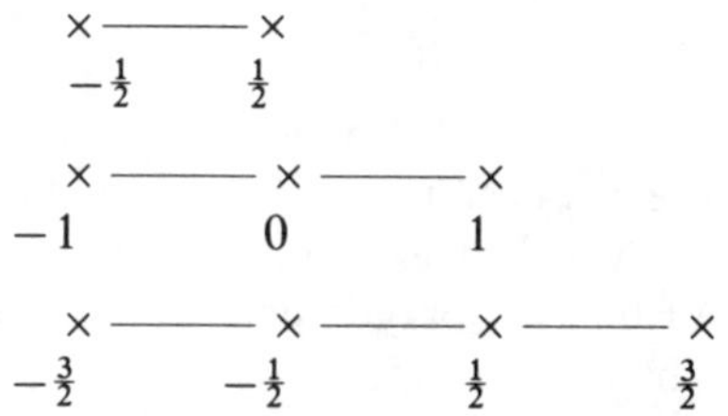

In product representation the eigenvalues of diagonal operators are additive. This addition of T_3-values can be represented graphically by putting the center point of the multiplet ($T_3 = 0$) on each point of the other multiplet.

We demonstrate this for the case of the product $[2] \otimes [2]$

$$= \times \!\!-\!\!-\!\!-\!\! \times \!\!-\!\!-\!\!-\!\! \times \quad \oplus \quad \times \quad .$$

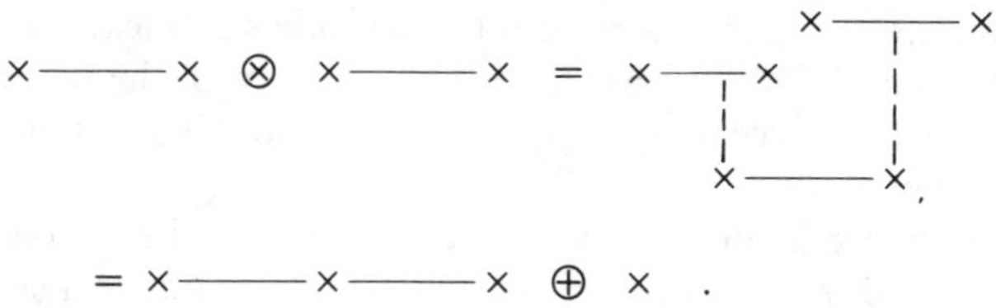

The product can by split up in the form

$$[2] \otimes [2] = [3] \oplus [1] \quad . \tag{2}$$

We use this result in order to reduce $[2] \otimes [2] \otimes [2]$ further:

$$[2] \otimes [2] \otimes [2] = [2] \otimes ([3] \oplus [1]) \quad . \tag{3}$$

Example 8.12

We apply the graphical method to the product $[2] \otimes [3]$

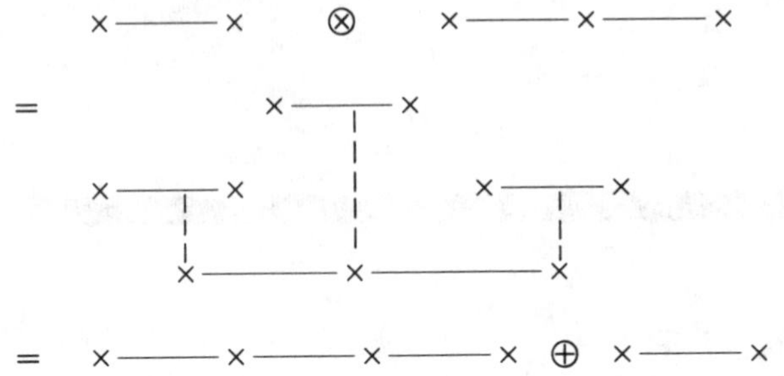

This yields

$$[2] \otimes [3] = [4] \oplus [2] \quad , \tag{4}$$

and therefore we finally obtain

$$[2] \otimes [2] \otimes [2] = [4] \oplus [2] \oplus [2] \quad . \tag{5}$$

This graphical method seems to be somewhat awkward but it is very useful for more complicated groups, as in the case of SU(3).

8.7 *U*-spin Invariance

Up to now the most extensive use has been made of the *T*-spin (isospin) group – an SU(2) subgroup of SU(3) – while investigating the SU(3) multiplets in detail. Thereby, it has turned out that the baryon octet can be associated with isospin multiplets by means of the hypercharge:

$$\begin{aligned} T=\tfrac{1}{2}&: \mathrm{n},\mathrm{p} && (\Delta M = 2\mathrm{MeV})\ Y=1 \quad ,\\ T=1&: \ \Sigma^-,\Sigma^0,\Sigma^+ && (\Delta M = 8\mathrm{MeV})\ Y=0 \quad ,\\ T=\tfrac{1}{2}&: \ \Xi^-,\Xi^0 && (\Delta M = 7\mathrm{MeV})\ Y=-1 \quad . \end{aligned} \tag{8.34}$$

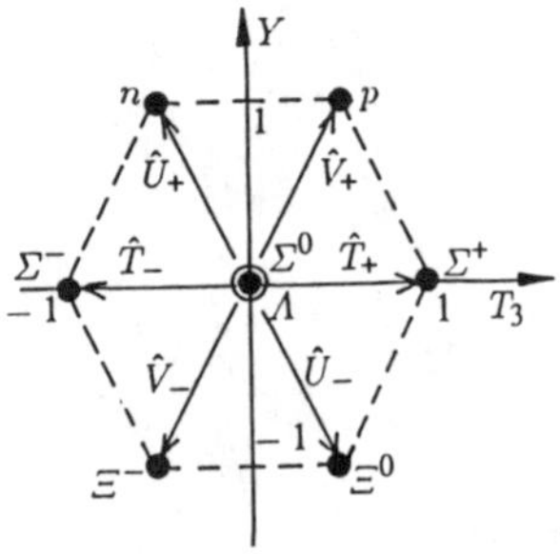

Fig. 8.11. The baryon octet as a function of the hypercharge Y and the isospin component T_3

States of a given multiplet are transformed into each other by means of the operators $\hat{T}_\pm$. Within a multiplet the mass differences ΔM are of the order of magnitude of 10 MeV, i.e. one percent of the particle mass, and it is said that the *isospin symmetry is weakly broken.*

Besides the *T*-spin we are acquainted with the *U*-spin and the *V*-spin as SU(2) subgroups of SU(3). First we consider the *U*-spin: The states of a single *U*-spin multiplet are transformed into each other by applying the operators $\hat{U}_\pm$. Thus one reads from Fig.8.11 for the multiplet:

$$\begin{aligned} &1.\ \Sigma^-,\Xi^- && \text{with } \Delta M = 124\ \text{MeV}, && \text{charge: } Q=-1 \quad ,\\ &2.\ \mathrm{n},\Sigma^0,\Xi^0 && \text{with } \Delta M = 374\ \text{MeV}, && \text{charge: } Q=0 \quad ,\\ &3.\ \mathrm{p},\Sigma^+ && \text{with } \Delta M = 251\ \text{MeV}, && \text{charge: } Q=+1 \quad . \end{aligned} \tag{8.35}$$

Obviously the states of a U-spin multiplet all have the same charge, i.e. one can draw in the (Y, T_3) diagram the analogous (Q, U_3) graph of the octet (see Fig. 8.12).

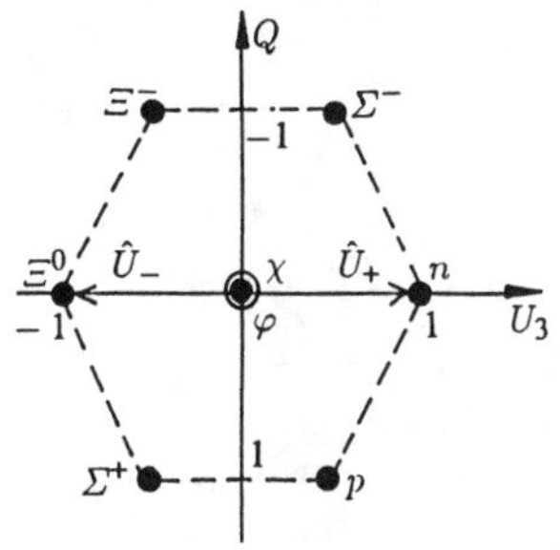

Fig. 8.12. The baryon octet as function of the charge Q and the U-spin component U_3

The states on the edges all have unit multiplicity, which means they are singly occupied, whereas the center of the diagram has the multiplicity two, indicating occupancy by two particles (the Σ^0 and the Λ). The Σ^0 and the Λ are eigenstates of the isospin (at least to a good approximation). *A state with multiplicity one must be simultaneously an eigenstate of T-, U- and V-spin, but if the multiplicity of a state is not one, then an isospin eigenstate with this multiplicity is in general no eigenstate of the U- or V-spin.* This is true for the Σ^0 and the Λ.

We can construct the eigenstates of U very easily, and we therefore start from the fundamental isospinors α and β. The wave functions of $\Sigma^\pm$, Σ^0 and Λ are then given by

$$\Sigma^+ = \alpha\alpha \quad , \quad \Sigma^0 = \frac{1}{\sqrt{2}}(\alpha\beta + \beta\alpha) \quad , \quad \Sigma^- = \beta\beta \quad , \quad \text{(triplet)}$$

$$\Lambda = \frac{1}{\sqrt{2}}(\alpha\beta - \beta\alpha) \quad . \quad \text{(singlet)} \tag{8.36}$$

Since T-, U- and V-spin are SU(2) algebras, their generators must be the Pauli matrices. We denote them in the case of the T-spin by $\hat{\tau}_i$ and in the case of the U-spin by $\hat{\mu}_i$, which means that

$$\hat{\mu}_3 = \hat{\tau}_3 = \begin{pmatrix} 1 & 0 \\ 0 & -1 \end{pmatrix} \quad , \quad \hat{\mu}_- = \hat{\tau}_- = \begin{pmatrix} 0 & 0 \\ 1 & 0 \end{pmatrix} , \quad \hat{\mu}_+ = \hat{\tau}_+ = \begin{pmatrix} 0 & 1 \\ 0 & 0 \end{pmatrix} \tag{8.37a}$$

and, for states with more than one particle,

$$\hat{T}_3 = \frac{1}{2}\sum_i \hat{\tau}_3(i) \quad , \quad \hat{T}_- = \sum_i \hat{\tau}_-(i) \quad , \quad \hat{T}_+ = \sum_i \hat{\tau}_+(i) \quad ,$$

$$\hat{U}_3 = \frac{1}{2}\sum_i \hat{\mu}_3(i) \quad , \quad \hat{U}_- = \sum_i \hat{\mu}_-(i) \quad , \quad \hat{U}_+ = \sum_i \hat{\mu}_+(i) \quad , \tag{8.37b}$$

We are searching now for the triplet eigenstates of the U-spin, which we denote by n, χ, Ξ^0 (let the singlet be ϕ), where χ and ϕ are orthogonal linear combinations of the isospin eigenstates Λ and Σ^0:

$$\chi = a\Sigma^0 + b\Lambda \quad , \quad \phi = b\Sigma^0 - a\Lambda \quad . \tag{8.38}$$

Let us recall the action of the isospin-operators on the isospin eigenstates, e.g.

$$\hat{T}_-\Sigma^+ = \sum_i \hat{\tau}_-(i)[\alpha\alpha] = (\hat{\tau}_-\,\alpha)\alpha + \alpha(\hat{\tau}_-\alpha) = \beta\alpha + \alpha\beta \quad , \tag{8.39}$$

i.e.

$$\hat{T}_-\Sigma^+ = \sqrt{2}\Sigma^0 \quad .$$

As $\hat{U}_-$ applied to the U-spin triplet eigenstate n has to act in the same way as $\hat{T}_-$ on the isospin eigenstate Σ^+, we conclude that

$$\hat{U}_- \mathrm{n} = \sqrt{2}\chi \quad . \tag{8.40}$$

Since furthermore $\hat{T}_+ \mathrm{n} = \mathrm{p}$, then

$$[\hat{U}_-, \hat{T}_+]_- \mathrm{n} = \hat{U}_- \hat{T}_+ \mathrm{n} - \hat{T}_+ \hat{U}_- \mathrm{n} = \hat{U}_- \mathrm{p} - \hat{T}_+ \sqrt{2}\chi \quad ,$$

or

$$[\hat{U}_-, \hat{T}_+]_- \mathrm{n} = \hat{U}_- \mathrm{p} - \sqrt{2}\hat{T}_+ (a\Sigma^0 + b\Lambda) \quad . \tag{8.41}$$

In addition, with

$$\hat{U}_- \mathrm{p} = \Sigma^+, \quad \hat{T}_+ \Sigma^0 = \sqrt{2}\Sigma^+ \quad \text{and} \quad \hat{T}_+ \Lambda = 0 \quad , \quad \text{results}$$

$$[\hat{U}_-, \hat{T}_+]_- \mathrm{n} = \Sigma^+ - 2a\Sigma^+ = (1 - 2a)\Sigma^+ \quad . \tag{8.42}$$

We know already [Chap. 7, Eq (7.28b)] that $[\hat{U}_-, \hat{T}_+] = 0$, i.e.

$$1 - 2a = 0 \quad \text{or} \quad a = \tfrac{1}{2} \quad . \tag{8.43}$$

Normalization of χ implies that

$$\chi^2 = |a\Sigma^0 + b\Lambda|^2 = |a|^2 + |b|^2 = 1$$

(since $|\Sigma^0|^2 = |\Lambda|^2 = 1$ and $\langle \Sigma^0 | \Lambda \rangle = 0$). With this follows, provided b is real,

$$b^2 = 1 - a^2 = \tfrac{3}{4} \quad , \quad b = \pm \tfrac{1}{2}\sqrt{3} \quad .$$

Usually one chooses b with positive sign, that is

$$\chi = \tfrac{1}{2}\Sigma^0 + \tfrac{1}{2}\sqrt{3}\Lambda \quad \text{and} \quad \phi = \tfrac{1}{2}\sqrt{3}\Sigma^0 - \tfrac{1}{2}\Lambda \quad . \tag{8.44}$$

Therefore ϕ and χ are the eigenstates of $\hat{U}_3$ (eigenvalue 0) corresponding to the eigenstates of $\hat{T}_3(\Sigma^0, \Lambda)$.

The center of the diagram must be doubly occupied, since from $[\hat{U}_-, \hat{T}_+] = 0$ follows $a^2 = \frac{1}{4}$. If, as well, $b = 0$, then χ would not be normalizable to 1: $\frac{1}{4} \neq 1$. Therefore, two states must exist in the center, something we were already aware of from our general discussion concerning the structure of the SU(3) multiplets [see Chap. 7, "The number of states on lattice points of inner shells"].

8.8 Test of U-spin Invariance

We know that the electromagnetic interaction violates isospin symmetry since it splits the T-multiplets, e.g. for the isodoublet the mass splitting is $\Delta M \simeq 2$ MeV. As argued earlier, this might be caused by the electromagnetic interaction. This kind of symmetry breaking should not occur for the U-spin, since we have seen that the members one U-spin multiplet all have the same charge ($[\hat{Q}, \hat{U}_3]_- = 0$). But this is only so if the U-spin symmetry is a symmetry of the strong interaction. However, this is only true to a very limited degree ($\Delta M_{\max} \simeq 100$ MeV, i.e. about 10% of the mass of the particle), whereas the isospin symmetry is broken only by about 1%. Therefore, we can at best expect that U-spin conservation influences the electromagnetic properties of the particles. If one *postulates* U-spin conservation, it might be expected, for example, that *p and* Σ^+ (since they belong to the same U-spin multiplet) have the same magnetic moment:

$$\mu(\Sigma^+) = \mu(\mathrm{p}) \quad . \tag{8.45}$$

Analogously we expect $\mu(\Sigma^-) = \mu(\Xi^-)$ and, for the triplet n, χ, Ξ^0, χ is a linear combination of Σ^0 and Λ and, therefore, not a physical particle. Experimentally one has found ($\mu_0 = e\hbar/2m_\mathrm{p}c$)

$$\begin{aligned} &\mu_\mathrm{p} = 2.79\,\mu_0 \quad , && \mu_{\Sigma^+} = (2.33 \pm 0.13)\mu_0 \quad , \\ &\mu_{\Xi^-} = (-0.69 \pm 0.04)\mu_0 \quad , && \mu_{\Sigma^-} = (-1.41 \pm 0.25)\mu_0 \quad , \\ &\mu_\mathrm{n} = -1.91\,\mu_0 \quad , && \mu_{\Xi^0} = (-1.253 \pm 0.014)\mu_0 \quad . \end{aligned} \tag{8.47}$$

Thus the predictions of U-spin invariance are only roughly confirmed; a fact which will be better understood in the framework of the quark model.

We can make further comments about the electromagnetic mass splitting: We assume therefore that the mass of a baryon on the one hand results from the strong interaction, which conserves isospin, and on the other from electromagnetic interaction, which conserves U-spin, i.e. the electromagnetic contributions to the mass must be equal within one U-spin multiplet. This implies, roughly, that the radii of these particles have to be the same. We find

$$\begin{aligned} &\delta M_\mathrm{p} = \delta M_{\Sigma^+} \quad , \quad \delta M_\mathrm{n} = \delta M_{\Xi^0} \quad , \quad \delta M_{\Sigma^-} = \delta M_{\Xi^-} \quad , \quad \text{or} \\ &\delta M_\mathrm{n} - \delta M_\mathrm{p} + \delta M_{\Xi^-} - \delta M_{\Xi^0} = \delta M_{\Sigma^-} - \delta M_{\Sigma^+} \quad . \end{aligned} \tag{8.48}$$

If, besides electromagnetic interaction, there exists no other interaction which destroys the degeneracy of an isomultiplet, then we must have

$$\begin{aligned} &\delta M_\mathrm{n} - \delta M_\mathrm{p} = M_\mathrm{n} - M_\mathrm{p} \quad , \quad \delta M_{\Xi^-} - \delta M_{\Xi^0} = M_{\Xi^-} - M_{\Xi^0} \quad , \\ &\delta M_{\Sigma^-} - \delta M_{\Sigma^+} = M_{\Sigma^-} - M_{\Sigma^+} \quad . \end{aligned}$$

Inserting this in (8.48) yields

$$M_\mathrm{n} - M_\mathrm{p} + M_{\Xi^-} - M_{\Xi^0} = M_{\Sigma^-} - M_{\Sigma^+} \quad . \tag{8.49}$$

This simple relation is called the "Coleman-Glashow" relation[15] and is well-confirmed experimentally by

$$\begin{aligned} &M_\mathrm{n} - M_\mathrm{p} = 1.3\ \mathrm{MeV} \quad , \quad M_{\Xi^-} - M_{\Xi^0} = (6.5 \pm 0.7)\ \mathrm{MeV} \quad , \\ &M_{\Sigma^-} - M_{\Sigma^+} = 8.0\ \mathrm{MeV} \quad . \end{aligned} \tag{8.50}$$

This shows that U-spin invariance is to some extend a useful hypothesis in the context of the SU(3) multiplets.

[15] S. Coleman, S.L. Glashow: Phys. Rev. Lett. **6**, 423 (1961).

Finally we study the V-spin multiplets,

Σ^-,n : $\Delta M \simeq 257$ MeV ,

Ξ^-,ϱ,p : $\Delta M \simeq 383$ MeV as triplet; and σ as singlet

Ξ^0, Σ^+ : $\Delta M \simeq 125$ MeV .

Here ϱ and σ are linear combinations of Σ^0 and Λ. For the U-spin we have seen that all members of one U-spin multiplet carry the same charge. The members of a V-spin multiplet, however, share no common property. Therefore the $V-$ symmetry, which is strongly broken ($\Delta M/M \geqslant 20\%$), is not useful for the classification of the states.

8.9 The Gell-Mann-Okubo Mass Formula

If the SU(3) symmetry was an exact symmetry of the strong interaction, then all states of one SU(3) multiplet would be energetically degenerate, which means they should have the same mass. Experiment, however, shows that (e. g. for the baryon octet the mass splitting is of the order $\Delta M/M \mathrel{\hat=} 100\,\mathrm{MeV}/1000\,\mathrm{MeV} = 10\%$) SU(3) symmetry is broken considerably more strongly than the isospin symmetry by the electromagnetic interaction: For an isospin multiplet we had $\Delta M/M \mathrel{\hat=} 10\,\mathrm{MeV}/1000\,\mathrm{MeV} = 1\%$.

Therefore we have to start from the idea that the Hamiltonian of the strong interaction splits into one part $\hat{H}_{ss}$ (ss for "superstrong"), which is SU(3) invariant, and one part $\hat{H}_{ms}$ (ms for "medium strong") which breaks SU(3) invariance. Since the mass splitting within a multiplet amounts only to about 10% of the mean mass, we can assume that the contribution to the mass which originates from $\hat{H}_{ms}$ is small, that is

$$M = \langle \hat{H}_{ss} \rangle + \langle \hat{H}_{ms} \rangle \quad , \quad \langle \hat{H}_{ss} \rangle \gg \langle \hat{H}_{ms} \rangle \quad . \tag{8.51}$$

According to $\hat{H}_{ss}$ the multiplets are degenerate, and $\hat{H}_{ms}$ removes this degeneracy. One should recognize that $\hat{H}_{ms}$ must be constructed from generators and not from Casimir operators since the former break SU(3) symmetry, whereas the latter do not. To be precise, the generators, if applied to the states of a multiplet, yield in general different values, depending on the state, while the Casimir operator does not.

From here on we will neglect the electromagnetic mass splitting, which means that all members of an isospin multiplet will have the same mass. This implies, of course, that the symmetry breaking part of the strong interaction hamiltonian $\hat{H}_{\mathrm{strong}} = \hat{H}_{ss} + \hat{H}_{ms}$ has to commute with $\hat{T}_3$; therefore $[\hat{H}_{ms}, \hat{T}_3] = 0$ and $\hat{H}_{ms}$ can only contain such generators of SU(3) which commute with $\hat{T}_3$. This is only the case for $\hat{F}_8 = \sqrt{3}\,\frac{1}{2}\,\hat{Y}$, and the simplest assumption is therefore

$$\hat{H}_{ms} = b\,\hat{Y} \quad , \tag{8.52}$$

where b is a constant. Calculating the expectation value of $\hat{H}_{ms}$ between the unperturbed wave functions $|TT_3Y\rangle$, we obtain by first order perturbation theory ($\langle\hat{H}_{ms}\rangle \ll \langle\hat{H}_{ss}\rangle$):

$$\langle TT_3 Y|\hat{H}_{ms}|TT_3 Y\rangle = bY \quad , \quad \text{i.e.} \quad M = a + bY \quad , \tag{8.53}$$

where a is the mean mass of the multiplet (for $Y = 0$) and a and b are constant for all members of one multiplet. Looking at the decuplet of the baryon resonances (Chap. 6, Example 6.5), it follows that

$$M_{\Omega^-} - M_{\Xi^*} = M_{\Xi^*} - M_{\Sigma^*} = M_{\Sigma^*} - M_{\Delta} \quad . \tag{8.54}$$

It is difficult to compare this prediction with experiment (because of the electromagnetic mass splitting). We can, however, minimize this effect when we consider only particles with the same charge (U-spin multiplets) for which we find:

$$\begin{aligned} M_{\Omega^-} - M_{\Xi^{*-}} &= (137 \pm 1)\,\text{MeV} \quad , \\ M_{\Xi^{*-}} - M_{\Sigma^{*-}} &= (148 \pm 1)\,\text{MeV} \quad , \\ M_{\Sigma^{*-}} - M_{\Delta^-} &= (148 \pm 5)\,\text{MeV} \quad . \end{aligned} \tag{8.55}$$

The agreement of theory with experiment is obviously very good, but our mass formula fails for the baryon octet since, because of $Y_{\Sigma^0} = Y_{\Lambda}$, we should also have $M_{\Sigma^0} = M_{\Lambda}$. In reality the mass difference is $M_{\Sigma^0} - M_{\Lambda} = 77$ MeV. This experience forces us to modify (8.52) for $\hat{H}_{ms}$: As well as $\hat{Y}$, both $\hat{T}^2$ and $\hat{Y}^2$ commute with $\hat{T}_3$, and we try therefore

$$\hat{H}_{ms} = b\hat{Y} + c\hat{T}^2 + d\hat{Y}^2 \tag{8.56}$$

as the next simplest assumption. Its expectation value leads to the formula

$$M = a + bY + cT(T+1) + dY^2 \quad . \tag{8.57}$$

Here again the coefficients c and d are constant in an SU(3) multiplet. But (8.57) leads to a problem, namely that this relation does not yield a constant mass splitting for the decuplet, i.e. the good agreement [(8.54, 55)] with experiment is lost.

We can correct this disadvantage by demanding that the decuplet $cT(T+1) + dY^2$ has the same form as $a + bY$, in other words like

$$cT(T+1) + dY^2 = x + yY \quad , \tag{8.57a}$$

where x and y are constants. Inserting the values of Δ^-, Ξ^{*-} and Ω^- results in (see adjacent table).

	T	Y
Δ^-	$\frac{3}{2}$	1
Ξ^{*-}	$\frac{1}{2}$	-1
Ω^-	0	-2

One gets from (8.57a)
for Δ^-: $15/4c + d = x + y$,
for Ξ^{*-}: $3/4c + d = x - y$,
for Ω^-: $4d = x - 2y$.

The last equation is consistent with the first two if one chooses $d = -\frac{1}{4}c$. Indeed, multiplying the first equation by $-\frac{1}{2}$, the second one by $\frac{3}{2}$ and adding both of them yields

$$-\tfrac{6}{8}c + d = x - 2y \quad .$$

This is identical with the last equation for $d = -\frac{1}{4}c$, as already stated.

$$M = a + bY + c[T(T+1) - 1/4Y^2] \quad . \tag{8.58}$$

This is the *mass formula of Gell-Mann and Okubo*[16], first deduced by Okubo, and then applied by Gell-Mann to predict the mass of the Ω^-. We wish to test it now in the case of the baryon octet and hence obtain

$$\tfrac{1}{2}(M_N + M_\Xi) = \tfrac{1}{4}(3M_\Lambda + M_\Sigma) \quad .$$

To minimize electromagnetic effects we consider the neutral particles

$$\tfrac{1}{2}M_n + \tfrac{1}{2}M_{\Xi^0} = (1127.1 \pm 0.7)\,\mathrm{MeV} \quad , \quad \tfrac{3}{4}M_\Lambda + \tfrac{1}{4}M_{\Sigma^0} = (1134.8 \pm 0.2)\,\mathrm{MeV} \quad ,$$

The energy difference between both values amounts to 7.7 MeV (much smaller than the mean mass splitting of around 100 MeV). This indicates that the Gell-Mann-Okubo mass formula is quite well satisfied; it also holds for the other SU(3) multiplets (for baryons as well as for mesons), although in the case of the mesons the mass formula must be modified since it is not valid for the masses, but for the squared masses. We can justify this – as we will see later – by the fact that baryons are fermions obeying the Dirac equation [which contains linear terms in energy (mass)], while mesons are bosons, obeying the Klein-Gordon equation [which contains the energy (mass) squared].

8.10 The Clebsch-Gordan Coefficients of the SU(3)

In the case of the isospin group we have found that the relative probabilities for two reactions taking place within an isospin multiplet are just given by their Clebsch-Gordan coefficients (CGC) (see Example 5.8). Therefore, in order to examine reactions within different SU(3) multiplets, we require knowledge of the CGC of the SU(3) group. We already know how to reduce Kronecker products of irreducible representation, for example:

$$[3] \otimes [3] = [6] \oplus [\bar{3}] \text{ or}$$

$$[3] \otimes [3] \otimes [3] = [1] \oplus [8] \oplus [8] \oplus [10] \quad , \quad \text{etc.}$$

To construct the CGC of the group SU(3), let us consider two irreducible representations $\alpha = D(p_1, q_1)$ and $\beta = D(p_2, q_2)$ which are generated by the two basis functions $\psi_\mu^{(\alpha)}$ and $\psi_\nu^{(\beta)}$. Here α and β denote the corresponding representations and μ, ν stand for all other quantum numbers (T, T_3 and Y). We use the abbreviations:

$$\mu = \{y, t, t_3\} \quad , \quad \nu = \{y', t', t'_3\} \quad , \quad m = \{Y, T, T_3\} \quad .$$

Let N be the dimension of the representation $D(p, q)$. In the well-known manner we first construct eigenfunctions of the isospin from $\psi_{ytt_3}^{(\alpha)}$ and $\psi_{y't't'_3}^{(\beta)}$,

$$\chi_{yty't'}^{(T,T_3)} = \sum_{t_3 t'_3} (tt'T|t_3 t'_3 T_3)\psi_{ytt_3}^{(\alpha)} \psi_{y't't'_3}^{(\beta)} \quad , \tag{8.59}$$

[16] S. Okubo: Prog. Theor. Phys. **27**, 949 (1962).

where $(tt'T|t_3t'_3T_3)$ are the usual SU(2) Clebsch-Gordan coefficients. Now we construct similarly the eigenstates of the irreducible representation γ of dimension N,

$$\psi^{(N\gamma)}_{YTT_3} = \sum_{yty't'} (\alpha\beta yty't'|NYT\gamma)\chi^{(T,T_3)}_{yty't'} \quad . \tag{8.60}$$

The coefficients $(\alpha\beta yty't'|NYT\gamma)$ only depend on T, not on T_3; therefore they are called *isoscalar factors*. Of course, we must have $y + y' = Y$. Only then will we have contributions to the sum in (8.60). Thus, by inserting $\chi^{(T,T_3)}_{yty't'}$, we obtain the CGC of SU(3) as

$$(\alpha\beta ytt_3y't't'_3|Nm\gamma) = (\alpha\beta\mu\nu|Nm\gamma) = (\alpha\beta yty't'|NYT\gamma)(tt'T|t_3t'_3T_3) \quad ,$$

$$\text{SU(3)} - \text{CGC} = \text{isoscalar factor} \times \text{SU(3)} - \text{CGC}. \tag{8.61}$$

To determine a CGC of the SU(3), we only need the SU(2)-CGC and the T_3-independent factors[17] $(\alpha\beta yty't'|NYT_\gamma)$. As an example let us construct SU(3)-symmetric wave functions [wave functions with well-defined SU(3) symmetry] for the simple quark model. We know that [3] and [$\bar{3}$] are two fundamental (smallest non-trivial) representations of the SU(3): [3] for the quarks and [$\bar{3}$] for the antiquarks. Consequently, these representations correspond to the three quark states q_1, q_2, q_3 (or also u, d, s) and the three antiquark states $\bar{q}_1, \bar{q}_2, \bar{q}_3$ (or also $\bar{u}, \bar{d}, \bar{s}$), the quantum numbers of which are listed in Table 8.1. The physically most significant Kronecker products of these representations are

Table 8.1. The quantum numbers of the quark and antiquark states

	T	T_3	Q	Y
$q_1(u)$	$\frac{1}{2}$	$\frac{1}{2}$	$\frac{2}{3}$	$\frac{1}{3}$
$q_2(d)$	$\frac{1}{2}$	$-\frac{1}{2}$	$-\frac{1}{3}$	$\frac{1}{3}$
$q_3(s)$	0	0	$-\frac{1}{3}$	$-\frac{2}{3}$
$\bar{q}_1(\bar{u})$	$\frac{1}{2}$	$-\frac{1}{2}$	$-\frac{2}{3}$	$-\frac{1}{3}$
$\bar{q}_2(\bar{d})$	$\frac{1}{2}$	$\frac{1}{2}$	$\frac{1}{3}$	$-\frac{1}{3}$
$\bar{q}_3(\bar{s})$	0	0	$\frac{1}{3}$	$\frac{2}{3}$

$$[3] \otimes [\bar{3}] = [8] \oplus [1] \quad , \tag{8.62}$$

which is the meson octet + singlet. Similarly, the reduction of the product

$$[3] \otimes [3] \otimes [3] = [1] \oplus [8]_1 \oplus [8]_2 \oplus [10] \tag{8.63}$$

yields a singlet, two octets with different symmetry, and a decuplet. This product reproduces the baryons and their resonances, respectively. As done previously (see the section about meson multiplets), we conclude from (8.62) that the mesons consist of a quark and an antiquark, while the nucleons are composed of three quarks. This is also the reason for expecting the spins $I = 0$ or $I = 1$ for mesons and $I = \frac{1}{2}$ or $I = \frac{3}{2}$ for baryons, remembering that the quarks have spin $I = \frac{1}{2}$.

Next we discuss the meson octet and, in particular, we construct the pion triplet: π^+ can only be a coupled $u\bar{d}$ state with an isospin function given by

$$\chi^{(1,1)}_{\frac{1}{3}\frac{1}{2},-\frac{1}{3}\frac{1}{2}} = (\tfrac{1}{2}\,\tfrac{1}{2}\,1\,|\,\tfrac{1}{2}\,\tfrac{1}{2}\,1)\psi^{(3)}_{\frac{1}{3}\frac{1}{2}\frac{1}{2}}\,\psi^{(\bar{3})}_{-\frac{1}{3}\frac{1}{2}\frac{1}{2}} \quad . \tag{8.64}$$

In this case the CGC is given by $(\frac{1}{2}\,\frac{1}{2}\,1\,|\,\frac{1}{2}\,\frac{1}{2}\,1) = 1$, and furthermore the isoscalar factor is found to be $(3\bar{3}\frac{1}{3}\frac{1}{2} - \frac{1}{3}\frac{1}{2}|8\,0\,1\,8) = 1$. We note that this result is valid for all pions, because the isoscalar factors are independent of T_3. The π^+ wave function is then given by $|\pi^+\rangle = ud$, i.e. the corresponding SU(3)-CGC equals 1. For the π^0 state we find similarly that $|\pi^0\rangle = (\frac{1}{2})^{1/2}(u\bar{u} - d\bar{d})$ and for the π^-

[17] One finds a detailed table of isoscalar factors, for example, in J.J. de Swart: Rev. Mod. Phys. **35**, 916 (1963).

state$|\pi^-\rangle = d\bar{u}$. There are no other possible ways to construct the $T = 1$-triplet than through coupling of the $t = \frac{1}{2}$-doublet of [3] and the $t = \frac{1}{2}$-doublet of $[\bar{3}]$. Similarly $T = \frac{1}{2}$-doublet states of the meson octet [8] with $Y = 1$ can only be coupled by the $T = \frac{1}{2}$-doublet ($Y = \frac{1}{3}$) of [3] and the $T = 0$-singlet ($Y = \frac{2}{3}$) of $[\bar{3}]$. Analogous reasoning holds for the $T = \frac{1}{2}$-doublet of the meson octet [8] with $Y = -1$. So there appear only ordinary SU(2) CGC [for example $\pm(\frac{1}{2})^{1/2}$]. For completeness we note the quark content of all pseudoscalar mesons in Table 8.2.

Table 8.2. The quark content of the pseudoscalar mesons

SU(3)-multiplet	Quark content	Y	T	T_3	Name
[8]	$u\bar{d}$	0	1	1	π^+
[8]	$\frac{1}{\sqrt{2}}(u\bar{u} - d\bar{d})$	0	1	0	π^0
[8]	$d\bar{u}$	0	1	-1	π^-
[8]	$u\bar{s}$	1	$\frac{1}{2}$	$+\frac{1}{2}$	K^+
[8]	$d\bar{s}$	1	$\frac{1}{2}$	$-\frac{1}{2}$	K^0
[8]	$s\bar{d}$	-1	$\frac{1}{2}$	$+\frac{1}{2}$	$\bar{K}^0$
[8]	$s\bar{u}$	-1	$\frac{1}{2}$	$-\frac{1}{2}$	K^-
[8]	$\frac{1}{\sqrt{6}}(u\bar{u} + d\bar{d} - 2s\bar{s})$	0	0	0	η, η'
[1]	$\frac{1}{\sqrt{3}}(u\bar{u} + d\bar{d} - 2s\bar{s})$	0	0	0	η, η'

In addition we have constructed isospin multiplets:

with $T = \frac{1}{2}$ and $Y = 1$: $\quad d\bar{s}, u\bar{s}$,

with $T = \frac{1}{2}$ and $Y = 0$: $\quad d\bar{u}, \frac{1}{\sqrt{2}}(u\bar{u} - d\bar{d}), u\bar{d}$,

with $T = \frac{1}{2}$ and $Y = -1$: $\quad s\bar{u}, s\bar{d}$.

As already mentioned, here all occurring CGC are SU(2)-CGC.

Thus we have built up seven independent states $\psi_\alpha, \psi_\beta, \ldots$ from the nine states $q_i q_j$ $(i,j = 1,2,3)$. The states are orthonormal, i.e.

$$\langle\psi_\alpha|\psi_\beta\rangle = \delta_{\alpha\beta} \quad , \quad \text{with} \quad \psi_\alpha = \sum_{ij} \alpha_{ij} q_i \bar{q}_j \quad ,$$

and, further, we can construct the state

$$\psi_1 = \frac{1}{\sqrt{3}}(u\bar{u} + d\bar{d} + s\bar{s}) \quad ,$$

which can also be written as

$$\psi_1 = \frac{1}{\sqrt{3}} \sum_i q_i \bar{q}_i \quad .$$

The invariance of this wave function with respect to SU(3) transformations is easily recognized with the aid of the results of Exercises 8.1 and 8.2. Namely, we have

$$\sum_i \bar{q}'_i q'_i = \sum_{ijk} \bar{q}_j U^\dagger_{ji} q_k U_{ki} = \sum_{ijk} \bar{q}_j q_k U_{ki} (\hat{U}^{-1})_{ij}$$
$$= \sum_{jk} \bar{q}_j q_k \delta_{jk} = \sum_i \bar{q}_i q_i \quad , \tag{8.65}$$

and consequently ψ_1 is completely invariant under SU(3) transformations since it will be transformed into itself. Thus ψ_1 *is an* SU(3) *singlet (scalar)*. The eighth state $\psi_8 = \frac{1}{\sqrt{6}}(u\bar{u} + d\bar{d} - 2s\bar{s})$ follows from the requirement of orthonormality. In simpler cases the SU(3) Clebsch-Gordan coefficients can be determined by symmetry arguments or normalization conditions, and in general there exist detailed tables on this subject.[18] To gain additional, and experimentally verifiable insight we continue the discussion of the quark models.

8.11 Quark Models with Inner Degrees of Freedom

So far we have discussed a naive quark model by neglecting all internal degrees of freedom of the hadrons (like quark spin, angular momentum etc.). Now we want to extend our discussion of these subjects. Taking the quark spin into account we have six independent quark states $q = \{u_1, u_2, d_1, d_2, s_1, s_2\}$ where, for example, u_1 means the u quark with spin up, u_2 the u quark with spin down, etc. In the case of SU(3) the quark triplet is the "simplest" representation of the group. Demanding our quark sextet q to be the smallest nontrivial representation of a symmetry group, we will be lead directly to the application of the SU(6) group of which [6] and $[\bar{6}]$ are the fundamental representations. Again we can form Kronecker products and reduce them (graphically), so that it seems reasonable to introduce a notation simplifying the reduction. This is realized by classifying the representations of the SU(6) group by means of the quantum numbers in relation to the *subgroup* SU(3) × SU(2) ⊂ SU(6). We denote

$$^{[6]}\mathrm{SU}(6) \rightarrow {}^{[\{3\},\frac{1}{2}]}\mathrm{SU}(3)\times\mathrm{SU}(2) \quad ,$$
$$^{[\bar{6}]}\mathrm{SU}(6) \rightarrow {}^{[\{\bar{3}\},\frac{1}{2}]}\mathrm{SU}(3)\times\mathrm{SU}(2) \quad , \tag{8.66}$$

where the first number marks the SU(3) content and the second the total spin. Again mesons should be composed of a quark and an antiquark pair. Hence they have to be imbedded into the direct product $[6] \otimes [\bar{6}]$. Because the product contains the spin, we should obtain both pseudoscalar mesons with spin 0 and also vector mesons with spin 1. Reduction of the product yields

$$[6] \otimes [\bar{6}] = [1] \oplus [35] \quad , \tag{8.67a}$$

or, by denoting the spin explicitly,

$$[\{3\},\tfrac{1}{2}] \otimes [\{\bar{3}\},\tfrac{1}{2}] = [\{1\},0] \oplus [\{1\},1] \oplus [\{8\},1] \oplus [\{8\},0] \quad . \tag{8.67b}$$

[18] See, for example, P. McNamee, F. Chilton: Rev. Mod. Phys. **36**, 1005 (1964).

Here we have used that $[3] \otimes [\bar{3}] = [1] \oplus [8]$ is valid in the case of SU(3). Furthermore we have coupled the individual spins $S_1 = \frac{1}{2}$ and $S_2 = \frac{1}{2}$ to the total spin $S = 0$ or $S = 1$ and, since $q\bar{q}$ is the quark content of the mesons, we have obtained a better classification for mesons: A singlet and an octet with spin 0 and a singlet and an octet with spin 1. Indeed, these are just the pseudoscalar and vector mesons.

To summarize, we first have to determine the SU(6) multiplets (in the case $[6] \otimes [\bar{6}]$ these are one singlet and one 35-plet) and then decompose them into SU(3) multiplets. Let us do this again taking the baryons as an example:

$$[6] \otimes [6] \otimes [6] = [20] \oplus [56] \oplus [70] \oplus [70] \quad . \tag{8.68a}$$

Instead of splitting up the baryons according to fundamental SU(6)-multiplets, we can also classify them by means of SU(3) × SU(3) subgroups. For this purpose we reduce the direct product

$$[\{3\}, \tfrac{1}{2}] \otimes [\{3\}, \tfrac{1}{2}] \otimes [\{3\}, \tfrac{1}{2}]$$

of fundamental quark triplets with spin.

$$\underbrace{([3] \otimes [3] \otimes [3])}_{\text{flavour SU(3)}} \otimes \underbrace{([2] \otimes [2] \otimes [2])}_{\text{spin SU(3)}}$$

Reduction of the products yields

$$\begin{aligned}
&([10] \oplus [8] \oplus [8] \oplus [1]) \otimes ([4] \oplus [2] \oplus [2]) \\
&\quad = [\{10\}, \tfrac{3}{2}] \oplus [\{8\}, \tfrac{3}{2}] \oplus [\{8\}, \tfrac{3}{2}] \oplus [\{1\}, \tfrac{3}{2}] \\
&\qquad \oplus [\{10\}, \tfrac{1}{2}] \oplus [\{8\}, \tfrac{1}{2}] \oplus [\{8\}, \tfrac{1}{2}] \oplus [\{1\}, \tfrac{1}{2}] \\
&\qquad \oplus [\{10\}, \tfrac{1}{2}] \oplus [\{8\}, \tfrac{1}{2}] \oplus [\{8\}, \tfrac{1}{2}] \oplus [\{1\}, \tfrac{1}{2}] \quad .
\end{aligned}$$

In this way we get a large number of states, which are far more numerous than the number of states observed in nature. We already know the baryon decuplet with spin $\frac{3}{2}$ and the baryon octet with spin $\frac{1}{2}$.

$$[\{8\}, \tfrac{1}{2}] \oplus [\{10\}, \tfrac{3}{2}] \quad .$$

The octet with spin $\frac{1}{2}$ contains $8 \times (2s + 1) = 16$ states, the decuplet with spin $\frac{3}{2}$ $10 \times (2s + 1) = 40$ states. In total these are 56 states of the SU(6) 56-plet. We will understand the fact that it is the 56-plet that is put into reality by nature in Chapter 9, where we will deal with the group of permutations and the Young tableaux. There we will show that the 56 states are the 56 completely symmetric combinations of three spin-$\frac{1}{2}$ quarks.

Apparently only those wavefunctions exist which are totally symmetric according to an arbitrary permutation of the quarks. Together with the total antisymmetric colour wave function of the quarks this leads to a total wave function which is antisymmetric, as required by the Pauli principle. The quark wave function of the triplet, $\psi^{(3)}_{ytt_3}$, is supplemented by the spin wave function. Hence

$$\psi^{(3)}_{ytt_3} \chi_{\frac{1}{2}\mu} \quad , \tag{8.70a}$$

and similarly for the antitriplet

$$\psi^{(\bar{3})}_{ytt_3}\chi_{\frac{1}{2}\nu} \quad , \tag{8.70b}$$

From these quark sextets we now construct the π-meson isotriplet ($T = 1$), which is embedded into the meson octet and has $Y = 0$ (cf. figure in Example 8.6 and Sect. 8.10). According to (8.64) the isoscalar factor is known to be equal to one. Therefore the states of the $T = 1$ isotriplet of the pions (spin 0) can be written as

$$\begin{aligned}\psi^{8}_{Y=0,T=1,T_3,I=0,M=0} = \sum_{t_3,t'_3,\mu,\nu} &(\tfrac{1}{2}\,\tfrac{1}{2}1\,|\,t_3 t'_3 T_3)(\tfrac{1}{2}\,\tfrac{1}{2}\,0\,|\,\mu\nu 0)\\ &\times \psi^{(3)}_{\frac{1}{3}\frac{1}{2}t_3}(1)\psi^{(\bar{3})}_{-\frac{1}{3}\frac{1}{2}t'_3}(2)\chi_{\frac{1}{2}\mu}(1)\chi_{\frac{1}{2}\nu}(2) \quad .\end{aligned} \tag{8.71}$$

The arguments (1) and (2) denote the coordinates of the particle 1(quark) and particle 2 (antiquark), respectively. We explicitly evaluate (8.71) for different values of T_3:

a) $T_3 = 1$. In this case both t_3 and t'_3 can only take the value $\frac{1}{2}$. Hence the wave function reads

$$|\pi^+\rangle = (\tfrac{1}{2}\,\tfrac{1}{2}1\,|\,\tfrac{1}{2}\,\tfrac{1}{2}\,1)\sum_{\mu}(\tfrac{1}{2}\,\tfrac{1}{2}\,0\,|\,\mu-\mu 0)\psi^{(3)}_{\frac{1}{3}\frac{1}{2}\frac{1}{2}}(1)\psi^{(\bar{3})}_{-\frac{1}{3}\frac{1}{2}\frac{1}{2}}(2)\chi_{\frac{1}{2}\mu}(1)\chi_{\frac{1}{2}-\mu}(2) \quad . \tag{8.72}$$

Since the Clebsch-Gordan coefficients are

$$(j,j,2j\,|\,j,j,2j) = 1 \quad , \tag{8.73a}$$

$$(j,j,0\,|\,\mu-\mu 0) = \frac{(-1)^{j-\mu}}{\sqrt{2j+1}} \quad , \tag{8.73b}$$

it follows that

$$\begin{aligned}|\pi^+\rangle &= \psi^{(3)}_{\frac{1}{3}\frac{1}{2}\frac{1}{2}}(1)\psi^{(\bar{3})}_{-\frac{1}{3}\frac{1}{2}\frac{1}{2}}(2)\frac{1}{\sqrt{2}}\left[\chi_{\frac{1}{2}-\frac{1}{2}}(1)\chi_{\frac{1}{2}\frac{1}{2}}(2) - \chi_{\frac{1}{2}\frac{1}{2}}(1)\,\chi_{\frac{1}{2}-\frac{1}{2}}(2)\right]\\ &= u(1)\bar{d}(2)\frac{1}{\sqrt{2}}\left[\chi_{\frac{1}{2}-\frac{1}{2}}(1)\chi_{\frac{1}{2}\frac{1}{2}}(2) - \chi_{\frac{1}{2}\frac{1}{2}}(1)\,\chi_{\frac{1}{2}-\frac{1}{2}}(2)\right]\\ &= \frac{1}{\sqrt{2}}[u(1)_\downarrow \bar{d}(2)_\uparrow - u(1)_\uparrow \bar{d}(2)_\downarrow]\\ &= -\frac{1}{\sqrt{2}}(u_\uparrow \bar{d}_\downarrow - u_\downarrow \bar{d}_\uparrow) \quad .\end{aligned} \tag{8.74}$$

In the final step wave functions and spin up and spin down are indicated with the indices $\uparrow$ and $\downarrow$, respectively. At the same time the position of the wave function denotes the argument; thus $u_\uparrow \bar{d}_\downarrow$ actually implies $u_\uparrow(1), \bar{d}_\downarrow(2)$, and so on.

b) $T_3 = 0$. According to (8.71) the $|\pi^0\rangle$ wave function reads

$$|\pi^0\rangle = [(\tfrac{1}{2}\,\tfrac{1}{2}1\,|\,\tfrac{1}{2}\,-\tfrac{1}{2}0)\psi^{(3)}_{\frac{1}{3}\frac{1}{2}\frac{1}{2}}(1)\psi^{(\bar{3})}_{-\frac{1}{3}\frac{1}{2}-\frac{1}{2}}(2) + (\tfrac{1}{2}\,\tfrac{1}{2}1\,|\,-\tfrac{1}{2}\,\tfrac{1}{2}0)$$
$$\times\chi^{(3)}_{\frac{1}{3}\frac{1}{2}-\frac{1}{2}}(1)\psi^{(\bar{3})}_{-\frac{1}{3}\frac{1}{2}\frac{1}{2}}(2)](-\tfrac{1}{2}\,\tfrac{1}{2}0\,|\,\mu-\mu 0)\chi_{\frac{1}{2}\mu}(1)\chi_{\frac{1}{2}-\mu}(2) \quad . \qquad (8.75)$$

The SU(2) Clebsch–Gordan coefficients take the values

$$(\tfrac{1}{2}\,\tfrac{1}{2}1\,|\,\tfrac{1}{2}-\tfrac{1}{2}0) = \frac{1}{\sqrt{2}} \qquad (8.76a)$$

$$(\tfrac{1}{2}\,\tfrac{1}{2}1\,|\,-\tfrac{1}{2}\,\tfrac{1}{2}0) = -\frac{1}{\sqrt{2}} \qquad (8.76b)$$

$$(\tfrac{1}{2}\,\tfrac{1}{2}0\,|\,\mu-\mu 0) = \frac{(-)^{\frac{1}{2}+\mu}}{\sqrt{2}} \quad , \qquad (8.76c)$$

which lead to

$$|\pi^0\rangle = \frac{1}{\sqrt{2}}(u(1)\bar{u}(2) + d(1)\bar{d}(2))\left[-\frac{1}{\sqrt{2}}\right](\chi_{\frac{1}{2}\frac{1}{2}}(1)\chi_{\frac{1}{2}-\frac{1}{2}}(2) - \chi_{\frac{1}{2}-\frac{1}{2}}(1)\chi_{\frac{1}{2}\frac{1}{2}}(2))$$
$$= \tfrac{1}{2}(u(1)_\uparrow\bar{u}(2)_\downarrow + d(1)_\uparrow\bar{d}(2)_\downarrow - u(1)_\downarrow\bar{u}(2)_\uparrow - d(1)_\downarrow\bar{d}(2)_\uparrow)$$
$$= \tfrac{1}{2}(u_\uparrow\bar{u}_\downarrow + d_\uparrow\bar{d}_\downarrow - u_\downarrow\bar{u}_\uparrow - d_\downarrow\bar{d}_\uparrow) \quad . \qquad (8.77)$$

This continues in the same way for the π^- meson ($T_3 = -1$). In the following table we summarize the wave functions for the (pseudo-)scalar (pions, spin 0) and the vector particles (ϱ mesons, spin 1) which belong to the isospin triplet of the SU(3) octet:

$$|\pi^+\rangle = \tfrac{1}{\sqrt{2}}(u_\uparrow\bar{d}_\downarrow - u_\downarrow\bar{d}_\uparrow)$$
$$|\pi^0\rangle = \tfrac{1}{2}(u_\uparrow\bar{u}_\downarrow - u_\downarrow\bar{u}_\uparrow - d_\uparrow\bar{d}_\downarrow + d_\downarrow\bar{d}_\uparrow)$$
$$|\pi^-\rangle = \tfrac{1}{\sqrt{2}}(d_\uparrow\bar{u}_\downarrow - d_\downarrow\bar{u}_\uparrow)$$
$$|\varrho^+_{\pm 1}\rangle = (u_{\uparrow\downarrow}\bar{d}_{\uparrow\downarrow})$$
$$|\varrho^+_0\rangle = \tfrac{1}{\sqrt{2}}(u_\uparrow\bar{d}_\downarrow + u_\downarrow\bar{d}_\uparrow)$$
$$|\varrho^0_{\pm 1}\rangle = \tfrac{1}{\sqrt{2}}(u_{\uparrow,\downarrow}\bar{u}_{\uparrow,\downarrow} - d_{\uparrow,\downarrow}\bar{d}_{\uparrow,\downarrow})$$
$$|\varrho^0_0\rangle = \tfrac{1}{2}(u_\uparrow\bar{u}_\downarrow + u_\downarrow\bar{u}_\uparrow - d_\uparrow\bar{d}_\downarrow - d_\downarrow\bar{d}_\uparrow)$$
$$|\varrho^-_{\pm 1}\rangle = (d_{\uparrow\downarrow}\bar{u}_{\uparrow\downarrow})$$
$$|\varrho^-_0\rangle = \tfrac{1}{\sqrt{2}}(d_\uparrow\bar{u}_\downarrow + d_\downarrow\bar{u}_\uparrow) \quad . \qquad (8.78)$$

The upper index refers to the charge and the lower one the z component of the spin: $u_\uparrow$ means $u|\uparrow\rangle$, $u_\downarrow$ means $u|\downarrow\rangle$, etc. The pions have spin 0, the ϱ mesons spin 1 (vector mesons). The latter have therefore $j_z = 0, \pm 1$ as indicated in (8.78). The wave functions of the baryons can be constructed similarly, which is quite easy for the decuplet $[\{10\}, \tfrac{3}{2}]$. Let us start, e.g. with the state of maximal weight Δ^{++}:

$$\Delta^{++}_{jz} = \begin{cases} \Delta^{++}_{+\frac{3}{2}} = u_\uparrow u_\uparrow u_\uparrow \\ \Delta^{++}_{+\frac{1}{2}} = \frac{1}{\sqrt{3}}(u_\uparrow u_\uparrow u_\downarrow + u_\uparrow u_\downarrow u_\uparrow + u_\downarrow u_\uparrow u_\uparrow) \end{cases} \quad . \tag{8.79}$$

In connection with the wave function for $\Delta^{++}_{+3} = u_\uparrow u_\uparrow u_\uparrow$ we come into contact with a new problem. Apparently, this wave function violates the Pauli principle. (The same problem occurs in the second wave function because all three quarks are in the same state $u_\uparrow$.)

In order to restore the Pauli principle in the realm of quarks one has to introduce a new quantum number for quarks. This quantum number is called colour or colour charge. Three colours are necessary (red, green, blue) so that the first $u_\uparrow$ quark is red, the second green and the third blue.

Later, in volumes 8 and 9 of these lectures (Gauge Theory of Weak Interactions and Quantum Chromodynamics) we will discuss this phenomenon more precisely.

We now explain the construction of the second wave function in (8.79) in greater detail. The SU(3) component for Δ^{++} is $u(1)u(2)u(3)$; the spin component of these $j = \frac{3}{2}$-particles reads as

$$\sum_{\mu\nu m\tau} (\tfrac{1}{2}\tfrac{1}{2}1\,|\,\mu\nu m)\,(1\tfrac{1}{2}\tfrac{3}{2}\,|\,m\tau M)\chi_{\frac{1}{2}\mu}(1)\chi_{\frac{1}{2}\nu}(2)\chi_{\frac{1}{2}\tau}(3) \quad . \tag{8.80}$$

This corresponds to the only possible coupling for $j = \frac{3}{2}$,

$$\left[[\tfrac{1}{2}\otimes\tfrac{1}{2}]^{[1]}\otimes\tfrac{1}{2}\right]^{[\frac{3}{2}]} \quad . \tag{8.81}$$

Now we evaluate (8.80) for $M = \frac{3}{2}$ and $M = \frac{1}{2}$.

a) $M = \frac{3}{2}$. Accordingly one must have $\tau = \frac{1}{2}$ and $m = 1$, which leads to $\mu = \nu = \frac{1}{2}$. For (8.80) this results in

$$(\tfrac{1}{2}\tfrac{1}{2}1\,|\,\tfrac{1}{2}\tfrac{1}{2}1)\,(1\tfrac{1}{2}\tfrac{3}{2}\,|\,1\tfrac{1}{2}\tfrac{3}{2})\chi_{\frac{1}{2}\frac{1}{2}}(1)\chi_{\frac{1}{2}\frac{1}{2}}(2)\chi_{\frac{1}{2}\frac{1}{2}}(3) \quad . \tag{8.82}$$

The two Clebsch-Gordan coefficients are equal to 1 (maximal coupling, no other possibility!) and hence we get for the total wave function,

$$\begin{aligned} \Delta^{++}_{j_3=\frac{3}{2}} &= \frac{1}{\sqrt{3}}\left[u(1)u(2)u(3)\,\chi_{\frac{1}{2}\frac{1}{2}}(1)\chi_{\frac{1}{2}\frac{1}{2}}(2)\chi_{\frac{1}{2}\frac{1}{2}}(3)\right] \\ &= u_\uparrow(1)u_\uparrow(2)u_\uparrow(3) \\ &= u_\uparrow u_\uparrow u_\uparrow \quad . \end{aligned} \tag{8.83}$$

b) $M = \frac{1}{2}$. For (8.80) must hold

$$m = 1 \quad , \quad \tau = -\tfrac{1}{2} \quad \text{or} \quad m = 0 \quad , \quad \tau = \tfrac{1}{2} \quad .$$

Further, for

$$m=1:\quad \mu=\tfrac{1}{2}\ ,\quad \nu=\tfrac{1}{2}$$

$$m=0:\quad \mu=\tfrac{1}{2}\ ,\quad \nu=-\tfrac{1}{2}$$

$$m=0:\quad \mu=-\tfrac{1}{2}\ ,\quad \nu=\tfrac{1}{2}\ .$$

For the spin wave function (8.80) this yields in all:

$$\begin{aligned} a &= (\tfrac{1}{2}\tfrac{1}{2}1\,|\,\tfrac{1}{2}\tfrac{1}{2}1)\,(1\tfrac{1}{2}\tfrac{3}{2}\,|\,1-\tfrac{1}{2}\tfrac{1}{2})\chi_{\frac{1}{2}\frac{1}{2}}(1)\chi_{\frac{1}{2}\frac{1}{2}}(2)\chi_{\frac{1}{2}-\frac{1}{2}}(3)\\ &\quad+(\tfrac{1}{2}\tfrac{1}{2}1\,|\,\tfrac{1}{2}\tfrac{1}{2}0)(1-\tfrac{1}{2}\tfrac{3}{2}\,|\,0\tfrac{1}{2}\tfrac{1}{2})\chi_{\frac{1}{2}\frac{1}{2}}(1)\chi_{\frac{1}{2}-\frac{1}{2}}(2)\chi_{\frac{1}{2}\frac{1}{2}}(3)\\ &\quad\times(\tfrac{1}{2}\tfrac{1}{2}1\,|-\tfrac{1}{2}\tfrac{1}{2}0)\,(1\tfrac{1}{2}\tfrac{3}{2}\,|\,0\tfrac{1}{2}\tfrac{1}{2})\chi_{\frac{1}{2}-\frac{1}{2}}(1)\chi_{\frac{1}{2}\frac{1}{2}}(2)\chi_{\frac{1}{2}\frac{1}{2}}(3)\ . \end{aligned} \tag{8.84}$$

The values for these Clebsch-Gordan coefficients can be found in various books[19], and we quote

$$(\tfrac{1}{2}\tfrac{1}{2}1\,|\,\tfrac{1}{2}\tfrac{1}{2}1)=1 \qquad (1\tfrac{1}{2}\tfrac{3}{2}\,|\,0\tfrac{1}{2}\tfrac{1}{2})=\sqrt{\tfrac{2}{3}}$$

$$(1\tfrac{1}{2}\tfrac{3}{2}\,|\,1-\tfrac{1}{2}\tfrac{1}{2})=\frac{1}{\sqrt{3}} \qquad (\tfrac{1}{2}\tfrac{1}{2}1\,|-\tfrac{1}{2}\tfrac{1}{2}0)=\frac{1}{\sqrt{2}}$$

$$(\tfrac{1}{2}\tfrac{1}{2}1\,|\,\tfrac{1}{2}-\tfrac{1}{2}0)=\frac{1}{\sqrt{2}} \qquad (1\tfrac{1}{2}\tfrac{3}{2}\,|\,0\tfrac{1}{2}\tfrac{3}{2})=\sqrt{\tfrac{2}{3}}\ .$$

This results in the total SU(6) wave function

$$\begin{aligned} \Delta^{++}_{j3=\frac{3}{2}} &= \frac{1}{\sqrt{3}}[u(1)u(2)u(3)]\cdot[\chi_{\frac{1}{2}\frac{1}{2}}(1)\chi_{\frac{1}{2}\frac{1}{2}}(2)\chi_{\frac{1}{2}-\frac{1}{2}}(3)+\chi_{\frac{1}{2}\frac{1}{2}}(1)\chi_{\frac{1}{2}-\frac{1}{2}}(2)\chi_{\frac{1}{2}\frac{1}{2}}(3)\\ &\quad+\chi_{\frac{1}{2}-\frac{1}{2}}(1)\,\chi_{\frac{1}{2}\frac{1}{2}}(2)\chi_{\frac{1}{2}\frac{1}{2}}(3)]\\ &= \frac{1}{\sqrt{3}}[u_\uparrow(1)u_\uparrow(2)u_\downarrow(3)+u_\uparrow(1)u_\downarrow(2)u_\uparrow(3)+u_\downarrow(1)u_\uparrow(2)u_\uparrow(3)]\\ &= \frac{1}{\sqrt{3}}[u_\uparrow u_\uparrow u_\downarrow+u_\uparrow u_\downarrow u_\uparrow+u_\downarrow u_\uparrow u_\uparrow]\ . \end{aligned} \tag{8.85a}$$

Using (2.18b) for $\hat{S}-\Delta^{++}_{\frac{3}{2}}=\sqrt{3}\,\Delta^{++}_{\frac{1}{2}}$ one can also obtain this result from (8.83) by means of spin ladder operators:

$$\hat{S}_-=\hat{S}_-(1)+\hat{S}_-(2)+\hat{S}_-(3)$$

Here again we have, according to (2.18b),

$$\hat{S}_-(1)u_\uparrow(1)=u_\downarrow(1)\ \text{etc.}$$

By means of acting with the lowering operators $\hat{T}_-$, $\hat{U}_-$, and $\hat{V}_-$ of the SU(3) symmetry group or with the spin lowering operators $\hat{S}_-$ on the state (8.79), all

[19] e.g. M. Rotenberg et al.: *The 3j- and the 6j-Symbols* (Technology Press, Cambridge, Mass. 1959).

the other wave functions of the SU(3) octet can be constructed, e.g. for the Σ^{*+} particle with the spin projection $j_z = +\frac{1}{2}$ we obtain the wave function

$$\Sigma^{*+}_{+\frac{1}{2}} = \text{const.}\, \hat{V}_- \Delta^{++}_{+\frac{1}{2}} = \frac{1}{\sqrt{9}}(s_\uparrow u_\uparrow u_\downarrow + s_\uparrow u_\downarrow u_\uparrow + s_\downarrow u_\uparrow u_\uparrow + \text{permutations}) \quad .$$

The permutations arise from the fact that $\hat{V}_- = \hat{V}_-(1) + \hat{V}_-(2) + \hat{V}_-(3)$, i.e. $\hat{V}_-$ is built up by the $\hat{V}_-$ operators of the three involved particles (quarks). Notice that the states of the decuplets $[\{10\}, \frac{3}{2}]$ automatically reveal symmetry under exchange of any quark pair. This can be immediately seen in (8.79 and 80).

To construct the states of the baryon octets with $j = \frac{1}{2}$ we combine two quarks to form an intermediate angular momentum and then couple them with the third one. Here we require that the total wave function is completely symmetric under exchange of two quarks, just as for the spin-$\frac{3}{2}$ baryons [cf. (8.79 and 80)]. If, for example, we want to obtain the wave function of the proton $p_{+\frac{1}{2}}$, we combine two u quarks to the (symmetric) triplet state $(uu)^{j=1}_{j3}$ and then couple the third quark, the d quark, to form a good total spin $\frac{1}{2}$:

$$\begin{aligned} |p_\uparrow\rangle &= \text{const.}\,[(1\tfrac{1}{2}\tfrac{1}{2}|1-\tfrac{1}{2}\tfrac{1}{2})\,(uu)^1_1 d_\downarrow + (1\tfrac{1}{2}\tfrac{1}{2}|0\tfrac{1}{2}\tfrac{1}{2})\,(uu)^1_0 d_\uparrow + \text{permut.}] \\ &= \text{const.}\,[\sqrt{\tfrac{2}{3}}u_\uparrow u_\uparrow d_\downarrow - \tfrac{1}{\sqrt{6}}(u_\uparrow u_\downarrow + u_\downarrow u_\uparrow)d_\uparrow + \text{permut.}] \\ &= \frac{1}{\sqrt{18}}[2u_\uparrow u_\uparrow d_\downarrow - u_\uparrow u_\downarrow d_\uparrow - u_\downarrow u_\uparrow d_\uparrow + \text{permut.}] \quad . \end{aligned}$$

In detail the wave function (showing all permutations) reads as

$$\begin{aligned} |p_\uparrow\rangle = \frac{1}{\sqrt{18}}[&2|u_\uparrow d_\downarrow u_\uparrow\rangle + 2|u_\uparrow u_\uparrow d_\downarrow\rangle + 2|d_\downarrow u_\uparrow u_\uparrow\rangle - |u_\uparrow u_\downarrow d_\uparrow\rangle - |u_\uparrow d_\uparrow u_\downarrow\rangle \\ &- |u_\downarrow d_\uparrow u_\uparrow\rangle - |d_\uparrow u_\downarrow u_\uparrow\rangle - |d_\uparrow u_\uparrow u_\downarrow\rangle - |u_\downarrow u_\uparrow d_\uparrow\rangle] \quad . \end{aligned} \tag{8.85b}$$

One must consider that here the position of the wave functions denotes the index of the position operator. For example

$$udu \leftrightarrow u(1)d(2)u(3)$$

By coupling the proton wave function via the intermediate spin $j = 1$ one assures the symmetry between the two u-quarks. The coupling via an intermediate spin $j = 0$ (which, at first, seems to be possible) is ruled out, because the wave function of the first two u-quarks would then be antisymmetric.

Still another remark on the uud-structure of the proton: Why is its structure not uus whose charge $Q = (\frac{2}{3} + \frac{2}{3} - \frac{1}{3}) = 1$, too? The answer is simply that the proton in [8] has to have hypercharge $Y = 1$ while the uus-configuration has hypercharge $Y = (\frac{1}{3} + \frac{1}{3} - \frac{2}{3}) = 0$. Similarly as for the proton above one obtains for the wave function of the neutron,

$$|n_\downarrow\rangle = \frac{1}{\sqrt{18}}[2d_\uparrow d_\uparrow u_\downarrow - d_\uparrow d_\downarrow u_\uparrow - d_\downarrow d_\uparrow u_\uparrow + \text{permut.}] \quad . \tag{8.85c}$$

The other states of the octet look quite similar, except the Λ^0 particle, which has to be determined by orthogonalization with respect to the Σ^0. One finds in this case that

$$|\Lambda^0_\uparrow\rangle = \frac{1}{\sqrt{12}}[u_\uparrow d_\downarrow s_\uparrow - u_\downarrow d_\uparrow s_\uparrow - d_\uparrow u_\downarrow s_\uparrow + d_\downarrow u_\uparrow s_\uparrow + \text{permut.}] \quad . \tag{8.85d}$$

In the following exercises the structure of the neutron wave function and especially also that of the wave functions of the baryon resonances is worked out in detail.

EXERCISE

8.13 Construction of the Neutron Wave Function

Problem. Calculate the wave function of the neutron by applying the shift operator $\hat{T}_-$ to the proton function.

Solution. We know that the wave function of the neutron results from the wave function of the proton by applying the $\hat{T}_-$ operator, which lowers the T_3 component by one unit. Since the wave functions of proton and neutron are three-body functions, we have to use the three-body operators $\hat{T}_- = \hat{T}_-(1) + \hat{T}_-(2) + \hat{T}_-(3)$. This operator consists of three parts, each of which acts in a different Hilbert space, though in a similar manner. The wave function of the proton is known from (8.85b) as

$$
\begin{aligned}
|\mathrm{p}_\uparrow\rangle = \frac{1}{\sqrt{18}}\cdot\{ & 2\psi_{\frac{1}{2}\frac{1}{2}}(1)\,\psi_{-\frac{1}{2}-\frac{1}{2}}(2)\,\psi_{\frac{1}{2}\frac{1}{2}}(3) \\
& + 2\psi_{\frac{1}{2}\frac{1}{2}}(1)\,\psi_{\frac{1}{2}\frac{1}{2}}(2)\psi_{-\frac{1}{2}-\frac{1}{2}}(3) + 2\psi_{-\frac{1}{2}-\frac{1}{2}}(1)\,\psi_{\frac{1}{2}\frac{1}{2}}(2)\,\psi_{\frac{1}{2}\frac{1}{2}}(3) \\
& - \psi_{\frac{1}{2}\frac{1}{2}}(1)\,\psi_{\frac{1}{2}-\frac{1}{2}}(2)\psi_{-\frac{1}{2}\frac{1}{2}}(3) - \psi_{\frac{1}{2}\frac{1}{2}}(1)\psi_{-\frac{1}{2}\frac{1}{2}}(2)\psi_{\frac{1}{2}-\frac{1}{2}}(3) \\
& - \psi_{\frac{1}{2}-\frac{1}{2}}(1)\psi_{-\frac{1}{2}\frac{1}{2}}(2)\psi_{\frac{1}{2}\frac{1}{2}}(3) - \psi_{-\frac{1}{2}\frac{1}{2}}(1)\psi_{\frac{1}{2}-\frac{1}{2}}(2)\psi_{\frac{1}{2}\frac{1}{2}}(3) \\
& - \psi_{-\frac{1}{2}\frac{1}{2}}(1)\psi_{\frac{1}{2}\frac{1}{2}}(2)\psi_{\frac{1}{2}-\frac{1}{2}}(3) - \psi_{\frac{1}{2}-\frac{1}{2}}(1)\psi_{\frac{1}{2}\frac{1}{2}}(2)\psi_{-\frac{1}{2}\frac{1}{2}}(3)\} \ . \qquad (1)
\end{aligned}
$$

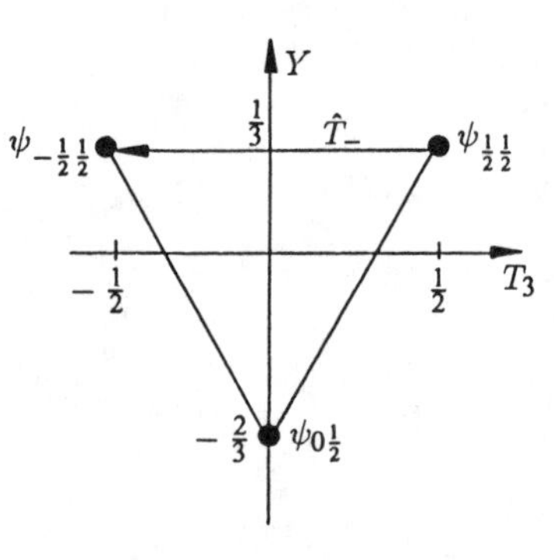

Elementary triplet of SU(3) for spin $+\frac{1}{2}$

Here $\psi_{t_3 j_3}$ denotes the wave function of a particle with spin component j_3 and isospin component t_3. The number given in parentheses specifies the corresponding Hilbert space, which is important because an operator applied to a function in a different Hilbert space does not affect that function [e.g. the notation $\hat{T}_-(1)$ should be understood as an abbreviation for the tensor product operator $\hat{T}_-(1)\hat{I}(2)\hat{I}(3)$, where $\hat{I}$ is the identity operator, etc.] We know that $\hat{T}_-$ lowers the isospin component t_3 by one unit, all other quantum numbers remaining unchanged. The $\hat{T}_3$ operator can not lead out of the multiplet of which the function is a member. Since the functions $\psi_{t_3 j_3}(i)$ are all functions of the elementary SU(3) triplet, it follows that $\hat{T}_-(i)\psi_{-\frac{1}{2}j_3}(i) = 0$ for all i and j_3 (see figure below for illustration). The only allowed shift which remains is

$$
\hat{T}_-(i)\psi_{\frac{1}{2}j_3}(i) = N\psi_{-\frac{1}{2}j_3}(i) \qquad (2)
$$

Exercise 8.13

for all i and j_3, where N is a normalization factor. Thus follows:

$$\begin{aligned}
|\mathrm{n}_\uparrow\rangle = N\hat{T}_-|\mathrm{p}_\uparrow\rangle = \frac{N}{\sqrt{18}}
&\cdot\{2(\hat{T}_-(1)\psi_{\frac{1}{2}\frac{1}{2}}(1))\psi_{-\frac{1}{2}-\frac{1}{2}}(2)\psi_{\frac{1}{2}\frac{1}{2}}(3) + 2\psi_{\frac{1}{2}\frac{1}{2}}(1)(\hat{T}_-(2)\psi_{-\frac{1}{2}-\frac{1}{2}}(2))\psi_{\frac{1}{2}\frac{1}{2}}(3) \\
&+ 2\psi_{\frac{1}{2}\frac{1}{2}}(1)\psi_{-\frac{1}{2}-\frac{1}{2}}(2)(\hat{T}_-(3)\psi_{\frac{1}{2}\frac{1}{2}}(3)) + 2(\hat{T}_-(1)\psi_{\frac{1}{2}\frac{1}{2}}(1))\psi_{\frac{1}{2}\frac{1}{2}}(2)\psi_{-\frac{1}{2}-\frac{1}{2}}(3) \\
&+ 2\psi_{\frac{1}{2}\frac{1}{2}}(1)(\hat{T}_-(2)\psi_{\frac{1}{2}\frac{1}{2}}(2))\psi_{-\frac{1}{2}-\frac{1}{2}}(3) + 2\psi_{\frac{1}{2}\frac{1}{2}}(1)\psi_{\frac{1}{2}\frac{1}{2}}(2)(\hat{T}_-(3)\psi_{-\frac{1}{2}-\frac{1}{2}}(3)) \\
&+ 2(\hat{T}_-(1)\psi_{-\frac{1}{2}-\frac{1}{2}}(1))\psi_{\frac{1}{2}\frac{1}{2}}(2)\psi_{\frac{1}{2}\frac{1}{2}}(3) + 2\psi_{-\frac{1}{2}-\frac{1}{2}}(1)(\hat{T}_-(2)\psi_{\frac{1}{2}\frac{1}{2}}(2))\psi_{\frac{1}{2}\frac{1}{2}}(3) \\
&+ 2\psi_{-\frac{1}{2}-\frac{1}{2}}(1)\psi_{\frac{1}{2}\frac{1}{2}}(2)(\hat{T}_-(3)\psi_{\frac{1}{2}\frac{1}{2}}(3)) - (\hat{T}_-(1)\psi_{\frac{1}{2}\frac{1}{2}}(1))\psi_{\frac{1}{2}-\frac{1}{2}}(2)\psi_{-\frac{1}{2}\frac{1}{2}}(3) \\
&- \psi_{\frac{1}{2}\frac{1}{2}}(1)(\hat{T}_-(2)\psi_{\frac{1}{2}-\frac{1}{2}}(2))\psi_{-\frac{1}{2}\frac{1}{2}}(3) - \psi_{\frac{1}{2}\frac{1}{2}}(1)\psi_{\frac{1}{2}-\frac{1}{2}}(2)(\hat{T}_-(3)\psi_{-\frac{1}{2}\frac{1}{2}}(3)) \\
&- (\hat{T}_-(1)\psi_{\frac{1}{2}\frac{1}{2}}(1))\psi_{-\frac{1}{2}\frac{1}{2}}(2)\psi_{\frac{1}{2}-\frac{1}{2}}(3) - \psi_{\frac{1}{2}\frac{1}{2}}(1)(\hat{T}_-(2)\psi_{-\frac{1}{2}\frac{1}{2}}(2))\psi_{\frac{1}{2}-\frac{1}{2}}(3) \\
&- \psi_{\frac{1}{2}\frac{1}{2}}(1)\psi_{-\frac{1}{2}\frac{1}{2}}(2)(\hat{T}_-(3)\psi_{\frac{1}{2}-\frac{1}{2}}(3)) - (\hat{T}_-(1)\psi_{\frac{1}{2}-\frac{1}{2}}(1))\psi_{-\frac{1}{2}\frac{1}{2}}(2)\psi_{\frac{1}{2}\frac{1}{2}}(3) \\
&- \psi_{\frac{1}{2}-\frac{1}{2}}(1)(\hat{T}_-(2)\psi_{-\frac{1}{2}\frac{1}{2}}(2))\psi_{\frac{1}{2}\frac{1}{2}}(3) - \psi_{\frac{1}{2}-\frac{1}{2}}(1)\psi_{-\frac{1}{2}\frac{1}{2}}(2)(\hat{T}_-(3)\psi_{\frac{1}{2}\frac{1}{2}}(3)) \\
&- (\hat{T}_-(1)\psi_{-\frac{1}{2}\frac{1}{2}}(1))\psi_{\frac{1}{2}-\frac{1}{2}}(2)\psi_{\frac{1}{2}\frac{1}{2}}(3) - \psi_{-\frac{1}{2}\frac{1}{2}}(1) - (\hat{T}_-(2)\psi_{\frac{1}{2}-\frac{1}{2}}(2))\psi_{\frac{1}{2}\frac{1}{2}}(3) \\
&- \psi_{-\frac{1}{2}\frac{1}{2}}(1)\psi_{\frac{1}{2}-\frac{1}{2}}(2)(\hat{T}_-(3)\psi_{\frac{1}{2}\frac{1}{2}}(3)) - (\hat{T}_-(1)\psi_{-\frac{1}{2}\frac{1}{2}}(1))\psi_{\frac{1}{2}\frac{1}{2}}(2)\psi_{\frac{1}{2}-\frac{1}{2}}(3) \\
&- \psi_{-\frac{1}{2}\frac{1}{2}}(1)(\hat{T}_-(2)\psi_{\frac{1}{2}\frac{1}{2}}(2))\psi_{\frac{1}{2}-\frac{1}{2}}(3) - \psi_{-\frac{1}{2}\frac{1}{2}}(1)\psi_{\frac{1}{2}\frac{1}{2}}(2)(\hat{T}_-(3)\psi_{\frac{1}{2}-\frac{1}{2}}(3)) \\
&- (\hat{T}_-(1)\psi_{\frac{1}{2}-\frac{1}{2}}(1))\psi_{\frac{1}{2}\frac{1}{2}}(2)\psi_{-\frac{1}{2}\frac{1}{2}}(3) - \psi_{\frac{1}{2}-\frac{1}{2}}(1)(\hat{T}_-(2)\psi_{\frac{1}{2}\frac{1}{2}}(2))\psi_{-\frac{1}{2}\frac{1}{2}}(3) \\
&- \psi_{\frac{1}{2}-\frac{1}{2}}(1)\psi_{\frac{1}{2}\frac{1}{2}}(2)(\hat{T}_-(3)\psi_{-\frac{1}{2}\frac{1}{2}}(3))\}
\end{aligned}$$

$$\begin{aligned}
|\mathrm{n}_\uparrow\rangle = \frac{N}{\sqrt{18}}&\cdot\{2\psi_{-\frac{1}{2}\frac{1}{2}}(1)\psi_{-\frac{1}{2}-\frac{1}{2}}(2)\psi_{\frac{1}{2}\frac{1}{2}}(3) + 0 \\
&+ 2\psi_{\frac{1}{2}\frac{1}{2}}(1)\psi_{-\frac{1}{2}-\frac{1}{2}}(2)\psi_{-\frac{1}{2}\frac{1}{2}}(3) + 2\psi_{-\frac{1}{2}\frac{1}{2}}(1)\psi_{\frac{1}{2}\frac{1}{2}}(2)\psi_{-\frac{1}{2}-\frac{1}{2}}(3) \\
&+ 2\,\psi_{\frac{1}{2}\frac{1}{2}}(1)\psi_{-\frac{1}{2}\frac{1}{2}}(2)\psi_{-\frac{1}{2}-\frac{1}{2}}(3) + 0 + 0 \\
&+ 2\psi_{-\frac{1}{2}-\frac{1}{2}}(1)\psi_{-\frac{1}{2}\frac{1}{2}}(2)\psi_{\frac{1}{2}\frac{1}{2}}(3) + 2\psi_{-\frac{1}{2}-\frac{1}{2}}(1)\psi_{-\frac{1}{2}\frac{1}{2}}(2)\psi_{\frac{1}{2}\frac{1}{2}}(3) \\
&- \psi_{-\frac{1}{2}\frac{1}{2}}\psi_{\frac{1}{2}-\frac{1}{2}}(2)\psi_{-\frac{1}{2}\frac{1}{2}}(3) - \psi_{\frac{1}{2}\frac{1}{2}}(1)\psi_{-\frac{1}{2}-\frac{1}{2}}(2)\psi_{-\frac{1}{2}\frac{1}{2}}(3) - 0 \\
&- \psi_{-\frac{1}{2}\frac{1}{2}}(1)\psi_{-\frac{1}{2}\frac{1}{2}}(2)\psi_{\frac{1}{2}-\frac{1}{2}}(3) - 0 - \psi_{\frac{1}{2}\frac{1}{2}}(1)\psi_{-\frac{1}{2}\frac{1}{2}}(2)\psi_{-\frac{1}{2}-\frac{1}{2}}(3) \\
&- \psi_{-\frac{1}{2}-\frac{1}{2}}(1)\psi_{\frac{1}{2}-\frac{1}{2}}(2)\psi_{-\frac{1}{2}\frac{1}{2}}(3) - 0 - \psi_{\frac{1}{2}-\frac{1}{2}}(1)\psi_{-\frac{1}{2}\frac{1}{2}}(1)\psi_{-\frac{1}{2}\frac{1}{2}}(2)\psi_{-\frac{1}{2}\frac{1}{2}}(3) \\
&- 0 - \psi_{-\frac{1}{2}\frac{1}{2}}(1)\psi_{-\frac{1}{2}-\frac{1}{2}}(2)\psi_{\frac{1}{2}\frac{1}{2}}(3) - \psi_{-\frac{1}{2}\frac{1}{2}}(1)\psi_{\frac{1}{2}-\frac{1}{2}}(2)\psi_{-\frac{1}{2}\frac{1}{2}}(3) \\
&- 0 - \psi_{-\frac{1}{2}\frac{1}{2}}(1)\psi_{-\frac{1}{2}\frac{1}{2}}(2)\psi_{\frac{1}{2}-\frac{1}{2}}(3) - \psi_{-\frac{1}{2}\frac{1}{2}}(1)\psi_{\frac{1}{2}\frac{1}{2}}(2)\psi_{-\frac{1}{2}-\frac{1}{2}}(3) \\
&- \psi_{-\frac{1}{2}-\frac{1}{2}}(1)\psi_{\frac{1}{2}\frac{1}{2}}(2)\psi_{-\frac{1}{2}\frac{1}{2}}(3) - \psi_{\frac{1}{2}-\frac{1}{2}}(1)\psi_{-\frac{1}{2}\frac{1}{2}}(2)\psi_{-\frac{1}{2}\frac{1}{2}}(3) - 0\}
\end{aligned}$$

Exercise 8.13

$$
\begin{aligned}
= -\frac{N}{\sqrt{18}}\cdot\{&2\psi_{-\frac12\frac12}(1)\psi_{\frac12-\frac12}(2)\psi_{-\frac12\frac12}(3)\\
&+2\psi_{-\frac12\frac12}(1)\psi_{-\frac12\frac12}(2)\psi_{\frac12-\frac12}(3)+2\psi_{\frac12-\frac12}(1)\psi_{-\frac12\frac12}(2)\psi_{-\frac12\frac12}(3)\\
&-\psi_{-\frac12\frac12}(1)\psi_{-\frac12-\frac12}(2)\psi_{\frac12\frac12}(3)-\psi_{\frac12\frac12}(1)\psi_{-\frac12-\frac12}(2)\psi_{-\frac12\frac12}(3)\\
&-\psi_{-\frac12\frac12}(1)\psi_{\frac12\frac12}(2)\psi_{-\frac12-\frac12}(3)-\psi_{\frac12\frac12}(1)\psi_{-\frac12\frac12}(2)\psi_{-\frac12-\frac12}(3)\\
&-\psi_{-\frac12-\frac12}(1)\psi_{-\frac12\frac12}(2)\psi_{\frac12\frac12}(3)-\psi_{-\frac12-\frac12}(1)\psi_{\frac12\frac12}(2)\psi_{-\frac12\frac12}(3)\}\quad . \qquad (3)
\end{aligned}
$$

The normalization condition is $\langle \mathrm{n}_\uparrow | \mathrm{n}_\uparrow \rangle = 1$, and the integral which has to be calculated consists of a sum of individual integrals which satisfy the elementary orthonormality conditions

$$
\begin{aligned}
&\langle \psi_{t_1 j_1}(1)\psi_{t_2 j_2}(2)\psi_{t_3 j_3}(3) \,|\, \psi_{t'_1 j'_1}(1)\psi_{t'_2 j'_2}(2)\psi_{t'_3 j'_3}(3)\rangle\\
&= \delta_{t_1 t'_1}\delta_{j_1 j'_1}\delta_{t_2 t'_2}\delta_{j_2 j'_2}\delta_{t_3 t'_3}\delta_{j_3 j'_3}\quad . \qquad (4)
\end{aligned}
$$

(We suppressed here the index 3 on both t and j.)
Making repeated use of this relation yields

$$
\begin{aligned}
1 &= \langle \mathrm{n}_\uparrow | \mathrm{n}_\downarrow \rangle\\
&= N^2\cdot\tfrac{1}{18}\cdot\{4+0+0+0+0+0+0+0+0+0+4+0+0+0\\
&\quad +0+0+0+0+0+0+4+0+0+0+0+0+0+0+0+0\\
&\quad +1+0+0+0+0+0+0+0+0+0+1+0+0+0+0+0\\
&\quad +0+0+0+0+1+0+0+0+0+0+0+0+0+0+1+0\\
&\quad +0+1+0+0+0+0+0+0+0+1+1\}\\
&= N^2\cdot\tfrac{1}{18}\cdot(4+4+4+1+1+1+1+1+1)\\
&= N^2\cdot\tfrac{18}{18} = N'^2 \quad . \qquad (5)
\end{aligned}
$$

Hence $|N^2| = 1$, i.e. $N = \mathrm{e}^{\mathrm{i}\phi}$, where the phase ϕ is arbitrary, because a wave function is only determined up to a phase. To obtain a result similar to the initial wave function, we choose $\phi = \pi$, i.e. $N = -1$. Hence the wave function becomes

$$
\begin{aligned}
|\mathrm{n}_\uparrow\rangle = \frac{1}{\sqrt{18}}\cdot\{&2\psi_{-\frac12\frac12}(1)\psi_{\frac12-\frac12}(2)\psi_{-\frac12\frac12}(3)\\
&+2\psi_{-\frac12\frac12}(1)\psi_{-\frac12\frac12}(2)\psi_{\frac12-\frac12}(3)+2\psi_{\frac12-\frac12}(1)\psi_{-\frac12\frac12}(2)\psi_{-\frac12\frac12}(3)\\
&-\psi_{-\frac12\frac12}(1)\psi_{-\frac12-\frac12}(2)\psi_{\frac12\frac12}(3)-\psi_{\frac12\frac12}(1)\psi_{-\frac12-\frac12}(2)\psi_{-\frac12\frac12}(3)\\
&-\psi_{-\frac12\frac12}(1)\psi_{\frac12\frac12}(2)\psi_{-\frac12-\frac12}(3)-\psi_{\frac12\frac12}(1)\psi_{-\frac12\frac12}(2)\psi_{-\frac12-\frac12}(3)\\
&-\psi_{-\frac12-\frac12}(1)\psi_{-\frac12\frac12}(2)\psi_{\frac12\frac12}(3)-\psi_{-\frac12-\frac12}(1)\psi_{\frac12\frac12}(2)\psi_{-\frac12\frac12}(3)\}\quad .
\end{aligned}
$$

With the abbreviation $d_\uparrow$ for $\psi_{-\frac12\frac12}$ and $u_\downarrow$ for $\psi_{\frac12-\frac12}$, respectively, it is possible to rewrite the equation. Bearing in mind that, for instance, *duu* stands for

$duu \equiv d(1)u(2)u(3)$, we obtain

$$|\mathrm{n}_\uparrow\rangle = \frac{1}{\sqrt{18}}\cdot\{2d_\uparrow u_\downarrow d_\uparrow + 2d_\uparrow d_\uparrow u_\downarrow + 2u_\downarrow d_\uparrow d_\uparrow - d_\uparrow d_\downarrow u_\uparrow - u_\uparrow d_\downarrow d_\uparrow - d_\uparrow u_\uparrow d_\downarrow - u_\uparrow d_\uparrow d_\downarrow - d_\downarrow d_\uparrow u_\uparrow - d_\downarrow u_\uparrow d_\downarrow\} \quad . \tag{6}$$

EXERCISE

8.14 Construction of the Wave Functions of the Baryon Decuplet

Problem. Construct the wave functions of the decuplet of baryonic resonances, starting from Δ^{++} which we already know to be [see (8.85a)]

$$|\Delta^{++}\rangle = \psi^{[3]}_{\frac{1}{3}\frac{1}{2}}(1)\cdot\psi^{[3]}_{\frac{1}{3}\frac{1}{2}}(2)\cdot\psi^{[3]}_{\frac{1}{3}\frac{1}{2}}(3)\,\chi_{\frac{1}{2}\frac{1}{2}}(1)\cdot\chi_{\frac{1}{2}\frac{1}{2}}(2)\cdot\chi_{\frac{1}{2}\frac{1}{2}}(3) \quad . \tag{1}$$

Solution. The wave function of the Δ^{++} is given by the product of the SU(3) wave functions $\psi^{[m]}_{yt_3}$ (in this case the functions are from the elementary triplet, thus $[m] = [3]$) and the spin functions χ_{jm} of the quarks constituting the particle. The number in parentheses specifies the Hilbert space of the respective function, so for example $\psi^{[3]}_{\frac{1}{3}\frac{1}{2}}(1)$ is the wave function of particle 1 which is part of the elementary triplet having the quantum numbers $y = \frac{1}{3}$, $t_3 = \frac{1}{2}$. Spin, isospin and hypercharge are additive quantum numbers, i.e. for Δ^{++} we have the values $Y = 1$, $T_3 = \frac{3}{2}$, $J = \frac{3}{2}$. Here we consider only states with maximal spin projection $M = \frac{3}{2}$. In general this is not necessary, we may also consider states with $M = \frac{1}{2}, = -\frac{1}{2}$ or $M = -\frac{3}{2}$. But the isospin-hypercharge part of the wave function is not affected by the spin-projection value, hence our choice of $M = \frac{3}{2}$ only.

Starting from the Δ^{++} function, the other wave functions of the baryon decuplet can be constructed by successively applying the shift operators. All observed baryonic resonances have spin $J = \frac{3}{2}$.

We know the effect of the shift operator acting on a state in $Y - T_3$ space (see figure below) and care has to be taken that it does not move the state beyond the boundaries of the SU(3) multiplet, in which case the result would be zero. In our case the wave function denotes states of the elementary triplet (see following figure).

Hence, for example $\hat{T}_+\psi^{[3]}_{\frac{1}{3}\frac{1}{2}} = 0$, because the state $\psi^{[3]}_{\frac{1}{3}\frac{1}{2}}$ does not exist. Similarly,

$$\hat{U}_+\psi^{[3]}_{\frac{1}{3}\frac{1}{2}} = \hat{U}_-\psi^{[3]}_{\frac{1}{3}\frac{1}{2}} = \hat{V}_+\psi^{[3]}_{\frac{1}{3}\frac{1}{2}} = 0 \quad .$$

Nevertheless the state

$$\hat{T}_+\psi^{[15]}_{\frac{1}{3}\frac{1}{2}} = \mathrm{const.}\psi^{[15]}_{\frac{1}{3}\frac{3}{2}}$$

exists, because in the case of the 15-plet the $\hat{T}_+$ operator acting on the state with $Y = \frac{1}{3}$, $T_3 = \frac{1}{2}$ does not result in an exit from the multiplet. The operations

Exercise 8.14

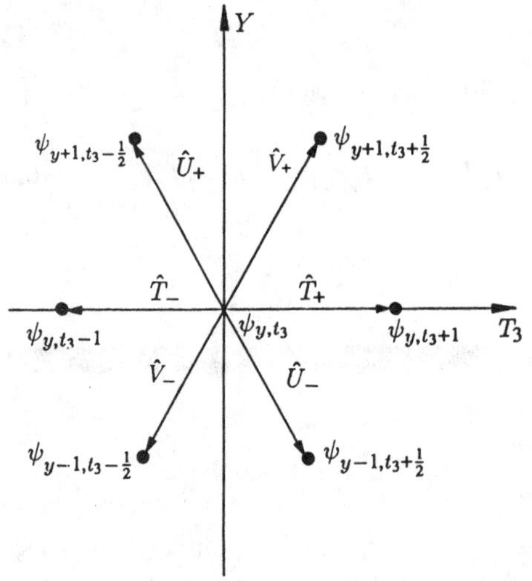

The effect of the shift operators

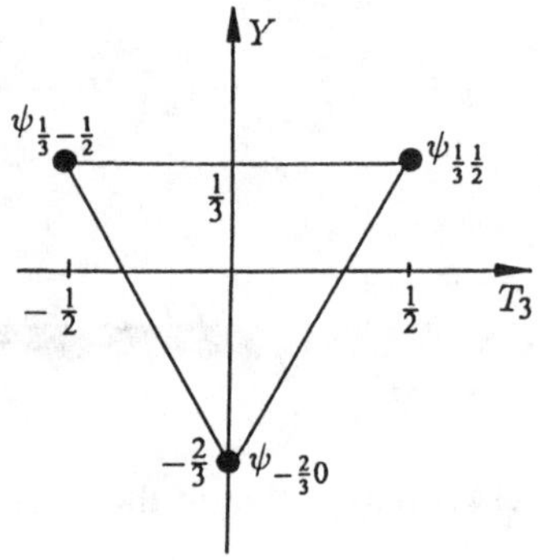

The elementary quark triplet

which are possible within the elementary triplet are

$$\hat{T}_-\psi^{[3]}_{\frac{1}{3}\frac{1}{2}} = \psi^{[3]}_{\frac{1}{3}-\frac{1}{2}} \quad , \quad \hat{V}_-\psi^{[3]}_{\frac{1}{3}\frac{1}{2}} = \psi^{[3]}_{-\frac{2}{3}0} \quad , \quad \hat{T}_+\psi^{[3]}_{\frac{1}{3}-\frac{1}{2}} = \psi^{[3]}_{\frac{1}{3}\frac{1}{2}}$$
$$\hat{U}_-\psi^{[3]}_{\frac{1}{3}-\frac{1}{2}} = \psi^{[3]}_{-\frac{2}{3}0} \quad , \quad \hat{V}_+\psi^{[3]}_{-\frac{2}{3}0} = \psi^{[3]}_{\frac{1}{3}\frac{1}{2}} \quad , \quad \hat{U}_+\psi^{[3]}_{-\frac{2}{3}0} = \psi^{[3]}_{\frac{1}{3}-\frac{1}{2}} \quad . \tag{2}$$

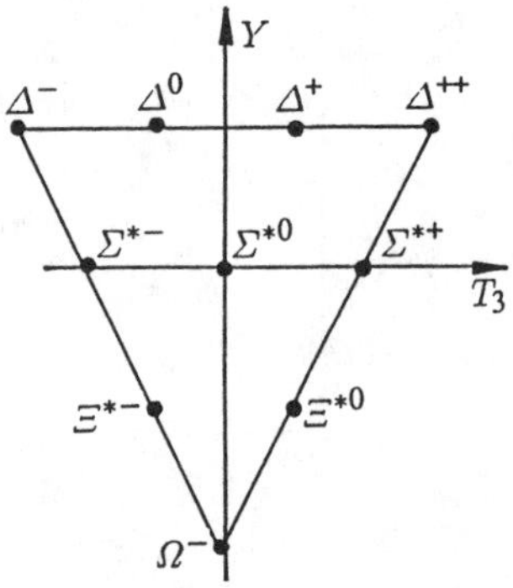

The decuplet of baryon resonances

Before we start to construct new states of the decuplet, (see figure below) we have to realize that we are here concerned with many-body states, and consequently the operators have to be many-body operators:

$$\hat{O}\phi \text{ with } \phi = \phi(1)\phi(2)\phi(3) \quad .$$

Therefore,

$$\{\hat{O}(1) + \hat{O}(2) + \hat{O}(3)\}\,\phi = (\hat{O}(1)\phi(1))\,\phi(2)\phi(3) + \phi(1)(\hat{O}(2)\phi(2))\,\phi(3)$$
$$+ \phi(1)\phi(2)(\hat{O}(3)\phi(3)) \quad . \tag{3}$$

Now we can start to construct the first new wave functions:

As we can see in the left figure, $|\Delta^+\rangle = N\cdot\hat{T}_-|\Delta^{++}\rangle$, where N is a normalization factor which ensures that $\langle\Delta^+|\Delta^+\rangle = 1$. Since we exclusively consider functions ψ belonging to the elementary triplet (which we assume to be normalized), we omit the superscript [3] in the following. Moreover we set

$$\chi_{\frac{1}{2}\frac{1}{2}}(1)\,\chi_{\frac{1}{2}\frac{1}{2}}(2)\,\chi_{\frac{1}{2}\frac{1}{2}}(3) \equiv \chi \quad ,$$

because the spin part of the wave function remains unchanged. Next we obtain

$$\begin{aligned}
|\Delta^+\rangle &= N\hat{T}_-|\Delta^{++}\rangle \\
&= N(\hat{T}_-(1) + \hat{T}_-(2) + \hat{T}_-(3))\,[\psi_{\frac{1}{3}\frac{1}{2}}(1)\psi_{\frac{1}{3}\frac{1}{2}}\psi_{\frac{1}{3}\frac{1}{2}}(3)]\cdot\chi \\
&= N\cdot\chi\cdot\{(\hat{T}_-(1)\psi_{\frac{1}{3}\frac{1}{2}}(1))\psi_{\frac{1}{3}\frac{1}{2}}(2)\psi_{\frac{1}{3}\frac{1}{2}}(3) \\
&\quad + \psi_{\frac{1}{3}\frac{1}{2}}(1)\,(\hat{T}_-(2)\psi_{\frac{1}{3}\frac{1}{2}}(2))\psi_{\frac{1}{3}\frac{1}{2}}(3) + \psi_{\frac{1}{3}\frac{1}{2}}(1)\psi_{\frac{1}{3}\frac{1}{2}}(2)\,(\hat{T}_-(3)\psi_{\frac{1}{3}\frac{1}{2}}(3))\} \\
&= N\cdot\chi\cdot\{\psi_{\frac{1}{3}-\frac{1}{2}}(1)\psi_{\frac{1}{3}\frac{1}{2}}(2)\psi_{\frac{1}{3}\frac{1}{2}}(3) + \psi_{\frac{1}{3}\frac{1}{2}}(1)\psi_{\frac{1}{3}-\frac{1}{2}}(2)\psi_{\frac{1}{3}\frac{1}{2}}(3) \\
&\quad + \psi_{\frac{1}{3}\frac{1}{2}}(1)\psi_{\frac{1}{3}\frac{1}{2}}(2)\psi_{\frac{1}{3}-\frac{1}{2}}(3)\} \quad .
\end{aligned} \tag{4}$$

Exercise 8.14

The factor N is determined with the help of the normalization condition

$$
\begin{aligned}
1 &\stackrel{!}{=} \langle \Delta^+ | \Delta^+ \rangle \\
&= \int \Bigg\{ N^* \cdot \bar{\chi} \cdot \Big[\underbrace{\bar{\psi}_{\frac{1}{3}-\frac{1}{2}}(1)\bar{\psi}_{\frac{1}{3}\frac{1}{2}}(2)\bar{\psi}_{\frac{1}{3}\frac{1}{2}}(3)}_{\psi_I} \\
&\quad + \underbrace{\bar{\psi}_{\frac{1}{3}\frac{1}{2}}(1)\bar{\psi}_{\frac{1}{3}-\frac{1}{2}}(2)\bar{\psi}_{\frac{1}{3}\frac{1}{2}}(3)}_{\psi_{II}} + \underbrace{\bar{\psi}_{\frac{1}{3}\frac{1}{2}}(1)\bar{\psi}_{\frac{1}{3}\frac{1}{2}}(2)\bar{\psi}_{\frac{1}{3}-\frac{1}{2}}(3)}_{\psi_{III}} \Big] \\
&\quad \times N \cdot \chi \cdot [\psi_{\frac{1}{3}-\frac{1}{2}}(1)\psi_{\frac{1}{3}\frac{1}{2}}(2)\psi_{\frac{1}{3}\frac{1}{2}}(3) \\
&\quad + \psi_{\frac{1}{3}\frac{1}{2}}(1)\psi_{\frac{1}{3}-\frac{1}{2}}(2)\psi_{\frac{1}{3}\frac{1}{2}}(3) + \psi_{\frac{1}{3}\frac{1}{2}}(1)\psi_{\frac{1}{3}\frac{1}{2}}(2)\psi_{\frac{1}{3}-\frac{1}{2}}(3)] \Bigg\} \, d^3x \\
&= |N|^2 \cdot \int |\chi|^2 \cdot \{ |\psi_I|^2 + \bar{\psi}_I \psi_{II} + \bar{\psi}_I \psi_{II} + \bar{\psi}_{II}\psi_I + |\psi_{II}|^2 \\
&\quad + \bar{\psi}_{II}\psi_{III} + \bar{\psi}_{III}\psi_I + \bar{\psi}_{III}\psi_{II} + |\psi_{III}|^2 \} \, d^3x_1 \, d^3x_2 \, d^3x_3 \quad . \qquad (5)
\end{aligned}
$$

Now the $\psi_{\tilde{\alpha}}$ are orthonormal, for example

$$
\begin{aligned}
&\int |\psi_I|^2 \cdot |\chi_{\frac{1}{2}\frac{1}{2}}|^2 d^3x_1 \, d^3x_2 \, d^3x_3 \\
&\quad = \int \bar{\psi}_{\frac{1}{3}-\frac{1}{2}}(1)\psi_{\frac{1}{3}-\frac{1}{2}}(1)|\chi_{\frac{1}{2}\frac{1}{2}}(1)|^2 \, d^3x_1 \int \bar{\psi}_{\frac{1}{3}\frac{1}{2}}(2)\psi_{\frac{1}{3}\frac{1}{2}}(2)|\chi_{\frac{1}{2}\frac{1}{2}}(2)|^2 d^3x_2 \\
&\quad \times \int \bar{\psi}_{\frac{1}{3}\frac{1}{2}}(3)\psi_{\frac{1}{3}\frac{1}{2}}(3)|\chi_{\frac{1}{2}\frac{1}{2}}(3)|^2 \, d^3x_3 \\
&\quad = 1 \cdot 1 \cdot 1 = 1 \\
&\int |\chi_{\frac{1}{2}\frac{1}{2}}|^2 \, \bar{\psi}_I \psi_{II} d^3x_1 \, d^3x_2 \, d^3x_3 \\
&\quad = \int \bar{\psi}_{\frac{1}{3}-\frac{1}{2}}(1)\psi_{\frac{1}{3}\frac{1}{2}}(1)|\chi_{\frac{1}{2}\frac{1}{2}}(1)|^2 \, d^3x_1 \int \bar{\psi}_{\frac{1}{3}\frac{1}{2}}(2)\psi_{\frac{1}{3}-\frac{1}{2}}(2)|\chi_{\frac{1}{2}\frac{1}{2}}(2)|^2 d^3x_2 \\
&\quad \times \int \bar{\psi}_{\frac{1}{3}\frac{1}{2}}(3)\psi_{\frac{1}{3}\frac{1}{2}}(3)|\chi_{\frac{1}{2}\frac{1}{2}}(3)|^2 \, d^3x_3 \\
&\quad = 0 \cdot 0 \cdot 1 = 0 \quad . \qquad (6)
\end{aligned}
$$

With this, one obtains

$$
\begin{aligned}
1 &= |N|^2 \cdot (1 \cdot 1 \cdot 1 + 0 \cdot 0 \cdot 1 + 0 \cdot 1 \cdot 0 \\
&\quad + 0 \cdot 0 \cdot 1 + 1 \cdot 1 \cdot 1 + 1 \cdot 0 \cdot 0 + 0 \cdot 1 \cdot 0 + 1 \cdot 0 \cdot 0 + 1 \cdot 1 \cdot 1) \\
&= |N|^2 \cdot 3 \ \Rightarrow \ N = \frac{1}{\sqrt{3}} \quad ,
\end{aligned}
$$

and hence,

$$
\begin{aligned}
|\Delta^+\rangle &= \frac{1}{\sqrt{3}} \cdot \{ \psi_{\frac{1}{3}-\frac{1}{2}}(1)\psi_{\frac{1}{3}\frac{1}{2}}(2)\psi_{\frac{1}{3}\frac{1}{2}}(3) + \psi_{\frac{1}{3}\frac{1}{2}}(1)\psi_{\frac{1}{3}-\frac{1}{2}}(2)\psi_{\frac{1}{3}\frac{1}{2}}(3) \\
&\quad + \psi_{\frac{1}{3}\frac{1}{2}}(1)\psi_{\frac{1}{3}\frac{1}{2}}(2)\psi_{\frac{1}{3}-\frac{1}{2}}(3) \} \chi_{\frac{1}{2}\frac{1}{2}}(1)\chi_{\frac{1}{2}\frac{1}{2}}(2)\chi_{\frac{1}{2}\frac{1}{2}}(3) \quad . \qquad (7)
\end{aligned}
$$

Exercise 8.14

Now we may calculate $|\Sigma^{*+}\rangle$ in the same way:

$$
\begin{aligned}
|\Sigma^{*+}\rangle &= N\cdot(\hat{V}_-(1)+\hat{V}_-(2)+\hat{V}_-(3))\psi_{\frac{1}{3}\frac{1}{2}}(1)\psi_{\frac{1}{3}\frac{1}{2}}(2)\psi_{\frac{1}{3}\frac{1}{2}}(3)\chi_{\frac{1}{2}\frac{1}{2}}(1)\chi_{\frac{1}{2}\frac{1}{2}}(2)\chi_{\frac{1}{2}\frac{1}{2}}(3)\\
&= N\cdot\{(\hat{V}_-(1)\psi_{\frac{1}{3}\frac{1}{2}})(1)\psi_{\frac{1}{3}\frac{1}{2}}(2)\psi_{\frac{1}{3}\frac{1}{2}}(3)\\
&\quad+\psi_{\frac{1}{3}\frac{1}{2}}(1)(\hat{V}_-(2)\psi_{\frac{1}{3}\frac{1}{2}}(2))\psi_{\frac{1}{3}\frac{1}{2}}(3)+\psi_{\frac{1}{3}\frac{1}{2}}(1)\psi_{\frac{1}{3}\frac{1}{2}}(2)(\hat{V}_-(3)\psi_{\frac{1}{3}\frac{1}{2}}(3))\}\cdot\chi\\
&= N\cdot\{\psi_{-\frac{2}{3}0}(1)\psi_{\frac{1}{3}\frac{1}{2}}(2)\psi_{\frac{1}{3}\frac{1}{2}}(3)\\
&\quad+\psi_{\frac{1}{3}\frac{1}{2}}(1)\psi_{-\frac{2}{3}0}(2)\psi_{\frac{1}{3}\frac{1}{2}}(3)+\psi_{\frac{1}{3}\frac{1}{2}}(1)\psi_{\frac{1}{3}\frac{1}{2}}(2)\psi_{-\frac{2}{3}0}(3)\}\cdot\chi\quad.
\end{aligned}
\tag{8}
$$

Again N arises from the normalization condition

$$
\begin{aligned}
1 &\overset{!}{=} |N|^2\cdot\langle[\psi_{-\frac{2}{3}0}(1)\psi_{\frac{1}{3}\frac{1}{2}}(2)\psi_{\frac{1}{3}\frac{1}{2}}(3)+\psi_{\frac{1}{3}\frac{1}{2}}(1)\psi_{-\frac{2}{3}0}(2)\psi_{\frac{1}{3}\frac{1}{2}}(3)\\
&\quad+\psi_{\frac{1}{3}\frac{1}{2}}(1)\psi_{\frac{1}{3}\frac{1}{2}}(2)\psi_{-\frac{2}{3}0}(3)]\cdot\chi\,|\,[\psi_{-\frac{2}{3}0}(1)\psi_{\frac{1}{3}\frac{1}{2}}(2)\psi_{\frac{1}{3}\frac{1}{2}}(3)\\
&\quad+\psi_{\frac{1}{3}\frac{1}{2}}(1)\psi_{-\frac{2}{3}0}(2)\psi_{\frac{1}{3}\frac{1}{2}}(3)+\psi_{\frac{1}{3}\frac{1}{2}}(1)\psi_{\frac{1}{3}\frac{1}{2}}(2)\psi_{-\frac{2}{3}0}(3)]\chi\rangle\\
&= |N|^2\cdot(1+0+0+0+1+0+0+0+1)=3|N|^2
\end{aligned}
\tag{9}
$$

Thus, $N=1/\sqrt{3}$ and we obtain

$$
\begin{aligned}
|\Sigma^{*+}\rangle &= \frac{1}{\sqrt{3}}\cdot\{\psi_{-\frac{2}{3}0}(1)\psi_{\frac{1}{3}\frac{1}{2}}(2)\psi_{\frac{1}{3}\frac{1}{2}}(3)+\psi_{\frac{1}{3}\frac{1}{2}}(1)\psi_{-\frac{2}{3}0}(2)\psi_{\frac{1}{3}\frac{1}{2}}(3)\\
&\quad+\psi_{\frac{1}{3}\frac{1}{2}}(1)\psi_{\frac{1}{3}\frac{1}{2}}(2)\psi_{-\frac{2}{3}0}(3)\}\chi_{\frac{1}{2}\frac{1}{2}}(1)\chi_{\frac{1}{2}\frac{1}{2}}(2)\chi_{\frac{1}{2}\frac{1}{2}}(3)\quad.
\end{aligned}
\tag{10}
$$

Similarly one can determine the wave functions of the particles Δ^0, Δ^-, Ξ^*, Ω:

$$
\begin{aligned}
|\Delta^0\rangle &= N\hat{T}_-|\Delta^+\rangle\\
&= N(\hat{T}_-(1)+\hat{T}_-(2)+\hat{T}_-(3))\cdot\left\{\frac{1}{\sqrt{3}}[\psi_{\frac{1}{3}-\frac{1}{2}}(1)\psi_{\frac{1}{3}\frac{1}{2}}(2)\psi_{\frac{1}{3}\frac{1}{2}}(3)\right.\\
&\quad\left.+\psi_{\frac{1}{3}\frac{1}{2}}(1)\psi_{\frac{1}{3}-\frac{1}{2}}(2)\psi_{\frac{1}{3}\frac{1}{2}}(3)+\psi_{\frac{1}{3}\frac{1}{2}}(1)\psi_{\frac{1}{3}\frac{1}{2}}(2)\psi_{\frac{1}{3}-\frac{1}{2}}(3)]\chi(1)\chi(2)\chi(3)\right\}\\
&= \frac{N}{\sqrt{3}}\cdot\{(\hat{T}_-(1)\psi_{\frac{1}{3}-\frac{1}{2}}(1))\psi_{\frac{1}{3}\frac{1}{2}}(2)\psi_{\frac{1}{3}\frac{1}{2}}(3)\\
&\quad+(\hat{T}_-(1)\psi_{\frac{1}{3}\frac{1}{2}}(1))\psi_{\frac{1}{3}-\frac{1}{2}}(2)\psi_{\frac{1}{3}\frac{1}{2}}(3)+(\hat{T}_-(1)\psi_{\frac{1}{3}\frac{1}{2}}(1))\psi_{\frac{1}{3}\frac{1}{2}}(2)\psi_{\frac{1}{3}-\frac{1}{2}}(3)\\
&\quad+\psi_{\frac{1}{3}-\frac{1}{2}}(1)(\hat{T}_-(2)\psi_{\frac{1}{3}\frac{1}{2}}(2))\psi_{\frac{1}{3}\frac{1}{2}}(3)+\psi_{\frac{1}{3}\frac{1}{2}}(1)(\hat{T}_-(2)\psi_{\frac{1}{3}-\frac{1}{2}}(2))\psi_{\frac{1}{3}\frac{1}{2}}(3)\\
&\quad+\psi_{\frac{1}{3}\frac{1}{2}}(1)(\hat{T}_-(2)\psi_{\frac{1}{3}\frac{1}{2}}(2))\psi_{\frac{1}{3}-\frac{1}{2}}(3)+\psi_{\frac{1}{3}-\frac{1}{2}}(1)\psi_{\frac{1}{3}\frac{1}{2}}(2)(\hat{T}_-(3)\psi_{\frac{1}{3}\frac{1}{2}}(3))\\
&\quad+\psi_{\frac{1}{3}\frac{1}{2}}(1)\psi_{\frac{1}{3}-\frac{1}{2}}(2)(\hat{T}_-(3)\psi_{\frac{1}{3}\frac{1}{2}}(3))+\psi_{\frac{1}{3}\frac{1}{2}}(1)\psi_{\frac{1}{3}\frac{1}{2}}(2)(\hat{T}_-(3)\psi_{\frac{1}{3}-\frac{1}{2}}(3))\}\cdot\chi\\
&= \frac{N}{\sqrt{3}}\cdot\chi\cdot\{0+\psi_{\frac{1}{3}-\frac{1}{2}}(1)\psi_{\frac{1}{3}-\frac{1}{2}}(2)\psi_{\frac{1}{3}\frac{1}{2}}(3)\\
&\quad+\psi_{\frac{1}{3}-\frac{1}{2}}(1)\psi_{\frac{1}{3}\frac{1}{2}}(2)\psi_{\frac{1}{3}-\frac{1}{2}}(3)+\psi_{\frac{1}{3}-\frac{1}{2}}(1)\psi_{\frac{1}{3}-\frac{1}{2}}(2)\psi_{\frac{1}{3}\frac{1}{2}}(3)+0\\
&\quad+\psi_{\frac{1}{3}\frac{1}{2}}(1)\psi_{\frac{1}{3}-\frac{1}{2}}(2)\psi_{\frac{1}{3}-\frac{1}{2}}(3)+\psi_{\frac{1}{3}-\frac{1}{2}}(1)\psi_{\frac{1}{3}\frac{1}{2}}(2)\psi_{\frac{1}{3}-\frac{1}{2}}(3)\\
&\quad+\psi_{\frac{1}{3}\frac{1}{2}}(1)\psi_{\frac{1}{3}-\frac{1}{2}}(2)\psi_{\frac{1}{3}-\frac{1}{2}}(3)+0\}
\end{aligned}
$$

Exercise 8.14

$$
\begin{aligned}
&= \frac{2}{\sqrt{3}}\cdot N\cdot\chi\cdot\{\psi_{\frac{1}{3}-\frac{1}{2}}(1)\psi_{\frac{1}{3}-\frac{1}{2}}(2)\psi_{\frac{1}{3}\frac{1}{2}}(3)\\
&\quad + \psi_{\frac{1}{3}-\frac{1}{2}}(1)\psi_{\frac{1}{3}\frac{1}{2}}(2)\psi_{\frac{1}{3}-\frac{1}{2}}(3) + \psi_{\frac{1}{3}\frac{1}{2}}(1)\psi_{\frac{1}{3}-\frac{1}{2}}(2)\psi_{\frac{1}{3}-\frac{1}{2}}(3)\} \quad .
\end{aligned}
$$

From $1 = \langle\Delta^0|\Delta^0\rangle$ follows

$$
1 = \tfrac{4}{3}\cdot|N|^2\cdot(1+0+0+0+1+0+0+0+1) = 4|N|^2 \Rightarrow N = \tfrac{1}{2} \quad ,
$$

leading to

$$
\begin{aligned}
|\Delta^0\rangle &= \frac{1}{\sqrt{3}}\cdot\{\psi_{\frac{1}{3}-\frac{1}{2}}(1)\psi_{\frac{1}{3}-\frac{1}{2}}(2)\psi_{\frac{1}{3}\frac{1}{2}}(3) + \psi_{\frac{1}{3}-\frac{1}{2}}(1)\psi_{\frac{1}{3}\frac{1}{2}}(2)\psi_{\frac{1}{3}-\frac{1}{2}}(3)\\
&\quad + \psi_{\frac{1}{3}\frac{1}{2}}(1)\psi_{\frac{1}{3}-\frac{1}{2}}(2)\psi_{\frac{1}{3}-\frac{1}{2}}(3)\}\,\chi_{\frac{1}{2}\frac{1}{2}}(1)\chi_{\frac{1}{2}\frac{1}{2}}(2)\chi_{\frac{1}{2}\frac{1}{2}}(3) \qquad (11)\\
|\Delta^-\rangle &= N\hat{T}_-|\Delta^0 = \frac{N}{\sqrt{3}}\cdot\{(\hat{T}_-(1)\psi_{\frac{1}{3}-\frac{1}{2}}(1))\psi_{\frac{1}{3}-\frac{1}{2}}(2)\psi_{\frac{1}{3}\frac{1}{2}}(3)\\
&\quad + (\hat{T}_-(1)\psi_{\frac{1}{3}-\frac{1}{2}}(1))\psi_{\frac{1}{3}\frac{1}{2}}(2)\psi_{\frac{1}{3}-\frac{1}{2}}(3) + (\hat{T}_-(1)\psi_{\frac{1}{3}\frac{1}{2}}(1))\psi_{\frac{1}{3}-\frac{1}{2}}(2)\psi_{\frac{1}{3}-\frac{1}{2}}(3)\\
&\quad + \psi_{\frac{1}{3}-\frac{1}{2}}(1)(\hat{T}_-(2)\psi_{\frac{1}{3}-\frac{1}{2}}(2))\psi_{\frac{1}{3}\frac{1}{2}}(3) + \psi_{\frac{1}{3}-\frac{1}{2}}(1)(\hat{T}_-(2)\psi_{\frac{1}{3}\frac{1}{2}}(2))\psi_{\frac{1}{3}-\frac{1}{2}}(3)\\
&\quad + \psi_{\frac{1}{3}\frac{1}{2}}(1)(\hat{T}_-(2)\psi_{\frac{1}{3}-\frac{1}{2}}(2))\psi_{\frac{1}{3}-\frac{1}{2}}(3) + \psi_{\frac{1}{3}-\frac{1}{2}}(1)\psi_{\frac{1}{3}-\frac{1}{2}}(2)(\hat{T}_-(3)\psi_{\frac{1}{3}\frac{1}{2}}(3))\\
&\quad + \psi_{\frac{1}{3}-\frac{1}{2}}(1)\psi_{\frac{1}{3}\frac{1}{2}}(2)(\hat{T}_-(3)\psi_{\frac{1}{3}-\frac{1}{2}}(3)) + \psi_{\frac{1}{3}\frac{1}{2}}(1)\psi_{\frac{1}{3}\frac{1}{2}}(2)(\hat{T}_-(3)\psi_{\frac{1}{3}-\frac{1}{2}}(3))\}\cdot\chi\\
|\Delta^-\rangle &= \frac{N}{\sqrt{3}}\cdot\chi\{0+0+\psi_{\frac{1}{3}-\frac{1}{2}}(1)\psi_{\frac{1}{3}-\frac{1}{2}}(2)\psi_{\frac{1}{3}-\frac{1}{2}}(3) + 0 + \psi_{\frac{1}{3}-\frac{1}{2}}(1)\psi_{\frac{1}{3}-\frac{1}{2}}(2)\psi_{\frac{1}{3}-\frac{1}{2}}(3)\\
&\quad + 0 + \psi_{\frac{1}{3}-\frac{1}{2}}(1)\psi_{\frac{1}{3}-\frac{1}{2}}(2)\psi_{\frac{1}{3}-\frac{1}{2}}(3) + 0 + 0\}\\
&= N\chi\cdot\sqrt{3}\cdot\psi_{\frac{1}{3}-\frac{1}{2}}(1)\psi_{\frac{1}{3}-\frac{1}{2}}(2)\psi_{\frac{1}{3}-\frac{1}{2}}(3) \quad .
\end{aligned}
$$

From $\langle\Delta^-|\Delta^-\rangle = 1$ follows $1 = 3\cdot|N|^2$, that is $N = 1/\sqrt{3}$. With this,

$$
\begin{aligned}
|\Delta^-\rangle &= \psi_{\frac{1}{3}-\frac{1}{2}}(1)\psi_{\frac{1}{3}-\frac{1}{2}}(2)\psi_{\frac{1}{3}-\frac{1}{2}}(3)\chi_{\frac{1}{2}\frac{1}{2}}(1)\chi_{\frac{1}{2}\frac{1}{2}}(2)\chi_{\frac{1}{2}\frac{1}{2}}(3) \qquad (12)\\
|\Xi^{*0}\rangle &= N\cdot\hat{V}_-|\Sigma^{*+}\rangle = \frac{N}{\sqrt{3}}\cdot\chi\cdot[(\hat{V}_-(1)\psi_{-\frac{2}{3}0}(1))\psi_{\frac{1}{3}\frac{1}{2}}(2)\psi_{\frac{1}{3}\frac{1}{2}}(3)\\
&\quad + (\hat{V}_-(1)\psi_{\frac{1}{3}\frac{1}{2}}(1))\psi_{-\frac{2}{3}0}(2)\psi_{\frac{1}{3}\frac{1}{2}}(3) + (\hat{V}_-(1)\psi_{\frac{1}{3}\frac{1}{2}}(1))\psi_{\frac{1}{3}\frac{1}{2}}(2)\psi_{-\frac{2}{3}0}(3)\\
&\quad + \psi_{-\frac{2}{3}0}(1)(\hat{V}_-(2)\psi_{\frac{1}{3}\frac{1}{2}}(2))\psi_{\frac{1}{3}\frac{1}{2}}(3) + \psi_{\frac{1}{3}-\frac{1}{2}}(1)(\hat{V}_-(2)\psi_{-\frac{2}{3}0}(2))\psi_{\frac{1}{3}\frac{1}{2}}(3)\\
&\quad + \psi_{\frac{1}{3}\frac{1}{2}}(1)(\hat{V}_-(2)\psi_{\frac{1}{3}\frac{1}{2}}(2))\psi_{-\frac{2}{3}0}(3) + \psi_{-\frac{2}{3}0}(1)\psi_{\frac{1}{3}\frac{1}{2}}(2)(\hat{V}_-(3)\psi_{\frac{1}{3}\frac{1}{2}}(3))\\
&\quad + \psi_{\frac{1}{3}\frac{1}{2}}(1)\psi_{-\frac{2}{3}0}(2)(\hat{V}_-(3)\psi_{\frac{1}{3}\frac{1}{2}}(3)) + \psi_{\frac{1}{3}\frac{1}{2}}(1)\psi_{\frac{1}{3}\frac{1}{2}}(2)(\hat{V}_-(3)\psi_{-\frac{2}{3}0}(3))]\\
&= \frac{N}{\sqrt{3}}\cdot 2\cdot\{\psi_{-\frac{2}{3}0}(1)\psi_{-\frac{2}{3}0}(2)\psi_{\frac{1}{3}\frac{1}{2}}(3)\\
&\quad + \psi_{-\frac{2}{3}0}(1)\psi_{\frac{1}{3}\frac{1}{2}}(2)\psi_{-\frac{2}{3}0}(3) + \psi_{\frac{1}{3}\frac{1}{2}}(1)\psi_{-\frac{2}{3}0}(2)\psi_{-\frac{2}{3}0}(3)\}\cdot\chi \quad . \qquad (13)
\end{aligned}
$$

Exercise 8.14

The normalization condition yields

$$|\Xi^{*0}\rangle = \frac{1}{\sqrt{3}}\cdot\{\psi_{-\frac{2}{3}0}(1)\psi_{-\frac{2}{3}0}(2)\psi_{\frac{1}{3}\frac{1}{2}}(3) + \psi_{-\frac{2}{3}0}(1)\psi_{\frac{1}{3}\frac{1}{2}}(2)\psi_{-\frac{2}{3}0}(3)$$
$$+ \psi_{\frac{1}{3}\frac{1}{2}}(1)\psi_{-\frac{2}{3}0}(2)\psi_{-\frac{2}{3}0}(3)\}\chi_{\frac{1}{2}\frac{1}{2}}(1)\chi_{\frac{1}{2}\frac{1}{2}}(2)\chi_{\frac{1}{2}\frac{1}{2}}(3) \quad , \tag{14}$$

and in the same way,

$$|\Omega^-\rangle = N\cdot\hat{V}_-|\Xi^{*0}\rangle$$
$$= \psi_{-\frac{2}{3}0}(1)\psi_{-\frac{2}{3}0}(2)\psi_{-\frac{2}{3}0}(3)\chi_{\frac{1}{2}\frac{1}{2}}(1)\chi_{\frac{1}{2}\frac{1}{2}}(2)\chi_{\frac{1}{2}\frac{1}{2}}(3) \quad . \tag{15}$$

Now several possibilities exist for deriving $|\Sigma^{*0}\rangle$ from the previously calculated states, as can be seen in the $Y - T_3$ diagram. On one hand we have

$$|\Sigma^{*0}\rangle = \hat{T}_-|\Sigma^{*+}\rangle \quad , \tag{16}$$

other possibilities being

$$|\Sigma^{*0}\rangle = \hat{V}_-|\Delta^+\rangle \quad \text{and} \quad |\Sigma^{*0}\rangle = \hat{U}_-|\Delta^0\rangle \quad .$$

We prove that all possibilities yield the same function (as one would expect), with

$$|\Sigma^{*0}\rangle = \hat{T}_-|\Sigma^{*+}\rangle$$
$$= N\cdot\hat{T}_-\left\{\frac{1}{\sqrt{3}}[\psi_{-\frac{2}{3}0}(1)\psi_{\frac{1}{3}\frac{1}{2}}(2)\psi_{\frac{1}{3}\frac{1}{2}}(3) + \psi_{\frac{1}{3}\frac{1}{2}}(1)\psi_{-\frac{2}{3}0}(2)\psi_{\frac{1}{3}\frac{1}{2}}(3)\right.$$
$$\left. + \psi_{\frac{1}{3}\frac{1}{2}}(1)\chi_{\frac{1}{3}\frac{1}{2}}(2)\psi_{-\frac{2}{3}0}(3)]\,\chi_{\frac{1}{2}\frac{1}{2}}\chi_{\frac{1}{2}\frac{1}{2}}(2)\chi_{\frac{1}{2}\frac{1}{2}}(3)\right\} \tag{17}$$
$$= \frac{N}{\sqrt{3}}\cdot\chi\cdot\{(\hat{T}_-(1)\psi_{-\frac{2}{3}0}(1))\psi_{\frac{1}{3}\frac{1}{2}}(2)\psi_{\frac{1}{3}\frac{1}{2}}(3)$$
$$+ (\hat{T}_-(1)\psi_{\frac{1}{3}\frac{1}{2}}(1))\psi_{-\frac{2}{3}0}(2)\psi_{\frac{1}{3}\frac{1}{2}}(3) + (\hat{T}_-(1)\psi_{\frac{1}{3}\frac{1}{2}}(1))\psi_{\frac{1}{3}\frac{1}{2}}(2)\psi_{-\frac{2}{3}0}(3)$$
$$+ \psi_{-\frac{2}{3}0}(1)(\hat{T}_-(2)\psi_{\frac{1}{3}\frac{1}{2}}(2))\psi_{\frac{1}{3}\frac{1}{2}}(3) + \psi_{\frac{1}{3}\frac{1}{2}}(1)(\hat{T}_-(2)\psi_{-\frac{2}{3}0}(2))\psi_{\frac{1}{3}\frac{1}{2}}(3)$$
$$+ \psi_{\frac{1}{3}\frac{1}{2}}(1)(\hat{T}_-(2)\psi_{\frac{1}{3}\frac{1}{2}}(2))\psi_{-\frac{2}{3}0}(3) + \psi_{-\frac{2}{3}0}(1)\psi_{\frac{1}{3}\frac{1}{2}}(2)(\hat{T}_-(3)\psi_{\frac{1}{3}\frac{1}{2}}(3))$$
$$+ \psi_{\frac{1}{3}\frac{1}{2}}(1)\psi_{-\frac{2}{3}0}(2)(\hat{T}_-(3)\psi_{\frac{1}{3}\frac{1}{2}}(3)) + \psi_{\frac{1}{3}\frac{1}{2}}(1)\psi_{\frac{1}{3}\frac{1}{2}}(2)(\hat{T}_-(3)\psi_{-\frac{2}{3}0}(3))\}$$
$$= \frac{N}{\sqrt{3}}\cdot\chi\cdot\{0 + \psi_{\frac{1}{3}-\frac{1}{2}}(1)\psi_{-\frac{2}{3}0}(2)\psi_{\frac{1}{3}\frac{1}{2}}(3) + \psi_{\frac{1}{3}-\frac{1}{2}}(1)\psi_{\frac{1}{3}\frac{1}{2}}(2)\psi_{-\frac{2}{3}0}(3)$$
$$+ \psi_{-\frac{2}{3}0}(1)\psi_{\frac{1}{3}-\frac{1}{2}}(2)\psi_{\frac{1}{3}\frac{1}{2}}(3) + 0 + \psi_{\frac{1}{3}\frac{1}{2}}(1)\psi_{\frac{1}{3}-\frac{1}{2}}(2)\psi_{-\frac{2}{3}0}(3)$$
$$+ \psi_{-\frac{2}{3}0}(1)\psi_{\frac{1}{3}\frac{1}{2}}(2)\psi_{\frac{1}{3}-\frac{1}{2}}(3) + \psi_{\frac{1}{3}\frac{1}{2}}(1)\psi_{-\frac{2}{3}0}(2)\psi_{\frac{1}{3}-\frac{1}{2}}(3) + 0\} \quad . \tag{18}$$

The normalization condition leads to

$$1 = \langle\Sigma^{*0}|\Sigma^{*0}\rangle = \langle[\psi_{\frac{1}{3}-\frac{1}{2}}(1)\psi_{-\frac{2}{3}0}(2)\psi_{\frac{1}{3}\frac{1}{2}}(3) + \psi_{\frac{1}{3}-\frac{1}{2}}(1)\psi_{\frac{1}{3}\frac{1}{2}}(2)\psi_{-\frac{2}{3}0}(3)$$
$$+ \psi_{-\frac{2}{3}0}(1)\psi_{\frac{1}{3}-\frac{1}{2}}(2)\psi_{\frac{1}{3}\frac{1}{2}}(3) + \psi_{\frac{1}{3}\frac{1}{2}}(1)\psi_{\frac{1}{3}-\frac{1}{2}}(2)\psi_{-\frac{2}{3}0}(3) + \psi_{-\frac{2}{3}0}(1)\psi_{\frac{1}{3}\frac{1}{2}}(2)\psi_{\frac{1}{3}-\frac{1}{2}}(3)$$
$$+ \psi_{\frac{1}{3}\frac{1}{2}}(1)\psi_{-\frac{2}{3}0}(2)\psi_{\frac{1}{3}-\frac{1}{2}}(3)]\cdot\chi\,|\,[\psi_{\frac{1}{3}-\frac{1}{2}}(1)\psi_{-\frac{2}{3}0}(2)\psi_{\frac{1}{3}\frac{1}{2}}(3)$$

Exercise 8.14

$$
\begin{aligned}
&+ \psi_{\frac{1}{3}-\frac{1}{2}}(1)\psi_{\frac{1}{3}\frac{1}{2}}(2)\psi_{-\frac{2}{3}0}(3) + \psi_{-\frac{2}{3}0}(1)\psi_{\frac{1}{3}-\frac{1}{2}}(2)\psi_{\frac{1}{3}\frac{1}{2}}(3)\\
&+ \psi_{\frac{1}{3}\frac{1}{2}}(1)\psi_{\frac{1}{3}-\frac{1}{2}}(2)\psi_{-\frac{2}{3}0}(3) + \psi_{-\frac{2}{3}0}(1)\psi_{\frac{1}{3}\frac{1}{2}}(2)\psi_{\frac{1}{3}-\frac{1}{2}}(3)\\
&+ \psi_{\frac{1}{3}\frac{1}{2}}(1)\psi_{-\frac{2}{3}0}(2)\psi_{\frac{1}{3}-\frac{1}{2}}(3)]\cdot\chi\rangle\cdot\frac{|N|^2}{3}\\
&= (1+0+0+0+0+0+0+1+0+0+0+0+0+0+1\\
&\quad +0+0+0+0+0+0+1+0+0+0+0+0+0+1\\
&\quad +0+0+0+0+0+0+1)\,\frac{|N|^2}{3} = 2|N|^2 \quad .
\end{aligned}
$$

Thus $N = 1/\sqrt{2}$ and therefore

$$
\begin{aligned}
|\Sigma^{*0}\rangle = \frac{1}{\sqrt{6}}\cdot\{&\psi_{\frac{1}{3}-\frac{1}{2}}(1)\psi_{-\frac{2}{3}0}(2)\psi_{\frac{1}{3}\frac{1}{2}}(3) + \psi_{\frac{1}{3}-\frac{1}{2}}(1)\psi_{\frac{1}{3}\frac{1}{2}}(2)\psi_{-\frac{2}{3}0}(3)\\
&+ \psi_{-\frac{2}{3}0}(1)\psi_{\frac{1}{3}-\frac{1}{2}}(2)\psi_{\frac{1}{3}\frac{1}{2}}(3) + \psi_{\frac{1}{3}\frac{1}{2}}(1)\psi_{\frac{1}{3}-\frac{1}{2}}(2)\psi_{-\frac{2}{3}0}(3)\\
&+ \psi_{-\frac{2}{3}0}(1)\psi_{\frac{1}{3}\frac{1}{2}}(2)\psi_{\frac{1}{3}-\frac{1}{2}}(3) + \psi_{\frac{1}{3}\frac{1}{2}}(1)\psi_{-\frac{2}{3}0}(2)\psi_{\frac{1}{3}-\frac{1}{2}}(3)\}\chi_{\frac{1}{2}\frac{1}{2}}(1)\chi_{\frac{1}{2}\frac{1}{2}}(2)\chi_{\frac{1}{2}\frac{1}{2}}(3)
\end{aligned}
$$

On the other hand,

$$
\begin{aligned}
|\Sigma^{*0}\rangle &= \tilde{N}\hat{V}_-|\Delta^+\rangle = \tilde{N}\hat{V}_-\Big[\frac{1}{\sqrt{3}}\cdot\{\psi_{\frac{1}{3}-\frac{1}{2}}(1)\psi_{\frac{1}{3}\frac{1}{2}}(2)\psi_{\frac{1}{3}\frac{1}{2}}(3)\\
&\quad + \psi_{\frac{1}{3}\frac{1}{2}}(1)\psi_{\frac{1}{3}-\frac{1}{2}}(2)\psi_{\frac{1}{3}\frac{1}{2}}(3) + \psi_{\frac{1}{3}\frac{1}{2}}(1)\psi_{\frac{1}{3}\frac{1}{2}}(2)\psi_{\frac{1}{3}-\frac{1}{2}}(3)\}\cdot\chi\Big]\\
&= \frac{\tilde{N}}{\sqrt{3}}\cdot\chi\cdot\{0 + \psi_{-\frac{2}{3}0}(1)\psi_{\frac{1}{3}-\frac{1}{2}}(2)\psi_{\frac{1}{3}\frac{1}{2}}(3) + \psi_{-\frac{2}{3}0}(1)\psi_{\frac{1}{3}\frac{1}{2}}(2)\psi_{\frac{1}{3}-\frac{1}{2}}(3)\\
&\quad + \psi_{\frac{1}{3}-\frac{1}{2}}(1)\psi_{-\frac{2}{3}0}(2)\psi_{\frac{1}{3}\frac{1}{2}}(3) + 0 + \psi_{\frac{1}{3}\frac{1}{2}}(1)\psi_{-\frac{2}{3}0}(2)\psi_{\frac{1}{3}-\frac{1}{2}}(3)\\
&\quad + \psi_{\frac{1}{3}-\frac{1}{2}}(1)\psi_{\frac{1}{3}\frac{1}{2}}(2)\psi_{-\frac{2}{3}0}(3) + \psi_{\frac{1}{3}\frac{1}{2}}(1)\psi_{\frac{1}{3}-\frac{1}{2}}(2)\psi_{-\frac{2}{3}0}(3) + 0\} \quad .
\end{aligned} \tag{19}
$$

From this it follows that $1 = 6\cdot|\tilde{N}|^2/3$ too, that is $\tilde{N} = 1/\sqrt{2}$; thus the function is the same as the one above. In the same way one makes sure that

$$
\begin{aligned}
|\Sigma^{*0}\rangle = \tilde{\tilde{N}}\hat{U}_-|\Delta^0\rangle &= \frac{\tilde{\tilde{N}}}{\sqrt{3}}\cdot\chi\cdot\{(\hat{U}_-(1)\psi_{\frac{1}{3}-\frac{1}{2}}(1))\chi_{\frac{1}{3}-\frac{1}{2}}(2)\psi_{\frac{1}{3}\frac{1}{2}}(3)\\
&\quad + (\hat{U}_-(1)\psi_{\frac{1}{3}-\frac{1}{2}}(1))\psi_{\frac{1}{3}\frac{1}{2}}(2)\psi_{\frac{1}{3}-\frac{1}{2}}(3) + (\hat{U}_-(1)\psi_{\frac{1}{3}\frac{1}{2}}(1))\psi_{\frac{1}{3}-\frac{1}{2}}(2)\psi_{\frac{1}{3}-\frac{1}{2}}(3)\\
&\quad + \psi_{\frac{1}{3}-\frac{1}{2}}(1)(\hat{U}_-(2)\psi_{\frac{1}{3}-\frac{1}{2}}(2))\psi_{\frac{1}{3}\frac{1}{2}}(3) + \psi_{\frac{1}{3}-\frac{1}{2}}(1)(\hat{U}_-(2)\psi_{\frac{1}{3}\frac{1}{2}}(2))\psi_{\frac{1}{3}-\frac{1}{2}}(3)\\
&\quad + \psi_{\frac{1}{3}\frac{1}{2}}(1)(\hat{U}_-(2)\psi_{\frac{1}{3}-\frac{1}{2}}(2))\psi_{\frac{1}{3}-\frac{1}{2}}(3) + \psi_{\frac{1}{2}-\frac{1}{2}}(1)\psi_{\frac{1}{3}-\frac{1}{2}}(2)(\hat{U}_-(3)\psi_{\frac{1}{3}\frac{1}{2}}(3))\\
&\quad + \psi_{\frac{1}{3}-\frac{1}{2}}(1)\psi_{\frac{1}{3}\frac{1}{2}}(2)(\hat{U}_-(3)\psi_{\frac{1}{3}-\frac{1}{2}}(3)) + \psi_{\frac{1}{3}\frac{1}{2}}(1)\psi_{\frac{1}{3}-\frac{1}{2}}(2)(\hat{U}_-(3)\psi_{\frac{1}{3}-\frac{1}{2}}(3))\}\\
&= \frac{\tilde{\tilde{N}}}{\sqrt{3}}\cdot\chi\cdot\{\psi_{-\frac{2}{3}0}(1)\psi_{\frac{1}{3}-\frac{1}{2}}(2)\psi_{\frac{1}{3}\frac{1}{2}}(3) + \psi_{-\frac{2}{3}0}(1)\psi_{\frac{1}{3}\frac{1}{2}}(2)\psi_{\frac{1}{3}-\frac{1}{2}}(3)\\
&\quad + \psi_{\frac{1}{3}-\frac{1}{2}}(1)\psi_{-\frac{2}{3}0}(2)\psi_{\frac{1}{3}\frac{1}{2}}(3) + \psi_{\frac{1}{3}\frac{1}{2}}(1)\psi_{-\frac{2}{3}0}(2)\psi_{\frac{1}{3}-\frac{1}{2}}(3)\\
&\quad + \psi_{\frac{1}{3}-\frac{1}{2}}(1)\psi_{\frac{1}{3}\frac{1}{2}}(2)\psi_{-\frac{2}{3}0}(3) + \psi_{\frac{1}{3}\frac{1}{2}}(1)\psi_{\frac{1}{3}-\frac{1}{2}}(2)\psi_{-\frac{2}{3}0}(3)\} \quad ,
\end{aligned} \tag{20}
$$

Exercise 8.14

which, after normalization, yields the same function. Now we calculate $|\Sigma^{*-}\rangle$ from

$$\begin{aligned}|\Sigma^{*-}\rangle = N\hat{U}_-|\Delta^-\rangle = N\cdot\{&(\hat{U}_-(1)\psi_{\frac{1}{3}-\frac{1}{2}}(1))\psi_{\frac{1}{3}-\frac{1}{2}}(2)\psi_{\frac{1}{3}-\frac{1}{2}}(3)\\ &+\psi_{\frac{1}{3}-\frac{1}{2}}(1)(\hat{U}_-(2)\psi_{\frac{1}{3}-\frac{1}{2}}(2))\psi_{\frac{1}{3}-\frac{1}{2}}(3)\\ &+\psi_{\frac{1}{3}-\frac{1}{2}}(1)\psi_{\frac{1}{3}-\frac{1}{2}}(2)(\hat{U}_-(3)\psi_{\frac{1}{3}-\frac{1}{2}}(3))\}\chi_{\frac{1}{2}\frac{1}{2}}(1)\chi_{\frac{1}{2}\frac{1}{2}}(2)\chi_{\frac{1}{2}\frac{1}{2}}(3)\\ = N\cdot\{&\psi_{-\frac{2}{3}0}(1)\psi_{\frac{1}{3}-\frac{1}{2}}(2)\psi_{\frac{1}{3}-\frac{1}{2}}(3)+\psi_{\frac{1}{3}-\frac{1}{2}}(1)\psi_{-\frac{2}{3}0}(2)\psi_{\frac{1}{3}-\frac{1}{2}}(3)\\ &+\psi_{\frac{1}{3}-\frac{1}{2}}(1)\psi_{\frac{1}{3}-\frac{1}{2}}(2)\psi_{-\frac{2}{3}0}(3)\}\cdot\chi\quad,\end{aligned}\tag{21}$$

normalization yielding

$$1=\langle\Sigma^{*-}|\Sigma^{*-}\rangle=N^2\cdot 3\quad\text{i.e.:}\quad N=\frac{1}{\sqrt{3}}$$

$$\begin{aligned}|\Sigma^{*-}\rangle=\frac{1}{\sqrt{3}}\cdot\{&\psi_{-\frac{2}{3}0}(1)\psi_{\frac{1}{3}-\frac{1}{2}}(2)\psi_{\frac{1}{3}-\frac{1}{2}}(3)+\psi_{\frac{1}{3}-\frac{1}{2}}(1)\psi_{-\frac{2}{3}0}(2)\psi_{\frac{1}{3}-\frac{1}{2}}(3)\\ &+\psi_{\frac{1}{3}-\frac{1}{2}}(1)\psi_{\frac{1}{3}-\frac{1}{2}}(2)\psi_{-\frac{2}{3}0}(3)\}\chi_{\frac{1}{2}\frac{1}{2}}(1)\chi_{\frac{1}{2}\frac{1}{2}}(2)\chi_{\frac{1}{2}\frac{1}{2}}(3)\quad.\end{aligned}\tag{22}$$

Calculating $\hat{T}_-|\Sigma^{*0}\rangle$ or $\hat{V}_-|\Delta^0\rangle$ gives the same result.

Finally there is $|\Xi^{*-}\rangle$ to be calculated. The easiest way to obtain this function is by

$$\begin{aligned}|\Xi^{*-}\rangle &= N\hat{U}_+|\Omega^-\rangle\\ &= N\cdot\{(\hat{U}_+(1)\psi_{-\frac{2}{3}0}(1))\psi_{-\frac{2}{3}0}(2)\psi_{-\frac{2}{3}0}(3)+\psi_{-\frac{2}{3}0}(1)(\hat{U}_+(2)\psi_{-\frac{2}{3}0}(2))\psi_{-\frac{2}{3}0}(3)\\ &\quad+\psi_{-\frac{2}{3}0}(1)\psi_{-\frac{2}{3}0}(2)(\hat{U}_+(3)\psi_{-\frac{2}{3}0}(3))\}\chi_{\frac{1}{2}\frac{1}{2}}(1)\chi_{\frac{1}{2}\frac{1}{2}}(2)\chi_{\frac{1}{2}\frac{1}{2}}(3)\\ &= N\cdot\{\psi_{\frac{1}{3}-\frac{1}{2}}(1)\psi_{-\frac{2}{3}0}(2)\psi_{-\frac{2}{3}0}(3)+\psi_{-\frac{2}{3}0}(1)\psi_{\frac{1}{3}-\frac{1}{2}}(2)\psi_{-\frac{2}{3}0}(3)\\ &\quad+\psi_{-\frac{2}{3}0}(1)\psi_{-\frac{2}{3}0}(2)\psi_{\frac{1}{3}-\frac{1}{2}}(3)\}\chi_{\frac{1}{2}\frac{1}{2}}(1)\chi_{\frac{1}{2}\frac{1}{2}}(2)\chi_{\frac{1}{2}\frac{1}{2}}(3)\quad.\end{aligned}\tag{23}$$

On normalizing,

$$1=\langle\Xi^{*-}|\Xi^{*-}\rangle=|N|^2\cdot 3\quad,\quad N=\frac{1}{\sqrt{3}}\quad\text{and}$$

$$\begin{aligned}|\Xi^{*-}\rangle=\frac{1}{\sqrt{3}}\cdot\{&\psi_{\frac{1}{3}-\frac{1}{2}}(1)\psi_{-\frac{2}{3}0}(2)\psi_{-\frac{2}{3}0}(3)+\psi_{-\frac{2}{3}0}(1)\psi_{\frac{1}{3}-\frac{1}{2}}(2)\psi_{-\frac{2}{3}0}(3)\\ &+\psi_{-\frac{2}{3}0}(1)\psi_{-\frac{2}{3}0}(2)\psi_{\frac{1}{3}-\frac{1}{2}}(3)\}\chi_{\frac{1}{2}\frac{1}{2}}(1)\chi_{\frac{1}{2}\frac{1}{2}}(2)\chi_{\frac{1}{2}\frac{1}{2}}(3)\quad.\end{aligned}\tag{24}$$

Now we combine all the wave functions, using the abbreviations.

$$\psi_{\frac{1}{3}\frac{1}{2}}\chi_{\frac{1}{2}\frac{1}{2}}=u_\uparrow\quad;\quad\psi_{\frac{1}{3}-\frac{1}{2}}\chi_{\frac{1}{2}\frac{1}{2}}=d_\uparrow\quad;\quad\psi_{-\frac{2}{3}0}\chi_{\frac{1}{2}\frac{1}{2}}=s_\uparrow\quad,\tag{25}$$

and, with regard to the the implicit order, i.e. $udu=u(1)d(2)u(3)$ etc., we write

$$|\Delta^{++}\rangle=u_\uparrow u_\uparrow u_\uparrow$$

$$|\Delta^{+}\rangle=\frac{1}{\sqrt{3}}(d_\uparrow u_\uparrow u_\uparrow+u_\uparrow d_\uparrow u_\uparrow+u_\uparrow u_\uparrow d_\uparrow)$$

Exercise 8.14

$$|\Delta^0\rangle = \frac{1}{\sqrt{3}}(d_\uparrow d_\uparrow u_\uparrow + d_\uparrow u_\uparrow d_\uparrow + u_\uparrow d_\uparrow d_\uparrow)$$

$$|\Delta^-\rangle = d_\uparrow d_\uparrow d_\uparrow$$

$$|\Sigma^{*+}\rangle = \frac{1}{\sqrt{3}}(s_\uparrow u_\uparrow u_\uparrow + u_\uparrow s_\uparrow u_\uparrow + u_\uparrow u_\uparrow s_\uparrow)$$

$$|\Sigma^{*0}\rangle = \frac{1}{\sqrt{6}}(s_\uparrow d_\uparrow u_\uparrow + s_\uparrow u_\uparrow d_\uparrow + d_\uparrow s_\uparrow u_\uparrow + d_\uparrow u_\uparrow s_\uparrow + u_\uparrow s_\uparrow d_\uparrow + u_\uparrow d_\uparrow s_\uparrow)$$

$$|\Sigma^{*-}\rangle = \frac{1}{\sqrt{3}}(s_\uparrow d_\uparrow d_\uparrow + d_\uparrow s_\uparrow d_\uparrow + d_\uparrow d_\uparrow s_\uparrow)$$

$$|\Xi^{*0}\rangle = \frac{1}{\sqrt{3}}(s_\uparrow s_\uparrow u_\uparrow + s_\uparrow u_\uparrow s_\uparrow + u_\uparrow s_\uparrow s_\uparrow)$$

$$|\Xi^{*-}\rangle = \frac{1}{\sqrt{3}}(s_\uparrow s_\uparrow d_\uparrow + s_\uparrow d_\uparrow s_\uparrow + d_\uparrow s_\uparrow s_\uparrow)$$

$$|\Omega^-\rangle = s_\uparrow s_\uparrow s_\uparrow \quad . \tag{26}$$

To construct particles with different spin projection j_3, it is only necessary just to recalculate the different spin part of the wave function.

EXERCISE

8.15 Construction of the Spin-Flavour Wave Functions of the Baryon Octet

Problem. Construct the wave functions of the baryon octet in the frame of $\mathrm{SU(3)} \times \mathrm{SU(2)}$ (i.e. the baryonic wave functions as eigenfunctions of isospin, hypercharge *and* spin.)

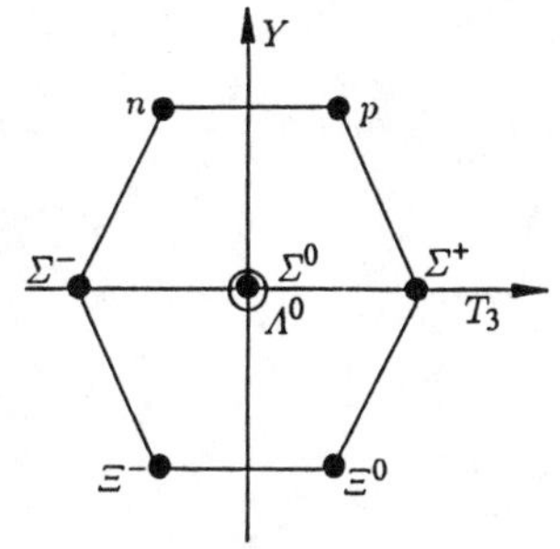

The baryon octet

Solution. I. The baryon octet and the effect of the shift operators. The baryon octet in the $Y - T_3$ plane is known from Example 6.3 (see figure) and we know the effect of the shift operators from Chap. 7, in particular from the consideration of the subalgebras of the SU(3). This is shown once again in the figure on page 276.

A shift operator applied to an arbitrary state of the baryon octet yields the adjacent state with respect to the shift direction of the operator. If it is a state of the octet, one obtains the wave function of the corresponding particle. But if the shift crosses the boundaries of the octet, the wave function vanishes. The construction of functions of the doubly occupied points of the octet (this occurs only at the origin of the $Y - T_3$ plane) will be considered separately. We know here (from our investigation concerning the number of states belonging to lattice

Exercise 8.15

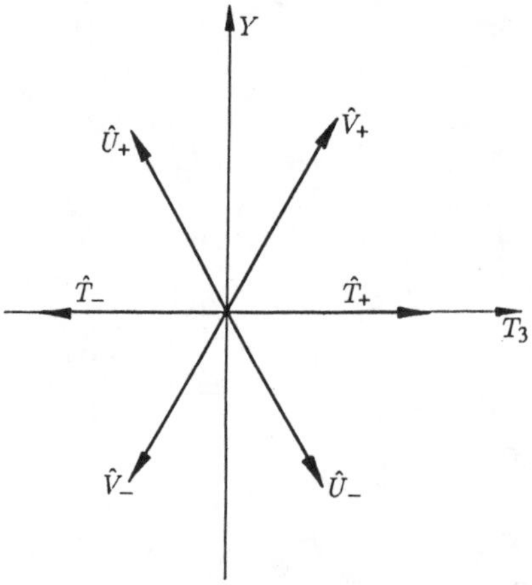

The effect of the shift operators

points of inner shells, Chap. 7) that we obtain two different states if, starting from Σ^+, we apply $\hat{T}_-$ or the operator combination $\hat{U}+\hat{V}_-$ to Σ^+: In the first case we generate a Σ^0 and in the second case a Λ^0. Before we start to construct the wave functions there are still two important things to be considered:

II. The explicit form of the shift operators: Each baryon consists of three quarks; in general its wave function consists of a linear combination of products of quark wave functions which obey the restrictions of the possible baryon quantum number Y, T, T_3 (initially we will not concern ourselves with the spin):

$$\psi_{\mathrm{B}}(\boldsymbol{r}) = \sum_i \lambda_i (q_{t^1,t_3^1,y^1}(\boldsymbol{r}_1)\cdot q_{t^2,t_3^2,y^2}(\boldsymbol{r}_2)\cdot q_{t^3,t_3^3,y^3}(\boldsymbol{r}_3))_i \quad . \tag{1}$$

The parentheses $(\dots)_i$ indicate the ith of the possible three-quark wave functions; $q_{t^jt_3^jy^j}(j)$ is the wave function of the jth quark with quantum numbers t^j, t_3^j, y^j (in the following we frequently use the common abbreviations $|u\rangle$ for $q_{\frac{1}{2}\frac{1}{2}\frac{1}{3}}$, $|d\rangle$ for $q_{\frac{1}{2}-\frac{1}{2}\frac{1}{3}}$ and $|s\rangle$ for $q_{00-\frac{2}{3}}$). In the first instance the λ_i, are arbitrary complex numbers.

Looking at (1) we recognize that in reality the baryon wave function is a many-body function. Therefore an operator (for example one of the shift operators) which acts in the space of the baryon wave functions is a many-body operator, too, and

$$\hat{O}_{\mathrm{B}} = \hat{O}_1 + \hat{O}_2 + \hat{O}_3 \quad . \tag{2}$$

$\hat{O}_j$ are operators which act only in the space of the corresponding quark wave function $q_{t^jt_3^jy^j}(j)$.[20] This means that

$$\begin{aligned} \hat{O}_{\mathrm{B}}\psi_{\mathrm{B}}(\boldsymbol{r}) &= \sum_i \lambda_i \hat{O}_{\mathrm{B}}(q_1(\boldsymbol{r}_1)q_2(\boldsymbol{r}_2)q_3(\boldsymbol{r}_3))_i \\ &= \sum_i \lambda_i\{(\hat{O}_1 q_1(\boldsymbol{r}_1))q_2(\boldsymbol{r}_1)q_3(\boldsymbol{r}_3) + q_1(\boldsymbol{r}_1)(\hat{O}_2 q_2(\boldsymbol{r}_2))q_3(\boldsymbol{r}_3) \\ &\quad + q_1(\boldsymbol{r}_1)q_2(\boldsymbol{r}_2)(\hat{O}_3 q_3(\boldsymbol{r}_3))\}_i \end{aligned} \tag{3}$$

[here we have abbreviated "(t^i, t_3^i, y^i)" by "i"].

[20] See Vol. 1 of this series, *Quantum Mechanics I-An Introduction* (Springer, Berlin, Heidelberg 1989).

Finally we have to consider the functions on which the operators act:

Exercise 8.15

III. The form of a baryon wave function with spin component: Neglecting spin the wave function has the form (1). If we include spin, at first sight it may seem that the function will be very complicated. As has been shown previously (see footnote 12), the total wave function is the direct product of the wave function (1) and the spin wave function and is defined as

$$\begin{aligned}\psi_{\mathrm{B}}(\boldsymbol{r})\otimes\chi(\boldsymbol{r}) &= \sum_i \lambda_i(q_1(\boldsymbol{r}_1)q_2(\boldsymbol{r}_2)q_3(\boldsymbol{r}_3))_i \otimes \sum_j \lambda_j(\chi_{\frac{1}{2}m_s^1}(1)\chi_{\frac{1}{2}m_s^2}(2)\chi_{\frac{1}{2}m_s^3}(3)) \\ &= \sum_k \lambda_k(q_1\otimes\chi_{\frac{1}{2}m_s^1}(1)\cdot q_2\otimes\chi_{\frac{1}{2}m_s^2}(2)\cdot q_3\otimes\chi_{\frac{1}{2}m_s^3}(3))_k \quad . \end{aligned} \tag{4}$$

Here ψ_{B} is a function in the space of eigenfunctions of isospin and hypercharge, χ is a function in the space of spin eigenfunctions.

The point which significantly simplifies the calculation is that the SU(3) shift operators do not influence the spin functions. Therefore, applying one of the shift operators to an arbitrary function belonging to the octet will not change the spin part. Thus one can define the spin projection at the outset and leave it unchanged throughout the calculation.

Since the quarks have spin $s=\frac{1}{2}$, with the projections $m_s=\pm\frac{1}{2}$, the coupling of three quarks provides the total spin projections $J_z=\pm\frac{3}{2},\pm\frac{1}{2}$ for the baryon. Yet we known from Example 6.3 that all baryons of the octet have spin $J=\frac{1}{2}$, i.e. $J_z=\pm\frac{1}{2}$ (on the other hand the decuplet of baryonic resonances has spin $J=\frac{3}{2}$). Hence we only have to consider the two cases $J_z=+\frac{1}{2}$ and $J_z=-\frac{1}{2}$ and, since the spin part of the wave function is not changed by applying the shift operators, we may restrict ourselves to $J_z=+\frac{1}{2}$ from the beginning. The other case can be treated similarly.

In order to construct a wave function one has first to calculate an initial state. Then the other states can be generated by applying the shift operators.

IV. The construction of the first baryonic state: We choose the proton function as the initial state. We know the quark content of the proton to be two u quarks and one d quark in the framework of SU(3) (or a linear combination of these if the function has to have specific symmetry properties). In the framework of SU(3) × SU(2) the quark content has to be a linear combination of the product wave functions of three of the quarks

$$|u_\uparrow\rangle := |u\rangle\otimes\chi_{\frac{1}{2}\frac{1}{2}} \quad ; \quad |u_\downarrow\rangle := |u\rangle\otimes\chi_{\frac{1}{2}-\frac{1}{2}} \quad ; \quad |d_\uparrow\rangle \text{ and } |d_\downarrow\rangle$$

(where two u quarks with spin are always coupled to one d quark with spin. Furthermore, $J=J_z=+\frac{1}{2}$).

To generate a wave function of this kind from these quarks, one has to apply the formalism of angular momentum coupling of Chap. 2. First we couple $|u_\uparrow\rangle$ and $|u_\uparrow\rangle$ to $(uu)^{J=1}_{J_z=1}\equiv|u_\uparrow u_\uparrow\rangle$ and then the resulting function is coupled with $|d_\downarrow\rangle$ (so that it yields a $J_z=\frac{1}{2}$ state). In the same way $|u_\uparrow\rangle$ and $|u_\downarrow\rangle$ are coupled to $(uu)^{J=1}_{J_z=0}$ and the result is coupled to $|d_\uparrow\rangle$. Since we have omitted the arguments $\boldsymbol{r}_1,\boldsymbol{r}_2,\boldsymbol{r}_3$ for the sake of greater simplicity, the order of the quarks corresponds to the order of arguments; thus being of crucial importance. The quark placed first has the quantum numbers t^1, t^1_3, y^1 and the argument $\boldsymbol{r}_1$ of the

Exercise 8.15

corresponding wave function, the second quark is specified by t^2, t_3^2, y^2 etc. Hence,

$$(uu)_0^1 = \sum_{m_s, m_s'} (\tfrac{1}{2}\tfrac{1}{2}1|m_s m_s' 0)\,(|u(\boldsymbol{r}_1)\rangle \otimes \chi_{\frac{1}{2}m_s}(1))\,(|u(\boldsymbol{r}_2)\rangle \otimes \chi_{\frac{1}{2}m_s}(2))$$
$$= (\tfrac{1}{2}\tfrac{1}{2}1|\tfrac{1}{2} - \tfrac{1}{2}0)\,|u_\uparrow u_\downarrow\rangle + (\tfrac{1}{2}\tfrac{1}{2}1| - \tfrac{1}{2}\tfrac{1}{2}0)\,|u_\uparrow u_\downarrow\rangle$$
$$= \frac{1}{\sqrt{2}}(|u_\uparrow u_\downarrow\rangle + |u_\downarrow u_\uparrow\rangle) \quad ,$$

leading to

$$|\mathrm{p}_\uparrow\rangle = (1\tfrac{1}{2}\tfrac{1}{2}|1 - \tfrac{1}{2}\tfrac{1}{2})\,|u_\uparrow u_\uparrow d_\downarrow\rangle + (1\tfrac{1}{2}\tfrac{1}{2}|0\tfrac{1}{2}\tfrac{1}{2})\,\frac{1}{\sqrt{2}}(|u_\uparrow u_\downarrow d_\uparrow\rangle + |u_\downarrow u_\uparrow d_\uparrow\rangle)$$
$$= \sqrt{\frac{2}{3}}|u_\uparrow u_\uparrow d_\downarrow\rangle - \frac{1}{\sqrt{6}}(|u_\uparrow u_\downarrow d_\uparrow\rangle + |u_\downarrow u_\uparrow d_\uparrow\rangle) \quad . \tag{5}$$

The wave function is correctly normalized but, since baryons have even parity (cf. Example 6.3), we must additionally demand that the wave function is symmetric with respect to the exchange of the quarks. Therefore we have to symmetrize the function.

An arbitrary wave function can be symmetrized applying the *symmetrization operator* (cf. Chap. 9). The total symmetrizing operator for a three-body system is defined as the sum of all permutations of the three particles

$$\hat{S}_{123} = \mathbb{1} + \hat{P}_{12} + \hat{P}_{13} + \hat{P}_{23} + \hat{P}_{13}\hat{P}_{12} + \hat{P}_{12}\hat{P}_{13} \quad , \tag{6}$$

where the $\hat{P}_{ij}$ denote the operators which exchange the ith with the jth particle. Hence the total symmetrical proton wave function is

$$|\mathrm{p}_\uparrow\rangle = N \cdot \frac{1}{\sqrt{6}} \cdot \{2|u_\uparrow u_\uparrow d_\downarrow\rangle - |u_\uparrow u_\downarrow d_\uparrow\rangle - |u_\downarrow u_\uparrow d_\uparrow\rangle + 2|u_\uparrow u_\uparrow d_\downarrow\rangle$$
$$- |u_\downarrow u_\uparrow d_\uparrow\rangle - |u_\uparrow u_\downarrow d_\uparrow\rangle + 2|d_\downarrow u_\uparrow u_\uparrow\rangle - |d_\uparrow u_\downarrow u_\uparrow\rangle - |d_\uparrow u_\uparrow u_\downarrow\rangle$$
$$+ 2|u_\uparrow d_\downarrow u_\uparrow\rangle - |u_\uparrow d_\uparrow u_\downarrow\rangle - |u_\downarrow d_\uparrow u_\uparrow\rangle + 2|d_\downarrow u_\uparrow u_\uparrow\rangle - |d_\uparrow u_\uparrow u_\downarrow\rangle$$
$$- |d_\uparrow u_\downarrow u_\uparrow\rangle + 2|u_\uparrow d_\downarrow u_\uparrow\rangle - |u_\downarrow d_\uparrow u_\uparrow\rangle - |u_\uparrow d_\uparrow u_\downarrow\rangle\}$$

$$|\mathrm{p}_\uparrow\rangle = \frac{1}{\sqrt{18}} \cdot \{2|u_\uparrow u_\uparrow d_\downarrow\rangle + 2|d_\downarrow u_\uparrow u_\uparrow\rangle + 2|u_\uparrow d_\downarrow u_\uparrow\rangle - |u_\uparrow u_\downarrow d_\uparrow\rangle$$
$$- |d_\uparrow u_\downarrow u_\uparrow\rangle - |u_\uparrow d_\uparrow u_\downarrow\rangle - |u_\downarrow u_\uparrow d_\uparrow\rangle - |d_\uparrow u_\uparrow u_\downarrow\rangle - |u_\downarrow d_\uparrow u_\uparrow\rangle\} \quad . \tag{7}$$

If we choose $\langle \mathrm{p}_\uparrow | \mathrm{p}_\uparrow \rangle = 1$, $N = 1/\sqrt{12}$ results by normalization [for normalization of the wave function see Sect. V of this exercise].

It is easy to prove that this wave function is totally symmetric with respect to the exchange of two arbitrary quarks.

V. The construction of the other baryonic states: Starting from the wave function of the proton we construct the other baryonic wave functions according to

Sect. I. The neutron wave function is hence

Exercise 8.15

$$
\begin{aligned}
|\mathrm{n}_\uparrow\rangle = N\cdot\hat{T}_-|\mathrm{p}_\uparrow\rangle = N\cdot\sum_{i=1}^{3}\hat{T}_-\cdot\frac{1}{\sqrt{18}}&\{2|u_\uparrow u_\uparrow d_\downarrow\rangle + 2|d_\downarrow u_\uparrow u_\uparrow\rangle\\
&+ 2|u_\uparrow d_\downarrow u_\uparrow\rangle - |u_\uparrow u_\downarrow d_\uparrow\rangle - |d_\uparrow u_\downarrow u_\uparrow\rangle - |u_\uparrow d_\uparrow u_\downarrow\rangle\\
&- |u_\downarrow u_\uparrow d_\uparrow\rangle - |d_\uparrow u_\uparrow u_\downarrow\rangle - |u_\downarrow d_\uparrow u_\uparrow\rangle\}\\
=\frac{N}{\sqrt{18}}\cdot&\{2|(\hat{T}_-(1)u)_\uparrow u_\uparrow d_\downarrow\rangle + 2|u_\uparrow(\hat{T}_-(2)u)_\uparrow d_\downarrow\rangle + 2|u_\uparrow u_\uparrow(\hat{T}_-(3)d)_\downarrow\rangle\\
&+ 2|(\hat{T}_-(1)d)_\downarrow u_\uparrow u_\uparrow\rangle + 2|d_\downarrow(\hat{T}_-(2)u)_\uparrow u_\uparrow\rangle + 2|d_\downarrow u_\uparrow(\hat{T}_-(3)u)_\uparrow\rangle\\
&+ 2|(\hat{T}_-(1)u)_\uparrow d_\downarrow u_\uparrow\rangle + 2|u_\uparrow(\hat{T}_-(2)d)_\downarrow u_\uparrow\rangle + 2|u_\uparrow d_\downarrow(\hat{T}_-(3)u)_\uparrow\rangle\\
&- |(\hat{T}_-(1)u)_\uparrow u_\downarrow d_\uparrow\rangle - |u_\uparrow(\hat{T}_-(2)u)_\downarrow d_\uparrow\rangle - |u_\uparrow u_\downarrow(\hat{T}_-(3)d)_\uparrow\rangle\\
&- |(\hat{T}_-(1)d)_\uparrow u_\downarrow u_\uparrow\rangle - |d_\uparrow(\hat{T}_-(2)u)_\downarrow u_\uparrow\rangle - |d_\uparrow u_\downarrow(\hat{T}_-(3)u)_\uparrow\rangle\\
&- |(\hat{T}_-(1)u)_\uparrow d_\uparrow u_\downarrow\rangle - |u_\uparrow(\hat{T}_-(2)d)_\uparrow u_\downarrow\rangle - |u_\uparrow d_\uparrow(\hat{T}_-(3)u)_\downarrow\rangle\\
&- |(\hat{T}_-(1)u)_\downarrow u_\uparrow d_\uparrow\rangle - |u_\downarrow(\hat{T}_-(2)u)_\uparrow d_\uparrow\rangle - |u_\downarrow u_\uparrow(\hat{T}_-(3)d)_\uparrow\rangle\\
&- |(\hat{T}_-(1)d)_\uparrow u_\uparrow u_\downarrow\rangle - |d_\uparrow(\hat{T}_-(2)u)_\uparrow u_\downarrow\rangle - |d_\uparrow u_\uparrow(\hat{T}_-(3)u)_\downarrow\rangle\\
&- |(\hat{T}_-(1)u)_\downarrow d_\uparrow u_\uparrow\rangle - u_\downarrow(\hat{T}_-(2)d)_\uparrow u_\uparrow\rangle - |u_\downarrow d_\uparrow(\hat{T}_-(3)u)_\uparrow\rangle\} \quad . \qquad (8)
\end{aligned}
$$

From the structure of the elementary triplet (see figure below) and the effect of the shift operators we may read off terms like $\hat{T}_-(i)q_i$:

$$
\hat{T}_-(i)|u_i\rangle = |d_i\rangle \quad , \quad \hat{T}_-(i)|d_i\rangle \equiv 0 \quad .
$$

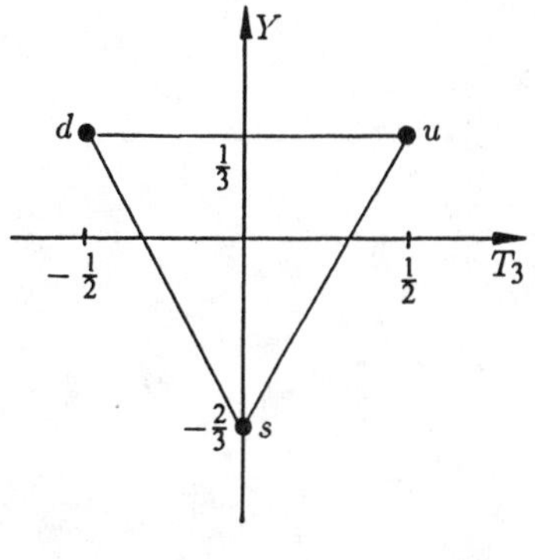

The elementary triplet

It follows that

$$
\begin{aligned}
|\mathrm{n}_\uparrow\rangle = \frac{N}{\sqrt{18}}\cdot&\{2|d_\uparrow u_\uparrow d_\downarrow\rangle + 2|u_\uparrow d_\uparrow d_\downarrow\rangle + 2|d_\downarrow d_\uparrow u_\uparrow\rangle + 2|d_\downarrow u_\uparrow d_\uparrow\rangle\\
&+ 2|d_\uparrow d_\downarrow u_\uparrow\rangle + 2|u_\uparrow d_\downarrow d_\uparrow\rangle - |d_\uparrow u_\downarrow d_\uparrow\rangle - |u_\uparrow d_\downarrow d_\uparrow\rangle - |d_\uparrow d_\downarrow u_\uparrow\rangle\\
&- |d_\uparrow u_\downarrow d_\uparrow\rangle - |d_\uparrow d_\uparrow u_\downarrow\rangle - |u_\uparrow d_\uparrow d_\downarrow\rangle - |d_\downarrow u_\uparrow d_\uparrow\rangle - |u_\downarrow d_\uparrow d_\uparrow\rangle\\
&- |d_\uparrow d_\uparrow u_\downarrow\rangle - |d_\uparrow u_\uparrow d_\downarrow\rangle - |d_\downarrow d_\uparrow u_\uparrow\rangle - |u_\downarrow d_\uparrow d_\uparrow\rangle\} \quad . \qquad (9)
\end{aligned}
$$

Altogether we have

$$
\begin{aligned}
|\mathrm{n}_\uparrow\rangle = \frac{N}{\sqrt{18}}\cdot&\{-2|d_\uparrow d_\uparrow u_\downarrow\rangle - 2|u_\downarrow d_\uparrow d_\uparrow\rangle - 2|d_\uparrow u_\downarrow d_\uparrow\rangle + |d_\uparrow d_\downarrow u_\uparrow\rangle\\
&+ |u_\uparrow d_\downarrow d_\uparrow\rangle + |d_\uparrow u_\uparrow d_\downarrow\rangle + |d_\downarrow d_\uparrow u_\uparrow\rangle + |u_\uparrow d_\uparrow d_\downarrow\rangle + |d_\downarrow u_\uparrow d_\uparrow\rangle\} \qquad (10)
\end{aligned}
$$

and the normalization factor N is obtained from the condition

$$
\langle\mathrm{n}_\uparrow|\mathrm{n}_\uparrow\rangle = \int \psi_{\mathrm{n}}^\dagger\psi_{\mathrm{n}}d^3\boldsymbol{r}_1 d^3\boldsymbol{r}_2 d^3\boldsymbol{r}_3 \overset{!}{=} 1 \quad . \qquad (11)
$$

Exercise 8.15

Since

$$\int q^{*}_{t'^1,t_3'^1,y'^1}\,\chi^{\dagger}_{\frac{1}{2}m'^1_s}(\boldsymbol{r}_1)\,q^{*}_{t'^2,t_3'^2,y'^2}\,\chi^{\dagger}_{\frac{1}{2}m'^2_s}(\boldsymbol{r}_2)\,q^{*}_{t'^3,t_3'^3,y'^3}\,\chi^{\dagger}_{\frac{1}{2}m'^3_s}(\boldsymbol{r}_3)\,q_{t^1,t_3^1,y^1}\,\chi_{\frac{1}{2}m^1_s}(\boldsymbol{r}_1)$$
$$\times\, q_{t^2,t_3^2,y^2}\,\chi_{\frac{1}{2}m^2_s}(\boldsymbol{r}_2)\,q_{t^3,t_3^3,y^3}\,\chi_{\frac{1}{2}m^3_s}(\boldsymbol{r}_3)\,d^3\boldsymbol{r}_1\,d^3\boldsymbol{r}_2\,d^3\boldsymbol{r}_3$$
$$= \prod_{i=1}^{3} \int q^{*}_{t'^i,t_3'^i,y'^i}\,\chi^{\dagger}_{\frac{1}{2}m'^i_s}(\boldsymbol{r}_i)\,q_{t^i,t_3^i,y^i}\,\chi_{\frac{1}{2}m^i_s}(\boldsymbol{r}_i)\,d^3\boldsymbol{r}_i$$
$$= \prod_{i=1}^{3} \delta_{y^i,y'^i}\,\delta_{t^i,t'^i}\,\delta_{t_3^i,t_3'^i}\,\delta_{m_s^i,m_s'^i} \quad , \tag{12}$$

then

$$1 = \langle \mathrm{n}_\uparrow | \mathrm{n}_\uparrow \rangle = N^2 \cdot \frac{1}{18}\{3\cdot 4 + 6\cdot 1\} = N^2 \quad ,$$

i.e. $N = \pm 1$.

If we choose $N = -1$, then the neutron wave function is of a form similar to the proton wave function:

$$|\mathrm{n}_\uparrow\rangle = \frac{1}{\sqrt{18}}\cdot\{2|d_\uparrow d_\uparrow u_\downarrow\rangle + 2|u_\downarrow d_\uparrow d_\uparrow\rangle + 2|d_\uparrow u_\downarrow d_\uparrow\rangle - |d_\uparrow d_\downarrow u_\uparrow\rangle - |u_\uparrow u_\downarrow d_\uparrow\rangle$$
$$- |d_\uparrow u_\uparrow d_\downarrow\rangle - |d_\downarrow d_\uparrow u_\uparrow\rangle - |u_\uparrow d_\uparrow d_\downarrow\rangle - |d_\downarrow u_\uparrow d_\uparrow\rangle\} \quad . \tag{13}$$

We proceed similarly with $|\Sigma^+_\uparrow\rangle$: By replacing the operator $\hat{T}_-$ by $\hat{U}_-$ in (8) and using the corresponding relations $\hat{U}_-|d\rangle = |s\rangle$, $\hat{U}_-|u\rangle = 0$), we immediately write

$$|\Sigma^+_\uparrow\rangle = U_-|\mathrm{p}_\uparrow\rangle = \frac{1}{\sqrt{18}}\{2|u_\uparrow u_\uparrow s_\downarrow\rangle + 2|s_\downarrow u_\uparrow u_\uparrow\rangle + 2|u_\uparrow s_\downarrow u_\uparrow\rangle$$
$$- |u_\uparrow u_\downarrow s_\uparrow\rangle - |s_\uparrow u_\downarrow u_\uparrow\rangle - |u_\uparrow s_\uparrow u_\downarrow\rangle$$
$$- |u_\downarrow u_\uparrow s_\uparrow\rangle - |s_\uparrow u_\uparrow u_\downarrow\rangle - |u_\downarrow s_\uparrow u_\uparrow\rangle\} \quad . \tag{14}$$

We choose the normalization factor in such a way that $\langle\Sigma^+|\Sigma^+\rangle = 1$.
Furthermore,

$$|\Sigma^0_\uparrow\rangle = \hat{T}_-|\Sigma^+_\uparrow\rangle = \tfrac{1}{6}\cdot\{2|d_\uparrow u_\uparrow s_\downarrow\rangle + 2|u_\uparrow d_\uparrow s_\downarrow\rangle + 2|s_\downarrow d_\uparrow u_\uparrow\rangle + 2|s_\downarrow u_\uparrow d_\uparrow\rangle$$
$$- |d_\uparrow u_\downarrow s_\uparrow\rangle - |u_\uparrow d_\downarrow s_\uparrow\rangle - |s_\uparrow d_\downarrow u_\uparrow\rangle - |s_\uparrow u_\downarrow d_\uparrow\rangle$$
$$- |d_\downarrow u_\uparrow s_\uparrow\rangle - |u_\downarrow d_\uparrow s_\uparrow\rangle - |s_\uparrow d_\uparrow u_\downarrow\rangle - |s_\uparrow u_\uparrow d_\downarrow\rangle$$
$$+ 2|d_\uparrow s_\downarrow u_\uparrow\rangle + 2|u_\uparrow s_\downarrow d_\uparrow\rangle - |d_\uparrow s_\uparrow u_\downarrow\rangle - |u_\uparrow s_\uparrow d_\downarrow\rangle$$
$$- |d_\downarrow s_\uparrow u_\uparrow\rangle - |u_\downarrow s_\uparrow d_\uparrow\rangle\} \tag{15}$$

since

$$\hat{T}_-|s\rangle = \hat{T}_-|d\rangle \equiv 0, \quad \hat{T}_-|u\rangle = |d\rangle \quad .$$

$$\begin{aligned}|\Sigma_\uparrow^-\rangle = \hat{T}_-|\Sigma_\uparrow^0\rangle = \frac{N}{6}\cdot\{&2|d_\uparrow d_\uparrow s_\downarrow\rangle + 2|d_\uparrow d_\uparrow s_\downarrow\rangle + 2|s_\downarrow d_\uparrow d_\uparrow\rangle + 2|s_\downarrow d_\uparrow d_\uparrow\rangle\\ &- |d_\uparrow d_\downarrow s_\uparrow\rangle - |d_\uparrow d_\downarrow s_\uparrow\rangle - |s_\uparrow d_\downarrow d_\uparrow\rangle - |s_\uparrow d_\downarrow d_\uparrow\rangle\\ &- |d_\downarrow d_\uparrow s_\uparrow\rangle - |d_\downarrow d_\uparrow s_\uparrow\rangle - |s_\uparrow d_\uparrow d_\downarrow\rangle - |s_\uparrow d_\uparrow d_\downarrow\rangle\\ &+ 2|d_\uparrow s_\downarrow d_\uparrow\rangle + 2|d_\uparrow s_\downarrow d_\uparrow\rangle - |d_\uparrow s_\uparrow d_\downarrow\rangle - |d_\uparrow s_\uparrow d_\downarrow\rangle\\ &- |d_\downarrow s_\uparrow d_\uparrow\rangle - |d_\downarrow s_\uparrow d_\uparrow\rangle\}\end{aligned}$$

$$\begin{aligned}|\Sigma_\uparrow^-\rangle = \frac{1}{\sqrt{18}}\cdot\{&2|d_\uparrow d_\uparrow s_\downarrow\rangle + 2|s_\downarrow d_\uparrow d_\uparrow\rangle + 2|d_\uparrow s_\downarrow d_\uparrow\rangle\\ &- |d_\uparrow d_\downarrow s_\uparrow\rangle - |s_\uparrow d_\downarrow d_\uparrow\rangle - |d_\uparrow s_\uparrow d_\downarrow\rangle - |d_\downarrow d_\uparrow s_\uparrow\rangle\\ &- |s_\uparrow d_\uparrow d_\downarrow\rangle - |d_\downarrow s_\uparrow d_\uparrow\rangle\} \quad ,\end{aligned} \tag{16}$$

with $N = 1/\sqrt{2}$ and

$$\hat{T}_-|u\rangle = |d\rangle, \quad \hat{T}_-|s\rangle = \hat{T}_-|d\rangle \equiv 0 \quad .$$

$$\begin{aligned}|\Xi_\uparrow^-\rangle = \hat{U}_-|\Sigma_\uparrow^-\rangle = \frac{N}{\sqrt{18}}\cdot\{&2|s_\uparrow d_\uparrow s_\downarrow\rangle + 2|d_\uparrow s_\uparrow s_\downarrow\rangle + 2|s_\downarrow s_\uparrow d_\uparrow\rangle\\ &+ 2|s_\downarrow d_\uparrow s_\uparrow\rangle - |s_\uparrow d_\downarrow s_\uparrow\rangle - |d_\uparrow s_\downarrow s_\uparrow\rangle - |s_\uparrow s_\downarrow d_\uparrow\rangle\\ &- |s_\uparrow d_\downarrow s_\uparrow\rangle - |s_\downarrow d_\uparrow s_\uparrow\rangle - |d_\downarrow s_\uparrow s_\uparrow\rangle - |s_\uparrow s_\uparrow d_\downarrow\rangle\\ &- |s_\uparrow d_\uparrow s_\downarrow\rangle + 2|s_\uparrow s_\downarrow d_\uparrow\rangle + 2|d_\uparrow s_\downarrow s_\uparrow\rangle - |s_\uparrow s_\uparrow d_\downarrow\rangle\\ &- |s_\uparrow s_\uparrow d_\downarrow\rangle - |d_\uparrow s_\uparrow s_\downarrow\rangle - |s_\downarrow s_\uparrow d_\uparrow\rangle - |d_\downarrow s_\uparrow s_\uparrow\rangle\}\end{aligned}$$

$$\begin{aligned}|\Xi_\uparrow^-\rangle = \frac{1}{\sqrt{18}}\cdot\{&2|s_\uparrow s_\uparrow d_\downarrow\rangle + 2|d_\downarrow s_\uparrow s_\uparrow\rangle + 2|s_\uparrow d_\downarrow s_\uparrow\rangle\\ &- |s_\uparrow s_\downarrow d_\uparrow\rangle - |d_\uparrow s_\downarrow s_\uparrow\rangle - |s_\downarrow d_\uparrow s_\uparrow\rangle - |s_\downarrow s_\uparrow d_\uparrow\rangle\\ &- |d_\uparrow s_\uparrow s_\downarrow\rangle - |s_\uparrow d_\uparrow s_\downarrow\rangle\} \quad ,\end{aligned} \tag{17}$$

setting $N = -1$ to produce formal similarity to the other wave functions. Similarly,

$$\begin{aligned}|\Xi_\uparrow^0\rangle = \hat{T}_+|\Xi_\uparrow^-\rangle = \frac{1}{\sqrt{18}}\cdot\{&2|s_\uparrow s_\uparrow u_\downarrow\rangle + 2|u_\downarrow s_\uparrow s_\uparrow\rangle\\ &+ 2|s_\uparrow u_\downarrow s_\uparrow\rangle - |s_\uparrow s_\downarrow u_\uparrow\rangle - |u_\uparrow s_\downarrow s_\uparrow\rangle - |s_\downarrow u_\uparrow s_\uparrow\rangle\\ &- |s_\downarrow s_\uparrow u_\uparrow\rangle - |u_\uparrow s_\uparrow s_\downarrow\rangle - |s_\uparrow u_\uparrow s_\downarrow\rangle\end{aligned} \tag{18}$$

because

$$T_+|u\rangle = \hat{T}_+|s\rangle \equiv 0, \quad \hat{T}_+|d\rangle = |u\rangle \quad .$$

Exercise 8.15

In order to construct $|\Lambda^0_\uparrow\rangle$, we first construct the second state at the centre of the $Y-T_3$ diagram as explained in Sect. I:

$$\begin{aligned}|\tilde{\Lambda}^0_\uparrow\rangle &= \hat{U}_+\hat{V}_-|\Sigma^+_\uparrow\rangle = \hat{U}_+|\Xi^0_\uparrow\rangle \\ &= \tfrac{1}{6}\{2|d_\uparrow s_\uparrow u_\downarrow\rangle + 2|s_\uparrow d_\uparrow u_\downarrow\rangle + 2|u_\downarrow d_\uparrow s_\uparrow\rangle + 2|u_\downarrow s_\uparrow d_\uparrow\rangle \\ &\quad - |d_\uparrow s_\downarrow u_\uparrow\rangle - |s_\uparrow d_\downarrow u_\uparrow\rangle - |u_\uparrow d_\downarrow s_\uparrow\rangle - |u_\uparrow s_\downarrow d_\uparrow\rangle \\ &\quad - |d_\downarrow s_\uparrow u_\uparrow\rangle - |s_\downarrow d_\uparrow u_\uparrow\rangle - |u_\uparrow d_\uparrow s_\downarrow\rangle - |u_\uparrow s_\uparrow d_\downarrow\rangle \\ &\quad + 2|d_\uparrow u_\downarrow s_\uparrow\rangle + 2|s_\uparrow u_\downarrow d_\uparrow\rangle - |d_\downarrow u_\uparrow s_\uparrow\rangle - |s_\downarrow u_\uparrow d_\uparrow\rangle \\ &\quad - |d_\uparrow u_\uparrow s_\downarrow\rangle - |s_\uparrow u_\uparrow d_\downarrow\rangle\} \quad ,\end{aligned} \tag{19}$$

since

$$\hat{U}_+|d\rangle = \hat{U}_+|u\rangle \equiv 0, \quad \hat{U}|s\rangle = |d\rangle \quad .$$

This wave function is similar to that of Σ^0; nevertheless, the two functions are linearly independent. To get the physical Λ^0 wave function we have to orthogonalize the function (19) with respect to Σ^0 (15).

According to the usual orthogonalization procedure of Schmidt the orthogonal wave function $|\Lambda^0_\uparrow\rangle$ is given by

$$|\Lambda^0_\uparrow\rangle = N\cdot(|\tilde{\Lambda}^0_\uparrow\rangle - \langle\Sigma^0_\uparrow|\tilde{\Lambda}^0_\uparrow\rangle|\Sigma^0_\uparrow\rangle) \quad .$$

Then

$$\langle\Sigma^0_\uparrow|\Lambda^0_\uparrow\rangle = N\cdot\Big(\langle\Sigma^0_\uparrow|\tilde{\Lambda}^0_\uparrow\rangle - \langle\Sigma^0_\uparrow|\tilde{\Lambda}^0_\uparrow\rangle\underbrace{\langle\Sigma^0_\uparrow|\Sigma^0_\uparrow\rangle}_{=1}\Big) = 0 \quad , \tag{20}$$

explicitly [because of (7)]

$$\langle\Sigma^0_\uparrow|\tilde{\Lambda}^0_\uparrow\rangle = \tfrac{1}{36}\{-2-2+1\}\cdot 6 = -\tfrac{18}{36} = -\tfrac{1}{2} \quad .$$

Thus we find

$$\begin{aligned}|\Lambda^0_\uparrow\rangle = \frac{N}{6}\cdot\{&2|d_\uparrow s_\uparrow u_\downarrow\rangle + 2|s_\uparrow d_\uparrow u_\downarrow\rangle + 2|u_\downarrow d_\uparrow s_\uparrow\rangle + 2|u_\downarrow s_\uparrow d_\uparrow\rangle \\ &+ 2|d_\uparrow u_\downarrow s_\uparrow\rangle + 2|s_\uparrow u_\downarrow d_\uparrow\rangle + \tfrac{1}{2}(-|d_\uparrow s_\uparrow u_\downarrow\rangle - |s_\uparrow d_\uparrow u_\downarrow\rangle \\ &- |u_\downarrow d_\uparrow s_\uparrow\rangle - |u_\downarrow s_\uparrow d_\uparrow\rangle - |d_\uparrow u_\downarrow s_\uparrow\rangle - |s_\uparrow u_\downarrow d_\uparrow\rangle - |d_\uparrow s_\downarrow u_\uparrow\rangle \\ &- |s_\uparrow d_\downarrow u_\uparrow\rangle - |u_\uparrow d_\downarrow s_\uparrow\rangle - |u_\uparrow s_\downarrow d_\uparrow\rangle - |d_\downarrow u_\uparrow s_\uparrow\rangle - |s_\downarrow u_\uparrow d_\uparrow\rangle \\ &+ \tfrac{1}{2}(2|d_\uparrow s_\downarrow u_\uparrow\rangle) + \tfrac{1}{2}(-|s_\uparrow d_\downarrow u_\uparrow\rangle) + \tfrac{1}{2}(-|u_\uparrow d_\downarrow s_\uparrow\rangle) + \tfrac{1}{2}(2|u_\uparrow s_\downarrow d_\uparrow\rangle) \\ &+ \tfrac{1}{2}(-|d_\downarrow u_\uparrow s_\uparrow\rangle) + \tfrac{1}{2}(2|s_\downarrow u_\uparrow d_\uparrow\rangle) - |d_\downarrow s_\uparrow u_\uparrow\rangle - |s_\downarrow d_\uparrow u_\uparrow\rangle \\ &- |u_\uparrow d_\uparrow s_\downarrow\rangle - |u_\uparrow s_\uparrow d_\downarrow\rangle - |d_\uparrow u_\uparrow s_\downarrow\rangle - |s_\uparrow u_\uparrow d_\downarrow\rangle + \tfrac{1}{2}(-|d_\downarrow s_\uparrow u_\uparrow\rangle) \\ &+ \tfrac{1}{2}(2|s_\downarrow d_\uparrow u_\uparrow\rangle) + \tfrac{1}{2}(2|u_\uparrow d_\uparrow s_\downarrow\rangle) + \tfrac{1}{2}(-|u_\uparrow s_\uparrow d_\downarrow\rangle) + \tfrac{1}{2}(2|d_\uparrow u_\uparrow s_\downarrow\rangle) \\ &+ \tfrac{1}{2}(-|s_\uparrow u_\uparrow d_\downarrow\rangle)\}\end{aligned}$$

Exercise 8.15

$$= \frac{N}{6} \cdot \{\tfrac{3}{2}(|d_\uparrow s_\uparrow u_\downarrow\rangle + |s_\uparrow d_\uparrow u_\downarrow\rangle + |u_\downarrow d_\uparrow s_\uparrow\rangle + |u_\downarrow s_\uparrow d_\uparrow\rangle + |d_\uparrow u_\downarrow s_\uparrow\rangle$$
$$+ |s_\uparrow u_\downarrow d_\uparrow\rangle) - \tfrac{3}{2}(|d_\downarrow s_\uparrow u_\uparrow\rangle + |s_\uparrow d_\downarrow u_\uparrow\rangle + |u_\uparrow d_\downarrow s_\uparrow\rangle$$
$$+ |u_\uparrow s_\uparrow d_\downarrow\rangle + |d_\downarrow u_\uparrow s_\uparrow\rangle + |s_\uparrow u_\uparrow d_\downarrow\rangle)\}$$

and, after normalization,

$$|\Lambda^0_\uparrow\rangle = \frac{1}{\sqrt{12}} \cdot \{|d_\uparrow s_\uparrow u_\downarrow\rangle + |s_\uparrow d_\uparrow u_\downarrow\rangle + |u_\downarrow d_\uparrow s_\uparrow\rangle + |u_\downarrow s_\uparrow d_\uparrow\rangle + |d_\uparrow u_\downarrow s_\uparrow\rangle$$
$$+ |s_\uparrow u_\downarrow d_\uparrow\rangle - |d_\downarrow s_\uparrow u_\uparrow\rangle - |s_\uparrow d_\downarrow u_\uparrow\rangle - |u_\uparrow d_\downarrow s_\uparrow\rangle$$
$$- |u_\uparrow s_\uparrow d_\downarrow\rangle - |d_\downarrow u_\uparrow s_\uparrow\rangle - |s_\uparrow u_\uparrow d_\downarrow\rangle\} \quad . \tag{21}$$

$|\Lambda^0_\uparrow\rangle$ is the orthogonal wave function constructed from $|\tilde{\Lambda}^0_\uparrow\rangle$, normalized to unity. It is also totally symmetric with respect to the exchange of two arbitrary quarks.

8.12 The Mass Formula in SU(6)

It is easy to generalize the Gell-Mann-Okubo mass formula in the framework of SU(6). To this end we assume the symmetry-breaking part of the Hamiltonian to be a scalar in spin space, e.g. $[\hat{H}_{\mathrm{ms}}, \boldsymbol{S}^2] = 0$. Thus $\hat{H}_{\mathrm{ms}}$ leads to a mass splitting of particles having different spin within a SU(6)-multiplet. The simplest assumption is then $\hat{H}_{\mathrm{ms}} \propto \hat{\boldsymbol{S}}^2$, with which we obtain the mass formula

$$M = a + bY + c[T(T+1) - \tfrac{1}{4}Y^2] + dS(S+1) \quad .$$

This relation is called the *Gürsey-Radicati mass formula.*[21] It is valid for the baryon octet and the decuplet (for all particles of the 56-plet). Fitting the constants for the masses of N, Λ, Σ, Σ^* and Ξ^* we obtain

$$a = 1066.6\ \mathrm{MeV}/c^2 \quad , \quad b = -196.1\ \mathrm{MeV}/c^2 \quad ,$$
$$c = 38.8\ \mathrm{MeV}/c^2 \quad , \quad d = 65.3\ \mathrm{MeV}/c^2 \quad .$$

With these constants the calculated masses of the other particles compare well with the experimental masses (see Table 8.3).

Table 8.3. Comparison of masses calculated by the Gürsey-Radicati mass formula with their experimental values

Name	Calcul. mass [MeV/c^2]	Exper. mass [MeV/c^2]
Ξ	1331	1318
Δ	1251.2	1232
Ω^-	1664.9	1672.4

[21] F. Gürsey, L.A. Radicati: Phys. Rev. Lett. **13**, 173 (1964).

8.13 Magnetic Moments in the Quark Model

Up to now we have assumed that the spatial wave function of the quarks' has orbital angular momentum $l = 0$, i.e. they are bound in an s state. This means, however, that the magnetic moment of ground state hadrons should be sum of the magnetic moments of the *constituent quarks*. Therefore we assign to each quark a magnetic moment μ_q (with $q = u, d, s$) with as yet undetermined value, and make an ansatz for the operator of the total magnetic moment:

$$\hat{\boldsymbol{\mu}} = \sum_i \mu_q(i)\hat{\boldsymbol{\sigma}}(i) \quad , \tag{8.86}$$

where $\frac{1}{2}\hat{\boldsymbol{\sigma}}(i)$ is the spin operator of the ith quark within the hadron. The measured (static) magnetic moment of a hadron $|h\rangle$ is the expectation value of the z component of $\boldsymbol{\mu}$, i.e.

$$\mu_h = \left\langle h \left| \sum_i \mu_1(i)\hat{\sigma}_z(i) \right| h \right\rangle \quad . \tag{8.87}$$

By definition $|h\rangle$ is always taken as the state of maximal spin projection. This relation can be directly evaluated by use of the hadron wave functions derived previously. Let us look at the magnetic moment of a proton described by the SU(6) wave function [see Exercise 8.13, Eq. (7)]

$$|p_\downarrow\rangle = \frac{1}{\sqrt{18}}|2u_\uparrow u_\uparrow d_\downarrow - u_\uparrow u_\downarrow d_\uparrow - u_\downarrow u_\uparrow d_\uparrow + \text{permut.}\rangle \quad .$$

Inserting this we get

$$\begin{aligned}
\mu_p = \frac{3}{18}\Bigg[& 4\left\langle u_\uparrow u_\uparrow d_\downarrow \left| \sum_i \mu_q(i)\hat{\sigma}_z(i) \right| u_\uparrow u_\uparrow d_\downarrow \right\rangle \\
& + \left\langle u_\uparrow u_\downarrow d_\uparrow \left| \sum_i \mu_q(i)\hat{\sigma}_z(i) \right| u_\uparrow u_\downarrow d_\uparrow \right\rangle + \left\langle u_\downarrow u_\uparrow d_\uparrow \left| \sum_i \mu_q(i)\hat{\sigma}_z(i) \right| u_\downarrow u_\uparrow d_\uparrow \right\rangle \Bigg] \\
= {} & \tfrac{3}{18}[4(\mu_u + \mu_u - \mu_d) + (\mu_u - \mu_u + \mu_d) + (-\mu_u + \mu_u + \mu_d)] \\
= {} & \tfrac{3}{18}(8\mu_u - 2\mu_d) = \tfrac{1}{3}(4\mu_u - \mu_d) \quad .
\end{aligned}$$

The magnetic moment of the neutron follows in a similar manner (interchange u and d), with

$$\mu_n = \tfrac{1}{3}(4\mu_d - \mu_u) \quad .$$

From these results μ_u and μ_d may be determined by comparison with the experimental values of μ_p and μ_n. We can derive an independent relation if we assume that the magnetic moments of quarks are proportional to their charge [see (8.14)] and if we neglect mass differences, i.e.

$$\mu_d = \frac{-1/3}{2/3}\mu_u = -\frac{1}{2}\mu_u \quad .$$

We thus get the prediction that

$$\left(\frac{\mu_n}{\mu_p}\right)_{\text{theor.}} = \frac{4\mu_d - \mu_u}{4\mu_u - \mu_d} = -\frac{2}{3} \quad ,$$

while $(\mu_n/\mu_p)_{\text{exp}} = -0.685$. The agreement is obviously quite good in view of the simplicity of the assumption and is one of the main successes of the nonrelativistic additive quark model.

Conversely, we can calculate the magnetic moments of quarks in units of the nuclear magneton from the experimental values; namely

$$\mu_u = \tfrac{1}{5}(\mu_n + 4\mu_p) = 1.852\mu_0 \quad ,$$

$$\mu_d = \tfrac{1}{5}(\mu_p + 4\mu_n) = -0.972\mu_0 \quad ,$$

$$\mu_s = \mu_\Lambda = -0.613\mu_0 \quad ,$$

$$\mu_0 = \frac{e\hbar}{2Mc} \tag{8.88a}$$

where M is the proton mass. The last value was obtained by use of the wave function of the Λ^0 baryon. With these values predictions can be made for the other baryons which have a sufficiently long lifetime. In Table 8.4 the results are compared to the available experimental data.

Table 8.4 Comparison of predicted and measured values of the magnetic moments of some baryons

Baryon	μ_h	Prediction $[\mu_0]$	Experimental $[\mu_0]$*
Σ^+	$\frac{4}{3}\mu_u - \frac{1}{3}\mu_s$	2.67	2.42 ± 0.05
Σ^0	$\frac{2}{3}(\mu_u + \mu_d) - \frac{1}{3}\mu_s$	0.76	–
Σ^-	$\frac{4}{3}\mu_d - \frac{1}{3}\mu_s$	– 1.09	– 1.160 ± 0.025
Ξ^0	$\frac{4}{3}\mu_s - \frac{1}{3}\mu_u$	– 1.44	– 1.250 ± 0.014
Ξ^-	$\frac{4}{3}\mu_s - \frac{1}{3}\mu_d$	– 0.49	– 0.6507 ± 0.0025

* From the *Review of Particle Properties* Phys. Rev. D45, Part 2 (1992)

Obviously the agreement is satisfactory. The remaining deviations may probably be attributed to admixtures of a d wave contribution to the orbital angular momentum. Furthermore, if we assume that quarks are elementary fermions having a g-factor of 2, the relation

$$\mu_q = \frac{Qe\hbar}{2M_q c} \tag{8.88b}$$

should hold for their magnetic moment. From this the "*masses*" of the quarks follow as

$$M_u = 338\ \text{MeV} \quad , \quad M_d = 322\ \text{MeV} \quad , \quad M_s = 510\ \text{MeV} \quad . \tag{8.89}$$

These masses are surprisingly small, especially since even in the largest accelerators no particles with masses below 100 GeV/c^2 that could be interpreted as quarks have been found. In fact, particles with fractional charges ($\frac{1}{3}$, $\frac{2}{3}$) have *never* been found. It has therefore become a major theoretical problem to explain why quarks exist only inside hadrons, the so-called problem of *quark confinement*. However, it is quite possible that sooner or later free quarks may be found and their mass could well be much larger than 100 GeV/c^2, a region not yet accessible to today's accelerators. Such a large mass for free quarks is not a contradiction of our statement that $M_q \approx 300$ MeV/c^2, since quarks inside hadrons are bound particles and therefore their mass is reduced by the binding energy. (More precisely, one should call the values M_q energy eigenvalues E_q, not masses.) The reader will note that the quark masses (8.89) roughly add up to the masses of hadrons according to their quark content. This fact, which represents a further success of the nonrelativistic quark model, can be utilized to derive mass formulae for hadrons which are not based on the flavour SU(3) symmetry group. E_q is commonly called the "*mass of the valence quark*".

8.14 Excited Meson and Baryon States

We have seen that in the framework of SU(6) based on the quark triplet with spin, the 0^- and 1^- mesons can be explained as members of a singlet and a [35] representation, because these representations can be decomposed into the SU(3) representations [1] (spin 0), [8] (spin 0), [1] (spin 1) and [8] (spin 1). Similarly for the baryons, the [56]-plet decomposes into an octet and a decuplet.

In the following we want to concern ourselves with the excited mesonic and baryonic states. There are two ways to do this:

(a) one imagines the excited states to consist of more than 2 or 3 quarks (for example $q\bar{q}q\bar{q}$ for mesons or $qqqq\bar{q}$ for baryons), so-called "*exotic*" *quark states*, or
(b) one keeps the structure of $q\bar{q}$ and qqq and interpretes the excited states as states with an orbital angular momentum $l \neq 0$.

To begin with, we want to examine the first possibility.

8.14.1 Combinations of More Than Three Quarks

The parity of a bound quark-antiquark pair is given by

$$\hat{P}\,|q\bar{q}\rangle = -(-1)^l\,|q\bar{q}\rangle$$

where l denotes the orbital angular momentum of the system. In the groundstate ($l = 0$) the parity of a quark-antiquark pair is always negative.

An example for this are the pseudoscalar mesons (Example 8.6) $J^P = 0^-$, and the vector mesons $J^P = 1^-$ (Example 8.9).

If we adhere to the concept that quarks are bound in s states ($l = 0$), the combination $q\bar{q}q\bar{q}$ allows for the construction of mesons with positive parity

($0^- \otimes 0^- \to 0^+$; $0^- \otimes 1^- \to 1^+$; $1^- \otimes 1 \to 0^+, 1^+, 2^+$). If each $q\bar{q}$ is an SU(6) representation, we obtain the following possible combinations:

$$[6] \otimes [\bar{6}] = [1] \oplus [35] \quad , \quad [1] \otimes [1] = [1] \quad , \quad [1] \otimes [35] = [35] \quad ,$$

$$[35] \otimes [35] = [1] \oplus [35] \oplus [35'] \oplus [189] \oplus [280] \oplus [\overline{280}] \oplus [405] \quad .$$

Again one can decompose these SU(6) multiplets with respect to their SU(3) content. We obtain multiplets of different spin (all of positive parity) which are shown in Table 8.5.

Table 8.5 Decomposition of SU(6) multiplets

		SU(6) multiplets decompose into				
		[1]	[35]	[189]	[280]	[405]
	[27]	–	–	0^+	1^+	$2^+, 1^+, 0^*$
SU(3)	$[\overline{10}]$	–	–	1^+	$1^+, 0^+$	1^+
multiplets	[10]	–	–	1^+	$2^+, 0^+$	1^+
	[8]	–	$0^+, 1^+$	$2^+, 1^+, 1^+, 0^+$	$2^+, 1^+, 1^+, 0^+$	$2^+, 1^+, 1^+, 0^+$
	[1]	0^+	1^+	$2^+, 0^+$	1^+	$2^+, 0^+$

Besides the 2^+ nonet (A_2, f, f', K*, cf. Example 8.10), out of this multitude of states only a few states of the 0^+ nonet (e.g. the σ-meson cf. Example 8.8) are known. There is no (or only very weak) experimental evidence for all the other states

Now let us consider the baryon combination $qqqq\bar{q}$. If we assume that pairs are always bound in s states ($l = 0$), all possible states have negative parity ($\frac{1}{2}^+ \otimes 0^- \to \frac{1}{2}^-$; $\frac{1}{2}^+ \otimes 1^- \to \frac{3}{2}^-, \frac{1}{2}^-$; $\frac{3}{2}^+ \otimes 0^- \to \frac{3}{2}^-$; $\frac{3}{2}^+ \otimes 1^- \to \frac{5}{2}^-, \frac{3}{2}^-, \frac{1}{2}^-$). In these case we couple a (qqq) [56]-plet to a ($q\bar{q}$) [35]-plet:

$$[35] \otimes [56] = [56] \oplus [70] \oplus [700] \oplus [1134] \quad .$$

These SU(6) multiplets have the SU(3) decompositions with negative parity and spin between $\frac{1}{2}$ and $\frac{5}{2}$ that are shown in Table 8.6.

Table 8.6 Decomposition of SU(6) multiplets into SU(3) multiplets with negative parity

SU(3) \ SU(6)	[56]	[70]	[700]	[1134]
[35]	–	–	$\frac{5}{2}^-, \frac{3}{2}^-$	$\frac{3}{2}^-, \frac{1}{2}^-$
[27]	–	–	$\frac{3}{2}^-, \frac{1}{2}^-$	$\frac{5}{2}^-, \frac{3}{2}^-, \frac{3}{2}^-, \frac{1}{2}^-, \frac{1}{2}$
$[\overline{10}]$	–	–	$\frac{1}{2}^-$	$\frac{3}{2}^-, \frac{1}{2}^-$
[10]	$\frac{3}{2}^-$	$\frac{1}{2}^-$	$\frac{5}{2}^-, \frac{3}{2}^-, \frac{1}{2}^-$	$\frac{5}{2}^-, \frac{3}{2}^-, \frac{3}{2}^-, \frac{1}{2}^-, \frac{1}{2}^-$
[8]	$\frac{1}{2}^-$	$\frac{1}{2}^-, \frac{3}{2}^-$	$\frac{1}{2}^-, \frac{3}{2}^-$	$\frac{5}{2}^-, \frac{3}{2}^-, \frac{3}{2}^-, \frac{3}{2}^-, \frac{1}{2}^-, \frac{1}{2}^-, \frac{1}{2}^-$
[1]	–	$\frac{1}{2}^-$		$\frac{3}{2}^-, \frac{1}{2}^-$

This decomposition yields far too many states, most of which have not been experimentally verified. Since this is the case for the mesons too, we have to conclude that this model is not suitable for a description of the physical situation and we turn to the alternative, (b).

8.15 Excited States with Orbital Angular Momentum

We first have a look at the excited states of mesons:

We have seen that in the framework of the multi-quark model, where we interpreted excited meson states as combinations of more than two valence quarks, we were only able to construct states with positive parity.

If we remain in the $q\bar{q}$ model and interpret the excited states as states with orbital angular momentum, we are led, using $P = -(-1)^l$, to the following table of excited meson multiplets, which again we classify by means of the quantum number J^{P} and l:

Meson multiplets with angular momentum

J^{P}	l
0^-	0
0^+	1
1^-	0, 2
1^+	1

J^{P}	l
2^-	2
2^+	1, 3
3^-	2, 4
3^+	3

For $l = 0$ we have here the pseudoscalar and vector mesons. Additionally, we get the scalar mesons, which we know from Example 8.8. We see that they are a result of the coupling of the total spin of the quarks $s = 1$ and the orbital angular momentum $l = 1$ to $J = 0$. The second possible coupling of $s = 1$ together with $l = 1$ to $J = 2$ yields the tensor mesons of Example 8.10. The third possibility, namely the coupling of $s = 1$ and $l = 1$ to $J = 1$ gives the pseudo vector mesons 1^+. All these particles have been observed in nature. One denotes multiplets with orbital angular momentum as supermultiplets. The intrinsic orbital angular momentum gives the parity of the particles in the multiplet.

For the baryons we stated that the $qqqq\bar{q}$ model produces too many states. Furthermore all of these states have negative parity, whereas states of positive parity are observed too. Hence we will try the qqq model with internal angular momentum. Therefore we assume an attractive central force acting between quark pairs, for which the ground state is completely symmetric (having $l = 0$). It belongs to the $[56]^+$ multiplet (positive parity) that has the SU(3) decomposition $[8]_2 \oplus [10]_4$ (the index means $2s + 1$). Of course the sequence of the excited states of a quark triplet depends on the particular quark-quark interaction, and to illustrate this we adopt a simple model: We place the quarks in a harmonic oscillator, solve the Schrödinger equation and classify the states in terms of SU(6) multiplets. The obtained energy level scheme is shown in Fig. 8.13.

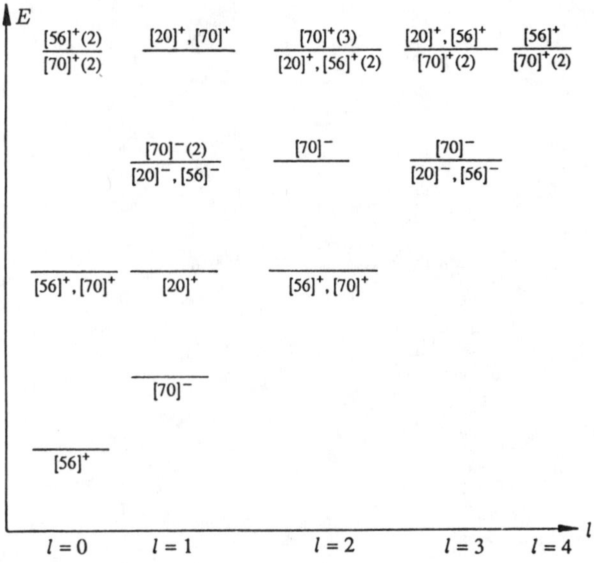

Fig. 8.13. Energy levels of quarks in a harmonic oscillator potential classified in terms of SU(6) multiplets

The ground state is the $[56]^+$ multiplet with $l = 0$, and so the first excited state is the $l = 1$ state with negative parity that belongs to the [70] representation of SU(6). The SU(3) decomposition of the [70]-plet is

$$[70] = [1]_2 \oplus [8]_2 \oplus [10]_2 \oplus [8]_4 \quad ,$$

and this means that we expect the following SU(3) multiplets with negative parity that are connected to the first excited qqq state:

$^4P_{\frac{5}{2},\frac{3}{2},\frac{1}{2}}$ (three octets) and

$^2P_{\frac{3}{2},\frac{1}{2}}$ (two singlets, two octets and two decuplets) .

Experiments show that there is at least one (in most cases several) verified state for each of these nine multiplets, and moreover they have found up to now exclusively supermultiplets of the kind $[56]^+$ (with even l) and $[70]^-$ (with odd l). This fact can only be explained by the use of the quark's internal quantum number "colour".

The generalized mass formula which has been derived in the framework of this ($l = 1$) supermultiplet model is very successful for the N*, Δ, Λ*, Σ*, Ξ* and Ω* resonances. The fine-structure can also be well explained in the framework of a spin-orbit interaction.[22] These and other experimentally verified predictions have led to the present conviction that the SU(6) quark model [SU(3) with spin] is the correct scheme to classify consistently mesons and baryons as well as their resonances.

[22] B.T. Feld: *Models for Elementary Particles* (Blaisdell, London 1969) p. 333.

9. Representations of the Permutation Group and Young Tableaux

9.1 The Permutation Group and Identical Particles

Up to now we have studied the unitary groups SU(N), especially those with $N = 2$, 3, 4 and 6 dimensions. Now we want to discuss some properties of the permutation group S_N, which is also called the symmetric group. The group S_N is important if we have to deal with several identical particles. In this section we will aquaint ourselves with the concept of Young diagrams, which in turn is useful for the construction of the basis functions of the unitary irreducible representations of SU(N).

The *transpositions* are those permutations which exchange two quantities (objects) with each other. For example, starting with N objects O_n ($n = 1, 2, \ldots, N$) in the order $O_1, O_2, O_3, \ldots, O_N$ a transposition of O_3 and O_1 will yield the objects in the order $O_3, O_2, O_1, \ldots, O_N$. They are of special importance, because an arbitrary permutation can always be represented as a product of transpositions. Let us imagine a number of boxes which contain certain objects, and which are labelled $1, 2, \ldots, N$. Then the exchange operator $\hat{P}_{ij}$ describes transpositions which exchange the objects in the boxes i and j. It is easily seen that

$$\hat{P}_{ij} = \hat{P}_{ji} \qquad \hat{P}^2_{ij} = 1 \quad . \tag{9.1}$$

A permutation is called *even* or *odd* depending on whether it can be produced by an even or an odd number of transpositions, respectively. For example, the above mentioned ordering of the objects

$$(O_3, O_2, O_1, O_4, \ldots, O_N) = \hat{P}_{13}(O_1, O_2, O_3, O_4, \ldots, O_N) \tag{9.2}$$

is odd while the permutation

$$(O_4, O_2, O_1, O_3, O_5, \ldots, O_N) = \hat{P}_{14}\hat{P}_{13}(O_1, O_2, O_3, O_4, \ldots, O_N) \tag{9.3}$$

is even.

In the following we define the permutations mathematically, especially in view of applications in quantum mechanics. For this purpose we imagine *non-identical objects* which are placed in the individual *identical* boxes; then we denote a permutation by

$$\begin{pmatrix} 1, 2. 3, \ldots, N \\ b_1, b_2, b_3, \ldots, b_N \end{pmatrix} \quad , \tag{9.4}$$

meaning that the object which initially was in the first box has now been moved into box b_1, the one in the second box is transferred to b_2, etc. If we have N identical particles labelled by $1, 2, \ldots, N$ and each of them is in a certain distinct quantum mechanical state, the identical particles correspond to the identical boxes and the non-identical states correspond to the non-identical objects. At first, this seems somewhat puzzling since the objects are inside the boxes, whereas the particles (which correspond to the boxes) occupy specific states (which correspond to the objects). However, eventually we get used to this confusing game!

There are several different ways to express the distribution of objects into boxes or that of particles into states. Let α, β, γ, δ be four nonidentical objects or four nonidentical one-particle states (wavefunctions). Then

$$\alpha(1)\beta(2)\gamma(3)\delta(4) \tag{9.5}$$

means that object α is situated in box 1, object β in box 2, object γ in box 3, and object δ in box 4. With our physical interpretation, expression (9.5) also means that particle 1 is in state (wavefunction) α, particle 2 is in state β, particle 3 is in state γ, and particle 4 in state δ. In the following we will always use the latter – quantum mechanical – interpretation of (9.5). If we apply the permutation $\left(\begin{smallmatrix}1&2&3&4\\2&3&4&1\end{smallmatrix}\right)$ to the four-particle wavefunction $\alpha(1)\beta(2)\gamma(3)\delta(4)$, we get the four-particle wavefunction $\alpha(2)\beta(3)\gamma(4)\delta(1)$. We write this as

$$\begin{pmatrix}1&2&3&4\\2&3&4&1\end{pmatrix}\alpha(1)\beta(2)\gamma(3)\delta(4) = \alpha(2)\beta(3)\gamma(4)\delta(1) \quad . \tag{9.6}$$

This notation can further be simplified if we agree to the procedure that the state of particle 1 is always noted first, followed by the state of particle 2 and finally by the state of particle 3. Using this notation (9.6) simply reads:

$$\begin{pmatrix}1&2&3&4\\2&3&4&1\end{pmatrix}\alpha\beta\gamma\delta = \delta\alpha\beta\gamma \quad . \tag{9.7}$$

We also remark that the permutation $\left(\begin{smallmatrix}1&2&3&4\\2&3&4&1\end{smallmatrix}\right)$ can the represented by a product of three transpositions, namely

$$\begin{aligned}\hat{P}_{14}\hat{P}_{13}\hat{P}_{12} &= \begin{pmatrix}1&2&3&4\\4&2&3&1\end{pmatrix}\begin{pmatrix}1&2&3&4\\3&2&1&4\end{pmatrix}\begin{pmatrix}1&2&3&4\\2&1&3&4\end{pmatrix}\\ &= \begin{pmatrix}1&2&3&4\\4&2&3&1\end{pmatrix}\begin{pmatrix}2&1&3&4\\2&3&1&4\end{pmatrix}\begin{pmatrix}1&2&3&4\\2&1&3&4\end{pmatrix}\\ &= \begin{pmatrix}1&2&3&4\\4&2&3&1\end{pmatrix}\begin{pmatrix}1&2&3&4\\2&3&1&4\end{pmatrix} = \begin{pmatrix}1&2&3&4\\2&3&4&1\end{pmatrix} \quad .\end{aligned} \tag{9.8}$$

Here we have to take care that we must always start with the right-most permutation. For example, the product $\left(\begin{smallmatrix}1&2&3&4\\3&2&1&4\end{smallmatrix}\right)\left(\begin{smallmatrix}1&2&3&4\\2&1&3&4\end{smallmatrix}\right)$ appearing in the first step means that 1 is changed into 2, followed by 2 changing to 2; 2 is changed into 1, followed by 1 changing to 3; 3 changing to 3, followed by 3 changing to 1; and

4 changing to 4, followed by 4 changing to 4. Hence

$$\begin{pmatrix}1&2&3&4\\3&2&1&4\end{pmatrix}\begin{pmatrix}1&2&3&4\\2&1&3&4\end{pmatrix}=\begin{pmatrix}1&2&3&4\\2&3&1&4\end{pmatrix}\quad, \tag{9.9}$$

which explains the last but one step in (9.8), etc. Sometimes the state is simply written as $\psi(1,2,3,4)$ instead of $\alpha(1)\beta(2)\gamma(3)\delta(4)$. This is more general in so far as $\psi(1,2,3,4)$ need not necessarily be a product of one-particle states $(\alpha\beta\gamma\delta)$. In this notation the correspondence to $\psi(2,3,4,1)$ is:

$$\psi(2,3,4,1) \Leftrightarrow \alpha(2)\beta(3)\gamma(4)\delta(1) \quad \text{or} \quad \delta\alpha\beta\gamma \quad . \tag{9.10}$$

As an example let us consider for simplicity states of 2 and 3 identical particles. We start with the permutation group S_2 and choose a two-particle state $\psi(1,2)$. The numbers 1 and 2 comprise all coordinates (position, spin, isospin) of the particles 1 and 2 respectively, i.e. $1 \Leftrightarrow x_1, y_1, z_1, s_1, \tau_1; 2 \Leftrightarrow x_2, y_2, z_2, s_2, \tau_2$. In order to examine the symmetry under exchange of particles, we note that in general $\psi(1,2)$ does not have a peculiar symmetry of this kind. But we can always build a symmetric (ψ_s) and an antisymmetric (ψ_a) state out of $\psi(1,2)$, namely

$$\psi_s = \psi(1,2) + \psi(2,1) \quad , \quad \psi_a = \psi(1,2) - \psi(2,1) \quad . \tag{9.11}$$

Here

$$\psi(2,1) = \hat{P}_{12}\psi(1,2) \tag{9.12}$$

denotes that two-particle wave function which is obtained from $\psi(1,2)$ by exchanging the coordinates 1 with the coordinates 2. Both, ψ_s as well as ψ_a, are eigenstates of the permutation operator $\hat{P}_{12}$. This is evident and it means that ψ_s as well as ψ_a, each one individually, represent a non-degenerate multiplet, i.e. *a singlet of the permutation group* S_2. We also say: Both ψ_s and ψ_a are separately basis functions of a one-dimensional representation (multiplet) of the permutation group.

Now, the two-particle states ψ_s and ψ_a can be graphically represented in the following way:

$\psi_s =$ 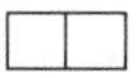 , $\psi_a =$ 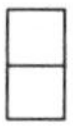 (9.13)

Here each particle is associated with a box; two boxes in one row describe a symmetric state; two boxes in one column describe an antisymmetric state. These representations of the symmetry of states are called *Young diagrams or Young tableaux*[1]. Obviously there is only one box for one-particle states; so there is only one Young diagram for one-particle states. This is rather obvious. The situation is more interesting for the three particle states, since in this case

[1] In the literature one sometimes distinguishes between Young diagrams and Young tableaux: the nomenclature says that a tableau is a diagram containing a positive number in each box.

there are three Young diagrams:

$\psi_s =$ 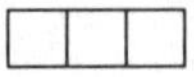 (fully symmetric state)

$\psi_a =$ 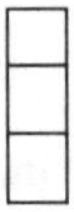 (fully antisymmetric state) ,

$\psi_m =$ 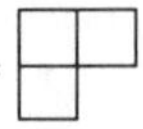 (mixed symmetric state) . (9.14)

The first two schemes and their interpretations are clear. The last diagram represents *all states* which are symmetric against exchange of two particles but antisymmetric with respect to the exchange of one of these (symmetrized) particles and the third particle. We can state this more precisely if we introduce the identity operator $\mathscr{I}$, the *symmetrization operator* (symmetrizer) $\hat{S}_{ij}$ and the antisymmetrization operator (antisymmetrizer) $\hat{A}_{ij}$. Then

$$\begin{aligned} &\hat{S}_{ij} = \mathscr{I} + \hat{P}_{ij} \quad , \quad \hat{A}_{ij} = \mathscr{I} - \hat{P}_{ij} \quad , \quad \text{and} \\ &\hat{S}_{ij}\hat{A}_{ij} = \mathscr{I} + \hat{P}_{ij} - \hat{P}_{ij} - \hat{P}_{ij}^2 = \mathscr{I} - \mathscr{I} = 0 \quad . \end{aligned} \tag{9.15}$$

If we start with the state $\psi(1,2,3)$, clearly the states of mixed symmetry ψ_m $(m = 1,2,3,4)$ read as

$$\begin{aligned} &\psi_1 = \hat{A}_{13}\hat{S}_{12}\psi(1,2,3) \quad , \quad \psi_2 = \hat{A}_{23}\hat{S}_{12}\psi(1,2,3) \quad , \\ &\psi_3 = \hat{A}_{23}\hat{S}_{13}\psi(1,2,3) \quad , \quad \psi_4 = \hat{A}_{12}\hat{S}_{13}\psi(1,2,3) \quad . \end{aligned} \tag{9.16}$$

As an example we denote ψ_1 explicitly, i.e.

$$\begin{aligned} \psi_1 &= \hat{A}_{13}[\psi(1,2,3) + \psi(2,1,3)] \\ &= [\psi(1,2,3) + \psi(2,1,3)] - [\psi(3,2,1) + \psi(2,3,1)] \\ &= \psi(1,2,3) + \psi(2,1,3) - \psi(3,2,1) - \psi(2,3,1) \quad . \end{aligned} \tag{9.17}$$

These four states (9.16) of mixed symmetry, together with the fully symmetric state and the fully antisymmetric state, form six linearly independent combinations of the six permutations of $\psi(1,2,3)$ including the identity. However, two additional states may be constructed by use of the symmetrizer $\hat{S}_{23}$. It is easy to check by explicit calculation, similar to that in (9.17), that these two additional states $\hat{A}_{12}\hat{S}_{23}\psi(1,2,3)$ and $\hat{A}_{13}\hat{S}_{23}\psi(1,2,3)$ are linearly dependent on the states (9.16). Hence we can forget them. One can also interpret the mixed symmetric Young diagram

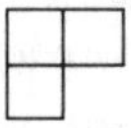 (9.18)

as a representation of the states

$$\psi_n = \hat{S}_{ij}\hat{A}_{ik}\psi(1,2,3) \quad , \tag{9.19}$$

which are constructed in such a way that the antisymmetrization is done first, followed by a symmetrization. The states obtained, however, are linearly dependent on the ones of (9.16). One should note that if a state is first symmetrized in i and j and thereafter antisymmetrized in j and k, then in general the symmetry concerning the exchange of i and j is lost. One says: *The symmetrizer (or antisymmetrizer) which is applied last controls the result.*

Let us summarize the six possible states which can be constructed by application of the permutation operators on $\psi(1,2,3)$. They are

$$\psi_{\mathrm{s}} \equiv \hat{S}_{123}\psi(1,2,3)$$
$$= [\mathcal{I} + \hat{P}_{12} + \hat{P}_{13} + \hat{P}_{23} + \hat{P}_{12}\hat{P}_{13} + \hat{P}_{13}\hat{P}_{12}]\psi(1,2,3) \equiv$$ 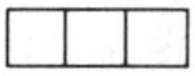

(9.20a)

$$\psi_{\mathrm{a}} = \hat{A}_{123}\psi(1,2,3) = \hat{A}_{12}\hat{A}_{13}\hat{A}_{23}\psi(1,2,3)$$
$$= [\mathcal{I} - \hat{P}_{12} - \hat{P}_{13} - \hat{P}_{23} + \hat{P}_{12}\hat{P}_{13} + \hat{P}_{13}\hat{P}_{12}]\psi(1,2,3) \equiv$$  (9.20b)

and

$$\psi_n = \hat{A}_{ik}\hat{S}_{ij}\psi(1,2,3) \equiv$$ 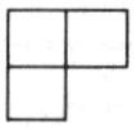 . (9.20c)

The four states (9.20c) are denoted in (9.16). In the final step of (9.20a) and (9.20b) we made use of the identities

$$\hat{P}_{12}\hat{P}_{13}\hat{P}_{23} = \hat{P}_{13} \quad , \quad \hat{P}_{13}\hat{P}_{23} = \hat{P}_{12}\hat{P}_{13} \quad , \quad \hat{P}_{13}\hat{P}_{12} = \hat{P}_{12}\hat{P}_{23} \tag{9.21}$$

and kept identical terms only once. These six states form the *basis of irreducible representations of the permutation group* S_3. Here all those states with well-defined symmetry properties are called *basis functions.* As there is just one fully symmetric and one fully antisymmetric state, the remaining four basis functions (9.19) have to be mixed-symmetric. One could imagine that these four states yield a four-dimensional irreducible representation of the permutation group S_3. This is incorrect, however; instead the four mixed-symmetric states are linear combinations of two two-dimensional irreducible representations. We will derive the dimension of the irreducible unitary representations of the permutation group S_N more clearly in the following section.

9.2 The Standard Form of Young Diagrams

One can easily convince oneself that in general both the totally symmetric state and the totally antisymmetric state form the basis function of a one-dimensional irreducible unitary representation of S_N. Namely, if one permutes the fully symmetric state $\psi_{\mathrm{s}}(1,2,\dots,N)$ by applying any permutation operator $\hat{P}_{ij}$ to ψ_{s}, the state ψ_{s} is transformed into itself. Application of group operators $\hat{P}_{ij}$ does not lead out of ψ_{s}. Thus ψ_{s} is an invariant subspace. A similar statement holds for the fully antisymmetric state $\psi_{\mathrm{a}}(1,2,\dots,n)$. All non-equivalent mixed-symmetric states, however, decompose into multi-dimensional irreducible representations of S_N. As we shall see, the dimensions of these representations can be obtained by use of the Young diagrams. This is a very powerful technique. To understand it we have first to explain the so-called *standard arrangement or standard form* of the Young diagrams.

Consider a Young tableau with various rows. Let q_i denote the number of boxes in the jth row. We now agree to the convention that $q_i \geq q_{i+1}$ should hold, which can always be achieved by rearranging the rows. In other words, the columns of a Young diagram should be arranged so that the number of boxes decreases with increasing row number. The total number of boxes gives the number of particles whose wave functions have a certain permutation symmetry expressed in the diagram. For example, a typical Young diagram is of the form

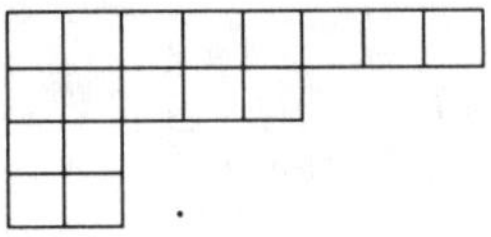 (9.22)

Here $q_1 = 8$, $q_2 = 5$, $q_3 = 2$, $q_4 = 2$, $q_5 = 0$. The symmetry of the 17-particle wave function represented by this Young diagram corresponds to permutations where the rows are symmetrized first and the columns are antisymmetrized afterwards, or vice versa. In particular scheme (9.22) describes all possible states of 17 particles that possess the symmetry expressed by the diagram.

It is often suitable to characterize the diagram not by the q_i's, but by the integer numbers p_i,

$$p_i = q_i - q_{i+1} \quad , \tag{9.23}$$

given by the differences of succeeding q_i's. In (9.22) this is

$$p_1 = 3 \quad , \quad p_2 = 3 \quad , \quad p_3 = 0 \quad , \quad p_4 = 2 \quad .$$

We recognize that p_1 is the number of columns with one box, p_2 the number of columns with 2 boxes, p_3 the number of columns with three boxes, and p_4 the number of columns with four boxes. Hence (9.22) can be described either by

$$q = (q_1, q_2, q_3, q_4) = (8, 5, 2, 2) \tag{9.24}$$

or by

$$p = (p_1, p_2, p_3, p_4) = (3, 3, 0, 2) \quad . \tag{9.25}$$

From now on we shall choose the *p characterization* of Young diagrams.

The *standard form* of a Young diagram is defined in the way that the boxes are labelled by positive integers obeying the rules that the numbers *in one row* of the diagram (read from left to right) *do not decrease* and those *in one column* (read from top to bottom) *increase* steadily. Each number represents a possible state of a single particle (one-particle state). If n states are available to a single particle, these states are ordered in some way and labelled by numbers between 1 and n. Thus the number j in a box has to be between 1 and n, i.e.

$$1 \leq j \leq n \quad . \tag{9.26}$$

We explain this by using the standard form of a three-particle Young diagram having four allowed one-particle states: the symmetric diagram is given by $\begin{array}{|c|c|c|}\hline &&\\ \hline\end{array}$ and the standard forms are

$$\begin{array}{llll}
\begin{array}{|c|c|c|}\hline 1&1&1\\ \hline\end{array} & \begin{array}{|c|c|c|}\hline 1&1&2\\ \hline\end{array} & \begin{array}{|c|c|c|}\hline 1&1&3\\ \hline\end{array} & \begin{array}{|c|c|c|}\hline 1&1&4\\ \hline\end{array} \\
\begin{array}{|c|c|c|}\hline 1&2&2\\ \hline\end{array} & \begin{array}{|c|c|c|}\hline 1&2&3\\ \hline\end{array} & \begin{array}{|c|c|c|}\hline 1&2&4\\ \hline\end{array} & \begin{array}{|c|c|c|}\hline 1&3&3\\ \hline\end{array} \\
\begin{array}{|c|c|c|}\hline 1&3&4\\ \hline\end{array} & \begin{array}{|c|c|c|}\hline 2&2&2\\ \hline\end{array} & \begin{array}{|c|c|c|}\hline 2&2&3\\ \hline\end{array} & \begin{array}{|c|c|c|}\hline 2&2&4\\ \hline\end{array} \\
\begin{array}{|c|c|c|}\hline 2&3&3\\ \hline\end{array} & \begin{array}{|c|c|c|}\hline 2&3&4\\ \hline\end{array} & \begin{array}{|c|c|c|}\hline 3&3&3\\ \hline\end{array} & \begin{array}{|c|c|c|}\hline 3&3&4\\ \hline\end{array} \\
\begin{array}{|c|c|c|}\hline 3&4&4\\ \hline\end{array} & \begin{array}{|c|c|c|}\hline 4&4&4\\ \hline\end{array} & . &
\end{array} \tag{9.27}$$

The antisymmetric diagram is given by $\begin{array}{|c|}\hline \\ \hline \\ \hline \\ \hline\end{array}$ and posesses the following standard forms:

$$\begin{array}{|c|}\hline 1\\ \hline 2\\ \hline 3\\ \hline\end{array}\quad \begin{array}{|c|}\hline 1\\ \hline 2\\ \hline 4\\ \hline\end{array}\quad \begin{array}{|c|}\hline 1\\ \hline 3\\ \hline 4\\ \hline\end{array}\quad \begin{array}{|c|}\hline 2\\ \hline 3\\ \hline 4\\ \hline\end{array}\quad . \tag{9.28}$$

Immediately one notices that a number (e.g. 2 or 3 or 4) can occur several times in one row, but not in a single column: all numbers in a column have to be different from each other. This means nothing more than that two particles can be symmetrized in the same state (there is a symmetrized two-particle state with the same one-particle function), but that two particles can never be antisymmetrized in the same one-particle state (these two-particle states are identically zero).

As another example we list the symmetric and antisymmetric standard arrangements for a two-particle Young tableau with three allowed single-particle states:

a) symmetric arrangements:

$$\begin{array}{|c|c|}\hline 1&1\\ \hline\end{array}\quad \begin{array}{|c|c|}\hline 1&2\\ \hline\end{array}\quad \begin{array}{|c|c|}\hline 1&3\\ \hline\end{array}\quad \begin{array}{|c|c|}\hline 2&2\\ \hline\end{array}\quad \begin{array}{|c|c|}\hline 2&3\\ \hline\end{array}\quad \begin{array}{|c|c|}\hline 3&3\\ \hline\end{array}\quad . \tag{9.29}$$

b) antisymmetric arrangements:

$$\begin{array}{|c|}\hline 1\\ \hline 2\\ \hline\end{array}\quad \begin{array}{|c|}\hline 1\\ \hline 3\\ \hline\end{array}\quad \begin{array}{|c|}\hline 2\\ \hline 3\\ \hline\end{array}\quad . \tag{9.30}$$

9.3 Standard Form and Dimension of Irreducible Representations of the Permutation Group S_N

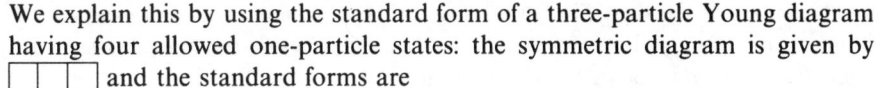

The dimensions of irreducible representations of the symmetric group S_N can be determined by use of the standard forms (arrangements). For this purpose all Young diagrams with N boxes have to be considered; each distinct diagram represents another permutation symmetry and thus another representation of the permutation group. The dimension of a representation is obtained by counting the standard arrangements for N available one-particle states. Here, each particle should be in a *different* state.

We explain this in the following example for three identical particles in different single-particle states: The standard arrangements of the Young tableaux read

$$\begin{array}{|c|c|c|}\hline 1&2&3\\\hline\end{array}\quad \begin{array}{|c|}\hline 1\\\hline 2\\\hline 3\\\hline\end{array}\quad \begin{array}{|c|c|}\hline 1&2\\\hline 3\\\cline{1-1}\end{array}\quad \begin{array}{|c|c|}\hline 1&3\\\hline 2\\\cline{1-1}\end{array}\quad . \tag{9.31}$$

Obviously there is only one possibility for standard arrangement of the diagram $\begin{array}{|c|c|c|}\hline \;&\;&\;\\\hline\end{array}$ with *different* one-particle states (1, 2, 3). The same holds for the diagram $\begin{array}{|c|}\hline \;\\\hline \;\\\hline \;\\\hline\end{array}$. This means that there is only one symmetric and one antisymmetric combination of three particles, which is rather plausible. It is also obvious that this statement holds for any number of particles. The non-standard arrangements correspond to the same states as the standard arrangements do after a corresponding symmetrization. Therefore, they must not be counted. We illustrate this with the non-standard arrangement $\begin{array}{|c|c|c|}\hline 1&3&2\\\hline\end{array}$, which yields the same total state as the standard arrangement $\begin{array}{|c|c|c|}\hline 1&2&3\\\hline\end{array}$, because we symmetrize with respect to all three particles.

As noted above there is only one symmetric and one antisymmetric standard arrangement. Each of these two arrangements represents a basis vector (the totally symmetric and the totally antisymmetric wavefunction, respectively) of a one-dimensional, irreducible unitary representation of S_3. Each of these representations is a non-degenerate multiplet, namely a singlet. Indeed, the generators of the permutation group, i.e. the operators $\hat{P}_{ij}$ of (9.1), reproduce the symmetric and antisymmetric state (many particle wavefunction) if applied to the symmetric or antisymmetric state, respectively. Since all members of a multiplet can be reached if the operators of the group are applied to any one state of the multiplet [see Chap. 3, (3.25)], the multiplet contains only a single member in these cases, and one deals therefore with singlet representations.

Consider now the mixed-symmetric Young diagram

$$\begin{array}{|c|c|}\hline \;&\;\\\hline \;\\\cline{1-1}\end{array}\quad , \tag{9.32}$$

for which there are two standard forms (arrangements), namely the tableaux

$$\begin{array}{|c|c|}\hline 1&2\\\hline 3\\\cline{1-1}\end{array}\quad \text{and}\quad \begin{array}{|c|c|}\hline 1&3\\\hline 2\\\cline{1-1}\end{array}\quad . \tag{9.33}$$

The two wavefunctions represented by these tableaux cannot be transformed into each other by permutations of the group S_3. Hence they must constitute basis vectors of two different irreducible representations of S_3 which are not singlets. Up to now we have considered four basis functions, namely those corresponding to the tableaux (9.31). However, we know that in total we obtain $3! = 6$ basis functions from $\psi(1, 2, 3)$ by permutation. Where have the other two basis functions gone? The remaining two functions combine with the two functions (9.33) to form two doublets of S_3 which are represented by the same scheme (9.32). The associated basis functions are given by the nonstandard tableaux.

$$\begin{array}{|c|c|}\hline 1 & 2 \\ \hline \multicolumn{1}{c|}{} & 3 \\ \cline{2-2}\end{array} \quad \text{and} \quad \begin{array}{|c|c|}\hline 1 & 3 \\ \hline \multicolumn{1}{c|}{} & 2 \\ \cline{2-2}\end{array} \quad . \tag{9.34}$$

They are different from the functions (9.33), e.g. $\begin{array}{|c|c|}\hline 2 & 1 \\ \hline 3 \\ \cline{1-1}\end{array}$ is antisymmetric against exchange of 2 and 3, which none of the functions (9.33) are. Application of the group operators

$$\hat{P}_{ij} = \{\hat{P}_{12}, \hat{P}_{13}, \hat{P}_{23}\} \tag{9.35}$$

transform the states (9.33) and (9.34) into each other, therefore they constitute two invariant subspaces, i.e. two multiplets (namely doublets). These two invariant subspaces are formed by the linear combinations

$$\lambda \begin{array}{|c|c|}\hline 1 & 2 \\ \hline 3 \\ \cline{1-1}\end{array} + \mu \begin{array}{|c|c|}\hline 2 & 1 \\ \hline 3 \\ \cline{1-1}\end{array} \tag{9.36}$$

and

$$\lambda' \begin{array}{|c|c|}\hline 1 & 3 \\ \hline 2 \\ \cline{1-1}\end{array} + \mu' \begin{array}{|c|c|}\hline 3 & 1 \\ \hline 2 \\ \cline{1-1}\end{array} \tag{9.37}$$

respectively.

Any permutations of the numbers 1, 2, 3 in the wave function represented by (9.36) again yields a linear combination of the basic vectors and the same is the case for (9.37). We will demonstrate this in a simple example:

$$\begin{aligned}\lambda \begin{array}{|c|c|}\hline 1 & 2 \\ \hline 3 \\ \cline{1-1}\end{array} + \mu \begin{array}{|c|c|}\hline 2 & 1 \\ \hline 3 \\ \cline{1-1}\end{array} &= \lambda(\psi(1,2,3) + \psi(2,1,3) - \psi(3,2,1) - \psi(2,3,1)) \\ &\quad + \mu(\psi(1,2,3) + \psi(2,1,3) - \psi(1,3,2) - \psi(3,1,2)) \quad .\end{aligned} \tag{9.38}$$

Acting with e.g. the permutation operator $P\left(\begin{smallmatrix}1&2&3\\2&3&1\end{smallmatrix}\right)$ on this, one gets

$$\begin{aligned}&P\begin{pmatrix}1&2&3\\2&3&1\end{pmatrix}\left(\lambda \begin{array}{|c|c|}\hline 1 & 2 \\ \hline 3 \\ \cline{1-1}\end{array} + \mu \begin{array}{|c|c|}\hline 2 & 1 \\ \hline 3 \\ \cline{1-1}\end{array}\right) \\ &\quad = \lambda(\psi(2,3,1) + \psi(3,2,1) - \psi(1,3,2) - \psi(3,1,2)) \\ &\qquad + \mu(\psi(2,3,1) + \psi(3,2,1) - \psi(2,1,3) - \psi(1,2,3)) \quad .\end{aligned} \tag{9.39}$$

The expression on the right-hand side can be written as a linear combination of the two basis vectors, i.e.

$$\begin{aligned}&P\begin{pmatrix}1&2&3\\2&3&1\end{pmatrix}\left(\lambda \begin{array}{|c|c|}\hline 1 & 2 \\ \hline 3 \\ \cline{1-1}\end{array} + \mu \begin{array}{|c|c|}\hline 2 & 1 \\ \hline 3 \\ \cline{1-1}\end{array}\right) \\ &\quad = -(\lambda + \mu) \begin{array}{|c|c|}\hline 1 & 2 \\ \hline 3 \\ \cline{1-1}\end{array} + \lambda \begin{array}{|c|c|}\hline 2 & 1 \\ \hline 3 \\ \cline{1-1}\end{array} \\ &\quad = -(\lambda + \mu)(\psi(1,2,3) + \psi(2,1,3) - \psi(3,2,1) - \psi(2,3,1)) \\ &\qquad + \lambda(\psi(1,2,3) + \psi(2,1,3) - \psi(1,3,2) - \psi(3,1,2)) \quad .\end{aligned} \tag{9.40}$$

Also, all other permutations of the numbers 1, 2, 3 with stay in this irreducible subspace. The same is valid for the subspace defined through (9.37). However, the basis vectors defined in such a way have the disadvantage that they are not orthonormalized, i.e., they form a non-orthogonal basis with reference to the scalar product of the wave functions. In Example 9.1, we will orthonormalize the basis vectors according to the orthonormalization procedure of Schmidt. In general it can be shown that if an irreducible representation of S_N is N dimensional, then there are exactly N such representations.[2]

In general a Young diagram with N boxes represents the basis functions of an irreducible representation of S_N. The basis functions are also often called *basis tensors* (instead of basis vectors), because they depend on N indices.

EXERCISE

9.1 Basis Functions of S_3

Problem. Determine the basis functions of the permutation group S_3 in terms of the single-particle wave functions $\alpha(i)\beta(j)\gamma(k)$.

Solution. According to the previous results the total symmetric wavefunction is given by

$$\psi_s = \boxed{1}\boxed{2}\boxed{3} = \hat{S}_{123}\alpha\beta\gamma \quad . \tag{1}$$

The total symmetry operator is the following combination of transpositions $\hat{P}_{ij}$ [see (9.20a)]:

$$\hat{S}_{123} = \mathbb{1} + \hat{P}_{12} + \hat{P}_{13} + \hat{P}_{23} + \hat{P}_{13}\hat{P}_{12} + \hat{P}_{12}\hat{P}_{13} \quad . \tag{2}$$

Normalizing ψ_s we get

$$\psi_s = 1/\sqrt{6}(\alpha\beta\gamma + \beta\alpha\gamma + \gamma\beta\alpha + \alpha\gamma\beta + \gamma\alpha\beta + \beta\gamma\alpha) \quad . \tag{3}$$

Similarly for the antisymmetric wavefunction:

$$\psi_a = \begin{array}{|c|}\hline 1 \\ \hline 2 \\ \hline 3 \\ \hline\end{array} = \hat{A}_{123}\alpha\beta\gamma \quad , \tag{4}$$

where

$$\hat{A}_{123} = \mathbb{1} - \hat{P}_{12} - \hat{P}_{13} - \hat{P}_{23} + \hat{P}_{13}\hat{P}_{12} + \hat{P}_{12}\hat{P}_{13} \tag{5}$$

---

[2] See, e.g., W.K. Tung: *Group Theory in Physics* (World Scientific, Singapore 1985), Chap. 5.

Exercise 9.1

is the total antisymmetrizer [see (9.20b)]. Therefore we find the normalized ψ_a to be

$$\psi_a = 1/\sqrt{6}(\alpha\beta\gamma - \beta\alpha\gamma - \gamma\beta\alpha - \alpha\gamma\beta + \gamma\alpha\beta + \beta\gamma\alpha) \quad . \tag{6}$$

However, the basis functions of the representations with mixed symmetry are not unambiguously determined, but it is possible to derive alternative basis functions with the help of unitary transformations. It is already known to us that the S_3 contains two doublets with mixed symmetry. To obtain suitable basis functions, we start by constructing four wavefunctions with mixed symmetry with the help of the symmetrizers $\hat{S}_{ij}$ and the antisymmetrizers $\hat{A}_{ij}$:

$$\psi_1 = \begin{array}{|c|c|}\hline 1 & 2 \\ \hline 3 & \multicolumn{1}{c}{} \\ \cline{1-1}\end{array} = \hat{A}_{13}\hat{S}_{12}\alpha\beta\gamma = \alpha\beta\gamma + \beta\alpha\gamma - \gamma\beta\alpha - \gamma\alpha\beta \quad , \tag{7a}$$

$$\psi_2 = \begin{array}{|c|c|}\hline 2 & 1 \\ \hline 3 & \multicolumn{1}{c}{} \\ \cline{1-1}\end{array} = \hat{A}_{23}\hat{S}_{12}\alpha\beta\gamma = \alpha\beta\gamma + \beta\alpha\gamma - \alpha\gamma\beta - \beta\gamma\alpha \quad , \tag{7b}$$

$$\psi_3 = \begin{array}{|c|c|}\hline 1 & 3 \\ \hline 2 & \multicolumn{1}{c}{} \\ \cline{1-1}\end{array} = \hat{A}_{12}\hat{S}_{13}\alpha\beta\gamma = \alpha\beta\gamma + \gamma\beta\alpha - \beta\alpha\gamma - \beta\gamma\alpha \quad , \tag{7c}$$

$$\psi_4 = \begin{array}{|c|c|}\hline 3 & 1 \\ \hline 2 & \multicolumn{1}{c}{} \\ \cline{1-1}\end{array} = \hat{A}_{23}\hat{S}_{13}\alpha\beta\gamma = \alpha\beta\gamma + \gamma\beta\alpha - \alpha\gamma\beta - \gamma\alpha\beta \quad . \tag{7d}$$

These four wavefunctions represent both mixed doublets, yet they are not orthogonal. To get an orthogonalized basis we form a linear combination of ψ_1 and ψ_2 that is orthogonal to ψ_1. Therefore we set

$$\psi'_2 = \psi_2 + \alpha\psi_1 \quad , \tag{8}$$

with the condition

$$\begin{aligned} 0 = \langle\psi_1|\psi'_2\rangle = \langle\alpha\beta\gamma + \beta\alpha\gamma - \gamma\beta\alpha - \gamma\alpha\beta||(1+\alpha)(\alpha\beta\gamma + \beta\alpha\gamma) - \alpha\gamma\beta \\ - \beta\gamma\alpha - \alpha(\gamma\beta\alpha + \gamma\alpha\beta)\rangle = 2(1+a) + 2a = 0 \quad . \end{aligned} \tag{9}$$

Here we used the orthonormality of the single particle states α, β and γ, which has the effect that nonidentical product wavefunctions are always orthogonal. With the help of (9) we find $a = -1/2$, i.e.

$$\psi'_2 = \psi_2 - \tfrac{1}{2}\psi_1 = \tfrac{1}{2}(\alpha\beta\gamma + \beta\alpha\gamma + \gamma\beta\alpha + \gamma\alpha\beta) - \alpha\gamma\beta - \beta\gamma\alpha \quad . \tag{10}$$

By applying the six permutations of S_3 to ψ_1 and ψ'_2 explicitly, we can express the result as linear combinations of ψ_1 and ψ'_2. Thus the two normalized vectors

$$\tfrac{1}{2}\psi_1 \quad , \quad \frac{1}{\sqrt{3}}\psi'_2 \tag{11}$$

are an orthonormal basis of a two-dimensional irreducible representation of S_3, i.e. they form a doublet.

We proceed to construct a basis function from ψ_3, ψ_1 and ψ'_2, required to be orthogonal to ψ_1 and ψ'_2. Repeating the Schmidt orthogonalization process we find

$$\psi'_3 = \psi_3 + \tfrac{1}{4}\psi_1 - \tfrac{1}{2}\psi'_2 = \alpha\beta\gamma - \beta\alpha\gamma + \tfrac{1}{2}(\gamma\beta\alpha + \alpha\gamma\beta - \gamma\alpha\beta - \beta\gamma\alpha) \quad . \tag{12}$$

Exercise 9.1

Finally, the fourth orthogonal function is given by the linear combination

$$\psi'_4 = \psi_4 - \tfrac{1}{4}(\psi_1 + 2\psi'_2 + 2\psi'_3) = \tfrac{3}{4}(\beta\gamma\alpha + \gamma\beta\alpha - \alpha\gamma\beta - \gamma\alpha\beta) \quad . \tag{13}$$

Here S_3 transforms the functions ψ'_3 and ψ'_4 into linear combinations of these same functions again. Hence we obtain the functions

$$\frac{1}{\sqrt{3}}\psi'_3 \quad , \quad \tfrac{2}{3}\psi'_4 \tag{14}$$

as an orthonormalized basis of a second doublet.

EXERCISE

9.2 Irreducible Representations of S_4

Problem. Discuss the irreducible representations of the permutation group S_4.

Solution. The irreducible representations of S_4 are given by the Young diagrams

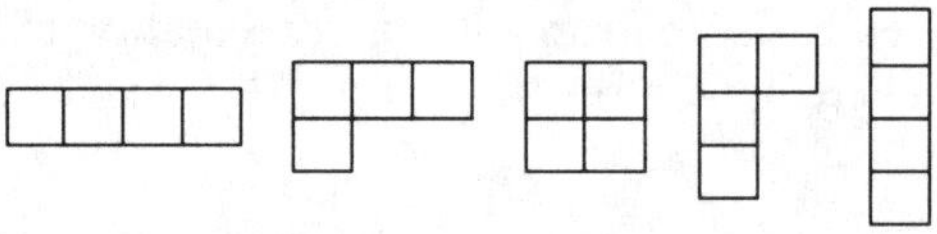

(1)

and the associated standard arrangements read:

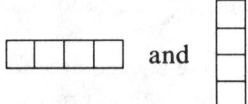

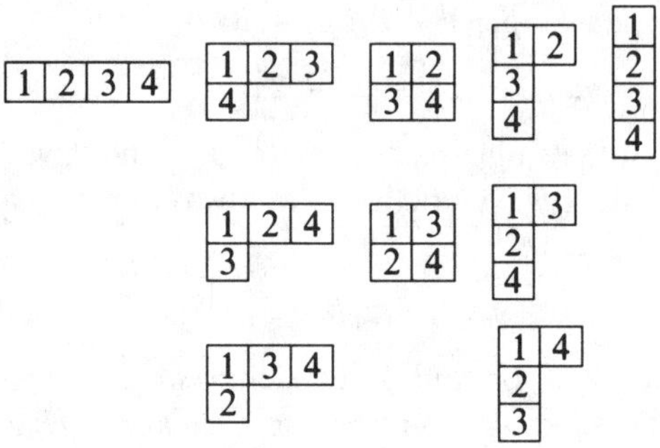

(2)

a) b) c) d) e)

The diagrams

 and 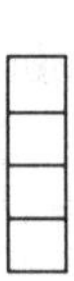

characterize the basis vectors to two one-dimensional representatives. Namely, each one of the group operators $\hat{P}_{ij} = \{\hat{P}_{12}, \hat{P}_{13}, \hat{P}_{14}, \hat{P}_{23}, \hat{P}_{24}, \hat{P}_{34}\}$ transforms

Exercise 9.2

every one of these vectors into itself. On the other hand the Young diagrams

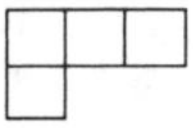 and 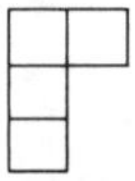 (3)

each represent three three-dimensional representations. To clarify this we consider the left diagram and its standard arrangements that are found in the second column of (2).

The wavefunction represented by the first tableau in (2b) is given by

$$\begin{aligned}
\psi_1 &= \hat{A}_{14}\hat{S}_{123}\alpha\beta\gamma\delta \\
&= [\alpha(1)\beta(2)\gamma(3) + \alpha(1)\beta(3)\gamma(2) + \alpha(2)\beta(1)\gamma(3) + \alpha(2)\beta(3)\gamma(1) \\
&\quad + \alpha(3)\beta(1)\gamma(2) + \alpha(3)\beta(2)\gamma(1)]\delta(4) - [\alpha(4)\beta(2)\gamma(3) + \alpha(4)\beta(3)\gamma(2) \\
&\quad + \alpha(2)\beta(4)\gamma(3) + \alpha(2)\beta(3)\gamma(4) + \alpha(3)\beta(4)\gamma(2) + \alpha(3)\beta(2)\gamma(4)]\delta(1) \\
&= \alpha\beta\gamma\delta + \alpha\gamma\beta\delta + \beta\alpha\gamma\delta + \gamma\alpha\beta\delta + \beta\gamma\alpha\delta + \gamma\beta\alpha\delta - \delta\beta\gamma\alpha - \delta\gamma\beta\alpha \\
&\quad - \delta\alpha\gamma\beta - \delta\alpha\beta\gamma - \delta\gamma\alpha\beta - \delta\beta\alpha\gamma \quad .
\end{aligned} \tag{4a}$$

Similarly, the other two tableaux in (2b) yield

$$\begin{aligned}
\psi_2 &= \hat{A}_{13}\hat{A}_{124}\alpha\beta\gamma\delta \\
&= \alpha\beta\gamma\delta + \alpha\delta\gamma\beta + \beta\alpha\gamma\delta + \delta\alpha\gamma\beta + \beta\delta\gamma\alpha + \delta\beta\gamma\alpha - \gamma\beta\alpha\delta - \gamma\delta\alpha\beta \\
&\quad - \gamma\alpha\beta\delta - \gamma\alpha\delta\beta - \gamma\delta\beta\alpha - \gamma\beta\delta\alpha
\end{aligned} \tag{4b}$$

and

$$\begin{aligned}
\psi_3 &= \hat{A}_{12}\hat{A}_{134}\alpha\beta\gamma\delta \\
&= \alpha\beta\gamma\delta + \alpha\beta\delta\gamma + \gamma\beta\alpha\delta + \delta\beta\alpha\gamma + \gamma\beta\delta\alpha + \delta\beta\gamma\alpha - \beta\alpha\gamma\delta - \alpha\beta\delta\gamma \\
&\quad - \beta\gamma\alpha\delta - \beta\delta\alpha\gamma - \beta\gamma\delta\alpha - \beta\delta\gamma\alpha \quad .
\end{aligned} \tag{4c}$$

These three wavefunctions serve as generating basis vectors for the three equivalent three-dimensional representations of the group S_4, which are denoted by the second Young diagram in (1). The complete representations are obtained by operating on ψ_1, ψ_2, and ψ_3 with all 24 operators of the group S_4. Since the representations are three-dimensional, only three of each 24 wavefunctions obtained in this way are linearly independent. One can take, e.g., the permutations of two numbers in the top row. For the first representation one has the three possible choices

$$\hat{P}_{12}\psi_1, \hat{P}_{13}\psi_1, \quad \text{and} \quad \hat{P}_{23}\psi_1 \quad .$$

The last of these is identical to ψ_1, because the symmetry in 2 and 3 is not destroyed by antisymmetrization between 1 and 4. Therefore only two choices remain; so that for the first representation one can take the wavefunctions

$$\psi_1, \hat{P}_{12}\psi_1, \hat{P}_{13}\psi_1 \quad . \tag{5a}$$

Exercise 9.2

For the other two representations one may take similarly:

$$\psi_2,\ \hat{P}_{12}\psi_2,\ \hat{P}_{14}\psi_2 \quad , \quad \text{and} \tag{5b}$$

$$\psi_3,\ \hat{P}_{13}\psi_3,\ \hat{P}_{14}\psi_3 \quad . \tag{5c}$$

These wavefunctions can be represented by non-standard Young tableaux. By explicitly writing out the permutations one easily checks that

$$\hat{P}_{12}\psi_1 = \hat{P}_{12}\hat{A}_{14}\hat{S}_{123}\alpha\beta\gamma\delta = \hat{A}_{24}\hat{S}_{123}\alpha\beta\gamma\delta \quad , \tag{6}$$

or in diagrammatic notation:[3]

$$\hat{P}_{12}\begin{array}{|c|c|c|}\hline 1&2&3\\ \hline 4& & \\ \hline\end{array} = \begin{array}{|c|c|c|}\hline 1&2&3\\ \hline &4& \\ \hline\end{array} \quad .$$

Similarly, one finds:

$$\hat{P}_{13}\begin{array}{|c|c|c|}\hline 1&2&3\\ \hline 4& & \\ \hline\end{array} = \begin{array}{|c|c|c|}\hline 1&2&3\\ \hline & &4\\ \hline\end{array} \quad .$$

The three three-dimensional representations (5a–c) can therefore be denoted by the Young tableaux with "shifted lower box".

$$\begin{array}{|c|c|c|}\hline 1&2&3\\ \hline 4& & \\ \hline\end{array} \quad \begin{array}{|c|c|c|}\hline 1&2&3\\ \hline &4& \\ \hline\end{array} \quad \begin{array}{|c|c|c|}\hline 1&2&3\\ \hline & &4\\ \hline\end{array} \tag{7a}$$

$$\begin{array}{|c|c|c|}\hline 1&2&4\\ \hline 3& & \\ \hline\end{array} \quad \begin{array}{|c|c|c|}\hline 1&2&4\\ \hline &3& \\ \hline\end{array} \quad \begin{array}{|c|c|c|}\hline 1&2&4\\ \hline & &3\\ \hline\end{array} \tag{7b}$$

$$\begin{array}{|c|c|c|}\hline 1&3&4\\ \hline 2& & \\ \hline\end{array} \quad \begin{array}{|c|c|c|}\hline 1&3&4\\ \hline &2& \\ \hline\end{array} \quad \begin{array}{|c|c|c|}\hline 1&3&4\\ \hline & &2\\ \hline\end{array} \tag{7c}$$

Of course, these wavefunctions are not yet orthogonal, but they can be orthogonalized as demonstrated in Problem 9.1 for the case of the group S_3.

Now let us discuss the two square tableaux (2c). The wavefunction represented by the first tableau reads explicitly:

$$\begin{aligned}\begin{array}{|c|c|}\hline 1&2\\ \hline 3&4\\ \hline\end{array} &= \hat{A}_{24}\hat{A}_{13}\hat{S}_{34}\hat{S}_{12}\alpha\beta\gamma\delta \\ &= [\alpha(1)\beta(2)+\beta(1)\alpha(2)][\gamma(3)\delta(4)+\gamma(4)\delta(3)] \\ &\quad - [\alpha(3)\beta(2)+\beta(3)\alpha(2)][\gamma(1)\delta(4)+\gamma(4)\delta(1)] \\ &\quad - [\alpha(1)\beta(4)+\beta(1)\alpha(4)][\gamma(3)\delta(2)+\gamma(2)\delta(3)] \\ &\quad + [\alpha(3)\beta(4)+\beta(3)\alpha(4)][\gamma(1)\delta(2)+\gamma(2)\delta(1)] \quad ,\end{aligned} \tag{8}$$

[3] Note that this rule of exchanging numbers in the tableau does not hold for numbers in different rows and columns, e.g.

$$\hat{P}_{23}\begin{array}{|c|c|}\hline 1&2\\ \hline 3& \\ \hline\end{array} = \hat{P}_{23}(\hat{A}_{13}\hat{S}_{12}\alpha\beta\gamma) \neq \hat{A}_{12}\hat{S}_{13}\alpha\beta\gamma = \begin{array}{|c|c|}\hline 1&3\\ \hline 2& \\ \hline\end{array}\ !$$

Exercise 9.2

whereas that of the second tableau is:

$$\begin{array}{|c|c|}\hline 1&3\\\hline 2&4\\\hline\end{array} = \hat{A}_{34}\hat{A}_{12}\hat{S}_{24}\hat{S}_{13}\alpha\beta\gamma\delta$$

$$\begin{aligned} &= [\alpha(1)\gamma(3)+\alpha(3)\gamma(1)][\beta(2)\delta(4)+\beta(4)\delta(2)] \\ &\quad -[\alpha(2)\gamma(3)+\alpha(3)\gamma(2)][\beta(1)\delta(4)+\beta(4)\delta(1)] \\ &\quad -[\alpha(1)\gamma(4)+\alpha(4)\gamma(1)][\beta(2)\delta(3)+\beta(3)\delta(2)] \\ &\quad +[\alpha(2)\gamma(4)+\alpha(4)\gamma(2)][\beta(1)\delta(3)+\beta(3)\delta(1)] \quad . \end{aligned} \tag{9}$$

Again, the complete representations are obtained by applying all 24 group operators on the generating vectors (8) and (9). Since the representations are two-dimensional, only one other wavefunction obtained in this way is linearly independent. As before, this can be obtained by applying a permutation within a row of the Young tableau. In the first case, Eq. (8), one obtains the two possible wavefunctions

$$\hat{P}_{12}\begin{array}{|c|c|}\hline 1&2\\\hline 3&4\\\hline\end{array} = \begin{array}{|c|c|}\hline 2&1\\\hline 3&4\\\hline\end{array} \quad \text{and} \quad \hat{P}_{34}\begin{array}{|c|c|}\hline 1&2\\\hline 3&4\\\hline\end{array} = \begin{array}{|c|c|}\hline 1&2\\\hline 4&3\\\hline\end{array} \quad . \tag{10}$$

These are identical states due to the fact that the wavefunction remains unchanged if we interchange entire rows or columns of a tableau. For the second representation (9) one has the two wavefunctions

$$\hat{P}_{13}\begin{array}{|c|c|}\hline 1&3\\\hline 2&4\\\hline\end{array} = \begin{array}{|c|c|}\hline 3&1\\\hline 2&4\\\hline\end{array} \quad \text{and} \quad \hat{P}_{24}\begin{array}{|c|c|}\hline 1&3\\\hline 2&4\\\hline\end{array} = \begin{array}{|c|c|}\hline 1&3\\\hline 4&2\\\hline\end{array} \tag{11}$$

which again represent identical states. From the four states

$$\begin{array}{|c|c|}\hline 1&2\\\hline 3&4\\\hline\end{array} \,,\quad \begin{array}{|c|c|}\hline 1&2\\\hline 4&3\\\hline\end{array} \quad \text{and} \quad \begin{array}{|c|c|}\hline 1&3\\\hline 2&4\\\hline\end{array} \,,\quad \begin{array}{|c|c|}\hline 1&3\\\hline 4&2\\\hline\end{array} \tag{12}$$

one can construct two doublets of orthogonal wavefunctions which transform only among themselves by application of $\hat{P}_{ij}$. In Exercise 9.1 we have shown this explicitly for the simple basis functions of S_3, but for now we omit this.

We conclude: the standard diagrams of the permutation group S_4 lead to a total number of basis vectors of

$$1 + 3^2 + 2^2 + 3^2 + 1 = 24 \quad . \tag{13}$$

Indeed, S_4 has exactly $4! = 24$ elements, and therefore, starting with one given vector $\psi(1,2,3,4)$, exactly 24 vectors can be formed by permutation. These are reducible and can be decomposed into the singlets, doublets and triplets given in (2) and (13). By decomposition of the reducible 24 basis vectors into the irreducible representations just mentioned [see (13)] the symmetry properties of S_4 can be exposed even more clearly.

Next we shall discuss the concept of the *conjugate Young diagram*. There are two possible ways to define it: one is for the symmetric group and the other is for

SU(N). Here we devote ourselves only to the conjugation of the symmetric group; the conjugation of SU(N) is discussed later (Chap. 12). Let us take a certain Young diagram with n boxes, which represents n identical particles (each one in a distinct state). Then for the symmetric group *the conjugate diagram is given by changing every column to a row and every row to a column.* We explain this for the following four-particle states:

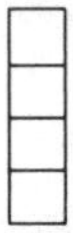 is conjugate to 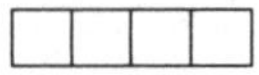 (9.41)

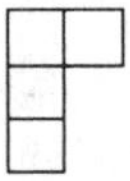 is conjugate to 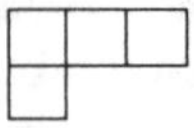 and (9.42)

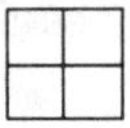 is self-conjugate . (9.43)

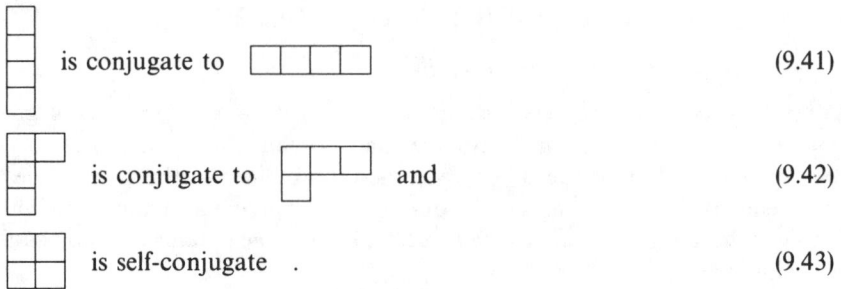

Conjugate diagrams belong to different non-equivalent representations of the symmetric group with the same dimension.

We stress the importance of the dimension of irreducible representations once again: all basis states of the same irreducible representation of a group (in our case of S_N) have equal energy if the Hamiltonian of the system is invariant with respect to this group (in our case it is invariant under permutations of identical particles). Thus the *dimension of an irreducible representation tells us the degeneracy of an energy level,* of course omitting accidental degeneracies. Therefore we also say: the dimension of an irreducible representation of a symmetry group is equal to the *essential* degeneracy.

Now let us assume that the Hamiltonian of a system is invariant under two or more different groups. The transformations of all these groups together form a larger group that can be equal to the direct product of the original groups, or not, depending on whether all group transformations of each single group commute with all transformations of all the other groups, or not. In any case the degenerate multiplets are basis vectors of the irreducible representations of the larger group. If the larger group is a direct product, the multiplicities are simply the products of the multiplicities of the groups that form the direct product.

All particles observed in nature are fermions or bosons. State vectors of N identical bosons belong to the fully symmetric representation of S_N; state vectors of fermions, however, belong to the fully antisymmetric representation of S_N. We may also express this important fact in the following way: All particles which have been found in nature up to now belong to the one-dimensional representation of the symmetric group; they are singlets with respect to the symmetric group.

The reader may wonder why we study the higher multiplets of S_N at all? One reason is that particles could be discovered in the future that are neither fermions nor bosons. There is, however, a much more important reason: Despite the fact that states of several identical particles are singlets under any permutation of all coordinates of the identical particles (more precisely, singlets with respect to the *complete* exchange of two particles), there are states of higher multiplicities (thus belonging to higher multiplets of S_N) if only permutations of some (not all) coordinates of the particles are considered. This is best explained by an example: Let us consider three identical spin-1 particles. The quantum state of these particles is symmetric against exchange of any two particles as a whole, i.e. the simultaneous exchange of the spin and space coordinates of two particles. But if one considers the exchange of spin or space coordinates alone, the state can very well possess a mixed symmetry under such a permutation. Then this mixed symmetry is of such a type that the state is overall symmetric under exchange of all coordinates (that means space coordinates *and* spin).

9.4 The Connection Between SU(2) and S_2

To begin with, we consider the connection between the irreducible representations of the permutation groups S_N and the unitary groups $SU(N)$, for SU(2). The fundamental representation of SU(2) is spanned by the basis vectors

$$\psi_1 = \begin{pmatrix} 1 \\ 0 \end{pmatrix} \quad , \quad \psi_2 = \begin{pmatrix} 0 \\ 1 \end{pmatrix} \quad . \tag{9.44}$$

These vectors may, for instance, represent the two states of a particle with spin $\frac{1}{2}$. In the following, these basis vectors are represented by the Young tableaux consisting of one box, i.e.

$$\psi_1 = \boxed{1} \quad , \quad \psi_2 = \boxed{2} \quad . \tag{9.45}$$

If we erase the contents of the box, the Young diagram $\square$ symbolizes both members of the doublet.

We know from (9.13) that both irreducible representations of the permutation group S_2 are represented by the Young diagrams

$$\begin{array}{|c|c|}\hline & \\ \hline \end{array} \ \text{symmetric representation} \quad , \quad \begin{array}{|c|}\hline \\ \hline \\ \hline \end{array} \ \text{antisymmetric representation} \tag{9.46}$$

By numbering the boxes we perceive that these diagrams also symbolize irreducible representations of the SU(2). We start with the symmetrical two-particle states. If both particles are in the state ψ_1, we have the tableau $\begin{array}{|c|c|}\hline 1 & 1 \\ \hline \end{array}$; if both are in state ψ_2, we have $\begin{array}{|c|c|}\hline 2 & 2 \\ \hline \end{array}$. There is a possible third state

$$\psi = \psi_1(1)\psi_2(2) + \psi_1(2)\psi_2(1) \quad , \tag{9.47}$$

symbolized by the tableau $\begin{array}{|c|c|}\hline 1 & 2 \\ \hline \end{array}$. Because of the symmetry, the tableau $\begin{array}{|c|c|}\hline 2 & 1 \\ \hline \end{array}$ describes the same state. Therefore we can restrict ourselves to the enumeration of the standard Young tableaux. From this argument we learn that the Young diagram $\begin{array}{|c|c|}\hline & \\ \hline \end{array}$ represents three different standard configurations, i.e. the corresponding multiplets have three dimensions.

If we proceed with the antisymmetric representation of S_2, however, there is only one possible way to construct an antisymmetric two-particle state:

$$\psi = \psi_1(1)\psi_2(2) - \psi_1(2)\psi_2(1) \quad . \tag{9.48}$$

This state is represented by the tableau

$$\begin{array}{|c|}\hline 1 \\ \hline 2 \\ \hline \end{array} \ . \tag{9.49}$$

Thus the related irreducible representation is a singlet. The other Young tableaux that in principle can be constructed,

$$\begin{array}{|c|}\hline 1 \\ \hline 1 \\ \hline \end{array}\ ,\quad \begin{array}{|c|}\hline 2 \\ \hline 2 \\ \hline \end{array}\ ,\quad \begin{array}{|c|}\hline 2 \\ \hline 1 \\ \hline \end{array}\ , \tag{9.50}$$

do not form standard configurations, because the numbers in the boxes do not increase from top to bottom of a column. The first two states in (9.37) vanish. The last one is identical to (9.36). We conclude that we also obtain the correct multiplicity by restriction to standard tableaux for the antisymmetric representation.

EXERCISE

9.3 Multiplets of a System of Three Spin-$\frac{1}{2}$ Particles

Problem. Determine the multiplets of a system of three spin-$\frac{1}{2}$-particles.

Solution. Three diagrams with three boxes are possible.

$$\underset{\text{a)}}{\begin{array}{|c|c|c|}\hline \ & \ & \ \\ \hline \end{array}}\ ,\quad \underset{\text{b)}}{\begin{array}{|c|c|}\hline \ & \ \\ \hline \ & \multicolumn{1}{c}{} \\ \cline{1-1} \end{array}}\quad \text{and}\quad \underset{\text{c)}}{\begin{array}{|c|}\hline \ \\ \hline \ \\ \hline \ \\ \hline \end{array}}\ . \tag{1}$$

In the case of SU(2), (1c) can be omitted, because no totally antisymmetrical state with three particles can be constructed if there are only two states (this is the Pauli principle). Diagram (1a) allows four standard configurations,

$$\begin{array}{|c|c|c|}\hline 1 & 1 & 1 \\ \hline \end{array}\ ,\quad \begin{array}{|c|c|c|}\hline 1 & 1 & 2 \\ \hline \end{array}\ ,\quad \begin{array}{|c|c|c|}\hline 1 & 2 & 2 \\ \hline \end{array}\ ,\quad \begin{array}{|c|c|c|}\hline 2 & 2 & 2 \\ \hline \end{array}\ , \tag{2}$$

i.e. the symmetrical states form a quartet describing a multiplet of spin $\frac{3}{2}$. Diagram (1b) generates only two standard tableaux,

$$\begin{array}{|c|c|}\hline 1 & 1 \\ \hline 2 & \multicolumn{1}{c}{} \\ \cline{1-1} \end{array}\quad \text{and}\quad \begin{array}{|c|c|}\hline 1 & 2 \\ \hline 2 & \multicolumn{1}{c}{} \\ \cline{1-1} \end{array}\ . \tag{3}$$

The states with mixed symmetry hence form a doublet, describing states of total spin $\frac{1}{2}$.

In order to clarify the connection between SU(2) and S_2, we normalize the wavefunctions represented by the Young tableaux. The singlet representation reads

$$\begin{array}{|c|}\hline 1\\\hline 2\\\hline\end{array} \mathrel{\hat{=}} \frac{1}{\sqrt{2}}(\psi(1)\psi(2) - \psi(2)\psi(1))$$

and for the triplet representation we find

$$\begin{array}{|c|c|}\hline 1 & 1\\\hline\end{array} \mathrel{\hat{=}} \psi(1)\psi(1)$$

$$\begin{array}{|c|c|}\hline 2 & 2\\\hline\end{array} \mathrel{\hat{=}} \psi(2)\psi(2)$$

$$\begin{array}{|c|c|}\hline 1 & 2\\\hline\end{array} \mathrel{\hat{=}} \frac{1}{\sqrt{2}}(\psi(1)\psi(2) + \psi(2)\psi(1)) \quad .$$

These are exactly the possible states which result from the coupling of two spin-$\frac{1}{2}$-particles to spin 0 or spin 1, respectively. The spin-0 singlet reads

$$|JM\rangle = |00\rangle = \frac{1}{\sqrt{2}}(\uparrow\downarrow - \downarrow\uparrow),$$

the spin-1 triplet is given by

$$|1\,1\rangle = \uparrow\uparrow$$

$$|1\,-1\rangle = \downarrow\downarrow$$

$$|1\,0\rangle = \frac{1}{\sqrt{2}}(\uparrow\downarrow + \downarrow\uparrow) \quad .$$

We perceive that in general a column cannot have more than two boxes if only two different states are allowed [i.e., in the case of SU(2)]; otherwise antisymmetrization would automatically yield zero. A column with two boxes necessarily has the numbering $\begin{array}{|c|}\hline 1\\\hline 2\\\hline\end{array}$. There is no other choice. Therefore we can omit all columns with two boxes if we determine the dimension of a SU(2) multiplet. Hence the Young diagrams

$$\begin{array}{|c|}\hline \\\hline\end{array}\ , \quad \begin{array}{|c|c|}\hline & \\\hline & \multicolumn{1}{|c}{}\\\cline{1-1}\end{array}\ , \quad \begin{array}{|c|c|c|}\hline & & \\\hline & & \multicolumn{1}{|c}{}\\\cline{1-2}\end{array}\ , \quad \text{etc.} \tag{9.51}$$

all represent a doublet. From the group-theoretical point of view it is the same irreducible representation of the SU(2), for a one-, two-, three- or many-particle system, respectively.

Consequently we obtain all irreducible representations of the SU(2) by constructing all possible Young diagrams with only one horizontal row; in each case the dimension is given by the number of possible distinct standard configurations with box indices 1 and 2. In a particular sense the singlet presents an exception, since if we omit the column with two boxes from the diagram $\begin{array}{|c|}\hline \\\hline \\\hline\end{array}$ we get a diagram with no box at all. To keep track of the "no-box" diagram we symbolize the singlet by ①. Thus we obtain successively:

①	singlet	
□	doublet	[1] , [2]
□□	triplet	[1|1] , [1|2] , [2|2]
□□□	quartet	[1|1|1] , [1|1|2] , [1|2|2] , [2|2|2]

and so on.

The diagram with p boxes can contain the number 2 not at all, once, twice, ... up to p times, i.e. its dimension is $(p + 1)$. This corresponds to our earlier result: for every positive integer p there exists exactly one irreducible SU(2) representation.

9.5 The Irreducible Representations of SU(n)

We can treat the irreducible representation of the group SU(n) similarly by numbering the boxes of the Young diagrams by 1 to n. The generalization relies upon the following theorem, which we quote without proof[4]:

Theorem. Every N-particle state that both belongs to an irreducible representation of the permutation group S_N and is built up from single particle states of an n-dimensional SU(n) multiplet transforms under an irreducible representation of the group SU(n).

From this theorem we conclude that each Young diagram with at most n rows represents an irreducible representation of the group SU(n). A diagram with more than n rows must necessarily include a column with at least $(n + 1)$ boxes. Hence the state must be antisymmetrized with respect to these $(n + 1)$ particles; but this cannot be done, since we have only n different single-particle basis states of the SU(n). To obtain the dimension of an irreducible representation related to a certain Young diagram, we count all possible standard configurations of this diagram built up by inserting the numbers 1 to n. Each box represents a particle and the number in the box marks one of the different states which the particle can occupy. Therefore we state the following rule: First, write one of the numbers 1 to n into each box by noting that in each column the values of the numbers increase downwards and do not decrease from left to right within a row.

---

[4] A proof of this theorem is found in B.G. Wyborne: *Classical Groups for Physicists* (Wiley, New York 1974), Chap. 22.

EXERCISE

9.4 Multiplets of a Two-Particle System in the Group SU(3)

Problem. Construct the possible SU(3) multiplets of a two-particle system.

Solution. The possible Young diagrams are:

$$\square\square \quad \text{and} \quad \begin{array}{c}\square\\ \square\end{array} . \tag{1}$$

For these diagrams we find the following standard configurations:

$$\boxed{1|1}\,,\ \boxed{1|2}\,,\ \boxed{1|3}\,,\ \boxed{2|2}\,,\ \boxed{2|3}\,,\ \boxed{3|3}\quad \text{, as well as}\quad \begin{array}{|c|}\hline 1\\ \hline 2\\ \hline\end{array}\,,\ \begin{array}{|c|}\hline 1\\ \hline 3\\ \hline\end{array}\,,\ \begin{array}{|c|}\hline 2\\ \hline 3\\ \hline\end{array}\ . \tag{2}$$

Therefore we have a sextet and a triplet. In fact, we are dealing here with the antitriplet, because the two particle wave functions represented by $\begin{array}{c}\square\\ \square\end{array}$ are antisymmetric. The triplet is represented by the Young-diagram $\square$.

Previously we represented the basic states of an SU(3) multiplet as points in the T_3-Y-plane. The coordinates of a point were called the "weight" of the related state. Now we can identify the points with the corresponding Young tableaux. In the case of the fundamental representation we obtain the following diagram (Fig. 9.1). From this we derive Table 9.1.

Since the weights behave as additive numbers, the weights of many-particle states are obtained by summing up the components. In this way we obtain the basis states of the sextet (from Exercise 9.4) built up by two particles (see Table 9.2).

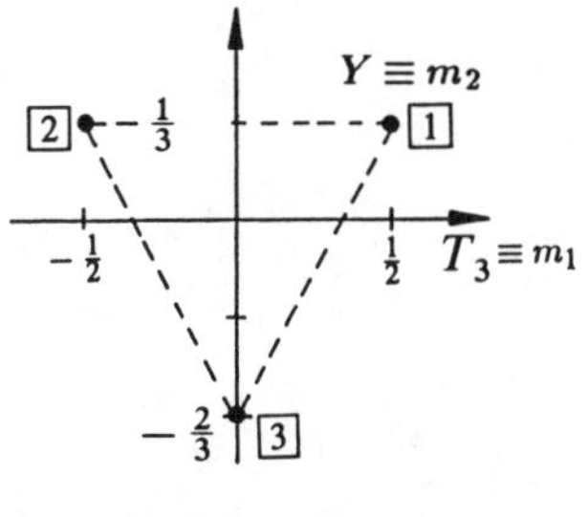

Fig. 9.1. Fundamental representation of SU(3)

Table 9.1. Weight of the triplet states

State	Weight (T_3, Y)
[1]	$(\frac{1}{2}, \frac{1}{3})$
[2]	$(-\frac{1}{2}, \frac{1}{3})$
[3]	$(0, -\frac{2}{3})$

Table 9.2. Weight of the symmetric states of the sextet

State	Weight
[1|1]	$(1, \frac{2}{3})$
[1|2]	$(0, \frac{2}{3})$
[2|2]	$(-1, \frac{2}{3})$
[1|3]	$(\frac{1}{2}, -\frac{1}{3})$
[2|3]	$(-\frac{1}{2}, -\frac{1}{3})$
[3|3]	$(0, -\frac{4}{3})$

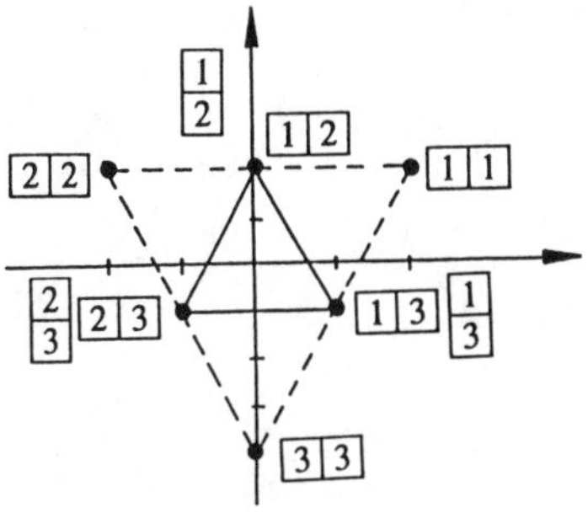

Fig. 9.2. The two particle states (sextet- and triplet-states) in SU(3).

The antisymmetric states of the two-particle triplet have the same weights, because the numbers within the boxes are only required during the construction of the multiplet, but their special configuration does not influence the result (see Table 9.3).

Table 9.3. Weight of the antisymmetric states of the two-particle triplet

State	Weight
$\begin{matrix}1\\2\end{matrix}$	$(0, \frac{2}{3})$
$\begin{matrix}1\\3\end{matrix}$	$(\frac{1}{2}, -\frac{1}{3})$
$\begin{matrix}2\\3\end{matrix}$	$(-\frac{1}{2}, -\frac{1}{3})$

If we draw these points into Fig. 9.2, the sextet and the anti-triplet representation, which are already known to us are produced.

EXERCISE

9.5 Multiplets of the SU(3) Constructed from Three Particles

Problem. Construct the SU(3) multiplets built up from three particles.

Solution. We get the following Young diagrams, consisting of three boxes:

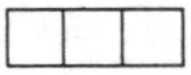 , 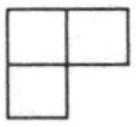 and  . (1)

Since the third diagram only has the one possible configuration $\begin{matrix}1\\2\\3\end{matrix}$, it represents the singlet. The second one, which permits eight configurations, represents the octet (see following table):

Weight of the three particle octet-states

State	Weight
$\begin{matrix}1&1\\2&\end{matrix}$	$(\frac{1}{2}, 1)$
$\begin{matrix}1&2\\2&\end{matrix}$	$(-\frac{1}{2}, 1)$
$\begin{matrix}1&3\\2&\end{matrix}$	$(0, 0)$
$\begin{matrix}1&1\\3&\end{matrix}$	$(1, 0)$
$\begin{matrix}1&2\\3&\end{matrix}$	$(0, 0)$
$\begin{matrix}1&3\\3&\end{matrix}$	$(\frac{1}{2}, -1)$
$\begin{matrix}2&2\\3&\end{matrix}$	$(-1, 0)$
$\begin{matrix}2&3\\3&\end{matrix}$	$(-\frac{1}{2}, -1)$

(2)

Exercise 9.5

Finally the first diagram, illustrating the totally symmetrical representation, contains ten states:

$$\boxed{1|1|1}\left(\frac{3}{2},1\right),\quad \boxed{2|2|2}\left(-\frac{3}{2},1\right)\quad,\quad \boxed{3|3|3}\,(0,-2)\quad, \tag{3}$$

as well as

$$\boxed{1|1|2}\ ,\quad \boxed{1|2|2}\ ,\quad \boxed{1|1|3}\ ,\quad \boxed{1|3|3}\ ,$$
$$\boxed{1|2|3}\ ,\quad \boxed{2|2|3}\ ,\quad \boxed{2|3|3}\ , \tag{4}$$

whose weights are already contained within the octet states (2). By drawing the states in the T_3-Y coordinate frame we obtain the well-known diagrams (see Figure below). It is interesting that the central state of the octet automatically acquires the correct multiplicity 2.

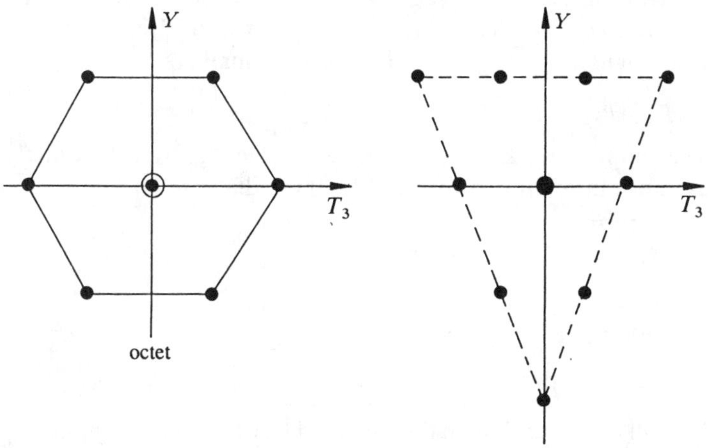

Octet and decuplet representation

Returning to the general SU(n), we already know that every column may consist of, at most, n boxes. Just one single way exists for n particles in n states to form an antisymmetric wavefunction. Hence we can omit all columns of the Young diagram with exactly n boxes if we want to determine the dimension of an irreducible representation of the SU(n). For example, we consider the diagram

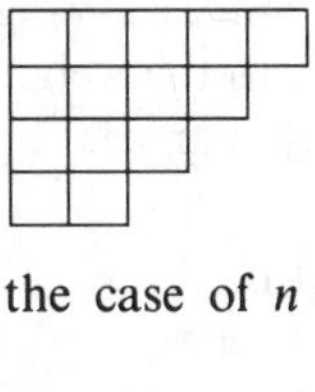

. (9.52)

In the case of $n = 4$, the same irreducible representation is illustrated more

simply by

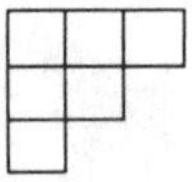
. (9.53)

Therefore (9.52) can be equated with (9.53):

SU(4): 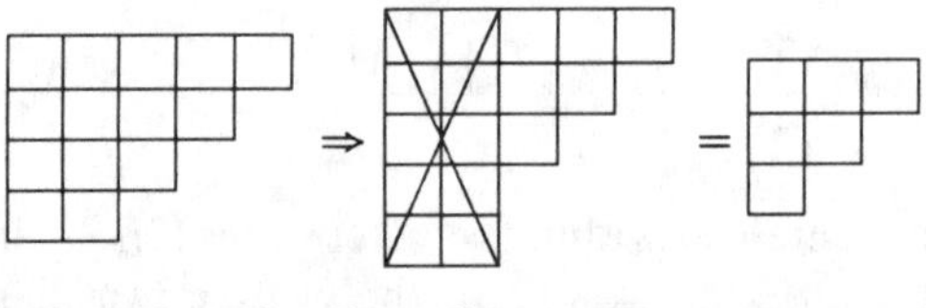 . (9.54)

Despite the fact that the first diagram represents a 14-particle state and the second one a 6-particle state, both irreducible representations are equivalent and have the same dimension. (But note that, for instance, in the framework of the SU(5) these Young diagrams represent different multiplets!)

Thus in the case of SU(n) we have to consider only diagrams with at most $(n-1)$ rows. We have already shown [see (9.22)ff.] that such a diagram is unambiguously characterized by $(n-1)$ numbers:

$$\boldsymbol{p} = (p_1, \dots, p_{n-1}) \quad . \tag{9.55}$$

The number p_k denotes the number of columns with exactly k boxes. For example, in the framework of SU(4) the diagram

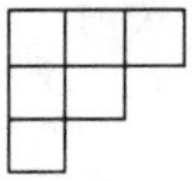
. (9.56)

is characterized by the vector

$$\boldsymbol{p} = (1, 1, 1) \quad . \tag{9.57}$$

We will find later that, in the case of SU(3), the numbers p_1 and p_2 correspond to the numbers p and q used previously to characterize the multiplets of SU(3).

Finally, we want to consider whether there may occur a change of the weights of the members within a multiplet when we omit all columns with n boxes. If we number the boxes of a Young tableau in the standard configuration, the numbers have to increase downwards within every column, i.e.

A column with n boxes has to contain all numbers from 1 to n. Since the weights of the single particle states are additive, the n boxes contribute to the weight of the tableau a number equal to the sum of the weights of all single-particle states,

$$\left(\sum_{i=1}^{n} W_1(\boxed{i}), \sum_{i=1}^{n} W_2(\boxed{i}), \sum_{i=1}^{n} W_3(\boxed{i}), \dots, \sum_{i=1}^{n} W_{N-1}(\boxed{i}) \right) \tag{9.58}$$

The weights of the single particle states are given by the diagonal elements of the diagonal commuting generators of the SU(n). The sum (9.58) of all n weights vanishes, because, in the case of SU(n), these matrices are traceless. Therefore we concluded that columns with n boxes do not contribute to the weights of a Young tableau and hence can be omitted [this is not necessarily valid for other groups like the U(n)].

Thus in general the multiplets of SU(n) are specified by the $(n - 1)$ integers $p_1, \ldots, p_{n-1}$. The simplest case,

$$p_1 = p_2 = \cdots = p_{n-1} = 0 \quad , \tag{9.59}$$

represents the singlet state of SU(n). We symbolize it by ① in analogy to SU(2). The following basic Young diagrams are characterized by one number p_i being equal to 1, whereas all other p_k vanish:

$$p_k = \delta_{ik} \quad , \quad k = 1, \ldots, n - 1 \quad . \tag{9.60}$$

i can take every value from 1 to $(n - 1)$ and these diagrams consist of a single column with exactly i boxes. These representations are called *fundamental representations*; in the case of SU(n) we have $(n - 1)$ diagrams of this kind so that the SU(2) has only *one* fundamental representation, namely the doublet

$$\square = (1) \quad . \tag{9.61}$$

In the case of SU(3) there exists the triplet and the antitriplet

$$\square = (1, 0) \qquad \begin{array}{c}\square\\\square\end{array} = (0, 1) \quad , \tag{9.62}$$

and for the SU(4) there are three fundamental representations:

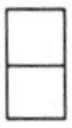

$$\square = (1, 0, 0) \qquad \begin{array}{c}\square\\\square\end{array} = (0, 1, 0) \qquad \begin{array}{c}\square\\\square\\\square\end{array} = (0, 0, 1) \quad . \tag{9.63}$$

Here the first one is the quartet and the last one is the antiquartet, whereas the second representation is a sextet

Now we introduce the concept of the *conjugate* Young diagram related to the SU(n). Consider a diagram which is characterized by the numbers

$$(p_1, \ldots, p_{n-1}) \quad . \tag{9.64}$$

If we invert the order of the numbers

$$(p_{n-1}, \ldots, p_1) \quad , \tag{9.65}$$

we call the diagram constructed in this way "conjugate" to the former one. (Note: The conjugate diagram related to the permutation group S_n, which we introduced in the chapter before, was defined differently!) The 6-particle Young diagram related to the group SU(6),

$$\begin{array}{c}\square\square\square\\\square\square\\\square\end{array} \quad , \tag{9.66}$$

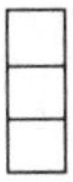

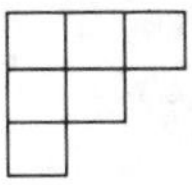

is denoted by the numbers $(1, 1, 1, 0, 0)$. The conjugate diagram $(0, 0, 1, 1, 1)$ has the form

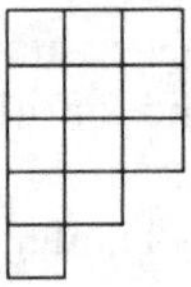

$$\tag{9.67}$$

If the original diagram is identical to the conjugate diagram, which is always the case for the SU(2), we call this diagram *self-conjugate*. The octet representation of the SU(3) is an example of a self-conjugate diagram in a higher group:

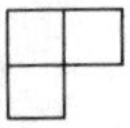

$$= (1, 1) \quad . \tag{9.68}$$

Another example is the second fundamental representation of SU(4),

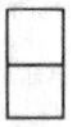

$$= (0, 1, 0) \quad . \tag{9.69}$$

In general, the dimension of the conjugate representation is identical to the dimension of the representation given by the original diagram.

9.6 Determination of the Dimension

In general, it can be very laborious to determine the dimension of an SU(n) multiplet by numbering the standard configurations of the related Young diagram. Hence it is very useful to have a universal formula at our disposal to calculate the dimension given the numbers $p_1, \dots, p_{n-1}$. We begin with the group SU(2), in which every multiplet (Young diagram, respectively) is characterized by a single number p. The diagram consists of one row with p boxes:

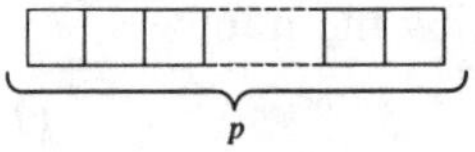

$$\underbrace{\boxed{}}_{p} \quad . \tag{9.70}$$

The standard configurations contain the numbers 1 or 2 in these boxes. In the diagram all numbers 2 have to be *on the rhs* of all numbers 1. Thus each standard configuration is given by the number of times q the index 2 occurs; q can take all integer values from 0 to p, i.e.

$$\underbrace{\boxed{1\,|\,1\,|\,1 \quad |\,1\,|\,1}}_{q=0} \quad , \quad \underbrace{\boxed{1\,|\,1\,|\,1 \quad |\,1\,|\,2}}_{q=1} \quad , \dots ,$$

$$\underbrace{\boxed{2\,|\,2\,|\,2 \quad |\,2\,|\,2}}_{q=p} \quad . \tag{9.71}$$

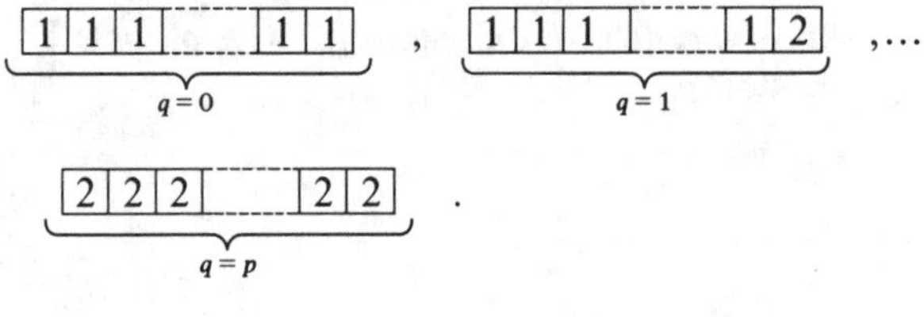

Each value of q is related to exactly one standard tableau. Therefore the total number of standard configurations, i.e. the dimension of the multiplet, is

$$D_2(p) = \sum_{q=0}^{p} 1 = p + 1 \quad . \tag{9.72}$$

If we calculate dimensions in higher groups SU(n), we have to use combinatorial methods, e.g.

$$D_n(p_1, \ldots, p_{n-1}) \quad . \tag{9.73}$$

We can derive a recursion formula by similar considerations, relating the dimension of a diagram of SU($n + 1$) to the dimension of the same diagram of SU(n). By passing from SU($n + 1$) down to SU(n) we note that all columns with n boxes can be omitted, simplifying the Young diagram considerably. This corresponds to the omission of the last number p_n, i.e.

$$(p_1, \ldots, p_{n-1}, p_n) \overset{\mathrm{SU}(n)}{\rightarrow} (p_1, \ldots, p_{n-1}) \quad . \tag{9.74}$$

The recursion formula mentioned above, for which a general proof will be given in Example 10.4, reads

$$\begin{aligned} D_{n+1}(p_1, \ldots, p_n) = {} & \frac{1}{n!}(p_n + 1)(p_n + p_{n-1} + 2) \cdots (p_n + \cdots + p_1 + n) \\ & \times D_n(p_1, \ldots, p_{n-1}) \quad . \end{aligned} \tag{9.75}$$

For the group SU(3) we obtain the formula

$$\begin{aligned} D_3(p_1, p_2) &= \tfrac{1}{2!}(p_2 + 1)(p_1 + p_2 + 2)D_2(p_1) \\ &= \tfrac{1}{2}(p_1 + 1)(p_2 + 1)(p_1 + p_2 + 2) \quad . \end{aligned} \tag{9.76}$$

If we identify the numbers (p_1, p_2) with the numbers (p, q), this is exactly the expression we obtained in Chap. 8 by performing geometrical considerations.

For the SU(4) we obtain similarly

$$\begin{aligned} D_4(p_1, p_2, p_3) &= \tfrac{1}{3!}(p_3 + 1)(p_2 + p_3 + 2)(p_1 + p_2 + p_3 + 3)D_3(p_1, p_2) \\ &= \tfrac{1}{12}(p_1 + 1)(p_2 + 1)(p_3 + 1)(p_1 + p_2 + 2) \\ &\quad \times (p_2 + p_3 + 2)(p_1 + p_2 + p_3 + 3) \quad . \end{aligned} \tag{9.77}$$

Clearly the number of factors is increasing rapidly with the rank of the group. For large values of n it is therefore often useful to use methods which yield the appropriate dimension in a simpler way; thus the dimension of an SU(n) multiplet is represented by a fraction[5],

$$D_n(p_1, \ldots, p_{n-1}) = \frac{a_n(\boldsymbol{p})}{b(\boldsymbol{p})} \quad , \tag{9.78}$$

whose numerator and denominator are determined as follows.

[5] H.J. Coleman: *Symmetry Groups made easy*, Adv. Quantum Chemistry **4**, 83 (1968).

To determine the numerator a_n, each box is marked by a number; in the case of SU(n) we start with the number n in the upper left box of the tableau. Within the first row the numbers have to increase by 1 from left to the right. In the second row the boxes contain the number of the corresponding boxes above, less one. For example, let us consider the following Young diagram for SU(3), i.e. $n = 3$.

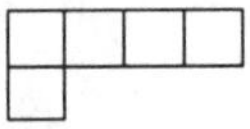 $= (3, 1) \quad .$ (9.79)

We adopt the rule stated above by filling the boxes with the numbers

$$\begin{array}{|c|c|c|c|} \hline 3 & 4 & 5 & 6 \\ \hline 2 \\ \cline{1-1} \end{array} \quad , \tag{9.80}$$

and the numerator results from the product of these numbers:

$$a_3(3, 1) = 3 \times 4 \times 5 \times 6 \times 2 \quad . \tag{9.81}$$

To obtain the denominator, again each box is numbered, but in a different way. For that purpose we draw a horizontal line from the centre of the box to the right and a descending vertical line. Then we count the number of boxes passed through by these lines, including the original box, and insert this number into the appropriate box. Multiplication of these numbers yields the value of b. Taking the first box from the left, in the first row of our example, we have

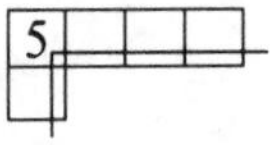 . (9.82)

and, altogether,

$$\begin{array}{|c|c|c|c|} \hline 5 & 3 & 2 & 1 \\ \hline 1 \\ \cline{1-1} \end{array} \quad . \tag{9.83}$$

From this numbering we can calculate b, e.g.

$$b(3, 1) \;= 5 \times 3 \times 2 \times 1 \times 1 \quad . \tag{9.84}$$

Hence the dimension of the representation (3,1) of SU(3) is

$$D_3(3, 1) = \frac{a_3(3, 1)}{b(3, 2)} = \frac{3 \times 4 \times 5 \times 6 \times 2}{5 \times 3 \times 2 \times 1 \times 1} = 24 \quad . \tag{9.85}$$

Formula (9.76) yields the same result:

$$D_3(3, 1) = \tfrac{1}{2}(3 + 1)(1 + 1)(3 + 1 + 2) = \tfrac{1}{2} \times 4 \times 2 \times 6 = 24 \quad . \tag{9.86}$$

The advantage of this new method becomes obvious when we consider a similar Young diagram as a multiplet of the group SU(9). We just have to recalculate the numerator, obtaining

$$\begin{array}{|c|c|c|c|c|} \hline 9 & 10 & 11 & 12 & 13 \\ \hline 8 & 9 \\ \cline{1-2} \end{array} \quad . \tag{9.87}$$

We now find for the dimension

$$D_9(3, 2, 0, \ldots, 0) = \frac{9 \times 10 \times 11 \times 12 \times 13 \times 8 \times 9}{6 \times 5 \times 3 \times 2 \times 1 \times 2 \times 1} = 30888 \quad . \tag{9.88}$$

Using the standard formula (9.75), in this example, we would have to multiply a product of $8! = 40320$ factors.

EXERCISE

9.6 Dimension Formula for the SU(3)

Problem. Derive the expression (9.76) for the dimension of SU(3) multiplets by using the result for the SU(2).

Solution. In a standard configuration of the SU(3) involving the numbers 1, 2 and 3, the second row may only include the numbers 2 and 3; also all the numbers 3 have to be on the rhs of the numbers 2 and only in those (single) boxes which do not have any box underneath. In the first row, boxes lying above a number 2 may only contain the value 1. In the first row the number 3 may only occur at the right end, in boxes with no box directly below them. Denoting by r the number of boxes in the first row which contain a 3, we must have $0 \leq r \leq p_1$. If q is the number of boxes in the second row containing a 3, the following relation holds: $0 \leq q \leq p_2$. Therefore we get a tableau of the form:

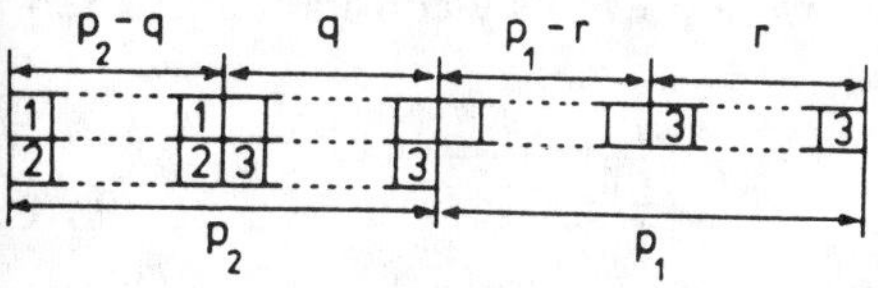

Up to now only the value of the $q + (p_1 - r)$ boxes in the centre of the first row has not been determined. Applying the standard method we can arbitrarily fill these boxes with the numbers 1 and 2. But the number of configurations is just equal to the dimension of the Young diagram with $q + (p_1 - r)$ boxes in SU(2):

$$D_2(q + p_1 - r) = q + p_1 - r + 1 \quad . \tag{1}$$

In the case of SU(3) we obtain the dimension of the complete diagram by summing over q and r in the allowed domains, giving

$$\begin{aligned} D_3(p_1; p_2) &= \sum_{q=0}^{p_2} \sum_{r=0}^{p_1} D_2(q + p_1 - r) \\ &= \sum_{q=0}^{p_2} \left[(q + p_1 + 1)(p_1 + 1) - \frac{1}{2} p_1 (p_1 + 1) \right] \\ &= \frac{1}{2}(p_1 + 1) \sum_{q=0}^{p_2} (2q + p_1 + 2) \\ &= \frac{1}{2}(p_1 + 1)[p_2(p_2 + 1) + (p_1 + 2)(p_2 + 1)] \\ &= \frac{1}{2}(p_1 + 1)(p_2 + 1)(p_1 + p_2 + 2) \quad , \end{aligned} \tag{2}$$

Exercise 9.6

where we have used the relations

$$\sum_{i=0}^{m} 1 = m + 1 \quad , \quad \sum_{i=0}^{m} i = \frac{1}{2} m(m+1) \quad . \tag{3}$$

9.7 The SU$(n-1)$ Subgroups of SU(n)

The weight diagram of an SU(3) multiplet contains a number of SU(2) multiplets. Because of the symmetry of the diagrams there is no difference between submultiplets related to isospin or others related to U- or V-spin. The quark triplet includes an isospin singlet (the s-quark) and an isospin doublet (the u- and d-quarks), for instance. The octet of the baryons contains two isospin doublets (p, n and Ξ^-, Ξ^0), a triplet ($\Sigma^-, \Sigma^0, \Sigma^+$) and a singlet ($\Lambda$).

Now we need to perform the decomposition into subgroups with the help of Young diagrams. To illustrate the procedure we take the octet of SU(3) represented by

$$\begin{array}{|c|c|}\hline & \\ \hline & \multicolumn{1}{c}{} \\ \cline{1-1}\end{array} = (1,1) \quad . \tag{9.89}$$

The eight corresponding Young tableaux were previously considered in Exercise 9.5 and are

$$\begin{array}{|c|c|}\hline 1 & 1 \\ \hline 2 & \multicolumn{1}{c}{} \\ \cline{1-1}\end{array}\,,\ \begin{array}{|c|c|}\hline 1 & 1 \\ \hline 3 & \multicolumn{1}{c}{} \\ \cline{1-1}\end{array}\,,\ \begin{array}{|c|c|}\hline 1 & 2 \\ \hline 2 & \multicolumn{1}{c}{} \\ \cline{1-1}\end{array}\,,\ \begin{array}{|c|c|}\hline 1 & 2 \\ \hline 3 & \multicolumn{1}{c}{} \\ \cline{1-1}\end{array}\,,\ \begin{array}{|c|c|}\hline 1 & 3 \\ \hline 2 & \multicolumn{1}{c}{} \\ \cline{1-1}\end{array}\,,\ \begin{array}{|c|c|}\hline 1 & 3 \\ \hline 3 & \multicolumn{1}{c}{} \\ \cline{1-1}\end{array}\,,\ \begin{array}{|c|c|}\hline 2 & 2 \\ \hline 3 & \multicolumn{1}{c}{} \\ \cline{1-1}\end{array}\,,\ \begin{array}{|c|c|}\hline 2 & 3 \\ \hline 3 & \multicolumn{1}{c}{} \\ \cline{1-1}\end{array} \,. \tag{9.90}$$

In the case of SU(2), boxes containing the number 3 do not exist. Thus we divide the tableaux into groups according to the respective position of boxes containing a 3. First we find two tableaux that don't contain the number 3, i.e.

$$\begin{array}{|c|c|}\hline 1 & 1 \\ \hline 2 & \multicolumn{1}{c}{} \\ \cline{1-1}\end{array} \quad \text{and} \quad \begin{array}{|c|c|}\hline 1 & 2 \\ \hline 2 & \multicolumn{1}{c}{} \\ \cline{1-1}\end{array} \quad . \tag{9.91}$$

In SU(2) the column $\begin{array}{|c|}\hline 1 \\ \hline 2 \\ \hline\end{array}$ can be omitted [because it represents a singlet under SU(2)], so we obtain a doublet,

$$\begin{array}{|c|}\hline 1 \\ \hline\end{array} \quad , \quad \begin{array}{|c|}\hline 2 \\ \hline\end{array} \quad . \tag{9.92}$$

Now we look for a tableau with a single number 3 on the right. There is only one of this kind,

$$\begin{array}{|c|c|}\hline 1 & 3 \\ \hline 2 & \multicolumn{1}{c}{} \\ \cline{1-1}\end{array} \quad . \tag{9.93}$$

Since the number of 3 is meaningless in the case of SU(2), we can neglect this box and obtain an SU(2) singlet

$$\begin{array}{|c|}\hline 1 \\ \hline 2 \\ \hline\end{array} = ① \quad . \tag{9.94}$$

There are three tableaux with the number 3 at the bottom,

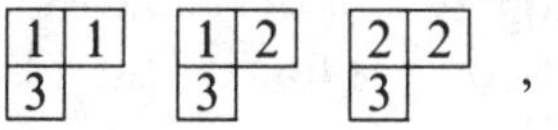

$$\boxed{\begin{array}{cc}1&1\\3&\end{array}} \quad \boxed{\begin{array}{cc}1&2\\3&\end{array}} \quad \boxed{\begin{array}{cc}2&2\\3&\end{array}} \;, \tag{9.95}$$

and, once again neglecting the boxes containing the number 3, we obtain an SU(2) triplet,

$$\boxed{1\,1} \;, \quad \boxed{1\,2} \;, \quad \boxed{2\,2} \;. \tag{9.96}$$

Finally we have two tableaux containing the number 3 twice, i.e.

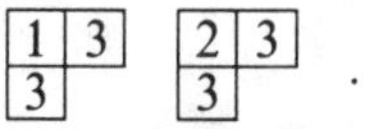

$$\boxed{\begin{array}{cc}1&3\\3&\end{array}} \quad \boxed{\begin{array}{cc}2&3\\3&\end{array}} \;. \tag{9.97}$$

By erasing the three boxes containing a "3", a doublet results,

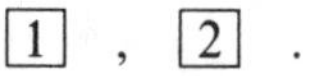

$$\boxed{1} \;, \quad \boxed{2} \;. \tag{9.98}$$

Thus altogether we have recovered one singlet, two doublets and a triplet.

We would have obtained the same result in a simpler way by filling boxes with a "3" for allowed configurations only:

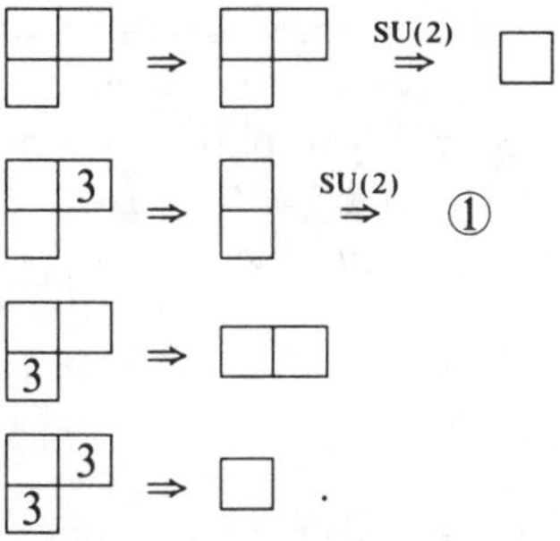

$$\tag{9.99}$$

In the first step we remove all boxes containing the number 3 and in the second step we eliminate all columns with two boxes.

We can easily generalize this method to the group SU(n). Consider all allowed positions of the boxes in the diagram containing the number n and remove these boxes. The SU($n-1$) submultiplets are then given by the remaining boxes. For example, we consider the Young diagram

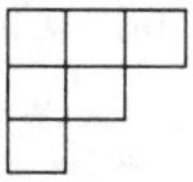

$$\tag{9.100}$$

for which there are eight different ways of labelling the boxes with the number "n", given by

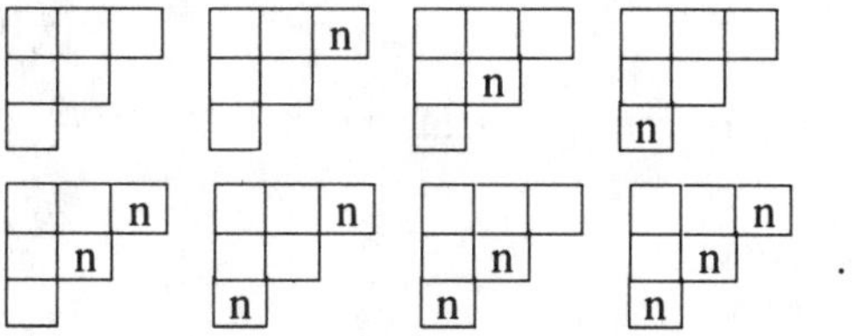

The rule here is that "n" can occur only at the bottom of a column, because the numbers in the squares have to increase from top to bottom, and cannot decrease from right to left. We thus obtain the decomposition of the SU(n) multiplet into submultiplets of the group SU($n-1$), i.e.

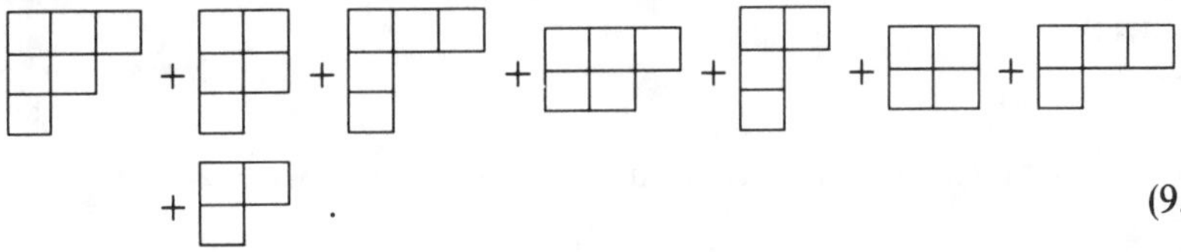

$$(9.102)$$

We may then further decompose these multiplets into multiplets of SU($n-2$), and so on.

9.8 Decomposition of the Tensor Product of Two Multiplets

It is an important problem in group theory to decompose the states of a many-particle system belonging to a certain multiplet. We have already discussed the example for which there are two possible ways to couple two spin-$\frac{1}{2}$-particles to states with good total angular momentum (to $j=0$ and $j=1$, respectively). Stating this in the language of group theory, the tensor product of two SU(2) doublets decomposes into a singlet and triplet

$$[2] \otimes [2] = [3] \oplus [1] \quad . \tag{9.103}$$

Expressing this result with the help of Young diagrams we have

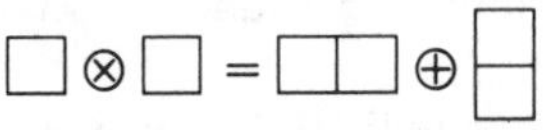

$$\square \otimes \square = \square\square \oplus \begin{array}{c}\square\\\square\end{array} \quad . \tag{9.104}$$

Obviously both new multiplets are obtained by adjoining the boxes of both original diagrams in an appropriate way. We notice that the combination of the Young diagrams (9.104) has a much more general validity than relation (9.103). While (9.103) is only valid in the case of SU(2), equation (9.104) is valid for any group SU(n). For the example of SU(3) the diagram $\square$ represents a fundamental triplet (quark). Now (9.104) states that two quarks can be coupled to a sextet $\square\square$ and to an antitriplet $\begin{array}{c}\square\\\square\end{array}$, i.e.

$$[3] \otimes [3] = [6] \oplus [\bar{3}] \quad . \tag{9.105}$$

In the framework of SU(4), (9.104) represents the relation

$$[4] \otimes [4] = [\underline{10}] \oplus [\underline{6}] \quad , \tag{9.106}$$

and so on.

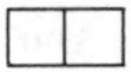

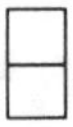

We now proceed, without proof, to state the general rules on how to couple two multiplets of the SU(n).[6] We start by drawing the Young diagrams representing the two multiplets and marking each box of the second diagram with the corresponding number of its row. We continue by adding all the boxes of the second diagram to the first one. The boxes may only be added to the right or the bottom of the first diagram observing the following rules:

a) Each resulting diagram has to be an allowed configuration, i.e. no row must be longer than the row above.
b) In the framework of SU(n) no column must contain more than n boxes.
c) Within a row the numbers in the boxes originating from the second diagram must not decrease from left to right.
d) The numbers must increase from top to bottom within one column.
e) We draw a path though all boxes of the resulting Young diagram which crosses each row from right to left, beginning at the top. Along the path the number i must at no time have occurred more often than the number $(i-1)$, i.e.

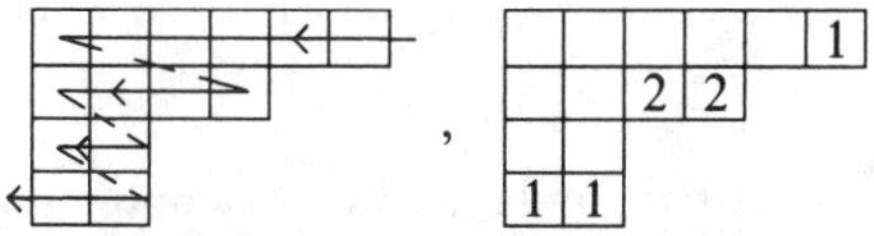

path , forbidden scheme (9.107)

Among other things, this rule states that a box of the i-th row of the second Young diagram must not be attached to the first $(i-1)$ rows of the first Young diagram.

As an example we consider the tensor product of the two fundamental representations of the SU(n), $n \geq 3$.

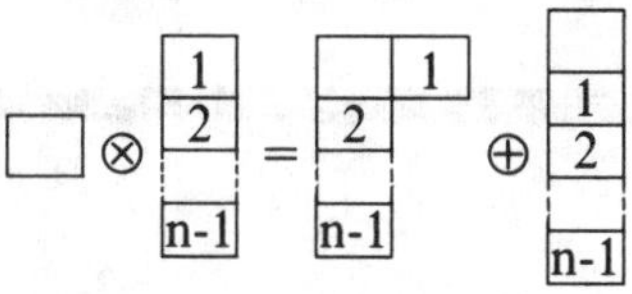

$= (n^2 - 1) \oplus \underline{1}$. (9.108)

The combinations

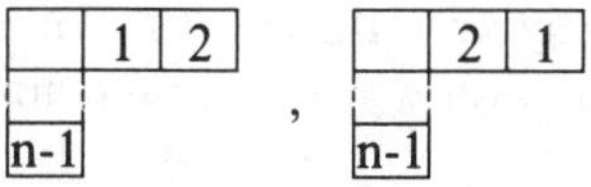

etc. (9.109)

are forbidden by rules (e) and (c), respectively. In the context of the group SU(3), the relation (9.108) means that

$$[\underline{3}] \otimes [\underline{\bar{3}}] = [\underline{8}] \oplus [\underline{1}] \quad . \qquad (9.110)$$

[6] D.B. Lichtenberg: Unitary symmetry and elementary particles (Academic Press, New York, London 1970) Chap. 7.4.

EXERCISE

9.7 Decomposition of a Tensor Product

Problem. Find the decomposition of the tensor product $\underline{8} \otimes \underline{8}$ for the SU(3).

Solution.

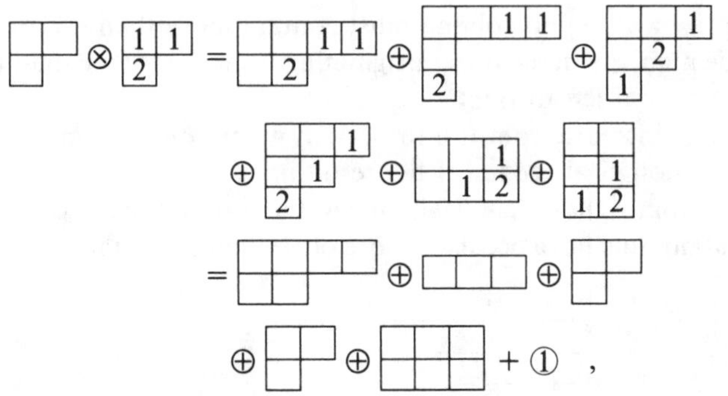

where in the last step we have omitted redundant columns containing three boxes. Thus, denoting the multiplets by their dimension, we obtain

$$[\underline{8}] \otimes [\underline{8}] = [\underline{27}] \oplus [\underline{10}] \oplus [\underline{8}] \oplus [\underline{8}] \oplus [\underline{10}] \oplus [\underline{1}] \quad . \tag{1}$$

Here it is worth noting that the octet occurs twice in the decomposition. Due to a general rule this may happen if the original diagrams do not *both* have the form of a rectangle.

If we consider products of representations, an arrangement into classes is often useful. This is seen as follows. Combining two Young diagrams with b_1 and b_2 boxes, respectively, initially yields a diagram with $(b_1 + b_2)$ boxes. However, since for an SU(n) we can omit columns with n boxes, the remaining diagrams can have

$$b_1 + b_2 \quad , \quad b_1 + b_2 - n \quad , \quad b_1 + b_2 - 2n \quad , \quad \text{etc.} \tag{9.111}$$

boxes. To unify all these possibilities it is useful to classify the number of boxes according to the residue with respect to n. Thus we say that an irreducible representation is of class m (with $0 \leq m \leq n-1$) if the number of boxes of the according Young diagram can be written in the form

$$b = in + m \quad , \tag{9.112}$$

with some integer i. If we form the tensor product of two representations of the classes m_1 and m_2 according to what we have said above, it follows that every resulting Young diagram must be a member of the class

$$m \equiv m_1 + m_2 \pmod{n} \quad . \tag{9.113}$$

So the representations of the SU(n) fall into classes which, when tensor products are formed, behave like an additive group. This group of the *residue classes* (mod n) is labelled Z_n and is called the "*centre of the group*". Instead of the residue classes m, one can also use the unit roots $\exp(2\mathrm{i}m\pi/n)$ to describe the different classes of representations of SU(n), because

$$\mathrm{e}^{2\mathrm{i}m_1\pi/n}\mathrm{e}^{2\mathrm{i}m_2\pi/n} = \mathrm{e}^{2\mathrm{i}m\pi/n} \tag{9.114}$$

holds with $m \equiv m_1 + m_2 (\mathrm{mod}\, n)$. This classification has the advantage that the multiplicative character of the formation of tensor products becomes apparent.

EXAMPLE

9.8 Representations of the SU(2) and Spin

Within the representation of the SU(2) there are representations with an even number of boxes and others with an odd number. Those having an even number of boxes correspond to the representations of integer spin or isospin, those with an odd number represent multiplets with half-integer (iso-) spin.

EXAMPLE

9.9 Triality and Quark Confinement

In the case of the SU(3) one distinguishes multiplets with *triality*; $\tau = 0$ (for $k = 0$), $\tau = 1$ (for $k = 1$) and $\tau = -1$ (for $k = 2$). For the quarks $\tau = 1$, for the antiquarks $\tau = -1$. The trialties of several particles are additive (modulo 3), and the baryons, containing three quarks, have triality $\tau = 0$, and the mesons, consisting of one quark and one antiquark, have $\tau = 0$, It is conjectured that in nature only particles with $\tau = 0$ occur as free particles (*quark confinement*).

10. Mathematical Excursion. Group Characters

10.1 Definition of Group Characters

In this chapter a special group theoretical concept is introduced which has many applications. It describes the main properties of representations and is therefore called "group character". It solves the problem of how to describe the invariant properties of a group representation in a simple way. If we denote an element of a group G by $\hat{G}_a$, a representation $\hat{D}(\hat{G}_a)$ is not unambiguous, because every similarity transformation $\hat{A}\hat{D}(\hat{G}_a)\hat{A}^{-1}$, $\hat{A} \in D(G)$ yields an equivalent form. One possibility for the description of the invariant properties would be to use the eigenvalues of the representation matrix, which do not change under a similarity transformation. This leads to the construction of the Casimir operators, the eigenvalues of which classify the representations. The construction of the Casimir operators and their eigenvalues is in general a very difficult nonlinear problem. Fortunately, in many cases it is sufficient to use a simpler invariant, namely the trace of the representation matrix

$$\chi(\hat{G}_a) = \sum_{i=1}^{d} D_{ii}(\hat{G}_a) \quad , \tag{10.1}$$

where d is the dimension of the matrix representation. Equation (10.1) is in fact invariant under similarity transformations, because

$$\begin{aligned} \psi'(\hat{G}_a) &= \sum_{i} D'_{ii}(\hat{G}_a) = \sum_{ijk} A_{ij} D_{jk}(\hat{G}_a)(\hat{A}^{-1})_{ki} \\ &= \sum_{jk} D_{jk}(\hat{A}^{-1}\hat{A})_{kj} = \sum_{j} D_{jj}(\hat{G}_a) = \chi(\hat{G}_a) \quad . \end{aligned} \tag{10.2}$$

$\chi(\hat{G}_a)$ is called the "*group character*" of the representation.

For the present we restrict ourselves to finite groups. These consist only of a finite number of elements and are therefore easier to handle. Later we generalize the results to continuous compact groups, i.e. compact Lie groups, like, for instance, U(N) and SU(N). First, however, we introduce some fundamental concepts which will be necessary later on.

10.2 Schur's Lemmas

10.2.1 Schur's First Lemma

Let $\hat{D}(\hat{G}_a)$ be an irreducible representation of a group G defined in the vector space $\mathscr{R}$ (e.g. the three-dimensional physical space), and let $\hat{A}$ be a fixed operator in $\mathscr{R}$. The *first lemma of Schur states*:

If $\hat{A}$ commutates with $\hat{D}(\hat{G}_a)$ for all $\hat{G}_a$, i.e.

$$[\hat{A}, \hat{D}(\hat{G}_a)]_- = 0 \quad , \tag{10.3}$$

then $\hat{A}$ is proportional to the unity operator (unit matrix), or

$$\hat{A} = \lambda \mathbb{1} \quad . \tag{10.4}$$

This has already been seen in Chap. 4, (4.6) and (4.9), but in a somewhat different form. What is new and, for important, the following discussion is that $\hat{D}(\hat{G}_a)$ is not just a general operator of the group, but a representation matrix. If the group is just the space $\mathscr{R}$, then this reduces to the former case.

10.2.2. Schur's Second Lemma

Let $\hat{D}^1(\hat{G}_a)$ and $\hat{D}^2(\hat{G}_a)$ be two nonequivalent irreducible representations of the group G in two spaces $\mathscr{R}_1$ and $\mathscr{R}_2$ with the dimensions d_1 and d_2, respectively (the case $\mathscr{R}_1 = \mathscr{R}_2$ or $d_1 = d_2$ is not excluded); and let $\hat{A}$ be an operator which maps from $\mathscr{R}_1$ into $\mathscr{R}_2$. *Schur's second lemma states*:

If

$$\hat{D}^1(\hat{G}_a)\hat{A} = \hat{A}\hat{D}^2(\hat{G}_a) \tag{10.5}$$

for all $\hat{G}_a$ of $\hat{G}$, then $\hat{A}$ is the zero operator, $\hat{A} = \hat{0}$.

Proof of Schur's Second Lemma. Since Schur's first lemma was explicitly treated in Chap. 4, it is sufficient to prove the second lemma. This can be achieved in the form of proof by contradiction:

(i) First we consider the case $d_2 \leq d_1$. Then $\hat{A}$ applied to $\mathscr{R}_2$ generates a subspace $\mathscr{R}_a$ of $\mathscr{R}_1$, with the dimension $d_a \leq d_2 \leq d_1$. This subspace is composed of all vectors $\hat{A}\boldsymbol{r}$, with $\boldsymbol{r} \in \mathscr{R}_2$. From the assumption (10.5), it follows at once that $\mathscr{R}_a$ is invariant with respect to an application of the representation of the group G:

$$\hat{D}^1(\hat{G}_a)\hat{A}\boldsymbol{r} = \hat{A}\hat{D}^2(\hat{G}_a)\boldsymbol{r} \equiv \hat{A}\boldsymbol{r}_a \quad . \tag{10.6}$$

This belongs to $\mathscr{R}_a$, too, because $\boldsymbol{r}_a = \hat{D}^2(\hat{G}_a)\boldsymbol{r}$ belongs to $\mathscr{R}_2$. It was assumed, however, that $\hat{D}^1$ is an irreducible representation, i.e. $\mathscr{R}_1$ has no true invariant subspace. Otherwise $\hat{D}^1(\hat{G}_a)$ could be brought into block form. Since $\mathscr{R}_a$ is an invariant subspace, there is a contradiction unless $\mathscr{R}_a$ is either the null space ($d_A = 0$), i.e. $\hat{A} = \hat{0}$, or when $\mathscr{R}_a$ is the full space $\mathscr{R}_1$ i.e. ($d_a = d_1 = d_2$). The last possibility is excluded, because $\hat{D}^1(\hat{G}_a)$ and $\hat{D}^2(\hat{G}_a)$ have been assumed to be different, i.e. non-equivalent, representations. If both have the same dimension, we could invert $\hat{A}$, so that

$$\hat{D}^1(\hat{G}_a) = \hat{A}\hat{D}^2(\hat{G}_a)\hat{A}^{-1} \quad ,$$

i.e. $\hat{D}^1(\hat{G}_a)$ would be similar to $\hat{D}^2(\hat{G}_a)$; therefore the single possibility $\hat{A} = \hat{0}$ remains.

$d_2 > 1$. Since $\hat{A}$ maps the full space $\mathscr{R}_2$ onto $\mathscr{R}_1$, $d_a < d_2$ follows necessarily because of $d_2 > d_1$. This means that there are vectors $\boldsymbol{r}$ in $\mathscr{R}_2$ which are mapped onto 0 ($\hat{A}\boldsymbol{r} = 0$). We denote this subspace of $\mathscr{R}_2$ which has the dimension $d_b = d_2 - d_a$, by $\mathscr{R}_b$. $\mathscr{R}_b$ is also an invariant subspace, because, if $\boldsymbol{r}_a = \hat{D}^2(\hat{G}_a)\boldsymbol{r}$, then

$$\hat{A}\boldsymbol{r}_a = \hat{A}\hat{D}^2(\hat{G}_a)\boldsymbol{r} = \hat{D}^1(\hat{G}_a)\hat{A}\boldsymbol{r} = 0 \quad , \tag{10.7}$$

i.e. $\boldsymbol{r}_a$ belongs to $\mathscr{R}_b$, too. If $d_b < d_2$, this is a contradiction of the irreducibility of D^2. Hence it follows that $d_b = d_2$, i.e $d_a = 0$, and again $\hat{A} = \hat{0}$ must hold. Thus Schur's second lemma is proven.

10.3 Orthogonality Relations of Representations and Discrete Groups

First we will show that for two arbitrary, irreducible representations $\hat{D}^\alpha$ and $\hat{D}^\beta$, the matrix

$$\hat{A} = \sum_b \hat{D}^\alpha(\hat{G}_b)\hat{X}\hat{D}^\beta(\hat{G}_b^{-1}) \tag{10.8}$$

has the properties which are demanded by Schur's lemma. Here $\hat{X}$ is an arbitrary matrix of dimension $d_\alpha \times d_\beta$ and the sum is over all elements $\hat{G}_b$ of the group. Remember that we are still dealing with finite groups and therefore the sum is finite.

Now we show by explicit calculation that (10.5) holds:

$$\begin{aligned}\hat{D}^\alpha(\hat{G}_a)\hat{A} &= \sum_b \hat{D}^\alpha(\hat{G}_a)\hat{D}^\alpha(\hat{G}_b)\hat{X}\hat{D}^\beta(\hat{G}_b^{-1}) \\ &= \sum_b \hat{D}^\alpha(\hat{G}_a\hat{G}_b)\hat{X}\hat{D}^\beta(\hat{G}_b^{-1})\hat{D}^\beta(\hat{G}_a^{-1})\hat{D}^\beta(\hat{G}_a) \quad .\end{aligned} \tag{10.9}$$

Here we have used $\hat{D}^\alpha(\hat{G}_a)\hat{D}^\alpha(\hat{G}_b) = \hat{D}^\alpha(\hat{G}_a\hat{G}_b)$, as well as $\hat{D}^\beta(\hat{G}_a^{-1})\hat{D}^\beta(\hat{G}_a) = D^\beta(E) = \mathbb{1}$ (E being the neutral element of the group). The above expression can be written as

$$\begin{aligned}\sum_b \hat{D}^\alpha(\hat{G}_a\hat{G}_b)\hat{X}\hat{D}^\beta((\hat{G}_a\hat{G}_b)^{-1})\hat{D}^\beta(\hat{G}_a) &= \sum_c \hat{D}^\alpha(\hat{G}_c)\hat{X}\hat{D}^\beta(\hat{G}_c^{-1})\hat{D}^\beta(\hat{G}_a) \\ &= \hat{A}\hat{D}^\beta(\hat{G}_a) \quad ,\end{aligned} \tag{10.10}$$

where $\hat{G}_c = \hat{G}_a\hat{G}_b$. This last step is justified, because the summation runs over all group elements $\hat{G}_b$, and $\hat{G}_a\hat{G}_b$ just causes a permutation of elements; at the end the sum again includes all elements and we regain the operator $\hat{A}$. The last equation implies, however, that

$$\hat{D}^\alpha(\hat{G}_a)\hat{A} = \hat{A}\hat{D}^\beta(\hat{G}_a) \quad , \tag{10.11}$$

and therefore $\hat{A}$ has the property (10.5) demanded by Schur's lemma. In the matrix representation $\hat{A}$ as the form

$$\hat{A} = \lambda\delta_{\alpha\beta}\mathbb{1} \quad . \tag{10.12}$$

In (10.8) we can choose $\hat{X}$ arbitrarily because we made no assumptions concerning $\hat{X}$ in the derivation. If we choose $x_{km} = \delta_{kp}\delta_{mq}$, i.e. if $\hat{X}$ is represented by a matrix which has only zeros except for the qth row and the pth column, then from (10.8) follows

$$\sum_{a=1}^{g}\sum_{m=1}^{d_\beta}\sum_{k=1}^{d_\alpha} D^\alpha_{ik}(\hat{G}_a)X_{km}D^\beta_{mj}(\hat{G}_a^{-1})$$

$$= A_{ij} = \lambda^{(p,q)}\delta_{\alpha\beta}\delta_{ij} = \sum_{a=1}^{g} D^\alpha_{ip}(\hat{G}_a)D^\beta_{qj}(\hat{G}_a^{-1}) \quad . \tag{10.13}$$

Here g is the order of the group, i.e. the number of elements in G, and d_α, d_β are the dimensions of the irreducible representations $\hat{D}^\alpha$, $\hat{D}^\beta$. Thus only λ is left to be determined. λ is of importance only for the cases $i = j$ and $\alpha = \beta$. If (10.13) is used and summation over all i is performed, one finds that

$$\sum_{i=1}^{d_\alpha}\sum_{a=1}^{g} D^\alpha_{ip}(\hat{G}_a)D^\alpha_{qi}(\hat{G}_a^{-1}) = \lambda^{(p,q)}\sum_{i=1}^{d_\alpha} 1 = d_\alpha\lambda^{(p,q)} \quad . \tag{10.14}$$

On the other hand,

$$\sum_{i=1}^{d_\alpha} D^\alpha_{ip}(\hat{G}_a)D^\alpha_{qi}(\hat{G}_a^{-1}) = D^\alpha_{qp}(\hat{E})$$

holds, i.e.

$$\sum_{a=1}^{g} D^\alpha_{qp}(\hat{E}) = gD^\alpha_{qp}(\hat{E}) = \lambda^{(p,q)}d_\alpha \quad . \tag{10.15}$$

The representation matrix of the neutral element, however, is $D^\alpha_{qp}(E) = \delta_{qp}$, and thus from (10.15) it follows that

$$\lambda^{(p,q)} = \delta_{qp}\frac{g}{d_\alpha} \quad . \tag{10.16}$$

Hence the final form of the orthogonality relation for representations of finite groups is found to be

$$\sum_{a=1}^{g} D^\alpha_{ip}(\hat{G}_a)D^\beta_{qj}(\hat{G}_a^{-1}) = \frac{g}{d_\alpha}\delta_{\alpha\beta}\delta_{ij}\delta_{pq} \quad . \tag{10.17}$$

Later we will need this form for the derivation of the orthogonality relations for group characters. If the representation is unitary, which is the case to which we want to restrict ourselves from now on, then (10.17) can be simplified because of

$D^\beta_{qj}(\hat{G}_a^{-1}) = D^\beta_{jq}(\hat{G}_a)$ and therefore assumes the form

$$\sum_{a=1}^{g} D^\alpha_{ip}(\hat{G}_a) D^\beta_{jq}(\hat{G}_a)^* = \frac{g}{d_\alpha}\delta_{\alpha\beta}\delta_{ij}\delta_{pq} \quad . \tag{10.18}$$

10.4 Equivalence Classes

Before we introduce the notion of equivalence classes, we need the definition of the "conjugate" group elements: $\hat{G}_a$ is called conjugate to the element $\hat{G}_b$ if there exists an element $\hat{G}_n \in G$ with the property

$$\hat{G}_a = \hat{G}_n \hat{G}_b \hat{G}_n^{-1} \quad . \tag{10.19}$$

If $\hat{G}_b$ and $\hat{G}_c$ are conjugate to $\hat{G}_a$, then (10.19) holds for $\hat{G}_b$ and $\hat{G}_c$, as well, i.e.

$$\hat{G}_a = \hat{G}_n \hat{G}_b \hat{G}_n^{-1} \quad \text{and} \quad \hat{G}_a = \hat{G}_m \hat{G}_c \hat{G}_m^{-1} \quad .$$

This yields

$$\hat{G}_b = \hat{G}_n^{-1}\hat{G}_a\hat{G}_n = \hat{G}_n^{-1}\hat{G}_m\hat{G}_c\hat{G}_m^{-1}\hat{G}_n = (\hat{G}_n^{-1}\hat{G}_m)\hat{G}_c(\hat{G}_n^{-1}\hat{G}_m)^{-1} \quad . \tag{10.20}$$

This is exactly the property of an equivalence relation. All elements which are pair-wise conjugate form an equivalence class. The following examples serve to illustrate this concept.

EXAMPLE

10.1 The Group D_3

The symmetry group D_3 of an equilateral triangle (see figure) is composed of 6 elements $\hat{E}$, $\hat{R}_1$, $\hat{R}_2$, $\hat{R}_3$, $\hat{R}_4$, $\hat{R}_5$. $\hat{E}$ is the unity transformation, the application of which does not change the triangle. $\hat{R}_1$ and $\hat{R}_2$ rotate the triangle by the angle

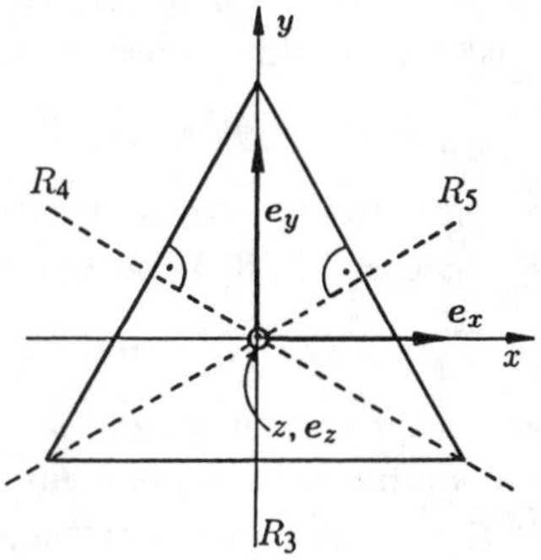

The symmetry group D_3 of an equilateral triangle

Example 10.1

$2\pi/3$ and $4\pi/3$ respectively, about the z-axis. The remaining operations $\hat{R}_3$, $\hat{R}_4$ and $\hat{R}_5$ denote reflections at the axes. With the help of these definitions we can construct a multiplication table, shown below.

Multiplication table of the symmetry group D_3 of an equilateral triangle

$\hat{G}_b$ \ $\hat{G}_a$	$\hat{E}$	$\hat{R}_1$	$\hat{R}_2$	$\hat{R}_3$	$\hat{R}_4$	$\hat{R}_5$
$\hat{E}$	$\hat{E}$	$\hat{R}_1$	$\hat{R}_2$	$\hat{R}_3$	$\hat{R}_4$	$\hat{R}_5$
$\hat{R}_1$	$\hat{R}_1$	$\hat{R}_2$	$\hat{E}$	$\hat{R}_4$	$\hat{R}_5$	$\hat{R}_3$
$\hat{R}_2$	$\hat{R}_2$	$\hat{E}$	$\hat{R}_1$	$\hat{R}_5$	$\hat{R}_3$	$\hat{R}_4$
$\hat{R}_3$	$\hat{R}_3$	$\hat{R}_5$	$\hat{R}_4$	$\hat{E}$	$\hat{R}_2$	$\hat{R}_1$
$\hat{R}_4$	$\hat{R}_4$	$\hat{R}_3$	$\hat{R}_5$	$\hat{R}_1$	$\hat{E}$	$\hat{R}_2$
$\hat{R}_5$	$\hat{R}_5$	$\hat{R}_4$	$\hat{R}_3$	$\hat{R}_2$	$\hat{R}_1$	$\hat{E}$

The notation is as follows: First one element of the upper row, e.g. the second one, $\hat{R}_1$, is applied to the triangle. This is followed by application of an element of the left column, e.g. the fifth one, $\hat{R}_4$. The result is given by the intersection of both elements, which, in this example, is in the fifth row and the second column.

We now look at these operations in more detail, by numbering the corners of the triangle.

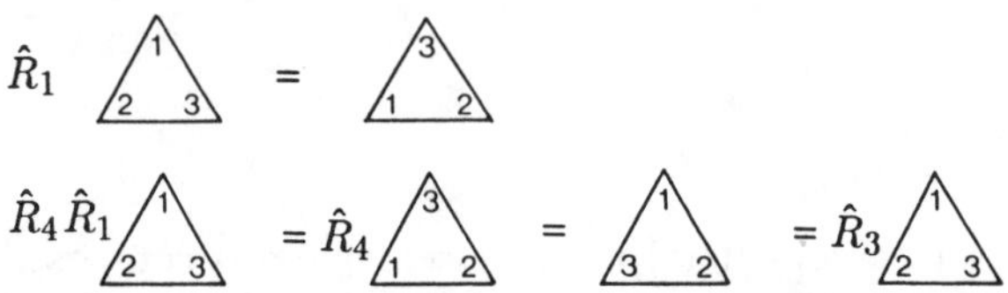

(1)

This leads to the result $\hat{R}_4\hat{R}_1 = \hat{R}_3$. In a similar manner, one can calculate all multiplications. The result is shown in the table.

Now what are the equivalence classes of this group? One finds that there exist exactly three classes:

$$\phi_1 = \hat{E} \qquad \phi_2 = \{\hat{R}_1, \hat{R}_2\} \qquad \phi_3 = \{\hat{R}_3, \hat{R}_4, \hat{R}_5\} \quad . \tag{2}$$

$\hat{E}$, the unity transformation, is always conjugate to itself, because $\hat{R}_i^{-1}\hat{E}\hat{R}_i = \hat{R}_i^{-1}\hat{R}_i = \hat{E}$. After simple calculation it follows, for example, that

$$\hat{R}_2 = \hat{R}_3\hat{R}_1\hat{R}_3^{-1} \quad \text{and} \quad \hat{R}_3 = \hat{R}_1\hat{R}_4\hat{R}_1^{-1} = \hat{R}_2\hat{R}_5\hat{R}_2^{-1} \quad , \tag{3}$$

which shows that $\hat{R}_2$ and $\hat{R}_1$ are conjugate to each other, as are $\hat{R}_3$, $\hat{R}_4$ and $\hat{R}_5$. To complete the assignment of all elements to classes, other combinations, e.g. $\hat{R}_4\hat{R}_1\hat{R}_4^{-1}$, must be constructed, and this is left as an exercise for the reader.

EXAMPLE

10.2 The Rotation Group O(3)

The O(3) consists of the elements

$$\hat{R}(\phi_1, \phi_2, \phi_3) = \exp\left(\mathrm{i}\sum \phi_k \hat{L}_k\right) \quad ,$$

where the $\hat{L}_k$ are the generators (angular momentum components) of the group SO(3). In order to be able to classify these elements, we must apply all possible similarity transformations on $\hat{R}(\phi_1, \phi_2, \phi_3)$, i.e.

$$\hat{R}'\hat{R}(\phi_1\,\phi_2\,\phi_3)\hat{R}'^{-1} = \hat{\tilde{R}}(\phi_1\,\phi_2\,\phi_3) \tag{1}$$

for arbitrary $\hat{R}'$. The interpretation of (1) is simple: First, with the help of $\hat{R}'^{-1}$, we rotate into a new system of axes, in which we perform a rotation with the Euler angles (ϕ_1, ϕ_2, ϕ_3), before rotating back into the old system. Therefore $\hat{\tilde{R}}(\phi_1, \phi_2, \phi_3)$ denotes the same rotation defined in other coordinate system. From (1) we see which elements constitute a class. All rotations by the same angle belong to the same class. To elucidate this point we consider the rotation by the angle ϕ about the axis $\boldsymbol{k}$ (e.g. the z-axis):

$$R_k(\phi) = \begin{pmatrix} \cos\phi & -\sin\phi & 0 \\ \sin\phi & \cos\phi & 0 \\ 0 & 0 & 1 \end{pmatrix} \quad . \tag{2}$$

Equation (1) transforms this into a rotation around an axis $\boldsymbol{k}'$, which can have an arbitrary direction in space.

$$\hat{R}'\hat{R}_k(\phi)\hat{R}'^{-1} = \hat{\tilde{R}}_{k'}(\phi) \quad . \tag{3}$$

$\hat{\tilde{R}}_{k'}(\phi)$ describes a rotation about the *same* angle ϕ. All such rotations by this angle about some arbitrary axis in three-dimensional space belong to the same class.

With this knowledge we can easily understand the result obtained for the group D_3: $\phi_1 = \{E\}$ describes a rotation by the angle 0, $\phi_2 = \{R_1, R_2\}$ describes a rotation by the angle $2\pi/3$ ($\phi = 4\pi/3$ is equivalent to $\phi = -2\pi/3$), and the elements in $\phi_3 = \{R_3, R_4, R_5\}$ each represent a rotation by π. Thus the group D_3 is a discrete subgroup of O(3), for which only distinct axes of rotation (each with fixed values of ϕ) are allowed. Groups like D_3 play an important role in crystallography.

10.5 Orthogonality Relations of the Group Characters for Discrete Groups and Other Relations

After these examples we return to the group characters which were defined in (10.1). First we prove the orthogonality relation for group characters by applying (10.18), setting $p = 1$ and $q = j$, and summing over all values of i and j.

$$\sum_{a=1}^{g} \sum_{i} D_{ii}^{\alpha}(\hat{G}_a) \sum_{j} D_{jj}^{\beta}(\hat{G}_a)^* = g\delta_{\alpha\beta} \quad . \tag{10.21}$$

Above we have used $\sum_i 1 = d_\alpha$; and α and β denote the two irreducible representations. If we use the definition of the group characters, (10.1), the orthogonality relation for group characters follows from (10.21) as

$$\sum_{a=1}^{g} \chi^{(\alpha)}(\hat{G}_a)\chi^{(\beta)}(\hat{G}_a)^* = g\delta_{\alpha\beta} \quad . \tag{10.22}$$

Since the characters are invariant under similarity transformations [cf. (10.2)], i.e. since all elements of one class have the same character, (10.22) simplifies to

$$\sum_{p=1}^{n} c_p \chi_p^{(\alpha)} \chi_p^{(\beta)} = g\delta_{\alpha\beta} \quad , \tag{10.23}$$

where c_p denotes the number of elements in class P and n is the total number of equivalence classes. We obtain an interesting result for the case of two identical representations, $\alpha = \beta$:

$$\sum_{a=1}^{g} |\chi^{(\alpha)}(\hat{G}_a)|^2 = \sum_{p=1}^{n} c_p |\chi_p^{(\alpha)}|^2 = g \quad . \tag{10.24}$$

We can also interpret the characters in (10.23) as follows: The character $\chi^{(\alpha)}$ can be understood as a vector with the components $\sqrt{c_p}\chi_p^{(\alpha)}$ in an n-dimensional space in which the irreducible characters form a set of orthogonal vectors. Thus it is obvious that there cannot exist more than n different irreducible representations!

10.6 Orthogonality Relations of the Group Characters for the Example of the Group D_3

First we try to construct a matrix representation for the group D_3 previously introduced in Example 10.1. For this purpose we consider the transformations of the triangle as rotations in three-dimensional space combined with reflections. We choose the axes $\boldsymbol{e}_x$, $\boldsymbol{e}_y$, and $\boldsymbol{e}_z$, as shown in the figure of Example 10.1, which, for instance, yields

$$\begin{aligned}
\hat{D}(\hat{R}_1)\boldsymbol{e}_x &= -\tfrac{1}{2}\boldsymbol{e}_x + \sqrt{\tfrac{3}{4}}\boldsymbol{e}_y \quad ,\\
\hat{D}(\hat{R}_1)\boldsymbol{e}_y &= -\sqrt{\tfrac{3}{4}}\boldsymbol{e}_x - \tfrac{1}{2}\boldsymbol{e}_y \quad ,\\
\hat{D}(\hat{R}_1)\boldsymbol{e}_z &= \boldsymbol{e}_z
\end{aligned} \tag{10.25}$$

for R_1 (rotation by $2\pi/3$ around the z-axis). For the other group elements we obtain

$$\hat{D}(\hat{R}_1) = \begin{pmatrix} -\frac{1}{2} & -\sqrt{\frac{3}{4}} & 0 \\ \sqrt{\frac{3}{4}} & -\frac{1}{2} & 0 \\ 0 & 0 & 1 \end{pmatrix} \qquad \hat{D}(\hat{R}_4) = \begin{pmatrix} \frac{1}{2} & -\sqrt{\frac{3}{4}} & 0 \\ -\sqrt{\frac{3}{4}} & -\frac{1}{2} & 0 \\ 0 & 0 & -1 \end{pmatrix}$$

$$\hat{D}(\hat{R}_2) = \begin{pmatrix} -\frac{1}{2} & -\sqrt{\frac{3}{4}} & 0 \\ \sqrt{\frac{3}{4}} & -\frac{1}{2} & 0 \\ 0 & 0 & 1 \end{pmatrix} \qquad \hat{D}(\hat{R}_5) = \begin{pmatrix} \frac{1}{2} & \sqrt{\frac{3}{4}} & 0 \\ \sqrt{\frac{3}{4}} & -\frac{1}{2} & 0 \\ 0 & 0 & -1 \end{pmatrix} \ . \tag{10.26}$$

$$\hat{D}(\hat{R}_3) = \begin{pmatrix} -1 & 0 & 0 \\ 0 & 1 & 0 \\ 0 & 0 & -1 \end{pmatrix} \qquad \hat{D}(\hat{E}) = \begin{pmatrix} 1 & 0 & 0 \\ 0 & 1 & 0 \\ 0 & 0 & 1 \end{pmatrix}$$

It is easy to verify that these matrices satisfy the rules of multiplication given in the table previously. Further, (10.26) reveals that all matrices have block structure, allowing the decomposition into a (2×2) and a (1×1) submatrix; therefore, the representation given in (10.26) is reducible. However, the submatrices form an irreducible representation because they cannot be further decomposed. The (2×2) submatrices form the irreducible representation $\hat{D}^{(3)}$ and the (1×1) submatrices form the representation $\hat{D}^{(2)}$. Finally there exists a trivial representation, namely the identical one, that is characterized by $\hat{D}^{(1)}(\hat{R}_i) = 1$ for all $\hat{R}_i$. The results are summarized in Table 10.1.

Table 10.1. The irreducible representations of the group element of D_3

Representation \ Group element	$\hat{E}$	$\hat{R}_1$	$\hat{R}_2$	$\hat{R}_3$	$\hat{R}_4$	$\hat{R}_5$
$\hat{D}^{(1)}$	1	1	1	1	1	1
$\hat{D}^{(2)}$	1	1	1	-1	-1	-1
$\hat{D}^{(3)}$	$\begin{pmatrix} 1 & 0 \\ 0 & 1 \end{pmatrix}$	$\begin{pmatrix} -\frac{1}{2} & -\sqrt{\frac{3}{4}} \\ \sqrt{\frac{3}{4}} & -\frac{1}{2} \end{pmatrix}$	$\begin{pmatrix} -\frac{1}{2} & \sqrt{\frac{3}{4}} \\ -\sqrt{\frac{3}{4}} & -\frac{1}{2} \end{pmatrix}$	$\begin{pmatrix} -1 & 0 \\ 0 & 1 \end{pmatrix}$	$\begin{pmatrix} \frac{1}{2} & -\sqrt{\frac{3}{4}} \\ -\sqrt{\frac{3}{4}} & -\frac{1}{2} \end{pmatrix}$	$\begin{pmatrix} \frac{1}{2} & \sqrt{\frac{3}{4}} \\ \sqrt{\frac{3}{4}} & -\frac{1}{2} \end{pmatrix}$

That they satisfy the orthogonality relations for irreducible representations (10.18) is left to the reader as an exercise.

After performing the trace of the matrices in Table 10.1, we obtain a character diagram, which is given in Table 10.2.

Table 10.2. Character table of the group D_3

Class / Representation	$\phi_1(\hat{E})$	$\phi_2(\hat{R}_1, \hat{R}_2)$	$\phi_3(\hat{R}_3, \hat{R}_4, \hat{R}_5)$
$\hat{D}^{(1)}$	1	1	1
$\hat{D}^{(2)}$	1	1	-1
$\hat{D}^{(3)}$	2	-1	0

The characters also satisfy the orthogonality relations (10.23). We consider, say, the orthogonality between the representation $\hat{D}^{(2)}$ and $\hat{D}^{(3)}$ (number of classes $n = 3$) as an exercise:

$$\sum_{p=1}^{n} c_p \chi_p^{(2)} \chi_p^{(3)*} = 1 \times 2 \times 1 + 2 \times 1 \times (-1) + 3 \times (-1) \times 0 = 0 \qquad (10.27)$$

$$\sum_{p=1}^{n} c_p \chi_p^{(2)} \chi_p^{(2)*} = 1 \times 1^2 + 2 \times 1^2 + 3 \times (-1)^2 = 6 = g \quad . \qquad (10.28)$$

Equation (10.28) just yields the number of group elements!

10.7 Reduction of a Representation

We now consider a general reducible representation $\hat{D}(\hat{G}_a)$. The group characters can help us to decompose this representation. We know that the matrices of the matrix representation $\hat{D}(\hat{G}_a)$ can be cast into block form. Then the character is the sum over the diagonal elements of each block. If $\hat{D}(\hat{G}_a)$ belongs to the class p, then

$$\chi_p = \sum_p m_\alpha \chi_p^{(\alpha)} \quad , \qquad (10.29)$$

where m_α is the number of equivalent irreducible representations, i.e. the number of times $D^\alpha(G_a)$ appears in $D(G_a)$. With the help of the orthogonality relation (10.23) we can determine m_α. Therefore we multiply with $(c_p/g)\chi_p^{(\beta)*}$ and sum over all classes p, obtaining

$$\frac{1}{g}\sum_p c_p \chi_p^{(\beta)*} \chi_p = \frac{1}{g}\sum_\alpha m_\alpha \sum_p c_p \chi_p^{(\beta)*} \chi_p^{(\alpha)} = \frac{1}{g}\sum_\alpha m_\alpha g \delta_{\alpha\beta} = m_\beta \quad . \qquad (10.30)$$

As an example we again consider D_3 and the representation introduced above. We decompose this representation (10.26) into the irreducible components (see Table 10.1)

$$\hat{D} = m_1\hat{D}^{(1)} \oplus m_2\hat{D}^{(2)} \oplus m_3\hat{D}^{(3)} \quad . \qquad (10.31)$$

"$\oplus$" denotes the direct sum, i.e. the formation of an extended matrix by joining the single matrices. With the help of (10.31) we get

$$m_1 = \tfrac{1}{6}(3+0-3) = 0 \quad , \quad m_2 = \tfrac{1}{6}(3+0+3) = 1 \quad ,$$
$$m_3 = \tfrac{1}{6}(6+0+0) = 1 \quad , \qquad (10.32)$$

i.e. the representations $\hat{D}^{(2)}$ and $\hat{D}^{(3)}$ are both included in $\hat{D}$, but $\hat{D}^{(1)}$ is not. This is the trivial result which has previously been given.

10.8 Criterion for Irreducibility

From the knowledge of the character we can conclude at once whether a given representation is irreducible or not. We then know from (10.24) that

$$\sum_p c_p|\chi_p|^2 = g \qquad (10.33)$$

holds. Equation (10.33) is a sufficient condition for irreducibility, which may be proved by considering (10.23) and (10.29);

$$\sum_p c_p|\chi_p|^2 = \sum_{\alpha,\beta,p} c_p m_\alpha m_\beta \chi^{(\alpha)}\chi_p^{(\beta)*} = g\sum_\alpha m_\alpha^2 \quad . \qquad (10.34)$$

Thus (10.33) holds if $\sum_\alpha m_\alpha^2 = 1$. Since all m_α are positive numbers or zero, however, this condition can only be fulfilled if all numbers m_α except one, e.g. $m_{\gamma'}$, are equal to zero, and $m_{\gamma'} = 1$. We consider the representation $D^{(3)}(R_i)$ of the group D_3 as an example and show that this is irreducible since

$$\sum_p c_p|\chi_p|^2 = (1\times 4 + 2\times 1 + 3\times 0) = 6 = g \quad . \qquad (10.35)$$

10.9 Direct Product of Representations

In the following we determine the character of a direct product of two irreducible representations. Let $\{D^\alpha_{im}(G_a)\}$ and $\{D^\beta_{im}(G_a)\}$ be the representation matrices of the irreducible representations α and β. Then the representation matrix of the direct product is given by the expression

$$\{D^{(\alpha\times\beta)}_{(ij)(mn)}(\hat{G}_a)\} = \{D^\alpha_{im}(\hat{G}_a)D^\beta_{jn}(\hat{G}_a)\} \qquad (10.36)$$

The trace is the sum over all diagonal elements

$$\chi^{(\alpha\times\beta)} = \sum_{ij} D^{(\alpha\times\beta)}_{(ii)(jj)}(\hat{G}_a) = \sum_i D^{(\alpha)}_{ii}(\hat{G}_a)\sum_j D^{(\beta)}_{jj}(\hat{G}_a) \quad . \tag{10.37}$$

The character of the representation of the direct product is equal to the product of the characters of the original representations α and β, which implies that

$$\chi^{(\alpha\times\beta)} = \chi^{(\alpha)}\cdot\chi^{(\beta)} \quad . \tag{10.38}$$

The representation resulting from (10.36) is in general reducible, i.e.

$$\hat{D}^{(\alpha\times\beta)}(\hat{G}_a) = \bigoplus_\gamma m_\gamma \hat{D}^\gamma(\hat{G}_a) \tag{10.39}$$

is the direct sum of irreducible representations. With the help of (10.30) and (10.38) we can determine m_γ to be

$$m_\gamma = \frac{1}{g}\sum_p c_p \chi_p^{(\gamma)*}\chi_p^{(\alpha)}\chi_p^{(\beta)} \quad . \tag{10.40}$$

Again we consider the group D_3 as an example: We can determine the character of the direct product $\hat{D}^{(3)}\otimes\hat{D}^{(3)} = \hat{D}^{(3\times 3)}$ with the help of Table 10.2 and (10.38). For the three classes the character takes the values

$$\chi^{(3\times 3)} = (4,1,0) = \chi^{(3)}\cdot\chi^{(3)} \quad , \tag{10.41}$$

with which, inserting this into (10.9.5), we obtain

$$\begin{aligned} m_1 &= \tfrac{1}{6}(1\times 1\times 4 + 2\times 1\times 1 + 3\times 1\times 0) = 1 \quad , \\ m_2 &= \tfrac{1}{6}(1\times 1\times 4 + 2\times 1\times 1 + 3\times(-1)\times 0) = 1 \quad , \\ m_3 &= \tfrac{1}{6}(1\times 2\times 4 + 2\times(-1)\times 1 + 3\times 0\times 0) = 1 \quad . \end{aligned} \tag{10.42}$$

Thus the representation $\hat{D}^{(3\times 3)}$ can be decomposed into

$$\hat{D}^{(3\times 3)} = \hat{D}^{(3)}\otimes\hat{D}^{(3)} = \hat{D}^{(1)}\oplus\hat{D}^{(2)}\oplus\hat{D}^{(3)} \quad . \tag{10.43}$$

10.10 Extension to Continuous, Compact Groups

Up to now we have only considered discrete groups because they are easier to handle. Yet groups with a continuous range of parameters have achieved great importance in physics, for example the groups SU(2) or SU(3), which play an important role in the elementary particle physics (see Chaps. 5 and 8). The difficulty resides in the fact that we now have an infinite number of group elements, because the group parameters are continuous. The sums appearing in the formulae derived above change into integrals, and the number of group elements is replaced by an appropriate normalization factor. First however, we need several new concepts, which shall be introduced in the following sections.

10.11 Mathematical Excursion: Group Integration

First we consider the expression

$$\sum_{a=1}^{g} f(\hat{G}_a) \tag{10.44}$$

for the case of a discrete finite-dimensional group. One important property of these groups was that in (10.44) $\hat{G}_a$ can be replaced by $\hat{G}_c = \hat{G}_b\hat{G}_a$ (with fixed but arbitrary $\hat{G}_b$) without changing the value. If the values of $\hat{G}_a$ cover the whole group, then the corresponding set $\hat{G}_c$ also covers the whole group exactly once. Otherwise some $\hat{G}_{a'}$, would exist which yields the same element $\hat{G}_c$:

$$\hat{G}_c = \hat{G}_b\hat{G}_a = \hat{G}_b\hat{G}_{a'} \quad , \tag{10.45}$$

Multiplication by $\hat{G}_b^{-1}$ from the left yields

$$\hat{G}_b^{-1}\hat{G}_c = \hat{G}_a = \hat{G}_{a'} \quad , \tag{10.46}$$

i.e. the elements $\hat{G}_a$ and $\hat{G}_{a'}$ must be identical!

In the case of a continuous group, (10.44) turns into an integral $\int \mu(\underline{a}) f(\underline{a}) d\underline{a}$, where $\underline{a}$ denotes the parameters of the group; the measure function $|\mu(\underline{a})|$ is chosen in order to fulfil

$$\int_{\underline{a}\in G} f(\underline{a})\mu(\underline{a})d\underline{a} = \int_{\underline{a}\in G} f(\underline{c})\mu(\underline{a})d\underline{a} \quad , \tag{10.47}$$

where c is defined by

$$\hat{G}(\underline{c}) = \hat{G}(\underline{a})\hat{G}(\underline{b}) \quad . \tag{10.48}$$

Here an arbitrary, but fixed, group element $\hat{G}(\underline{b})$ means that the group property for finite groups is conserved.

The integration only makes sense if the range of integration is compact (i.e. finite and closed). For the case of non-compact groups, the integral could possibly be divergent. In (10.47), $\underline{a}$ is the abbreviation for $\underline{a} = (a_1, a_2, \dots, a_r)$, and $d\underline{a} = da_1 d_2 \dots da_r$; r being the number of parameters of the group. Besides (10.47) there exists the trivial condition

$$\int_{\underline{a}\in G} f(\underline{a})\mu(\underline{a})d\underline{a} = \int_{\underline{c}\in G} f(\underline{c})\mu(\underline{c})d\underline{c} \quad . \tag{10.49}$$

If we change the integration variable on the rhs of (10.49) from $\underline{c}$ to $\underline{a}$ we get

$$\int_{\underline{a}\in G} f(\underline{a})\mu(\underline{a})d\underline{a} = \int_{\underline{a}\in G} f(\underline{c})\mu(\underline{c})\frac{\partial \underline{c}}{\partial \underline{a}}d\underline{a} \quad , \tag{10.50}$$

$\partial\underline{c}/\partial\underline{a}$ being the functional determinant (Jacobi determinant)

$$\frac{\partial \underline{c}}{\partial \underline{a}} = \begin{vmatrix} \dfrac{\partial c_1}{\partial a_1} & \dfrac{\partial c_1}{\partial a_2} & \cdots & \dfrac{\partial c_1}{\partial a_r} \\ \vdots & \vdots & \ddots & \vdots \\ \dfrac{\partial c_r}{\partial a_1} & \dfrac{\partial c_r}{\partial a_2} & \cdots & \dfrac{\partial c_r}{\partial a_r} \end{vmatrix} \quad . \tag{10.51}$$

If we compare (10.50) with the relation (10.44), which was required to hold, then the condition for μ is

$$\mu(\underset{\sim}{c})\frac{\partial \underset{\sim}{c}}{\partial \underset{\sim}{a}} = \mu(\underset{\sim}{a}) \quad , \tag{10.52}$$

valid for all $\underset{\sim}{a}$. Assuming that such a measure μ exists, we can calculate $\mu(\underset{\sim}{c})$ by assigning a value for an arbitrary element $\underset{\sim}{a}$, e.g. $\underset{\sim}{a} = \underset{\sim}{0}$ (this accords with the unity element of the Lie group). With this,

$$\mu(\underset{\sim}{c}) = \mu(0) \Big/ \left(\frac{\partial \underset{\sim}{c}}{\partial \underset{\sim}{a}}\right)_{\underset{\sim}{a}=\underset{\sim}{0}} \quad , \tag{10.53}$$

for which the constant factor $\mu(\underset{\sim}{0})$ is unimportant, so that it is sufficient to set $\mu(\underset{\sim}{0}) = 1$ and to calculate $(\partial \underset{\sim}{c}/\partial \underset{\sim}{a})_{\underset{\sim}{a}=\underset{\sim}{0}}$. Later we will explicitly carry out this procedure for the case of unitary groups, e.g. SU(2) and SU(3).

In the case of a finite, discrete group we have an expression of the general form

$$\frac{1}{g}\sum_{a=1}^{g} f(\hat{G}_a) \quad , \tag{10.54}$$

where g plays the role of the "volume" of the group and is therefore replaced by

$$g \to V(G) = \int_{\underset{\sim}{a} \in G} \mu(\underset{\sim}{a}) d\underset{\sim}{a} \quad . \tag{10.55}$$

For a continuous group, the expression

$$\frac{1}{V(G)} \int_{\underset{\sim}{a} \in G} f(\underset{\sim}{a}) \mu(\underset{\sim}{a}) d\underset{\sim}{a} \tag{10.56}$$

may be considered as a generalization of (10.54). In the case of the trivial representation $f(\underset{\sim}{a}) = 1$, then (10.56) has the value "one". *All formulae* of this previous section can be transformed into compact groups by replacing the expression (10.54) by (10.56), though before we consider the practical calculation of the measure function $\mu(\underset{\sim}{a})$, we must study the unitary groups in greater detail.

10.12 Unitary Groups

The unitary group $U(N)$ has N^2 generators $\hat{C}_{im}$ $(i, m = 1, \ldots, N)$, which satisfy the commutation rules

$$[\hat{C}_{im}, \hat{C}_{jm}]_- = \delta_{jm}\hat{C}_{in} - \delta_{in}\hat{C}_{im} \quad . \tag{10.57}$$

For example, we can take $\hat{C}_{im} = \hat{b}_i^+ \hat{b}_m$ as generators if $\hat{b}_i^+$, $\hat{b}_m$ are creation and annihilation operators with the commutation rules $[\hat{b}_m, \hat{b}_i^+] = \delta_{im}$, for which

$$\begin{aligned}[][\hat{b}_i^+ \hat{b}_m, \hat{b}_j^+ \hat{b}_n] &= \hat{b}_i^+ [\hat{b}_m, \hat{b}_j^+] \hat{b}_n + \hat{b}_i^+ \hat{b}_j^+ [\hat{b}_m, \hat{b}_n] + [\hat{b}_i^+, \hat{b}_j^+] \hat{b}_n \hat{b}_m + \hat{b}_j^+ [\hat{b}_i^+, \hat{b}_n] \hat{b}_m \\ &= \hat{b}_i^+ \delta_{mj} \hat{b}_n - \hat{b}_j^+ \delta_{in} \hat{b}_m = \delta_{jm} \hat{b}_i^+ \hat{b}_n - \delta_{in} \hat{b}_j^+ \hat{b}_m \quad ,\end{aligned}$$

is required. Another possible representation is

$$(\hat{C}_{im}) = (\delta_{im}) \quad , \tag{10.58}$$

where (δ_{im}) denotes the matrix which has a "1" at the intersection of the ith row and the mth column and "zeros" everywhere else. This matrix also satisfies the commutation rules (10.57). An arbitrary group element of $U(N)$ is given by

$$\exp\left(-\mathrm{i} \sum_{kl} \theta_{kl} \hat{C}_{kl} \right) \quad , \tag{10.59}$$

where the "angles" θ_{kl} parametrize the group transformation.

The transition to SU(N) can be made by constructing traceless matrices from the $\hat{C}_{im}$, which, for $i \neq m$, is automatically fulfilled. For $i = m$ one only has to subtract N^{-1} from each diagonal element, made possible because the unit matrix commutes with each matrix and therefore does not disturb the commutation rules:

$$\hat{\hat{C}}_{ii} = \hat{C}_{ii} - \frac{1}{N} \mathbb{1} \quad , \quad \hat{\hat{C}}_{ij} = \hat{C}_{ij} \quad , \quad i \neq j \quad , \tag{10.60}$$

i.e. $\hat{\hat{C}}_{ii}$ in (10.60) has the form

$$\hat{\hat{C}}_{ii} = \underbrace{\left(\begin{array}{ccccccc} -\frac{1}{N} & & & & & & \\ & \ddots & & & 0 & & \\ & & -\frac{1}{N} & & & & \\ & & & \underset{\substack{\uparrow \\ i}}{1 - \frac{1}{N}} & & & \longleftarrow i \\ & & & & -\frac{1}{N} & & \\ & & 0 & & & \ddots & \\ & & & & & & -\frac{1}{N} \end{array} \right)}_{N} \left. \begin{array}{c} \\ \\ \\ \\ \\ \\ \\ \end{array} \right| N \quad , \tag{10.61}$$

from which it is easy to calculate that $\mathrm{Tr}(\hat{\hat{C}}_{ii}) = 0$. Equation (10.60) gives us a condition for the $\hat{C}_{ii}$'s, because

$$\sum_i \hat{\hat{C}}_{ii} = 0 \tag{10.62}$$

holds, i.e. just $(N-1)$ out of N creation operators are linearly independent. All in all we obtain exactly N^2-1 generators for the $\mathrm{SU}(N)$.

10.13 The Transition from U(N) to SU(N) for the example SU(3)

Up to now we have used other matrices as generators of SU(3). These are connected with the matrices (10.58) by the following relations:

$$\hat{\lambda}_1 = \begin{pmatrix} 0 & 1 & 0 \\ 1 & 0 & 0 \\ 0 & 0 & 0 \end{pmatrix} = \hat{C}_{21} + \hat{C}_{12} \quad ,$$

$$\hat{\lambda}_2 = \begin{pmatrix} 0 & -\mathrm{i} & 0 \\ \mathrm{i} & 0 & 0 \\ 0 & 0 & 0 \end{pmatrix} = \mathrm{i}(\hat{C}_{21} - \hat{C}_{12}) \quad ,$$

$$\hat{\lambda}_3 = \begin{pmatrix} 1 & 0 & 0 \\ 0 & -1 & 0 \\ 0 & 0 & 0 \end{pmatrix} = \hat{C}_{11} - \hat{C}_{22} \quad ,$$

$$\hat{\lambda}_4 = \begin{pmatrix} 0 & 0 & 1 \\ 0 & 0 & 0 \\ 1 & 0 & 0 \end{pmatrix} = \hat{C}_{31} + \hat{C}_{13} \quad ,$$

$$\hat{\lambda}_5 = \begin{pmatrix} 0 & 0 & -\mathrm{i} \\ 0 & 0 & 0 \\ \mathrm{i} & 0 & 0 \end{pmatrix} = \mathrm{i}(\hat{C}_{31} - \hat{C}_{13}) \quad ,$$

$$\hat{\lambda}_6 = \begin{pmatrix} 0 & 0 & 0 \\ 0 & 0 & 1 \\ 0 & 1 & 0 \end{pmatrix} = \hat{C}_{32} + \hat{C}_{23} \quad ,$$

$$\hat{\lambda}_7 = \begin{pmatrix} 0 & 0 & 0 \\ 0 & 0 & -\mathrm{i} \\ 0 & \mathrm{i} & 0 \end{pmatrix} = \mathrm{i}(\hat{C}_{32} - \hat{C}_{23}) \quad ,$$

$$\hat{\lambda}_8 = \frac{1}{\sqrt{3}} \begin{pmatrix} 1 & 0 & 0 \\ 0 & 1 & 0 \\ 0 & 0 & -2 \end{pmatrix} = \frac{1}{\sqrt{3}}(\hat{C}_{11} + \hat{C}_{22} - 2\hat{C}_{33}) \quad . \tag{10.63}$$

We will repeatedly make use of (10.63) in the following.

The connection between the groups U(3) and SU(3) is as follows: An element of U(3) is given by

$$\exp\left(-\mathrm{i}\sum_{k,l=1}^{3}\theta_{kl}\hat{C}_{kl}\right)=\exp\left(-\mathrm{i}\sum_{k\neq l=1}^{3}\theta_{kl}\hat{C}_{kl}-\mathrm{i}\sum_{k=1}^{3}\theta_{kk}\hat{C}_{kk}\right) \quad , \tag{10.64}$$

which can be further transformed by using (10.60) to obtain

$$\exp\left(-\mathrm{i}\sum_{k\neq l=1}^{3}\theta_{kl}\hat{\hat{C}}_{kl}-\mathrm{i}\sum_{k=1}^{3}\theta_{kk}\hat{\hat{C}}_{kk}\right)\exp\left(-\frac{\mathrm{i}}{N}\mathbb{1}\sum_{k=1}^{3}\theta_{kk}\right) \quad . \tag{10.65}$$

The term $-\frac{\mathrm{i}}{N}(\sum_{k=1}^{3}\theta_{kk})\,\mathbb{1}$ in the exponent cancels the $\hat{C}_{kk}$, and the left hand part of (10.65) represents a group element which has just the form of an SU(3) element; the right hand part is simply a phase factor. Now the transition from U(3) to SU(3) takes place by requiring

$$\mathrm{U}(3)\rightarrow\mathrm{SU}(3):\ \sum_{k=1}^{3}\theta_{kk}=0 \tag{10.66}$$

in (10.65). In the general case of $U(N)$, one proceeds in the same manner, with the difference that the index k runs from 1 to N. If, however, we want to use the generators in (10.63), we must apply the following method. First we confine ourselves to the diagonal matrices $\hat{C}_{11}$, $\hat{C}_{22}$, $\hat{C}_{33}$, returning to the remaining matrices later. Then group element generated only by $\hat{C}_{11}$, $\hat{C}_{22}$, and $\hat{C}_{33}$ has the expression

$$\theta_{11}\hat{C}_{11}+\theta_{22}\hat{C}_{22}+\theta_{33}\hat{C}_{33}=\phi\hat{\lambda}_3+(\psi/\sqrt{3})\hat{\lambda}_8+1/3(\theta_{11}+\theta_{22}+\theta_{33})\hat{N} \tag{10.67}$$

in the exponent. We are interested in the relation between (ϕ,ψ) and $(\theta_{11},\theta_{22},\theta_{33})$ for the unit matrix $N[\hat{N}=(\hat{C}_{11}+\hat{C}_{22}+\hat{C}_{33})=\mathbb{1}]$. If we make use of (10.63), $\hat{\lambda}_3$, $\hat{\lambda}_8$ and $\hat{N}$ can be expressed by $\hat{C}_{11}$, $\hat{C}_{22}$ and $\hat{C}_{33}$ and vice versa. This yields the relations between the angles

$$\theta_{11}-\theta_{22}=\phi \quad , \quad \theta_{11}-\theta_{33}=\psi+\tfrac{1}{2}\phi \quad , \quad \theta_{22}-\theta_{33}=\psi-\tfrac{1}{2}\phi \quad . \tag{10.68}$$

In the case of SU(3) we have to satisfy the auxiliary condition $\theta_{11}+\theta_{22}+\theta_{33}=0$.

Finally we consider the form of a finite transformation which consists of diagonal elements only. The infinitesimal transformation is given by

$$1-\mathrm{i}\sum_{k}\theta_{kk}\hat{C}_{kk}-\begin{pmatrix}1-\mathrm{i}\theta_{11} & 0 & \cdots & 0\\ 0 & 1-\mathrm{i}\theta_{22} & \cdots & 0\\ \vdots & \vdots & \ddots & \vdots\\ 0 & 0 & \cdots & 1-\mathrm{i}\theta_{NN}\end{pmatrix} \quad , \tag{10.69}$$

from which the finite transformation follows as

$$\exp\left(-\mathrm{i}\sum_k \theta_{kk}\hat{C}_{kk}\right) = \begin{pmatrix} \mathrm{e}^{-\mathrm{i}\theta_{11}} & 0 & \cdots & 0 \\ 0 & \mathrm{e}^{-\mathrm{i}\theta_{22}} & \cdots & 0 \\ \vdots & \vdots & \ddots & \vdots \\ 0 & 0 & \cdots & \mathrm{e}^{-\mathrm{i}\theta_{NN}} \end{pmatrix}$$

$$= \begin{pmatrix} \varepsilon_1 & 0 & \cdots & 0 \\ 0 & \varepsilon_2 & \cdots & 0 \\ \vdots & \vdots & \ddots & \vdots \\ 0 & 0 & \cdots & \varepsilon_N \end{pmatrix} , \tag{10.70}$$

[introducing $\varepsilon_k = \exp(-\mathrm{i}\theta_{kk})$]. Later on we need the most general form of a unitary transformation; therefore we construct other generators as the $\hat{C}_{ij}(i \neq j)$ so that, in analogy to SU(3) [see (10.63)],

$$\hat{\lambda}_{ij} = \hat{C}_{ij} + \hat{C}_{ji} \quad , \quad \hat{\lambda}_{ji} = \mathrm{i}(\hat{C}_{ij} - \hat{C}_{ji}) \quad , \quad i < j \ . \tag{10.71}$$

By this we obtain for an infinitesimal transformation:

$$1 - \mathrm{i}\sum_{kl} \theta_{kl}\hat{C}_{kl} = 1 - \mathrm{i}\sum_k \tilde{\theta}_{kk}\hat{C}_{kk} - \mathrm{i}\sum_{k<l} \tilde{\theta}_{kl}\hat{\lambda}_{kl} - \mathrm{i}\sum_{k<l} \tilde{\theta}_{lk}\hat{\lambda}_{lk}$$

$$= \begin{pmatrix} 1 - \mathrm{i}\tilde{\theta}_{11} & \tilde{\theta}_{12} - \mathrm{i}\tilde{\theta}_{21} & \cdots \\ \tilde{\theta}_{12} + \mathrm{i}\tilde{\theta}_{21} & 1 - \mathrm{i}\tilde{\theta}_{22} & \cdots \\ \vdots & \vdots & \ddots \end{pmatrix} \Bigg\updownarrow N \tag{10.72}$$

with

$$\tilde{\theta}_{kl} + \mathrm{i}\tilde{\theta}_{lk} = \theta_{kl} \quad , \quad \tilde{\theta}_{kl} - \mathrm{i}\tilde{\theta}_{lk} = \theta_{lk} \quad , \quad k < l \quad \text{and} \quad \tilde{\theta}_{kk} = \theta_{kk} \ . \tag{10.73}$$

10.14 Integration over Unitary Groups

The unitary group U(N) depends on N^2 parameters so that an arbitrary transformation written in terms of operators is given by

$$\exp\left(-\mathrm{i}\sum_{k,l=1}^{N} \theta_{kl}\hat{C}_{kl}\right) \ . \tag{10.74}$$

In order to determine $\mu(\theta_{kl})$ we have to calculate the product $\hat{U}(c) = \hat{U}(a)\hat{U}(b)$, where $\hat{U}(b)$ is chosen to be diagonal matrix, i.e. in (10.74) there occur only operators of the Cartan sub-algebra which commute with each other (see Chap. 12). $\hat{U}(b)$ is of the form (10.72) implying that for $\hat{U}(c)$ we obtain

$$\hat{U}(c) = \begin{pmatrix} (1 - \mathrm{i}a_{11})\varepsilon_1 & (a_{12} - \mathrm{i}a_{21})\varepsilon_2 & \cdots \\ (a_{12} + \mathrm{i}a_{21})\varepsilon_1 & (1 - \mathrm{i}a_{22})\varepsilon_2 & \cdots \\ \vdots & \vdots & \ddots \end{pmatrix} \ . \tag{10.75}$$

This matrix can be diagonalized by the transformation

$$\hat{U} = \hat{V}^{-1}\hat{W}\hat{V} \quad , \quad \text{with} \tag{10.76}$$

$$V(c) = \begin{pmatrix} 1 & \dfrac{a_{12} - \mathrm{i}a_{21}}{\varepsilon_1/\varepsilon_2 - 1} & \cdots \\ \dfrac{a_{12} + \mathrm{i}a_{12}}{\varepsilon_2/\varepsilon_1 - 1} & 1 & \cdots \\ \vdots & \vdots & \ddots \end{pmatrix} \quad \text{and} \tag{10.77}$$

$$W(c) = \begin{pmatrix} (1 - \mathrm{i}a_{11})\varepsilon_1 & 0 & \cdots \\ 0 & (1 - \mathrm{i}a_{22}) & \cdots \\ \vdots & \vdots & \ddots \end{pmatrix} \quad , \tag{10.78}$$

where we have neglected terms of second order ($a_{ik} \ll 1$). The explicit calculation of $\hat{V}\hat{U} = \hat{W}\hat{V}$ is the best way to prove the validity of (10.76), and so the transformation is separated into three parts; $\hat{W}$ is now a diagonal matrix and depends on N parameters only! If we write $(1 - \mathrm{i}a_{kk}) \simeq \exp(-\mathrm{i}a_{kk})$ and take into account that $\varepsilon_k = \exp(-\mathrm{i}\theta_{kk})$, we can immediately specify the transformation of N angles by

$$\hat{C}_{kk} = \theta_{kk} = \theta_{kk} + a_{kk} \quad . \tag{10.79}$$

The transformation matrix $\hat{V}$ depends on the remaining $N^2 - N$ parameters and contains the angles which describe the transformation into a system in which $\hat{U}$ is diagonal. We choose the transformation of the parameters [c.f. the element (12) of $\hat{V}$ with the corresponding one of $\hat{U}(c)$]:

$$\begin{aligned} \hat{C}_{12} &= a_{12}\,\mathrm{Re}(\varepsilon_1/\varepsilon_2 - 1)^{-1} + a_{21}\,\mathrm{Im}(\varepsilon_1/\varepsilon_2 - 1)^{-1} \quad , \\ \hat{C}_{21} &= a_{12}\,\mathrm{Im}(\varepsilon_1/\varepsilon_2 - 1)^{-1} - a_{21}\,\mathrm{Re}(\varepsilon_1/\varepsilon_2 - 1)^{-1} \quad , \quad \text{etc.} \end{aligned} \tag{10.80}$$

Thus we achieve the following structure of the Jacobi determinant that has, with regard to the first N parameters, a diagonal form:

$$\partial\hat{C}_{kk}/\partial a_{ij} = \delta_{ki}\delta_{kj} \quad . \tag{10.81}$$

The remaining part is a product of $\frac{1}{2}N(N-1)$ factors, i.e.

$$\begin{vmatrix} \partial\hat{C}_{12}/\partial a_{12} & \partial\hat{C}_{12}/\partial a_{21} \\ \partial\hat{C}_{21}/\partial a_{12} & \partial\hat{C}_{21}/\partial a_{21} \end{vmatrix} = |\varepsilon_1 - \varepsilon_2|^{-2} \quad , \quad \text{etc.} \tag{10.82}$$

Finally for $\mu(\underline{c})$ we obtain

$$\mu(\underline{c}) = (\partial\underline{c}/\partial\underline{a})^{-1}_{\underline{a}=\underline{0}} = \prod_{i<j}^{N} |\varepsilon_i - \varepsilon_j|^2 \quad . \tag{10.83}$$

Note, that $\mu(c = \theta)$ *depends only on* N *of the parameters*!

a) U(2), SU(2). As an example let us have a look at the group U(2). The measure is given by (10.83); hence

$$\begin{aligned}\mu(\mathrm{U}(2)) &= |\varepsilon_1 - \varepsilon_2|^2 = |\exp(-\mathrm{i}\theta_{11}) - \exp(-\mathrm{i}\theta_{22})|^2 \\ &= |\exp(-\tfrac{1}{2}\mathrm{i}(\theta_{11}+\theta_{22}))|^2 \exp(-\tfrac{1}{2}\mathrm{i}(\theta_{11}-\theta_{22})) \\ &\quad - |\exp(\tfrac{1}{2}\mathrm{i}(\theta_{11}-\theta_{22}))|^2 = 4\sin^2(\tfrac{1}{2}(\theta_{11}-\theta_{22})) \quad .\end{aligned} \tag{10.84}$$

The restriction to SU(2) yields $\theta_{11} + \theta_{22} = 0$. With $\theta_{11} - \theta_{22} = \phi$ then

$$\mu(\mathrm{SU}(2)) = 4\sin^2\tfrac{1}{2}\phi \quad . \tag{10.85}$$

According to (10.56) this result must be divided by

$$V = \int_0^{4\pi} \mu(\mathrm{SU}(2))\, d\phi = 8\pi \quad ,$$

with the consequence that

$$V^{-1}\int \mu(\phi) f(\phi)\, d\phi = \frac{1}{2\pi}\int_0^{4\pi} (\sin^2\tfrac{1}{2}\phi) f(\phi)(d\phi) \quad . \tag{10.86}$$

b) U(3), SU(3). In this case we proceed similarly by looking first at U(3). According to (10.83),

$$\mu(\mathrm{U}(3)) = |\varepsilon_1 - \varepsilon_2|^2 |\varepsilon_1 - \varepsilon_3|^2 |\varepsilon_2 - \varepsilon_3|^2 \tag{10.87}$$

is valid. Each factor is treated in the same way as in the example U(2) → SU(2) and

$$\begin{aligned}\mu(\mathrm{U}(3)) &= 4\left|\exp\left[-\frac{\mathrm{i}(\theta_{11}+\theta_{22})}{2}\right]\right|^2 \sin^2\left(\frac{\theta_{11}-\theta_{22}}{2}\right) \\ &\quad \times 4\left|\exp\left[-\frac{\mathrm{i}(\theta_{11}+\theta_{33})}{2}\right]\right|^2 \sin^2\left(\frac{\theta_{11}-\theta_{33}}{2}\right) \\ &\quad \times 4\left|\exp\left[-\frac{\mathrm{i}(\theta_{22}+\theta_{33})}{2}\right]\right|^2 \sin^2\left(\frac{\theta_{22}-\theta_{33}}{2}\right) \\ &= 64\sin^2\left(\frac{\theta_{11}-\theta_{22}}{2}\right)\sin^2\left(\frac{\theta_{11}-\theta_{33}}{2}\right)\sin^2\left(\frac{\theta_{22}-\theta_{33}}{2}\right) \quad .\end{aligned} \tag{10.88}$$

Finally for SU(3) we get, according to (10.66) and (10.68),

$$\mu(\mathrm{SU}(3)) = 64\sin^2(\tfrac{1}{2}\phi)\sin^2[\tfrac{1}{2}(\tfrac{1}{2}\phi + \psi)]\sin^2[\tfrac{1}{2}(-\tfrac{1}{2}\phi + \psi)] \quad , \tag{10.89}$$

where ϕ is the parameter for isospin and ψ for hypercharge, respectively. According to (10.56) we have to divide by

$$V = \int_{-\pi}^{+\pi}\int_{-\pi}^{+\pi} d(\tfrac{1}{2}\phi)\, d(\tfrac{1}{3}\psi)\, \mu(\mathrm{SU}(3)) \quad . \tag{10.90}$$

The limits above arise because in (10.88) the limits for the arguments of the sine are given by

$$0 \le \theta_{11} - \theta_{22} \quad , \quad \theta_{11} - \theta_{33} \le 2\pi \quad , \quad \theta_{22} - \theta_{33} \le 2\pi \quad , \tag{10.91}$$

i.e. over a total period. For $\frac{1}{2}\phi = \frac{1}{2}(\theta_{11} - \theta_{22})$ this yields the limits

$$-\pi \le \tfrac{1}{2}\phi \le +\pi \quad . \tag{10.92}$$

Finally the limits of ψ have to be determined. If we replace $\psi + \frac{1}{2}\phi = \theta_{11} - \theta_{33}$ by the corresponding maximal and minimal values for $\theta_{11} - \theta_{33}$ and $\frac{1}{2}\phi$, respectively, we obtain

$$-\pi \le \tfrac{1}{3}\psi \le +\pi \quad . \tag{10.93}$$

The result of the integration (10.90) is

$$V = 64 \times 3\pi^2/8 = 24\pi^2 \quad , \tag{10.94}$$

and thereby all integrations of interest are of the form

$$\begin{aligned}
&\frac{1}{V}\int_{\mathrm{SU(3)}} \mu(\mathrm{SU(3)})\,f(\phi,\psi)\,d(\mathrm{SU(3)}) \\
&= \frac{8}{3\pi^2}\int_{-\pi}^{+\pi}\int_{-\pi}^{+\pi} d(\tfrac{1}{2}\phi)\,d(\tfrac{1}{3}\psi)\sin^2\tfrac{1}{2}\phi\,\sin^2[\tfrac{1}{2}(\tfrac{1}{2}\phi + \psi)] \\
&\quad \times \sin^2[\tfrac{1}{2}(-\tfrac{1}{2}\phi + \psi)]\,f(\phi,\psi) \quad .
\end{aligned} \tag{10.95}$$

Here we assume that "f" only depends on ϕ and ψ. If f depended on other angles, too, we would have to integrate over them and multiply V by $V_0 = \int d(\theta_{ij})(i \ne j)$. We do not get into trouble with the volume element $\mu(\mathrm{SU}(3))$, because it does not depend on these angles!

10.15 Group Characters of Unitary Groups

If an element of a group is given, e.g.

$$\hat{G}_a = \exp\left(-\mathrm{i}\sum_{ij}\theta_{ij}\hat{C}_{ij}\right) \quad , \tag{10.96}$$

then the character is defined by

$$\chi(\hat{G}_a) = \mathrm{Tr}[\hat{G}_a] \quad . \tag{10.97}$$

However, we know that the trace formation is invariant under similarity transformations, i.e. $\hat{A}\hat{G}_\alpha\hat{A}^{-1}$ has the same trace and the same character, and every unitary matrix can be transformed into diagonal form. But since all representation matrices in diagonal form can be constructed solely by the generators of the Cartan sub-algebra

$$\exp\left(-\sum_{k=1}^{N}\theta_{kk}\hat{C}_{kk}\right) \quad , \tag{10.98}$$

it is sufficient to consider only group elements of this form. If we denote the eigenvalues (weights) of $\hat{C}_{kk}$ $(k = 1, \ldots, N)$ by $(r_1, \ldots, r_N)$, then (10.97) finally assumes the form

$$\chi^{(\alpha)} = \sum_{r_1,\ldots,r_N} \exp\left(-\mathrm{i}\sum_{k=1}^{N} \theta_{kk} r_k\right) \quad , \tag{10.99}$$

in which we sum over all possible weights. The term "(α)" denotes the representation and $(\theta_{11}, \ldots, \theta_{NN})$ the class.

There exists a general form of the group characters of the unitary group, which we will prove, having first presented the result and some examples. If $\varepsilon_j = \exp(-\mathrm{i}\theta_{jj})$, then the group character is given by the following ratios of two determinants:

$$\chi^{(\alpha)} = \begin{vmatrix} \varepsilon_1^{h_{1N}+N-1} & \varepsilon_1^{h_{2N}+N-2} & \ldots & \varepsilon_1^{h_{NN}} \\ \varepsilon_2^{h_{1N}+N-1} & \varepsilon_2^{h_{2N}+N-2} & \ldots & \varepsilon_2^{h_{NN}} \\ \vdots & & & \\ \varepsilon_N^{h_{1N}+N-1} & \varepsilon_N^{h_{2N}+N-2} & \ldots & \varepsilon_N^{h_{NN}} \end{vmatrix} \cdot \begin{vmatrix} \varepsilon_1^{N-1} & \varepsilon_1^{N-2} & \ldots & 1 \\ \varepsilon_2^{N-1} & \varepsilon_2^{N-2} & \ldots & 1 \\ \vdots & & & \\ \varepsilon_N^{N-1} & \varepsilon_N^{N-2} & \ldots & 1 \end{vmatrix}^{-1} \tag{10.100}$$

Here the values $(h_{1N}, \ldots, h_{NN})$ characterize the representation of the group U(N), being the maximal weights $(r_1 = h_{1N}, r_2 = h_{2N}, \ldots, r_N = h_{NN})$. In the case of SU$(N)$ we set $\sum_k \theta_{kk} = 0$, so that the representation of the SU(N) is then characterized by the numbers

$$(h_{1N} - h_{2N} \quad , \quad h_{2N} - h_{3N}, \ldots \quad , \quad h_{(N-1)N} - h_{NN}, 0) \quad . \tag{10.101}$$

a) U(1). This is trivial, because U(1) consists only of a single element

$$\chi^{(\alpha)} = \exp(-\mathrm{i}h_{11}\theta_{11}) \quad . \tag{10.102}$$

b) U(2). This representation is characterized by h_{12}, h_{22}, and the group character is

$$\chi^{(\alpha)} = \begin{vmatrix} \varepsilon_1^{h_{12}+1} & \varepsilon_1^{h_{22}} \\ \varepsilon_2^{h_{12}+1} & \varepsilon_2^{h_{22}} \end{vmatrix} \cdot \begin{vmatrix} \varepsilon_1 & 1 \\ \varepsilon_2 & 1 \end{vmatrix}^{-1} = \frac{\varepsilon_1^{h_{12}+1}\varepsilon_2^{h_{22}} - \varepsilon_1^{h_{22}}\varepsilon_2^{h_{12}+1}}{\varepsilon_1 - \varepsilon_2} \quad . \tag{10.103}$$

With $\varepsilon_i = \exp(-\mathrm{i}\theta_{ii})$ we obtain

$$\begin{aligned} \chi^{(\alpha)} &= \frac{\exp[-\mathrm{i}((h_{12}+1)\theta_{11} + h_{22}\theta_{22})] - \exp[-\mathrm{i}(h_{22}\theta_{11} + (h_{12}+1)\theta_{22})]}{\exp(-\mathrm{i}\theta_{11}) - \exp(-\mathrm{i}\theta_{22})} \\ &= \exp\left(\mathrm{i}\frac{(\theta_{11}+\theta_{22})(h_{12}+1+h_{22})}{2}\right) \\ &\quad \times \left(\frac{\exp\{-\mathrm{i}\frac{1}{2}[(h_{12}+1)\theta_{11} + \frac{1}{2}h_{22}\theta_{22} - \frac{1}{2}\theta_{11}h_{22} - \frac{1}{2}\theta_{22}(h_{12}+1)]\}}{\exp(-\mathrm{i}\theta_{11}) - \exp(-\mathrm{i}\theta_{22})}\right. \\ &\quad \left. - \frac{\exp\{-\mathrm{i}[\frac{1}{2}h_{22}\theta_{11} + \frac{1}{2}(h_{12}+1)\theta_{22} - \frac{1}{2}\theta_{11}(h_{12}+1) - \frac{1}{2}\theta_{22}h_{22}]\}}{\exp(-\mathrm{i}\theta_{11}) - \exp(-\mathrm{i}\theta_{22})}\right) \end{aligned}$$

$$
\begin{aligned}
&= \exp\left(-\mathrm{i}\,\frac{(\theta_{11}+\theta_{22})(h_{12}+h_{22}+1)}{2}\right)\\
&\quad\times\left(\frac{\exp\{-\mathrm{i}[\frac{1}{2}(h_{12}+1-h_{22})](\theta_{11}-\theta_{22})\}}{\exp(-\mathrm{i}\theta_{11})-\exp(-\mathrm{i}\theta_{22})}\right.\\
&\quad\left.-\frac{\exp\{\mathrm{i}[\frac{1}{2}(h_{12}+1-h_{22})](\theta_{11}-\theta_{22})\}}{\exp(-\mathrm{i}\theta_{11})-\exp(-\mathrm{i}\theta_{22})}\right)\\
&= \exp\left(-\mathrm{i}\,\frac{(\theta_{11}+\theta_{22})(h_{12}+h_{22})}{2}\right)\\
&\quad\times\left\{\sin\left[\frac{(h_{12}+1-h_{22})}{2}(\theta_{11}-\theta_{22})\right]\right\}\Big/\left(\sin\frac{\theta_{11}-\theta_{22}}{2}\right) \quad . \qquad (10.104)
\end{aligned}
$$

In the last step we made use of

$$
\begin{aligned}
&\exp(-\mathrm{i}\theta_{11})-\exp(-\mathrm{i}\theta_{22})\\
&\quad= \exp(-\tfrac{1}{2}\mathrm{i}(\theta_{11}+\theta_{22}))[\exp(-\tfrac{1}{2}\mathrm{i}(\theta_{11}-\theta_{22}))-\exp(\tfrac{1}{2}\mathrm{i}(\theta_{11}-\theta_{22}))] \quad .
\end{aligned} \qquad (10.105)
$$

When we make the transition to SU(2), the representation is characterized by

$$
(h_{12}-h_{22},0)=(2j,0) \quad , \qquad (10.106)
$$

according to (10.102). Furthermore, we set $\theta_{11}+\theta_{22}=0$ [according to (10.66)] and thus obtain

$$
\chi(\mathrm{SU}(2)) = \sin((2j+1)\phi/2)/\sin\phi/2 \qquad (10.107)
$$

for the character of SU(2), with $\phi=\theta_{11}-\theta_{22}$.

c) U(3). The representation is characterized by three numbers (h_{11},h_{22},h_{33}), with group character

$$
\chi^{(\alpha)}(\mathrm{U}(3)) = \begin{vmatrix} \varepsilon_1^{h_{13}+2} & \varepsilon_1^{h_{23}+1} & \varepsilon_1^{h_{33}} \\ \varepsilon_2^{h_{13}+2} & \varepsilon_2^{h_{23}+1} & \varepsilon_2^{h_{33}} \\ \varepsilon_3^{h_{13}+2} & \varepsilon_3^{h_{23}+1} & \varepsilon_3^{h_{33}} \end{vmatrix} \cdot \begin{vmatrix} \varepsilon_1^2 & \varepsilon_1 & 1 \\ \varepsilon_2^2 & \varepsilon_2 & 1 \\ \varepsilon_3^2 & \varepsilon_3 & 1 \end{vmatrix}^{-1} \quad . \qquad (10.108)
$$

The denominator can be simplified further, i.e.

$$
\begin{aligned}
&(\varepsilon_1-\varepsilon_2)(\varepsilon_1-\varepsilon_3)(\varepsilon_2-\varepsilon_3)\\
&\quad= -\exp[-\tfrac{1}{2}\mathrm{i}(\theta_{11}+\theta_{22})]\exp[-\tfrac{1}{2}\mathrm{i}(\theta_{11}+\theta_{33})]\exp[-\tfrac{1}{2}\mathrm{i}(\theta_{22}+\theta_{33})]\\
&\qquad\times 8\sin[\tfrac{1}{2}(\theta_{11}-\theta_{22})]\sin[\tfrac{1}{2}(\theta_{11}-\theta_{33})]\sin[\tfrac{1}{2}(\theta_{22}-\theta_{33})] \quad , \qquad (10.109)
\end{aligned}
$$

but the numerator cannot be transformed to a simpler form. Nevertheless we shall derive an expression which we will later need for the general proof of

(10.100):

$$\chi^{(\alpha)}(U(3)) = \sum_{h'_{12}=h_{23}}^{h_{13}} \sum_{h'_{22}=h_{33}}^{h_{23}} \frac{\begin{vmatrix} \varepsilon_1^{h'_{12}+1} & \varepsilon_1^{h'_{22}} \\ \varepsilon_2^{h'_{12}+1} & \varepsilon_2^{h'_{22}} \end{vmatrix}}{\begin{vmatrix} \varepsilon_1 & 1 \\ \varepsilon_2 & 1 \end{vmatrix}} \times \varepsilon_3^{h_{13}+h_{23}+h_{33}-h'_{12}-h'_{22}} \quad . \tag{10.110}$$

We still have to prove that this expression is equivalent to (10.108)!

To that end we multiply the first column in (10.110) by $\varepsilon_3^{-h'_{12}-1}$ and the second one by $\varepsilon_3^{h'_{22}}$, which yields for each term

$$\varepsilon_3^{h_{13}+h_{23}+h_{33}+1} \begin{vmatrix} \left(\frac{\varepsilon_1}{\varepsilon_3}\right)^{h'_{12}+1} & \left(\frac{\varepsilon_1}{\varepsilon_3}\right)^{h'_{22}} \\ \left(\frac{\varepsilon_2}{\varepsilon_3}\right)^{h'_{12}+1} & \left(\frac{\varepsilon_2}{\varepsilon_3}\right)^{h'_{22}} \end{vmatrix} \cdot \begin{vmatrix} \varepsilon_1 & 1 \\ \varepsilon_2 & 1 \end{vmatrix}^{-1} \quad . \tag{10.111}$$

Using this result, with the help of the addition theorem of determinants, equation (10.110) becomes

$$\varepsilon_3^{h_{13}+h_{23}+h_{33}+1}$$

$$\times \frac{\begin{vmatrix} \left(\frac{\varepsilon_1}{\varepsilon_3}\right)^{h'_{13}+1} + \left(\frac{\varepsilon_1}{\varepsilon_3}\right)^{h_{13}} + \cdots + \left(\frac{\varepsilon_1}{\varepsilon_3}\right)^{h_{23}+1} & \left(\frac{\varepsilon_1}{\varepsilon_3}\right)^{h_{23}} + \left(\frac{\varepsilon_1}{\varepsilon_3}\right)^{h_{23}-1} + \cdots + \left(\frac{\varepsilon_1}{\varepsilon_3}\right)^{h_{33}} \\ \left(\frac{\varepsilon_2}{\varepsilon_3}\right)^{h_{13}+1} + \left(\frac{\varepsilon_2}{\varepsilon_3}\right)^{h_{13}} + \cdots + \left(\frac{\varepsilon_2}{\varepsilon_3}\right)^{h_{23}+1} & \left(\frac{\varepsilon_2}{\varepsilon_3}\right)^{h_{23}} + \left(\frac{\varepsilon_2}{\varepsilon_3}\right)^{h_{23}-1} + \cdots + \left(\frac{\varepsilon_2}{\varepsilon_3}\right)^{h_{33}} \end{vmatrix}}{\begin{vmatrix} \varepsilon_1 & 1 \\ \varepsilon_2 & 1 \end{vmatrix}} \tag{10.112}$$

We multiply the first row by $(\varepsilon_1/\varepsilon_3 - 1)$, the second by $(\varepsilon_2/\varepsilon_3 - 1)$ and do the same in the denominator. For the denominator we now have

$$\begin{vmatrix} \varepsilon_1 & 1 \\ \varepsilon_2 & 1 \end{vmatrix} \left(\frac{\varepsilon_1}{\varepsilon_3} - 1\right)\left(\frac{\varepsilon_2}{\varepsilon_3} - 1\right) = \varepsilon_3^{-2}(\varepsilon_1 - \varepsilon_2)(\varepsilon_1 - \varepsilon_3)(\varepsilon_2 - \varepsilon_3)$$

$$= \varepsilon_3^{-2} \begin{vmatrix} \varepsilon_1^2 & \varepsilon_1 & 1 \\ \varepsilon_2^2 & \varepsilon_2 & 1 \\ \varepsilon_3^2 & \varepsilon_3 & 1 \end{vmatrix} \quad , \tag{10.113}$$

i.e. the denominator is already in the correct form. The numerator is given by

$$\begin{vmatrix} \left[\left(\frac{\varepsilon_1}{\varepsilon_3}\right)^{h_{13}+2} - \left(\frac{\varepsilon_1}{\varepsilon_3}\right)^{h_{23}+1}\right] & \left[\left(\frac{\varepsilon_1}{\varepsilon_3}\right)^{h_{23}+1} - \left(\frac{\varepsilon_1}{\varepsilon_3}\right)^{h_{33}}\right] & \left(\frac{\varepsilon_1}{\varepsilon_3}\right)^{h_{33}} \\ \left[\left(\frac{\varepsilon_2}{\varepsilon_3}\right)^{h_{13}+2} - \left(\frac{\varepsilon_2}{\varepsilon_3}\right)^{h_{23}+1}\right] & \left[\left(\frac{\varepsilon_2}{\varepsilon_3}\right)^{h_{23}+1} - \left(\frac{\varepsilon_2}{\varepsilon_3}\right)^{h_{33}}\right] & \left(\frac{\varepsilon_2}{\varepsilon_3}\right)^{h_{33}} \\ 0 & 0 & 1 \end{vmatrix} \quad , \tag{10.114}$$

where in the last step we added the column

$$\begin{vmatrix} \left(\dfrac{\varepsilon_1}{\varepsilon_3}\right)^{h_{33}} \\ \left(\dfrac{\varepsilon_2}{\varepsilon_3}\right)^{h_{33}} \\ 1 \end{vmatrix}$$

and inserted zeros into the third row. The value of the determinant does not change by this operation! Now we add the second and the third columns to the first one and add the third column to the second one. This yields

$$\begin{vmatrix} \left(\dfrac{\varepsilon_1}{\varepsilon_3}\right)^{h_{13}+2} & \left(\dfrac{\varepsilon_1}{\varepsilon_3}\right)^{h_{23}+1} & \left(\dfrac{\varepsilon_1}{\varepsilon_3}\right)^{h_{33}} \\ \left(\dfrac{\varepsilon_2}{\varepsilon_3}\right)^{h_{13}+2} & \left(\dfrac{\varepsilon_2}{\varepsilon_3}\right)^{h_{23}+1} & \left(\dfrac{\varepsilon_2}{\varepsilon_3}\right)^{h_{33}} \\ 1 & 1 & 1 \end{vmatrix}$$

$$= \varepsilon_3^{-(h_{13}+2)}\varepsilon_3^{-(h_{23}+1)}\varepsilon_3^{-h_{33}} \begin{vmatrix} \varepsilon_1^{h_{13}+2} & \varepsilon_1^{h_{23}+1} & \varepsilon_1^{h_{33}} \\ \varepsilon_2^{h_{13}+2} & \varepsilon_2^{h_{23}+1} & \varepsilon_2^{h_{33}} \\ \varepsilon_3^{h_{13}+2} & \varepsilon_3^{h_{23}+1} & \varepsilon_3^{h_{33}} \end{vmatrix} \quad . \tag{10.115}$$

Here we have removed the factors

$$\varepsilon_3^{-(h_{13}+2)} \quad , \quad \varepsilon_3^{-(h_{23}+1)} \quad \text{and} \quad \varepsilon_3^{-(h_{33})}$$

from the first, second and third columns, respectively. Finally, by application of (10.113) and (10.112) we obtain the result (10.110), the interpretation of which is easy. If we only consider the subgroup U(2), all angles θ_{33}, θ_{i3}, θ_{ij} $(i, j = 1, \ldots, N-1)$ must be fixed with zero value. Thus, only $(N-1)^2 = 2^2 = 4$ angles remain in the case of $N = 3$. The character of the representation can be written as the sum of the characters of the representation of the subgroup, which are included in the representation of the full group. With regard to U(3) this means that we set $\theta_{33} = 0$, and (10.110) follows as the sum of the characters of the representations of the subgroup U(2) [see section on U(3)]:

$$\chi^{(\alpha)}(\mathrm{U}(3))|_{\theta_{33}=0} = \sum_{h'_{12}=h_{23}}^{h_{13}} \; \sum_{h'_{22}=h_{33}}^{h_{23}} \chi^{(h'_{12}, h'_{22})}(\mathrm{U}(2)) \quad . \tag{10.116}$$

Equation (10.116) states that a representation (h_{13}, h_{23}, h_{33}) of the group U(3) contains only those representations (h_{12}, h_{22}) of U(2) which comply with the inequality

$$h_{13} \geq h_{12} \geq h_{23} \geq h_{22} \geq h_{33} \quad . \tag{10.117}$$

Here every representation of U(2) is contained exactly once!

Finally, to get the character of SU(3) we can use (10.110) as well as the results of the former example

$$\chi^{(h_{13},h_{23},h_{33})} = \chi^{(\alpha)} = \sum_{h'_{12}=h_{23}}^{h_{13}} \sum_{h'_{22}=h_{33}}^{h_{23}} \exp[\tfrac{i}{2}(\theta_{11}+\theta_{22})(h'_{12}+h'_{22})]$$
$$\times \exp[-\tfrac{i}{2}\theta_{33}(h_{13}+h_{23}+h_{33}-h'_{12}-h'_{22})]$$
$$\times \sin\left[\frac{(h'_{12}+1-h'_{22})}{2}(\theta_{11}-\theta_{22})\right]\left(\sin\frac{\theta_{11}-\theta_{22}}{2}\right)^{-1} \quad . \tag{10.118}$$

Consider the first two factors here, which can be rewritten in the form

$$\exp[-\tfrac{i}{3}(\theta_{11}+\theta_{22}+\theta_{33})(h_{13}+h_{23}+h_{33})]$$
$$\times \exp[-\tfrac{i}{2}(\theta_{11}+\theta_{22}-2\theta_{33})(h'_{12}+h'_{22})]$$
$$\times \exp\left[\frac{i}{3}(\theta_{11}+\theta_{22}-2\theta_{33})(h_{13}+h_{23}+h_{33})\right]$$
$$= \exp\left[-\frac{i}{3}(\theta_{11}+\theta_{22}+2\theta_{33})(h_{13}+h_{23}+h_{33})\right]$$
$$\times \exp\left[-\frac{i}{2}[(\theta_{11}-\theta_{33})+(\theta_{22}-\theta_{33})](h'_{12}+h'_{22})\right]$$
$$\times \exp\left[\frac{i}{3}[(\theta_{11}-\theta_{33})+(\theta_{22}-\theta_{33})](h_{13}+h_{23}+h_{33})\right] \quad . \tag{10.119}$$

Thus, for (10.110) or (10.118) we have

$$\chi_{\mathrm{U}(3)}^{(h_{13},h_{23},h_{33})}$$
$$= \exp\left[-\frac{i}{3}(\theta_{11}+\theta_{22}+\theta_{33})(h_{13}+h_{23}+h_{33})\right]$$
$$\times \exp\left[\frac{i}{3}[(\theta_{11}-\theta_{33})+(\theta_{22}-\theta_{33})](h_{13}+h_{23}+h_{33})\right]$$
$$\times \sum_{h'_{12}=h_{23}}^{h_{13}} \sum_{h'_{22}=h_{33}}^{h_{23}} \exp\left[-\frac{i}{2}[(\theta_{11}-\theta_{33})+(\theta_{22}-\theta_{33})](h'_{12}+h'_{22})\right]$$
$$\times \sin\left[\left(\frac{h'_{12}+1-h'_{22}}{2}\right)(\theta_{11}-\theta_{22})\right]\left(\sin\frac{\theta_{11}-\theta_{22}}{2}\right)^{-1} \quad . \tag{10.120}$$

In the case of SU(3) we have $\theta_{11}+\theta_{22}+\theta_{33}=0$, which with (10.68) leads to

$$\chi_{\mathrm{SU}(3)}^{(h_{13}-h_{23},h_{23}-h_{33},0)}$$
$$= \exp\left[\frac{2i\varphi}{3}(h_{13}+h_{23}+h_{33})\right] \sum_{h'_{12}=h_{23}}^{h_{13}} \sum_{h'_{22}=h_{33}}^{h_{23}} \exp[-i\psi(h'_{12}+h'_{22})]$$
$$\times \sin\left[\left(\frac{h'_{12}+1-h'_{22}}{2}\right)\varphi\right]\left(\sin\frac{\varphi}{2}\right)^{-1} \quad . \tag{10.121}$$

Here the representation of SU(3) is given by

$$(p,q,0) = (h_{13}-h_{23},h_{23}-h_{33},0) \quad . \tag{10.122}$$

A special case is the scalar representation. Here we have $h_{13} = h_{23} = h_{33} = 0$ and therefore $h_{12} = h_{22} = 0$, too; the character function is identically equal to 1, which is also obvious from (10.108).

In the last example, we outlined how in general to calculate the character of a unitary representation. The general proof is done by induction from a basis given by U(1), which is trivial. Now we assume that the assumption is valid for U(N) and verify it for U($N + 1$). The procedure is identical to the one in the last performed example, except that now more than two sums and more indices have to be considered.

Finally, some formulae which are useful for further reference in connection with unitary groups are presented once again.

From (10.22), (10.23) and (10.56), respectively, we obtain

$$V^{-1} \int_{\mathrm{U}(N)} \chi^{(\alpha)}\chi^{(\beta)*}\mu(\mathrm{U}(N))\,d\theta = \delta_{\alpha\beta} \quad , \tag{10.123}$$

where $d\theta$ is an abbreviation of $d\theta_{11}d\theta_{22}\ldots d\theta_{NN}$, and V is given by

$$V = \int_{\mathrm{U}(N)} \mu(\mathrm{U}(N))\,d\theta \quad . \tag{10.124}$$

Since the characters only depend on $\theta_{11}, \ldots, \theta_{NN}$, integration over these angles is sufficient. $\chi^{(\alpha)}$ is given by (10.100) and $\mu(\mathrm{U}(N))$ by (10.83). Furthermore, as in (10.29) we can express the character χ of a reducible representation in the form

$$\chi = \sum m_\alpha \chi^{(\alpha)} \quad , \tag{10.125}$$

where $\chi^{(\alpha)}$ are characters of irreducible representations only.

For a product representation we have [see (10.38)]

$$\chi^{(\alpha\times\beta)} = \chi^{(\alpha)}\cdot\chi^{(\beta)} \quad . \tag{10.126}$$

By using these formulae we are able to calculate the irreducible representations which are contained in an arbitrary representation. If we restrict our considerations to a single subalgebra, we can further calculate the representations into which the original representation can be decomposed. An example of this is given by the reduction U(3) → U(2) in (10.116).

EXAMPLE

10.3 Application of Group Characters: Partition Function for the Colour Singlet Quark-Gluon Plasma with Exact SU(3) Symmetry

In the past few years the so-called quark-gluon plasma[1] has been studied intensely by theoretical physicists. The name "quark-gluon plasma" describes a new state of nuclear matter in which quarks and gluons move freely. Supposedly this state existed during the first 20 microseconds after the "big-bang" of

[1] See e.g.: B. Müller: *The Physics of the Quark-Gluon Plasma*, Lecture Notes in Physics 225 (Springer, Berlin, Heidelberg 1985).

Example 10.3

our universe. Today one hopes to create this plasma in heavy-ion reactions at very high energies. In order to be able to recognize a quark-gluon plasma when it is formed, one must study its properties theoretically and obtain predictions for characteristic signatures, e.g. increased abundance of strange particles. Since the excitations of the plasma can have high energy and the surrounding nuclear matter can act as a "heat bath", then we are dealing with a thermodynamical problem. However, due to colour confinement, the total colour of the plasma state must always be neutral, which means that only states with an exact SU(3) singlet symmetry are realized.

In the following we shall state some definitions of thermodynamics and then show that, using the group characters, one can be restricted to states which belong to a certain representation of an SU(3) group. As an example we discuss the conservation of isospin, since in nuclear reactions, in which the electromagnetic interaction can be neglected, isospin is conserved, so that only states of a given isospin can be reached. As a last example we return to the quark-gluon plasma.

(1) Some Thermodynamical Expressions. The partition function is of central importance in the thermodynamical description of statistical systems. It is defined as

$$Z = \mathrm{Tr}[\exp(-\beta\hat{H})] \quad , \tag{1}$$

with $\beta = 1/kT$, where k is Boltzmann's constant, T is the temperature and $\hat{H}$ is the Hamiltonian. The trace operation means a sum over all diagonal matrix elements in Hilbert space. If a noninteracting system of particles has discrete states with energies ε_i, then the expectation value of $\hat{H}$ is given by

$$\sum_{i=1}^{\infty} n_i\varepsilon_i \quad , \tag{2}$$

where n_i are the occupation numbers, specifying how many particles are in a given state ε_i. If there is a continuous energy spectrum, the sum in (2) is replaced by an integral. Once the partition function is known, physical quantities such as pressure P and intrinsic energy U can be easily calculated by use of the formulae:

$$P = T\partial \log Z/\partial V \quad , \tag{3a}$$

$$U = -\partial \log Z/\partial\beta \quad , \tag{3b}$$

The partition function is given by different expressions for boson and fermion systems, and to show this we first abbreviate the exponentials in the partition function:

$$Z_i \equiv \exp(-\beta\varepsilon_i) \quad . \tag{4}$$

Thus (1) takes the form

$$Z_i \equiv \sum_{(n_i)} \exp\left(-\beta\sum_i n_i\varepsilon_i\right) = \sum_{(n_i)} \prod_{i=1}^{\infty} (Z_i)^{n_i} \quad , \tag{5}$$

where (n_i) denotes all possible sets of occupation numbers.

For bosons (e.g. gluons) this yields

Example 10.3

$$Z_{\mathrm{B}} = \left(\sum_{n_1=0}^{\infty} Z_1^{n_1}\right)\left(\sum_{n_2=0}^{\infty} Z_2^{n_2}\right)\cdots\left(\sum_{n_i=0}^{\infty} Z_i^{n_i}\right)\cdots \quad . \tag{6}$$

Each state can be occupied by an arbitrary number of bosons. Since every sum in (6) is a geometrical series, one gets

$$Z_B = \prod_{i=1}^{\infty} \frac{1}{1-Z_i} = \prod_{i=1}^{\infty} \frac{1}{1-\exp(-\beta\varepsilon_i)} \quad . \tag{7}$$

Considering fermions (e.g. quarks, nucleons), every state can be occupied by either one or no particle, i.e.

$$\sum_{n_i} Z_i = 1 + Z_i = 1 + \exp(-\beta\varepsilon_i) \quad . \tag{8}$$

The first term corresponds to the case that no particle is in the state i, whereas the second term corresponds to one particle in that state. Hence the fermion partition function is given by

$$Z_F = \prod_{i=1}^{\infty} [1 + \exp(-\beta\varepsilon_i)] \quad . \tag{9}$$

For the calculation the products are raised into the exponential

$$Z_{\mathrm{B}} = \exp\left(\ln \prod_i \frac{1}{1-Z_i}\right) = \exp\left[-\sum_{i=1}^{\infty} \ln(1-\mathrm{e}^{-\beta\varepsilon_i})\right] \quad , \tag{10a}$$

$$Z_{\mathrm{F}} = \exp\left[\ln \prod_i (1+Z_i)\right] = \exp\left[+\sum_{i=1}^{\infty} \ln(1+\mathrm{e}^{-\beta\varepsilon_i})\right] \quad . \tag{10b}$$

For $\beta\varepsilon_i \gg 1$ the logarithm can be expanded, an approximation that is justified in many cases, and the partition functions in (10a) and (10b) obey the respective statistics. But what happens if we are concerned with an exact global symmetry? Z_{B} and Z_{F} do *not* take into account an exactly conserved symmetry, i.e. they also include states which are not allowed. In the case of the quark-gluon plasma one has a product of Z_B and Z_F which takes into account coloured states, too.

In the following we shall first consider a general, exact SU(n) symmetry. We shall show how to project out states which belong to a certain representation of SU(n). First we add a sum over the weight operators $\hat{C}_{kk}$ to the exponential. Thus

$$\tilde{Z} = \mathrm{Tr}\left[\exp\left(-\beta\hat{H} - \mathrm{i}\sum_k \theta_{kk}\hat{C}_{kk}\right)\right] \quad , \tag{11}$$

where the tilde over Z denotes that $\tilde{Z}$ differs from the old partition function Z. The trace is defined as the sum of the diagonal matrix elements over all possible states:

$$\tilde{Z} = \sum_{(p)}\sum_{(m)} \langle (p)(m)|\exp\left(-\beta\hat{H} - \mathrm{i}\sum_k \theta_{kk}\hat{C}_{kk}\right)|(p)(m)\rangle \quad . \tag{12}$$

Example 10.3

Here (m) is an abbreviation for all quantum numbers labelling the states in the irreducible representation (p). In the case of SU(3) we have $(m) = (y, T_z)$. Inserting a complete set of states yields:

$$\tilde{Z} = \sum_{\substack{(p)\\(p')}} \sum_{\substack{(m)\\(m')}} \langle (p)(m)|\exp(-\beta\hat{H})|(p')(m')\rangle$$

$$\times \langle (p')(m')|\exp\left(-\mathrm{i}\sum_{k=1}^{n}\theta_{kk}\hat{C}_{kk}\right)|(p)(m)\rangle \quad . \tag{13}$$

We now note that the matrix elements of $\hat{H}$ are independent of, and diagonal in, (m), because $\hat{H}$ exactly conserves the colour symmetry. Realizing that C_{kk} changes neither the representation (p) nor the quantum numbers (m), we get

$$\tilde{Z} = \sum_{(p)} \langle (p)|\mathrm{e}^{-\beta\hat{H}}|(p)\rangle \sum_{(m)} \langle (p)(m)|\exp\left(-\mathrm{i}\sum_{k}\theta_{kk}\hat{C}_{kk}\right)|(p)(m)\rangle \quad . \tag{14}$$

The latter sum is equal to the definition of a character of the representation (p):

$$\chi^{(p)}_{(\theta_{kk})} = \sum_{(m)} \langle (p)(m)|\exp\left(-\mathrm{i}\sum_{k}\theta_{kk}\hat{C}_{kk}\right)|(p)(m)\rangle \quad . \tag{15}$$

The first factor in (14) can be rewritten as

$$\langle (p)|\mathrm{e}^{-\beta\hat{H}}|(p)\rangle = \frac{1}{\dim(p)}\sum_{\overline{m}} \langle (p)(\overline{m})|\mathrm{e}^{-\beta\hat{H}}|(p)(\overline{m})\rangle \equiv \frac{1}{\dim(p)} Z_{(p)} \quad . \tag{16}$$

Since the matrix element in the sum is independent of (m), we obtain $\dim(p)$ times the same value [$\dim(p)$ is the dimension of the representation (p)]. The sum in (16) is thus equal to

$$Z_{(p)} = \mathrm{Tr}_{(p)}[\exp(-\beta\hat{H})] \quad , \tag{17}$$

i.e. the trace, which is performed only on the states in the (p) representation. Equation (17) is the partition function which we finally want to obtain. Summarizing we get for $\tilde{Z}$:

$$\tilde{Z} = \sum_{(p)} \frac{Z_{(p)}}{\dim(p)} \chi^{(p)}(\theta_{kk}) \quad . \tag{18}$$

Z is calculated in the same way as in (10a) and (10b), with the addition of the weight operators $\hat{C}_{kk}$. To obtain $Z(p)$ we apply the orthogonality relation (10.123) for group characters:

$$Z_{(p)} = \dim(p) \int d\mu(\theta_{kk})\chi^{(p)}(\theta_{kk})\tilde{Z}(\beta,\hat{H},\theta_{kk}) \quad ,$$

with

$$d\mu(\theta_{kk}) = \frac{1}{V}\mu(\mathrm{SU}(n))d(\mathrm{SU})(n)) \quad . \tag{19}$$

Equation (19) is the relation used for the computation of $Z(p)$.

Example 10.3

Example (a): Conservation of isospin in nuclear reactions. Earlier in the chapter, the character of SU(2) was defined in (10.107) as

$$\chi^{(p)}_{\mathrm{SU}(2)} = \frac{\sin(\frac{1}{2}(p+1)\phi)}{\sin\frac{1}{2}\phi} \quad , \tag{20}$$

with $p = 2j$, $\phi = \phi_{11} - \phi_{22}$ and $\phi_{11} + \phi_{22} = 0$. The measure was shown in (10.86) to be given by

$$d\mu(\phi) = \frac{1}{2\pi}\sin^2\left(\frac{\phi}{2}\right)d\phi \quad , \quad \text{with} \quad 0 \le \phi \le 4\pi \quad . \tag{21}$$

Thus we get an expression for the partition function which takes into account only states with certain isospin I [$\dim(p) = 2I + 1$]:

$$Z_{(I)} = \frac{(2I+1)}{2\pi}\int_0^{4\pi} d\phi \cdot \sin^2\tfrac{1}{2}\phi \cdot \tilde{Z}(\beta, \hat{H}, \phi) \quad . \tag{22}$$

Due to (11), $\tilde{Z}$ has the following form:

$$\tilde{Z}(\phi) = \mathrm{Tr}\left\{\exp\left[-\beta\hat{H} - \frac{\mathrm{i}}{2}\phi(\hat{C}_{11} - \hat{C}_{22})\right]\right\} \quad . \tag{23}$$

In moving from U(n) to SU(n), we have used [see (10.66) the case U(3) → SU(3)]

$$\sum_{k=1}^{n} \theta_{kk} = 0 \quad . \tag{24}$$

Equation (23) defines a very strongly oscillating function, with values between 0 and some powers of ten. Analytical solutions for (22) are not available; however methods of approximation or numerical approaches exist.

Example (b): Conservation of the SU(3) colour charge. If a quark-gluon plasma is formed in heavy-ion reactions, the total colour must remain zero at all times. The character of an SU(3) representation has the form given by (10.118), and we consider a scalar (colourless) representation [singlet$(p) = (p_1, p_2) = (0,0)$]. The singlet character is $\chi^{(0,0)} = 1$. The measure is given by (10.95):

$$\begin{aligned} d\mu(\phi,\psi) &= \frac{8}{3\pi^2}\sin^2\frac{\phi}{2}\sin^2\left[\frac{1}{2}\left(\frac{\phi}{2}+\psi\right)\right] \\ &\quad\times \sin^2\left[\frac{1}{2}\left(-\frac{\phi}{2}+\psi\right)\right] d\left(\frac{\phi}{z}\right) d\left(\frac{\psi}{3}\right) \end{aligned} \tag{25}$$

with

$$\begin{aligned} &\phi = \theta_{11} - \theta_{22}, \quad -\pi \le \tfrac{1}{2}\phi \le \pi \\ &\psi = \tfrac{1}{2}(\theta_{11} + \theta_{22} - 2\theta_{33}) \quad , \quad -\pi \le \frac{\psi}{3} \le \pi \quad \text{and} \\ &\theta_{11} + \theta_{22} + \theta_{33} = 0 \quad . \end{aligned} \tag{26}$$

Example 10.3

Thus we get for the partition function, which allows only colourless states, the integral

$$Z_{(0,0)} = \frac{8}{3\pi^2} \int_{-\pi}^{\pi} \int_{-\pi}^{\pi} d\left(\frac{\phi}{2}\right) d\left(\frac{\psi}{3}\right) \sin^2\frac{\phi}{2} \sin^2\frac{1}{2}\left(\frac{\phi}{2}+\psi\right)$$
$$\times \sin^2\frac{1}{2}\left(-\frac{\phi}{2}+\psi\right) \tilde{Z}(\beta, \hat{H}, \phi, \psi) \quad . \tag{27}$$

Again $\tilde{Z}$ is a strongly oscillating function with values between zero and a large power of ten; thus, numerical integration is necessary. A useful way to plot the result is to take the ratio of the energy $E_{(0,0)}$ calculated by $Z_{(0,0)}$ and the energy $E^{(0)}$ obtained without any symmetry restriction. The quantity $D_{\text{eff}} = E_{(0,0)}/E^{(0)}$ describes the deviation from the Stefan-Boltzmann ideal-gas behaviour and can be understood as an effective number of degrees of freedom. The result is shown in the figure below, where D_{eff} is shown as a function of the dimensionless quantity $TV^{1/2}/(\hbar c)$. We see that there is a transition region where the effective number of degrees of freedom decreases rapidly. This behaviour is reminiscent of a phase transition, but an inspection of the specific heat shows that there is no discontinuity involved.

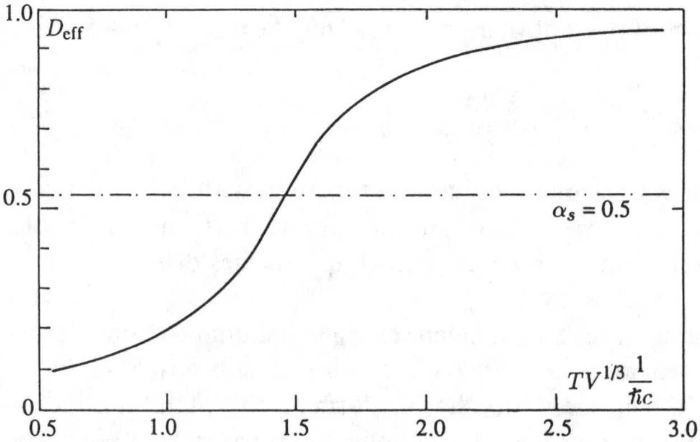

Effective reduction in the number of degrees of freedom in SU(3) gauge theory

EXAMPLE

10.4 Proof of the Recursion Formula for the Dimensions of the SU(n) Representations

Earlier in this chapter the general formula (10.100) for a character of a representation $[h_1, h_2, \ldots, h_n]$ in the group U(n) was given. The numbers $[h_1, h_2, \ldots, h_n]$ denote the number of boxes in a Young diagram, i.e. h_1 boxes in the first row, h_2 boxes in the second row and so on. A young diagram in SU(n) is obtained by cancelling the last row ($h_n = 0$). The numbers $p_1, p_2, \ldots, p_{n-1}$, which characterize the representation in SU(n) are related to the numbers h_i ($i = 1,2,\ldots,n$) by *Example 10.4*

$$p_i = h_i - h_{i+1} \quad . \tag{1}$$

For p_{n-1} this leads to

$$p_{n-1} = h_{n-1} - h_n = h_{n-1} \quad \text{if} \quad h_n = 0 \quad . \tag{2}$$

To calculate the dimension of the representation $(p_1, p_2, \ldots, p_{n-1})$ in SU(n), it is advantageous to calculate the dimension of the representation $[h_1, h_2, \ldots, h_{n-1}, 0]$ in U(n). For further calculations one needs the inversion of (1), which reads:

$$\begin{aligned}
h_{n-1} &= p_{n-1} \quad ,\\
h_{n-2} &= p_{n-1} + p_{n-2} \quad ,\\
h_{n-3} &= p_{n-1} + p_{n-2} + p_{n-3} \quad ,\\
&\vdots \\
h_2 &= p_{n-1} + p_{n-2} + p_{n-3} + \ldots + p_2 \quad ,\\
h_1 &= p_{n-1} + p_{n-2} + p_{n-3} + \ldots + p_2 + p_1 \quad .
\end{aligned} \tag{3}$$

The dimension of the representation is given by the value of the character of the identity (all angles $\theta_{kk} = 0$). In this case we have a diagonal matrix with value one everywhere, and the character of this transformation is the sum over all "1"'s i.e. the dimension. This becomes obvious through (10.99), where the exponent becomes equal to zero, and the summation is performed over all states. Before we show the general approach, we illustrate it by several examples.

(a) SU(2). According to (10.100) the character is given by

$$\chi^{[h_1 0]}_{\mathrm{U}(2)} = \frac{\begin{vmatrix} \varepsilon^{h_1+1} & 1 \\ \varepsilon_2^{h_1+1} & 1 \end{vmatrix}}{\begin{vmatrix} \varepsilon_1 & 1 \\ \varepsilon_2 & 1 \end{vmatrix}} \quad . \tag{4}$$

ε_1, ε_2 are exponentials $[\varepsilon_k = \exp(-\mathrm{i}\theta_{kk})]$, but they can be considered as variables. For the case ($\theta_{kk} = 0$) we have

$$\varepsilon_k \xrightarrow[\theta_{kk}\to 0]{} 1 \quad . \tag{5}$$

Since both numerator and denominator approach zero, we have to apply the rule of l'Hospital; i.e. we have to differentiate both numerator and denominator with respect to one of the variables. If the ratio of the latter quantities exists, then the limit of $\chi^{[h_1 0]}$, also exists and both values are equal. We choose $\partial/\partial\varepsilon_1$:

$$\frac{\partial/\partial\varepsilon_1 \begin{vmatrix} \varepsilon_1^{h_1+1} & 1 \\ \varepsilon_2^{h_2+1} & 1 \end{vmatrix}}{\partial/\partial\varepsilon_1 \begin{vmatrix} \varepsilon_1 & 1 \\ \varepsilon_2 & 1 \end{vmatrix}} = \frac{\begin{vmatrix} (h_1+1)\varepsilon_1^{h_1} & 0 \\ \varepsilon_2^{h_2+1} & 1 \end{vmatrix}}{\begin{vmatrix} 1 & 0 \\ \varepsilon_2 & 1 \end{vmatrix}} \theta_{kk} \to 0 \frac{\begin{vmatrix} (h_1+1) & 0 \\ 1 & 1 \end{vmatrix}}{\begin{vmatrix} 1 & 0 \\ 1 & 1 \end{vmatrix}} = h_1 + 1 \quad . \tag{6}$$

Since in the case of SU(2) $h_1 = p_1$, this result agrees with (9.59).

Example 10.4

(b) SU(3). According to (10.100) the character is given by

$$\chi_{U(3)}^{[h_1 h_2 0]} = \begin{vmatrix} \varepsilon_1^{h_1+2} & \varepsilon_1^{h_2+1} & 1 \\ \varepsilon_2 & \varepsilon_2^{h_2+1} & 1 \\ \varepsilon_3^{h_1+2} & \varepsilon_3^{h_2+1} & 1 \end{vmatrix} \cdot \begin{vmatrix} \varepsilon_1^2 & \varepsilon_1 & 1 \\ \varepsilon_2^2 & \varepsilon_2 & 1 \\ \varepsilon_3^2 & \varepsilon_3 & 1 \end{vmatrix}^{-1} \quad . \tag{7}$$

We shall now differentiate three times, once with respect to ε_2 and twice with respect to ε_1. (The converse would not suffice because numerator and denominator would yield zero.) The result is

$$\left(\frac{\partial^2}{\partial \varepsilon_1^2} \frac{\partial}{\partial \varepsilon_2} \begin{vmatrix} \varepsilon_1^{h_1+2} & \varepsilon_1^{h_2+1} & 1 \\ \varepsilon_2^{h_1+2} & \varepsilon_2^{h_2+1} & 1 \\ \varepsilon_3^{h_1+2} & \varepsilon_3^{h_2+1} & 1 \end{vmatrix} \right) \cdot \left(\frac{\partial^2}{\partial \varepsilon_1^2} \frac{\partial}{\partial \varepsilon_2} \begin{vmatrix} \varepsilon_1^2 & \varepsilon_1 & 1 \\ \varepsilon_2^2 & \varepsilon_2 & 1 \\ \varepsilon_3^2 & \varepsilon_3 & 1 \end{vmatrix} \right)^{-1}$$

$$= \begin{vmatrix} (h_1+2)(h_1+1)\varepsilon_1^{h_1} & (h_2+1)h_2\varepsilon_1^{h_2-1} & 0 \\ (h_1+2)\varepsilon_2^{h_1+1} & (h_2+1)\varepsilon_2^{h_2} & 0 \\ \varepsilon_3^{h_1+2} & \varepsilon_3^{h_2+1} & 1 \end{vmatrix} \cdot \begin{vmatrix} 2 & 0 & 0 \\ 2\varepsilon_2 & 1 & 0 \\ \varepsilon_3^2 & \varepsilon_3 & 1 \end{vmatrix}^{-1} \quad . \tag{8a}$$

In the limit $\theta_{kk} \to 0$, we obtain

$$\frac{1}{2 \cdot 1 \cdot 1} \begin{vmatrix} (h_1+2)(h_1+1) & (h_2+1)h_2 & 0 \\ (h_1+2) & (h_2+1) & 0 \\ 1 & 1 & 1 \end{vmatrix} = \tfrac{1}{2}(h_1+2)(h_2+1)(h_1-h_2+1) \quad , \tag{8b}$$

and then, using (3), which relates h_1, h_2 and p_1, p_2, we get the dimension formula for SU(3):

$$\tfrac{1}{2}(h_1-h_2+1)(h_2+1)(h_1+2) = \tfrac{1}{2}(p_1+1)(p_2+1)(p_1+p_2+2) \quad . \tag{8c}$$

These two examples indicate the general procedure.

It is interesting to express the character of U(3) through that of U(2), since this yields another hint how to obtain the recursion formula (9.6.2). For that purpose we consider numerator and denominator in (8a) and (8b) separately.

Numerator:

$$\begin{vmatrix} (h_1+2)(h_1+1)\varepsilon_1^{h_1} & (h_2+1)h_2\varepsilon_1^{h_2-1} & 0 \\ (h_1+2)\varepsilon_2^{h_1+1} & (h_2+1)\varepsilon_2^{h_2} & 0 \\ \varepsilon_3^{h_1+2} & \varepsilon_3^{h_2+1} & 1 \end{vmatrix}$$

$$= \begin{vmatrix} (h_1+2)(h_1+1)\varepsilon_1^{h_1} & (h_2+1)h_2\varepsilon_1^{h_2-1} \\ (h_1+2)\varepsilon_2^{h_1+1} & (h_2+1)\varepsilon_2^{h_2} \end{vmatrix}$$

$$= (h_1+2)(h_2+1) \begin{vmatrix} (h_1+1)\varepsilon_1^{h_1} & h_2\varepsilon_1^{h_2-1} \\ \varepsilon_1^{h_1+1} & \varepsilon_1^{h_2} \end{vmatrix} \quad . \tag{9}$$

Now we invert the rule of l'Hospital and write instead of (9): *Example 10.4*

$$(h_1 + 2)(h_2 + 1)\frac{\partial}{\partial \varepsilon_1}\begin{vmatrix} \varepsilon_1^{h_1+1} & \varepsilon_1^{h_2} \\ \varepsilon_2^{h_1+1} & \varepsilon_2^{h_2} \end{vmatrix} \quad . \tag{10a}$$

Denominator: Proceeding analogously we end up with

$$\begin{vmatrix} 2 & 0 & 0 \\ 2\varepsilon_2 & 1 & 0 \\ \varepsilon_3^2 & \varepsilon_3 & 1 \end{vmatrix} = \begin{vmatrix} 2 & 0 \\ 2\varepsilon_2 & 1 \end{vmatrix} = 2\begin{vmatrix} 1 & 0 \\ \varepsilon_2 & 1 \end{vmatrix} = 2\frac{\partial}{\partial \varepsilon_1}\begin{vmatrix} \varepsilon_1 & 1 \\ \varepsilon_2 & 1 \end{vmatrix} \quad . \tag{10b}$$

If we put (10a) and (10b) together, we obtain

$$\tfrac{1}{2}(h_2 + 1)(h_1 + 2)\left(\frac{\partial}{\partial \varepsilon_1}\begin{vmatrix} \varepsilon_1^{h_1+1} & \varepsilon_1^{h_2} \\ \varepsilon_2^{h_1+1} & \varepsilon_2^{h_2} \end{vmatrix}\right)\left(\frac{\partial}{\partial \varepsilon_1}\begin{vmatrix} \varepsilon_1 & 1 \\ \varepsilon_2 & 1 \end{vmatrix}\right)^{-1} \quad . \tag{10c}$$

Now, if we drop the derivatives, we get the character of the representation $[hh_1h_2]$ of U(2) and $(p_1) = (h_1 - h_2)$ of SU(2), respectively. Since (10c) exists as a limit for $\theta_{kk} \to 0$, the limit of (10c) exists without derivatives, as well. This means that

$$\dim\,[h_1h_20]_{U(3)} = \tfrac{1}{2!}(h_2 + 1)(h_1 + 2)\dim\,[h_1h_2]_{U(2)} \quad , \tag{11a}$$

or, due to (3),

$$\dim\,(p_1p_2)_{SU(3)} = \tfrac{1}{2!}(p_2 + 1)(p_1 + p_2 + 2)\dim\,(p_1)_{SU(2)} \quad . \tag{11b}$$

This is just the recursion formula (9.62) for the case $(n + 1) = 3$.

It should now be clear how to proceed in general. For the case of the group U(n), the character is given by (10.100). We shall consider this for the denominator and numerator separately.

Denominator:

$$\begin{aligned} &\frac{\partial^{n-1}}{\partial \varepsilon_1^{n-1}}\frac{\partial^{n-2}}{\partial \varepsilon_2^{n-2}}\cdots\frac{\partial}{\partial \varepsilon_{n-1}}\begin{vmatrix} \varepsilon_1^{n-1} & \varepsilon_1^{n-2} & \cdots & \varepsilon_1 & 1 \\ \varepsilon_2^{n-1} & \varepsilon_2^{n-2} & \cdots & \varepsilon_2 & 1 \\ \vdots & \vdots & & \vdots & \vdots \\ \varepsilon_{n-1}^{n-1} & \varepsilon_{n-1}^{n-2} & \cdots & \varepsilon_{n-1} & 1 \\ \varepsilon_n^{n-1} & \varepsilon_n^{n-2} & \cdots & \varepsilon_n & 1 \end{vmatrix} \\ &= \begin{vmatrix} (n-1)(n-2)\cdots 1 & 0 & \cdots & & 0 \\ (n-1)(n-2)\cdots 2\varepsilon_2 & (n-2)(n-3)\cdots 1 & 0 & \cdots & 0 \\ \vdots & \vdots & \vdots & \vdots & \vdots \\ (n-1)\varepsilon_{n-1}^{n-2} & (n-2)\varepsilon_{n-1}^{n-3} & \cdots & 1 & 0 \\ \varepsilon_n^{n-1} & \varepsilon_n^{n-2} & \cdots & \varepsilon_n & 1 \end{vmatrix} \quad . \end{aligned} \tag{12}$$

Example 10.4

Again we drop the nth row and column, respectively, and separate the factor $(n-1)(n-2)\cdots 1 = (n-1)!$:

$$\begin{vmatrix} (n-2)\cdots 1 & 0 & \cdots & 0 \\ (n-2)\cdots 2\varepsilon_2 & (n-3)\cdots 1 & \cdots & 0 \\ \vdots & \vdots & \vdots & \\ \varepsilon_{n-1}^{n-2} & \varepsilon_{n-1}^{n-3} & \cdots & 0 \end{vmatrix}$$

$$= (n-1)! \frac{\partial^{n-2}}{\partial\varepsilon_1^{n-2}} \frac{\partial^{n-3}}{\partial\varepsilon_2^{n-3}} \frac{\partial}{\partial\varepsilon_{n-2}} \begin{vmatrix} \varepsilon_1^{n-2} & \varepsilon_1^{n-3} & \cdots & \varepsilon_1 & 1 \\ \varepsilon_2^{n-2} & \varepsilon_2^{n-3} & \cdots & \varepsilon_2 & 1 \\ \vdots & \vdots & & \vdots & \vdots \\ \varepsilon_{n-1}^{n-2} & \varepsilon_{n-1}^{n-3} & \cdots & \varepsilon_{n-1} & 1 \end{vmatrix} . \tag{13}$$

Thus we have obtained the denominator of the character of the representation of $\mathrm{U}(n-1)$.

Numerator: Application of $(\partial^{n-1}/\partial\varepsilon_1^{n-1})\cdots(\partial/\partial\varepsilon_{n-1})$ yields

$$\begin{vmatrix} (h_1+n-1)\cdots(h_1+1)\varepsilon_1^{h_1} & (h_2+n-2)\cdots h_2\varepsilon_1^{h_2-1} & \cdots & 0 \\ (h_1+n-1)\cdots(h_1+2)\varepsilon_2^{h_1+1} & (h_2+n-2)\cdots(h_2+1)\varepsilon_1^{h_2} & \cdots & 0 \\ \vdots & \vdots & & \vdots \\ (h_1+n-1)\varepsilon_{n-1}^{h_1+n-2} & (h_2+n-2)\varepsilon_{n-1}^{h_2+n-3} & \cdots & 0 \\ \varepsilon_n^{h_1+n-1} & \varepsilon_n^{h_2+n-2} & \cdots & 1 \end{vmatrix} . \tag{14}$$

Omitting the last row and column, and extracting the factor

$$(h_1+n-1)(h_2+n-2)\cdots(h_{n-1}+1) \quad ,$$

we get from (14) that the numerator becomes

$$(h_{n-1}+1)(h_{n-2}+2)\cdots(h_2+n-2)(h_1+n-1)$$
$$\times \begin{vmatrix} (h_1+n-2)\cdots(h_1+1)\varepsilon_1^{h_1} & (h_2+n-3)\cdots h_2\varepsilon_1^{h_2-1} & \cdots & (h_{n-1})\cdots(h_{n-1}-n+3)\varepsilon_1^{h_{n-1}-n+2} \\ \vdots & \vdots & & \vdots \\ \varepsilon_{n-1}^{h_1+n-2} & \varepsilon_{n-1}^{h_2+n-3} & \cdots & \varepsilon_{n-1}^{h_{n-1}} \end{vmatrix}$$

or

$$(h_{n-1}+1)(h_{n-2}+2)\cdots(h_2+n-2)(h_1+n-1)$$
$$\times \frac{\partial^{n-2}}{\partial\varepsilon_1^{n-2}} \frac{\partial^{n-3}}{\partial\varepsilon_2^{n-3}} \cdots \frac{\partial}{\partial\varepsilon_{n-2}} \begin{vmatrix} \varepsilon_1^{h_1+(n-1)-1} & \varepsilon_1^{h_2+(n-1)-2} & \cdots & \varepsilon_1^{h_{n-1}} \\ \vdots & \vdots & & \vdots \\ \varepsilon_{n-1}^{h_1+(n-1)-1} & \varepsilon_{n-1}^{h_2+(n-1)-2} & \cdots & \varepsilon_{n-1}^{h_{n-1}} \end{vmatrix} . \tag{15}$$

Putting together (13) and (15) and omitting the derivatives, we obtain the recursion formula in which the character of $\mathrm{U}(n-1)$ appears. In the case of

Example 10.4

$\theta_{kk} \to 0$ we have

$$\begin{aligned}\dim[h_1 \ldots h_{n-1} 0]_{\mathrm{U}(n)} \\ &= (h_{n-1}+1)(h_{n-2}+2)\cdots(h_1+n-1)\dim[h_1 \ldots h_{n-1}]_{\mathrm{U}(n-1)}\end{aligned} \tag{16a}$$

or with (13):

$$\begin{aligned}\dim(p_1, \ldots, p_{n-1})_{\mathrm{SU}(n)} \\ &= \frac{1}{(n-1)!}(p_{n-1}+1)(p_{n-1}+p_{n-2}+2) \\ &\quad \times (p_n + \cdots + p_1 + n - 1)\dim(p_1 \ldots p_{n-2})_{\mathrm{SU}(n-1)} \quad .\end{aligned} \tag{16b}$$

This is just the recursion formula (9.62) if we replace (n) by $(n+1)$ and dim $(\ldots)$ by $N(\ldots)$:

$$\begin{aligned}N_{n+1}(p_1, \ldots, p_n) \\ &= \frac{1}{n!}(p_n+1)(p_n+p_{n-1}+2)\cdots(p_n+p_{n-1}+\ldots+p_1+n) \\ &\quad \times N_n(p_1, \ldots, p_{n-1}) \quad .\end{aligned} \tag{17}$$

Solution of the recursion formula. The dimension formula for the lowest groups reads:

$$\begin{aligned}&\mathrm{SU}(2)\colon \frac{1}{1!}(p_1+1) \quad , \\ &\mathrm{SU}(3)\colon \frac{1}{2!}(p_1+1)(p_2+1)(p_1+p_2+2) \quad , \\ &\mathrm{SU}(4)\colon \frac{1}{2!3!}(p_1+1)(p_2+1)(p_3+1)(p_1+p_2+2)(p_2+p_3+2) \\ &\qquad \times (p_1+p_2+p_3+3) \quad .\end{aligned} \tag{18}$$

This systematically leads us to the assumption that the general solution is given by

$$N_{n+1}(p_1 \ldots p_n) = \frac{\prod\limits_{l=1}^{n}\prod\limits_{k=0}^{n}\left(\sum\limits_{m=k-l+1}^{k} p_m + l\right)}{\prod\limits_{k=1}^{n} k!} \quad . \tag{19}$$

The proof is done by induction. In the cases of $n = 1, 2, 3$ we already have the basis for the proof. We now suppose the formula to be valid for some $\mathrm{SU}(n)$.

Example 10.4

With the help of the recursion formula (17) we obtain:

$$N_{n+1}(p_1 \dots p_n) = \frac{1}{n!}(p_n + 1)(p_n + p_{n-1} + 2)\cdots(p_n + p_{n-1} + \cdots + p_1 + n)$$
$$\times \frac{\prod\limits_{l=1}^{n-1} \prod\limits_{k=l}^{n-1} \left(\sum\limits_{m=k-l+1}^{k} p_m + l \right)}{\prod\limits_{k=1}^{n-1} k!} \quad . \tag{20}$$

The denominator can at once be written as $\prod_{k=1}^{n} k!$, and the factors can be put into the terms in the product in the numerator, i.e.

$$(p_n + 1) \quad \text{for} \quad k = n,\ l = 1 \quad , \quad (p_n + p_{n-1} + 2) \quad \text{for} \quad k = n,\ l = 2 \quad , \tag{21}$$

and so on. The last factor $(p_n + p_{n-1} + \cdots + p_1 + n)$ is put into the term $k = n$, $l = n$. Thus we get the ansatz (19).

11. Charm and SU(4)

Up to now we have discussed quark models with *three* constituent quark flavours (quark triplet). In November 1974 a new vector meson was discovered by two groups in Brookhaven and Stanford (USA),[1] which was denoted by J or Ψ, respectively, and is now generally called the J/Ψ meson. The discovery of other particles followed. Later these findings were experimentally confirmed at the Deutsches Elektronen-Synchroton (DESY) in the colliding storage ring DORIS at Hamburg and more particles have been discovered. Figure 11.1 gives an impression of the complicated construction of the accelerator in Hamburg, where such experiments have been performed.

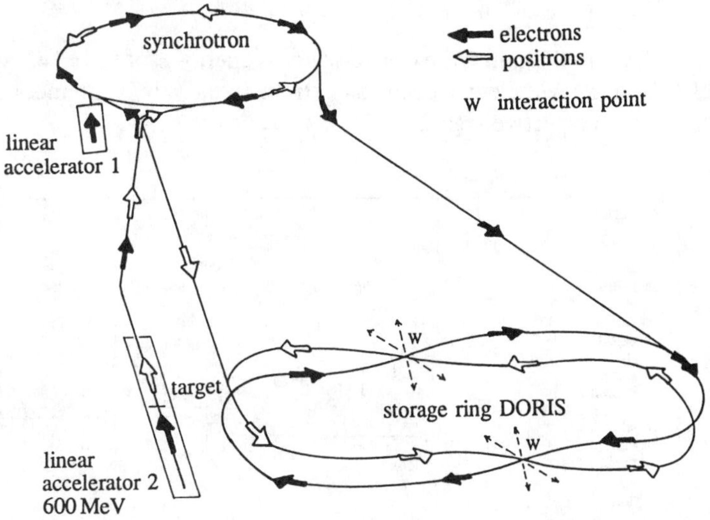

Fig. 11.1. Plan of the accelerator construction at DESY, Hamburg

The experiment involved electron (e^-)-positron (e^+) collisions, in which muons ($e^+e^- \to \mu^+\mu^-$) as well as hadrons ($e^+e^- \to$ hadrons) are produced. The J/Ψ meson has an unusually high mass of $M=3097$ MeV and an extremely small resonance width of $\Gamma=0.063$ MeV (see Table 11.1). In 1976[2] S.C.C. Ting and B. Richter were awarded the Nobel prize in physics for the discovery of the J/Ψ.

[1] J.J. Aubert et al. (15 authors): Phys. Rev. Lett. **33**, 1404 (1974);
J.-E. Augustin et al. (35 authors): Phys. Rev. Lett. **33**, 1406 (1974).

[2] Samuel C.C. Ting: Rev. Mod. Phys. **49**, 235 (1977);
Burton Richter: Rev. Mod. Phys. **49**, 251 (1977).

Table 11.1. The properties of the newly discovered vector mesons

Particle	Spin parity	Iso-spin	Mass [MeV]	Width [MeV]	Decays	[%]
$\psi(3100)$, J	1^-	0	3096.9 ± 0.1	0.063 ± 0.009	e^+e^-	7.5
					$\mu^+\mu^-$	7.5
					hadrons	85
$\psi(3700)$, ψ'	1^-	0	3686.0 ± 0.1	0.215 ± 0.040	e^+e^-	0.9
					$\mu^+\mu^-$	0.9
					hadrons	98.1
$\psi(4030)$, ψ''	1^-	?	4030 ± 5	51 ± 10	e^+e^-	0.0014
					hadrons	rest
$\psi(4415)$, ψ'''	1^-	?	4415 ± 6	43 ± 20	e^+e^-	0.0010
					hadrons	rest

The ratio of the cross-sections of the two reactions is calculated by

$$R = \frac{\sigma(e^+e^- \to \text{hadrons})}{\sigma(e^+e^- \to \mu^+\mu^-)} \quad .$$

If R is plotted against the centre-of-mass energy E of the e^+e^- system, then the J/Ψ and Ψ' will clearly occur as extremely narrow resonances at 3.1 GeV and 3.7 GeV, respectively (Fig. 11.2).

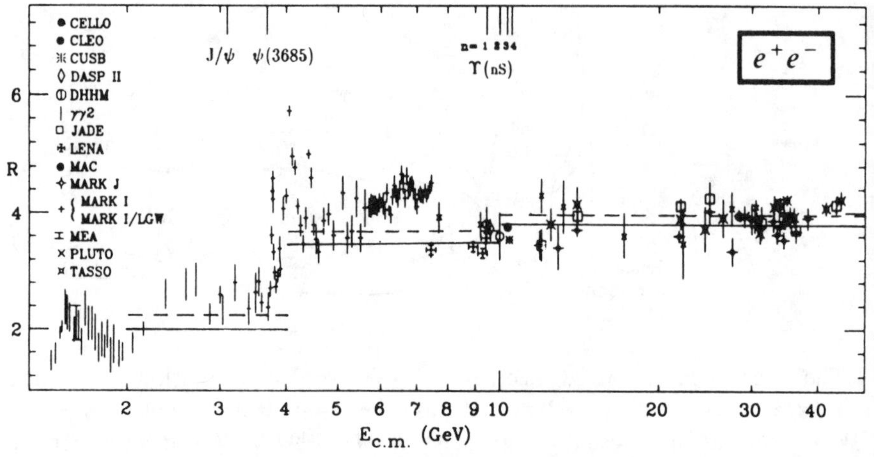

Fig. 11.2. The ratio of the cross-sections in the reaction $e^+e^- \to$ hadrons to that in the reaction $e^+e^- \to \mu^+\mu^-$

In addition, around $E \sim 4.1$ GeV and $E \sim 4.4$ GeV several resonances can be observed which, however, have a significantly larger width than Ψ at 3.1 GeV; we therefore first take a closer look at this state.

Since this vector meson is strongly interacting, which is well confirmed, its lifetime τ should be of the order of magnitude of the decay times of other high-lying hadronic resonances; thus $\tau \sim 5 \times 10^{-24}$ s, which is equivalent to a width of $\Gamma \sim 100$ MeV. However the measured value is $\Gamma_{\text{exp}} \sim 0.063$ MeV. This is a hint for a new quantum number which is conserved, strongly suppressing the decay of Ψ induced by strong interactions. The simplest method of adopting this new quantum number into the framework of the quark model is to introduce a fourth quark *c* for (for "*charm*") which differs from u, d and s by the *charm quantum number* C; u, d and s have $C=0$, while the fourth quark has $C=1$. Charm is an additive (charge-like) quantum number like T_3 and Y, i.e. the T_3-Y diagram is extended to a third axis. There also exists the antiquark $\bar{c}$ with $C=-1$. Thus, instead of the fundamental triplet (3) we have a quartet (4) as the smallest representation of the symmetry group. This leads to the group SU(4).

11.1 Particles with Charm and the SU(4)

From now on we assume the existence of a fourth quark c, which has an additional quantum number denoted by "charm C", i.e. the c quark has $C=1$. The other quantum numbers of the c quark are $T=T_3=0$ and $Y=\frac{1}{3}$. Therefore c is a singlet with respect to the usual flavour SU(3), and we now have the quartet q = u, d, s, c instead of the triplet of SU(3) as the fundamental representation of the symmetry group. Furthermore, in addition to the additive quantum numbers T_3 and Y of the SU(3) (group of rank 2) we have to consider the charm quantum number C, i.e. the states of a multiplet of the new symmetry group are given by $|T_3 YC\rangle$. Since we have three additive quantum numbers now, the symmetry group has to be of rank 3 and must also have the quartet (antiquartet) as fundamental representation. The group which naturally fulfills these requirements is the SU(4). Therefore we will base our examination of particles with charm on the group theoretical consideration of SU(4).

11.2 The Group Properties of SU(4)

Since the SU(3) multiplets have to be included in the SU(4) model, we want to preserve as much as possible of the structure of SU(3) in SU(4) [analogous to the imbedding of isospin SU(2) in SU(3)]. Again we use the matrix representation of the Lie algebra and its generators, i.e. its generators are 4×4 matrices. Analogously to the SU(3) matrices $\hat{\lambda}_i$, which were derived with the help of the $\hat{\tau}_i$ (Pauli matrices) of SU(2), we will derive the SU(4) matrices from the $\hat{\lambda}_i$ of SU(3).

Since the number of generators of the group SU(n) is in general given by n^2-1, we get fifteen 4×4 matrices for the SU(4). The first eight are created by the eight generators of SU(3). (We also denote the generators of the SU(4) by $\hat{\lambda}_i$

to stress the formal similarity.)

$$\hat{\lambda}_1=\begin{pmatrix}0&1&0&0\\1&0&0&0\\0&0&0&0\\0&0&0&0\end{pmatrix},\quad \hat{\lambda}_2=\begin{pmatrix}0&-i&0&0\\i&0&0&0\\0&0&0&0\\0&0&0&0\end{pmatrix},\quad \hat{\lambda}_3=\begin{pmatrix}1&0&0&0\\0&-1&0&0\\0&0&0&0\\0&0&0&0\end{pmatrix},$$

$$\hat{\lambda}_4=\begin{pmatrix}0&0&1&0\\0&0&0&0\\1&0&0&0\\0&0&0&0\end{pmatrix},\quad \hat{\lambda}_5=\begin{pmatrix}0&0&-i&0\\0&0&0&0\\i&0&0&0\\0&0&0&0\end{pmatrix},\quad \hat{\lambda}_6=\begin{pmatrix}0&0&0&0\\0&0&1&0\\0&1&0&0\\0&0&0&0\end{pmatrix},$$

$$\hat{\lambda}_7=\begin{pmatrix}0&0&0&0\\0&0&-i&0\\0&i&0&0\\0&0&0&0\end{pmatrix},\quad \hat{\lambda}_8=\frac{1}{\sqrt{3}}\begin{pmatrix}1&0&0&0\\0&1&0&0\\0&0&-2&0\\0&0&0&0\end{pmatrix}. \tag{11.1}$$

The following six $\hat{\lambda}_i$ are constructed by displacing the non-vanishing elements 1, -1 and $-i$, i respectively (such as $\hat{\lambda}_4$ to $\hat{\lambda}_7$ from $\hat{\lambda}_1$ and $\hat{\lambda}_2$):

$$\hat{\lambda}_9=\begin{pmatrix}0&0&0&1\\0&0&0&0\\0&0&0&0\\1&0&0&0\end{pmatrix},\quad \hat{\lambda}_{10}=\begin{pmatrix}0&0&0&-i\\0&0&0&0\\0&0&0&0\\i&0&0&0\end{pmatrix},$$

$$\hat{\lambda}_{11}=\begin{pmatrix}0&0&0&0\\0&0&0&1\\0&0&0&0\\0&1&0&0\end{pmatrix},\quad \hat{\lambda}_{12}=\begin{pmatrix}0&0&0&0\\0&0&0&-i\\0&0&0&0\\0&i&0&0\end{pmatrix},$$

$$\hat{\lambda}_{13}=\begin{pmatrix}0&0&0&0\\0&0&0&0\\0&0&0&1\\0&0&1&0\end{pmatrix},\quad \hat{\lambda}_{14}=\begin{pmatrix}0&0&0&0\\0&0&0&0\\0&0&0&-i\\0&0&0&0\end{pmatrix}. \tag{11.2}$$

$\hat{\lambda}_{15}$ is chosen in such a way that all $\hat{\lambda}_i$ are linearly independent. Usually one chooses $\hat{\lambda}_{15}$ analogously to $\hat{\lambda}_8$, i.e.

$$\hat{\lambda}_{15}=\frac{1}{\sqrt{6}}\begin{pmatrix}1&0&0&0\\0&1&0&0\\0&0&1&0\\0&0&0&-3\end{pmatrix}. \tag{11.3}$$

The matrices $\hat{\lambda}_i$ of the SU(4) fulfill similar commutation and trace relations as the SU(3) matrices:

$$[\hat{\lambda}_i, \hat{\lambda}_j]_- = 2\mathrm{i} f_{ijk} \hat{\lambda}_k \quad , \quad [\hat{\lambda}_i, \hat{\lambda}_j]_+ = \delta_{ij} \mathbb{1} + 2 d_{ijk} \hat{\lambda}_k \quad , \tag{11.4a}$$

$$\mathrm{Tr}(\hat{\lambda}_i) = 0 \quad , \quad \mathrm{Tr}(\hat{\lambda}_i \hat{\lambda}_j) = 2\delta_{ij} \quad . \tag{11.4b}$$

The first and the third equations are evident, while the second and fourth equations are proven in the following exercise for general SU(n). The structure constants f_{ijk} and d_{ijk} of SU(4) are given by the relations already shown (see Exercises 7.3 and 7.4 and also the following exercises 11.1 – 11.3)

$$f_{ijk} = \frac{1}{4\mathrm{i}} \mathrm{Tr}([\hat{\lambda}_i, \hat{\lambda}_j]_- \hat{\lambda}_k) \quad , \quad d_{ijk} = \frac{1}{4} \mathrm{Tr}([\hat{\lambda}_i, \hat{\lambda}_j]_+ \hat{\lambda}_k) \quad . \tag{11.5}$$

EXERCISE

11.1 Anticommutators of the Generators of SU(N)

Problem. Show that the matrix representation of an SU(N) with ($N^2 - 1$) traceless $\hat{\lambda}_i$-matrices satisfies

$$[\hat{\lambda}_i, \hat{\lambda}_j]_+ = \frac{4}{N} \delta_{ij} \mathbb{1}_{N \times N} + 2 d_{ijk} \hat{\lambda}_k \quad , \tag{1}$$

with

$$d_{ijk} := \frac{1}{4} \mathrm{Tr}\{[\hat{\lambda}_i, \hat{\lambda}_j]_+ \hat{\lambda}_k\} \quad . \tag{2}$$

Solution. Together with the unit matrix $\mathbb{1}_{N \times N}$, the $\hat{\lambda}_i$ form a basis of the $N \times N$ matrices. This allows the representation as

$$[\hat{\lambda}_i, \hat{\lambda}_j]_+ = \mathbb{1}_{N \times N} \alpha_{ij} + \beta_{ijk} \hat{\lambda}_k \quad , \tag{3}$$

with coefficients α_{ij} and β_{ijk} which have to be determined. The trace of (3) follows as

$$\mathrm{Tr}\{[\hat{\lambda}_i, \hat{\lambda}_j]_+\} = 2\mathrm{Tr}\{\hat{\lambda}_i \hat{\lambda}_j\} = N\alpha_{ij} + 0 \quad ,$$

$$\alpha_{ij} = \frac{4}{N} \delta_{ij} \quad , \tag{4}$$

where $\mathrm{Tr}\{\hat{\lambda}_k\} = 0$ and $\mathrm{Tr}\{\hat{\lambda}_i, \hat{\lambda}_j\} = 2\delta_{ij}$ have been used. The last relation will be proven later (see also exercise 11.2), but now we multiply by $\hat{\lambda}_l$ and calculate the trace again.

$$\mathrm{Tr}\{\hat{\lambda}_l [\hat{\lambda}_i, \hat{\lambda}_j]_+\} = \frac{4}{N} \mathrm{Tr}\{\hat{\lambda}_l\} \delta_{ij} + \beta_{ijk} \mathrm{Tr}\{\hat{\lambda}_l \hat{\lambda}_k\} \quad ,$$

$$4 d_{ijl} = 2\beta_{ijk} \delta_{lk} = 2\beta_{ijl} \quad , \quad \beta_{ijl} = 2 d_{ijl} \quad . \tag{5}$$

Exercise 11.1

Therefore

$$[\hat{\lambda}_i, \hat{\lambda}_j]_+ = \frac{4}{N}\delta_{ij} + 2d_{ijk}\hat{\lambda}_k$$

holds; q.e.d.

We still have to establish the equation $\mathrm{Tr}\{\hat{\lambda}_i\hat{\lambda}_j\} = 2\delta_{ij}$, which is simply a normalization condition. This is done by assuming that

$$\mathrm{Tr}\{\hat{\lambda}_i^2\} = 2 \text{ (normalization)} \tag{6}$$

and showing that $\mathrm{Tr}\{\hat{\lambda}_i\hat{\lambda}_j\} = 0$ for $i \neq j$.

The product $\hat{\lambda}_i\hat{\lambda}_j$, $i \neq j$, has diagonal elements only in two cases, namely:

1) $\hat{\lambda}_i$ and $\hat{\lambda}_j$ are non-diagonal matrices and belong to the same SU(2) subalgebra with the translation operators

$$\hat{\Lambda}_\pm = \tfrac{1}{2}(\hat{\lambda}_i \pm \mathrm{i}\hat{\lambda}_j) \quad . \tag{7}$$

Then

$$\hat{\Lambda}_+^2 = \tfrac{1}{4}(\hat{\lambda}_i^2 - \hat{\lambda}_j^2 + \mathrm{i}\hat{\lambda}_i\hat{\lambda}_j + \mathrm{i}\hat{\lambda}_j\hat{\lambda}_i) \quad . \tag{8}$$

The diagonal elements of $\hat{\Lambda}_+$ and $\hat{\Lambda}_+^2$ are zero. Therefore,

$$\mathrm{Tr}\{\hat{\Lambda}_+^2\} = \mathrm{Tr}\{\hat{\Lambda}_+\hat{\Lambda}_+\} = 0 \quad , \tag{9}$$

$$\mathrm{Tr}\{\hat{\lambda}_i^2 - \hat{\lambda}_j^2 + 2\mathrm{i}\hat{\lambda}_i\hat{\lambda}_j\} = 0 \quad , \tag{10}$$

or, combined with (1),

$$\mathrm{Tr}\{\hat{\lambda}_i\hat{\lambda}_j\} = 0 \quad , \quad i \neq j \ . \tag{11}$$

2) $\hat{\lambda}_i$ and $\hat{\lambda}_j$ are diagonal matrices, therefore

$$i, j \in \{3, 8, 15, \ldots, n^2 - 1, \ldots\} \quad .$$

Now the $\hat{\lambda}$-matrices are constructed in such a way that

$$\hat{\lambda}_l\hat{\lambda}_m \sim \hat{\lambda}_l \quad , \quad \text{for} \quad l < m \quad , \tag{12}$$

which leads, with (12) ($i \neq j$), to

$$\mathrm{Tr}\{\hat{\lambda}_i\hat{\lambda}_j\} \sim \mathrm{Tr}\{\hat{\lambda}_{\min(i,j)}\} = 0 \quad i \neq j \quad . \tag{13}$$

From (6), (11) and (13) we obtain

$$\mathrm{Tr}\{\hat{\lambda}_i\hat{\lambda}_j\} = 2\delta_{ij} \quad , \quad \text{q.e.d.} \tag{14}$$

Compare this result with the following Exercise 11.2!

EXERCISE

11.2 Trace of a Generator Product in the SU(N)

Problem. Show that for the $\hat{\lambda}_i$ matrix representation of any group SU(N),

$$\mathrm{Tr}\{\hat{\lambda}_i\hat{\lambda}_j\} = 2\delta_{ij}$$

is valid.

Solution. To solve this problem, one has to devise a general scheme to construct the $\hat{\lambda}_i$ of an arbitrary SU(N).

1) To every $i, j = 1, 2, \ldots, N$, $i < j$, we define the two $N \times N$ matrices

$$[\hat{\lambda}^{(1)}(i,j)]_{\mu\nu} = \delta_{j\mu}\delta_{i\nu} + \delta_{j\nu}\delta_{i\mu} \quad ,$$

$$[\hat{\lambda}^{(2)}(i,j)]_{\mu\nu} = -\mathrm{i}(\delta_{i\mu}\delta_{j\nu} - \delta_{i\nu}\delta_{j\mu}) \quad , \tag{1}$$

which form $N(N-1)/2 + N(N-1)/2 = N(N-1)$ linearly independent matrices.

2) We construct a further $N-1$ matrices according to

$$\hat{\lambda}_{n^2-1} = \sqrt{\frac{2}{n^2-n}}\begin{pmatrix} 1 & & & & & \\ & \ddots & & & & \\ & & 1 & & & 0 \\ & & & -(n-1) & & \\ & & & & 0 & \\ 0 & & & & & 0 \end{pmatrix} \quad ; \quad (\text{first } n-1 \text{ diagonal entries } 1)$$

$$n = 2, 3, \ldots, N \quad . \tag{2}$$

This gives $N-1$ additional linear independent matrices in such a way that we get a total number $N(N-1) + N - 1 = N^2 - 1$ matrices. The $\hat{\lambda}$ matrices defined this way from a basis of the vector space of the traceless $N \times N$ matrices and thus a representation of the SU(N) generators.

Now we can prove the required relation by simple calculation.

$$\mathrm{Tr}\{\hat{\lambda}_i^2\} = 2 \quad \text{for all} \quad i = 1, \ldots, N^2 - 1 \quad , \tag{3}$$

which is achieved by an appropriate choice of normalization. For $\mathrm{Tr}\{\hat{\lambda}_{n^2-1}\hat{\lambda}_{m^2-1}\}$, with $n < m$, equation (2) yields

$$\hat{\lambda}_{n^2-1}\hat{\lambda}_{m^2-1} = \sqrt{\frac{2}{m^2-m}}\,\hat{\lambda}_{n^2-1} \quad \text{for} \quad n < m \quad , \tag{4}$$

hence

$$\mathrm{Tr}\{\hat{\lambda}_{n^2-1}\hat{\lambda}_{m^2-1}\} = \sqrt{\frac{2}{m^2-m}}\,\mathrm{Tr}\{\hat{\lambda}_{n^2-1}\} = 0 \quad . \tag{5}$$

Furthermore

$$\begin{aligned}\mathrm{Tr}\{\hat{\lambda}_{n^2-1}\hat{\lambda}^{(1)}(i,j)\} &= \Sigma_\mu \alpha_\mu^{(n)}\delta_{\mu\nu}[\hat{\lambda}^{(1)}(i,j)]_{\nu\mu} \\ &= \Sigma_\mu \alpha_\mu^{(n)}[\delta_{j\mu}\delta_{i\nu}\delta_{\mu\nu} + \delta_{j\nu}\delta_{i\mu}\delta_{\mu\nu}] = 0 \quad ,\end{aligned} \tag{6}$$

Exercise 11.2

since $i < j$, and in the same way we get

$$\mathrm{Tr}\{\hat{\lambda}_{n^2-1}\hat{\lambda}^{(2)}(i,j)\} = 0 \quad , \tag{7}$$

Now we calculate

$$\mathrm{Tr}\{\hat{\lambda}^{(1)}(i,j)\hat{\lambda}^{(1)}(k,l)\} \quad \text{for} \quad (i,j) \neq (k,l) \quad , \quad \text{i.e. } i \neq k \quad \text{or} \quad j \neq l \quad ,$$

$$\begin{aligned}\mathrm{Tr}\{\hat{\lambda}^{(1)}(i,j)\hat{\lambda}^{(1)}(k,l)\} &= \sum_{\mu,\nu} (\delta_{j\mu}\delta_{i\nu} + \delta_{j\nu}\delta_{i\mu})(\delta_{k\nu}\delta_{l\mu} + \delta_{k\mu}\delta_{l\nu}) \quad . \\ &= \delta_{jl}\delta_{ik} + \delta_{jk}\delta_{il} + \delta_{jk}\delta_{il} + \delta_{jl}\delta_{ik} \\ &= 2\delta_{jk}\delta_{il} = 0 \quad ,\end{aligned} \tag{8}$$

since $i < j$ and $k < l$ (assuming $i = l$ and $j = k$ yields $i < j \Rightarrow l < k$, and hence a contradiction to $k < l$). Analogously one finds that

$$\mathrm{Tr}\{\hat{\lambda}^{(2)}(i,j)\hat{\lambda}^{(2)}(k,l)\} = 0 \quad . \tag{9}$$

The final quantity required is:

$$\begin{aligned}\mathrm{Tr}\{\hat{\lambda}^{(1)}(i,j)\hat{\lambda}^{(2)}(k,l)\} &= \sum_{\mu,\nu} -\mathrm{i}[\delta_{j\mu}\delta_{i\nu} + \delta_{j\nu}\delta_{i\mu}][\delta_{l\mu}\delta_{k\nu} - \delta_{k\mu}\delta_{l\nu}] \\ &= -\mathrm{i}[\delta_{jl}\delta_{ik} + \delta_{jk}\delta_{il} - \delta_{kj}\delta_{il} - \delta_{jl}\delta_{ik}] = 0 \quad .\end{aligned} \tag{10}$$

Thus we have shown that

$$\mathrm{Tr}\{\hat{\lambda}_i\hat{\lambda}_j\} = 2\delta_{ij} \quad . \tag{11}$$

EXERCISE

11.3 The Completeness Relation for $\hat{F}_a$

Problem: Prove the completeness relation for the SU(N) generators $\hat{F}_a = \hat{\lambda}_a/2$.

$$\sum_a (\hat{F}_a)_{il}(\hat{F}_a)_{jk} = \tfrac{1}{2}\,\delta_{ik}\,\delta_{jl} - \frac{1}{2N}\delta_{il}\,\delta_{jk} \quad . \tag{1}$$

Derive the following useful identities

$$\text{a)} \quad \sum_a (\hat{F}_a)_{ik}(\hat{F}_a)_{jl} = \frac{N^2-1}{2N}\,\delta_{il}\delta_{jk} - \frac{1}{N}\sum_a (\hat{F}_a)_{il}(\hat{F}_a)_{jk} \quad , \tag{2}$$

and

$$\sum_a (\hat{F}_a)_{ij}(\hat{F}_a)_{jk} = \frac{N^2-1}{2N}\,\delta_{ik} \quad , \tag{3}$$

$$\text{b)} \quad f^{acd}f^{bcd} = N\,\delta^{ab} \quad . \tag{4}$$

Exercise 11.3

Solution: The $\hat{F}_a$ and the unity matrix are one possible basis of the hermitian $N \times N$ matrices. Hence, any hermitian $N \times N$ matrix can be represented by

$$(A)_{ij} = c_0\,\delta_{ij} + \sum_{a=1}^{N^2-1} c_a(\hat{F}_a)_{ij} \quad . \tag{5}$$

In the following we use Einstein's summation convention over identical indices. The coefficients c_0 and c_a ($a = 1, \dots, N^2 - 1$) are determined from the normalization conditions

$$\mathrm{Tr}\{\hat{F}_a\hat{F}_b\} = \tfrac{1}{2}\delta_{ab} \tag{6}$$

and

$$\mathrm{Tr}\{\mathbb{1}\} = \delta_{ii} = N \quad . \tag{7}$$

The multiplication of (5) by δ_{ij} leads to

$$A_{ij}\delta_{ij} = c_0\,\delta_{ij}\delta_{ij} + c_a(\hat{F}_a)_{ij}\delta_{ij}$$

$$A_{ii} = c_0\,\delta_{ii} + c_a(\hat{F}_a)_{ii} = c_0\,N + c_a\,\mathrm{Tr}\{\hat{F}_a\} = c_0\,N \quad . \tag{8}$$

Hence, the property $\mathrm{Tr}\{\hat{F}_a\} = 0$ leads to

$$c_0 = \frac{A_{ii}}{N} \quad . \tag{9}$$

The multiplication of (5) by $(\hat{F}_b)_{ji}$ leads to

$$A_{ij}(\hat{F}_b)_{ij} = c_a(\hat{F}_a)_{ij}(\hat{F}_b)_{ji} = c_a\,\tfrac{1}{2}\,\delta_{ab} = \tfrac{1}{2}c_b \tag{10}$$

and therefore

$$A_{ij} = \frac{A_{ll}}{N}\delta_{ij} + 2\,A_{lm}(\hat{F}_b)_{ml}(\hat{F}_b)_{ij} \quad . \tag{11}$$

We bring all terms to the left-hand side and take the factor A_{lm} outside the brackets

$$A_{lm}\left(\delta_{li}\,\delta_{jm} - \frac{1}{N}\delta_{ij}\,\delta_{lm} - 2(\hat{F}_b)_{ml}(\hat{F}_b)_{ij}\right) = 0 \quad .$$

This equation is valid for arbitrary matrices A_{lm}. Therefore, the expression within the brackets has to vanish and one obtains

$$(\hat{F}_b)_{ml}(\hat{F}_b)_{ij} = \frac{1}{2}\delta_{li}\,\delta_{mj} - \frac{1}{2N}\delta_{ij}\,\delta_{lm} \quad . \tag{12}$$

After renaming the indices this is just the desired relation (1).

a) This relation is a simple corollary of the completeness relation. Adding the expression

$$(\hat{F}_a)_{ik}(\hat{F}_a)_{jl} = \frac{1}{2}\delta_{il}\,\delta_{kj} - \frac{1}{2N}\delta_{ik}\delta_{jl} \quad . \tag{13}$$

Exercise 11.3

to

$$\frac{1}{N}(\hat{F}_a)_{il}(\hat{F}_a)_{ik} = -\frac{1}{2N^2}\delta_{il}\delta_{kj} + \frac{1}{2N}\delta_{ik}\delta_{jl} \quad . \tag{14}$$

leads immediately to the desired relation

$$(\hat{F}_a)_{ik}(\hat{F}_a)_{jl} = \frac{1}{2N^2}(N^2-1)\,\delta_{il}\,\delta_{jk} - \frac{1}{N}(\hat{F}_a)_{il}(\hat{F}_a)_{jk} \quad . \tag{15}$$

Multiplying by δ_{kj} yields

$$(\hat{F}_a)_{ik}(\hat{F}_a)_{kl} = \frac{N^2-1}{2N^2}\delta_{il} \quad . \tag{16}$$

With this expression we constructed one of the Casimir operators, because

$$\hat{F}_a \cdot \hat{F}_a = \frac{N^2-1}{2N^2}\mathbb{1} \quad . \tag{17}$$

commutes with all generators.

b) Now we show that

$$f^{acd} f^{bcd} = N\,\delta^{ab} \quad . \tag{18}$$

First, we express the f^{abc} *by* $\hat{F}^a$, $\hat{F}^b$ and $\hat{F}^c$

$$\begin{aligned} f^{abc} &= -2\mathrm{i}\,\mathrm{Tr}\{[\hat{F}^a,\hat{F}^b]\hat{F}^c\} \\ &= -2\mathrm{i}\,\mathrm{Tr}\{\hat{F}^a[\hat{F}^b,\hat{F}^c]\} \end{aligned}$$

(see Exercise 7.4).
Therefore, this gives for (18)

$$\begin{aligned} f^{acd} f^{bcd} &= -4\,\mathrm{Tr}\{\hat{F}^a[\hat{F}^c,\hat{F}^d]\}\,\mathrm{Tr}\{\hat{F}^b[\hat{F}^c,\hat{F}^d]\} \\ &= -4\,[(\mathrm{Tr}\{\hat{F}^a\hat{F}^c\hat{F}^d\} - \mathrm{Tr}\{\hat{F}^a\hat{F}^d\hat{F}^c\}) \\ &\quad \times(\mathrm{Tr}\{\hat{F}^b\hat{F}^c\hat{F}^d\} - \mathrm{Tr}\{\hat{F}^b\hat{F}^d\hat{F}^c\})] \\ &= -8\,(\mathrm{Tr}\{\hat{F}^a\hat{F}^c\hat{F}^d\}\,\mathrm{Tr}\{\hat{F}^b\hat{F}^c\hat{F}^d\} \\ &\quad - \mathrm{Tr}\{\hat{F}^a\hat{F}^c\hat{F}^d\}\,\mathrm{Tr}\{\hat{F}^b\hat{F}^d\hat{F}^c\}) \quad . \end{aligned} \tag{19}$$

Now we have to calculate the terms

$$\mathrm{Tr}\{\hat{F}^a\hat{F}^c\hat{F}^d\}\,\mathrm{Tr}\{\hat{F}^b\hat{F}^c\hat{F}^d\} \quad . \tag{20}$$

and

$$\mathrm{Tr}\{\hat{F}^a\hat{F}^c\hat{F}^d\}\,\mathrm{Tr}\{\hat{F}^b\hat{F}^d\hat{F}^c\} \quad . \tag{21}$$

This can be done with the help of the completeness relation (1)

$$\begin{aligned} \mathrm{Tr}\{\hat{F}^a\hat{F}^c\hat{F}^d\}\,\mathrm{Tr}\{\hat{F}^b\hat{F}^c\hat{F}^d\} &= (\hat{F}^a_{ij}\hat{F}^c_{jl}\hat{F}^d_{li})(\hat{F}^b_{\alpha\beta}\,\hat{F}^c_{\beta\gamma}\,\hat{F}^d_{\gamma\alpha}) \\ &= \hat{F}^a_{ij}\,\hat{F}^b_{\alpha\beta}\,(\hat{F}^c_{jl}\,\hat{F}^c_{\beta\gamma})(\hat{F}^d_{li}\,\hat{F}^d_{\gamma\alpha}) \quad . \end{aligned} \tag{22}$$

Now we insert (1)

$$= \hat{F}^a_{ij}\hat{F}^b_{\alpha\beta}\left(\frac{1}{2}\delta_{j\gamma}\delta_{l\beta} - \frac{1}{2N}\delta_{jl}\delta_{\beta\gamma}\right)\left(\frac{1}{2}\delta_{l\alpha}\delta_{i\gamma} - \frac{1}{2N}\delta_{li}\delta_{\gamma\alpha}\right)$$

$$= \frac{1}{4}\hat{F}^a_{ij}\hat{F}^b_{\alpha\beta}\left(\delta_{ji}\delta_{\alpha\beta} - \frac{1}{N}\delta_{i\beta}\delta_{j\alpha} - \frac{1}{N}\delta_{j\alpha}\delta_{i\beta} - \frac{1}{N^2}\delta_{jl}\delta_{\beta\gamma}\right)$$

$$= \frac{1}{4}\hat{F}^a_{ij}\hat{F}^b_{\alpha\beta}\left(-\frac{2}{N}\delta_{i\beta}\delta_{j\alpha}\right) = -\frac{1}{2N}\mathrm{Tr}\{\hat{F}^a\hat{F}^b\}$$

$$= -\frac{1}{4N}\delta^{ab} \quad . \tag{23}$$

Here, we used again $\mathrm{Tr}\{\hat{F}^\alpha\} = 0$.

The second term (21) is calculated analogously

$$\mathrm{Tr}\{\hat{F}^d\hat{F}^c\hat{F}^d\}\,\mathrm{Tr}\{\hat{F}^b\hat{F}^d\hat{F}^c\} = \hat{F}^a_{ij}\hat{F}^b_{\alpha\beta}(\hat{F}^c_{jl}\hat{F}^c_{\gamma\alpha})(\hat{F}^d_{li}\hat{F}^d_{\beta\gamma}) \quad . \tag{24}$$

After the insertion of (1) this gives

$$\mathrm{Tr}\{\hat{F}^a\hat{F}^c\hat{F}^d\}\,\mathrm{Tr}\{\hat{F}^b\hat{F}^d\hat{F}^c\} = \frac{1}{8}\delta^{ab}\left(N - \frac{2}{N}\right) \quad . \tag{25}$$

This expression together with (23) inserted into (19) finally leads to

$$f^{acd}f^{bcd} = -8\left\{-\frac{1}{4N}\delta^{ab} - \frac{1}{8}N\,\delta^{ab} + \frac{1}{4N}\delta^{ab}\right\} = N\,\delta^{ab} \quad .$$

EXERCISE

11.4 Eigenvalue of the Casimir Operator $\hat{C}_1$ of a Fundamental Representation of the SU(N)

Problem. Show that the eigenvalue of the Casimir operator of a fundamental representation of SU(N) has the value $(N^2-1)/2N$.

Solution. First of all we show that $\Sigma_i\, d_{iik} = 0$ for all k. $\hat{C}_1$ is a Casimir operator, and hence we have $[\hat{F}_\sigma, \hat{C}_1]_- = 0$. By using the matrix representation of the $\hat{\lambda}_i$ this gives

$$\sum_{i=1}^{N^2-1}[\tfrac{1}{2}\hat{\lambda}_\sigma, \tfrac{1}{4}\hat{\lambda}_i^2]_- = \frac{1}{8}\sum_{i=1}^{N^2-1}\left[\hat{\lambda}_\sigma, \left(\frac{2}{N}\delta_{ii}\mathbb{1}_{N\times N} + d_{iik}\hat{\lambda}_k\right)\right]$$

$$= \frac{\mathrm{i}}{4}\sum_k f_{\sigma kl}\hat{\lambda}_l\left\{\sum_{i=1}^{N^2-1} d_{iil}\right\} \quad , \tag{1}$$

where we have used the result of Exercise 11.1,

$$[\hat{\lambda}_i, \hat{\lambda}_j]_+ = \frac{4}{N}\delta_{ij}\mathbb{1}_{N\times N} + 2d_{ijk}\hat{\lambda}_k \quad . \tag{2}$$

Exercise 11.4

Now the following are valid:
1) For given σ and l there exists only one k in such a way that $f_{\sigma kl} \neq 0$.
2) The $\hat{\lambda}_l$'s are linearly independent.

It then follows that the expression in (1) is zero only if

$$\sum_{i=1}^{N^2-1} d_{iil} = 0 \quad \text{for all } l \quad . \tag{3}$$

The $\hat{\lambda}_i$ are just the matrix representation of the $\hat{F}_i$ of the fundamental representation. Therefore the value of $\hat{C}_1$ is directly calculable in the case of the fundamental representation:

$$\hat{G}_1^{\text{f.r}} = \frac{1}{4}\sum_i \hat{\lambda}_i\hat{\lambda}_i = \frac{1}{4}\sum_i \left(\frac{2}{N}\,\delta_{ii}\,\mathbb{1}_{N\times N} + d_{iik}\,\hat{\lambda}_k\right) = \frac{N^2-1}{2N}\,\mathbb{1}_{N\times N} \quad . \tag{4}$$

11.3 Tables of the Structure Constants f_{ijk} and the Coefficients d_{ijk} for SU(4)

The non-zero structure constants of the SU(4) are given in Tables 11.2 and 11.3:

We recognize the f_{ijk} to be completely antisymmetric, while the d_{ijk} are again symmetric with respect to all of their indices. This follows from the relations

Table 11.2. The non-vanishing structure constants f_{ijk}

i	j	k	f_{ijk}	i	j	k	f_{ijk}
1	2	3	1	4	10	13	$-1/2$
1	4	77	$1/2$	5	9	13	$1/2$
1	5	6	$-1/2$	5	10	14	$1/2$
1	9	12	$1/2$	6	7	8	$\sqrt{3}/2$
1	10	11	$-1/2$	6	11	14	$1/2$
2	4	6	$1/2$	6	12	13	$-1/2$
2	5	7	$1/2$	7	11	13	$1/2$
2	9	11	$1/2$	7	12	14	$1/2$
2	10	12	$1/2$	8	9	10	$1/(2\sqrt{3})$
3	4	5	$1/2$	8	11	12	$1/(2\sqrt{3})$
3	6	7	$-1/2$	8	13	14	$-1/(\sqrt{3})$
3	9	10	$1/2$	9	10	15	$\sqrt{2/3}$
3	11	12	$-1/2$	11	12	15	$\sqrt{2/3}$
4	5	8	$\sqrt{3}/2$	13	14	15	$\sqrt{2/3}$
4	9	14	$1/2$				

Table 11.3. The non-vanishing structure coefficients d_{ijk}

i	j	k	d_{ijk}	i	j	k	d_{ijk}
1	1	8	$1/\sqrt{3}$	5	5	8	$-1/(2\sqrt{3})$
1	1	15	$1/\sqrt{6}$	5	5	15	$1/\sqrt{6}$
1	4	6	$1/2$	5	9	14	$-1/2$
1	5	7	$1/2$	5	10	13	$1/2$
1	9	11	$1/2$	6	6	8	$-1/(2\sqrt{3})$
1	10	12	$1/2$	6	6	15	$1/\sqrt{6}$
2	2	8	$1/\sqrt{3}$	6	11	13	$1/2$
2	2	15	$1/\sqrt{6}$	6	12	14	$1/2$
2	4	7	$-1/2$	7	7	8	$-1/(2\sqrt{3})$
2	5	6	$1/2$	7	7	15	$1/\sqrt{6}$
2	9	12	$-1/2$	7	11	14	$-1/2$
2	10	11	$1/2$	7	12	13	$1/2$
3	3	8	$1/\sqrt{3}$	8	8	8	$-1/\sqrt{3}$
3	3	15	$1/\sqrt{6}$	8	8	15	$1/\sqrt{6}$
3	4	4	$1/2$	8	9	9	$1/(2\sqrt{3})$
3	5	5	$1/2$	8	10	10	$1/(2\sqrt{3})$
3	6	6	$-1/2$	8	11	11	$1/(2\sqrt{3})$
3	7	7	$-1/2$	8	12	12	$1/(2\sqrt{3})$
3	9	9	$1/2$	8	13	13	$-1/\sqrt{3}$
3	10	10	$1/2$	8	14	14	$-1/\sqrt{3}$
3	11	11	$-1/2$	9	9	15	$-1/\sqrt{6}$
3	12	12	$-1/2$	10	10	15	$-1/\sqrt{6}$
4	4	8	$-1/(2\sqrt{3})$	11	11	15	$-1/\sqrt{6}$
4	4	15	$1/\sqrt{6}$	12	12	15	$-1/\sqrt{6}$
4	9	13	$1/2$	13	13	15	$-1/\sqrt{6}$
4	10	14	$1/2$	14	14	15	$-1/\sqrt{6}$
				15	15	15	$-1/\sqrt{2/3}$

(11.4) and (11.5) in complete analogy to our conclusions in the earlier Exercises 7.3 und 7.4.

The usual generators of the SU(4) are

$$\hat{F}_i = \tfrac{1}{2}\hat{\lambda}_i \quad (i = 1, \ldots, 15) \quad , \tag{11.6}$$

which correspond to the $\hat{F}_i$ of SU(3). Furthermore we may introduce the isospin operators $\hat{T}_\pm = \hat{F}_1 \pm \mathrm{i}\hat{F}_2$, $\hat{T}_3 = \hat{F}_3$; the V-spin operators $\hat{V}_\pm = \hat{F}_4 \pm \mathrm{i}\hat{F}_5$; and the U-spin operators $\hat{U}_\pm = \hat{F}_6 \pm \mathrm{i}\hat{F}_7$, analogously to the SU(3). We do not use the usual SU(3) definition for the hypercharge $\hat{Y} = Z/\sqrt{3}\hat{F}_8$ because it leads to zero hypercharge $Y = 0$ for the charm quark. This definition would lead to non-integer values for the hypercharge of hadrons with open charm. Therefore, we extend the definition with the help of the third diagonal matrix $\hat{F}_{15}$

$$\hat{Y} = \frac{2}{\sqrt{3}}\hat{F}_8 + \frac{1}{12}\left(\mathbb{1} - 2\sqrt{6}\frac{1}{\sqrt{15}}\right).$$

This definition leads to the hypercharge $Y = \frac{1}{3}$ for the charm quark. We could now construct further quasi-spin operators with the help of the remaining operators (for example a W-spin, etc.); however, this does not seem to be sensible. While the isospin was very useful for the classification of states within an SU(3) multiplet (mass differences $\Delta M \leqslant 10$ MeV), already for the U-spin this holds only in a very restricted sense; the U-spin multiplets show mass splittings of some 100 MeV (10% of the mass).

Trying to use W-spin for classification of SU(4) multiplets would be completely inappropriate; the mass differences (e.g. a value of 2 GeV between ϱ^0 and Ψ, approximately 100% of the mass) are much higher, i.e. the SU(4) symmetry is broken much more severely than the SU(3) symmetry. Therefore the classification according to SU(2) subgroups is not practical. However, in the following we will see that the classification of the SU(4) multiplet according to SU(3) submultiplets of the quark triplets uds (i.e. a SU(3) group that acts on a triplet of three of the four quark states) still makes sense.

11.4 Multiplet Structure of SU(4)

We have already realized that [4] and [$\bar{4}$] are the fundamental representations of the SU(4); their quantum numbers are listed in Table 11.4. Analogously to SU(3) we have to form direct products of these fundamental representations in order to classify hadrons, i.e. [4] ⊗ [$\bar{4}$] for mesons (q$\bar{\text{q}}$), [4] ⊗ [4] ⊗ [4] for baryons (qqq).

Since there are three additive quantum numbers (T_3, Y, C) for each multiplet (for the quartet see Table 11.4), we get a three-dimensional $T_3 - Y - C$-representation of [4] and [$\bar{4}$], illustrated in Fig. 11.5. This compares to the one-dimensional representation of SU(2) shown in Fig. 11.3 and the two-dimensions of SU(3) (Fig. 11.4).

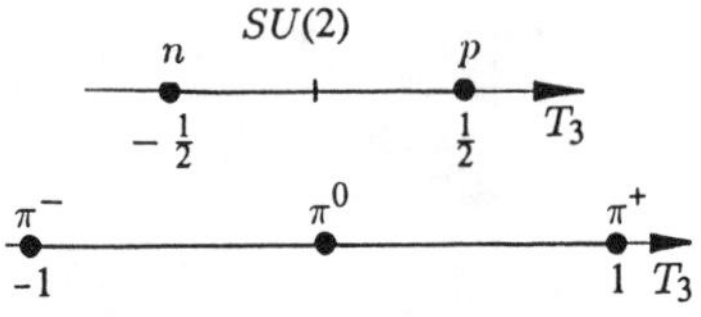

Fig. 11.3. The lowest non-trivial representations of SU(2).

Table 11.4. Quantum numbers of the quark quartet; S = "strangeness", B = baryon number

Symbol	Q [e]	T	T_3	S	B	Y	C
u_1, u	$\frac{2}{3}$	$\frac{1}{2}$	$\frac{1}{2}$	0	$\frac{1}{3}$	$\frac{1}{3}$	0
u_2, d	$-\frac{1}{3}$	$\frac{1}{2}$	$-\frac{1}{2}$	0	$\frac{1}{3}$	$\frac{1}{3}$	0
u_2, s	$-\frac{1}{3}$	0	0	-1	$\frac{1}{3}$	$-\frac{2}{3}$	0
u_4, c	$\frac{2}{3}$	0	0	0	$\frac{1}{3}$	$\frac{1}{3}$	1

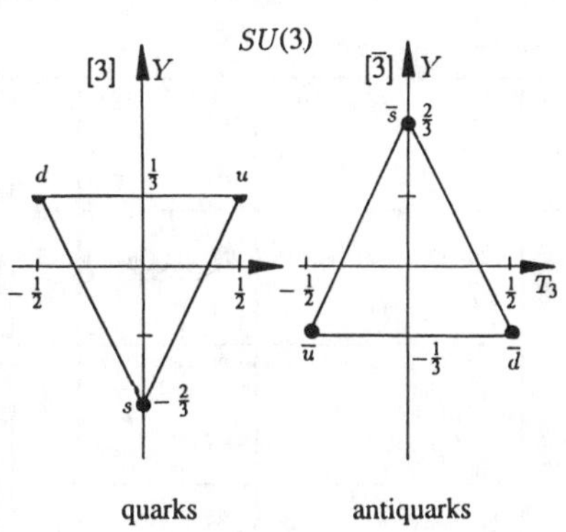

Fig. 11.4

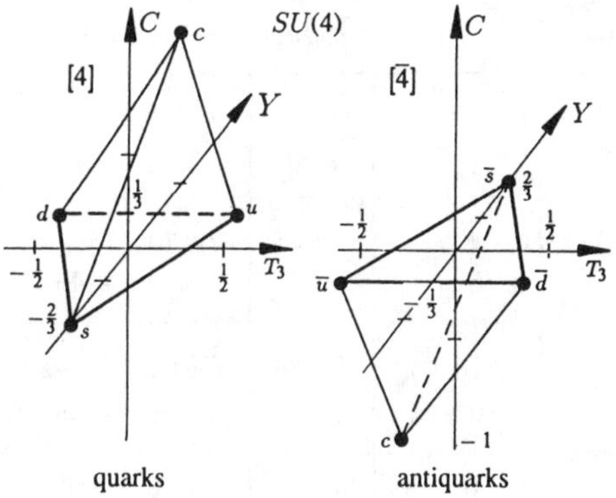

Fig. 11.5

Fig. 11.4. The lowest non-trivial representations of SU(3).

Fig. 11.5. The lowest non-trivial representations of SU(4).

Remark: In elementary particle physics the strangeness S is often used instead of the hypercharge Y, where $Y = B + S$ (see Fig. 11.6). The baryon number is $B = +1$ for all baryons and $B = -1$ for all antibaryons, and for mesons $Y = S$. The Gell-Mann-Nishijima relation then reads $Q = T_3 + 1/2(B + S)$. It will be necessary to generalize this relation to $Q = T_3 + \frac{1}{2}(Y + C) = T_3 + \frac{1}{2}(B + S + C)$, as we shall show later in Eq. (11.19). With that generalization the charge of the c-quark is calculated to the $Q_c = \frac{2}{3}$ (see Table 11.4).

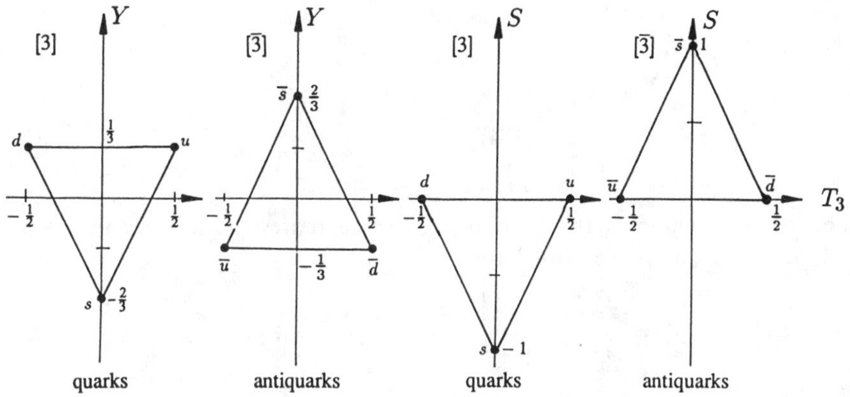

Fig. 11.6. Quarks and antiquarks in the representations of SU(3) as functions of the isospin component T_3 and the hypercharge Y or strangeness S.

While the graphical reduction of the Kronecker products is relatively simple in the case of SU(2) and SU(3), it is not so easily done in the case of SU(4) (especially for higher representations) because of the higher dimensionality. Therefore it is more appropriate to rely on group theoretical methods, e.g. the method of Young diagrams. We present the results (also for the representations [6], [10], [$\overline{10}$] etc.) in Table 11.5, some of which are derived in the solutions of Exercises 11.4–11.7.

As an example we select the Kronecker products [4] ⊗ [$\bar{4}$]. According to the rules for Young diagrams (see Chap. 9) the two fundamental representations are

Table 11.5. Table of the complete decomposition of the products $[X] \otimes [\bar{Y}]$

4	$\bar{4}$	6	10	$\overline{10}$	15	20	$\overline{20}$	20′	$\overline{20}′$	20″	x / y
6+10	1+15	$\bar{4}+20′$	20+20′	$\bar{4}+\overline{36}$	$4+\overline{20}′$ $+36$	35+45	$\overline{10}+70$	15+20″ $+45$	$6+\overline{10}$ $+64$	$\overline{20}′+60$	4
		1+15 $+20″$	15+45	$\overline{15}+\overline{45}$	6+10 $+\overline{10}+64$	36+84′	$\overline{36}+\overline{84}′$	$4+\overline{20}′$ $+36+60$	$\bar{4}+20′$ $+\overline{36}+\overline{60}$	6+50 $+64$	6
			20″+35 $+45$	1+15 $+84$	6+10 $+64+70$	56+60 $+84′$	$\bar{4}+\overline{36}$ $+\overline{160}$	$\overline{20}′+36$ $+60+84′$	$\bar{4}+20′$ $+\overline{36}+140″$	$\overline{10}+64$ $+126$	10
					1+15+15 $+20″+45$ $+\overline{45}+84$	20+20′ $+120$ $+140″$	$\overline{20}+\overline{20}′$ $+\overline{120}$ $+\overline{140}″$	$\bar{4}+20+20$ $+20′+36$ $+\overline{60}+140″$	$4+\overline{20}′$ $+20+\overline{20}$ $+36+60+\overline{140}″$	15+20″ $+45+\overline{45}$ $+175$	15

characterized by the diagrams

$$[4] = \square \quad , \quad [\bar{4}] = \begin{array}{c}\square \\ \square \\ \square\end{array} \quad . \tag{11.7}$$

The product yields

$$[4] \otimes [\bar{4}] = \square \otimes \begin{array}{c}\square \\ \square \\ \square\end{array} = \begin{array}{c}\square \\ \square \\ \square \\ \square\end{array} \oplus \begin{array}{c}\square\square \\ \square \\ \square\end{array} = [1] \oplus \begin{array}{c}\square\square \\ \square \\ \square\end{array} \quad , \tag{11.8}$$

and the dimension of this second irreducible representation is computed by adopting the rules given in Chap. 9:

$$\dim\left(\begin{array}{c}\square\square \\ \square \\ \square\end{array}\right) = \frac{\begin{array}{|c|c|}\hline 4 & 5 \\ \hline 3 \\ \hline 2 \\ \hline\end{array}}{\begin{array}{|c|c|}\hline 4 & 1 \\ \hline 2 \\ \hline 1 \\ \hline\end{array}} = \frac{4 \times 5 \times 3 \times 2}{4 \times 1 \times 2 \times 1} = 15 \quad . \tag{1.19}$$

For the meson combination $[4] \otimes [\bar{4}]$ we thus get

$$[4] \otimes [\bar{4}] = [1] \oplus [15] \quad , \tag{11.10}$$

i.e. an SU(4) singlet and a [15]-plet. We may decompose the [15]-plet into SU(3) multiplets:

$$[15] \xrightarrow{\text{SU(3)}} [1] \oplus [3] \oplus [\bar{3}] \oplus [8] \quad .$$

This follows from the SU(3) decomposition of $[4] = [3] + [1]$. In order to elucidate this circumstance we recall that SU(4) multiplets are constructed from SU(3) multiplets. The latter are levels (layers) with constant charms parallel to

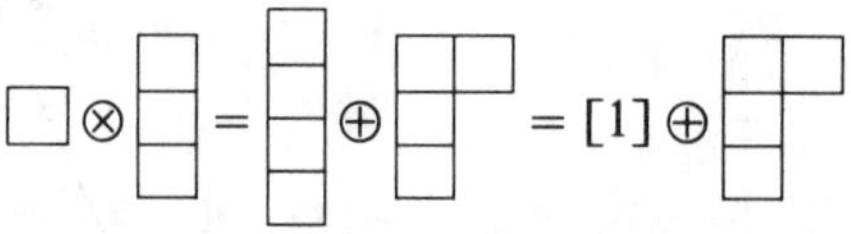

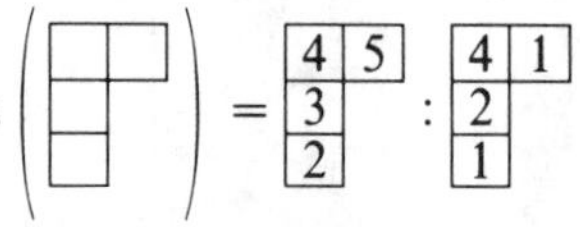

the $T_3 - Y$ plane. This is similar to the construction of SU(3) multiplets from SU(2) multiplets. In this case, the SU(2) multiplets (e.g. T-multiplets) are parallel to the T axis and belong to a different (but within every multiplet fixed) hypercharge Y. Hence, e.g. the SU(3) triplet $[3]=[2]^{1/3}\oplus[1]^{-2/3}$. The upper indices to the right give the hypercharge. The SU(3) antitriplet is decomposed as $[\bar{3}]=[2]^{-1/3}\oplus[1]^{2/3}$. Now one can easily calculate the SU(2) substructure of the SU(3) meson nonet:

$$\begin{aligned}[3]\otimes[\bar{3}] &= ([2]^{1/3}\oplus[1]^{-2/3})\otimes([2]^{-1/3}\oplus[1]^{2/3})\\ &= ([2]\otimes[2])^0\oplus[2]^{-1}\oplus[2]^1\oplus[1]^0\\ &= [3]^0\oplus[1]^0\oplus[2]^{-1}\oplus[2]^1\oplus[1]^0 \quad .\end{aligned}$$

This is exactly the structure of the meson nonet which was introduced in Chap. 8.

In SU(3) multiplets the upper indices denote the charm quantum number, e.g. $[1]^1$ means an SU(3)-singlet with $C=1$ or $[\bar{3}]^0$ indicates an SU(3)-triplet with $C=0$. Therefore according to (8.62) we have

$$\begin{aligned}[4]\otimes[\bar{4}] &= [[3]^0\oplus[1]^1]\otimes[[\bar{3}]^0\oplus[1]^{-1}]\\ &= [3]^0\otimes[\bar{3}]^0\oplus[3]^0\otimes[1]^{-1}\oplus[\bar{3}]^0\otimes[1]^1\oplus[1]^1\otimes[1]^{-1}\\ &= [8]^0\oplus[1]^0\oplus[3]^{-1}\oplus[\bar{3}]^1\oplus[1]^0 \quad . \qquad (11.11)\end{aligned}$$

This is illustrated in Figs. 11.7 and 11.8, which show the SU(3) multiplets (within every plane lying parallel to the $Y-T_3$ plane) and the two polyhedra of the SU(4). Altogether we get the well-known meson nonet $[1]^0+[8]^0$ (both lie

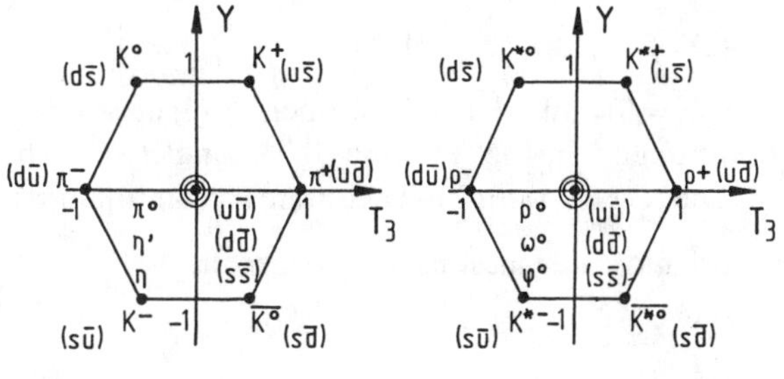

Fig. 11.7. SU(3)-multiples of pseudoscalar mesons (0^-) and vector mesons (1^-). The quark content of the analogous states is identical in the left and right multiplet, merely the coupling of the spin part of the wave function is different.

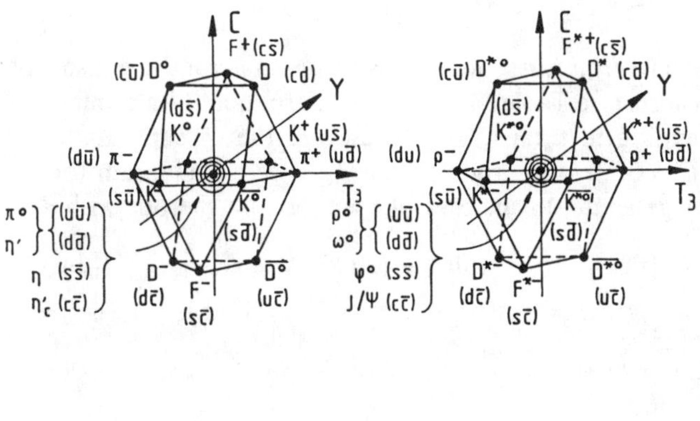

Fig. 11.8. SU(4)-multiplet of pseudoscalar mesons (0^-) and vector mesons (1^-). The quark content of the analogous states is the same in the left and in the right multiplets; merely the spin part of the wave function is coupled in a different way (to 0^- and 1^- respectively).

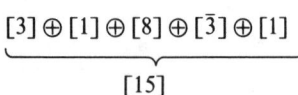

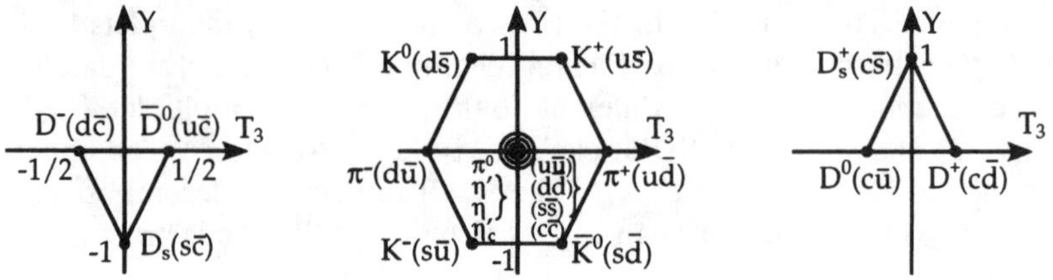

Fig. 11.9. The detailed decomposition of the pseudoscalar SU(4) multiplet $[15]\oplus[1]$ into SU(3) multiplets

$$\underbrace{[3]\oplus[1]\oplus[8]\oplus[\bar{3}]\oplus[1]}_{[15]}$$

clarifies the quantitative contents of the previous figure. The different C-levels are depicted separately.

exactly in the $Y-T_3$ plane of the polyhedron), but additionally a triplet $[3]^{-1}$ and an antitriplet $[\bar{3}]^1$. This SU(3) classification distinguishes between SU(3) multiplets with different charm: The triplet has $C=-1$, the nonet $C=0$ and the antitriplet $C=+1$ (see the SU(4) polyhedron in Fig. 11.8). The SU(3) decomposition of the $[4]\otimes[\bar{4}]$ product leads to an association of charm with the SU(3) triplet. Then the triplet results from coupling $[3]^0\otimes[1]^{-1}$, where $[1]^{-1}$ is the singlet from the decomposition of $[\bar{4}]$ with $C=-1$. $[\bar{3}]$ results from $[\bar{3}]^0\otimes[1]^1$, where now the singlet $[1]^1$ stems from the decomposition of $[4]$ with charm $C=+1$. Thus we construct spatial polygons from the basic tetrahedrons $[4]$ and $[\bar{4}]$, in analogy to the procedure of the construction of different SU(3) representations by using the basic triplets.

Clearly we obtain the following ($C=1$) states (containing one c quark), which we call the ($C=1$) SU(3) multiplet:

pseudoscalar mesons vector mesons

$$\begin{aligned}
&|\mathrm{D}^0\rangle=|\mathrm{c\bar{u}}\rangle\ , && |\mathrm{D}^{*0}\rangle=|\mathrm{c\bar{u}}\rangle\ ,\\
&|\mathrm{F}^+\rangle=|\mathrm{c\bar{s}}\rangle\ , && |\mathrm{F}^{*+}\rangle=|\mathrm{c\bar{s}}\rangle\ ,\\
&|\mathrm{D}^+\rangle=|\mathrm{c\bar{d}}\rangle\ , && |\mathrm{D}^{*+}\rangle=|\mathrm{c\bar{d}}\rangle\ .
\end{aligned}\tag{11.12}$$

The ($C=0$) multiplet is the former SU(3) meson nonet (it contains no c quarks). Additionally, in Fig. 11.8 the SU(4) singlet $\mathrm{c\bar{c}}$ is shown.

The ($C=-1$) multiplet contains a c antiquark $\bar{\mathrm{c}}$ within each wave function:

pseudoscalar mesons vector mesons

$$\begin{aligned}
&|\mathrm{D}^0\rangle=|\mathrm{u\bar{c}}\rangle\ , && |\bar{\mathrm{D}}^{*0}\rangle=|\mathrm{u\bar{c}}\rangle\ ,\\
&|\mathrm{F}^-\rangle=|\mathrm{s\bar{c}}\rangle\ , && |\mathrm{F}^{*-}\rangle=|\mathrm{s\bar{c}}\rangle\ ,\\
&|\mathrm{D}^-\rangle=|\mathrm{d\bar{c}}\rangle\ , && |\mathrm{D}^{*-}\rangle=|\mathrm{d\bar{c}}\rangle\ .
\end{aligned}\tag{11.13}$$

Apart from these mesons with *open charm* there also exist mesons with *hidden charm* (see Table 11.6), i.e. the charm quantum numbers of the two constituents adding to total $C=0$: These are the $\mathrm{c\bar{c}}$ combinations with $T_3=Y=0$ and $C=0$, i.e. the origin of the T_3-Y-C-diagram (or T_3-$S-C$) is four times degenerate. In the case of pseudoscalar mesons with $T_3=Y=0$, i.e.

$$\begin{aligned}
&|\pi^0\rangle=1/\sqrt{2}(|\mathrm{u\bar{u}}\rangle+|\mathrm{d\bar{d}}\rangle)\ , \quad |\eta\rangle\sim 1/\sqrt{2}(|\mathrm{u\bar{u}}\rangle-|\mathrm{d\bar{d}}\rangle)\ ,\\
&|\eta'\rangle\sim \mathrm{s\bar{s}}\ ,
\end{aligned}\tag{11.14}$$

Table 11.6. Quantum numbers of the pseudoscalar mesons and the vector mesons

Pseudosc. mesons	Vector mesons	Quark content	Q [e]	T	T_3	S	B	Y	C
π^-	ϱ^-	$\mathrm{d\bar{u}}$	-1	1	-1	0	0	0	0
$\pi^0(\eta',\eta,\eta_c)$	$\varrho^0(\omega^0,\phi^0,\psi)$	$\mathrm{u\bar{u}, d\bar{d}, s\bar{s}, dc\bar{c}\bar{d}}$	0	1	0	0	0	0	0
π^+	ϱ^+	$\mathrm{u\bar{d}}$	1	1	1	0	0	0	0
K^0	K^{*0}	$\mathrm{d\bar{s}}$	0	$\frac{1}{2}$	$-\frac{1}{2}$	1	0	1	0
K^+	K^{*+}	$\mathrm{u\bar{s}}$	1	$\frac{1}{2}$	$\frac{1}{2}$	1	0	1	0
K^-	K^{*-}	$\mathrm{s\bar{u}}$	-1	$\frac{1}{2}$	$-\frac{1}{2}$	-1	0	-1	0
$\overline{\mathrm{K}^0}$	$\overline{\mathrm{K}^{*0}}$	$\mathrm{s\bar{d}}$	0	$\frac{1}{2}$	$\frac{1}{2}$	-1	0	-1	0
D^0	D^{*0}	$\mathrm{c\bar{u}}$	0	$\frac{1}{2}$	$-\frac{1}{2}$	0	0	0	1
D^+	D^{*+}	$\mathrm{c\bar{d}}$	1	$\frac{1}{2}$	$\frac{1}{2}$	0	0	0	1
F^+	F^{*+}	$\mathrm{c\bar{s}}$	1	0	0	1	0	1	1
D^-	D^{*+}	$\mathrm{d\bar{c}}$	-1	$\frac{1}{2}$	$-\frac{1}{2}$	0	0	0	-1
$\overline{\mathrm{D}^0}$	$\overline{\mathrm{D}^{*0}}$	$\mathrm{u\bar{c}}$	0	$\frac{1}{2}$	$\frac{1}{2}$	0	0	0	-1
F^-	F^{*-}	$\mathrm{s\bar{c}}$	-1	0	0	-1	0	-1	-1

we may now add

$$|\eta_c\rangle = |\mathrm{c\bar{c}}\rangle \quad .$$

Obviously the strangeness S gives the number of strange quarks of the particle, and we note that the strangeness is $S = +1$ if an $\bar{\mathrm{s}}$ (anti-strange quark) and $S = -1$ if an s (strange quark) is contained within the configuration, i.e. an "anti-intuitive" mapping.

In the case of vector mesons the states in the centre of the multiplet with the quantum numbers $T_3 = 0$, $Y = 0$, $C = 0$ read as

$$|\varrho^0\rangle = 1/\sqrt{2}(|\mathrm{u\bar{u}}\rangle + |\mathrm{d\bar{d}}\rangle) \quad , \quad |\omega^0\rangle = 1/\sqrt{2}(|\mathrm{u\bar{u}}\rangle - |\mathrm{d\bar{d}}\rangle) \quad ,$$
$$|\phi\rangle = |\mathrm{s\bar{s}}\rangle \quad , \tag{11.15}$$

and now also as

$$|\Psi\rangle = |\mathrm{c\bar{c}}\rangle \quad .$$

We identify $|\Psi\rangle$ with $|\mathrm{c\bar{c}}\rangle$ since this state has exactly the quantum number which $\Psi(3,1\,\mathrm{GeV})$ shows in experiment. This concept will be explained in more detail in the following section. In analogy to the bound state $\mathrm{e^+e^-}$, which is called *positronium*, we call $\mathrm{c\bar{c}}$ *charmonium.* Later on we will see that each of the recently discovered vector mesons may be understood in the framework of the charmonium model.

The relative stability of the $c\bar{c}$ may be explained by the so-called *Okubo-Zweig-Iizuka-Rule* (OZI-*rule*). This rule, which was deduced from empirical considerations, states: The quarks contained in the incoming particles are distributed over the particles in the final state of the reaction. Otherwise the corresponding reaction is strongly suppressed. We want to make the OZI rule clearer by using the example of the ϕ meson ($|\phi\rangle = |s\bar{s}\rangle$). Therefore it is useful to represent each quark line by an arrow ($\rightarrow$) and each antiquark line by an arrow into the opposite direction ($\leftarrow$), i.e. a meson has the graphical representation

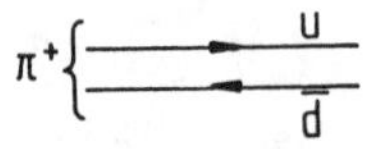

and a baryon has the graphical representation

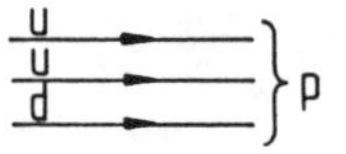

The creation and annihilation of a $q\bar{q}$ pair is denoted by

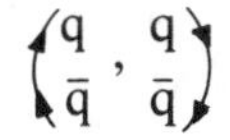

respectively. This completes the list of elements for the construction of our quark flow diagrams. It is now easy to see that the decay $\phi \rightarrow K^+K^-$ is Zweig-allowed.

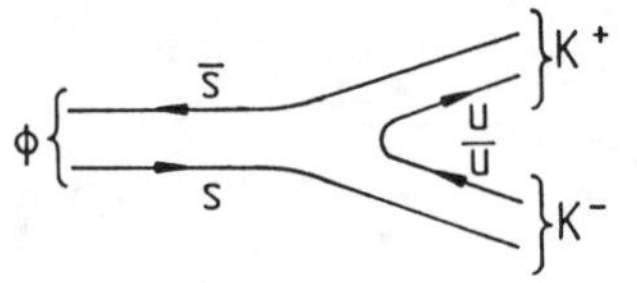

The quark components of the initial $\phi(s\bar{s})$ are divided to the two particles (K^+K^-) in the final state. Also the decay $\phi \rightarrow K^0\bar{K}^0$ is possible since

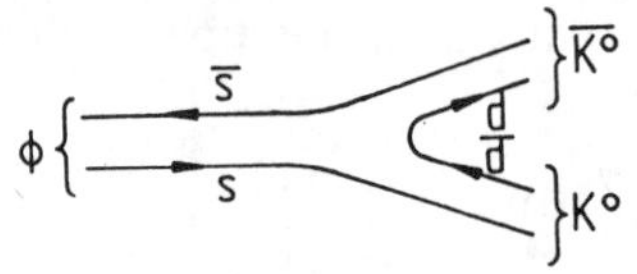

On the contrary the decay $\phi \rightarrow \pi^+\pi^-\pi^0$ is Zweig forbidden:

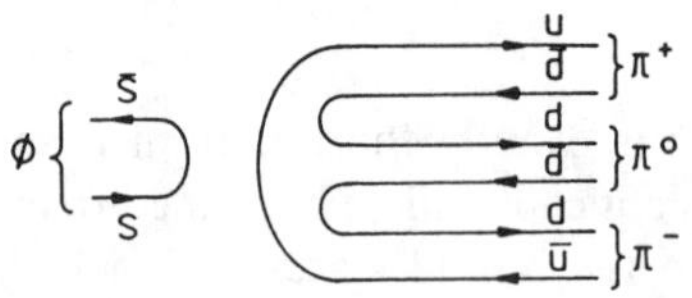

Here we have no possibility of dividing the s and $\bar{s}$ between any final state particles since pions contain no s quarks. In fact experiment shows the following

distribution of the decay channels:

$\phi \to K^+K^-$ 47%

$\phi\phi \to K^0K^0$ 35%

$\phi \to \pi^+\pi^-\pi^0$ 16% .

This is most remarkable since the energy release in the last reaction (605 MeV) is much higher than that in the first two decays (35 and 25 MeV, respectively). Consequently the Okubo-Zweig-Iizuka rule is remarkably well fulfilled. Field theoretical considerations indicate that the validity of the OZI rule should be even better for heavy mesons (see Sect. 11.5).

11.5 Advanced Considerations

11.5.1 Decay of Mesons with Hidden Charm

Nowadays the OZI rule is relatively well understood on the basis of quantum chromodynamics (QCD)[3]. Then the Zweig-forbidden decay of a spin-1 meson (ϕ, ψ etc.) takes place by pair annihilation of the quark-antiquark pair into three virtual gluons, which each create a quark-antiquark pair. This is illustrated in Fig. 11.10.

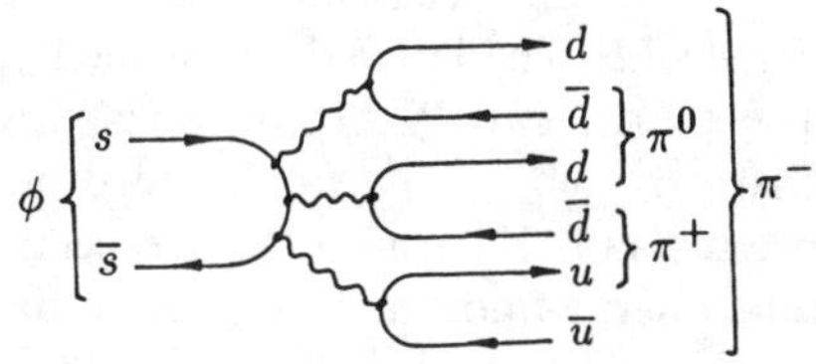

Fig. 11.10. The Zweig-forbidden decay of the ϕ meson.

Every virtual gluon contributes a factor α_s (the coupling constant of QCD) to the Feynman diagram, which is consequently proportional to α_s^3. For energies in the region of 1 GeV (approximately the mass of ϕ) $\alpha_s \sim 0.5$, i.e. the Zweig-forbidden diagram is suppressed by a factor of 0.1. For energies in the region of 3.5 GeV, $\alpha_s \sim 0.2$ as result of the "asymptotic freedom" of QCD gauge theory. Thus the suppression factor for the Zweig-forbidden decay is now ~ 0.01. With further increasing energy, α_s continues to decrease. This energy dependence of the coupling constant is called "running" of the coupling constant.

In Fig. 11.11 we show the Zweig diagrams for the decay of the ψ meson.

The decays $\Psi \to D^0\overline{D^0}$, D^+D^-, F^+F^- are allowed according to the OZI rule, but $\Psi \to \pi^+\pi^-\pi^0$ is forbidden. However the calculations which we will

[3] See also W. Greiner, A. Schäfer: *Quantum Chromodynamics* (Harri Deutsch, Thun, Frankfurt am Main 1989); F.E. Close: *An Introduction to Quarks and Partons* (Academic Press, New York 1979).

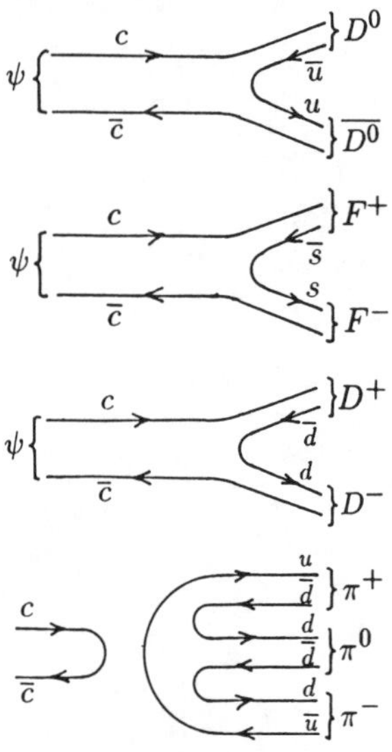

Fig. 11.11. The decays of the ψ meson OZI-forbidden and -allowed. The last graph describes the OZI-forbidden process.

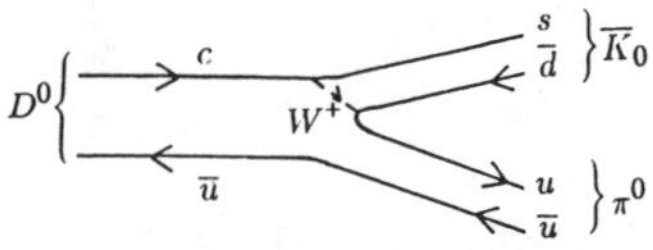

Fig. 11.12. Diagrammatic representation of the decay of the D^0 meson into $\bar{K}^0$ and π^0. The c quark has charge $Q_c = +2/3$ and decays into an s quark with charge $Q_s = -1/3$ by emission of a W^+. The W^+ decays into a $\bar{d}(Q_{\bar{d}} = 1/3)$ and a u quark $(Q_u = 2/3)$. Compare this with Table 11.6.

perform later in the framework of a potential model show that the mass of the Ψ is less than two times the mass of the D- or F-mesons, and so these decays are energetically not possible. They should occur though for highly excited $c\bar{c}$ states, and indeed they have actually been observed.

Thus only the *Zweig-forbidden* hadronic decays of the Ψ particle, such as $\Psi \to \pi^+\pi^-\pi^0$, remain. As a consequence the decay of Ψ is strongly suppressed. This explains the small width of the Ψ. Furthermore, we have seen that the widths of the other vector mesons (at higher energies) are often larger: In the case of $\Psi(4030)$, $\Gamma \sim 50$ MeV; and for $\Psi(4415)$, $\Gamma \sim 40$ MeV. These larger widths express the fact that decay into two D- or F-mesons, given in Fig. 11.11 is possible. Thus the masses of the Ψ-mesons at higher energies [e.g. $\Psi(4030)$ and $\Psi(4415)$] have to be at least as big as the sum of the masses of the particles occurring in the decay, i.e. at least twice the mass of the single D- and F-mesons respectively. From this observation we get an estimate for these masses of

$$\tfrac{1}{2}M(\Psi' = \Psi(3700)) < M(\mathrm{D\ or\ F}) \leqslant \tfrac{1}{2}M(\Psi(4030)) \quad , \tag{11.16}$$

i.e. $1800\ \mathrm{MeV} < M(\mathrm{D,F}) \leqslant 2000\ \mathrm{MeV}$.

11.5.2 Decay of Mesons with Open Charm

The D-mesons decay into π^+K^- or π^-K^+ mainly by the weak interaction (see Fig. 11.12). The latter decay is mediated by the exchange of W- and Z-bosons. There exist two W bosons: a positively charged one (W^+) and a negatively charged one (W^-)[4]. In Stanford and Hamburg[5] the yield of such reaction products (π^+K^- and π^-K^+ respectively) was measured as a function of their total energy and the experimental data are shown in Fig. 11.13.

One finds a distinct maximum at $M = (1865 \pm 1)$ MeV which has all the properties that are expected for the D^0- and $\bar{D}^0$-mesons. By this method the existence of a particle with *open charm* was established for the first time. Later D^+- and D^--mesons with a mass of 1869 MeV were also detected. This is a brilliant success of the quark model. However, before we consider the mesons in more detail (cf. Sect. 11.6. "The Potential Model of Charmonium"), we will first look at the baryon combinations of SU(4).

---

[4] These phenomena are discussed in Vol. 5 of this series, *Gauge Theory of Electro-Weak Interactions* (Springer, Berlin, Heidelberg) to be published.

[5] See G. Goldhaber et al. (41 authors): Phys. Rev. Lett. **37**, 255 (1976); I. Peruzzi et al. (40 authors): Phys. Rev. Lett. **37**, 569 (1976); I.E. Wiss et al. (40 authors): Phys. Rev. Lett. **37** 1531 (1976).

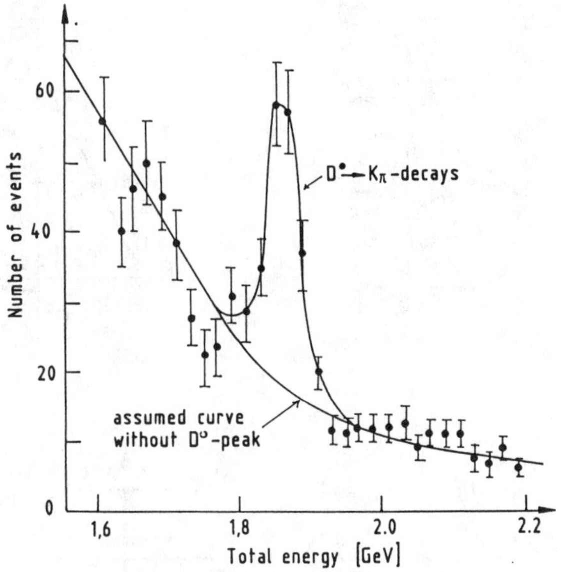

Fig. 11.13. The decay of D mesons into $K\pi$

11.5.3 Baryon Multiplets

Baryons are built as qqq combinations, i.e. out of the SU(4) representation.

$$\begin{aligned}[4] \otimes [4] \otimes [4] &= ([6] \oplus [10]) \otimes [4] \\ &= [\bar{4}] \oplus [20] \oplus [20] \oplus [20'] \quad . \end{aligned} \tag{11.17}$$

(See Table 11.5 and Exercises 11.5 and 11.6.) The SU(3) reduction of the SU(4) multiplets yields

$$\begin{aligned}[20] &= [3] \oplus [\bar{3}] \oplus [6] \oplus [8] \quad , \\ [20'] &= [1] \oplus [3] \oplus [6] \oplus [10] \quad , \end{aligned} \tag{11.18}$$

where [20] and [20′] differ by the composition of SU(3)-multiplets. Both contain 20 states, but different SU(3)-multiplets, which are responsible for the SU(3) classification (see Figs. 11.14, 11.15 and 11.16): Clearly the [20] representation

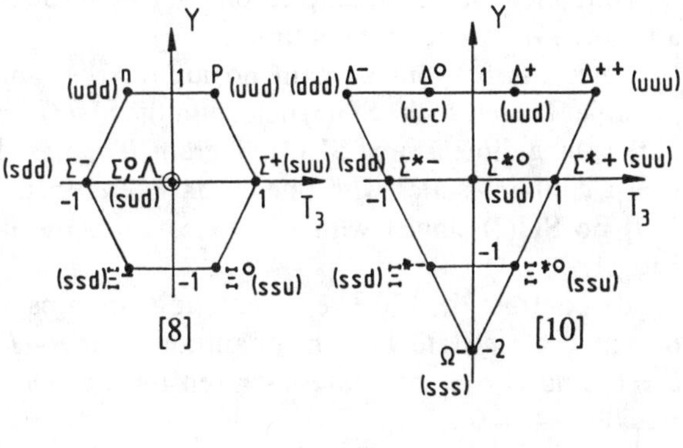

Fig. 11.14. The SU(3) baryon multiplets

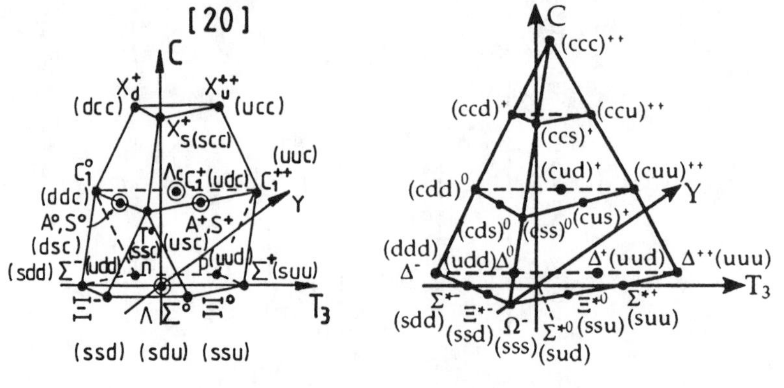

Fig. 11.15. The SU(4) baryon multiplets

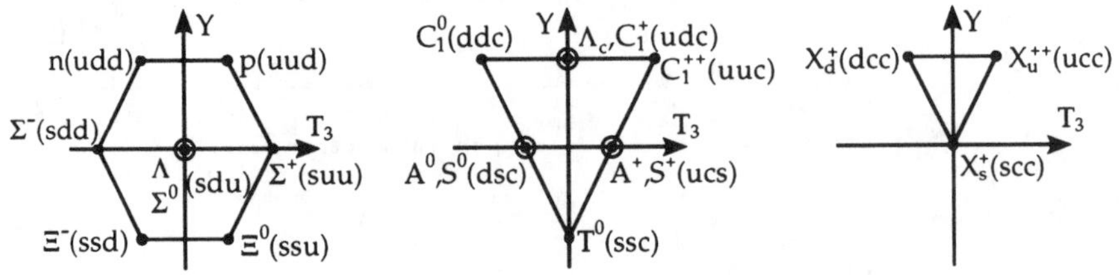

Fig. 11.16. The detailed decomposition of the SU(4)-multiplet [20] (Fig. 11.15) into SU(3) multiplets [8] ⊕ [6] ⊕ [3] ⊕ [3] clarifies the substructure of the three-dimensional figure. The [20′] multiplet can be decomposed similarly. See also Exercise 11.7.

of SU(4) is suitable for the classification of the baryons because it contains an SU(3) octet, while [20′] represents the baryon resonances because it contains a decouple SU(3)–[10] (all states have $C = 0$). The baryon sextet and the baryon antitriplet both have $C = 1$. The states of the antitriplet have the same quark content as some states of the sextet, i.e. three points of the sextet are occupied twice (indicated by rings around the points). Furthermore there also occurs a [3] (triplet) with $C = 2$. The reason why the baryon states with $C = 1$ correspond to two different SU(3) representations, but those of the baryon resonances correspond only to a single SU(3) representation, lies in the fact that they contain two light quarks u, d, and s. We know that the product of two SU(3) triplets decomposes into a sextet and an antitriplet. The sextet is symmetric, whereas the antitriplet is antisymmetric. Therefore, both representations can be contained within the baryon multiplet which has mixed symmetry, but only the sextet can be contained in the multiplet of baryon resonances which corresponds to a totally symmetric representation.

Since the baryons contain no antiquarks, only particles with $C \geqslant 0$ are possible. Therefore the SU(3) octet and the SU(3) decuplet always form the basis of the three-dimensional SU(4) diagram. The next higher SU(3) multiplet with $C = 1$ contains states with one c quark, etc. In the case of the representation [20] no SU(3) singlet with $C = 3$ exists, i.e. the figure is flat at the top (see Fig. 11.4).

By contrast the [20′] representation contains an SU(3) singlet with $C \neq 0$, namely the ccc state with the quantum numbers $T_3 = 0$, $Y = B + S = 1$ (since $S = 0$) and $C = 3$. Its charge is given by the *generalized Gell-Mann-Nishijima* relation, namely

$$Q = T_3 + \tfrac{1}{2}(B + S + C) = T_3 + \tfrac{1}{2}(Y + C) \quad , \tag{11.19}$$

i.e. the charge of the ccc state is $Q = 2$. The charge of the c quark, $Q_c = 2/3$, also follows from (11.19). This can be verified by the experimentally measured charges of the D- and F-mesons. The quantum numbers of the baryons are listed in Table 11.7.

Table 11.7. Quantum numbers of the baryons

Particles	Quark content	Q [e]	T	T_3	S	B	Y	C
n	udd	0	$\frac{1}{2}$	$-\frac{1}{2}$	0	1	1	0
p	uud	1	$\frac{1}{2}$	$\frac{1}{2}$	0	1	1	0
Σ^-	dds	-1	1	-1	-1	1	0	0
$\Sigma^0(\Lambda)$	uds	0	1(0)	0	-1	1	0	0
Σ^+	uus	1	1	1	-1	1	0	0
Ξ^-	dss	-1	$\frac{1}{2}$	$-\frac{1}{2}$	-2	1	-1	0
Ξ^0	uss	0	$\frac{1}{2}$	$\frac{1}{2}$	-2	1	-1	0
T^0	ssc	0	0	0	-2	1	-1	1
$\mathrm{A}^0, \mathrm{S}^0$	dsc	0	$\frac{1}{2}$	$-\frac{1}{2}$	-1	1	0	1
$\mathrm{A}^+, \mathrm{S}^+$	usc	1	$\frac{1}{2}$	$\frac{1}{2}$	-1	1	0	1
C_1^0	ddc	0	1	-1	0	1	1	1
$\mathrm{Ac}(\mathrm{C}_1^+)$	udc	1	0(1)	0	0	1	1	1
C_1^{++}	uuc	2	1	1	0	1	1	1
X_d^+	dcc	1	$\frac{1}{2}$	$-\frac{1}{2}$	0	1	1	2
X_u^{++}	ucc	2	$\frac{1}{2}$	$\frac{1}{2}$	0	1	1	2
X_s^+	scc	1	0	0	-1	1	0	2

Thus baryons which contain a c quark have *open charm* ($C \neq 0$). Since the decays of charmed baryons into particles without charm are Zweig- or Cabibbo-suppressed, the experimental observation of baryons with $C > 0$ is very difficult. Until now only one of the charmed baryons is firmly established; the antibaryon $\bar{\Lambda}_c$ (*quark combination* $\bar{u}\bar{d}\bar{c}$) decays through weak interactions into

$$\bar{\Lambda}_c \rightarrow \bar{\Lambda}\pi^+\pi^-\pi^- \quad ,$$

and this excitation is shown in Fig. 11.17.

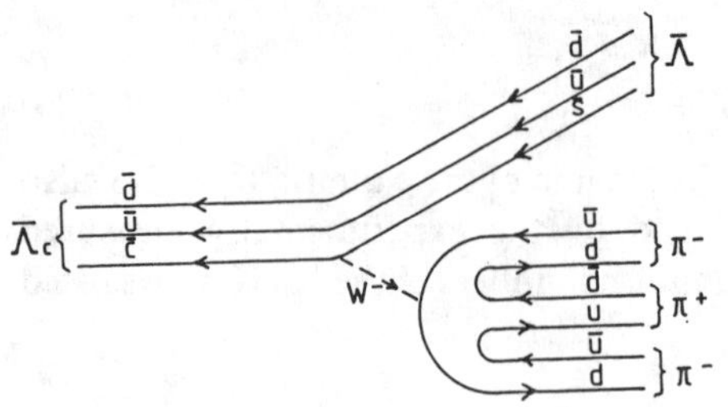

Fig. 11.17. The decay of $\bar{\Lambda}_c$: A $\bar{c}$ quark with charge $Q_{\bar{c}} = -2/3$ changes into an $\bar{s}$ quark ($Q_s = 1/3$) by emitting a W^- boson. The W^- decays into an $\bar{u}d$ pair which "pionizes"

At Fermilab (USA) a resonance at $M = (2282 \pm 3)$ MeV was seen in the mass distributions of $(\bar{\Lambda}\pi^+\pi^-\pi^-)$ systems[6], but not in the distribution of $(\bar{\Lambda}\pi^+\pi^+\pi^-)$ systems by bombarding a target with high energy photons in the 400 GeV accelerator (see Fig. 11.18). If the maximum is caused by the decay of $\bar{\Lambda}_c$ particles, this can be easily understood, because $\bar{\Lambda}_c$ has quantum numbers $T = T_3 = 0$. Therefore it is an isospin singlet and appears in only one charge state.

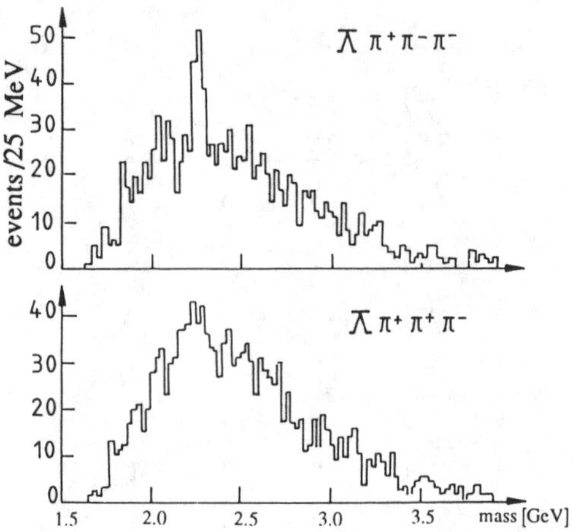

Fig. 11.18. Experimental excitation of the $\bar{\Lambda}_c$ decay at Batavia (Fermi Lab.). The excitation of $\bar{\Lambda}_c$ (i.e. the resonance which is interpreted that way) occurs in collisions with high-energy photons

The quantum numbers $T = 1$ and $T_3 = -1$ ($\bar{\Lambda}$ has $T_3 = 0$; for π^-, $T_3 = -1$; for π^+, $T_3 = +1$) follow from the quark configuration of the final state. If we try to explain the maximum by the strong decay of an SU(3) resonance, it would have to be an isospin triplet, i.e. there would have to be three charge states. Therefore one should also see a maximum in the $(\bar{\Lambda}\pi^+\pi^+\pi^-)$ channel. The missing of such a maximum indicates the (weak) decay of a charmed baryon.

Finally, note that the terms "flavour isospin", "flavour hypercharge" etc. in the sense associated with the framework of strong-interaction theory should be differentiated from the terms "weak isospin", "weak hypercharge" etc. in the

[6] B. Knapp et al. (19 authors): Phys. Rev. Lett. **37**, 882 (1976).

theory of weak interactions[7], which should be distinguished from those discussed here.

EXERCISE

11.5 SU(3) Content of the SU(4) Meson Multiplet

Problem. Calculate the direct product $[4]_{SU(4)} \otimes [\bar{4}]_{SU(4)}$ in SU(4) (a meson multiplet). Decompose the result into the representations of the SU(3) subgroup and determine the charm content of each SU(3) representation.

Solution. (a) Calculation of the direct product. First of all we represent [4] and $[\bar{4}]$ by their associated Young diagrams. Then we multiply both representations with the help of our multiplication rules. The graphical representation is

$[4]_{SU(4)} \otimes [\bar{4}]_{SU(4)} =$ $= [1]_{SU(4)} \oplus [15]_{SU(4)}$. (1)

(b) Reduction of the SU(4) representation to SU(3) representations. If we reduce an SU(4) representation to the group SU(3), we perform this by subtracting one box after another from the Young diagram. Here one uses the following rule[8]: If h_k is the number of boxes in the kth row of an SU(4) representation and h'_k the corresponding number of the SU(3) representation, then we have

$$h_1 \geqslant h'_1 \geqslant h_2 \geqslant h'_2 \geqslant h_3 \; . \tag{2}$$

Therefore it follows that

$$[1]_{SU(4)} = \text{①} \xrightarrow{SU(3)} \text{①} = [1]_{SU(3)} \quad ,$$

$[15]_{SU(4)} =$ $$= [3]_{SU(3)} \oplus [8]_{SU(3)} \oplus [1]_{SU(3)} \oplus [\bar{3}]_{SU(3)} \quad . \tag{3}$$

(c) Determination of the charm content. The SU(3) multiplets lie parallel to the $Y - T_3$-plane within the SU(4) multiplet. As a first step to determine the charm content of an SU(3) representation we decompose $[4]_{SU(4)}$ and $[\bar{4}]_{SU(4)}$ into their

---

[7] See Vol. 5 of this series, *Gauge Theory of Electro-Weak Interactions* (Springer, Berlin, Heidelberg) to be published.

[8] M. Hamermesh: *Group Theory and its Application to Physical Problems* (Addison-Wesley, Reading, MA 1962); J.P. Elliott, P.G. Dawber: *Symmetry in Physics* (Oxford University Press, Oxford 1979).

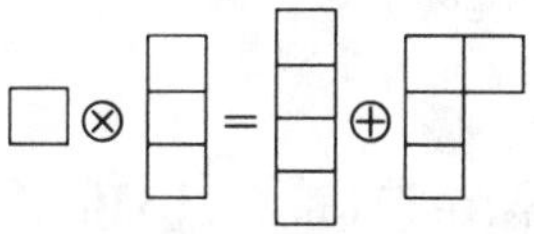

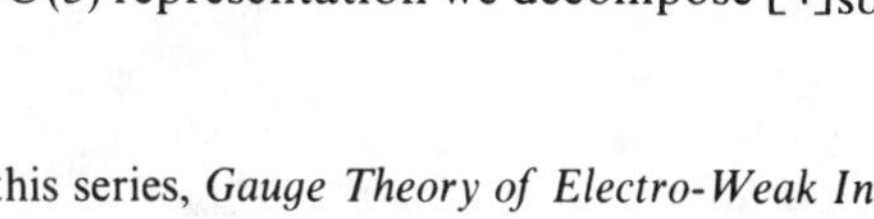

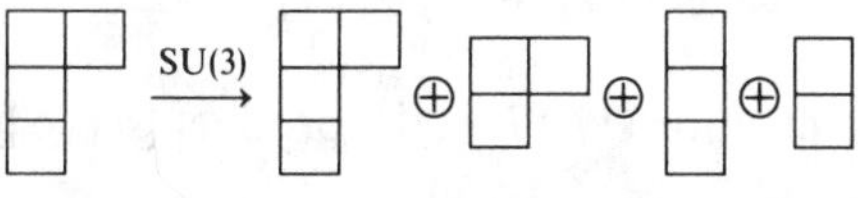

Exercise 11.5

SU(3) components:

$$[4]_{SU(4)} = [3]^{0}_{SU(3)} \oplus [1]^{1}_{SU(3)} \quad , \quad [\bar{4}]_{SU(4)} = [\bar{3}]^{0}_{SU(3)} \oplus [\bar{1}]^{-1}_{SU(3)} \quad . \tag{4}$$

The superscript 0 or 1 denotes the charm content of the SU(3) multiplet. Antiquarks are represented by $[\bar{4}]$. Now negative charm values may occur as a result of the presence of a charmed antiquark (in contrast to the baryon multiplet).

With (4) the product $[\bar{4}] \otimes [4]$ (in the following we drop the indices SU(3) and SU(4)) becomes

$$\begin{aligned}[4] \otimes [\bar{4}] &= ([3]^0 \oplus [1]^1) \otimes ([\bar{3}]^0 \oplus [1]^{-1}) \\ &= ([3]^0 \otimes [\bar{3}]^0) \oplus ([3]^0 \otimes [1]^{-1}) \oplus ([1]^1 \otimes [\bar{3}]^0) \\ &\quad \oplus ([1]^1 \otimes [1]^{-1}) \\ &= ([8]^0 \oplus [1]^0) \oplus ([3]^{-1}) \oplus ([\bar{3}]^1) \oplus ([1]^0) \quad . \end{aligned} \tag{5}$$

By comparison of (3) and (5) we get

$$\begin{aligned} &[1]_{SU(4)} = [1]^{0}_{SU(3)} \quad , \\ &[15]_{SU(4)} = [1]^{0}_{SU(3)} \oplus [8]^{0}_{SU(3)} \oplus [3]^{-1}_{SU(3)} \oplus [\bar{3}]^{1}_{SU(3)} \quad . \end{aligned} \tag{6}$$

This means that the SU(3) octet and the SU(3) singlet lie in the $Y - T_3$ plane, whereas the SU(3) triplet is shifted one unit downwards on the C axis and the antitriplet is shifted one unit upwards on the C axis. So we obtain Fig. 11.8.

EXERCISE

11.6 Decomposition of the Product $[4] \otimes [4] \otimes [4]$

Problem. Verify the decomposition (11.17) of the product $[4] \otimes [4] \otimes [4]$ using the method of Young diagrams.

Solution. First we calculate the product $[4] \otimes [4]$.

$$[4] \otimes [4] = \square \otimes \square = \begin{array}{|c|}\hline \ \\ \hline \ \\ \hline \end{array} \oplus \begin{array}{|c|c|}\hline \ & \ \\ \hline \end{array} = [6] \oplus [10] \quad , \tag{1}$$

since we have

$$\begin{aligned} &\dim \begin{array}{|c|}\hline \ \\ \hline \ \\ \hline \end{array} = \begin{array}{|c|}\hline 4 \\ \hline 3 \\ \hline \end{array} : \begin{array}{|c|}\hline 2 \\ \hline 1 \\ \hline \end{array} = \frac{4 \cdot 3}{2 \cdot 1} = 6 \quad , \\ &\dim \begin{array}{|c|c|}\hline \ & \ \\ \hline \end{array} = \begin{array}{|c|c|}\hline 4 & 5 \\ \hline \end{array} : \begin{array}{|c|c|}\hline 2 & 1 \\ \hline \end{array} = \frac{4 \cdot 5}{2 \cdot 1} = 10 \quad . \end{aligned} \tag{2}$$

If we multiply that product by [4] once again, we obtain on one hand

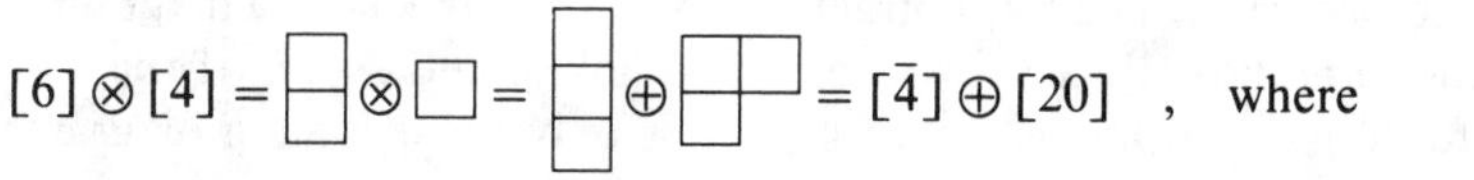

Exercise 11.6

$$[6] \otimes [4] = \square\!\square \otimes \square = \square\!\square \oplus \square\!\square\!\square = [\bar{4}] \oplus [20] \quad , \quad \text{where} \tag{3}$$

$$\dim \square\!\square = \boxed{\begin{array}{cc}4&5\\3\end{array}} : \boxed{\begin{array}{cc}3&1\\1\end{array}} = \frac{4 \cdot 5 \cdot 3}{3 \cdot 1 \cdot 1} = 20 \quad , \tag{4}$$

and on the other hand

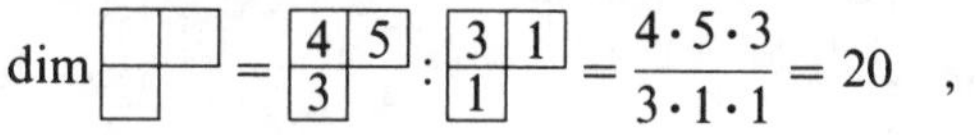

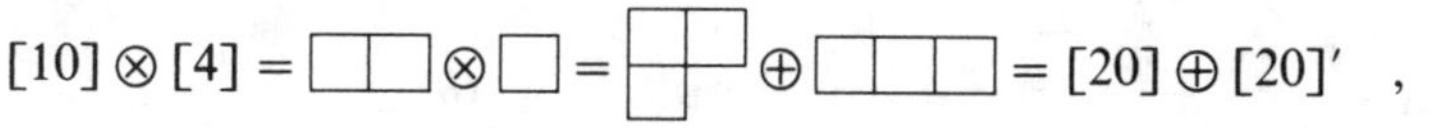

$$[10] \otimes [4] = \boxed{\ \ } \otimes \square = \square\!\square \oplus \boxed{\ \ \ } = [20] \oplus [20]' \quad , \tag{5}$$

where

$$\dim \boxed{\ \ \ } = \boxed{\begin{array}{ccc}4&5&6\end{array}} : \boxed{\begin{array}{ccc}3&2&1\end{array}} = \frac{4 \cdot 5 \cdot 6}{3 \cdot 2 \cdot 1} = 20 \quad . \tag{6}$$

The two 20-dimensional representations $\square\!\square$ and $\boxed{\ \ \ }$ are different irreducible representations, which differ in their symmetry properties. $\boxed{\ \ \ } = [20]'$ contains all 20 possible totally symmetric combinations of three SU(4)-quarks. $\square\!\square = [20]$ contains the combinations of mixed symmetry.

EXERCISE ▬▬▬▬▬▬▬▬▬▬▬

11.7 SU(3) Content of the SU(4) Baryon Multiplet

Problem. Calculate the direct product of three fundamental representations [4] of SU(4) (i.e. the baryon multiplet). Decompose the result into representations of the SU(3) subgroup and determine the charm contents of every SU(3) multiplet.

Solution. a) Calculation of the direct product. First we represent the fundamental representation [4] by the Young diagram $\square$ and take the product of two of these. That product is then multiplied with the third representation. The graphic representation of the result is

$$([4] \otimes [4]) \otimes [4] = (\square \otimes \square) \otimes \square = \left(\square\!\square \oplus \square\!\square \right) \otimes \square$$

$$= \left(\square\!\square \oplus \square\!\square\!\square \right) \oplus \left(\square\!\square \oplus \square\!\square \right)$$

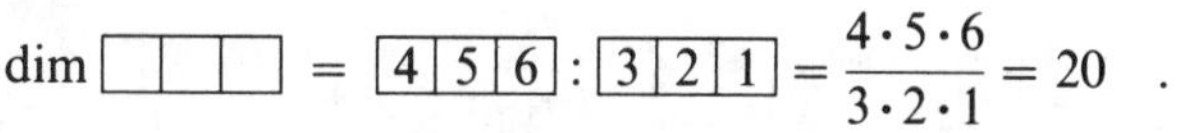

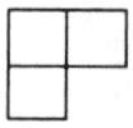

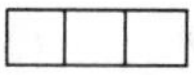

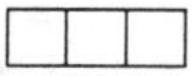

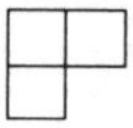

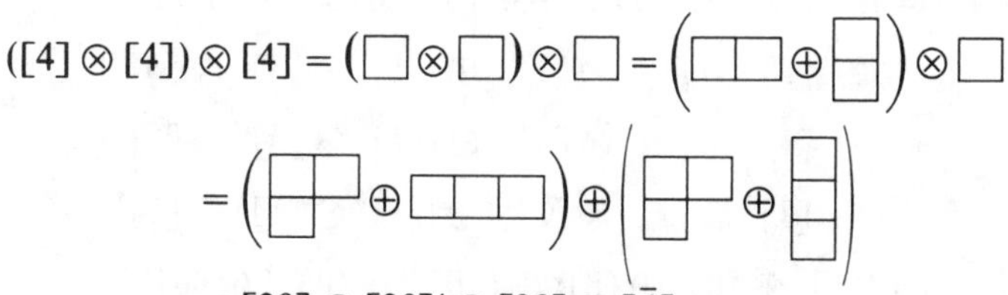

$$= [20] \oplus [20]' \oplus [20] \oplus [\bar{4}] \quad . \tag{1}$$

In the last line we have written down the dimensions of the representations. Note that there are several multiplets of the same dimension.

Exercise 11.7

b) Reduction of the SU(4) representations to SU(3) subrepresentations. This reduction can be performed by subtracting boxes from the Young diagram; first no box, then one box, then two boxes, three boxes, and so on. If the Young diagram to be reduced has h_k boxes in row k, only those Young diagrams of SU(3) are allowed that have h'_k boxes in row k with

$$h_1 \geqslant h'_1 \geqslant h_2 \geqslant \cdots \geqslant h_k \geqslant h'_k \geqslant \cdots \tag{2}$$

Thus we obtain

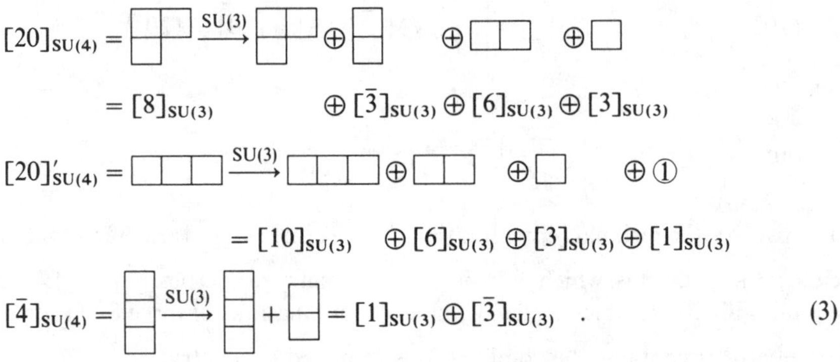

$$[20]_{SU(4)} = [8]_{SU(3)} \oplus [\bar{3}]_{SU(3)} \oplus [6]_{SU(3)} \oplus [3]_{SU(3)}$$

$$[20]'_{SU(4)} = [10]_{SU(3)} \oplus [6]_{SU(3)} \oplus [3]_{SU(3)} \oplus [1]_{SU(3)}$$

$$[\bar{4}]_{SU(4)} = [1]_{SU(3)} \oplus [\bar{3}]_{SU(3)} \quad . \tag{3}$$

c) Determination of the charm content. The SU(3) multiplets lie parallel to the $Y - T_3$ plane in the SU(4) multiplets, but their exact position, i.e. the charm content is not yet clear. In order to determine it we first decompose the fundamental representation $[4]_{SU(4)}$ into the SU(3) subrepresentations which are contained within it:

$$[4]_{SU(4)} = [3]^0_{SU(3)} \oplus [1]^1_{SU(3)} \quad . \tag{4}$$

The superscript on the upper right describes the charm content, i.e. the number of charmed quarks contained in the particles forming the submultiplet. The indices remind us which group we are actually concerned with, but in the following calculations they will be omitted since it is clear from the context which group is meant.

We repeat the direct product in order to compare it with the previous results (1) and (3).

$$\begin{aligned}
[4] \otimes [4] \otimes [4] &= ([3]^0 \oplus [1]^1) \otimes ([3]^0 \oplus [1]^1) \otimes ([3]^0 \oplus [1]^1) \\
&= ([3]^0 \otimes [3]^0 \otimes [3]^0) \oplus ([3]^0 \otimes [3]^0 \otimes [1]^1) \\
&\quad \oplus ([3]^0 \otimes [1]^1 \otimes [3]^0) \oplus ([1]^1 \otimes [3]^0 \otimes [3]^0) \\
&\quad \oplus ([1]^1 \otimes [1]^1 \otimes [3]^0) \oplus ([1]^1 \otimes [3]^0 \otimes [1]^1) \\
&\quad \oplus ([3]^0 \otimes [1]^1 \otimes [1]^1) \oplus ([1]^1 \otimes [1]^1 \otimes [1]^1) \\
&= ([1]^0 \oplus [8]^0 \oplus [8]^0 \oplus [10]^0) \oplus ([\bar{3}]^1 \oplus [6]^1) \\
&\quad \oplus ([\bar{3}]^1 \oplus [6]^1) \oplus ([\bar{3}]^1 \oplus [6]^1) \oplus [3]^2 \\
&\quad \oplus [3]^2 \oplus [3]^2 \oplus [1]^3 \quad .
\end{aligned} \tag{5}$$

Except for the singlets the relation between the representations is unique. Taking into account that the representation $[\bar{4}]$ is decomposed into $[1]^0$ and $[\bar{3}]^1$, we obtain the final result

$$
\begin{aligned}
[20] &= [8]^0 \oplus [6]^1 \oplus [\bar{3}]^1 \oplus [3]^2 \\
[20]' &= [10]^0 \oplus [6]^1 \oplus [3]^2 \oplus [1]^3 \\
[\bar{4}] &= [1]^0 \oplus [\bar{3}]^1 \quad ,
\end{aligned}
$$

and, as we know from (1), the representation [20] appears twice. Note that $[1]^0 \oplus [\bar{3}]^1$ is indeed a $[\bar{4}]$, but with the singlet $[1]^0$ at $C = 0$ and the antitriplet $[\bar{3}]^1$ at $C = 1$, i.e. shifted by one unit along the positive C-axis. This tells us that the SU(4) multiplets are not all symmetric with respect to the origin in the T_3-Y-C-coordinate system. This is in contrast to SU(3), where all multiplets contain the $T_3 - Y$ origin in a symmetric way.

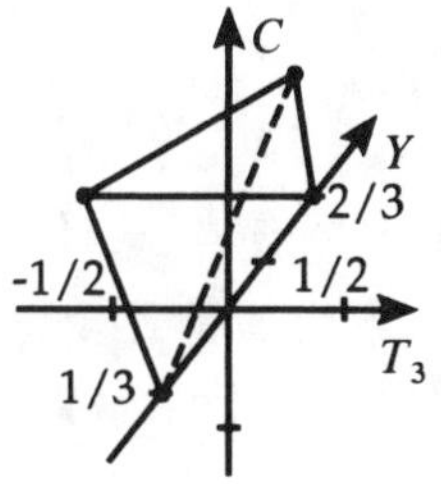

EXAMPLE

11.8 Decomposition and Dimension of Higher SU(4) Multiplets

We now calculate the direct product of several representations $[x]$ with $[4]$ in SU(4) and decompose them into SU(4) representations before determining the dimensions of some of these representations in SU(4). We use the rules for the decomposition of the direct product of two multiplets, which we studied in Chap. 9, in the following examples.

$$
[4] \otimes [4] = \Box \otimes ① = \boxed{}\boxed{1} \oplus \begin{array}{|c|}\hline \\ \hline 1 \\ \hline\end{array} = [10] \oplus [6] \quad , \tag{1}
$$

$$
\begin{aligned}
[\bar{4}] \otimes [4] &= \begin{array}{|c|}\hline \\ \hline \\ \hline \\ \hline\end{array} \otimes ① = \begin{array}{|c|c|}\hline & 1 \\ \hline \\ \cline{1-1} \\ \cline{1-1}\end{array} \oplus \begin{array}{|c|}\hline \\ \hline \\ \hline \\ \hline 1 \\ \hline\end{array} \\
&= [15] \oplus [1] \quad ,
\end{aligned} \tag{2}
$$

Example 11.8

$$[6] \otimes [4] = \square \otimes \textcircled{1} = \square \oplus \square$$

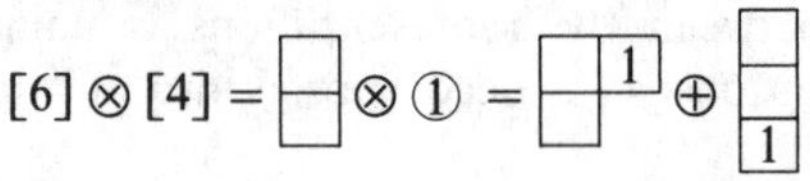

$$= [20] \oplus [\bar{4}] \quad , \tag{3}$$

$$[10] \otimes [4] = \square \otimes \textcircled{1} = \square \oplus \square$$

$$= [20]' \oplus [20] \quad , \tag{4}$$

$$[\overline{10}] \otimes [4] = \square \otimes \textcircled{1} = \square \oplus \square$$

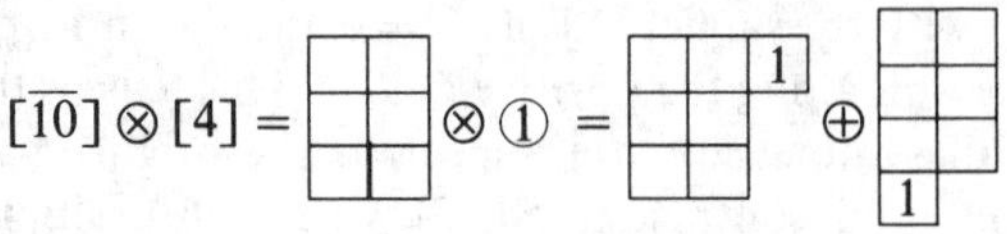

$$= [\overline{36}] \oplus [\bar{4}] \quad , \tag{5}$$

$$[15] \otimes [4] = \square \otimes \textcircled{1} = \square \oplus \square \oplus \square$$

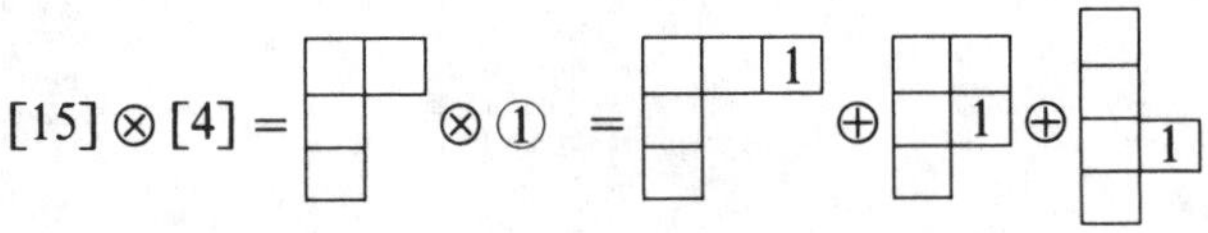

$$= [36] \oplus [\overline{20}] \oplus [4] \quad , \tag{6}$$

$$[20]' \otimes [4] = \square \otimes \textcircled{1} = \square \oplus \square$$

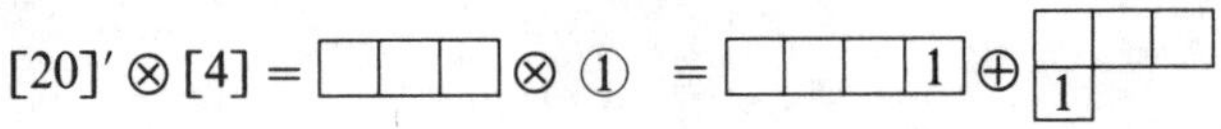

$$= [35] \oplus [45] \quad , \tag{7}$$

$$[\overline{20}]' \otimes [4] = \square \otimes \textcircled{1} = \square \oplus \square$$

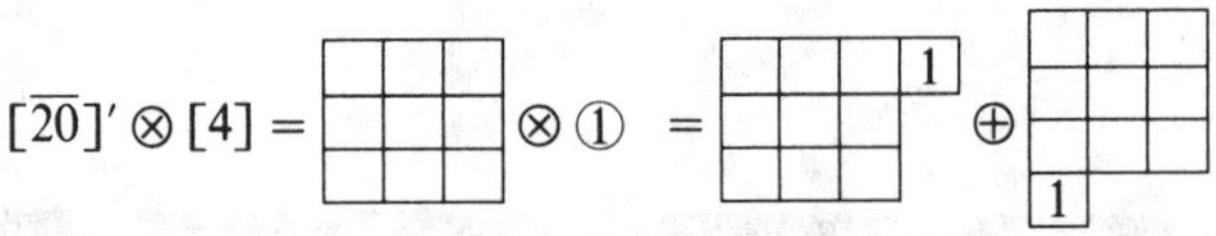

$$= [\overline{70}] \oplus [\overline{10}] \quad , \tag{8}$$

$$[20] \otimes [4] = \square \otimes \textcircled{1} = \square \oplus \square \oplus \square$$

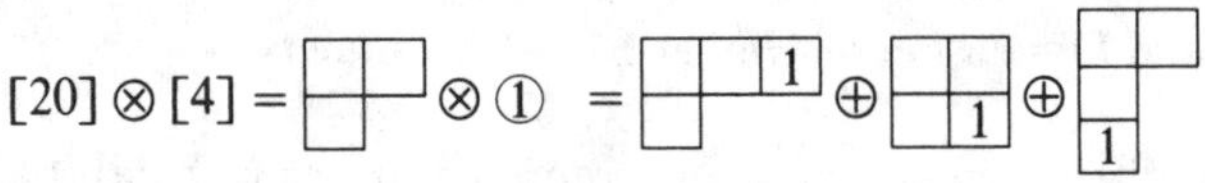

$$[45] \oplus [20]'' \oplus [15] \quad , \tag{9}$$

$$[\overline{20}] \otimes [4] = \square \otimes \textcircled{1} = \square \oplus \square \oplus \square$$

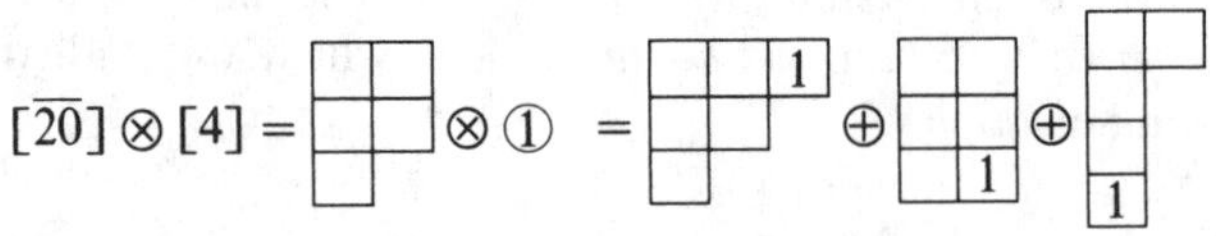

$$= [64] \oplus [\overline{10}] \oplus [6] \quad , \tag{10}$$

$$[20]'' \otimes [4] = \square \otimes \textcircled{1} = \square \oplus \square$$

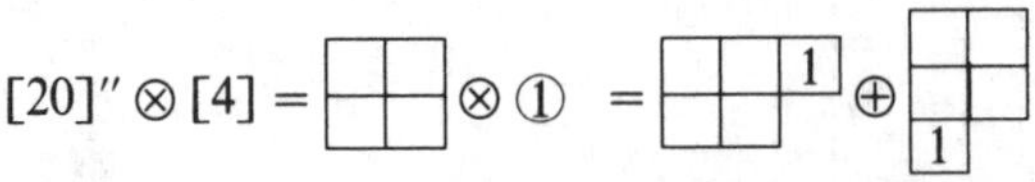

$$= [60] \oplus [\overline{20}] \quad . \tag{11}$$

Now we try a more complicated example *Example 11.8*

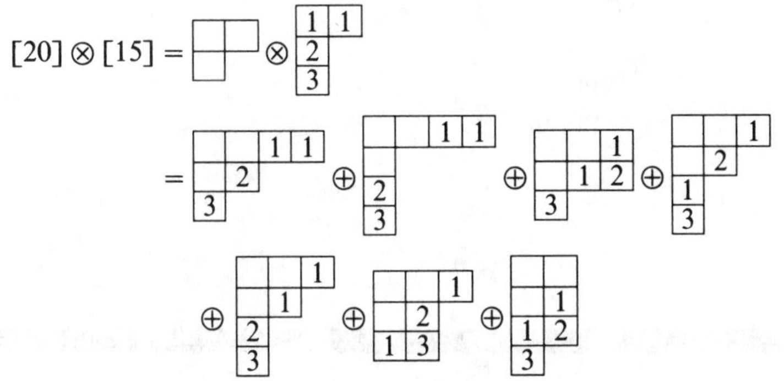

$$= [140]'' \oplus [20]' \oplus [\overline{60}] \oplus [20] \oplus [20] \oplus [\overline{36}] \oplus [\bar{4}] \quad . \quad (12)$$

According to the rules given in Chap 9 the following tableaux are not allowed:

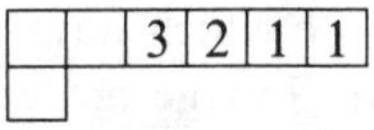 rule 3 (numbers descending in a row) ;

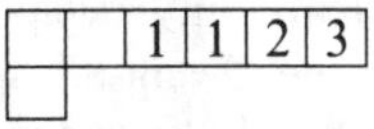 rule 5 ("3" appears on the path before "2") ;

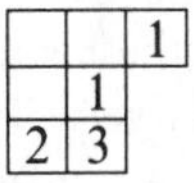 rule 5 ("3" appears on the path before "2") ;

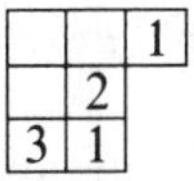 rules 3 and 4 (numbers descending in a row and in a column) .

Finally, we determine the dimensions of some of the representations for our example:

dim 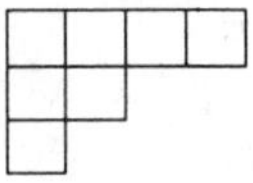 $= \dfrac{(4\cdot 5\cdot 6\cdot 7)(3\cdot 4)\cdot 2}{(6\cdot 4\cdot 2\cdot 1)(3\cdot 1)\cdot 1} = 140 \;;$ (13)

dim 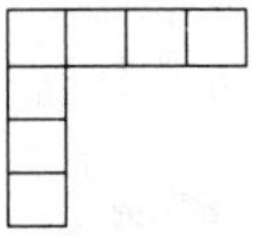 = dim 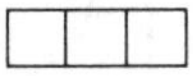 $= \dfrac{4\cdot 5\cdot 6}{3\cdot 2\cdot 1} = 20 \;;$ (14)

dim 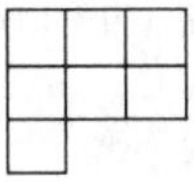 $= \dfrac{(4\cdot 5\cdot 6)(3\cdot 4\cdot 5)\cdot 2}{(5\cdot 3\cdot 2)(4\cdot 2\cdot 1)\cdot 1} = 60 \;;$ (15)

dim 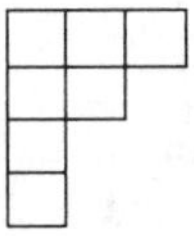 = dim 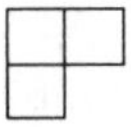 $= \dfrac{(4\cdot 5)\cdot 3}{(3\cdot 1)\cdot 1} = 20 \;;$ (16)

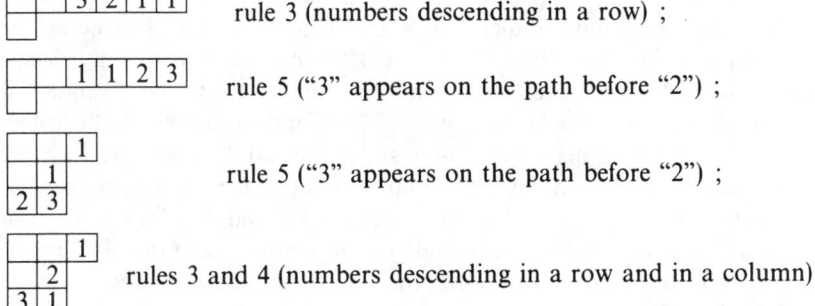

Example 11.8

dim 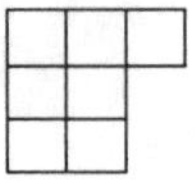 $= \dfrac{(4\cdot 5\cdot 6)(3\cdot 4)(2\cdot 3)}{(5\cdot 4\cdot 1)(3\cdot 2)(2\cdot 1)} = 36 \;;$ (17)

dim 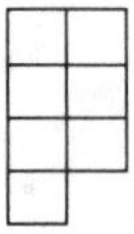 = dim 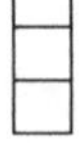 $= \dfrac{(4)(3)(2)}{(3)(2)(1)} = 4 \quad .$ (18)

11.6 The Potential Model of Charmonium

We have explained the Ψ particles as bound states of c and $\bar{c}$ which have the overall charm quantum number $C = 0$, i.e. "hidden" charm. Furthermore we have seen that for energetic reasons the J/Ψ and the Ψ' can only decay in a Zweig-forbidden way, such as $\Psi \to \pi^+\pi^0\pi^-$, but not into the Zweig-allowed channels $\Psi \to D^0 D^0$, $\Psi \to D^+ D^-$, $\Psi \to F^+ F^-$. On the other hand, those decays are possible for the higher charmonium states Ψ'' and Ψ''', because these states are sufficiently heavy to decay into D mesons and F mesons respectively. This explains their much larger with Γ with respect to Ψ and Ψ'. If we now consider all Ψ resonances as states of the bound system $c\bar{c}$, we see that the higher excited states are able to decay into D and F, but not the ground state and first excited state of spin 1. Since the $c\bar{c}$ system is formally similar to the bound states of an electron and a positron, which are called "positronium", we call the $c\bar{c}$ system "*charmonium*".[9]

We now try to describe the $c\bar{c}$ states by a suitable potential model, in which we assume the system to be spherically symmetric and to be described by the nonrelativistic Schrödinger equation. After a transformation of the coordinates to the centre of mass of the $c\bar{c}$ system, we get

$$\left[\frac{1}{m_c}\left(-\frac{1}{r}\frac{d^2}{dr^2}r + \frac{l(l+1)}{r^2}\right) + (V(r) - E)\right]R(r) = 0 \quad , \quad (\hbar = 1) \quad , \qquad (11.20)$$

where m_c is the mass of the charmed quark. From the fact that free quarks have not been observed, we conclude that the potential must cause quark confinement at large distances (more precisely, for distances r much larger than the diameter of a hadron). The simplest choice is to assume that the potential of the

[9] A. de Rujula, S.L. Glashow: Phys. Rev. Lett. **34**, 46 (1975); T. Appelquist, A. de Rujula, H.D. Politzer, S.L. Glashow: Phys. Rev. Lett. **34**, 365 (1975); E. Eichten, K. Gottfried, T. Kinoshita, J. Kogut, K.D. Lane, T.-M. Yan: Phys. Rev. Lett. **34**, 369 (1975).

quarks increases proportionally to the distance, i.e. $V(r) = kr$, and the constants k and m_c must be fitted to the experimental data.

For $l = 0$ we can solve the Schrödinger equation

$$-\frac{1}{m_c}\frac{d^2(rR)}{dr^2} + (kr - E)rR = 0 \tag{11.21}$$

exactly. We now introduce the notation $L^3 = 1/m_c \cdot k$ and $\lambda = L^2 m_c E$, and transform to a new variable x, defined by

$$x = \frac{r}{L} - \lambda \quad , \quad rR(r) = u(x) \quad ,$$

obtaining

$$\frac{d^2u(x)}{dx^2} - xu = 0 \quad . \tag{11.22}$$

Since the wave function should be regular at $r = 0$, i.e. $R(0) < \infty$, we get $\lim_{r\to 0} rR(r) = 0$, which requires

$$u(-\lambda) = 0 \quad . \tag{11.23}$$

Furthermore, we have to demand $u(\infty) = 0$, because the bound states must be normalizable. The solutions of the differential equation (11.22) can be expressed in terms of modified Bessel functions of order 1/3, and that solution satisfying the boundary condition $u(\infty) = 0$ is called the *Airy function* (see the Mathematical Supplement 11.8). Hence

$$u(x) = C \times \mathrm{Ai}(x) \quad , \quad \text{where} \tag{11.24}$$

$$\mathrm{Ai}(x) = \frac{1}{\pi}\sqrt{\frac{x}{3}}\,K_{1/3}\left(\frac{2}{3}x^{3/2}\right) \quad \text{for } x > 0 \quad .$$

The asymptotic behaviour is derived from the formula

$$K_\nu(z) \to \sqrt{\frac{\pi}{2z}}\,\mathrm{e}^{-z} \quad \text{for } z \to \infty \quad ,$$

so that

$$u(x) \to \frac{C}{2}\sqrt{\frac{3}{\pi}}\,x^{-1/4}\exp\left(-\frac{2}{3}x^{3/2}\right) \quad \text{for } x \to +\infty \quad . \tag{11.25}$$

In the range $x < 0$ the Airy function is expressed as

$$\mathrm{Ai}(-x) = \tfrac{1}{3}\sqrt{x}\{J_{1/3}(\tfrac{2}{3}x^{3/2}) + J_{-1/3}(\tfrac{2}{3}x^{3/2})\} \quad , \tag{11.26}$$

so that

$$u(x) = \frac{C}{3}\sqrt{|x|}\,\{J_{1/3}(\tfrac{2}{3}|x|^{3/2}) + J_{-1/3}(\tfrac{2}{3}|x|^{3/2})\} \quad \text{for } x < 0 \; .$$

Because of the second boundary condition we require $\mathrm{Ai}(-\lambda) = 0$. Note that λ is just the energy eigenvalue (in units of $L^2 m_c$), which means that the energy eigenvalues are determined by the roots of the Airy function

$$E_n = \left(\frac{k^2}{m_c}\right)^{1/3} \lambda_n \quad . \tag{11.27}$$

These energies have to be added to the rest mass of the $c\bar{c}$ pair, so that the mass of the nth excited state of charmonium is given by

$$M^{(n)} = 2m_c + \left(\frac{k^2}{m_c}\right)^{1/3} \lambda_n \quad , \tag{11.28}$$

where λ_n is the nth root of the Airy function: $\mathrm{Ai}(-\lambda) = 0$. Since we have the two input parameters m_c and k, we need two masses for the fit. We choose $M^{(1)} = 3.097$ GeV and $M^{(2)} = 3.686$ GeV and obtain from this $k = (0.458\ \mathrm{GeV})^2$ and for the mass of the c quark $m_c = 1.155$ GeV. We see that the *effective* mass of the c quark is much larger than that of u, d, or s quarks which are about 330 MeV, which explains why particles containing c quarks have very large masses. With the knowledge of the parameters we are now able to calculate the masses of the higher excited states, and these are listed in Table 11.8.

Table 11.8. Masses of radially excited states (*ns* states) of charmonium in the linear potential model

n	λ_n	M^n [GeV]	ΔM_{rel}	$M^{(n)}_{eff}$	M_{exp}
1	2.338	3.097 (fit)	—	3.097	3.097
2	4.088	3.686 (fit)	—	3.686	3.686
3	5.521	4.17	−0.15	4.02	4.040
4	6.787	4.59	−0.23	4.36	4.415

In this table we have taken into account that states with $n > 2$ can no longer be described by a nonrelativistic equation, and the relativistic mass correction ΔM_{rel}, is a consequence of this relativistic generalization of our model. Comparison with experiment shows that the excited spin-1 states of charmonium are well described by our simple model.

Next we have to solve the Schrödinger equation for arbitrary orbital angular momentum l. Furthermore, we have to consider the quark spin, for which there are two possibilities:

a) Spin antiparallel, "paracharmonium" ($s = 0$),
b) Spin parallel, "orthocharmonium" ($s = 1$).

Since the state with antiparallel spin has lower energy than that with parallel spin, it appears natural to identify the pseudoscalar mesons η_c and η'_c (masses 2.98 GeV and 3.59 GeV) with paracharmonium and the vector mesons Ψ, Ψ', etc. with orthocharmonium. Further consideration of the spin-orbit coupling (**L**·**S**) and the hyperfine structure (HFS) leads to the level scheme shown in Fig. 11.19.

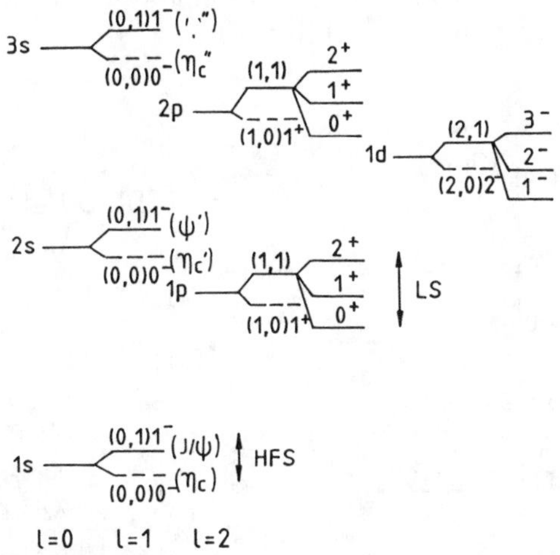

Fig. 11.19. Level scheme of charmonium. The paracharmonium states are characterized (--- ↑↓) and the orthocharmonium states (—— ↑↑). The numbers in parantheses denote orbital angular momentum and spin (l, s)

The paracharmonium states are characterized by dashed lines and those of the orthocharmonium by solid lines. The states are characterized by their total angular momentum J and the parity Π, written J^{Π} (e.g. 3^- means $J = 3$ and negative parity). The numbers in parentheses denote the orbital angular momentum and spin, (l, s). In analogy to its use in atomic physics, the spectroscopic notation $n^M l_J$ is employed, where n is the principal quantum number, M is the multiplicity (e.g. "3" is a triplet), l is the orbital and J the total angular momentum. In this notation, η_c' is the $2\,{}^1S_0$ state, etc. The presently (1985) accepted assignments are listed in Table 11.9. (A question mark indicates that

Table 11.9. Masses of experimentally observed states of charmonium

Name	State	Mass [MeV]	Name	State	Mass [MeV]
J/Ψ	1^3S_1	3097	η_c	1^1S_0	2980
Ψ'	2^3S_1	3686	η_c' (?)	2^1S_0 (?)	3590
			χ_0	2^1P_0	3415
			χ_1	1^3P_1	3510
			χ_2	1^5P_2	3556
Ψ''	3^3S_1	4040		1^3D_1	3770
	?	4160			
Ψ'''	4^3S_1	4415			

Table 11.10. Classification of mesons with hidden charm

Names	State	Mass [MeV]
D^0	$^1S_0$	1864.7 ± 0.6
D^+	$^1S_0$	1869.4 ± 0.6
D^{0*}	$^3S_1$	2007 ± 2
D^{+*}	$^3S_1$	2010.1 ± 0.7
F^+	$^1S_0$	1971 ± 6
F^{+*}(?)	$^3S_1$	2140 ± 60

the assignment is not unique.) The mesons with open charm can be classified analogously, as shown in Table 11.10.

Finally, Fig. 11.20 shows the experimental level scheme of the charmonium system. Here the decay modes are denoted along with their respective branching ratios. The states (particles) above the energy $2\,M_D = 3.73$ MeV (dashed line) decay into $D\bar{D}$, $D^*\bar{D}^*$, $F\bar{F}$.

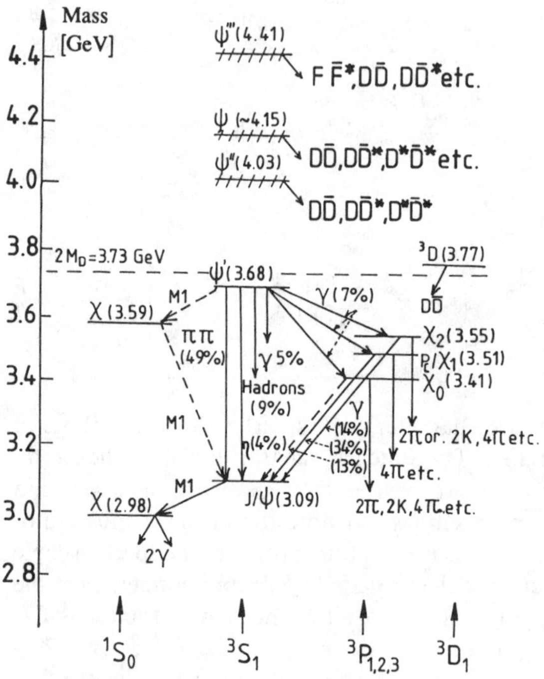

Fig. 11.20. Experimental level scheme of charmonium

EXAMPLE

11.9 Mathematical Supplement. Airy Functions

The Airy functions are solutions of the differential equation

$$\frac{d^2u(x)}{dx^2} - xu(x) = 0 \quad , \tag{1}$$

which has two linear independent solutions. The solution which decreases to zero as $x \to \infty$ is called $\mathrm{Ai}(x)$, and the one that diverges at infinity is called $\mathrm{Bi}(x)$. These functions can be represented by normal and modified Bessel functions, as we will see in the following. First we consider the range $x > 0$ and introduce the coordinate transformation

$$z = \beta x^{\gamma} \quad . \tag{2}$$

We try the ansatz

$$u(x) = z^{\alpha} f(z) \tag{3}$$

in order to solve (1), where the constants α, β and γ still have to be determined. We obtain

$$\frac{d^2u}{dx^2} = \gamma x^{-2} z^{\alpha}\{\alpha(\gamma\alpha - 1)f(z) + (2\alpha\gamma + \gamma - 1)zf'(z) + \gamma z^2 f''(z)\} \quad , \tag{4}$$

where a prime indicates a derivative with respect to z. Inserting (4) into (1) we obtain the equation

$$0 = f''(z) + \frac{2\alpha\gamma + \gamma - 1}{\gamma z} f'(z) - \frac{x^3}{\gamma^2 z^2} f(z) + (\alpha\gamma - 1)\frac{\alpha}{\gamma z^2} f(z) \quad . \tag{5}$$

We can eliminate the x dependence in the third term of the rhs by a suitable choice of γ, so that the coefficient of $f(z)$ becomes constant. We set

$$\frac{x^3}{\gamma^2 z^2} = 1 \tag{6}$$

and get, according to (2), the condition

$$\gamma = \tfrac{3}{2} \quad , \quad \beta = \tfrac{2}{3} \quad . \tag{7}$$

Thus, (5) is cast into the form

$$0 = f''(z) + \frac{6\alpha + 1}{3z} f'(z) - \left(1 - (\alpha - \tfrac{2}{3})\frac{\alpha}{z^2}\right) f(z) \quad , \tag{8}$$

and, with the choice of $\alpha = 1/3$, we obtain the standard form of Bessel's differential equation

$$0 = f''(z) + \frac{1}{z} f'(z) - \left(1 + (\tfrac{1}{3})^2 \frac{1}{z^2}\right) f(z) \quad , \tag{9}$$

the solutions of which are the modified Bessel functions of order $1/3$,

$$f(z) = \{I_{1/3}, I_{-1/3}, K_{1/3}\} \quad .$$

It is common to represent the Airy functions by

$$\begin{aligned}\mathrm{Ai}(x) &= \frac{1}{\pi}\sqrt{\frac{x}{3}} K_{1/3}(\tfrac{2}{3}x^{3/2}) \quad , \quad \mathrm{Bi}(x) = \sqrt{\frac{x}{3}}\{I_{1/3}(\tfrac{2}{3}x^{3/2}) \\ &\quad + I_{-1/3}(\tfrac{2}{3}x^{3/2})\} \quad , \qquad \text{for } x > 0\end{aligned} \tag{10}$$

where we have reintroduced the original variable x.

Next, to derive the solutions for negative values of x, we make the transformation

$$y = -x \quad , \quad u(x) = v(y) \quad , \tag{11}$$

Example 11.9

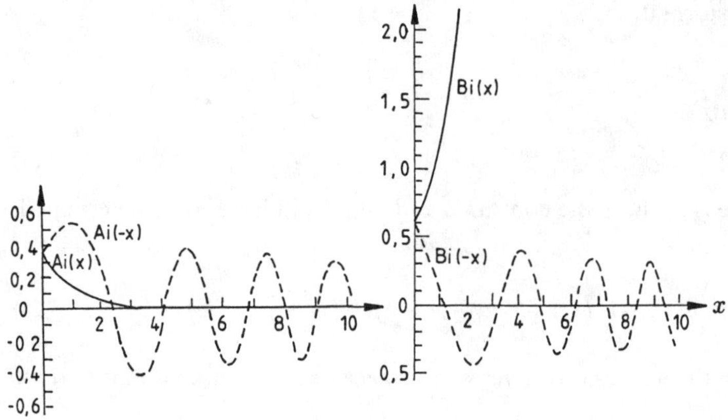

Airy functions

whereupon (1) takes the form

$$\frac{d^2v(y)}{dy^2} + yv(y) = 0 \quad . \tag{12}$$

Now we perform the same calculations as before, except with the variable y instead of x, using

$$z = \beta y^\gamma \quad , \quad v(y) = z^\alpha f(z) \quad . \tag{13}$$

We then get the already familiar form

$$0 = f''(z) + \frac{2\alpha\gamma + \gamma - 1}{\gamma z} f'(z) + \frac{y^3}{\gamma^2 z^2} f(z) + (\alpha\gamma - 1)\frac{\alpha}{\gamma z^2} f(z) \quad .$$

In analogy to (7) we choose

$$\gamma = \tfrac{3}{2} \quad , \quad \beta = \tfrac{2}{3} \quad , \tag{14}$$

and obtain

$$0 = f''(z) + \frac{6\alpha + 1}{3z} f'(z) + \left(1 + (a - \tfrac{2}{3})\frac{\alpha}{z^2}\right) f(z) \quad ,$$

which immediately reduces to a differential equation of the Bessel type if we set $\alpha = 1/3$.

$$0 = f''(z) + \frac{1}{z} f'(z) + \left(1 - (\tfrac{1}{3})^2 \frac{1}{z^2}\right) f(z) \quad . \tag{15}$$

The solutions are Bessel functions of order 1/3 or $-1/3$, i.e. the functions $J_{\pm 1/3}(z)$. Returning to the original variable x, we write the Airy-functions in the form

$$\begin{aligned} \mathrm{Ai}(x) &= \tfrac{1}{3}\sqrt{|x|}\,\{J_{1/3}(\tfrac{2}{3}|x|^{3/2}) + J_{+1/3}(\tfrac{2}{3}|x|^{3/2})\} \quad , \\ \mathrm{Bi}(x) &= \sqrt{\tfrac{1}{3}|x|}\,\{J_{1/3}(\tfrac{2}{3}|x|^{3/2}) - J_{+1/3}(\tfrac{2}{3}|x|^{3/2})\} \quad , \\ &\text{for } x < 0 \quad . \end{aligned} \tag{16}$$

Example 11.9

The definitions (10) and (16) determine the Airy functions uniquely in the full range of x [cf. (11.22)], and, using some well-known relations concerning Bessel functions, we try to investigate their behaviour for $x \to 0$. The following relations are taken from standard mathematical literature.[10]

For small values of $|z|$ we find

$$J_\nu(z) \simeq \frac{1}{\Gamma(\nu+1)}(\tfrac{1}{2}z)^\nu \left(1 - \frac{z^2}{4(\nu+1)}\right) \quad , \tag{17}$$

$$I_\nu(z) \simeq \frac{1}{\Gamma(\nu+1)}(\tfrac{1}{2}z)^\nu \left(1 + \frac{z^2}{4(\nu+1)}\right) \quad . \tag{18}$$

The behaviour of $K_\nu(z)$ can be calculated by means of the general formula

$$K_\nu(z) = \frac{\pi}{2}\frac{I_{-\nu} - I_\nu}{\sin(\nu\pi)} \quad , \tag{19}$$

and in particular we get

$$K_{1/3}(z) \simeq \frac{\pi}{\sqrt{3}}(1/\Gamma(\tfrac{2}{3}))(\tfrac{1}{2}z)^{-1/3} \times \{1 - (\tfrac{1}{2}z)^{2/3}\Gamma(\tfrac{2}{3})/\Gamma(\tfrac{4}{3})\} \quad . \tag{20}$$

Thus, for small positive values of x we obtain

$$\mathrm{Ai}(x) \simeq (\tfrac{1}{3})^{2/3}\left(\frac{1}{\Gamma(\tfrac{2}{3})}\right)\{ - (\tfrac{1}{3})^{2/3} x\Gamma(\tfrac{2}{3})/\Gamma(\tfrac{4}{3})\}$$

$$\mathrm{Bi}(x) \simeq (\tfrac{1}{3})^{1/6}\left(\frac{1}{\Gamma(\tfrac{2}{3})}\right)\{1 + (\tfrac{1}{3})^{2/3} x\Gamma(\tfrac{2}{3})/\Gamma(\tfrac{4}{3})\} \quad ,$$

and for negative values near zero,

$$\mathrm{Ai}(x) \simeq (\tfrac{1}{3})^{2/3}\left(\frac{1}{\Gamma(\tfrac{2}{3})}\right)\{1 + (\tfrac{1}{3})^{2/3} |x|\Gamma(\tfrac{2}{3})/\Gamma(\tfrac{4}{3})\} \quad ,$$

$$\mathrm{Bi}(x) \simeq (\tfrac{1}{3})^{1/6}\left(\frac{1}{\Gamma(\tfrac{2}{3})}\right)\{1 - (\tfrac{1}{3})^{2/3} |x|\Gamma(\tfrac{2}{3})/\Gamma(\tfrac{4}{3})\} \quad .$$

Thus we immediately see that the Airy functions are continuous at $x = 0$ and also have continuous derivatives at this point.

$$\mathrm{Ai}(0) = (\tfrac{1}{3})^{1/2} \; ; \quad \mathrm{Ai}'(0) = -(\tfrac{1}{3})^{1/2} \; ;$$

$$\mathrm{Bi}(0) = (\tfrac{1}{3})^{2/3}/\Gamma(\tfrac{2}{3}) \; ; \quad \mathrm{Bi}'(0) = -(\tfrac{1}{3})^{4/3}/\Gamma(\tfrac{4}{3}) \; . \tag{21}$$

For application to physical problems it is necessary to know the asymptotic behaviour. This is easily derived from the well-known behaviour of the regular-

[10] e.g. M. Abramowitz, J.A. Stegun: *Handbook of Mathematical Functions* (Dover, New York 1965).

Example 11.9

and modified-Bessel functions. For $|z| \to \infty$ we see that

$$J_\nu(z) \to \sqrt{\frac{2}{\pi z}} \cos\left(z - \frac{\pi}{2}(\nu + \tfrac{1}{2})\right) \quad , \tag{22}$$

$$I_\nu(z) \to \sqrt{\frac{1}{2\pi z}}\, e^z \quad , \tag{23}$$

$$K_\nu(z) \to \sqrt{\frac{\pi}{2z}}\, e^{-z} \quad . \tag{24}$$

Thus we obtain for $x \to \infty$

$$\mathrm{Ai}(x) \to \sqrt{\frac{1}{4\pi}}\, x^{-1/4}[\exp(-x^{3/2})]^{2/3} \quad , \tag{25}$$

$$\mathrm{Bi}(x) \to \sqrt{\frac{1}{\pi}}\, x^{-1/4}[\exp(x^{3/2})]^{2/3} \quad . \tag{26}$$

For negative values both functions are oscillating and exhibit asymptotic behaviour.

$$\mathrm{Ai}(x) \to \sqrt{\frac{1}{\pi}}\, |x|^{-1/4} \cos\left(\tfrac{2}{3}|x|^{3/2} - \frac{\pi}{4}\right) \quad , \tag{27}$$

$$\mathrm{Bi}(x) \to -\sqrt{\frac{1}{\pi}}\, |x|^{-1/4} \sin\left(\tfrac{2}{3}|x|^{3/2} - \frac{\pi}{4}\right) \quad . \tag{28}$$

11.7 The SU(4) [SU(8)] Mass Formula

If we want to extend the mass formula of SU(6) to hadrons containing charmed quarks, we obtain SU(4) multiplets with spin, i.e. SU(8) multiplets. The mass formula of SU(6) was (cf. Chap. 8)

$$M = M_0 + aY + b[T(T+1) - \tfrac{1}{4}Y^2] + cJ(J+1) \quad , \tag{11.29}$$

where J is the spin of the particle. Again, we assume the Hamiltonian of the strong interaction to be composed of two parts, i.e. $\hat{H}_{\mathrm{strong}} = \hat{H}_{\mathrm{ss}} + \hat{H}_{\mathrm{ms}}$. The Hamiltonian of the basic strong interaction $\hat{H}_{\mathrm{ss}}$ is invariant under the group SU(4), which means that all particles of an SU(4) multiplet have the same mass, $M_0 = \langle \hat{H}_{\mathrm{ss}} \rangle$. Furthermore, we assume the symmetry breaking part $\hat{H}_{\mathrm{ms}}$ to cause mass splitting only between different isospin multiplets, and not within such a multiplet, i.e. $[\hat{H}_{\mathrm{ms}}, \hat{T}_3] = 0$. Thus, we simply take the mass formula of SU(6) and consider how to describe the mass splitting between multiplets with different charm content.

Besides $\hat{\lambda}_8 = 2\hat{F}_8 = \sqrt{3}\,\hat{Y}$, only the generator $\hat{\lambda}_{15}$ commutes with $\hat{T}_3 = \frac{1}{2}\hat{\lambda}_3$. We now construct the operator from $\hat{\lambda}_{15}$

$$\hat{Z} = \sqrt{\frac{3}{8}}\hat{\lambda}_{15} = \frac{1}{4}\begin{pmatrix} 1 & 0 & 0 & 0 \\ 0 & 1 & 0 & 0 \\ 0 & 0 & 1 & 0 \\ 0 & 0 & 0 & -3 \end{pmatrix} \tag{11.30}$$

which obviously satisfies $[\hat{T}_3, \hat{Z}] = 0$. The operator $\hat{Z}$ acts on the quark states in the following way:

$$\hat{Z}|\mathrm{u}\rangle = \tfrac{1}{4}|\mathrm{u}\rangle \quad , \quad \hat{Z}|\mathrm{d}\rangle = \tfrac{1}{4}|\mathrm{d}\rangle \quad ,$$
$$\hat{Z}|\mathrm{s}\rangle = \tfrac{1}{4}|\mathrm{s}\rangle \quad , \quad \hat{Z}|\mathrm{c}\rangle = -\tfrac{3}{4}|\mathrm{c}\rangle \; . \tag{11.31}$$

It is now easy to construct the charm operator from $\hat{Z}$. We try the ansatz

$$\hat{C} = \tfrac{3}{4}\hat{B} - \hat{Z} \quad , \tag{11.32}$$

where $\hat{B}$ denotes the baryon-number operator. The relation

$$\hat{B}|\mathrm{q}_i\rangle = \tfrac{1}{3}|\mathrm{q}_i\rangle \quad \text{for} \quad i = 1,2,3,4$$

is valid for all quark states, because each quark has by definition the baryon number $B = 1/3$. This ensures that baryons, consisting of 3 quarks, have baryon number $B = 1$. Hence

$$\hat{C} = \tfrac{1}{4}\mathbb{1} - \hat{Z} \quad , \tag{11.33}$$

which can be checked explicitly, i.e.

$$\hat{C}|\mathrm{u}\rangle = 0|\mathrm{u}\rangle \quad , \quad \hat{C}|\mathrm{d}\rangle = 0|\mathrm{d}\rangle \quad , \quad \hat{C}|\mathrm{s}\rangle = 0|\mathrm{s}\rangle \quad , \quad \hat{C}|\mathrm{c}\rangle = 1|\mathrm{c}\rangle \quad .$$

We see that $\hat{C}$ is indeed the *charm operator* with the eigenvalues $C_i = 0, 0, 0, 1$ for the quark quartet. Since $\hat{Z}$ and $\mathbb{1}$ commute with $\hat{T}_3$, the commutator of $\hat{C}$ and $\hat{T}_3$ vanishes as well $[\hat{C}, \hat{T}_3] = 0$, i.e. $\hat{C}$ is the desired operator. The *mass formula for baryons* must therefore have the form

$$M = M_0 + aY + b[T(T+1) - \tfrac{1}{4}Y^2] + cJ(J+1) + dC \quad , \tag{11.34}$$

where the parameters M_0, a, b, c and d have to be fitted to the experimental masses. As up to now no charmed baryon with $J = 3/2$ has been found, it is sufficient to consider the flavour SU(4) mass formula,

$$M = M_0 + \alpha Y + \beta[T(T+1) - \tfrac{1}{4}Y^2] + \gamma C \quad ,$$

instead of the full SU(8) formula. To determine the parameters we use the data from Table 11.11.

With the values $M_0 = 1116$ MeV, $\alpha = -196$ MeV, $\beta = 38$ MeV and $\gamma = 1349.5$ MeV (compare with Sect. 8.12) the remaining masses can be calculated and the theoretical values are compared with the experimental data in Table 11.12.

Table 11.11. Quantum numbers of some baryons with $J=1/2$

Particle	Mass [MeV]	T	Y	C
Nucleon	939	$\frac{1}{2}$	1	0
Σ	1192	1	0	0
Λ	1116	0	0	0
Λ_c	2282	0	1	1

Table 11.12. Comparison of theoretical and experimental values for the masses of the baryons with $J = 1/2$

Particle C	Calculated Mass [MeV]	Observed Mass [MeV]	T	Y	C
Ξ	1331	1317	$\frac{1}{2}$	-1	0
Ξ_c	2494	2465?	$\frac{1}{2}$	0	1
Σ_c	2336	2450?	1	1	1
X_u, X_d	3638	?	$\frac{1}{2}$	1	2
X_s	3815	?	0	0	2

As the charmed baryons have not yet been clearly identified, it is difficult to determine the success of the extended mass formula.

Next we write down an *extended mass formula for mesons.* We have already noted, in the context of the SU(3) group, that the meson mass formula holds for the squared masses and not for the masses themselves. Since the mesons with $C = 1$ as well as those with $C = -1$ have larger masses than mesons with $C = 0$, the mass formula must not contain C as a linear term, because in that case either the states with $C = 1$ or those with $C = -1$ would have smaller masses than the states with $C = 0$. Therefore the simplest choice for the mass formula is a term proportional to C^2, so that

$$\mu^2 = \mu_0^2 + \beta C^2 + \gamma[T(T+1) - \tfrac{1}{4}Y^2] + \delta J(J+1) \quad . \tag{11.35}$$

This formula cannot explain the mass differences between ω^0, Φ^0 and Ψ, because these states are degenerate within the T_3-Y-C-diagram. These differences can only be explained by taking into account the different quark compositions of the states ($\omega^0 \sim$ uu, d$\bar{\text{d}}$, $\Phi^0 \sim$ s$\bar{\text{s}}$, $\Psi \sim$ c$\bar{\text{c}}$), i.e. different quarks should have different effective masses. Considering the magnetic moment, we obtain $m_{u,d} \approx 330$ MeV, $m_s \approx 470$ MeV, while the charmonium model yields $m_c \approx 1.15$ GeV. These mass differences can in fact be explained by these data.

In order to apply the mass formula to vector mesons ($J = 1$), we use the SU(4)-formula

$$\mu^2 = \mu_0^2 + \beta C^2 + \gamma[T(T+1) - \tfrac{1}{4}Y^2] \quad . \tag{11.36}$$

With the data from Table 11.13 we derive the following values for the parameters:

$$\mu_0^2 = (783)^2(\text{MeV})^2 \quad ,$$

$$\gamma = -7780\ \text{MeV}^2 \quad ,$$

$$\beta = 3432846\ \text{MeV}^2 \quad .$$

The masses of the other vector mesons are listed, in comparison with the observed values, in Table 11.14.

Table 11.13. Masses and quantum numbers of three vector mesons

Particle	Mass [MeV]	T	Y	C
ϱ	773	1	0	0
ω	783	0	0	0
D^{0*}	2010	$\frac{1}{2}$	0	1

Table 11.14. Comparison of the predicted and observed values for the masses of the K*, F* vector mesons

Particle	Calculated Mass [MeV]	Observed Mass [MeV]	T	Y	C
K*	790	892	$\frac{1}{2}$	1	0
F*	2012	2140	0	-1	-1

As in the case of the spin-flavour group SU(6), we note that the SU(4) meson mass formula is much worse than the baryon mass formula. Because β is very much larger than μ_0^2 and γ, the largest contribution to the mass is due to the charm term βC^2; therefore it is not surprising that the masses of F* and D* are nearly equal (the difference is 2 MeV).

To summarize, we can state that the SU(4) symmetry is far more strongly violated than the SU(3) symmetry. This can be seen especially in the mass formulae. Therefore the applicability of the SU(4) symmetry is much more restricted than that of SU(3), and one mostly tries to explain the fine structure of the level splittings of the charmonium states with a potential model. This also yields the transition probabilities and the decay widths from first principles without taking into account the SU(4) symmetry. Such a treatment may be successful, because the probably fundamental theory of interactions between quarks, the quantum chromodynamics (QCD), at energies much higher than 1000 MeV leads to weak coupling and allows perturbative calculations. In the case of "light" baryons and mesons (i.e. those without charm), perturbation theory cannot be applied and symmetry considerations are quite successful. The spectrum of charmonium can be reasonably well explained in analogy to Bohr's atomic model, because it can be understood as a nonrelativistic, bound state of two heavy particles.

11.8 The Υ Resonances

In 1977 a team of physicists[11] observed the reaction $p + N \to \mu^+\mu^- + X$ at the Fermi National Laboratory (Illinois, USA) another family of very heavy meson resonances. These so-called *upsilon resonances* Υ, Υ', Υ'', Υ''' are in a way the "big sisters" of the charmonium states. Shortly after, experiments at DESY in Hamburg[12] confirmed the Fermilab results with much higher resolution, and since then the masses of Υ, Υ' and Υ'' have been measured very accurately at

[11] S.W. Herb et al. (16 authors): Phys. Rev. Lett. **39**, 252 (1977).

[12] C.W. Darden et al. (15 authors): Phys. Lett. **76**B, 246 and **78**B, 364 (1978).

DESY:

$$M(\Upsilon) = \ \ 9.460\ \mathrm{GeV/c^2} \quad ,$$

$$M(\Upsilon') = 10.023\ \mathrm{GeV/c^2} \quad ,$$

$$M(\Upsilon'') = 10.355\ \mathrm{GeV/c^2} \quad .$$

The cross-section for e^+e^- scattering as a function of CM energy is depicted in Fig. 11.21. One concludes from the small leptonic decay widths[13] of the resonances that they must be interpreted as bound states of a new heavy quark and its antiquark. The new quark is called *b quark* ("*bottom*" or "*beauty*").

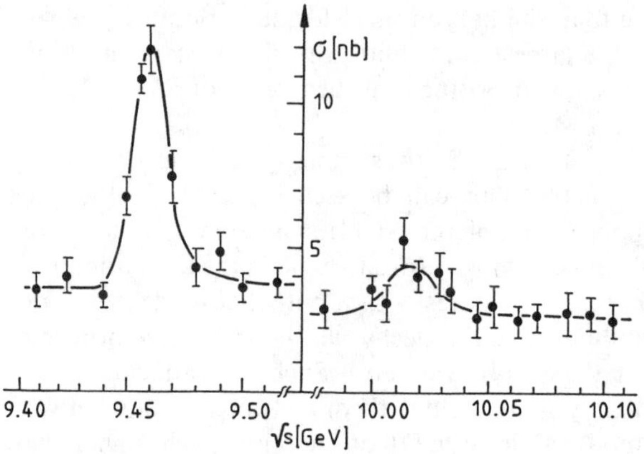

Fig. 11.21. Observed cross-sections $\sigma\,(e^+e^- \to \mathrm{hadrons})$ in the region of Υ (lhs) and Υ' (rhs). The total widths are $\Gamma_{\mathrm{tot}}(\Upsilon) \approx 7.8$ MeV and $\Gamma_{\mathrm{tot}}(\Upsilon') \approx 8.7$ MeV, while the decay widths into electrons are $\Gamma_e(\Upsilon) \approx 1$ keV and $\Gamma_e(\Upsilon') \approx 0.32$ keV

To explain the upsilon masses of about 10 GeV within the framework of a symmetry group SU(5) would require symmetry violations even larger than in the SU(4) theory, so that such symmetry considerations are abandoned. Instead this system is described within a potential model. Because of the high-energy scale, such a model represents an excellent approximation, and in this way the b quark mass is found to be

$$M_b = 4500\ \mathrm{MeV/c^2} \quad .$$

Since the relativistic corrections for the heavy b-quarks are smaller than for the light c-quarks, one hopes to get new information about the quark-antiquark potential from the measurement of the decay modes and further excited states of the $b\bar{b}$ system. In this way a better understanding of the fundamental interaction between quarks and antiquarks and the confinement problem is hoped for.

The new resonances are clearly apparent in a plot of the ratio R,

$$R = \frac{\sigma(e^+e^- \to \mathrm{Hadrons})}{\sigma(e^+e^- \to \mu^+\mu^-)} , \tag{11.37}$$

which plays, as we have seen, (cf. Fig. 11.2) a decisive role in determining the correct quark model. In the energy region $2\ \mathrm{GeV} \leqslant E_{\mathrm{cm}} \leqslant 3\ \mathrm{GeV}$, R is nearly

[13] J.L. Rosner, C. Quigg, H.B. Thacker: Phys. Lett. **74B**, 350 (1978).

constant at

$$R = 2 \quad . \tag{11.38}$$

At higher energies R increases and reaches the value 3.5 at an energy of 4.5 GeV. In this region all the charmonium resonances are found. At an energy of about 10 GeV, R runs through another threshold region, the region of the upsilon resonances. Here it rises to a value of nearly 4.

The elementary process which produces the cross-section $e^+e^- \to$ hadrons is the annihilation of an e^+e^- pair into a virtual photon followed by the creation of a quark-antiquark pair (see Fig. 11.22).

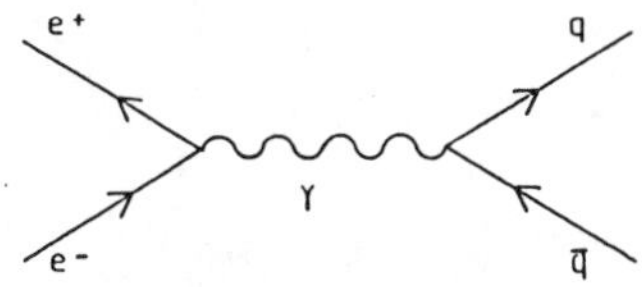

Fig. 11.22. Creation of a quark-antiquark pair

Quantum electrodynamical calculations show[14] that this cross-section is essentially given by the square of the charges of the created particles. Since all quarks are created with the same probability (given sufficient energy), we sum over all quark charges and find that

$$R = \sum_i Q_i^2 \quad . \tag{11.39}$$

By using the usual SU(3) quark model we obtain

$$R_{\mathrm{SU(3)}} = \tfrac{1}{9} + \tfrac{1}{9} + \tfrac{4}{9} = \tfrac{6}{9} = \tfrac{2}{3} \quad , \tag{11.40}$$

though obviously this value does not agree with the experiment. If we take into account that each quark appears in three colours, we get

$$R^{\mathrm{colour}} = 3 \sum_i Q_i^2 \quad , \tag{11.41}$$

and for the flavour SU(3) model follows

$$R^{\mathrm{colour}}_{\mathrm{SU(3)}} = 3 \times \tfrac{2}{3} = 2 \quad . \tag{11.42}$$

In fact this is the experimental value in the energy region $E_{\mathrm{cm}} \leqslant 3$ GeV. Hence we conclude that the $[\mathrm{SU(3)_{colour}} \times \mathrm{SU(3)_{flavour}}]$ model describes the low-energy behaviour of R correctly.

Evidently a new channel of particle production opens between 3 GeV and 4.5 GeV, and we cannot explain the detailed structure of this transition within our simple model. However, starting at $E_{\mathrm{cm}} = 5$ GeV, R is once again nearly constant. According to our considerations the value of R should now be described by the charm model. The charge of the c quark is $Q = 2/3$ and therefore we would get the wrong value, i.e.

$$R_{(\mathrm{u,d,s,c})} = \tfrac{10}{9} \quad , \tag{11.43}$$

if we counted each quark flavour only once. However, if we allow for colour, we obtain

$$R^{\mathrm{colour}}_{(\mathrm{u,d,s,c})} = 3 \times \tfrac{10}{9} = \tfrac{10}{3} \quad , \tag{11.44}$$

and once again this value agrees well with experiment.

[14] W. Greiner, A. Schäfer: *Quantum Chromodynamics* (Harri Deutsch, Thun 1989).

The recently discovered increase at 10 GeV can only be explained by a further quark flavour, the *b-quark* ("*bottom*" or "*beauty*"). Denoting the charge of the new quark by Q_b, the relation

$$R = \tfrac{10}{3} + 3Q_b^2 \qquad (11.45)$$

must hold beyond 10 GeV. If we restrict ourselves to charges which are one or two thirds of the elementary charge, the experimental value of R yields $Q_b = \pm 1/3$. The possibility $Q_b = +1/3$ can be excluded by a simple consideration: The so-called Λ_b particle, which consists of one u-, one d- and one b-quark, has the charge

$$Q(\Lambda_b) = (+\tfrac{2}{3}) + (-\tfrac{1}{3}) + Q_b = \tfrac{1}{3} + Q_b \quad . \qquad (11.46)$$

In the case of $Q_b = +1/3$, the charge of Λ_b would not be an integer. Consequently, because of charge conservation, the Λ_b particle could not decay into known hadrons which all have integer charges. The Λ_b would be absolutely stable. Since such a stable particle has not been observed, the b quark must have the charge $-1/3$, like the d- and the s-quarks.

Based on considerations of the theory of weak interactions[15], it has been conjectured that there is at least a sixth quark flavour, the *t-quark* ("*top*" or "*true*"). This quark, which has not yet been discovered experimentally, should have the charge $Q_t = +2/3$. Then all quark flavours can be arranged into doublets to give

$$\begin{pmatrix} \mathrm{u} \\ \mathrm{d} \end{pmatrix} , \quad \begin{pmatrix} \mathrm{c} \\ \mathrm{s} \end{pmatrix} , \quad \begin{pmatrix} \mathrm{t} \\ \mathrm{b} \end{pmatrix} , \quad \text{and so on (?)} \qquad (11.47)$$

It is not known how many of these doublets exist. It is quite possible that more and more quark flavours will be discovered until it may become questionable to consider the quarks as "elementary" constituents. Various models of such possible quark substructure have already been proposed.[16]

[15] See Vol. 5 of this series, W. Greiner, B. Müller: *Gauge Theory of Electro-Weak Interactions* (Springer, Berlin, Heidelberg) to be published.

[16] See e.g. R.N. Mohapatra: *Unification and Supersymmetry* (Springer, Berlin, Heidelberg 1986).

12. Mathematical Supplement

12.1 Introduction

Up to now, we have seen various examples of Lie groups, especially unitary (U(n) and SU(n)) groups. Let us see if we can find some common denominator in the structure of their algebras U(n) and SU(n) (algebras are denoted by lower case letters).

The first example we discussed extensively is the SU(2) algebra, which is identical to O(3). Various notations were used but here we are interested in the spherical form ($\hat{J}_+, \hat{J}_0, \hat{J}_-$) of the algebra, which then is given by

$$[\hat{J}_+, \hat{J}_-] = 2\hat{J}_0 \quad , \quad [\hat{J}_0, \hat{J}_\pm] = \pm \hat{J}_\pm . \tag{12.1}$$

The operators $\hat{J}_+$ and $\hat{J}_-$ can be interpreted as a raising and lowering operator respectively, simply because they raise and lower the magnetic quantum number by one. The operator $\hat{J}_0$ can be called "weight" operator because it gives just the "weight" M. Inspecting (12.1) more closely we can interpret $\hat{J}_+$ and $\hat{J}_-$ as operators of weight plus and minus one respectively, because $\hat{J}_0$ acting on $\hat{J}_\pm$, defined by the commutator, gives back $\hat{J}_\pm$ with a plus-minus sign. See also the definition of the regular representation of a Lie algebra, Section 5.3. There, the action of the generators on themselves defines a representation. In this sense, e.g. $\hat{J}_+$ can be viewed as a "state" of weight one, etc. Equations (12.1) tell us that $\hat{J}_+$ acting on $\hat{J}_-$, and vice versa, gives back the weight operator $\hat{J}_0$ and the action of $\hat{J}_0$ on $\hat{J}_\pm$ gives back $\hat{J}_\pm$ with the "weight" in front.

This division in terms of raising, lowering and weight operators turned out to be very useful in defining a suitable basis and in the construction of matrix elements (see Chapter 2 for the SU(2)-case). The question is now: Do other algebras show a similar structure? If yes we can copy the procedures proposed for the SU(2)-case and apply them to the other algebras. Let us consider next the SU(3) algebra as an example.

In Section 7.3, Eq. (7.28), we indeed find the same structure. We have operators of "positive" weight ($\hat{T}_+, \hat{V}_+, \hat{U}_+$), of negative weight ($\hat{T}_-, \hat{V}_-, \hat{U}_-$) and two weight operators ($\hat{T}_3$, $\hat{Y}$). The weight operators commute with each other and form the so-called "carton" subalgebra, given by the maximal possible number of generators which commute with each other. *Note* that with the definition of "highest" weight, given in Chapter 7, $\hat{T}_+$, $\hat{V}_+$ and $\hat{U}_-$ are the raising operators in SU(3). Nevertheless, it is equally justified to define $\hat{T}_+$, $\hat{V}_+$ and $\hat{U}_+$

as raising operators, resulting in a different highest weight state. All this is just a matter of definition and should not disturb us here.

The structure of the SU(3) algebra seems to be more complicated but in reality it shows the same structure as SU(2). For reasons of brevity we now denote all raising and lowering operators as $\hat{E}_\alpha$ with α defining the weight of the operators. For example, the action of $\hat{T}_3$ and $\hat{Y}$ on $\hat{T}_\pm$ is given by $[\hat{T}_3, \hat{T}_\pm] = \pm T_\pm$ and $[\hat{Y}, \hat{T}_\pm] = 0$, showing that the weight α of these operators is $\alpha = (\pm 1, 0)$. Analogously $\hat{V}_\pm$ and $\hat{U}_\pm$ have the weights (see Chapter 7, Eq. (7.28)) $\alpha = (\pm\frac{1}{2}, \pm 1)$ and $\alpha = (\mp\frac{1}{2}, \pm 1)$, respectively. The weight α uniquely defines the operator! Also the weight operators $\hat{T}_3$ and $\hat{Y}$ we shall now denote specially by $\hat{H}_i$ ($i = 1, 2$). With this new notation the algebra reads

$$[\hat{E}_\alpha, \hat{E}_\beta] = N_{\alpha\beta}\hat{E}_{\alpha+\beta} \quad , \quad \alpha \neq \beta \tag{12.2a}$$

$$[\hat{E}_\alpha, \hat{E}_{-\alpha}] = \alpha^i \hat{H}_i \tag{12.2b}$$

$$[\hat{H}_i, \hat{E}_\alpha] = \alpha_i \hat{E}_\alpha \tag{12.2c}$$

$$[\hat{H}_i, \hat{H}_j] = 0 \tag{12.2d}$$

with $N_{\alpha\beta}, \alpha^i, \alpha_i$ being constants and summing over repeated indices. As an example for (12.2a) there is the commutator $[\hat{T}_+, \hat{V}_-] = -\hat{U}_-$, which reads in the new notation $[\hat{E}_{(+1,0)}, \hat{E}_{(-\frac{1}{2},-1)}] = -\hat{E}_{(\frac{1}{2},-1)}$, i.e. the weights *simply add*. As an example for (12.2b) we have $[\hat{U}_+, \hat{U}_-] = \frac{3}{2}\hat{Y} - \hat{T}_3$, which in the new notation reads $[\hat{E}_{(-\frac{1}{2},+1)}, \hat{E}_{(+\frac{1}{2},-1)}] = -\hat{H}_1 + \frac{3}{2}\hat{H}_2$. For (12.2c) we have $[\hat{T}_3, \hat{U}_\pm] = \mp\frac{1}{2}\hat{U}_\pm$, which now reads $[\hat{H}_1, \hat{E}_{(\mp\frac{1}{2},\pm 1)}] = (\mp\frac{1}{2})\hat{E}_{(\mp\frac{1}{2},\pm 1)}$, giving the first component of the weight $\alpha = (\alpha_1, \alpha_2)$, i.e. α_i is a weight component. Finally (12.2d) is represented by e.g. $[\hat{T}_3, \hat{Y}] = 0$, which is $[\hat{H}_1, \hat{H}_2] = 0$. The two operators $\hat{T}_3$ and $\hat{Y}$ form the Cartan subalgebra. As will be shown later, the α^i in (12.2b) are related to the α_i. For the moment consider them to be different. Also some change will come from the redefinition of the generators via a simple multiplication by a constant, which can be different for each generator! This redefines the weights and other constants in (12.2) accordingly.

If it were possible to write an algebra in the form (12.2), then the advantage is obvious. The weight operators, together with the Casimir operators form a commuting set of Hermitian operators. Then states can be defined which are eigenstates with respect to these operators. Or in other words: *we can systematically define quantum numbers*! Furthermore, we can define raising and lowering operators whose action can be deduced from the algebra itself, in just the same way as in Chapter 2 for the SU(2) group. Also, analogously matrix elements and Clebsch-Gordan coefficients can be calculated.

In Exercise 7.1 we show that SU(4) also exhibits the structure of (12.2) and thus implies that possibly all known semi-simple algebras can be put to the form (12.2).

As a last example we discuss a non-compact algebra sp(2, R). It is related to the one-dimensional harmonic oscillator ($\hbar = m = \omega = 1$)

$$\hat{H} = \tfrac{1}{2}(\hat{p}^2 + x^2) \tag{12.3}$$

with $[x, \hat{p}] = \mathrm{i}$. The name "symplectic" comes from first redefining $X = \hat{Z}_1$ and $\hat{p} = \hat{Z}_2$, resulting in the commutator $[\hat{Z}_i, \hat{Z}_j] = \mathrm{i}g_{ij}$ with $(g_{ij}) = \left(\begin{smallmatrix} 0 & 1 \\ -1 & 0 \end{smallmatrix}\right)$. Only those transformations in the phase space $(\hat{Z}_1, \hat{Z}_2)$ are allowed which preserve the "metric" (g_{ij}), which is called symplectic. The number 2 is related to the dimension of the phase space.

Instead of coordinates and momenta we can introduce boson creation and annihilation operators via

$$\hat{X} = \frac{1}{\sqrt{2}}(\hat{b}^+ + \hat{b}) \qquad \hat{p} = \frac{\mathrm{i}}{\sqrt{2}}(\hat{b}^+ - \hat{b}) \quad . \tag{12.4}$$

Let us now construct all possible squares, i.e.

$$\hat{B}^+ = \hat{b}^+\hat{b}^+ \qquad \hat{B} = \hat{b}\hat{b} \qquad \hat{C} = \hat{b}^+\hat{b} + \tfrac{1}{2} \quad . \tag{12.5}$$

The $\hat{B}^+$ then raises a state of N-quanta by two quanta and $\hat{B}$ lowers it. The operator $\hat{C}$ stays within the same state or simply gives the weight $(N + \frac{1}{2})$. With (12.7) we can connect *all* states of the one-dimensional harmonic oscillator, dividing the states into two spaces with $N =$ even or odd. We can now define a lowest weight state (l.w.) by

$$\hat{B}|\mathrm{l.w.}\rangle = 0 \tag{12.6}$$

having two solutions $N = 0$ or 1. The $\hat{C}$ applied to this state then defines the irreducible representation, i.e. $(\frac{1}{2})$ or $(\frac{3}{2})$ depending on whether we act on the space with $N =$ even or odd. The representation is clearly of infinite dimension because we can apply $\hat{B}^+$ as often as we wish.

The operators in (12.5) form the algebra

$$[\hat{B}, \hat{B}^+] = 4\hat{C} \qquad [\hat{C}, \hat{B}^+] = +\hat{B}^+ \qquad [\hat{C}, \hat{B}] = -\hat{B} \quad . \tag{12.7}$$

This algebra is called sp(2, R). Now let us redefine the operators, i.e.

$$\hat{B}^+ \rightarrow \hat{E}_{+1} \qquad \hat{B} \rightarrow \hat{E}_{-1} \qquad \hat{C} \rightarrow \hat{H}_1 \quad . \tag{12.8}$$

With this the algebra reads

$$[\hat{E}_{+1}, \hat{E}_{-1}] = -4\hat{H}_1 \qquad [\hat{H}_1, \hat{E}_{\pm 1}] = \pm \hat{E}_{\pm 1} \quad .$$

In fact, multiplying $\hat{B}^+$ and $\hat{B}$ by $\frac{1}{\sqrt{2}}$ only changes the first commutator now with a 2 in front. Comparing to (12.1) we then obtain nearly the same algebra except for the sign in the first commutator. This is no coincidence. The algebra is the same as the one of SU(1, 1), i.e. sp(2, R)$\simeq$SU(1, 1). More important for our puspose is that *again* we have the same structure.

In conclusion, we have found various examples of algebras which can be put into a *standard* form, i.e. weight, raising and lowering operators. *This remarkable property deserves more attention and investigation.* In what follows, we will see that indeed all semi-simple Lie algebras can be put into the standard form and can be classified according to it. Though the algebra will have the same structure as indicated above, the constants $N_{\alpha\beta}$, α_i and α^i will differ from the previous examples due to a convenient multiplication by constant factors of the generators.

Unfortunately, we have had to go through a rather dry topic, which is however necessary. During the various steps one has to keep in mind the various examples. As a reward one will understand that all semi-simple, classical Lie algebras are related to each other in their structure and that there can be only a *finite* number of types of semi-simple algebras. This knowledge enormously simplifies the treatment and application of groups to various topics in physics!

EXAMPLE

12.1 Weight Operators of the SU(4)-Algebra

The algebra SU(4) was introduced in Chapter 11. It has SU(3) as a subalgebra, so we need only to see if the additional operators exhibit a continuation of (12.2). The answer is yes! For this use use the $\hat{\lambda}_i$-operators ($i = 1, \dots, 15$) as defined in Eqs. (11.1)–(11.3). For $\hat{\lambda}$ with $i = 1, \dots, 8$ we have done this already in Eq. (7.27) for SU(3). We do the same for the rest, i.e. we define $\hat{F}_i = \frac{1}{2}\hat{\lambda}_i$ and introduce the operators

$$\hat{W}_\pm = \hat{F}_9 \pm \mathrm{i}\hat{F}_{10} \qquad \hat{X}_\pm = \hat{F}_{11} \pm \mathrm{i}\hat{F}_{12}$$

$$\hat{Z}_\pm = \hat{F}_{13} \pm \mathrm{i}\hat{F}_{14} \qquad \hat{H}_3 = \sqrt{6}\hat{F}_{15} = \frac{\sqrt{6}}{2}\hat{\lambda}_{15} \tag{1}$$

We now have *three* weight operators, namely $\hat{H}_1 = \hat{T}_3, \hat{H}_2 = \hat{Y}$ and $\hat{H}_3 = \frac{6}{\sqrt{2}}\hat{\lambda}_{15}$. This implies that the weight of the operators carry three numbers related to the action of $\hat{H}_i$ ($i = 1, 2, 3$) on the operators.

In order to determine the weight of $W_\pm$, $\hat{X}_\pm$ and $\hat{Z}_\pm$ we act respectively with $\hat{H}_1$, $\hat{H}_2$ and $\hat{H}_3$ on them. We start with $\hat{W}_\pm$. Using Table 11.2 and Eq. (11.4a) we obtain

$$[\hat{H}_1, \hat{W}_\pm] = [\tfrac{1}{2}\hat{\lambda}_3, \tfrac{1}{2}(\hat{\lambda}_9 \pm \mathrm{i}\hat{\lambda}_{10})] = \pm \hat{W}_\pm \tag{2a}$$

$$[\hat{H}_2, \hat{W}_\pm] = \left[\frac{1}{\sqrt{3}}\hat{\lambda}_8, \tfrac{1}{2}(\hat{\lambda}_9 \pm \mathrm{i}\hat{\lambda}_{10})\right] = \pm \tfrac{1}{3}\hat{W}_+ \tag{2b}$$

$$[\hat{H}_3, \hat{W}_\pm] = \left[\frac{\sqrt{6}}{2}\hat{\lambda}_{15}, \tfrac{1}{2}(\hat{\lambda}_9 \pm \mathrm{i}\hat{\lambda}_{10})\right] = \pm 2\hat{W}_\pm \quad . \tag{2c}$$

This implies the weight $\alpha = (\alpha_1, \alpha_2, \alpha_3) = (\pm 1, \pm \frac{1}{3}, \pm 2)$ and the substitution of $\hat{W}_\pm$ by

$$\hat{W}_\pm \rightarrow \hat{E}_{(\pm 1, \pm \frac{1}{3}, \pm 2)} \quad . \tag{3a}$$

Similarly we have

$$\hat{X}_\pm \rightarrow \hat{E}_{(\mp \frac{1}{4}, \pm \frac{1}{3}, \pm 2)} \tag{3b}$$

and

$$\hat{Z}_\pm \rightarrow \hat{E}_{(0, \mp \frac{2}{3}, \pm 2)} \tag{3c}$$

Example 12.1

The weight of each generator of SU(3) with respect to the third weight operator is zero as can be deduced easily from Table 11.2, because the commutator $\hat{H}_3 = (\sqrt{6}/2)\hat{\lambda}_{15}$, i.e. $\hat{\lambda}_{15}$ with $\hat{\lambda}_k$ $(k = 1, \dots, 8)$, vanishes.

Note also that in this example we choose the weights in such a way that no square roots appear. By a suitable rescaling, i.e. multiplication of the generators, this is changed. In fact the final convenient form of weights will deviate from the example.

12.2 Root Vectors and Classical Lie Algebras

In this chapter we shall try to classify all semi-simple Lie algebras. Since we know that one basis of a Lie algebra can be transformed into another by a linear transformation, we first search for a standard form of the commutators of the elements (generators) $\hat{X}_\mu$ of a semi-simple Lie algebra. One arbitrary linear combination of these elements is denoted by $\hat{A}$, where

$$\hat{A} = a^\mu \hat{X}_\mu \quad , \tag{12.9}$$

and another by $\hat{X}$, defined by

$$\hat{X} = x^\mu \hat{X}_\mu \quad . \tag{12.10}$$

We now demand that these operators fulfil the eigenvalue equation

$$[\hat{A}, \hat{X}] = r\hat{X} \quad , \tag{12.11}$$

where r and $\hat{X}$ represent eigenvalue and eigenvector respectively. By using the commutation relations between the $\hat{X}_\mu$

$$[\hat{X}_\mu, \hat{X}_\nu]_- = C_{\mu\nu\sigma}\hat{X}_\sigma \quad , \tag{12.12}$$

equation (12.11) takes the form

$$a^\mu x^\nu C_{\mu\nu\sigma}\hat{X}_\sigma = r x^\sigma \hat{X}_\sigma \quad . \tag{12.13}$$

Since the $\hat{X}_\sigma$ are by definition linearly independent, then (12.13) is equivalent to

$$(a^\mu C_{\mu\nu\sigma} - r\delta_{\nu\sigma})x^\nu = 0 \quad ,$$

and the corresponding secular equation is

$$\det|a^\mu C_{\mu\nu\sigma} - r\delta_{\nu\sigma}| = 0 \quad . \tag{12.14}$$

If the Lie algebra consists of N elements, then (12.14) has N solutions, although some of them may be degenerate. As Cartan showed, one can choose $\hat{A}$ in such a way that the number of different solutions of (12.14) becomes maximal. In the case of semi-simple Lie algebras only the eigenvalue $r = 0$ remains degenerate. If $r = 0$ is l times degenerate we call l the *rank of the semi-simple algebra*, because (12.11) with $r = 0$ defines the commuting elements $\hat{H}_i$ of the Lie algebra.

The eigenvalue $r = 0$ has l corresponding linearly independent eigenvectors $\hat{H}_i$ which make up an l-dimensional subspace of the N-dimensional vector space, i.e.

$$[\hat{A}, \hat{H}_i]_- = 0 \quad (i = 1, 2, \dots, l) \quad . \tag{12.15}$$

The eigenvectors $\hat{E}_\alpha$, which correspond to the remaining $(N - l)$ solutions, form an $(N - l)$-dimensional subspace for which

$$[\hat{A}, \hat{E}_\alpha]_- = \alpha \hat{E}_\alpha \tag{12.16}$$

holds, where the eigenvalues are now denoted by α. The values α are called the roots of the Lie algebra. Next we write

$$[H_i, \hat{E}_\alpha]_- = \alpha \hat{E}_\alpha \quad , \quad [\hat{H}_i, \hat{H}_j]_- = 0 \quad .$$

The commutativity of the H_i may be understood by initially assuming that the H_i do not commute, so that

$$[H_1, H_i] = \tilde{C}^j_{1i} H_j \quad \text{with} \quad \tilde{C}^j_{1i} \neq 0 \quad ,$$

i.e. the equation

$$(\tilde{C}^j_{1i} - s\delta^j_i) H_j = 0$$

has at least one eigenvalue s that is non-zero. Now we replace A by

$$A' = A + \varepsilon H_1 \quad ,$$

where ε is chosen so small that the eigenvalues r of (12.14) which are non-zero are not decisively altered (i.e. none of the eigenvalues become zero). We also obtain the eigenvalue $\varepsilon \cdot s$ if we choose a suitable linear combination of H_i for $\hat{X}$ and thus the degree of degeneration of the eigenvalue zero becomes smaller. This is in contrast to the assumption that A has already been chosen so that (12.14) has the greatest number of *different* solutions. Any operator $\hat{A}$ that commutes with all $\hat{H}_i$ [as in (12.15)] can be expressed as a linear combination of the operators $\hat{H}_i$, i.e.

$$\hat{A} = a^i \hat{H}_i \quad . \tag{12.17}$$

To study the properties of the roots α, we calculate

$$\begin{aligned}
[\hat{A}, [\hat{H}_i, \hat{E}_\alpha]_-]_- &= [\hat{A}, \hat{H}_i \hat{E}_\alpha]_- - [\hat{A}, \hat{E}_\alpha, \hat{H}_i] \\
&= [\hat{A}, \hat{H}_i]_- \hat{E}_\alpha + \hat{H}_i [\hat{A}, \hat{E}_\alpha]_- - [\hat{A}, \hat{E}_\alpha]_- \hat{H}_i - \hat{E}_\alpha [\hat{A}, \hat{H}_i]_- \\
&= \alpha [\hat{H}_i, \hat{E}_\alpha]_- \quad .
\end{aligned} \tag{12.18}$$

Here we have taken into account (12.15) and (12.16), though (12.18) can be obtained more quickly and in a more elegant way using the Jacobi identity.

$$\begin{aligned}
[\hat{A}, [\hat{H}_i, \hat{E}_\alpha]_-]_- &= -[\hat{H}_i, [\hat{E}_\alpha, \hat{A}]_-]_- - [\hat{E}_\alpha, [\hat{A}, \hat{H}_i]_-]_- \\
&= \alpha [\hat{H}_i, \hat{E}_\alpha]_- \quad ,
\end{aligned}$$

once again using (12.15) and (12.16) in the last step. If $\hat{E}_\alpha$ is an eigenvector with eigenvalue α, then, because of (12.18), there exist l eigenvectors $[\hat{H}_i, \hat{E}_\alpha]_-$ with the same eigenvalue. But since α is not degenerate, all $[\hat{H}_i, \hat{E}_\alpha]_-$ must be proportional to $\hat{E}_\alpha$, i.e.

$$[\hat{H}_i, \hat{E}_\alpha]_- = \alpha_i \hat{E}_\alpha \quad , \quad \text{or} \tag{12.19}$$

$$C^\sigma_{i\alpha} = \alpha_i \delta^\sigma_\alpha \quad . \tag{12.20}$$

In other words the $\hat{H}_i$ and $\hat{E}_\alpha$ span the Lie algebra. The structure constants defined in (12.20) are denoted by $C^\sigma_{i\alpha}$ and have a very simple form because they are diagonal in α and σ. From (12.16), (12.17) and (12.18) we see

$$\alpha = a^i \alpha_i \quad , \quad (i = 1, 2, \ldots, l) \quad . \tag{12.21}$$

Here the α_i can be considered as the covariant components of a vector α in an l-dimensional space. This vector is called the *root vector*.

The Jacobi identity

$$[\hat{A}, [\hat{E}_\alpha, \hat{E}_\beta]_-]_- + [\hat{E}_\alpha, [\hat{E}_\beta, \hat{A}]_-]_- + [\hat{E}_\beta, [\hat{A}, \hat{E}_\alpha]_-]_- = 0 \tag{12.22}$$

in connection with (12.16) yields

$$[\hat{A}, [\hat{E}_\alpha, \hat{E}_\beta]_-]_- = (\alpha + \beta)[\hat{E}_\alpha, \hat{E}_\beta]_- \quad , \tag{12.23}$$

i.e. $[\hat{E}_\alpha, \hat{E}_\beta]_-$ is an eigenvector with eigenvalue $\alpha + \beta$, if $\alpha + \beta \neq 0$. Hence

$$[\hat{E}_\alpha, \hat{E}_\beta]_- = C^\sigma_{\alpha\beta} \delta^{\alpha+\beta}_\sigma \hat{E}_{\alpha+\beta} \quad .$$

If $\alpha + \beta = 0$, then $[\hat{E}_\alpha, \hat{E}_\beta]_-$ is a linear combination of the generators $\hat{H}_i$,

$$[\hat{E}_\alpha, \hat{E}_{-\alpha}]_- = C^i_{\alpha,-\alpha} \hat{H}_i \quad , \tag{12.24}$$

where $C^\sigma_{\alpha\beta} = 0$ in the case $\sigma \neq \alpha + \beta$. If $\alpha + \beta$ is a nonvanishing root, then

$$[\hat{E}_\alpha, \hat{E}_\beta]_- = N_{\alpha\beta} \hat{E}_{\alpha+\beta} \quad , \quad \text{i.e. } C^{\alpha+\beta}_{\alpha\beta} = N_{\alpha\beta} \tag{12.25}$$

holds. Next we construct the metric tensor

$$g_{\alpha\sigma} = C^\mu_{\alpha\nu} C^\nu_{\sigma\mu} \quad . \tag{12.26}$$

The summations over μ and ν are performed with respect to the restrictions (12.19), (12.20) and (12.24),

$$g_{\alpha\sigma} = C^\alpha_{\alpha\nu} C^\nu_{\sigma\alpha} + C^\mu_{\alpha,-\alpha} C^{-\alpha}_{\sigma\mu} + \sum_{\beta \neq -\alpha} C^{\alpha+\beta}_{\alpha\beta} C^\beta_{\sigma,\alpha+\beta} \quad .$$

α as well as σ is a fixed index, i.e. although α appears twice, one must not sum over it. In this equation only terms with $\sigma = -\alpha$ can exist [cf. (12.19) and (12.24)], i.e.

$$g_{\alpha\sigma} = 0 \quad , \quad \text{if } \sigma \neq -\alpha \quad . \tag{12.27}$$

If $-\alpha$ does not solve (12.14), then $\det|g_{\alpha\sigma}| = 0$, which means that Cartan's condition for a semi-simple Lie algebra is not satisfied. We thus conclude that for each nonvanishing root α of a semi-simple Lie algebra, $-\alpha$ is also a root.

By normalizing the $\hat{E}_\alpha$ in such a way that $g_{\alpha-\alpha} = 1$ and by reordering the basis elements, we obtain

$$g_{\mu\nu} = \left(\begin{array}{c|cccccc} g_{ik} & & & 0 & & & \\ \hline & 0 & 1 & & & & \\ & 1 & 0 & & & & \\ & & & \ddots & 0 & & \\ 0 & & & & & & \\ & & 0 & & & 0 & 1 \\ & & & & & 1 & 0 \end{array}\right)$$

From $\det|g_{\mu\nu}| \neq 0$ follows $\det|g_{ik}| \neq 0$ and we obtain, by using (12.19) and (12.20),

$$g_{ik} = \sum_\alpha C^\alpha_{i\alpha} C^\alpha_{k\alpha} = \sum_\alpha \alpha_i \alpha_k \quad . \tag{12.28}$$

This tensor g_{ik} can be considered as a metrical tensor of the l-dimensional space, which is spanned by the vectors α [cf. (12.21)]. It is only left to determine the coefficients in (12.24).

Since $C_{ikl} \equiv g_{lj} C^j_{ik}$ the indices may be cyclically permutated, i.e. $C_{ikl} = C_{kli} = C_{lik}$ as shown in Exercise 12.1, we find

$$C^i_{\alpha-\alpha} = g^{ik} C_{\alpha-\alpha,k} = g^{ik} C_{k,\alpha-\alpha} = g^{ik} C^\alpha_{k\alpha} = g^{ik}\alpha_k = \alpha^i \quad . \tag{12.29}$$

Hence

$$[\hat{E}_\alpha, \hat{E}_{-\alpha}]_- = \alpha^i \hat{H}_i \quad . \tag{12.30}$$

Here the α^i are the contravariant components of the vector α. Now we can write down the so-called *Cartan-Weyl basis* of the commutator relations of a semi-simple Lie algebra:

$$[\hat{H}_i, \hat{H}_k]_- = 0 \quad , \quad (i = 1, 2, \dots, l) \tag{12.31}$$

$$[\hat{H}_i, \hat{E}_\alpha]_- = \alpha_i \hat{E}_\alpha \quad , \tag{12.32}$$

$$[\hat{E}_\alpha, \hat{E}_\beta]_- = N_{\alpha\beta}\hat{E}_{\alpha+\beta} \quad , \quad (\alpha + \beta \neq 0) \tag{12.33}$$

$$[\hat{E}_\alpha, \hat{E}_{-\alpha}]_- = \alpha^i \hat{H}_i \quad . \tag{12.34}$$

Equation (12.31) may be used to define a commutative subalgebra, also called *Cartan subalgebra*, from the elements of the semi-simple Lie algebra. It represents the maximal abelian subalgebra of a given Lie algebra.

EXERCISE

12.2 Proof of a Relation for the Structure Constants C_{ikl}

Problem. Show that $C_{ikl} = C_{kli} = C_{lik}$.

Solution. By definition we have

$$C_{ikl} = g_{lj} C^{j}_{ik} = C^{n}_{lm} C^{m}_{jn} C^{j}_{ik} \tag{1}$$

so that, using the Jacobi-identity, we may write

$$C_{ikl} = - C^{n}_{lm} C^{m}_{ji} C^{j}_{kn} - C^{n}_{lm} C^{m}_{jk} C^{j}_{ni} = C^{m}_{ij} C^{j}_{kn} C^{n}_{lm} - C^{j}_{in} C^{m}_{kj} C^{n}_{lm} \quad . \tag{2}$$

We see immediately by suitably renaming the indices j, m, and n that it remains unchanged under a cyclic permutation, e.g.

$$C^{m}_{kj} C^{j}_{ln} C^{n}_{im} = C^{j}_{kn} C^{n}_{lm} C^{m}_{ij} \quad . \tag{3}$$

12.3 Scalar Products of Eigenvalues

In (12.21) and (12.29) we have seen that the eigenvalues α may be considered as vectors in an l-dimensional space. Now we define the scalar product of two root vectors as

$$(\alpha, \beta) = \alpha^{i} \beta_{i} \quad . \tag{12.35}$$

For these scalar products the following **Lemma 1** holds:
If α and β are roots of (12.14), then $2(\alpha, \beta)/(\alpha, \alpha)$ is an integer and $\{\beta - [2\alpha(\alpha, \beta)/(\alpha, \alpha)]\}$ is also a root.

Proof: We suppose that α and β are roots of (12.14) and choose a third root γ in such a way that $\alpha + \gamma$ does not solve (12.14). From (12.33) then follows

$$[\hat{E}_{-\alpha}, \hat{E}_{\gamma}]_{-} = N_{\alpha\gamma} \hat{E}_{\gamma-\alpha} \quad .$$

Since we do not care about normalization, we denote the rhs by $\hat{E}'_{\gamma-\alpha}$. Also adopting this definition in the following, we get

$$[\hat{E}_{-\alpha}, \hat{E}_{\gamma}]_{-} = \hat{E}_{\gamma-\alpha} \quad , \quad [\hat{E}_{-\alpha}, \hat{E}'_{\gamma-\alpha}]_{-} = \hat{E}'_{\gamma-2\alpha} \quad ,$$

and finally

$$[\hat{E}_{-\alpha}, \hat{E}'_{\gamma-j\alpha}]_{-} = \hat{E}'_{\gamma-(j+1)\alpha} \quad . \tag{12.36}$$

This series of equations has to terminate after n steps, because there only exist a finite number of $\hat{E}_\beta$, that is

$$[\hat{E}_{-\alpha}, \hat{E}'_{\gamma-n\alpha}]_- = \hat{E}'_{\gamma-(n+1)\alpha} = 0 \quad . \tag{12.37}$$

From (12.25) an analogous relation is also derived, i.e.

$$[\hat{E}_\alpha, \hat{E}'_{\gamma-(j+1)\alpha}]_- = \mu_{j+1}\hat{E}'_{\gamma-j\alpha} \quad . \tag{12.38}$$

By eliminating $\hat{E}'_{\gamma-(j+1)\alpha}$ from (12.36) and (12.38) and using the Jacobi identity, we deduce that

$$\begin{aligned}\mu_{j+1}\hat{E}'_{\gamma-j\alpha} &= [\hat{E}_\alpha,[\hat{E}_{-\alpha}, \hat{E}'_{\gamma-j\alpha}]_-]_- \\ &= [\hat{E}'_{\gamma-j\alpha}, [\hat{E}_\alpha, \hat{E}_{-\alpha}]_-]_- - [\hat{E}_{-\alpha}, [\hat{E}'_{\gamma-j\alpha}, \hat{E}_\alpha]_-]_-\end{aligned}$$

[due to (12.31 − 34)] and because of (12.38),

$$= -[\hat{E}'_{\gamma-j\alpha}, \alpha^i\hat{H}_i]_- + \mu_j[\hat{E}_{-\alpha}, \hat{E}'_{\gamma-(j-1)\alpha}]_- \quad ,$$

which with (12.36)

$$= \alpha^i[\hat{H}_i, \hat{E}'_{\gamma-j\alpha}]_- + \mu_j\hat{E}'_{\gamma-j\alpha} \quad ,$$

and finally using (12.32)

$$= \alpha_i(\gamma_i - j\alpha_i)\hat{E}'_{\gamma-j\alpha} + \mu_j\hat{E}'_{\gamma-j\alpha}$$

The $\hat{E}'_\beta$ are linearly independent (since the $\hat{E}_\beta$ are) and the last equation therefore yields the recursion relation

$$\mu_{j+1} = (\alpha,\gamma) - j(\alpha,\alpha) + \mu_j \quad . \tag{12.39}$$

μ_0 is not defined originally by (12.36), so that choosing $\mu_0 = 0$, we apply (12.31) for all values $j \geq 0$; therefore we obtain by recursion from (12.39)

$$\mu_j = j(\alpha,\gamma) - \tfrac{1}{2}j(j-1)(\alpha,\alpha) \quad . \tag{12.40}$$

But now (12.37) shows that $\mu_{n+1} = 0$; hence

$$(\alpha,\gamma) = \tfrac{1}{2}n(\alpha,\alpha) \quad , \quad \text{or} \tag{12.41}$$

$$\mu_j = \frac{j}{2}(n-j+1)(\alpha,\alpha)$$

If β is any root, there exists a certain $j \geqslant 0$ with the property that $\gamma = \beta + j\alpha$ is also a root, but not $\gamma + \alpha$. Inserting this into (12.41) yields

$$(\alpha,\beta) = \tfrac{1}{2}(n-2j)(\alpha,\alpha) \quad , \quad \text{or}$$

$$2\frac{(\alpha,\beta)}{(\alpha,\alpha)} = n - 2j \quad . \tag{12.42}$$

This equation proves the first part of our lemma, *because the rhs is certainly an integer.*

(α,α) must not be zero, because otherwise (12.41) would imply that α is orthogonal to all roots. Since these roots make up the whole l-dimensional space, this would be in contradiction to Cartan's criterion for a semi-simple algebra. Thus we can always divide by (α, α) and (12.41) also yields

$$n = 2\frac{(\alpha, \gamma)}{(\alpha, \alpha)} \quad .$$

To each pair α and γ for which $\alpha + \gamma$ is not a solution a series of roots corresponds

$$\gamma, \gamma - a, \ldots, \gamma - n\alpha \quad , \tag{12.43}$$

which is invariant under reflection at the hyperplane. This series also touches the origin and is orthogonal to the vector α. Since each β must be a member of one of these series, we can say that $\beta = \gamma - m\alpha$. $\delta = \gamma - (n - m)\alpha$ is also a root because of reflection symmetry, i.e. $\alpha = \beta - (n - 2m)\alpha$. Therefore (12.41) implies that

$$\delta = \beta - 2\alpha\frac{(\alpha, \beta)}{(\alpha, \alpha)} \tag{12.44}$$

is also a root, which proves the second part of Lemma 1.

It is almost trivial to prove **Lemma 2**. If α is a root vector, then α, 0 and $-\alpha$ are the only integer multiples (denoted by k) of α that are also root vectors.

Proof: From $[\hat{E}_\alpha, \hat{E}_\alpha]_- = 0$ and (12.33) we directly see that 2α cannot be a root. But each value $|k| > 1$ gives rise to a root series, which must contain 2α. Therefore $|k| > 1$ is impossible and Lemma 2 has been proved.

Lemma 3 is also important. A root series based on α which contains another root β consists of not more than four roots, which satisfy

$$2\frac{(\alpha, \beta)}{(\alpha, \alpha)} = 0, \ \pm 1, \ \pm 2, \ \pm 3 \quad . \tag{12.45}$$

Proof: Since the case $\beta = \pm\alpha$ was considered in Lemma 2, we can assume that $\beta \neq \pm\alpha$. We further assume that at least five roots exist, which we can denote by $\beta - 2\alpha$, $\beta - \alpha$, β, $\beta + \alpha$, $\beta + 2\alpha$.

Lemma 2 shows, however, that 2α and $2(\alpha + \beta)$ are not roots. On the other hand we have

$$2\alpha = (\beta + 2\alpha) - \beta \quad \text{and} \quad 2(\alpha + \beta) = (\beta + 2\alpha) + \beta \quad .$$

Therefore the β-root series, which contains $\beta + 2\alpha$, only consists of a single member, i.e. $\beta + 2\alpha$. Hence $(\beta + 2\alpha, \beta) = 0$. Similarly neither $\beta - 2\alpha - \beta$ nor $\beta - 2\alpha + \beta$ are roots, so that $(\beta - 2\alpha, \beta) = 0$ holds. Addition of both these equations yields $(\beta, \beta) = 0$, which is only possible for $\beta = 0$. Zero roots, however, are excluded from our considerations and therefore at most four roots are left.

We still have to prove (12.43). To that end we write down a root series analogous to (12.41). On one hand we have

$$k + j + 1 \leq 4 \tag{12.46}$$

and on the other, according to (12.40),

$$2\frac{(\alpha,\beta)}{(\alpha,\alpha)} = k - j \qquad (12.47)$$

if we set $n = k + j$. But (12.46) yields $k, j \leq 3$ and so (12.47) proves the assertion.

12.4 Cartan-Weyl Normalization

Before we consider the graphical representation of the root vectors, we make some remarks about the Cartan and Weyl normalization. The constants $N_{\alpha\beta}$ appearing in (12.33) have not yet been determined. Combining (12.36) and (12.38) and setting $\beta = \gamma - j\alpha$, we obtain

$$\begin{aligned}\mu_j \hat{E}_{\alpha+\beta} &= [\hat{E}_\alpha, [\hat{E}_{-\alpha}, \hat{E}_{\alpha+\beta}]_-]_- = N_{-\alpha,\alpha+\beta}[\hat{E}_\alpha, \hat{E}_\beta]_- \\ &= N_{-\alpha,\alpha+\beta} N_{\alpha\beta} \hat{E}_{\alpha+\beta} \quad ,\end{aligned} \qquad (12.48)$$

using (12.33) twice. Since μ_j is known from (12.40) and (12.41), then

$$N_{\alpha\beta} N_{-\alpha,\alpha+\beta} = \mu_j = \frac{j}{2}(n - j + 1)(\alpha,\alpha) \quad ,$$

and writing $n = k + j$, as above, we arrive at

$$N_{\alpha\beta} N_{-\alpha,\alpha+\beta} = \frac{j}{2}(k + 1)(\alpha,\alpha) \quad . \qquad (12.49)$$

The series of roots

$$\beta + j\alpha \quad , \quad \beta + (j-1)\alpha, \ldots, \beta, \ldots, \beta - k\alpha \qquad (12.50)$$

belongs to (12.49), for which the constants $N_{\alpha\beta}$ are determined only up to a phase. However, we reach a consistent choice of phase if we take into account the antisymmetry of the $N_{\alpha\beta}$. For that purpose we have to require the validity of the relations

$$N_{\alpha\beta} = -N_{\beta\alpha} = -N_{-\alpha,-\beta} = N_{-\beta,-\alpha} \quad .$$

12.5 Graphic Representation of the Root Vectors

In view of (12.21),

$$\alpha = \lambda^i \alpha_i (i = 1, \ldots, l)$$

can be considered as a root vector in an l-dimensional space with l covariant components α_i. If we draw the vectors, beginning at a certain point, an l-dimensional diagram arises. Van der Waerden showed that exactly one system of root vectors corresponds to each diagram and he used this to give a complete classification of the simple Lie algebras. His method is based upon the three lemmata which we have proved above. Here we once more repeat their relevant statements:

1. If α is a root vector, then $-\alpha$ is also a root.
2. If α and β are root vectors, then $2(\alpha,\beta)/(\alpha,\alpha)$ is an integer number.
3. If α and β are root vectors, then $\beta - [2\alpha(\alpha,\beta)/(\alpha,\alpha)]$ is also a root vector.

With help of the scalar product, defined in (12.35), we are able to introduce the angle ϕ between the root vectors α and β, where

$$\cos\phi = \frac{(\alpha,\beta)}{\sqrt{(\alpha,\alpha)(\beta,\beta)}} \quad . \tag{12.51}$$

With (12.19) and Lemma 2 this yields

$$\cos^2\phi = 0 \quad , \quad \tfrac{1}{4} \quad , \quad \tfrac{1}{2} \quad , \quad \tfrac{3}{4} \quad , \quad \text{or } 1 \quad . \tag{12.52}$$

Because of Lemma 1 it is sufficient to consider only positive angles, i.e. the angles

$$\phi = 0°, 30°, 45°, 60°, \text{ and } 90° \quad .$$

Now we connect these angles with the ratios of the scalar products:

1. $\phi = 0°$. This angle only appears in the case $\alpha = \beta$.
2. $\phi = 30°$. Then we have $(\alpha,\beta)/(\alpha,\alpha) = \frac{1}{2}$ or $\frac{3}{2}$ and $(\alpha,\beta)/(\beta,\beta) = \frac{3}{2}$ or $\frac{1}{2}$. Therefore we get $(\beta,\beta)/(\alpha,\alpha) = \frac{1}{3}$ or 3.
3. $\phi = 45°$. Then we have $(\alpha,\beta)/(\alpha,\alpha) = \frac{1}{2}$ or 1 and correspondingly $(\alpha,\beta)/(\beta,\beta) = 1$ or $\frac{1}{2}$. Therefore $(\beta,\beta)/(\alpha,\alpha) = \frac{1}{2}$ or 2.
4. $\phi = 60°$. Here $(\alpha,\beta)/(\alpha,\alpha) = \frac{1}{2}$ and $(\alpha,\beta)/(\beta,\beta) = \frac{1}{2}$. We obtain $(\beta,\beta) = (\alpha,\alpha)$.
5. $\phi = 90°$. $(\alpha,\beta) = 0$ and $(\beta,\beta)/(\alpha,\alpha)$ is not determined.

It is useful to introduce the ratio of the lengths of the vectors α and β,

$$k = \sqrt{\frac{(\alpha,\alpha)}{(\beta,\beta)}} \quad .$$

If β is the shorter vector, the following picture is obtained from the statements made above:

$$\begin{aligned} &\phi = 30°: \quad k^2 = 3 \qquad \phi = 60°: \quad k^2 = 1 \\ &\phi = 45°: \quad k^2 = 2 \qquad \phi = 90°: \quad k^2 \text{ not determined} \end{aligned} \tag{12.53}$$

Now we are able to construct vector diagrams for all simple Lie algebras.

12.6 Lie Algebra of Rank 1

In the case $l = 1$, (12.31–34) and Lemma 1 yield only two nonvanishing roots, $\pm\alpha$. Therefore we obtain $\phi = 0°$ and the diagram becomes quite simple, which is shown in Fig. 12.1. There is only one Lie algebra of rank one, which is the SU(2) [or the SO(3), which is isomorphic to SU(2)] and is commonly called A_1.

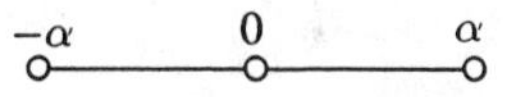

Fig. 12.1. Diagram of the Lie Algebra A_1

12.7 Lie Algebras of Rank 2

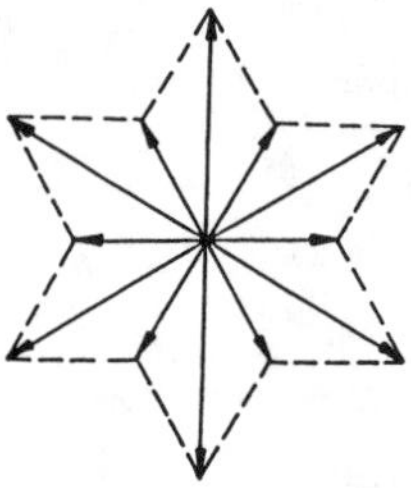

Fig. 12.2. Vector diagram of the Lie algebra G_2

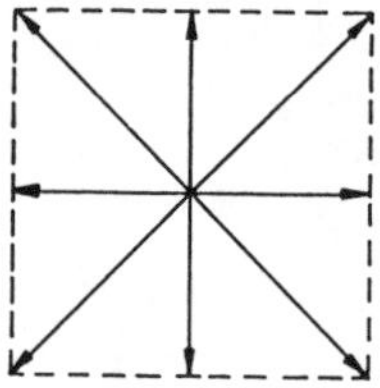

Fig. 12.3. Vector diagram of the Lie algebra B_2

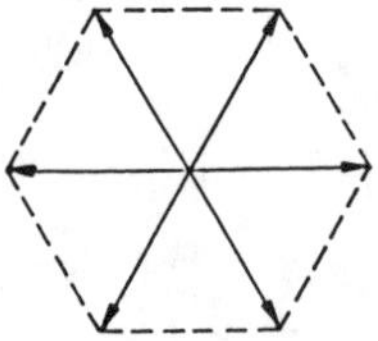

Fig. 12.4. Vector diagram of the Lie algebra A_2

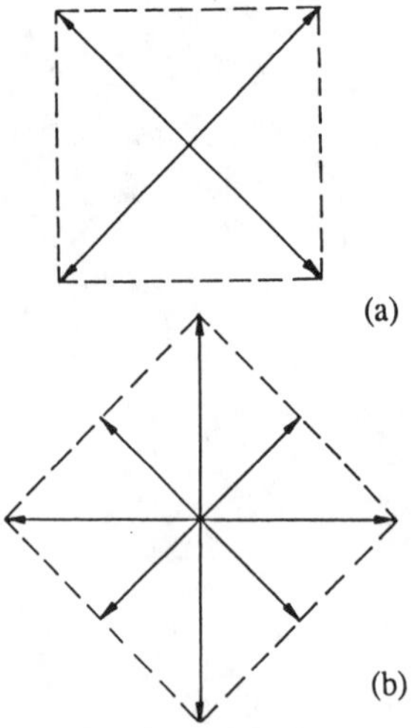

Fig. 12.5a. Vector diagram of the Lie algebra D_2. b Vector diagram of the Lie algebra C_2

The various diagrams of second rank Lie algebras are two-dimensional. We describe them separately according to the angle ϕ.

$\phi = 30°$:

The system of coordinates is chosen such that the vector α given by $(1,0)$ is shorter than the vector β. From (12.53) we have $(\beta, \beta) = 3$, and because the angle between α and β is 30°, the coordinates of β are $[3/2, \sqrt{3}/2]$. From (12.51) it follows that the vectors $-\alpha$ and $-\beta$ belong to the diagram, too. Since $(\alpha, \beta)/(\beta, \beta) = 1/2$, we see from Lemma 3 that $\alpha - \beta$ is a vector of the diagram with the coordinates $[-1/2, -\sqrt{3}/2]$. Thus it follows from Lemma 1 that $\beta - \alpha$, with the coordinates $[1/2, \sqrt{3}/2]$, also belongs to the diagram.

By successive application of this method we can find all 12 vectors which are nonzero. (Of course, for $l = 2$ there are two null vectors which belong to the Lie algebra). The diagram of this Lie algebra, labelled G_2 by Cartan, is represented in Fig. 12.2. Note that the long vectors pointing to the outer edges are each given by the sum of their two neighbouring vectors.

$\phi = 45°$:

Proceeding in the same way as above we obtain Fig. 12.3. This represents the B_2 Lie algebra according to Cartan's notation. It contains 10 vectors including the two null vectors and describes the Lie algebra of the group SO(5).

$\phi = 60°$:

In this case one obtains a hexagonal vector diagram that belongs to the A_2 Lie algebra.

This Lie algebra represents the group SU(3) and contains 8 vectors, including the null vectors.

$\phi = 90°$:

There are two different diagrams in this case (shown in Figs. 12.5a and b).

The vector diagram D_2 contains 6 vectors and can be separated into two groups of orthogonal vectors. Therefore it represents the Lie algebra SO(4), which is isomorphic to the direct sum of two SO(3) algebras. The second diagram belongs to the Lie algebra denoted as C_2 by Cartan. It results from a rotation of 45° of B_2, thus being isomorphic to it, and represents the algebra of the generators of the simplectic group Sp(4) in 4 dimensions.

12.8 Lie Algebras of Rank $l > 2$

The generalization of the vector diagrams for all possible simple Lie algebras of higher rank was given by Van der Waerden.

B$_l$: We introduce the vectors

$$e_1 = (1,0) \quad \text{and} \quad e_2 = (0,1)$$

and construct the vectors $\pm e_1$, $\pm e_2$ and $\pm e_1 \pm e_2$ with all possible combinations of the sign. Then we have

$$(\pm 1, 0), (0, \pm 1), (1, \pm 1) \text{ and } (-1, \pm 1)$$

in the coordinate representation. There are 8 vectors, which together with the two null vectors form the diagram B_2. For B_3 we consider the orthogonal vectors

$$e_1 = (1,0,0), \quad e_2 = (0,1,0) \quad \text{and} \quad e_3 = (0,0,1) \quad .$$

We build the vectors $\pm e_i$ and $\pm e_i \pm e_j$ and obtain 18 vectors, which represent the 21 roots of B_3 including the three null vectors. For the general diagram B_l we consider l orthogonal unit vectors and construct the vectors

$$\pm e_i \quad \text{and} \quad \pm e_1 \pm e_j \quad (i,j = 1, \dots, l) \tag{12.54}$$

of an l-dimensional space, which yields $2l^2$ vectors. Including the l null-vectors we obtain the vector diagram which represents the Lie algebra of rank $l(2l + 1)$ and belongs to the group SO$(2l + 1)$.

C$_l$: For C_l we can use the same unit vectors as in the case B_l, but we need to construct

$$\pm 2e_i \quad \text{and} \quad \pm e_i \pm e_j \quad (i,j = 1, \dots, l) \tag{12.55}$$

as vectors of the diagram. Evidently C_l is of the same order as B_l, but it belongs to the Lie algebra of the simplest group Sp$(2l)$.

D$_l$: For $l > 2$ we again use the same unit vectors and consider the vectors

$$\pm e_i \pm e_j \quad (i,j = 1, \dots, l) \quad . \tag{12.56}$$

$2l(l - 1)$ vectors of this kind exist, thus the algebra is of order $l(2l - 1)$ and therefore D_l represents the Lie algebra of the group SO$(2l)$.

A$_l$: We choose $l + 1$ orthogonal unit vectors of an $(l + 1)$-dimensional space and build all vectors of the form

$$e_i - e_j \quad (i,j = 1, \dots, l+1) \quad . \tag{12.57}$$

Now we project the vectors onto a suitable l-dimensional subspace, yielding $l(l + 1)$ vectors. Together with the l null vectors they correspond to a Lie algebra of order $l(l + 2)$, belonging to the group SU$(l + 1)$.

12.9 The Exceptional Lie Algebras

The four series of vector diagrams A_l, B_l, C_l and D_l correspond to the four *classical Lie algebras* of the groups SU$(l + 1)$, SO$(2l + 1)$, Sp$(2l)$ and SO$(2l)$. As

Van der Waerden showed, there are only five other diagrams which belong to the *exceptional Lie algebras* and there are G_2, F_4, E_6, E_7 and E_8 (in Cartan's notation).

G_2: This diagram has already been discussed.

F_4: For this diagram we add the 16 vectors

$$\tfrac{1}{2}(\pm e_1 \pm e_2 \pm e_3 \pm e_4) \tag{12.58}$$

to the vectors of B_4. These 48 vectors, together with the 4 null vectors, represent 52 roots and the algebra of F_4 is therefore of rank 52. Obviously B_4 is a subalgebra of F_4.

E_6: We add the following vectors to those of A_5:

$$\pm 2e_7 \quad \text{and} \quad \tfrac{1}{2}(\pm e_1 \pm e_2 \pm e_3 \pm e_4 \pm e_5 \pm e_6) \pm \frac{e_7}{\sqrt{2}} \quad , \tag{12.59}$$

where in the parenthesis we choose three positive and three negative signs. Then we have 72 nonvanishing roots, and the algebra is of order 78. Clearly the Lie algebra E_6 contains the algebra of the group $SU(6) \otimes SU(5)$ as a subalgebra.

E_7: Together with the vectors of A_7, we consider

$$\tfrac{1}{2}(\pm e_1 \pm e_2 \pm e_3 \pm e_4 \pm e_5 \pm e_6 \pm e_7 \pm e_8) \quad , \tag{12.60}$$

where we choose four positive and four negative signs and obtain 126 non-zero vectors belonging to the algebra E_7 of order 133. Evidently the algebra of $SU(8)$ is a subalgebra of E_7.

E_8: For this diagram we take the vectors of D_8 and add

$$\tfrac{1}{2}(\pm e_1 \pm e_2 \pm e_3 \pm e_4 \pm e_5 \pm e_6 \pm e_7 \pm e_8) \quad , \tag{12.61}$$

where we choose an even number of positive signs. Therefore the algebra E_8 is of order 248 and contains the algebra of the group $SO(16)$ as a subalgebra.

12.10 Simple Roots and Dynkin Diagrams

Van der Waerden's method for the construction of the vector diagrams is suitable only for groups of order $l \leq 2$, because for $l > 2$ a two-dimensional representation is not possible. Dynkin showed that most of the information about the roots of a semi-simple Lie algebra are contained in a small part σ of the entire set of root vectors Σ. These special root vectors are called *simple roots.* Dynkin also showed that the simple roots can be represented in two l-dimensional diagrams, the *Dynkin diagrams.* From these diagrams one can easily obtain the complete set of root vectors, including their lengths and the angles between them.

We call a root α a *positive root* if in some given basis the first non-zero coordinate is positive (of course, this depends on the basis choice). For example, we consider all non-zero roots in the diagram B_2, for which there are 8 vectors

$$(1,0);\ (1,1);\ (1,-1);\ (0,1);\ (0,-1);\ (-1,0);\ (-1,1);\ (-1,-1) \quad .$$

The first four of this set are positive root vectors. Generally speaking, one half of all non-zero vectors of a diagram are positive.

We call a root *simple* if it is positive and cannot be represented as the sum of two positive roots. For example we have

$$(1,0) = (1,-1) + (0,1) \quad \text{or} \quad (1,1) = (1,0) + (0,1) \quad ,$$

and therefore, (1, 0) and (1, 1) are not simple roots. A corresponding decomposition of (0, 1) and (1, − 1) is not possible, implying that the simple roots of B_2 are

$$\alpha = (0,1) \quad \text{and} \quad \beta = (1,-1) \quad . \tag{12.62}$$

Obviously, all simple roots are linearly independent and we denote the system of all simple roots by σ. We can represent every positive root as

$$\gamma = \sum_{\alpha_i \in \sigma} k_i \alpha_i \quad , \tag{12.63}$$

where the k_i are non-negative integers. A semi-simple Lie algebra of rank l has exactly l simple roots, which form a basis of the l-dimensional space of the root vectors.

Because the connection with the lemmata given above is evident, we give the three following lemmata without proof:

Lemma 4. If α and β are simple roots, then their difference is not a simple root, i.e.

$$\text{if} \quad \alpha, \beta \in \sigma \quad , \quad \text{then} \quad \alpha - \beta \notin \sigma \quad .$$

Lemma 5. If $\alpha, \beta \in \sigma$, then

$$2\frac{(\alpha,\beta)}{(\alpha,\alpha)} = -p \quad , \tag{12.64}$$

where p is a positive integer.

Lemma 6. If $\alpha, \beta \in \sigma$, the angle $\phi_{\alpha\beta}$ between them is either 90°, 120°, 135° or 150°. If $(\alpha,\alpha) \leq (\beta,\beta)$, then

$$\frac{(\beta,\beta)}{(\alpha,\alpha)} = \begin{cases} 1 & \text{for } \phi_{\alpha\beta} = 120° \quad , \\ 2 & \text{for } \phi_{\alpha\beta} = 135° \quad , \\ 3 & \text{for } \phi_{\alpha\beta} = 150° \quad , \\ \text{not determined} & \text{for } \phi_{\alpha\beta} = \ \ 90° \quad . \end{cases} \tag{12.65}$$

For illustration we consider B_2 and B_3. Due to (12.51), for B_2 the angle between the simple roots α and β given in (12.62) is found to be

$$\cos\phi_{\alpha\beta} = \frac{(\alpha, \beta)}{\sqrt{(\alpha, \alpha)(\beta, \beta)}} = -\sqrt{\tfrac{1}{2}} \quad ,$$

i.e.

$$\phi_{\alpha\beta} = 135^\circ \quad , \tag{12.66}$$

and due to (12.65) the relation between the lengths is

$$\frac{(\beta, \beta)}{(\alpha, \alpha)} = 2 \quad . \tag{12.67}$$

For B_3 there are three simple roots

$$\alpha = (0, 0, 1) \quad , \quad \beta = (0, 1, -1) \quad , \quad \gamma = (1, -1, 0) \quad , \tag{12.68}$$

which obey the relations

$$\begin{aligned}
&\cos\phi_{\alpha\beta} = -\sqrt{\tfrac{1}{2}} \quad , && \text{i.e. } \phi_{\alpha\beta} = 135^\circ \quad , && \frac{(\beta\beta)}{(\alpha\alpha)} = 2 \\
&\cos\phi_{\alpha\beta} = 0 \quad , && \text{i.e. } \phi_{\alpha\beta} = 90^\circ \quad , && \frac{(\gamma, \gamma)}{(\alpha, \alpha)} = 2 \\
&\cos\phi_{\alpha\beta} = -\tfrac{1}{2} \quad , && \text{i.e. } \phi_{\alpha\beta} = 120^\circ \quad , && \frac{(\beta, \beta)}{(\alpha, \alpha)} = 1 \quad .
\end{aligned} \tag{12.69}$$

A usual representation of the vectors (12.68) would be three-dimensional. Dynkin gave a prescription how to present the simple roots of every semi-simple Lie algebra in a two-dimensional space.

12.11 Dynkin's Prescription

Every simple root is represented by a small circle in the diagram. The circles are connected by one, two or three lines, according to whether the angle between the corresponding simple roots is 120°, 135° or 150°. Circles which belong to orthogonal roots remain unconnected. Circles corresponding to simple roots with the shortest length are filled, whereas circles corresponding to roots with the largest length remain unfilled. This prescription is unique because every simple Lie algebra contains simple roots with at most two different lengths.

EXAMPLE

12.3 Dynkin Diagrams for B_l

According to (12.66) and (12.67) the Dynkin diagram for B_2 is simply

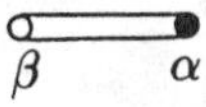

and due to (12.69) the Dynkin-diagram for B_3 is represented by

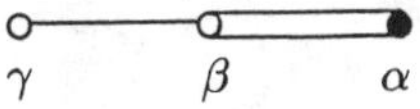

Order	Cartan's Notation	Group	Dynkin Diagram	Solutions
$l(l+2)$	A_l	$SU(l+1)$	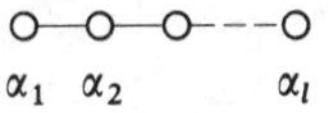	$e_i - e_j (i, j = 1, \ldots, l+1)$
$l(2l+1)$ $l \geq 2$	B_l	$SO(2l+1)$	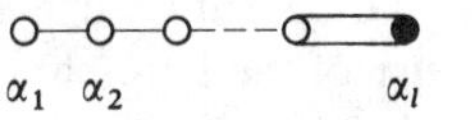	$\pm e_i$ and $\pm e_i \pm e_j$ $(i, j = 1, \ldots, l)$
$l(2l+1)$ $l \geq 3$	C_l	$Sp(2l)$		$\pm 2e_i$ and $\pm e_i \pm e_j$ $(i, j = 1, \ldots, l)$
$l(2l-1)$ $l \geq 4$	D_l	$SO(2l)$	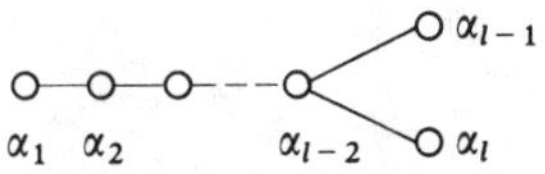	$\pm e_i \pm e_j$ $(i, j = 1, \ldots, l)$
14	G_2	G_2	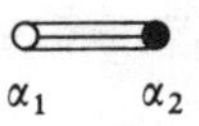	$e_i - e_j (i, j = 1, 2, 3; i \neq j)$ $\pm 2e_i \mp e_j \mp e_k (i, j, k = 1, 2, 3, i \neq j \neq k)$
52	F_4	F_4	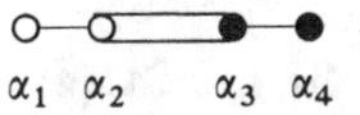	As for B_4 plus the 16 solutions $\frac{1}{2}(\pm e_1 \pm e_2 \pm e_3 \pm e_4)$
78	E_6	E_6	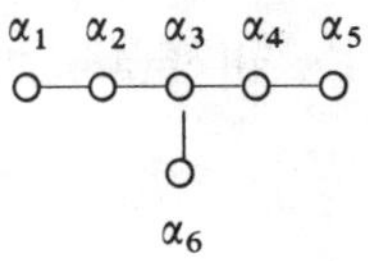	As for A_5 plus solutions $\pm\sqrt{2}e_7$ and $\frac{1}{2}(\pm e_1 \pm e_2 \pm e_3 \pm e_4 \pm e_5 \pm e_6) \pm e_7/\sqrt{2}$ (an arbitrary choice of 3 " + " and 3 " − " signs for the terms in parentheses)
133	E_7	E_7	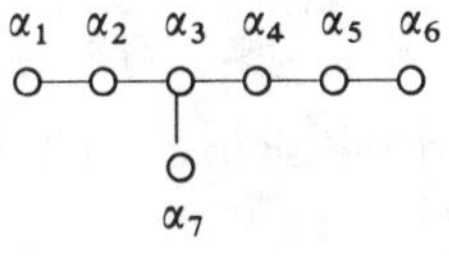	As for A_7 plus the solutions $\frac{1}{2}(\pm e_1 \pm e_2 \pm e_3 \pm e_4 \pm e_5 \pm e_6 \pm e_7 \pm e_8)$ (an arbitrary choice of 4 " + " and 4 " − " signs for the terms in parentheses)
248	E_8	E_8	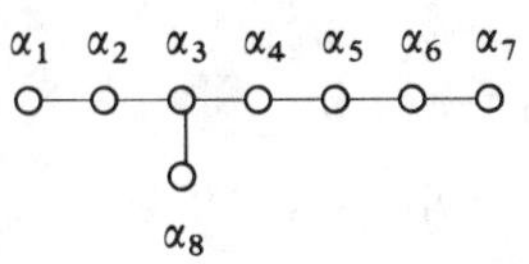	As for D_8 plus the solutions $\frac{1}{2}(\pm e_1 \pm e_2 \pm e_3 \pm e_4 \pm e_5 \pm e_6 \pm e_7 \pm e_8)$ with an even number of plus signs.

Example 12.3

Repeating this method we obtain the Dynkin diagram of every Lie algebra B_l, i.e.

Dynkin showed that to every sample Lie algebra there exists a unique representative diagram. These diagrams are given in the above table. Other diagrams are not possible.

12.12 The Cartan Matrix

The Cartan matrix is important for subsequent applications; hence we show the Cartan matrix (A_{ij}) with the elements given by

$$A_{ij} = 2(\alpha_i, \alpha_j)/(\alpha_i, \alpha_i) \tag{12.70}$$

for $\alpha_k \in \sigma$. For every Dynkin diagram we can easily calculate the corresponding Cartan matrix by making use of the Lemmata 4 to 6. From (12.70) the diagonal elements are always 2 and the non-diagonal terms are equal to -3, -2, -1, and 0.

EXAMPLE

12.4 The Cartan Matrices for SU(3), SU(4) and G_2

The Dynkin diagram for the Lie algebra of the group SU(3) is

From this we calculate that

$$(\alpha_1, \alpha_2) = (\alpha_2, \alpha_2) = 1 \quad \text{and} \quad (\alpha_1, \alpha_2) = -\tfrac{1}{2} \ ,$$

because the angle between the simple roots α_1 and α_2 is 120°. Therefore the Cartan matrix for SU(3) is

$$\begin{pmatrix} 2 & -1 \\ -1 & 2 \end{pmatrix} .$$

The Dynkin diagram for SU(4) is

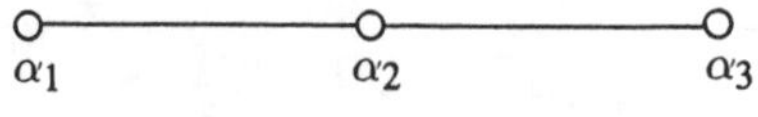

and thus

Example 12.4

$$(\alpha_1, \alpha_1) = (\alpha_2, \alpha_2) = (\alpha_3, \alpha_3) = 1, \quad (\alpha_1, \alpha_2) = (\alpha_2, \alpha_3) = -\tfrac{1}{2}$$

and $(\alpha_1, \alpha_3) = 0$. For SU(4) the Cartan matrix is

$$\begin{pmatrix} 2 & -1 & 0 \\ -1 & 2 & -1 \\ 0 & -1 & 2 \end{pmatrix} .$$

For G_2 we have the simple Dynkin diagram

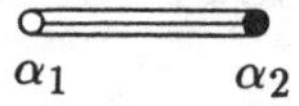

where $(\alpha_1, \alpha_1) = 3$, $(\alpha_2, \alpha_2) = 1$ and $(\alpha_1, \alpha_2) = -\frac{3}{2}$ because

$$\begin{aligned} (\alpha_1, \alpha_2) &= \sqrt{(\alpha_1, \alpha_1)(\alpha_2, \alpha_2)} \cos 150^\circ \\ &= -\sqrt{(\alpha_1, \alpha_1)(\alpha_2, \alpha_2)}\ \sqrt{\tfrac{3}{2}} \ . \end{aligned}$$

Therefore the corresponding Cartan matrix is given by

$$\begin{pmatrix} 2 & -1 \\ -3 & 2 \end{pmatrix} .$$

12.13 Determination of all Roots from the Simple Roots

We now show how to determine the complete Lie algebra from the system σ of the simple roots. For that purpose we search for a series $k_1, \dots, k_l$ of integers, such that $\sum_{\alpha_i \in \sigma} k_i \alpha_i$ is a root.

For practical applications it is sufficient to determine only the positive roots. If $\beta = \sum k_i \alpha_i$ is a root, then we call $|\beta| = \sum |k_i|$ the height of β. Obviously the height is a positive number and all simple roots have the height 1. To continue, we assume that we know the positive roots of height n. Then all the positive solutions of height $n+1$ have the form $\beta = \alpha + \alpha_j$, with $\alpha_j \in \sigma$. For a given positive root α of height n, we must therefore determine those $\alpha \in \sigma$ whose root is given by $\alpha + \alpha_j$. If $\alpha = \alpha_j$, then it follows that $\alpha + \alpha_j$ is no root and in that case we can assume that $\alpha = \sum k_i \alpha_i$, with some $k_i > 0$ $(i \neq j)$. These linear combinations $\alpha - \alpha_j, \alpha - 2\alpha_j, \dots$, being roots, are positive and are of height lower than n, so that we can deduce from our assumption which combinations yield roots. Thus the number s of the α_j-series which contains α,

$$\alpha - s\alpha_j, \dots, \alpha, \dots, \alpha + t\alpha_j \quad ,$$

is already known. For t the relation

$$t = s - 2(\alpha, \alpha_j)/(\alpha_j, \alpha_j) = s - \sum_{i=1}^{l} k_i A_{jl} \tag{12.71}$$

holds and it may be determined by using the Cartan matrix. Because $\alpha + \alpha_j$ is a root only if $t > 0$, we can easily determine whether a given $\alpha + \alpha_j$ is a root or not.

EXAMPLE

12.5 Determination of the Roots of G_2 Using the Corresponding Simple Roots

From Example 12.3 we know the Dynkin diagram

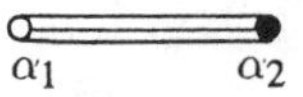

and the corresponding Cartan matrix

$$\begin{pmatrix} 2 & -1 \\ -3 & 2 \end{pmatrix} \tag{1}$$

for the Lie algebra G_2. Therefore we know from (1) that

$$2(\alpha_1, \alpha_2)/(\alpha_1, \alpha_1) = -1 \quad \text{and} \quad 2(\alpha_1, \alpha_2)/(\alpha_2, \alpha_2) = -3 \ . \tag{2}$$

Because $\alpha_1 - \alpha_2$ is not a root, it follows from these relations that the series α_1 containing α_2 and the series α_2 containing α_1 have the form

$$\alpha_2, \alpha_2 + \alpha_1 \quad ; \quad \alpha_1, \alpha_1 + \alpha_2, \alpha_1 + 2\alpha_2, \alpha_1 + 3\alpha_2 \ .$$

Here the only positive root of height 2 is $\alpha_1 + \alpha_2$. Also because $\alpha_2 + 2\alpha_1$ is no root, we conclude that $\alpha_1 + 2\alpha_2$ is the only root of height 3. $2a_1 + 2\alpha_2$ is also not a root; thus $\alpha_1 + 3\alpha_2$ is the single root of height 4. Finally, from (12.71), we find that

$$2(\alpha_1 + 3\alpha_2, \alpha_1)/(\alpha_1, \alpha_2) = 2 - 3 = -1 \ ,$$

which means that $(\alpha_1 + 3\alpha_2) + \alpha_1 = 2\alpha_1 + 3\alpha_2$ is a root. It is the only root of height 5, because $\alpha_1 + 4\alpha_2$ does not represent a root. Since the linear combinations $(2\alpha_1 + 3\alpha_2) + \alpha_1 = 3(\alpha_1 + \alpha_2)$ and $(2\alpha_1 + 3\alpha_2) + \alpha_2 = 2(\alpha_1 + 2\alpha_2)$ are not roots, there are none with a height greater than 5. Consequently the roots of G_2 that are non-zero are

$$\pm\alpha_1, \ \pm\alpha_2, \ \pm(\alpha_1 + \alpha_2), \ \pm(\alpha_1 + 2\alpha_2), \ \pm(\alpha_1 + 3\alpha_2), \ \pm(2\alpha_1 + 3\alpha_2) \ .$$

12.14 Two Simple Lie Algebras

As an illustration we want to determine the Lie algebras of the groups SO(3) and SU(3). The Dynkin diagram for SO(3) is simply one circle 0, meaning that there exists one simple root, which we mark by J_+. In Van der Waerden's representations the corresponding figure looks like

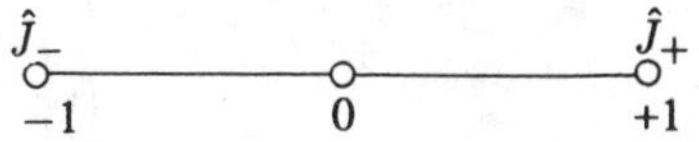

The corresponding commutator algebra follows from (12.31)–(12.34), i.e.

$$[\hat{H}_1, \hat{H}_1]_- = 0 \quad , \quad [\hat{H}_1, \hat{J}_\pm]_- = \pm \hat{J}_\pm \quad , \quad [\hat{J}_+, \hat{J}_-]_- = \hat{H}_1 \quad .$$

In order to obtain the above algebra, we express $\hat{H}_1$ and $\hat{J}$ by the infinitesimal operators of angular momentum $\hat{J}_x$, $\hat{J}_y$ and $\hat{J}_z$ for SO(3), so that

$$\hat{H}_1 = \hat{J}_z \quad \text{and} \quad \hat{J}_\pm = \tfrac{1}{2}(\hat{J}_x \pm \mathrm{i}\hat{J}_y) \quad .$$

Thus we obviously obtain the known commutator algebra for angular momenta. As we can see, the new definition deviates from the one of Chapter 2 by a factor of $\frac{1}{\sqrt{2}}$.

The Dynkin diagram of SU(3) is

This means that there exist 2 simple roots of equal height, enclosing an angle of 120°. Of course $-\alpha$ and $-\beta$ are also roots, as is $\pm(\alpha + \beta)$, so that adding these roots we obtain the diagram for A_2 shown in Fig. 12.6. We normalize the roots by using

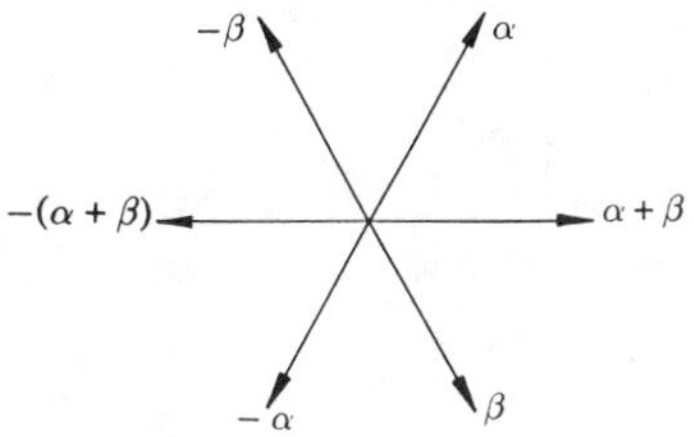

Fig. 12.6. Vector diagram for A_2

$$\sum_\alpha \alpha_i \alpha_j = \delta_{ij} \quad ,$$

and thus obtain

$$\alpha = \frac{1}{2\sqrt{3}}(1, 3), \quad \beta = \frac{1}{2\sqrt{3}}(1, -3), \quad \alpha + \beta = \frac{1}{\sqrt{3}}(1, 0) \quad .$$

For the commutator algebra (12.31)–(12.34) we obtain

$$\begin{aligned}
&[\hat{H}_1, \hat{E}_{\pm\alpha}]_- = 1/2\sqrt{3}\,\hat{E}_{\pm\alpha} && [\hat{H}_2, \hat{E}_{\pm\alpha}]_- = \pm\tfrac{1}{2}\hat{E}_{\pm\alpha}\\
&[\hat{H}_1, \hat{E}_{\pm\beta}]_- = 1/2\sqrt{3}\,\hat{E}_{\pm\beta} && [\hat{H}_2, \hat{E}_{\pm\beta}]_- = +\tfrac{1}{2}\hat{E}_{\pm\beta}\\
&[\hat{H}_1, \hat{E}_{\pm(\alpha+\beta)}]_- = 1/\sqrt{3}\,\hat{E}_{\pm(\alpha+\beta)} && [\hat{H}_2, \hat{E}_{\pm(\alpha+\beta)}]_- = 0\\
&[\hat{E}_\alpha, \hat{E}_{-\alpha}]_- = 1/2\sqrt{3}\,\hat{H}_1 + \tfrac{1}{2}\hat{H}_2 && [\hat{E}_\beta, \hat{E}_{-\beta}] = \tfrac{1}{2}\sqrt{3}\,\hat{H}_1 - \tfrac{1}{2}\hat{H}_2\\
&[\hat{E}_{\alpha+\beta}, \hat{E}_{-(\alpha+\beta)}]_- = 1/\sqrt{3}\,\hat{J}_1 && [\hat{E}_\alpha, \hat{E}_\beta]_- = 1/\sqrt{6}\,\hat{E}_{\alpha+\beta}\\
&[\hat{E}_\alpha, \hat{E}_{\alpha+\beta}]_- = 0 && [\hat{E}_\beta, \hat{E}_{\alpha+\beta}]_- = 0\\
&[\hat{E}_\alpha, \hat{E}_{-(\alpha+\beta)}]_- = -1/\sqrt{6}\,\hat{E}_{-\beta} && [\hat{E}_\beta, \hat{E}_{-(\alpha+\beta)}]_- = 1/\sqrt{6}\,\hat{E}_{-\alpha}\\
&[\hat{H}_1, \hat{H}_2]_- = 0 \quad .
\end{aligned}$$

12.15 Representations of the Classical Lie Algebras

There is a method which allows us to construct the irreducible representations of every Lie algebra using the above table. We want to illustrate this method without further proof. The basis vectors $|u\rangle$ of an irreducible representation are classified by their eigenvalues with respect to the operators H_i of the Cartan subalgebra, i.e.

$$H_i|u\rangle = \lambda_i(u)|u\rangle \quad . \tag{12.72}$$

One defines the *weight* of a state $|u\rangle$ to be the vector (rank for N of the Lie algebra)

$$\Lambda(u) := (\lambda_1(u), \lambda_2(u), \dots, \lambda_N(u)) \quad . \tag{12.73}$$

For the weights one introduces the following ordering relation: The weight Λ is called greater than the weight Λ' if the first non-zero number $\lambda_i - \lambda'_i$, $i = 1, 2, \dots, N$, is positive. This ordering relation determines uniquely a maximal weight $\Lambda_{\max}$ for each irreducible representation, which has the properties:

a) $\Lambda_{\max}$ is not degenerate, i.e. there exists only one state with the weight $\Lambda_{\max}$.
b) There is a one-to-one correspondence between the occurring irreducible representations and the corresponding $\Lambda_{\max}$.

One can prove these statements by generalizing the considerations of Chap. 7 in connection with SU(3). To justify the following procedure, we want to illustrate for the SU(3) triplet that one can deduce the entire irreducible representation from the largest weight $\Lambda_{\max}$. To this end, in analogy to Lemma 1, we need the following:

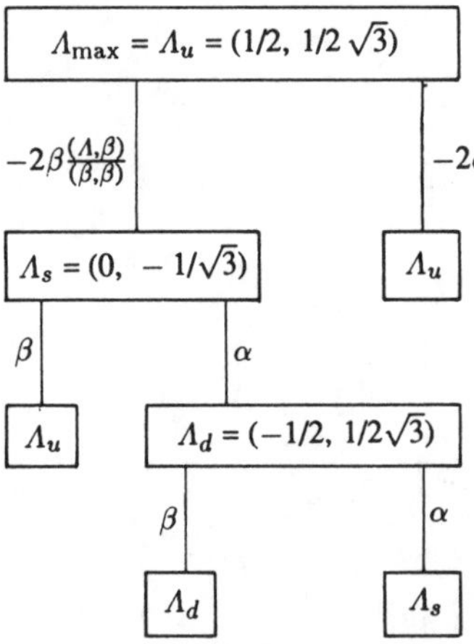

Fig. 12.7. Illustration of the derivation of the weights of all members of an irreducible representation from the largest weight $\Lambda_{\max}$

Lemma 1′. For every weight of an irreducible representation Λ and for every root α, $2(\Lambda, \alpha)/(\alpha, \alpha)$ is an integer and $\Lambda - 2\alpha[(\Lambda, \alpha)/(\alpha, \alpha)]$ is a weight of the irreducible representation.

The greatest weight of SU(3) is $\Lambda_{\max} = \Lambda_u = (1/2, 1/2\sqrt{3})$. The simple positive roots for the SU(3) are (see last section)

$$\alpha = \sqrt{2}(1/2, -\sqrt{3}/2) \quad ; \quad \beta = \sqrt{2}(1/2, \sqrt{3}/2) \quad .$$

Because of $(\Lambda, \alpha) = 0$, $(\Lambda, \beta) = 1/\sqrt{2}$ it follows from lemma 1′ that

$$\Lambda_u = (1/2, 1/2\sqrt{3}) \quad , \Lambda_d = (-1/2, 1/2\sqrt{3}) \quad , \Lambda_s = (0, -1/2)$$

The complete determination of an irreducible representation from $\Lambda_{\max}$ is generally more complicated because of the degeneracies, i.e. there are more states with the same weight. To obtain the general scheme for the complete determination of the irreducible representations of any simple Lie algebra, we must therefore:

1. be able to write down all the largest weights $\Lambda_{\max}$ and
2. know how often each state is occupied.

Problem 1 was solved by Dynkin by defining a new weight Λ.

$$\Lambda(\lambda_1, \ldots, \lambda_N) \to \tilde{\Lambda}(\tilde{\lambda}_1, \ldots, \tilde{\lambda}_N) \quad ; \quad \tilde{\lambda}_i = 2\frac{(\Lambda, \alpha_i)}{(\alpha_i, \alpha_i)} \quad , \tag{12.74}$$

where the positive roots are denoted by $\alpha_1 \ldots \alpha_N$. From lemma 1′ it follows that all $\tilde{\lambda}_i$ are integers and so the first problem is solved; Dynkin showed that all $\Lambda = (\lambda_1, \lambda_2, \ldots, \lambda_N)$, with $\lambda_i \geq 0$, $i = 1, 2, \ldots, N$, appear as the largest weights. The first problem thus reduces to writing down all N-dimensional vectors Λ with non-negative integer components.

By considering the series of roots one can derive the following extension of lemma 1′:

Lemma 7. If $\tilde{\Lambda}$ and $\tilde{\lambda}_j > 0$, then $\tilde{\Lambda} - \tilde{\alpha}_j$ is also a root, where $\tilde{\alpha}_j$ is the simple root α_j in the "Dynkin basis", and

$$(\tilde{\alpha}_j)_\lambda = 2\frac{(\alpha_j, \alpha_\lambda)}{(\alpha_\lambda, \alpha_\lambda)} = A_{\lambda j} = A_{j\lambda} \quad .$$

Lemma 7 provides the weights in the Dynkin basis. To transform this basis into the usual one we write

$$\Lambda = \sum_j b_j a_j \to a_i = \sum_j b_j A_{ij} \to b_j = \sum_i a_i (A^{-1})_{ij} \quad . \tag{12.75}$$

For the calculation of the degree of degeneracy one can use the so-called Freudenthal recursion formula, given by

$$\begin{aligned} &[(\Lambda_{\max} + \delta, \Lambda_{\max} + \delta) - (\Lambda + \delta, \Lambda + \delta)]n_\Lambda \\ &\quad = 2 \sum_{\substack{\alpha \text{ pos. root} \\ k > 0}} n_{\Lambda + k\alpha}(\Lambda + k\alpha, \alpha) \quad , \quad \text{with} \end{aligned} \tag{12.76}$$

$$\delta = \tfrac{1}{2} \sum_{\alpha \text{ pos. root}} \alpha \quad ;$$

$\tilde{\delta} = (1, 1, \ldots, 1)$ in the Dynkin basis ;
n_Λ is the number of states of weight Λ ;

and

$$
\begin{aligned}
(\Lambda, \Lambda') &= \sum_{ij} b_i b'_j (\alpha_i, \alpha_j) \\
&= \sum_{ijkl} a_k (A^{-1})_{ki} (\alpha_i, \alpha_j) a'_l (A^{-1})_{lj} \\
&= \sum_{jl} a'_l a_j \tfrac{1}{2} (\alpha_j, \alpha_j) (A^{-1})_{lj} \\
&= \sum_{jl} a'_l G_{lj} a_j
\end{aligned}
$$

$$G_{lj} = [(\alpha_i, \alpha_j)/2]\,[A^{-1}]_{lj} \ . \tag{12.77}$$

The metric G_{lj} is uniquely determined by normalizing the longest appearing roots to 2. Finally we give the so-called Weyl formula, whch allows us to compute the dimension of an irreducible representation from Λ_{max}.

$$N(\Lambda_{\text{max}}) = \prod_{\alpha \text{ pos. root}} (\Lambda_{\text{max}} + \delta, \alpha)/(\delta, \alpha) \ . \tag{12.78}$$

Concluding, we summarize the procedure for determining the irreducible representations of a given simple Lie algebra:

1. From the Dynkin diagram (see the table of Example 12.3) we obtain all positive roots α_i.
2. We calculate the Cartan matrix A and the metric G using (12.74) and (12.77).
3. From (12.78) we get all Λ_{max} of the irreducible representations with not too large dimension.
4. For the irreducible representations of interest we construct the diagram for the weights, making use of Lemma 7 and (12.78).

In Example 12.6 we shall perform this calculation for the octet representation of SU(3), though in practice one can use tables calculated by computers.

EXAMPLE

12.6 Analysis of SU(3)

1. We make use of the above table. Because the SU(3) is of second rank, with the two simple roots

$$\alpha_1 = \sqrt{2}(1, 0) \quad \alpha_2 = \sqrt{2}(-1/2, \sqrt{3}/2) \ ,$$

it follows that

$$\alpha_3 = \alpha_1 + \alpha_2 = \sqrt{2}(1/2, \sqrt{3}/2) \ ,$$

where $\alpha_1, \alpha_2, \alpha_3$ are the positive roots in the considered basis, which were fixed by $\alpha_1 = \sqrt{2}(1, 0)$.

Example 12.6

2. We calculate A and G. $A_{ij} = 2(\alpha_i, \alpha_j)/(\alpha_i, \alpha_i)$;

$$A = \begin{pmatrix} 2 & -1 \\ -1 & 2 \end{pmatrix} \quad \text{and} \quad G = \frac{1}{3}\begin{pmatrix} 2 & 1 \\ 1 & 2 \end{pmatrix} \quad \text{i.e.}$$

$$\tilde{\alpha}_1 = (2, -1), \tilde{\alpha}_2 = (-1, 2), \tilde{\alpha} = \tilde{\alpha}_1 + \tilde{\alpha}_2 = (1, 1) \quad .$$

3. We take $\tilde{\Lambda}_{\max}$ which has the general form $\tilde{\Lambda}_{\max} = (p, q)$, so that from (12.78) follows that

$$\begin{aligned} N(p, q) &= \prod_{\alpha=\alpha_1,\alpha_2,\alpha_3} \frac{1/3(p+1, q+1)\begin{pmatrix} 2 & 1 \\ 1 & 2 \end{pmatrix}\tilde{\alpha}^T}{1/3(1, 1)\begin{pmatrix} 2 & 1 \\ 1 & 2 \end{pmatrix}\tilde{\alpha}^T} \\ &= \prod_{\alpha=\alpha_1,\alpha_2,\alpha_3} \frac{1/3(2p+q+3, p+2q+3)\tilde{\alpha}^T}{(1, 1)\tilde{\alpha}^T} \\ &= (p+1)\times(q+1)\times\tfrac{1}{2}\times(p+q+2) \quad . \end{aligned}$$

This is exactly the relation which has been derived in Exercise 7.9. Thus for SU(N) the so-called Dynkin labels correspond to the labels derived from the Young diagrams.

4. We want to construct the octet $\Lambda_{\max} = (1, 1)$, for which the scheme below holds.

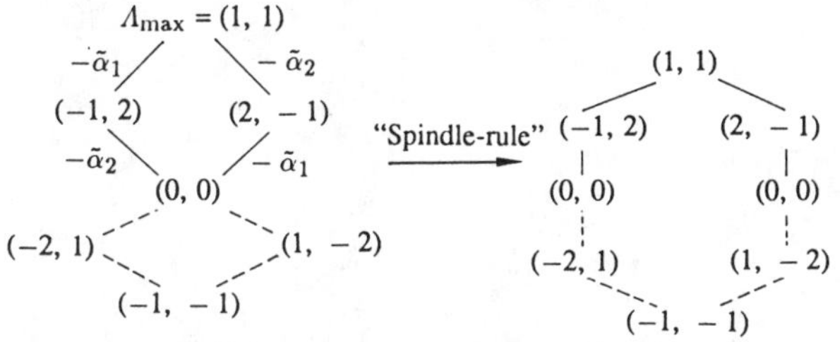

For $\Lambda_{\max} = (-1, 2)$, (12.78) yields

$$\left[(2, 2)\frac{1}{3}\begin{pmatrix} 2 & 1 \\ 1 & 2 \end{pmatrix}\begin{pmatrix} 2 \\ 2 \end{pmatrix} - (0, 3)\frac{1}{3}\begin{pmatrix} 2 & 1 \\ 1 & 2 \end{pmatrix}\begin{pmatrix} 0 \\ 2 \end{pmatrix}\right] n_{(-1, 2)}$$
$$= 2n_{(-1,1)}(1, 1)\frac{1}{3}\begin{pmatrix} 2 & 1 \\ 2 & -1 \end{pmatrix}\begin{pmatrix} 2 \\ -1 \end{pmatrix} \quad .$$

Because $\Lambda_{\max}$ is not degenerate if follows that

$$(8-6)n_{(-1, 2)} = 2n_{(1, 1)} = 2 \qquad n_{(-1, 2)} = 1 \quad .$$

Correspondingly, (2.68) results in

$$\tilde{\Lambda} = (2, -1) \quad , \quad n_{(2, -1)} = 1 \quad .$$

Example 12.6

So, for $\Lambda_{\max} = (0, 0)$ we have

$$\left[8 - (1,1)\frac{1}{3}\begin{pmatrix}2 & 1\\ 1 & 2\end{pmatrix}\begin{pmatrix}1\\ 1\end{pmatrix}\right] n_{(0,\,0)} = 2\left\{n_{(2,\,-1)}(2, -1)\frac{1}{3}\begin{pmatrix}2 & 1\\ 1 & 2\end{pmatrix}\begin{pmatrix}2\\ -1\end{pmatrix}\right.$$

$$\left. + n_{(-1,\,2)}(-1, 2)\frac{1}{3}\begin{pmatrix}2 & 1\\ 1 & 2\end{pmatrix}\begin{pmatrix}-1\\ 2\end{pmatrix} + n_{(1,\,1)}(1, 1)\frac{1}{3}\begin{pmatrix}2 & 1\\ 1 & 2\end{pmatrix}\begin{pmatrix}1\\ 1\end{pmatrix}\right\}$$

$$= 2(2 + 2 + 2) = 12 \quad ,$$

$$6n_{(0,\,0)} = 12 \rightarrow n_{(0,\,0)} = 2 \quad .$$

This result could also have been obtained by the "spindle rule", which states that the root diagrams are convex, i.e. the number of states which are obtained by the first, second, etc. subtraction, cannot decrease and then increase.

13. Special Discrete Symmetries

In the last two chapters of this book we return to symmetries which have a general significance in quantum mechanics. We shall begin with the discrete symmetries of space inversion and time reversal.

13.1 Space Reflection (Parity Transformation)

Parity plays an important role in quantum mechanics. One often finds[1] that the eigenfunctions of the Hamiltonian either stay the same or change sign when a parity transformation is performed, i.e. when $\boldsymbol{r}$ is replaced by $(-\boldsymbol{r})$, as long as the transformation does not effect the potential energy. A physical state is defined to have positive parity if the wave function does not change, whereas it is of negative parity if the wave function changes its sign.

At first we adopt a similar approach as in the case of rotations [cf. Chap. 1, Eq. (1.46)] and the transformation is written as

$$\boldsymbol{r}' = \hat{R}_1 \boldsymbol{r} = -\boldsymbol{r} \quad . \tag{13.1}$$

If we choose $\hat{R}_1$ to be

$$\hat{R}_1 = -\mathbb{1} = \begin{pmatrix} -1 & 0 & 0 \\ 0 & -1 & 0 \\ 0 & 0 & -1 \end{pmatrix} \quad , \tag{13.2}$$

this yields a space reflection. Obviously, $\hat{R}_1$ is real, orthonormal and, in addition, $\hat{R}_1^{-1} = \hat{R}_1$. But as the determinant of $\hat{R}_1$ is -1 this cannot be a transformation of the group SO(3), because the determinant of a rotation matrix is always 1. Every real orthonormal 3×3 matrix which has determinant -1, can be written as a product of $\hat{R}_1$ and a matrix that represents a rotation. The unit matrix $\mathbb{1}$ and $\hat{R}_1$ together form a discrete group of two elements $\hat{E}_i$ for which $\hat{E}_i^2 = 1$.

Now we consider a physical system in a state $|\alpha\rangle$ represented by the wave function $\psi_\alpha(\boldsymbol{r})$. Under a space reflection this state is transformed into $|\alpha'\rangle$, with

[1] See Vol. 1 of this series, *Quantum Mechanics, An Introduction* (Springer, Berlin, Heidelberg 1989).

$\psi_{\alpha'}(\boldsymbol{r})$. We assume that these two states are related by

$$\psi_{\alpha'}(\hat{R}_1 \boldsymbol{r}) = a\psi_\alpha(\boldsymbol{r}), \tag{13.3}$$

where we still have to determine a. The discrete nature of a space-inversion implies that a appears in this relation, in contrast to the analogous relations for translations or rotations [see (1.17) and (1.64) respectively].

Remark. For such continuous transformations one could also introduce factors a analogous to (13.3). These factors would be continuous functions of the respective transformation parameters, i.e. $a = a(\boldsymbol{a})$ for translations and $a = a(\boldsymbol{\phi})$ for rotations. These continuous functions must have the value 1 if the transformation parameters are zero, i.e. if the transformation is just the identity transformation. However, one can show that factors with this property have no physical meaning.

In analogy to former definitions we now introduce the *unitary space-reflection operators* $\hat{U}$ by the equation

$$\hat{U}_1 \psi_\alpha(\boldsymbol{r}) = \psi_{\alpha'}(\boldsymbol{r}) \quad , \tag{13.4}$$

which [with (13.3)] yields

$$\hat{U}_1 \psi_\alpha(\boldsymbol{r}) = a\psi_\alpha(\hat{R}_1^{-1} \boldsymbol{r}) = a\psi_\alpha(\hat{R}_1 \boldsymbol{r}) \quad . \tag{13.5}$$

Here we have made use of the fact that the inverse of a reflection is also a reflection; thus we apply this operation twice, we get

$$\hat{U}_1^2 \psi_\alpha(\boldsymbol{r}) = a\hat{U}_1 \psi_\alpha(\hat{R}_1 \boldsymbol{r}) = a^2 \psi_\alpha(\hat{R}_1^{-1} \hat{R}_1 \boldsymbol{r}) = a^2 \psi_\alpha(\boldsymbol{r}) \quad . \tag{13.6}$$

On the other hand, two consecutive space reflections transform coordinate space into itself, so that $\hat{U}_1^2$ must transform every state into itself. This means that $\hat{U}_i^2 \psi_\alpha$ and ψ_α can differ only by a phase factor, since $\hat{U}_1$ is a unitary operator. Therefore the absolute value of a^2 must be 1 and consequently the same holds for a. In addition we can impose a further restriction on a.

The physical states of a system of particles, which we denote by $\psi_\alpha(\boldsymbol{r})$, can be superposed to give

$$\psi(\boldsymbol{r}) = \sum_\alpha A_\alpha \psi_\alpha(\boldsymbol{r}) \quad , \tag{13.7}$$

leading to

$$\hat{U}_1^2 \psi(\boldsymbol{r}) = \sum a_\alpha^2 A_\alpha(\boldsymbol{r}) \quad . \tag{13.8}$$

The rhs of (13.8) describes a state which is different from $\psi(\boldsymbol{r})$ unless all factors a_α^2 are equal. Only then do two space reflections in sequence reproduce the original state up to a phase factor. This means that a_α^2 has to be the same for all states that can be superposed. Therefore it seems adequate to choose a certain value of a for each kind of particles. A rotation by 2π does not change the state vector of particles with integer spin; we assume the same for space reflections, and therefore $a^2 = 1$, which implies $a = \pm 1$. Particles with half-integer spin can be coupled in pairs, yielding states with integer spin. Therefore we expect that

a for particles with half-integer spin can take on the same values as the factor a^2 for particles with integer spin, so that a can be ± 1 and $\pm \mathrm{i}$, i.e. $a^4 = 1$. In Volume 3 of this series we have seen that the spinor of a spin-$\frac{1}{2}$ particle is reproduced by a rotation by 4π.

In order to determine a experimentally for different particles, one has to include their mutual interaction. Let us consider for example the π^0 meson, which has zero spin. It decays into two photons (see Example 8.6). Assuming that the parity of a system is conserved during the decay we can determine the *internal parity* of π^0 relative to that of the electromagnetic field. One finds $a = -1$, so that the π^0 meson has a negative parity. It is called a *pseudoscalar* particle, whereas a spin zero particle with $a = +1$ is called a *scalar particle*. The charged mesons $\pi^\pm$ also have spin zero, though it is not possible to determine their internal parity relative to the electromagnetic field, because they cannot decay into photons due to charge conservation. On the other hand these pions are created and annihilated in interactions between nucleons. Thus we can determine their parity relative to the nucleons, provided that parity is conserved in these interactions, too. According to such considerations, one assigns negative parity to pions and positive parity to nucleons.

13.2 Reflected States and Operators

The equation of motion of a space-reflected state is obtained by the application of $\hat{U}_1$ to the equation of motion of the original state. As expected, a reflected state fulfils the same Schrödinger equation as the original state, if $\hat{H}$ and $\hat{U}_1$ commute, i.e.

$$[\hat{U}_1, \hat{H}] = 0 \quad . \tag{13.9}$$

In this case $\hat{H}$ and $\hat{U}_1$ can be represented by diagonal matrices and the energy eigenstates can simultaneously be chosen to be states of good parity. In analogy to Example 1.6, (1) the matrix elements of a dynamical variable $\hat{A}$ for reflected states are equal to the matrix elements of $\hat{U}_1^\dagger \hat{A} \hat{U}_1$ for the original state, because

$$\langle \hat{U}_1 \psi_\alpha | \hat{A} | \hat{U}_1 \psi_\beta \rangle = \langle \psi_\alpha | \hat{U}_1^\dagger \hat{A} \hat{U}_1 | \psi_\beta \rangle \quad . \tag{13.10}$$

Since $\hat{U}_1$ was chosen to be a unitary operator, (13.5) becomes, on applying $\hat{U}_1^\dagger$,

$$\psi_\alpha(\boldsymbol{r}) = a \hat{U}_1^\dagger \, \psi_\alpha(\hat{R}_1^{-1} \boldsymbol{r})$$

or

$$\hat{U}_1^\dagger \psi_\alpha(\boldsymbol{r}) = \frac{1}{a} \psi_\alpha(\hat{R}_1 \boldsymbol{r}) \quad . \tag{13.11}$$

By means of this equation the expression $\hat{U}_1^\dagger \boldsymbol{r} \hat{U}_1$ can easily be calculated:

$$\hat{U}_1^\dagger \boldsymbol{r} \hat{U}_1 \psi_\alpha(\boldsymbol{r}) = a \hat{U}_1^\dagger \boldsymbol{r} \psi_\alpha(\hat{R}_1^{-1} \boldsymbol{r})$$

$$= a \frac{1}{a} (\hat{R}_1 \boldsymbol{r}) \psi_\alpha(\hat{R}_1 \hat{R}_1^{-1} \boldsymbol{r}) = (-\boldsymbol{r}) \psi_\alpha(\boldsymbol{r}) \quad . \tag{13.12}$$

Since (13.12) holds for every $\psi_\alpha(\boldsymbol{r})$, we conclude that

$$\hat{U}_1^\dagger \boldsymbol{r} \hat{U}_1 = -\boldsymbol{r} \quad , \tag{13.13}$$

so that instead of (13.4), the last equation could also be used to define the operator $\hat{U}$. Because of $\boldsymbol{p} = -\mathrm{i}\hbar\vec{\nabla}$ and $\boldsymbol{L} = \boldsymbol{r} \times \boldsymbol{p}$, one finds in the same way that

$$\hat{U}_1^\dagger \hat{\boldsymbol{p}} \hat{U}_1 = -\hat{\boldsymbol{p}} \quad \text{and} \tag{13.14}$$

$$\hat{U}_1^\dagger \hat{\boldsymbol{L}} \hat{U}_1 = \hat{\boldsymbol{L}} \quad . \tag{13.15}$$

Given that the operator $\hat{U}_1$ acts only on spatial coordinates, but not on the spin, it must commute with $\hat{\boldsymbol{S}}$. Because of (13.15) we therefore find that

$$\hat{U}_1^\dagger \hat{\boldsymbol{S}} \hat{U}_1 = \hat{\boldsymbol{S}} \quad , \tag{13.16}$$

$$\hat{U}_1^\dagger \hat{\boldsymbol{J}} \hat{U}_1 = \hat{\boldsymbol{J}} \quad , \tag{13.17}$$

with $\hat{\boldsymbol{J}} = \hat{\boldsymbol{L}} + \hat{\boldsymbol{S}}$. The results (13.13)–(13.17) show that coordinates, momenta, and angular momenta behave under space reflections exactly the way as expected classically. Accordingly, spatial coordinates and momenta are examples of vectors, also called *polar vectors*, whereas angular momentum is an example of an *axial vector* or *pseudovector*.

13.3 Time Reversal

The classical equations of motion for particles which move in a conservative potential are symmetrical if the direction of time propagation is reversed, since they only contain second derivatives with respect to time or squares of first time derivatives. In general we expect that this symmetry of time reversal also holds for quantum mechanical equations. However, we shall see that time reversal has some special properties in quantum mechanics. In part this is due to the fact that in quantum mechanics the time coordinate (contrary to the space coordinates) is not represented by an operator in Hilbert space, but is taken to be a parameter of the state. Therefore there is no analogue of (13.13). We require time reversal to transform the wave function $\psi_\alpha(\boldsymbol{r}, t)$ of state α into $\psi_{\alpha'}(\boldsymbol{r}, -t)$, i.e. we demand that $\psi_\alpha(\boldsymbol{r}, -t)$ propagates according to the reversed time direction. In this new state the *signs of momenta and angular momenta are reversed*, whereas all other quantities remain unchanged. Time reversal is described by the time-independent operator $\hat{T}$, which is defined by the equation

$$\hat{T}\psi_\alpha(\boldsymbol{r}, t) = \psi_{\alpha'}(\boldsymbol{r}, -t) \quad . \tag{13.18}$$

In the following we shall consider physical systems for which $\hat{T}$ is a symmetry transformation. This implies that, with u_k, $\hat{T}u_k$ is also an eigenstate to the Hamiltonian (which is supposed to be time-independent) with the same eigenvalue E_k.

Let us consider the Schrödinger equation

$$\mathrm{i}\hbar\frac{\partial}{\partial t}\psi_\alpha(\boldsymbol{r},t)=\hat{H}\psi_\alpha(\boldsymbol{r},t)\quad, \tag{13.19}$$

where $\hat{H}$ does not depend on time. For the time-reversed function $\psi_{\alpha'}(\boldsymbol{r},t)$, the Schrödinger equation therefore becomes

$$-\mathrm{i}\hbar\frac{\partial}{\partial t}\psi_{\alpha'}(\boldsymbol{r},t)=\hat{H}\psi_{\alpha'}(\boldsymbol{r},t)\quad, \tag{13.20}$$

where the negative sign on the rhs is due to the time derivative. For $\hat{T}$ to be a symmetry transformation it must commute with the Hamiltonian, which can be expressed by

$$\hat{T}\hat{H}=\hat{H}\hat{T}\quad,\quad\text{or}\quad\hat{T}\hat{H}\hat{T}^{-1}=\hat{H}\quad. \tag{13.21}$$

This can be verified in the following way: we multiply (13.19) with $\hat{T}$ from the lhs, which yields

$$(\hat{T}\mathrm{i}\hat{T}^{-1})\hbar\frac{\partial}{\partial t}\hat{T}\psi_\alpha(\boldsymbol{r},t)=(\hat{T}\hat{H}\hat{T}^{-1})\hat{T}\psi_\alpha(\boldsymbol{r},t)=\hat{H}\hat{T}\psi_\alpha(\boldsymbol{r},t)\quad. \tag{13.22}$$

This operation has been performed carefully, in the sense that we have included the imaginary unit i into this operation, i.e. we have allowed $\hat{T}\mathrm{i}\hat{T}^{-1}\neq\mathrm{i}$. Furthermore we have taken into account that the real parameter t commutes with $\hat{T}$. [Instead of (13.21) we could have required that $\hat{T}\hat{H}\hat{T}^{-1}=-\hat{H}$, so that i commutes with $\hat{T}$. In this case, however, time reversal would also change the spectrum of the Hamiltonian, which is then no longer positive definite. This would be physically unacceptable.] Comparing (13.22) with (13.20) shows that we have to postulate

$$\hat{T}\mathrm{i}\hat{T}^{-1}=-\mathrm{i}\quad. \tag{13.23}$$

This is a special case of the general relation

$$\hat{T}(a\psi)=a^*\hat{T}\psi\quad, \tag{13.24}$$

$$\hat{T}(a\psi+b\psi)=a^*\hat{T}\psi+b^*\hat{T}\psi\quad. \tag{13.25}$$

An operator which obeys these equations is called an *antilinear* operator. Accordingly the time reversal operator $\hat{T}$ is antilinear.

13.4 Antiunitary Operators

We introduce the operator of complex conjugation $\hat{K}$ by the definition

$$\hat{K}\psi=\psi^*\quad, \tag{13.26}$$

for any function ψ. Obviously $\hat{K}^2 = 1$, and $\hat{K}$ obeys the relations (13.24) and (13.25), i.e. it is antilinear. Now each antilinear operator can be represented by a product of $\hat{K}$ with a linear operator. Of special importance are products for which the linear operator is unitary. Such operators are called *antiunitary*, and clearly $\hat{K}$ itself is antiunitary.

EXERCISE

13.1 Effect of an Antiunitary Operator on Matrix Elements of Wavefunctions

Problem. Prove that the inner product of two states is equal to the complex conjugate of the inner product of the states which are obtained by applying an antiunitary operator $\hat{A}$ to the former states. What follows for the norm?

Solution. The two states are characterized by their wave functions, which we denote by ψ_α and ψ_β. From these states we obtain $\psi_{\alpha'} = \hat{A}\psi_\alpha$ and $\psi_{\beta'} = \hat{A}\psi_\beta$. Since $\hat{A}$ is antiunitary this can be written as

$$\hat{A} = \hat{U}\hat{K} \quad , \tag{1}$$

with unitary $\hat{U}$. This yields

$$\psi_{\alpha'} = \hat{U}\psi_\alpha^* \quad \text{and} \quad \psi_{\beta'} = \hat{U}\psi_\beta^* \quad . \tag{2}$$

Using this we can write the inner product as

$$\begin{aligned} \langle\alpha'|\beta'\rangle &\equiv \int d^3r \psi_{\alpha'}^\dagger \psi_{\beta'} = \int d^3r (\hat{U}\psi_\alpha^*)^\dagger (\hat{U}\psi_\beta^*) = \int d^3r \psi_\alpha^{*\dagger} \hat{U}^\dagger \hat{U} \psi_\beta^* \\ &= \left\{ \int d^3r \psi_\alpha^\dagger \psi_\beta \right\}^* = \langle\alpha|\beta\rangle^* = \langle\beta|\alpha\rangle \quad , \end{aligned} \tag{3}$$

where we have made use of the fact that $\hat{U}$ is unitary. For the norm we get

$$\langle\alpha'|\alpha'\rangle = \langle\alpha|\alpha\rangle^* = \langle\alpha|\alpha\rangle \quad , \tag{4}$$

since it is real number, which means that the norm remains unchanged.

According to the result of this problem, it is plausible to assume the time reversal operator to be antiunitary since it does not change the norm of the states and the absolute value of the inner product of two states also remains unchanged. Therefore we write $\hat{T}$ in the form

$$\hat{T} = \hat{U}\hat{K} \tag{13.27}$$

with unitary $\hat{U}$. Now we want to find an explicit expression for the operator $\hat{T}$ in such a way that in the time-reversed state the signs of all momenta and angular momenta are reversed, whereas all other quantities remain unchanged. First we examine – as the simplest case – the situation for a particle with spin zero. The corresponding states are characterized by a single component wave function. From any given wave function ψ_α we construct

$$\psi_\beta(\boldsymbol{r}, t) = \boldsymbol{r}\psi_\alpha(\boldsymbol{r}, t) \quad . \tag{13.28}$$

Denoting the corresponding time reversed states by $\psi_{\alpha'}(\boldsymbol{r}, t) = \hat{T}\psi_\alpha(\boldsymbol{r}, t)$ and $\psi_{\beta'} = \hat{T}\psi_\beta$ we get an analogous relation,

$$\psi_{\beta'}(\boldsymbol{r}, -t) = \boldsymbol{r}\psi_{\alpha'}(\boldsymbol{r}, t) \quad , \tag{13.29}$$

since $\boldsymbol{r}$ does not change sign under time reversal. From this we find that

$$\boldsymbol{r}\hat{T}\psi_\alpha = \boldsymbol{r}\psi_{\alpha'} = \psi_{\beta'} = \hat{T}\psi_\beta = \hat{T}\boldsymbol{r}\psi_\alpha \quad . \tag{13.30}$$

ψ_α can be chosen arbitrarily and so it follows that

$$\boldsymbol{r}\hat{T} = \hat{T}\boldsymbol{r} \quad . \tag{13.31}$$

We now write the state which is obtained from ψ_α by applying the momentum operator as ψ_γ, i.e. $\psi_\gamma(\boldsymbol{r}, t) = \hat{\boldsymbol{p}}\psi_\alpha(\boldsymbol{r}, t)$. Concerning the time-reversed states we expect

$$\psi_{\gamma'}(\boldsymbol{r}, -t) = -\hat{\boldsymbol{p}}\psi_{\alpha'}(\boldsymbol{r}, -t) \quad , \tag{13.32}$$

to hold, because the momentum changes its sign under time-reversal. This yields

$$\begin{aligned}\hat{\boldsymbol{p}}\hat{T}\psi_\alpha(\boldsymbol{r}, t) &= \hat{\boldsymbol{p}}\psi_{\alpha'}(\boldsymbol{r}, -t) = \psi_{\gamma'}(\boldsymbol{r}, t)\\ &= -\hat{T}\psi_\gamma(\boldsymbol{r}, t) = -\hat{T}\hat{\boldsymbol{p}}\psi_\alpha(\boldsymbol{r}, t) \quad \text{or}\end{aligned} \tag{13.33}$$

$$\hat{\boldsymbol{p}}\hat{T} = -\hat{T}\hat{\boldsymbol{p}} \quad . \tag{13.34}$$

In a similar way we find for $\hat{\boldsymbol{L}} = \boldsymbol{r} \times \hat{\boldsymbol{p}}$ that

$$\hat{\boldsymbol{L}}\hat{T} = -\hat{T}\hat{\boldsymbol{L}} \quad . \tag{13.35}$$

In the coordinate representation r is real operator and $\hat{\boldsymbol{p}} = -\mathrm{i}\hbar\nabla$ is purely imaginary. The simplest choice for $\hat{T}$ which fulfils (13.31), (13.34) and (13.35) is

$$\hat{U} = \mathbb{1} \quad ,$$

so that

$$\hat{T} = \hat{K} \quad , \tag{13.36}$$

though this conclusion depends on the representation chosen. In the momentum representation (where ψ_α is a function of $\boldsymbol{p}$, not $\boldsymbol{r}$) the momentum $\boldsymbol{p}$ is real multiplicative factor and $\boldsymbol{r}$ is purely imaginary $\boldsymbol{r} = \mathrm{i}\hbar\nabla_{\boldsymbol{p}}$). In this case the representation (13.27) for $\hat{T}$ is also valid, but now $\hat{U}$ has to be an operator which transforms $\hat{\boldsymbol{p}}$ into $-\hat{\boldsymbol{p}}$, i.e.

$$\hat{U}\psi_\alpha(\boldsymbol{p}) = \psi_\alpha(-\boldsymbol{p}) \quad .$$

We now return to the coordinate space representation, where for a particle with spin we expect

$$\hat{\boldsymbol{S}}\hat{T} = -\hat{T}\hat{\boldsymbol{S}} \quad , \tag{13.37}$$

[in analogy to (13.35)] as well as

$$\hat{J}\hat{T} = -\hat{T}\hat{J} \quad . \tag{13.38}$$

We have already seen that the form of $\hat{T}$ depends on the representation. Furthermore we have to choose a specific set of spin matrices if we consider particles with nonzero spin. We choose them in such a way that $\hat{S}_x$ and $\hat{S}_z$ are real operators and $\hat{S}_y$ is purely imaginary [see e.g. (2.21b) for particles with spin 1 and (2.21a) for particles with spin $\frac{1}{2}$]. This can always be achieved by a suitable unitary transformation.

The symbols $\boldsymbol{r}$, $\hat{S}_x$, and $\hat{S}_z$ are now real operators, whereas $\hat{\boldsymbol{p}}$, $\hat{\boldsymbol{L}}$, and $\hat{S}_y$ are purely imaginary. If $\hat{T}$ were equal to $\hat{K}$, then (13.31), (13.34), and (13.35) would be fulfilled as well as (13.37) for the component $\hat{S}_y$. On the other hand the equations for $\hat{S}_x$ and $\hat{S}_z$ would not be fulfilled. Therefore we have to choose a unitary operator $\hat{U}$ in the general representation (13.27) of $\hat{T}$ which commutes with $\hat{\boldsymbol{r}}$, $\hat{\boldsymbol{p}}$, $\hat{\boldsymbol{L}}$, and $\hat{S}_y$ and which obeys the relations

$$\hat{S}_x\hat{U} = -\hat{U}\hat{S}_x \quad \text{and} \quad \hat{S}_z\hat{U} = -\hat{U}\hat{S}_z \quad . \tag{13.39}$$

If we choose $\hat{U}$ to be a function of $\hat{S}_y$ alone, then the first three conditions

$$[\boldsymbol{r}, \hat{U}]_- = [\hat{\boldsymbol{p}}, \hat{U}]_- = [\hat{\boldsymbol{L}}, \hat{U}]_- = [\hat{S}_y, \hat{U}]_- = 0$$

are satisfied. On the other hand one can show (see the following problem) that (13.39) is also fulfilled if one chooses

$$\hat{U} = \exp(-\mathrm{i}\pi\hat{S}_y/\hbar) \quad . \tag{13.40}$$

Our considerations of rotations in Chap. 3 also led us to this expression. There we saw that the unitary operator $\exp(-\mathrm{i}\boldsymbol{\phi}\cdot\hat{\boldsymbol{S}}/\hbar)$ represents a rotation by an angle ϕ. Therefore a rotation by an angle π around the y-axis as in (13.40) transforms $\hat{S}_x$ into $-\hat{S}_x$, and $\hat{S}_z$ into $-\hat{S}_z$. For a particle with spin $\frac{1}{2}$ this takes on a simple form,

$$\hat{T} = -\mathrm{i}\hat{\sigma}_y\hat{K} \quad , \tag{13.41}$$

where $\hat{\sigma}_y$ is the Pauli matrix

$$\hat{\sigma}_y = \begin{pmatrix} 0 & -\mathrm{i} \\ \mathrm{i} & 0 \end{pmatrix} \quad .$$

EXERCISE

13.2 Commutation Relations Between $\hat{U}$ and $\hat{S}$

Problem. Prove, for $\hat{U} = \exp(-\mathrm{i}\pi\hat{S}_y/\hbar)$, the validity of the relations $\hat{U}\hat{S}_x = -\hat{S}_x\hat{U}$ and $\hat{U}\hat{S}_z = -\hat{S}_z\hat{U}$ by using the commutation relations between $\hat{S}_y$ and $\hat{S}_\pm = \hat{S}_z \pm \mathrm{i}\hat{S}_x$.

Exercise 13.2

Solution. Using the general commutator relations

$$[\hat{S}_i, \hat{S}_j]_- = \mathrm{i}\varepsilon_{ijk}\hat{S}_k \quad , \tag{1}$$

it follows directly that

$$[\hat{S}_y, \hat{S}_\pm]_- = \pm \hbar \hat{S}_\pm \quad . \tag{2}$$

These relations can then be cast into the form

$$\hat{S}_y\hat{S}_\pm = \hat{S}_\pm \hat{S}_y \pm \hbar\hat{S}_\pm = \hat{S}_\pm(\hat{S}_y \pm \hbar\mathbb{1}) \quad , \tag{3}$$

with the unit matrix $\mathbb{1}$. Now we assert that

$$\hat{S}_y^n\hat{S}_\pm = \hat{S}_\pm(\hat{S}_y \pm \hbar\mathbb{1})^n \tag{4}$$

holds for every n.

Proof. Obviously (4) is satisfied for $n = 1$. If we assume that it also holds for a certain $n > 1$, we only have to show that it is also true for $n + 1$, i.e.

$$\begin{aligned}\hat{S}_y^{n+1}\hat{S}_\pm &= \hat{S}_y\hat{S}_y^n\hat{S}_\pm \\ &= \hat{S}_y\{\hat{S}_\pm(\hat{S}_y \pm \hbar\mathbb{1})^n\} \\ &= \hat{S}_\pm(\hat{S}_y \pm \hbar\mathbb{1})(\hat{S}_y \pm \hbar\mathbb{1})^n \\ &= \hat{S}_\pm(\hat{S}_y \pm \hbar\mathbb{1})^{n+1} \quad .\end{aligned} \tag{5}$$

Now we can calculate $\hat{U}\hat{S}_\pm$:

$$\begin{aligned}\hat{U}\hat{S}_\pm &= \sum_n \left(\frac{-\mathrm{i}\pi}{\hbar}\right)^n \frac{1}{n!}\hat{S}_y^n\hat{S}_\pm \\ &= \hat{S}_\pm \sum_n \left(\frac{-\mathrm{i}\pi}{\hbar}\right)^n \frac{1}{n!}(\hat{S}_y \pm \hbar\mathbb{1})^n \\ &= \hat{S}_\pm \exp\left\{-\frac{\mathrm{i}\pi}{\hbar}(\hat{S}_y \pm \hbar\mathbb{1})\right\} \\ &= \hat{S}_\pm \hat{U} \exp\{\pm(-\mathrm{i}\pi)\mathbb{1}\} \quad .\end{aligned} \tag{6}$$

On the other hand, the last factor is equal to $-\mathbb{1}$, because

$$\exp(\pm\mathrm{i}\pi\mathbb{1}) = 1\exp(\pm\mathrm{i}\pi) = \mathbb{1}\cos\pi = -\mathbb{1} \quad , \tag{7}$$

and thus (6) becomes

$$\hat{U}\hat{S}_\pm = -\hat{S}_\pm\hat{U} \quad . \tag{8}$$

If we write down the real and imaginary parts separately, we get

$$\hat{U}\hat{S}_x = -\hat{S}_x\hat{U} \quad \text{and} \tag{9}$$

$$\hat{U}\hat{S}_z = -\hat{S}_z\hat{U} \quad . \tag{10}$$

This separation is possible because $\hat{U}$, $\hat{S}_x$ and $\hat{S}_z$ are real operators.

13.5 Many-Particle Systems

If we have a system of many particles, the time reversal operator $\hat{T}$ can be written as the product of unitary operators $\hat{U}$ belonging to the individual particles and the operators $\hat{K}$, i.e.

$$\hat{T} = \exp\{-\mathrm{i}\pi\hat{S}_{1y}/\hbar\} \cdots \exp\{-\mathrm{i}\pi\hat{S}_{ny}/\hbar\}\hat{K} \quad . \tag{13.42}$$

These operators may be placed in an arbitrary order, since each operator acts on the state of a different particle. Therefore they commute and the operator $\hat{T}$ obeys the relation (13.37) for each particle spin. Since all operators $\hat{S}_y$ are purely imaginary, all exponentials in (13.42) are real and therefore commute with $\hat{K}$. Knowing $\hat{K}^2 = 1$, we find that

$$\hat{T}^2 = \exp\{-2\mathrm{i}\pi\hat{S}_{1y}/\hbar\} \cdots \exp\{-2\mathrm{i}\pi\hat{S}_{ny}/\hbar\} \quad . \tag{13.43}$$

Each exponential represents a rotation by 2π, so that the corresponding coordinate space is transformed into itself. For particles with integer spin, the exponentials are equal to $+1$, whereas for particles with half-integer spin they are -1. Therefore $\hat{T}^2$ is $+1$ or -1, according to whether the number of particles in the system with half-integer spin is even or odd.

As we have already mentioned, if u_k is an energy eigenfunction, then so is $\hat{T}u_k$, and both have the same eigenvalue. Let us first assume that there is no degeneracy. In this case $\hat{T}u_k$ represents the same state as u_k, so that we can write

$$\hat{T}u_k = cu_k \quad , \tag{13.44}$$

with some complex number c, leading to

$$\hat{T}^2u_k = \hat{T}(cu_k) = c^*\hat{T}u_k = |c|^2u_k \quad . \tag{13.45}$$

$\hat{T}^2 = +1$, $|c|^2 = 1$ describes one possible situation. On the other hand, if $\hat{T}^2 = -1$ there is no c, in accordance with (13.46). This means that the application of $\hat{T}$ must yield a new state and the corresponding eigenvalue is degenerate.

In this case we can show that u_k and $\hat{T}u_k$ are orthogonal to each other. According to Exercise 13.1 we have $(\hat{T}\psi_1, \hat{T}\psi_2) = (\psi_1, \psi_2)^* = (\psi_2, \psi_1)$. If we now choose $\psi_1 = \hat{T}u_k$ and $\psi_2 = u_k$, we get

$$(\hat{T}^2u_k, \hat{T}u_k) = (u_k, Tu_k) \quad . \tag{13.46}$$

On the other hand, $\hat{T}^2 = -1$, so that the lhs of (13.46) is $-(u_k, \hat{T}u_k)$. Therefore both sides of (13.46) differ by a sign, and this is only possible if

$$(u_k, \hat{T}u_k) = 0 \quad , \tag{13.47}$$

which means that both states are orthogonal to each other. Hence for every state u_k we can find a well-defined degenerate state $\hat{T}u_k$, so that the number of degenerate states is always even.

13.6 Real Eigenfunctions

We want to consider a system which has no spin (or where the spin is irrelevant). In this case $\hat{U} = \mathbb{1}$ is the unit matrix, and $\hat{T} = \hat{K}$ if we choose to work in the space representation. Furthermore we assume that there is an operator $\hat{A}$, which commutes with $\hat{K}$ and whose eigenvalues A_μ are not degenerate. We shall denote its eigenfunctions by a_μ so that

$$\hat{A}a_r(\boldsymbol{r}, t) = A_\mu a_\mu(\boldsymbol{r}, t) \quad . \tag{13.48}$$

As we have explained above, $\hat{K}a_\mu$ represents the same state as a_μ, and we can write $\hat{K}a_\mu = a_\mu^* = ca_\mu$, where c is again complex. The function a_μ can be split up into its real and its imaginary parts, i.e. $a_\mu = v_\mu + \mathrm{i}w_\mu$, with two real functions v_μ and w_μ. The above equation now reads

$$v_\mu - \mathrm{i}w_\mu = c(v_\mu + \mathrm{i}w_\mu) \quad \text{or} \quad (1-c)v_\mu = \mathrm{i}(1+c)w_\mu \quad . \tag{13.49}$$

This shows that v_μ and w_μ are proportional to each other; therefore a_μ is real up to some complex factor. In this sense, in particular all nondegenerate eigenfunctions can be chosen real if the system is invariant under time-reversal.

Sometimes the above arguments can also be extended to the case of degenerate states. An interesting example is the real Hamiltonian which includes an arbitrary spherical potential, i.e. which fulfils the relations

$$[\hat{H}, \hat{L}^2]_- = 0 \quad , \quad [\hat{H}, \hat{K}]_- = 0 \quad \text{and} \quad [\hat{K}, \hat{L}^2]_- = 0 \quad . \tag{13.50}$$

It is well-known that the energy eigenfunctions (belonging to E_k) can be characterized by the quantum numbers k, l (angular momentum), and m (projection of the angular monentum), so that we can write these eigenfunctions as u_{klm}. The operator of the z-component of the angular momentum (with eigenvalues m) is purely imaginary: $\hat{L}_z = \mathrm{i}\hbar(\partial/\partial\phi)$. In accordance with (13.35) it does not commute with $\hat{K}$; therefore we cannot argue in the same way as above that $\hat{K}u_{klm} = cu_{klm}$, which would imply that u_{klm} is real (up to an irrelevant complex constant). This can also be seen directly by the fact that these eigenfunctions are proportional to the spherical harmonics $Y_{lm}(\theta, \phi)$, which are all complex for $m \neq 0$. On the other hand, if we restrict ourselves to functions with $m = 0$, then we only have eigenfunctions of $\hat{L}_z$ with the eigenvalue zero, i.e. $\hat{L}u_{kl0} = 0$. From the general relation (13.35) then follows, because of $\hat{L}_z\hat{K} = -\hat{K}\hat{L}_z$.

$$\hat{L}_z\hat{K}u_{kl0} = 0 \quad . \tag{13.51}$$

$\hat{K}u_{kl0}$ is therefore an eigenfunction of $\hat{L}_z$ with eigenvalue zero, too. If we do not have an additional degeneracy, we can be sure that the eigenfunctions of $\hat{H}$ and $\hat{L}$ with $m = 0$ are real functions in a generalized sense, i.e. up to a phase factor.

14. Dynamical Symmetries

We have seen in the preceding chapters that the symmetry and degeneracy of the states of a system are associated with each other. For example, a system that possesses rotational symmetry is usually degenerate with respect to the direction of the angular momentum, i.e. with respect to the eigenvalues of a particular component (usually J_z). An exception is the case $\boldsymbol{J} = 0$. In the case of the discrete symmetries of space-inversion and time-reversal discussed in the previous chapter, degeneracy is less common since the transformed states are quite frequently equal to the original states.

Subsequently we will see in more detail that beyond the degeneracies arising, say, in rotational symmetry there is the possibility of degeneracies of different origin. Such degeneracies are to be expected whenever the Schrödinger equation can be solved in more than one way, either in different coordinate systems, or in a single coordinate system which can be oriented in different directions. From our present considerations we should expect these degeneracies to be associated with some symmetry, too. These symmetries differ essentially from all the others considered so far, because their nature is not geometrical. They are called *dynamical symmetries*, since they are the consequence of particular forms of the Schrödinger equation or of the classical force law. We shall examine two relatively simple quantum mechanical examples which, starting from the corresponding classical systems, allow the derivation of the existence and then the general form of the dynamical symmetry in very nearly the same way as for geometrical symmetries. Of course this is not possible in general; indeed many quantum mechanical systems have no classical analogue.

14.1 The Hydrogen Atom

First we investigate the classical Kepler problem. In relative coordinates the Hamiltonian is given by

$$H = \boldsymbol{p}^2/2\mu - \kappa/r \quad . \tag{14.1}$$

Here μ is the reduced mass of the system and κ is a positive quantity (for the hydrogen atom $\kappa = Ze^2$). The bound solutions of the classical problem are ellipses, and the distance from perihelion P to aphelion A is called $2a$. If b denotes the length of the minor semiaxis, the eccentricity e is $e = (a^2 - b^2)^{1/2}/a$

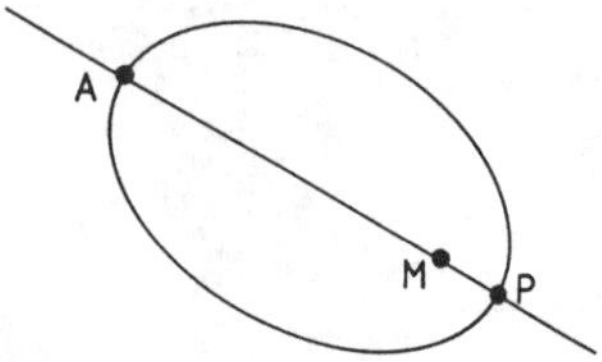

Fig. 14.1. Classical Kepler orbit with the centre of gravity M located in one of the foci of the ellipse

and the distance f from focus M to the geometrical centre is $f = a \cdot e$ (as shown in Fig. 14.1).

Since H is independent of time, the total energy E is a constant of motion; and, because H possesses rotational symmetry, the orbital angular momentum $\boldsymbol{L} = \boldsymbol{r} \times \boldsymbol{p}$ is a constant of motion. $\boldsymbol{L}$ is evidently an axial vector perpendicular to the plane of the orbit. It is easy to show that (Exercise 14.1)

$$E = -\kappa/2a \quad , \quad \text{and} \quad \boldsymbol{L}^2 = \mu\kappa a(1-e^2) \quad . \tag{14.2}$$

The rotational symmetry of H implies that the orbit lies in some plane through the centre of gravity M though it is not enough to ensure that the orbit is closed. A small deviation of the potential term from the Newtonian form $V(r) = -\kappa/r$ causes the major axis PA of the ellipse to precess slowly and so the orbit is not closed anymore. This suggests that there is some quantity other than H and $\boldsymbol{L}$ which is a constant of motion for potentials of the form $-\kappa/r$. It can be used to fix the orientation of the major axis in the orbital plane. We thus look for a constant vector $\boldsymbol{M}$ pointing from M to P or to A.

Such a vector has been known for a long time and is called the *Runge-Lenz vector*. It has the form

$$\boldsymbol{M} = \boldsymbol{p} \times \boldsymbol{L}/\mu - \kappa \boldsymbol{r}/r \quad , \tag{14.3}$$

for which $\boldsymbol{M}$ is easily seen to be a constant of motion. Since $\dot{\boldsymbol{L}} = 0$, we find, by differentiating $\boldsymbol{M}$ with respect to time, that

$$\dot{\boldsymbol{M}} = \dot{\boldsymbol{p}} \times \boldsymbol{L}/\mu - \kappa \dot{\boldsymbol{r}}/r + \kappa \boldsymbol{r}(\boldsymbol{r}\cdot\boldsymbol{p})/(\mu r^3) \quad , \tag{14.4}$$

using the relation $\boldsymbol{p} = \mu\dot{\boldsymbol{r}}$. If we take into account the relation $\boldsymbol{L} = \boldsymbol{r} \times \boldsymbol{p}$, we find

$$\dot{\boldsymbol{M}} = \boldsymbol{r}(\dot{\boldsymbol{p}}\cdot\boldsymbol{p})/\mu - \boldsymbol{p}(\dot{\boldsymbol{p}}\cdot\boldsymbol{r})/\mu - \boldsymbol{p}\kappa/(\mu r) + \boldsymbol{r}(\boldsymbol{r}\cdot\boldsymbol{p})\kappa/(\mu r^3) \quad .$$

Since Newton's equation is $\dot{\boldsymbol{p}} = -\kappa\boldsymbol{r}/r^3$ we see that

$$\dot{\boldsymbol{M}} = 0 \quad , \tag{14.5}$$

i.e. $\boldsymbol{M}$ is a constant of motion being a vector of length κe directed from M to perihelion P (see Exercise 14.2). There are two relations that are independent of the special choice of the orbital parameters a and e (see Exercise 14.3):

$$\boldsymbol{L}\cdot\boldsymbol{M} = 0 \quad \text{and} \quad \boldsymbol{M}^2 = 2E\boldsymbol{L}^2/\mu + \kappa^2 \quad . \tag{14.6}$$

In order to treat the hydrogen atom quantum mechanically we have to replace the classical functions by operators, which can be done easily for $\boldsymbol{r}, \boldsymbol{p}$ and $\boldsymbol{L}$. However, the vector products $\hat{\boldsymbol{p}} \times \hat{\boldsymbol{L}}$ and $-\hat{\boldsymbol{L}} \times \hat{\boldsymbol{p}}$ are not identical, because the components of $\hat{\boldsymbol{L}}$ and $\hat{\boldsymbol{p}}$ do not commute. Therefore the expression in (14.3) following from the replacement of the functions by operators is not Hermitian and we redefine $\hat{\boldsymbol{M}}$ as a symmetrized expression:

$$\hat{\boldsymbol{M}} = \frac{1}{2\mu}(\hat{\boldsymbol{p}} \times \hat{\boldsymbol{L}} - \hat{\boldsymbol{L}} \times \hat{\boldsymbol{p}}) - \kappa\frac{\boldsymbol{r}}{r} \quad . \tag{14.7}$$

By considering the commutation relations for $\boldsymbol{r}$ and $\hat{\boldsymbol{p}}$ we can show that

$$[\hat{\boldsymbol{M}}, \hat{H}]_- = 0 \quad , \tag{14.8}$$

$$\hat{\boldsymbol{L}} \cdot \hat{\boldsymbol{M}} = \hat{\boldsymbol{M}} \cdot \hat{\boldsymbol{L}} = 0 \quad \text{and} \tag{14.9}$$

$$\hat{\boldsymbol{M}}^2 = 2\hat{H}/\mu(\hat{\boldsymbol{L}}^2 + \hbar^2) + \kappa^2 \quad . \tag{14.10}$$

Relation (14.8) is the quantum-mechanical analogue to (14.5) and the expressions (14.9) and (14.10) correspond to (14.6).

The relations (14.7–10) were used by Pauli in 1926 to calculate the energy levels of the hydrogen atom. He regarded the three components of $\hat{\boldsymbol{M}}$ as generators of infinitesimal transformations. Following this method we deduce the algebra of the six generators $\hat{\boldsymbol{L}}$ and $\hat{\boldsymbol{M}}$, which consists of 15 commutation relations. Three of these are the known relations for the angular momentum operators:

$$[\hat{L}_i, \hat{L}_j]_- = \mathrm{i}\hbar\varepsilon_{ijk}\hat{L}_k \quad . \tag{14.11}$$

The commutators that include one component of $\hat{\boldsymbol{M}}$ and one component of $\hat{\boldsymbol{L}}$ yield nine further relations,

$$[\hat{M}_i, \hat{L}_j]_- = \mathrm{i}\hbar\varepsilon_{ijk}\hat{M}_k \quad . \tag{14.12}$$

After some additional calculations, we find the last three commutation relations,

$$[\hat{M}_i, \hat{M}_j]_- = -2\mathrm{i}\frac{\hbar}{\mu}\hat{H}\varepsilon_{ijk}\hat{L}_k \quad . \tag{14.13}$$

The components of $\hat{\boldsymbol{L}}$ constitute a closed algebra, as we have seen in Chap. 2, and generate the group O(3). The $\hat{\boldsymbol{L}}$ and $\hat{\boldsymbol{M}}$ together, however, do not form a closed algebra since, although the relations (14.12) involve only $\hat{\boldsymbol{M}}$ and $\hat{\boldsymbol{L}}$, relation (14.13) brings in the operator $\hat{H}$ as well. However, given that $\hat{H}$ is independent of time and commutes with $\hat{\boldsymbol{L}}$ and $\hat{\boldsymbol{M}}$, we can restrict ourselves to a subspace of the Hilbert space that corresponds to a particular eigenvalue E of $\hat{H}$. Then $\hat{H}$ in (14.13) can be replaced by its eigenvalue E. For bound states E has negative values, and it is convenient to replace $\hat{\boldsymbol{M}}$ by

$$\hat{\boldsymbol{M}}' = \sqrt{-\mu/2E}\,\hat{\boldsymbol{M}} \quad . \tag{14.14}$$

Evidently (14.12) and (14.13) are thus transformed into

$$[M'_i, L_j]_- = \mathrm{i}\hbar\varepsilon_{ijk}M'_k \quad \text{and} \tag{14.15}$$

$$[M'_i, M'_j]_- = \mathrm{i}\hbar\varepsilon_{ijk}L_k \quad . \tag{14.16}$$

14.2 The Group SO(4)

The six generators $\hat{\boldsymbol{L}}, \hat{\boldsymbol{M}}'$ constitute a closed algebra. To clarify this we relabel the indices of the components of $\hat{\boldsymbol{L}}$. First we write

$$\boldsymbol{r} = (r_1, r_2, r_3) \quad \text{and} \quad \hat{\boldsymbol{p}} = (\hat{p}_1, \hat{p}_2, \hat{p}_3) \quad , \tag{14.17}$$

for which

$$[r_i, \hat{p}_j]_- = \mathrm{i}\hbar\delta_{ij} \tag{14.18}$$

holds and we find, because of

$$\hat{L}_{ij} = r_i\hat{p}_j - r_j\hat{p}_i \quad , \tag{14.19}$$

the "natural indices"

$$\hat{\boldsymbol{L}} = (\hat{L}_{23}, \hat{L}_{31}, \hat{L}_{12}) \tag{14.20}$$

for the $\hat{L}$-operators. We now extend the indices to $i, j = 1, 2, 3, 4$ by introducing fourth components r_4 and $\hat{p}_4$ that fulfil (14.18) and (14.19) and for which

$$\hat{M}'_x = \hat{L}_{14} \quad , \quad \hat{M}'_y = \hat{L}_{24} \quad , \quad \hat{M}'_z = \hat{L}_{34} \tag{14.21}$$

is valid. It is easily verified that (14.18), (14.19) and (14.21) lead to the commutation relations (14.11), (14.15) and (14.16). The six generators $\hat{L}_{ij}$ obviously constitute a generalization of the three generators $\hat{\boldsymbol{L}}$ from three to four dimensions. The corresponding group can be shown to be the special orthogonal group or the proper rotation group in four dimensions, i.e. SO(4). It includes all real orthonormal 4×4 matrices with determinant equal to $+1$. This evidently does not represent a geometrical symmetry of the hydrogen atom since the fourth components r_4 and $\hat{p}_4$ are fictitious and cannot be identified with geometrical variables. For this reason SO(4) is said to describe a dynamical symmetry of the hydrogen atom. It contains the geometrical symmetry SO(3), generated by the angular momentum operators $\hat{L}_i$, as a subgroup.

It is essential to note that the SO(4) generators are obtained by restriction to bound states. For continuum states E is positive and the sign inside the square root of (14.14) has to be changed in order for $\hat{\boldsymbol{M}}'$ to be Hermitian. But then the sign on the rhs of (14.16) changes and the identifications of (14.21) are no longer possible. It turns out that the dynamical symmetry group in this case is isomorphic to the group of Lorentz transformations in one time- and three space-dimensions, rather than to the group of rotations in four space-dimensions. This is expressed by notation SO(3, 1).

14.3 The Energy Levels of the Hydrogen Atom

It now is comparatively simple to find the energy eigenvalues. We define the quantities

$$\hat{\boldsymbol{I}} = \tfrac{1}{2}(\hat{\boldsymbol{L}} + \hat{\boldsymbol{M}}') \quad \text{and} \quad \hat{\boldsymbol{K}} = \tfrac{1}{2}(\hat{\boldsymbol{L}} - \hat{\boldsymbol{M}}') \quad , \tag{14.22}$$

which satisfy the commutation relations

$$[\hat{I}_i, \hat{I}_j]_- = \mathrm{i}\hbar\varepsilon_{ijk}\hat{I}_k \quad , \tag{14.23}$$

$$[\hat{K}_i, \hat{K}_j]_- = \mathrm{i}\hbar\varepsilon_{ijk}\hat{K}_k \quad , \tag{14.24}$$

$$[\hat{I}_i, \hat{K}_j]_- = 0 \quad , \tag{14.25}$$

$$[\hat{\boldsymbol{I}}, \hat{H}]_- = [\hat{\boldsymbol{K}}, \hat{H}]_- = 0 \quad . \tag{14.26}$$

Because of (14.25) the algebras of the I_k and K_k are decoupled; therefore each $\hat{\boldsymbol{I}}$ and $\hat{\boldsymbol{K}}$ constitutes a SO(3)- or SU(2)-algebra, and we at once realize the eigenvalues to be

$$\hat{\boldsymbol{I}}^2 = i(i+1)\hbar^2 \quad , \quad i = 0, \tfrac{1}{2}, 1, \dots \quad , \tag{14.27}$$

$$\hat{\boldsymbol{K}}^2 = k(k+1)\hbar^2 \quad , \quad k = 0, \tfrac{1}{2}, 1, \dots \quad . \tag{14.28}$$

Relations (14.23)–(14.26) show that the SO(4) group is of rank 2, because one operator $\hat{I}_i$ and one operator $\hat{K}_j$ form a maximal system of commutating generators. Thus there are two Casimir operators which may evidently be chosen to be

$$\hat{\boldsymbol{I}}^2 = \tfrac{1}{4}(\hat{\boldsymbol{L}} + \hat{\boldsymbol{M}}')^2 \quad , \quad \text{and} \tag{14.29}$$

$$\hat{\boldsymbol{K}}^2 = \tfrac{1}{4}(\hat{\boldsymbol{L}} - \hat{\boldsymbol{M}}')^2 \quad . \tag{14.30}$$

Alternatively they may be chosen to be the sum and difference of $\hat{\boldsymbol{I}}^2$ and $\hat{\boldsymbol{K}}^2$:

$$\hat{C}_1 = \hat{\boldsymbol{I}}^2 + \hat{\boldsymbol{K}}^2 = \tfrac{1}{2}(\hat{\boldsymbol{L}}^2 + \hat{\boldsymbol{M}}'^2) \quad , \tag{14.31}$$

$$\hat{C}_2 = \hat{\boldsymbol{I}}^2 - \hat{\boldsymbol{K}}^2 = \hat{\boldsymbol{L}} \cdot \hat{\boldsymbol{M}}' \quad . \tag{14.32}$$

Equation (14.9) shows that $\hat{C}_2 = 0$, so that we should only deal with that part of SO(4) for which $\hat{\boldsymbol{I}}^2 = \hat{\boldsymbol{K}}^2$. Thus $i = k$, and the possible eigenvalues of $\hat{C}_1$ are

$$C_1 = 2k(k+1)\hbar^2 \quad , \quad k = 0, \tfrac{1}{2}, 1, \dots \quad . \tag{14.33}$$

Transforming (14.31) by taking into account (14.14) and (14.10) yields

$$\hat{C}_1 = \tfrac{1}{2}(\hat{\boldsymbol{L}}^2 - \tfrac{\mu}{2E}\hat{\boldsymbol{M}}^2) = -\mu\kappa^2/4E - \tfrac{1}{2}\hbar^2 \quad , \tag{14.34}$$

so that, using (14.33), the energy eigenvalues are found to be

$$E = -\mu\kappa^2/[2\hbar^2(2k+1)^2] \quad \text{with} \quad k = 0, \tfrac{1}{2}, 1, \dots \quad . \tag{14.35}$$

Note that one is allowed to use half odd-integer values for i and k; as we soon will see this does not yield any contradiction for the physical quantity $\hat{\boldsymbol{L}} = \hat{\boldsymbol{I}} + \hat{\boldsymbol{K}}$. Using the triangle rule we see that l [in $\hat{\boldsymbol{L}}^2 = l(l+1)\hbar^2$] can have any value in the interval $i + k = 2k, \dots, |i - k| = 0$ (subsequent values differ by steps of one unit). Obviously l is an integer, as it should be in the case of the orbital angular momentum. Also the degeneracy of states is reproduced correctly: $\hat{I}_z$ and $\hat{K}_z$ can each have $2k + 1$ independent eigenvalues, and therefore there are altogether $(2k+1)^2$ states. Also, in the preceding chapter we recognized that $\hat{\boldsymbol{L}}$ is an axial vector, not changing its sign on space inversion. But it is apparent from (14.7) that $\hat{\boldsymbol{M}}$ is a polar vector, which does change sign. We thus expect that states characterized by the symmetry generators $\hat{\boldsymbol{L}}$ and $\hat{\boldsymbol{M}}$ need not have well-defined parity. This is actually the case, since states of even and odd l are degenerate in the hydrogen atom.

14.4 The Classical Isotropic Oscillator

The three-dimensional isotropic harmonic oscillator is described by the Hamiltonian

$$H = \boldsymbol{p}^2/2m + \tfrac{1}{2}Kr^2 \quad . \tag{14.36}$$

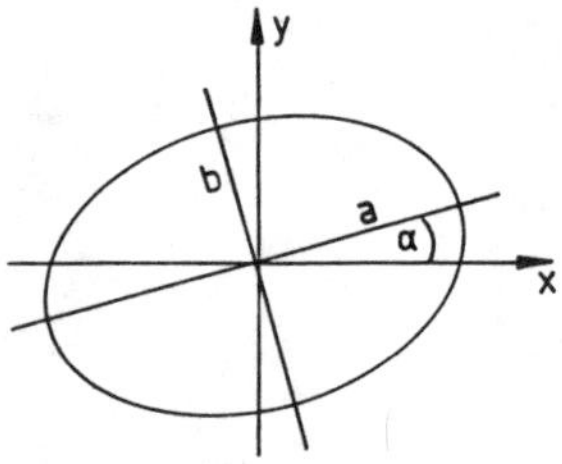

Fig. 14.2. Orbit of the classical harmonic oscillator. Centre of force in the middle of the ellipse

A particular classical orbit is an ellipse with semiaxes a and b. The major semiaxis forms an angle α with the x aixs (see Fig. 14.2). As in the Kepler problem, H and $\boldsymbol{L}$ are constants of motion with values given by

$$E = \tfrac{1}{2}K(a^2 + b^2) \quad \text{and} \tag{14.37}$$

$$\boldsymbol{L}^2 = mKa^2b^2 \quad . \tag{14.38}$$

The fact that the orbit is closed again suggests that there is some other constant of motion that can also be used to characterize the angle α. But comparing the corresponding figures, (Fig. 14.1 and Fig. 14.2) there is a striking difference: In the Kepler problem the centre of force is a focus of the ellipse, whereas in this problem it is at its centre.

Therefore the directions OA and OP are not equivalent in the Kepler problem, and the minor semiaxis is not a symmetry element. In contrast, in the oscillator problem both directions along the major semiaxis as well as the minor semiaxis are good symmetry elements. We thus expect that the additional constant of motion is not a vector (as in the Kepler problem), but rather a quadrupole tensor. We get

$$Q_{xy} = \tfrac{1}{2}a(a^2 - b^2)\sin 2\alpha \quad , \quad Q_{yz} = Q_{xz} = 0$$

$$Q_0 = \tfrac{1}{2}a/\sqrt{3}(a^2 + b^2) \quad \text{and} \quad Q_1 = \tfrac{1}{2}a(a^2 - b^2)\cos 2\alpha \quad . \tag{14.39}$$

As already expected from Fig. 14.2, the tensor components remain unchanged if α is replaced by $\alpha + \beta$, or if a and b are interchanged and α is replaced by $\alpha \pm \frac{1}{2}\pi p$ at the same time.

14.4.1 The Quantum Mechanical Isotropic Oscillator

Since the quantum mechanical problem separates in cartesian coordinates, the energy levels are easily found to be

$$E_n = (n + \tfrac{3}{2})\hbar\sqrt{K/m} \quad \text{with} \quad n = n_x + n_y + n_z \quad ,$$
$$n_x, n_y, n_z = 0, 1, 2, \ldots . \tag{14.40}$$

Obviously the states are $\frac{1}{2}(n+1)(n+2)$-fold-degenerate. The parity of the states is positive or negative, depending on n being even or odd. Thus the only possible values of l are just $n, n-2 \ldots$. down to 1 or 0, and each l occurs just once. It can be shown that in SU(3) it is possible to construct a Casimir operator from $\hat{\boldsymbol{L}}$ and the quadrupole tensor. Using the spatial representation one can find the relation

$$\begin{aligned}\hat{C} &= \hat{\boldsymbol{L}}^2/\hbar^2 + mK/(\hbar^2 a^2)(Q_{xy}^2 + Q_{yz}^2 + Q_{zx}^2 + Q_0^2 + Q_1^2) \\ &= -3 + \tfrac{1}{3}\hbar^2(\sqrt{mK}\boldsymbol{r}^2 + \hat{\boldsymbol{p}}^2/\sqrt{mK})^2 \quad . \end{aligned} \tag{14.41}$$

If we express the quantities in brackets in terms of the Hamiltonian, we find that

$$\hat{C} = -3 + 4m/(3\hbar^2 K)\hat{H}^2 \quad . \tag{14.42}$$

Substitution of the eigenvalues (14.40) yields

$$C = \tfrac{4}{3}(n^2 + 3n) \tag{14.43}$$

for the nth eigenvalue. Since SU(3) is of rank 2, there are two Casimir operators, which can be characterized by two parameters λ and μ, and which take on the values $0, 1, 2, \ldots$. The general expression for one of the Casimir operators in terms of these parameters is

$$C = \tfrac{4}{3}(\lambda^2 + \lambda\mu + \mu^2 + 3\lambda + 3\mu) \quad . \tag{14.44}$$

Comparison with (14.43) shows that only the representations of SU(3) with $(\lambda, \mu) = (n, 0)$ are realized by the isotropic oscillator. The situation here is somewhat analogous to that in the hydrogen atom, where only the representations of SO(4) with $i = k$ are realized. In contrast to the hydrogen atom we have seen that there is no parity mixing in the isotropic oscillator, since the l values in each degenerate state are either all even or odd. This is to be expected, because all eight generators, i.e. the three components of $\hat{\boldsymbol{L}}$ and the five components of the quadrupole tensor, do not change sign on space inversion. The group theoretical classification of the isotropic harmonic oscillator plays an important role in the explanation of the structure of light atomic nuclei in the framework of the shell model[1].

EXERCISE

14.1 Energy and Radial Angular Momentum of the Hydrogen Atom

Problem. Derive relation (14.2).

Solution. In spherical coordinates the Hamiltonian (14.1) is given by

$$H = \frac{\boldsymbol{p}^2}{2\mu} + \frac{\boldsymbol{L}^2}{2\mu r^2} - \frac{\kappa}{r} \quad . \tag{1}$$

[1] e.g. J.M. Eisenberg, W. Greiner: *Nuclear Theory, Vol. III: Microscopic Theory of the Nucleus*, 2nd ed. (North-Holland, Amsterdam 1972).

Exercise 14.1

At the aphelion $\boldsymbol{r}_a$ and the perihelion $\boldsymbol{r}_p$ the radial momentum is $p_r = 0$, so that we have two equations for the total energy:

$$E = \frac{L^2}{2\mu r_a^2} - \frac{\kappa}{r_a} \quad \text{(Aphelion)} \quad . \tag{2a}$$

$$E = \frac{L^2}{2\mu r_p^2} - \frac{\kappa}{r_p} \quad \text{(Perihelion)} \quad . \tag{2b}$$

If we divide (2a) by r_p^2 and (2b) by r_a^2 and then subtract them from each other, the unknown angular momentum is eliminated, yielding

$$E(r_a + r_p) = -\kappa \quad . \tag{3}$$

By subtracting (2) directly we have, in addition, that

$$\frac{L^2}{2\mu}(r_a + r_p) = \kappa r_a r_p \quad . \tag{4}$$

With the help of the geometrical relations $r_a = a + f$, $r_p = a - f$ and $f = (a^2 - b^2)^{1/2}$, we find that

$$r_p + r_a = 2a \quad \text{and} \quad r_a r_p = a^2 - f^2 = b^2 \quad . \tag{5}$$

Equations (3) and (4) thus yield the desired result.

$$E = -\frac{\kappa}{2a} \quad . \tag{6}$$

$$L^2 = 2\mu\kappa r_a r_p/(r_a + r_p) = \mu\kappa b^2/a = \mu\kappa a(1 - e^2) \quad . \tag{7}$$

EXERCISE

14.2 The Runge-Lenz Vector

Problem. Show that the Runge-Lenz vector can be written in the form

$$\hat{M} = \kappa e \boldsymbol{r}_p / r_p \quad , \tag{1}$$

where $\boldsymbol{r}_p$ is the vector directed from the centre of the ellipsoidal orbit to the perihelion, and e is the eccentricity.

Solution. We start with (14.3). Since $\hat{M}$ is a constant of motion, we can evaluate the rhs for $r = r_p$ and, because of the relation $\boldsymbol{L} = \boldsymbol{r} \times \boldsymbol{p}$, $\boldsymbol{r}, \boldsymbol{p}$ and $\boldsymbol{L}$ form a right-handed system of vectors at every point of the orbit. At the perihelion $\boldsymbol{r}$ and $\boldsymbol{p}$ are perpendicular, that is, $\boldsymbol{p} \times \boldsymbol{L}$ is oriented in the same way as $\boldsymbol{r}$. This can easily be seen using the general rule for double cross products:

$$\boldsymbol{p} \times \boldsymbol{L} = \boldsymbol{p} \times (\boldsymbol{r} \times \boldsymbol{p}) = \boldsymbol{r} p^2 - \boldsymbol{p}(\boldsymbol{r} \cdot \boldsymbol{p}) \quad , \tag{2}$$

yielding

$$(\boldsymbol{p} \times \boldsymbol{L})_{\mathrm{p}} = \boldsymbol{r}_{\mathrm{p}} p^2 = \boldsymbol{r}_{\mathrm{p}} L^2 / r_{\mathrm{p}}^2 \tag{3}$$

at the perihelion. With the help of (3) we have, using $r_{\mathrm{p}} = a - f = a(1 - e)$,

$$\frac{1}{\mu}(\boldsymbol{p} \times \boldsymbol{L})_{\mathrm{p}} = \boldsymbol{r}_{\mathrm{p}} \frac{\kappa a(1 - e^2)}{a(1 - e) r_{\mathrm{p}}} = \boldsymbol{r}_{\mathrm{p}} \kappa (1 + e)/r_{\mathrm{p}} \quad . \tag{4}$$

This enables us to write the desired equation for $\hat{M}$, i.e.

$$\hat{M} = \frac{1}{\mu}(\boldsymbol{p} \times \boldsymbol{L})_{\mathrm{p}} - \frac{\kappa \boldsymbol{r}_{\mathrm{p}}}{r_{\mathrm{p}}} = \kappa e \frac{\boldsymbol{r}_{\mathrm{p}}}{r_{\mathrm{p}}} \quad . \tag{5}$$

EXERCISE

14.3 Properties of the Runge-Lenz Vector $\hat{M}$

Problem. Derive relation (14.6).

Solution. First we prove (6a) by writing

$$\begin{aligned} \boldsymbol{L} \cdot \boldsymbol{M} &= \frac{1}{\mu} \boldsymbol{L} \cdot (\boldsymbol{p} \times \boldsymbol{L}) - \frac{\kappa}{r} \boldsymbol{L} \times \boldsymbol{r} \\ &= -\frac{\kappa}{r} (\boldsymbol{r} \times \boldsymbol{p}) \times \boldsymbol{r} = 0 \quad , \end{aligned} \tag{1}$$

because a mixed product including two identical vectors always vanishes. Next we establish (6b) through

$$\begin{aligned} \boldsymbol{M}^2 &= \frac{1}{\mu^2} (\boldsymbol{p} \times \boldsymbol{L})^2 - 2 \frac{\kappa}{\mu r} (\boldsymbol{p} \times \boldsymbol{L}) \cdot \boldsymbol{r} + \kappa^2 \\ &= \frac{1}{\mu^2} [\boldsymbol{p}^2 \boldsymbol{L}^2 - (\boldsymbol{p} \cdot \boldsymbol{L})^2] - 2 \frac{\kappa}{\mu r} (\boldsymbol{r} \times \boldsymbol{p}) \cdot \boldsymbol{L} + \kappa^2 \\ &= \frac{1}{\mu^2} (\boldsymbol{p}^2 \boldsymbol{L}^2) - 2 \frac{\kappa}{\mu r} \boldsymbol{L}^2 + \kappa^2 \quad , \end{aligned} \tag{2}$$

because again $\boldsymbol{p} \cdot \boldsymbol{L} = \boldsymbol{p} \cdot (\boldsymbol{r} \times \boldsymbol{p}) = 0$. The first two terms can be combined to produce

$$\begin{aligned} \boldsymbol{M}^2 &= \frac{2}{\mu} \boldsymbol{L}^2 \left(\frac{\boldsymbol{p}^2}{2\mu} - \frac{\kappa}{r} \right) + \kappa^2 \\ &= \frac{2}{\mu} \boldsymbol{L}^2 H + \kappa^2 = \frac{2}{\mu} E \boldsymbol{L}^2 + \kappa^2 \quad . \end{aligned} \tag{3}$$

EXERCISE

14.4 The Commutator Between $\hat{M}$ and $\hat{H}$

Problem. Prove the quantum mechanical relation

$$[\hat{M}, \hat{H}]_- = 0 \quad \text{for} \quad \hat{H} = \frac{\hat{p}^2}{2m} - \frac{c}{r} \quad \text{and}$$

$$\hat{M} = \frac{1}{2\mu}(\hat{p} \times \hat{L} - \hat{L} \times \hat{p}) - c\frac{\hat{r}}{r} \quad .$$

Solution. Since we only have to deal with operators in the rest of the exercises in this chapter, confusion is hardly possible so from now on for convenience we drop the operator sign. It is well known that the above Hamiltonian commutes with the angular momentum operators, $[H, L_i]_- = 0$, for all i. Next we calculate two auxiliary commutators, necessary for the subsequent calculations:

$$\begin{aligned}[p_i, H]_- &= -c\left[p_i, \frac{1}{r}\right]_- \\ &= \mathrm{i}\hbar c\left(\partial_i \frac{1}{r} - \frac{1}{r}\partial_i\right) = -\mathrm{i}\hbar c \frac{x_i}{r^3} \quad ,\end{aligned} \tag{1}$$

$$\left[\frac{x_i}{r}, \sum_k p_k^2\right]_- = -\hbar^2 \sum_k \left\{\frac{x_i}{r}\partial_k^2 - \partial_k^2 \frac{x_i}{r}\right\} \quad .$$

Now we have

$$\begin{aligned}\sum_k \partial_k \left\{\partial_k \frac{x_i}{r}\right\} &= \sum_k \partial_k \left\{\frac{\delta_{ik}}{r} - \frac{x_i x_k}{r^3} + \frac{x_i}{r}\partial_k\right\} \\ &= \sum_k \left\{-2\delta_{ik}\frac{x_k}{r^3} + 2\frac{\delta_{ik}}{r}\partial_k + 3\frac{x_i x_k^2}{r^5} - 2\frac{x_i x_k}{r^3}\partial_k - \frac{x_i}{r^3} + \frac{x_i}{r}\partial_k^2\right\} \quad ,\end{aligned}$$

and therefore

$$\left[\frac{x_i}{r}, \sum_k p_k^2\right]_- = 2\hbar^2\left(-\frac{x_i}{r^3} + \frac{1}{r}\partial_i - \frac{x_i}{r^3}\sum_k x_k \partial_k\right) \tag{2}$$

For the commutator in question we have

$$[M_i, H]_- = \frac{1}{2\mu}[(\boldsymbol{p} \times \boldsymbol{L} - \boldsymbol{L} \times \boldsymbol{p})_i, H]_- - c\left[\frac{x_i}{r}, H\right]_- \quad , \tag{3}$$

and the second commutator can be further reduced yielding

$$-\frac{c}{2m}\left[\frac{x_i}{r}, \sum_k p_k^2\right]_- \quad ,$$

Exercise 14.4

which just corresponds to (2). The first commutator is then treated in the following way:

$$[(\boldsymbol{p}\times\boldsymbol{L}-\boldsymbol{L}\times\boldsymbol{p})_i, H]_- = \sum_{jk}\varepsilon_{ijk}[(p_jL_k - L_jp_k), H]_- \quad . \tag{4}$$

Since H and $\boldsymbol{L}$ commute, H can be brought into the left-most position [note (1)] so that

$$p_jL_kH - L_jp_kH = p_jHL_k - L_jHp_k + \mathrm{i}\hbar cL_j\frac{x_k}{r^3}$$

$$= Hp_jL_k - HL_jp_k + \mathrm{i}\hbar c\left(L_j\frac{x_k}{r^3} - \frac{x_j}{r^3}L_k\right) \quad .$$

Now we insert $L_j = \sum_{m,n}\varepsilon_{jmn}x_mp_n$ and the analogous expression for L_k and find

$$[(\boldsymbol{p}\times\boldsymbol{L}-\boldsymbol{L}\times\boldsymbol{p})_i, H]_- = \mathrm{i}\hbar c\sum_{jkmn}\varepsilon_{ijk} \times\left\{\varepsilon_{jmn}x_mp_n\frac{x_k}{r^3} - \frac{x_j}{r^3}\varepsilon_{kmn}x_mp_n\right\} \quad . \tag{5}$$

To proceed further we use the simply derived auxiliary commutator

$$\left[p_n, \frac{x_k}{r^3}\right]_- = -\mathrm{i}\hbar\left(\frac{\delta_{kn}}{r^3} - 3\frac{x_kx_n}{r^5}\right)$$

and thus we have

$$[(\boldsymbol{p}\times\boldsymbol{L}-\boldsymbol{L}\times\boldsymbol{p})_i, H]_-$$

$$= -(\mathrm{i}\hbar)^2c\sum_{jkmn}\varepsilon_{ijk}\left\{\varepsilon_{jmn}\frac{x_mx_k}{r^3} - \varepsilon_{kmn}\frac{x_jx_m}{r^3}\right\}\partial_n$$

$$- (\mathrm{i}\hbar)^2c\sum_{jkm}\varepsilon_{ijk}\varepsilon_{jmk}\frac{x_m}{r^3} + 3(\mathrm{i}\hbar)^2c\sum_{jkmn}\varepsilon_{ijk}\varepsilon_{jmn}\frac{x_kx_mx_n}{r^5} \quad . \tag{6}$$

$\sum_{m,n}\varepsilon_{ijmn}x_mx_n$ vanishes for all j, as can easily be seen by explicit notation, since $[x_m, x_n]_- = 0$ in the last term of (6). The indices j and k may be interchanged in the second term of the first sum, yielding (by use of $\varepsilon_{ikj} = -\varepsilon_{ijk}$) the first term of this sum once again. Now we have

$$-2c(\mathrm{i}\hbar)^2\sum_{jkmn}\varepsilon_{ijk}\varepsilon_{jmn}\frac{x_mx_k}{r^3}\partial_n = -2c(\mathrm{i}\hbar)^2\sum_{kmn}(\delta_{km}\delta_{in} - \delta_{kn}\delta_{im})\frac{x_mx_k}{r^3}\partial_n$$

$$= 2c(\mathrm{i}\hbar)^2\left(\frac{x_i}{r^3}\sum_k x_k\partial_k - \frac{1}{r}\partial_i\right) \quad ,$$

using the relation

$$\sum_i\varepsilon_{ijk}\varepsilon_{imn} = \delta_{jm}\delta_{kn} - \delta_{jn}\delta_{km} \quad , \tag{7}$$

Exercise 14.4

which can easily be verified. The second sum in (6) can be simplified with the help of this relation, too, i.e.

$$-(\mathrm{i}\hbar)^2 c \sum_{jkm} \varepsilon_{ijk}\varepsilon_{jmk}\frac{x_m}{r^3} = -(\mathrm{i}\hbar)^2 c \sum_{jm}(\delta_{ij}\delta_{jm} - \delta_{im}\delta_{jj})\frac{x_m}{r^3}$$

$$= -(\mathrm{i}\hbar)^2 c\left(\frac{x_i}{r^3} - 3\frac{x_i}{r^3}\right) = 2c(\mathrm{i}\hbar)^2\frac{x_i}{r^3} \quad .$$

Collecting all the contributing terms, we find on the one hand, using (6), that

$$\frac{1}{2m}[(\boldsymbol{p}\times\boldsymbol{L} - \boldsymbol{L}\times\boldsymbol{p})_i, H]_- = \frac{c}{m}(\mathrm{i}\hbar)^2\left(\frac{x_i}{r^3} - \frac{1}{r}\partial_i + \frac{x_i}{r^3}\sum_k x_k\partial_k\right) \tag{8}$$

and on the other hand from (2) that

$$-c\left[\frac{x_i}{r}, H\right]_- = -\frac{c}{m}(\mathrm{i}\hbar)^2\left(\frac{x_i}{r^3} - \frac{1}{r}\partial_i + \frac{x_i}{r^3}\sum_k x_k\partial_k\right) \quad . \tag{9}$$

Adding Eqs. (8) and (9) yields $[M_i, H]_- = 0$.

EXERCISE

14.5 The Scalar Product $\hat{L}\cdot\hat{M}$

Problem. Prove the quantum mechanical relation $\hat{\boldsymbol{L}}\cdot\hat{\boldsymbol{M}} = 0$, using the definition of $\hat{\boldsymbol{M}}$ given in Exercise 14.4.

Solution. a) We have $\boldsymbol{L}\cdot\boldsymbol{r}/r = 0$ since

$$\sum_i L_i\frac{x_i}{r} = \sum_{imn}\varepsilon_{imn}x_m p_n\frac{x_i}{r} = \sum_{imn}\varepsilon_{imn}p_n x_m\frac{x_i}{r} + \mathrm{i}\hbar\sum_{imn}\varepsilon_{imn}\delta_{mn}\frac{x_i}{r} \quad . \tag{1}$$

Since $\sum_{i,m}\varepsilon_{imn}x_m x_i \equiv 0$, the first sum vanishes, as does the second because of $\varepsilon_{imm} \equiv 0$.

b) Furthermore we have $\boldsymbol{L}\cdot(\boldsymbol{p}\times\boldsymbol{L}) - \boldsymbol{L}\cdot(\boldsymbol{L}\times\boldsymbol{p}) = 0$ due to

$$\sum_{ijk}\varepsilon_{ijk}L_i(p_jL_k - L_jp_k) = \sum_i L_i\sum_{jkmn}\varepsilon_{ijk}(\varepsilon_{kmn}p_j m_m p_n - \varepsilon_{jmn}x_m p_n p_k)$$

$$= \sum_i L_i\sum_{jmn}(\delta_{im}\delta_{jn} - \delta_{in}\delta_{jm})p_j x_m p_n - \sum_i L_i\sum_{kmn}(\delta_{km}\delta_{in} - \delta_{kn}\delta_{im})x_m p_n p_k$$

$$= \sum_i L_i\left\{\sum_j(p_jx_ip_j - p_jx_jp_i - x_jp_ip_j + x_ip_jp_j)\right\} \tag{2}$$

(in the last transformation we have changed the summation index from k to j). Using the commutation relations for $\boldsymbol{x}$ and $\boldsymbol{p}$, $[p_j, x_i]_- = -\mathrm{i}\hbar\delta_{ij}$, we end up with

Exercise 14.5

$$\sum_i L_i\left\{\sum_j (p_jx_ip_j - p_jx_jp_j - x_jp_ip_j + x_ip_j^2)\right\}$$

$$= \sum_{ijmn} \varepsilon_{imn}x_m\Big(pn \underbrace{p_jx_i}_{x_ip_j - i\hbar\delta_{ij}} p_j - p_np_j \underbrace{x_jp_i}_{p_ix_j + i\hbar\delta_{ij}} - pn \underbrace{x_jp_i}_{p_ix_j + i\hbar\delta_{ij}} p_j + \underbrace{p_nx_i}_{x_ip_n - i\hbar\delta_{ni}} p_j^2\Big)$$

$$= \sum_{ijmn} \varepsilon_{imn}x_m\Big(\underbrace{p_nx_i}_{x_ip_n - i\hbar\delta_{in}} p_j^2 - i\hbar\delta_{ij}p_np_j - p_np_ip_jx_j - i\hbar\delta_{ij}p_np_i$$

$$- \underbrace{p_np_i}_{\substack{=0 \text{ after sum}\\ \text{over } n,i}} x_jp_j - i\hbar p_n\delta_{ij}p_j + \underbrace{x_i}_{\substack{=0 \text{ after sum}\\ \text{over } m,i}} p_np_j^2 - i\hbar \underbrace{\delta_{ni}}_{\substack{=0 \text{ since}\\ \varepsilon_{imi}=0}} p_j^2\Big)$$

$$= \sum_{ijmn} \varepsilon_{imn}x_m\Big(\underbrace{x_i}_{\substack{=0 \text{ after sum}\\ \text{over } i,m}} p_np_j^2 - i\hbar \underbrace{\delta_{ni}}_{\substack{=0 \text{ since}\\ \varepsilon_{imi}=0}} p_j^2 - i\hbar\delta_{ij}p_np_j$$

$$- \underbrace{p_n}_{\substack{=0 \text{ after sum}\\ \text{over } n,i}} p_ip_jx_j - i\hbar\delta_{ij}p_np_j - i\hbar p_n\delta_{ij}p_j\Big)$$

$$= \sum_{ijmn} \varepsilon_{imn}x_m(-3i\hbar\delta_{ij}p_np_j) = 0 \qquad (3)$$

since

$$\sum_j \delta_{ij}p_j = p_i \quad , \quad \sum_{imn} \varepsilon_{imn}x_mp_np_i = \sum_m x_m \sum_{in} \varepsilon_{imn}p_np_i = 0 \quad ,$$

Note that $\sum_k \delta_{kk} = 3$, and that the last sum vanishes for the same reason given for the case of the position operators in Exercise 14.4.

EXERCISE

14.6. Determination of $\hat{M}^2$

Problem. Prove the quantum mechanical relation $\hat{\boldsymbol{M}}^2 = 2\hat{H}/\mu(\hat{\boldsymbol{L}} + \hbar^2) + c^2$. For the definition of $\hat{\boldsymbol{M}}$ see Exercise 14.4.

Solution. First we prove the auxiliary relation

$$(\boldsymbol{L}\times\boldsymbol{p})_i = \sum jk\varepsilon_{ijk}L_jp_k = \sum_{jk} \varepsilon_{ijk}p_kL_j + i\hbar \sum_{jkm} \varepsilon_{ijk}\varepsilon_{jkm}p_m$$

$$= -(\boldsymbol{p}\times\boldsymbol{L})_i + 2i\hbar p_i \quad , \qquad (1)$$

Exercise 14.6

using $\sum_{j,k}\varepsilon_{jki}\varepsilon_{jkm} = 2\delta_{im}$ and the well-known commutation relation $[L_j, p_k]_- = \mathrm{i}\hbar\sum_m \varepsilon_{jkm}p_m$. Hence we find

$$\begin{aligned} \boldsymbol{M}^2 &= \left\{\frac{1}{2\mu}(\boldsymbol{p}\times\boldsymbol{L} - \boldsymbol{L}\times\boldsymbol{p}) - c\frac{\boldsymbol{r}}{r}\right\}^2 \\ &= \frac{1}{(2\mu)^2}4(\boldsymbol{p}\times\boldsymbol{L} - \mathrm{i}\hbar\boldsymbol{p})^2 - \frac{c}{\mu}(\boldsymbol{p}\times\boldsymbol{L} - \mathrm{i}\hbar\boldsymbol{p})\cdot\frac{\boldsymbol{r}}{r} \\ &\quad - \frac{c}{\mu}\frac{\boldsymbol{r}}{r}(\boldsymbol{p}\times\boldsymbol{L} - \mathrm{i}\hbar\boldsymbol{p}) + c^2\frac{\boldsymbol{r}^2}{r^2} \quad . \end{aligned} \tag{2}$$

The first quadratic expression on the rhs yields

$$(\boldsymbol{p}\times\boldsymbol{L} - \mathrm{i}\hbar\boldsymbol{p})^2 = (\boldsymbol{p}\times\boldsymbol{L})^2 - \mathrm{i}\hbar(\boldsymbol{p}\times\boldsymbol{L})\cdot\boldsymbol{p} - \mathrm{i}\hbar\boldsymbol{p}\cdot(\boldsymbol{p}\times\boldsymbol{L}) - \hbar^2\boldsymbol{p}^2 \quad .$$

These items are now evaluated separately [see Exercise 14.4, Eq. (7)]. First we obtain

$$\begin{aligned} (\boldsymbol{p}\times\boldsymbol{L})^2 &= \sum_{ijkmn}\varepsilon_{ijk}\varepsilon_{inm}p_jL_kp_nL_m \\ &= \sum_{jkmn}(\delta_{jn}\delta_{km} - \delta_{jm}\delta_{kn})p_jL_kp_nL_m \\ &= \sum_{jk}(p_jL_kp_jL_k - p_jL_kp_kL_j) \quad . \end{aligned}$$

Using $\sum_k L_kp_k = \sum_{k,m,n}\varepsilon_{kmn}x_mp_np_k = 0$, the second term vanishes, while in the first term the commutation of L_k and p_j yields

$$= \sum_{jk}p_j^2L_k^2 + \mathrm{i}\hbar\sum_{jkm}\varepsilon_{kjm}p_jp_mL_k = \boldsymbol{p}^2\boldsymbol{L}^2 \quad , \tag{3}$$

with the last sum again vanishing. Furthermore we have

$$\boldsymbol{p}\cdot(\boldsymbol{p}\times\boldsymbol{L}) = \sum_{ijk}\varepsilon_{ijk}p_ip_jL_k \equiv 0 \quad , \tag{4}$$

which is the same for the classical case, whereas

$$(\boldsymbol{p}\times\boldsymbol{L})\cdot\boldsymbol{p} = \sum_{ijk}\varepsilon_{ijk}p_jL_kp_i = \sum_{ijk}\varepsilon_{ijk}p_jp_iL_k + \mathrm{i}\hbar\sum_{ijkm}\varepsilon_{ijk}\varepsilon_{kim}p_jp_m = 2\mathrm{i}\hbar\boldsymbol{p}^2 \quad , \tag{5}$$

since the first sum vanishes and the relation $\sum_{k,i}\varepsilon_{kij}\varepsilon_{kim} = 2\delta_{jm}$ may be used in the second sum. Collecting the terms yields

$$(\boldsymbol{p}\times\boldsymbol{L} - \mathrm{i}\hbar\boldsymbol{p})^2 = \boldsymbol{p}^2\boldsymbol{L}^2 - 2(\mathrm{i}\hbar)^2\boldsymbol{p}^2 - \hbar^2\boldsymbol{p}^2 = \boldsymbol{p}^2(\boldsymbol{L}^2 + \hbar^2) \quad . \tag{6}$$

The further terms are treated as follows:

$$\begin{aligned} (\boldsymbol{p}\times\boldsymbol{L})\cdot\frac{\boldsymbol{r}}{r} &= \sum_{ijk}\varepsilon_{ijk}p_jL_k\frac{x_i}{r} \\ &= \sum_{ijk}\varepsilon_{ijk}p_j\frac{x_i}{r}L_k + \mathrm{i}\hbar\sum_{ijkm}\varepsilon_{ijk}\varepsilon_{kim}p_j\frac{x_m}{r} \\ &= \sum_{ijk}\varepsilon_{ijk}\frac{x_i}{r}p_jL_k - \mathrm{i}\hbar\sum_{ijk}\varepsilon_{ijk}\left(\frac{\delta_{ij}}{r} - \frac{x_ix_j}{r^3}\right)L_k + 2\mathrm{i}\hbar\boldsymbol{p}\cdot\frac{\boldsymbol{r}}{r} \end{aligned}$$

(using the commutator $[x_i/r, L_k]_- = \mathrm{i}\hbar\sum_m\varepsilon_{ikm}x_m/r$ in the penultimate line).

Exercise 14.6

The second sum, however, vanishes since $\varepsilon_{iik} \equiv 0$ and $\sum_{i,j,k} \varepsilon_{ijk} x_i x_j \equiv 0$. Substitution of $\sum_{i,j} \varepsilon_{ijk} x_i p_j$ by L_k yields

$$(\boldsymbol{p} \times \boldsymbol{L}) \cdot \frac{\boldsymbol{r}}{r} = \frac{1}{r} \sum_k L_k L_k + 2\mathrm{i}\hbar \boldsymbol{p} \cdot \frac{\boldsymbol{r}}{r} \quad . \tag{7}$$

Altogether we find that

$$\frac{\boldsymbol{r}}{r}(\boldsymbol{p} \times \boldsymbol{L} - \mathrm{i}\hbar \boldsymbol{p}) + (\boldsymbol{p} \times \boldsymbol{L} - \mathrm{i}\hbar \boldsymbol{p}) \cdot \frac{\boldsymbol{r}}{r} = \frac{1}{r} \boldsymbol{L}^2 - \mathrm{i}\hbar \frac{\boldsymbol{r}}{r} \cdot \boldsymbol{p} + \frac{1}{r} \boldsymbol{L}^2 + 2\mathrm{i}\hbar \boldsymbol{p} \frac{\boldsymbol{r}}{r} - \mathrm{i}\hbar \boldsymbol{p} \cdot \frac{\boldsymbol{r}}{r} \quad ,$$

and, using $\boldsymbol{p} \cdot \boldsymbol{r}/r = \boldsymbol{r}/r \cdot \boldsymbol{p} - 2\mathrm{i}\hbar/r$, we get

$$= \frac{2}{r} \boldsymbol{L}^2 - \frac{2}{r(\mathrm{i}\hbar)^2} = \frac{2}{r} (\boldsymbol{L}^2 + \hbar^2) \quad . \tag{8}$$

Collecting all these partial results we end up with

$$\boldsymbol{M}^2 = \frac{1}{(2\mu)^2} 4\boldsymbol{p}^2 (\boldsymbol{L}^2 + \hbar^2) - \frac{c}{\mu} \frac{2}{r} (\boldsymbol{L}^2 + \hbar^2) + c^2 \quad ,$$

and therefore

$$\boldsymbol{M}^2 = \frac{2}{\mu} \left(\frac{\boldsymbol{p}^2}{2\mu} - \frac{c}{r} \right) (\boldsymbol{L}^2 + \hbar^2) + c^2 \quad , \quad \boldsymbol{M}^2 = \frac{2}{\mu} H (\boldsymbol{L}^2 + \hbar^2) + c^2 \quad . \tag{9}$$

EXERCISE

14.7 Proof of the Commutation Relation for $[\hat{M}_i, \hat{L}_j]_-$

Problem. Prove the commutation relations

$$[\hat{M}_i, \hat{L}_j]_- = \mathrm{i}\hbar \sum_k \varepsilon_{ijk} \hat{M}_k \quad .$$

Solution. It is convenient to verify the relation explicitly for its components, e.g.

$$[M_x, L_y]_- = \mathrm{i}\hbar M_z = \left[\frac{1}{\mu} (\boldsymbol{p} \times \boldsymbol{L} - \mathrm{i}\hbar \boldsymbol{p})_x - c\frac{x}{r}, L_y \right]_-$$

$$= \frac{1}{\mu} (p_y L_z L_y - p_z L_y L_y - L_y p_y L_z + L_y p_z L_y) + \mathrm{i}\frac{\hbar}{\mu} (L_y p_x - p_x L_y) + c \left(L_y \frac{x}{r} - \frac{x}{r} L_y \right) \quad . \tag{1}$$

Exercise 14.7

We have

$$p_y L_z L_y = p_y L_y L_z - \mathrm{i}\hbar p_y L_x = L_y pIy L_z - \mathrm{i}\hbar p_y L_x \quad ,$$

and

$$p_z L_y L_y = L_y p_z L_y - \mathrm{i}\hbar p_x L_y \quad ,$$

and therefore

$$p_y L_z L_y - p_z L_y L_y - L_y p_y L_z + L_y p_z L_y = \mathrm{i}\hbar(p_x L_y - p_y L_x) \quad .$$

In addition,

$$L_y p_x - p_x L_y = -\mathrm{i}\hbar p_z \quad , \quad \text{and} \quad L_y \frac{x}{r} - \frac{x}{r} L_y = -\mathrm{i}\hbar \frac{z}{r} \quad ,$$

using the commutation relation $[L_i, p_k]_- = \mathrm{i}\hbar \sum_m \varepsilon_{ikm} p_m$ already mentioned in Exercise 14.6; thus we obtain

$$[M_x, L_y]_- = \mathrm{i}\frac{h}{\mu}(p_x L_y - p_y L_x - \mathrm{i}\hbar p_z) - \mathrm{i}\hbar c \frac{z}{r} = \mathrm{i}\hbar M_z \quad . \tag{2}$$

What we still have to prove is that, e.g., $[M_x, L_x] = 0$. Continuing, we write

$$\begin{aligned} &\left[\frac{1}{\mu}(\boldsymbol{p} \times \boldsymbol{L} - \mathrm{i}\hbar \boldsymbol{p})_x - c\frac{x}{r}, L_x\right]_- \\ &\quad = \frac{1}{\mu}(p_y L_z L_x - p_z L_y L_x - L_x p_y L_z + L_x p_z L_y) \\ &\qquad + \mathrm{i}\frac{\hbar}{\mu}(L_x p_x - p_x L_x) + c\left(L_x \frac{x}{r} - \frac{x}{r} L_x\right) \quad , \end{aligned} \tag{3}$$

which leads to

$$\begin{aligned} p_y L_z L_x &= p_y L_x L_z + \mathrm{i}\hbar p_y L_y = L_x p_y L_z - \mathrm{i}\hbar p_z L_z + \mathrm{i}\hbar p_y L_y \quad , \quad \text{and} \\ p_z L_y L_x &= p_z L_x L_y - \mathrm{i}\hbar p_z L_z = L_x p_z L_y + \mathrm{i}\hbar p_y L_y - \mathrm{i}\hbar p_z L_z \quad , \end{aligned}$$

yielding

$$p_y L_z L_x - p_z L_y L_x - L_x p_y L_z + L_x p_z L_y \equiv 0 \quad . \tag{4}$$

Furthermore $L_x p_x - p_x L_x = 0$, but also $L_x x/r - x/r L_x \equiv 0$ (since $[x_i/r, L_j]_- = \mathrm{i}\hbar \sum_k \varepsilon_{ijk} x_k/r$). So we end up with

$$[M_x, L_x] = 0 \quad . \tag{5}$$

The remaining commutation relations can be derived by cyclic permutation of x, y, and z.

EXERCISE

14.8 Proof of the Commutation Relation for $[\hat{M}_i, \hat{H}_j]_-$

Problem. Prove the commutation relations

$$[\hat{M}_i, \hat{M}_j]_- = \frac{2\mathrm{i}\hbar}{\mu}\hat{H}\sum_k \varepsilon_{ijk}\hat{L}_k \quad .$$

Solution. As in the preceding problem we proceed by proving it for its components.

$$\begin{aligned}
M_x M_y - M_y M_x &\\
= \left\{\frac{1}{\mu}(\boldsymbol{p}\times\boldsymbol{L} - \mathrm{i}\hbar\boldsymbol{p})_x - c\frac{x}{r}\right\}&\left\{\frac{1}{\mu}(\boldsymbol{p}\times\boldsymbol{L} - \mathrm{i}\hbar\boldsymbol{p})_y - c\frac{y}{r}\right\}\\
- \left\{\frac{1}{\mu}(\boldsymbol{p}\times\boldsymbol{L} - \mathrm{i}\hbar\boldsymbol{p})_y - c\frac{y}{r}\right\}&\left\{\frac{1}{\mu}(\boldsymbol{p}\times\boldsymbol{L} - \mathrm{i}\hbar\boldsymbol{p})_x - c\frac{x}{r}\right\} \qquad (1)
\end{aligned}$$

$$= \frac{1}{\mu^2}(p_y L_z - p_z L_y - \mathrm{i}\hbar p_x)(p_z L_x - p_x L_z - \mathrm{i}\hbar p_y) \qquad (2a)$$

$$- \frac{1}{\mu^2}(p_z L_x - p_x L_z - \mathrm{i}\hbar p_y)(p_y L_z - p_z L_y - \mathrm{i}\hbar p_x) \qquad (2b)$$

$$+ \frac{c}{\mu}\left[\frac{y}{r}(p_y L_z - p_z L_y - \mathrm{i}\hbar p_x) - (p_y L_z - p_z L_y - \mathrm{i}\hbar p_x)\frac{y}{r}\right] \qquad (2c)$$

$$+ \frac{c}{\mu}\left[(p_z L_x - p_x L_z - \mathrm{i}\hbar p_y)\frac{x}{r} - \frac{x}{r}(p_z L_x - p_x L_z - \mathrm{i}\hbar p_y)\right] \quad . \qquad (2d)$$

After performing the multiplications in (2a) and (2b) we end up with 18 terms. All terms created by multiplication with $\mathrm{i}\hbar p_x$ or $\mathrm{i}\hbar p_y$ cancel (maybe one has to commute terms). This leaves

$$\begin{aligned}
\frac{1}{\mu^2}(&p_y L_z p_z L_x - p_y L_z p_x L_z - p_z L_y p_z L_x + p_z L_y p_x L_z\\
&- p_z L_x p_y L_z + p_z L_x p_z L_y + p_x L_z p_y L_z - p_x L_z p_z L_y) \quad .
\end{aligned}$$

We have

$$\begin{aligned}
p_y L_z p_z L_x &= p_y p_z L_x L_z + \mathrm{i}\hbar p_y p_z L_y\\
&= p_z L_x p_y L_z + \mathrm{i}\hbar p_y p_z L_y - \mathrm{i}\hbar p_z p_z L_z \quad ,\\
p_z L_x p_z L_y &= p_z p_z L_x L_y + \mathrm{i}\hbar p_z p_y L_y\\
&= p_z p_z L_y L_x - \mathrm{i}\hbar p_z p_y L_y + \mathrm{i}\hbar p_z p_z L_z\\
&= p_z L_y p_z L_x - \mathrm{i}\hbar p_z p_y L_y + \mathrm{i}\hbar p_z p_y L_y - \mathrm{i}\hbar p_z p_x L_x \quad ,
\end{aligned}$$

Exercise 14.8

$$
\begin{aligned}
p_x L_z p_y L_z &= p_x p_y L_z L_z - \mathrm{i}\hbar p_x p_x L_z \\
&= p_y L_z p_x L_z - \mathrm{i}\hbar p_x p_x L_z - \mathrm{i}\hbar p_y p_y L_z \quad , \\
p_z L_y p_x L_z &= p_z p_x L_y L_z - \mathrm{i}\hbar p_z p_z L_z \\
&= p_x L_z p_z L_y - \mathrm{i}\hbar p_z p_z L_z + \mathrm{i}\hbar p_z p_x L_x \quad ,
\end{aligned}
$$

so we obtain

$$
\begin{aligned}
&\frac{\mathrm{i}\hbar}{\mu^2}(p_y p_z L_y - p_z^2 L_z - p_z p_y L_y + p_z^2 L_z \\
&\quad - p_z p_x L_x - p_x^2 L_z - p_y^2 L_z - p_z^2 L_z + p_z p_x L_x) \\
&= -\frac{\mathrm{i}\hbar}{\mu^2} \boldsymbol{p}^2 L_z
\end{aligned} \tag{3}
$$

for (2a) and (2b). Now consider the remaining terms of (2c) and (2d), which lead to

$$
\begin{aligned}
&p_z L_x \frac{x}{r} - p_x L_z \frac{x}{r} - \mathrm{i}\hbar p_y \frac{x}{r} - p_y L_z \frac{y}{r} + p_z L_y \frac{y}{r} + \mathrm{i}\hbar p_x \frac{y}{r} \\
&- \frac{x}{r} p_z L_x + \frac{x}{r} p_x L_z + \mathrm{i}\hbar \frac{x}{r} p_y + \frac{y}{r} p_y L_z - \frac{y}{r} p_z L_y - \mathrm{i}\hbar \frac{y}{r} p_x \quad .
\end{aligned} \tag{4}
$$

Here we find

$$
p_x \frac{y}{r} - p_y \frac{x}{r} + \frac{x}{r} p_y - \frac{y}{r} p_x \equiv 0 \quad ,
$$

since

$$
p_x \frac{y}{r} = \frac{y}{r} p_x + \mathrm{i}\hbar \frac{yx}{r^3} \quad \text{and} \quad -p_y \frac{x}{r} = -\frac{x}{r} p_y - \mathrm{i}\hbar \frac{xy}{r^3} \quad .
$$

The differences of the terms remaining in (4) may be simplified with the help of

$$
\begin{aligned}
p_z L_x \frac{x}{r} &= p_z \frac{x}{r} L_x = \frac{x}{r} p_z L_x + \mathrm{i}\hbar \frac{xz}{r^3} L_x \quad , \\
p_z L_y \frac{y}{r} &= p_z \frac{y}{r} L_y = \frac{y}{r} p_z L_y + \mathrm{i}\hbar \frac{yz}{r^3} L_y \quad , \\
-p_x L_z \frac{x}{r} &= -p_x \frac{x}{r} L_z - \mathrm{i}\hbar p_x \frac{y}{r} \\
&= -\frac{x}{r} p_x L_z + \mathrm{i}\hbar \left(\frac{1}{r} - \frac{x^2}{r^3} \right) L_z - \mathrm{i}\hbar p_x \frac{y}{r} \quad , \\
-p_y L_z \frac{y}{r} &= -p_y \frac{y}{r} L_z - \mathrm{i}\hbar p_y \frac{x}{r} \\
&= -\frac{y}{r} p_y L_z + \mathrm{i}\hbar \left(\frac{1}{r} - \frac{y^2}{r^3} \right) L_z + \mathrm{i}\hbar p_y \frac{x}{r} \quad .
\end{aligned}
$$

Exercise 14.8

This leaves the remaining terms as

$$p_z L_x \frac{x}{r} - p_x L_z \frac{x}{r} - \frac{x}{r} p_z L_x + \frac{x}{r} p_x L_z - p_y L_z \frac{y}{r} + p_z L_y \frac{y}{r} + \frac{y}{r} p_y L_z + \frac{y}{r} p_z L_y$$

$$= \mathrm{i}\hbar\left(\frac{xz}{r^3} L_x + \frac{yz}{r^3} L_y + \left(\frac{1}{r} - \frac{x^2}{r^3}\right) L_z + \left(\frac{1}{r} - \frac{y^2}{r^3}\right) L_z + \frac{x}{r} p_y - \frac{y}{r} p_x\right) \quad , \tag{5}$$

and we have

$$\begin{aligned} xL_x - yL_y &= -\mathrm{i}\hbar(x(y\partial_z z\partial_y) + y(z\partial_x - x\partial_z)) \\ &= -\mathrm{i}\hbar z(y\partial_x - x\partial_y) = -zL_z \quad . \end{aligned}$$

Substituting $xp_y - yp_x$ by L_z we can transform (5) into

$$\mathrm{i}\hbar\left\{-\frac{1}{r^3}(z^2 + x^2 + y^2)L_z + \frac{2}{r} L_z + \frac{1}{r} L_z\right\} = \mathrm{i}\hbar \frac{2}{r} L_z \tag{6}$$

Collecting the partial results (3) and (6), we end up with

$$\begin{aligned} [M_x, M_y]_- &= -2\mathrm{i}\frac{\hbar}{\mu}\left(\frac{\boldsymbol{p}^2}{2\mu}\right) L_z + \frac{c}{\mu}\frac{2\mathrm{i}\hbar}{r} L_z = -2\mathrm{i}\frac{\hbar}{\mu}\left(\frac{\boldsymbol{p}^2}{2m} - \frac{c}{r}\right) L_z \\ &= -2\mathrm{i}\frac{\hbar}{\mu} H L_z \quad . \end{aligned} \tag{7}$$

$[M_x, M_x]_- = 0$ is evident, and the remaining commutation relations are derived by cyclic permutation.

15. Mathematical Excursion: Non-compact Lie Groups

15.1 Definition and Examples of Non-compact Lie Groups

Compact and non-compact Lie groups differ from each other in their essential properties (which will be discussed later), so we should investigate the group quality "compactness" more fully.

A Lie group is *compact*, if its parameter range is compact. This non-topological definition is somewhat simplified, but is perfectly adequate for our purpose. In the case of parameter ranges that are subsets of $\mathbb{R}^n$, this means, owing to the theorem of Heine-Borel, that the parameter range has to be *closed* and *bounded*.

As an example we consider SO(3): every element of this Lie group can be written as $\exp(-\mathrm{i}\boldsymbol{\phi}\hat{\boldsymbol{L}})$, where $\boldsymbol{\phi} = (\phi_x, \phi_y, \phi_z)$ stands for three real parameters. As discussed in Sect. 1.9, the parameter range of SO(3) is the set of all rotation vectors $\boldsymbol{\phi}$ whose endpoints lie within a sphere of radius π, so that $|\boldsymbol{\phi}| \leq \pi$. A sphere is a compact set, and therefore SO(3) is compact.

The case of SU(2) is similar; its elements are given by

$$\exp\left(\frac{-\mathrm{i}}{2}\boldsymbol{\phi}\cdot\hat{\boldsymbol{\sigma}}\right) = \cos\left(\frac{\phi}{2}\right)\mathbb{1} - \mathrm{i}\sin\left(\frac{\phi}{2}\right)\boldsymbol{n}\cdot\hat{\boldsymbol{\sigma}} \tag{15.1}$$

(with the unit vector $\boldsymbol{n} = \boldsymbol{\phi}/|\boldsymbol{\phi}| = \boldsymbol{\phi}/\phi$). Because of the periodicity of $\sin x$ and $\cos x$, we need only a range $|\phi| \leq 2\pi$, and thus the rotation vectors $\boldsymbol{\phi}$ fill a sphere of radius 2π. Another proof of the compactness of SU(2) will be given in Exercise 15.1.

As the first example of a non-compact Lie group we take SU(1, 1). To understand the definition of this group better, we turn to the definition of the Lie groups SU(n) once more: these are unitary $n \times n$ matrices (n lines, n columns) with determinant 1. For $\hat{U} \in \mathrm{SU}(n)$ we thus have $\hat{U}^\dagger = \hat{U}^{-1}$.

Now let us consider the action of the transformation $\hat{U}$ on an n-dimensional vector $\boldsymbol{Z} = (z_1, \dots, z_n)^T$ with complex components z_i ($i = 1, \dots, n$). We get a vector $\boldsymbol{Z}' = \hat{U}\boldsymbol{Z}$ and

$$\|\boldsymbol{Z}'\|^2 = (\boldsymbol{Z}')^\dagger \boldsymbol{Z}' = \boldsymbol{Z}^\dagger \hat{U}^\dagger \hat{U} \boldsymbol{Z} = \boldsymbol{Z}^\dagger \boldsymbol{Z} = \|\boldsymbol{Z}\|^2$$

(because $\hat{U}^\dagger = \hat{U}^{-1}$). Thus every matrix of SU(n) transforms n-dimensional complex vectors in such a way that their *norm* is conserved, where this norm is defined as

$$\|\boldsymbol{Z}\| = \sqrt{|z_1|^2 + \cdots + |z_n|^2} \quad .$$

As we have

$$\|\boldsymbol{Z}\| = \sqrt{\boldsymbol{Z}^\dagger \hat{g} \boldsymbol{Z}} \quad , \tag{15.2}$$

where g is the so-called *metric tensor*, we get such a norm if the metric tensor is just the n-dimensional unit matrix, i.e.

$$\hat{g} = \mathbb{1}_{(n)} \quad .$$

As a generalization of SU(n) one can now define the Lie groups SU(p, q) (with $p + q = n$), which transform n-dimensional complex vectors in such a way that a norm is conserved, but this is now defined in a different way:

$$\|\boldsymbol{Z}\| = \sqrt{|z_1|^2 + \cdots + |z_p|^2 - |z_{p+1}|^2 - \cdots - |z_n|^2} \quad .$$

This norm corresponds to the metric tensor

$$\hat{g} = \begin{pmatrix} \mathbb{1}_{(p)} & 0 \\ 0 & \mathbb{1}_{(q)} \end{pmatrix} \quad . \tag{15.3}$$

For $\hat{U} \in \mathrm{SU}(1,1)$ we thus have

$$\begin{pmatrix} z'_1 \\ z'_2 \end{pmatrix} = \hat{U} \begin{pmatrix} z_1 \\ z_2 \end{pmatrix} \text{ with } |z'_1|^2 - |z'_2|^2 = |z_1|^2 - |z_2|^2 .$$

Furthermore, $\hat{U} \in \mathrm{SU}(p, q)$ has the determinant 1.

EXERCISE

15.1 Representation of SU(2) Matrices

Problem. Show that every $\hat{U} \in \mathrm{SU}(2)$ can be written as

$$\hat{U} = u_0 \hat{\sigma}_0 + \mathrm{i} \sum_{k=1}^{3} u_k \hat{\sigma}_k \quad (u_\nu \in \mathbb{R};\ \hat{\sigma}_0 = \mathbb{1}_{(2)})$$

with $u_0^2 + u_1^2 + u_2^2 + u_3^2 = 1$, and prove the compactness of SU(2).

Solution. We use (15.1):

$$\hat{U} = \mathbb{1}_{(2)} \cos(\phi/2) - \mathrm{i}\boldsymbol{n} \cdot \hat{\boldsymbol{\sigma}} \sin(\phi/2) = \hat{\sigma}_0 \cos(\phi/2) - \mathrm{i} \sum_{k+1}^{3} n_k \hat{\sigma}_k \sin(\phi/2) \quad .$$

Thus

$$\hat{U} = u_0 \hat{\sigma}_0 + \mathrm{i} \sum_{k=1}^{3} u_k \hat{\sigma}_k \quad , \quad \text{with} \quad u_0 = \cos(\phi/2), \quad u_k = -n_k \sin(\phi/2) \quad ,$$

$$\Rightarrow \sum_{k=1}^{3} u_k^2 = \sin^2(\phi/2) \sum_k n_k^2 = \sin^2(\phi/2) \quad ,$$

because $|\boldsymbol{n}| = 1$,

$$\Rightarrow u_0^2 + \sum_{k=1}^{3} u_k^2 = \cos^2(\phi/2) + \sin^2(\phi/2) = 1 \quad .$$

This is the surface (topologically, the boundary) of a 4-dimensional unit sphere and is therefore a compact set, so the SU(2) is compact.

EXERCISE

15.2 Representation of SU(1, 1) Matrices

Problem. Show that every $\hat{U} \in \mathrm{SU}(1,1)$ is a complex 2×2 matrix with

a) $\hat{U}^\dagger \hat{\sigma}_3 \hat{U} = \hat{\sigma}_3$ and $\det \hat{U} = 1$.

b) $\hat{U} = v_0\hat{\sigma}_0 + v_1\hat{\sigma}_1 + v_2\hat{\sigma}_2 + \mathrm{i}v_3\hat{\sigma}_3$, $(v_\nu \in \mathbb{R})$,
with $v_0^2 - v_1^2 - v_2^2 + v_3^2 = 1$.

Why is SU(1, 1) not compact?

Solution. a) As the norm is to be conserved, we have

$$\|Z'\| = \sqrt{Z'^\dagger \hat{g} Z'} = \sqrt{Z^\dagger \hat{U}^\dagger \hat{g} \hat{U} Z} = \sqrt{Z^\dagger \hat{g} Z} \quad ,$$

whence follows:

$$\hat{U}^\dagger \hat{g} \hat{U} = \hat{g} \quad .$$

With

$$\hat{g} = \begin{pmatrix} 1 & 0 \\ 0 & -1 \end{pmatrix} = \hat{\sigma}_3 \quad ,$$

it follows that

$$\hat{U}^\dagger \hat{\sigma}_3 \hat{U} = \hat{\sigma}_3 \quad .$$

$\det \hat{U} = 1$ is necessary, because $\hat{U} \in \mathrm{SU}(1,1)$.

b) Let $\hat{U} = \begin{pmatrix} a & b \\ c & d \end{pmatrix}$ with $a,b,c,d \in C$. This leads to

$$\hat{U}^\dagger \hat{g} \hat{U} = \begin{pmatrix} a^* & c^* \\ b^* & d^* \end{pmatrix} \begin{pmatrix} 1 & 0 \\ 0 & -1 \end{pmatrix} \begin{pmatrix} a & b \\ c & d \end{pmatrix} = \begin{pmatrix} aa^* - cc^* & a^*b - c^*d \\ ab^* - cd^* & bb^* - dd^* \end{pmatrix}$$

$$\stackrel{!}{=} \hat{g} = \begin{pmatrix} 1 & 0 \\ 0 & -1 \end{pmatrix} \quad .$$

Exercise 15.2

It follows that

$$a^*b - c^*d = 0 \quad \text{(A)}$$
$$|a|^2 - |c|^2 = 1 \quad \text{(B)}$$
$$|d|^2 - |b|^2 = 1 \quad \text{(C)}$$
$$1 = \det \hat{U} = ad - bc \quad \text{(D)} \quad .$$

(A) $\times\, a$ yields

$$0 = |a|^2 b - c^* ad \overset{\text{(D)}}{=} |a|^2 b - c^*(cb + 1)$$
$$= b(|a|^2 - |c|^2) - c^* \overset{\text{(B)}}{=} b - c^* \quad .$$

Thus

$$b = c^* \quad \text{(E)} \quad .$$

With (E) we get from (A)

$$b(a^* - d) = 0 \quad , \quad \text{thus} \quad a^* = d \quad .$$

Hence,

$$\hat{g} = \begin{pmatrix} a & b \\ b^* & a^* \end{pmatrix} \quad .$$

With $a = x + \mathrm{i}y$, $b = z + \mathrm{i}t$,

$$\hat{g} = \begin{pmatrix} x + \mathrm{i}y & z + \mathrm{i}t \\ z - \mathrm{i}t & x - \mathrm{i}y \end{pmatrix}$$
$$= x\begin{pmatrix} 1 & 0 \\ 0 & 1 \end{pmatrix} + \mathrm{i}y\begin{pmatrix} 1 & 0 \\ 0 & -1 \end{pmatrix} + z\begin{pmatrix} 0 & 1 \\ 1 & 0 \end{pmatrix} - t\begin{pmatrix} 0 & -\mathrm{i} \\ \mathrm{i} & 0 \end{pmatrix}$$
$$= x\hat{\sigma}_0 + z\hat{\sigma}_1 - t\hat{\sigma}_2 + \mathrm{i}y\hat{\sigma}_3 \quad ,$$

i.e. with $v_0 = x$, $v_1 = z$, $v_2 = -t$, $v_3 = y$.

$$v_0^2 - v_1^2 - v_2^2 + v_3^2 = x^2 + y^2 - (z^2 + t^2) = |a|^2 - |b|^2 \overset{\text{(E)}}{=} |a|^2 - |c|^2 \overset{\text{(B)}}{=} 1 \quad .$$

The range with $v_0^2 - v_1^2 - v_2^2 + v_3^2 = 1$ is a non-bounded set and therefore is not compact, so SU(1, 1) is not compact.

EXERCISE

15.3 Non-compactness of the Lorentz Group

Problem. Show that the proper Lorentz group L (Exercise 3.4) is not compact.

Solution. The 6 real parameters of L are the rotation angle ω and the rapidity $\boldsymbol{\xi} = (\boldsymbol{\beta}/|\boldsymbol{\beta}|)\tanh^{-1}|\boldsymbol{\beta}|$ with $\boldsymbol{\beta} = \boldsymbol{v}/c$.

Exercise 15.3

Now, the magnitude of the relative velocity $\boldsymbol{v}$ between two coordinate systems is always smaller than c, which leads to $0 \le |\boldsymbol{\beta}| < 1$. This yields $0 \le \tanh^{-1} |\boldsymbol{\beta}| < \infty$ and thus also $0 \le |\boldsymbol{\xi}| < \infty$. So the parameter range of L is not bounded and therefore is not compact.

Remark. Here one can see that it is very important to use the correct parameters to check compactness. In the single components of the rapidity $\boldsymbol{\xi}$, the Lorentz group is additive, by which the non-boundedness is immediately obvious. In the components of the velocity $\boldsymbol{\beta}$, however, we have a complicated law of addition, which makes the proof of the boundedness much more complicated. One can show that $|\boldsymbol{\beta}| = 1$ is not an element of the Lorentz group!

Now we come to another class of non-compact Lie groups, the groups $\mathrm{SO}(p,q)$ (with $p+q=n$), which are a generalization of the special n-dimensional rotation groups $\mathrm{SO}(n)$.

Whereas $\hat{U} \in \mathrm{SO}(n)$ transforms a vector $\boldsymbol{x} \in \mathbb{R}^n$ in such a way that $\|\boldsymbol{x}\| = \sqrt{\boldsymbol{x}^T\boldsymbol{x}}$ is conserved (i.e. $\hat{g} = \mathbb{1}_{(n)}$), $\mathrm{SO}(p,q)$ transformations leave invariant the norm

$$\|\boldsymbol{x}\| = \sqrt{|x_1|^2 + \cdots + |x_p|^2 - |x_{p+1}|^2 - \cdots - |x_n|^2} = \sqrt{\boldsymbol{x}^T \hat{g} \boldsymbol{x}}$$

with

$$\hat{g} = \begin{pmatrix} \mathbb{1}_{(p)} & 0 \\ 0 & \mathbb{1}_{(q)} \end{pmatrix} \quad .$$

The determinants of the $\mathrm{SO}(p,q)$ transformations matrices are $+1$.

As is well known, proper Lorentz transformations leave the quadratic form $x^2 + y^2 + z^2 - (ct)^2$ invariant and can thus be described by $\mathrm{SO}(3,1)$. The non-compactness of the Lorentz group (cf. Exercise 15.3) is therefore a special case of the non-compactness of all Lie groups $\mathrm{SO}(p,q)$ (with $p,q \neq 0$).

EXERCISE

15.4 Generators of SO(p, q)

Problem. How many real parameters has a Lie group $\hat{U} \in \mathrm{SO}(p,q)$ with $p+q=n$? Find a set of generators and show that the resulting infinitesimal operators (cf. Sect. 3.13) can be represented as

$$\hat{L}_{ij} = \mathrm{i}\left(x_i \frac{\partial}{\partial x_j} - x_j \frac{\partial}{\partial x_i}\right) \quad , (i<j) \quad \text{for } i,j = 1,\dots,p \text{ and } i,j = p+1,\dots,n$$

$$\hat{L}_{ij} = \mathrm{i}\left(x_i \frac{\partial}{\partial x_j} + x_j \frac{\partial}{\partial x_i}\right) \quad , \quad \text{for } i = 1,\dots,p \quad \text{and} \quad j = p+1,\dots,n \quad .$$

Exercise 15.4

Solution. For infinitesimal transformations we have

$$\hat{U} = \mathbb{1}_{(n)} + S\hat{U} \quad ,$$

where $S\hat{U}$ may contain only infinitesimal real parameters. With $\boldsymbol{x}' = \hat{U}\boldsymbol{x}$ ($\boldsymbol{x} \in \mathbb{R}^n$) it follows from norm-conservation that

$$(\boldsymbol{x}')^T \hat{g} \boldsymbol{x}' = \boldsymbol{x}^T \hat{U}^T \hat{g} \hat{U} \boldsymbol{x} = \boldsymbol{x}^T \hat{g} \boldsymbol{x} \quad ,$$

and thus

$$\hat{U}^T \hat{g} \hat{U} \boldsymbol{x} = \hat{g} = \begin{pmatrix} \mathbb{1}_{(p)} & 0 \\ 0 & -\mathbb{1}_{(q)} \end{pmatrix} \quad ,$$

$$\hat{U}^T \hat{g} \hat{U} \boldsymbol{x} = \hat{g} + \delta\hat{U}^T \hat{g} + \hat{g}\delta\hat{U} + \mathcal{O}(\delta\hat{U}^2) \stackrel{!}{=} \hat{g} \quad .$$

Because $\hat{g}^T = \hat{g}$, this leads to

$$\hat{g} S\hat{U} = -(\hat{g} S\hat{U})^T$$

The following ansatz is very helpful:

$$\delta\hat{U} = \begin{pmatrix} \hat{A} & \hat{B} \\ \hat{C} & \hat{D} \end{pmatrix}$$

(for example, $\hat{B}$ is also $p \times q$ matrix),

$$\Rightarrow \hat{g}\delta\hat{U} = \begin{pmatrix} \hat{A} & \hat{B} \\ -\hat{C} & -\hat{D} \end{pmatrix} \stackrel{!}{=} -(\hat{g}\delta\hat{U})^T = \begin{pmatrix} -\hat{A}^T & \hat{C}^T \\ -\hat{B}^T & \hat{D}^T \end{pmatrix} \quad .$$

It follows that

$$\hat{A}^T = -\hat{A} \quad , \quad \hat{D}^T = -\hat{D} \quad , \quad \hat{B} = -\hat{C}^T \quad .$$

Thus $S\hat{U}$ has the following form:

$$S\hat{U} = \begin{pmatrix} \hat{A} & \hat{B} \\ \hat{B}^T & \hat{C} \end{pmatrix} \quad ,$$

where $\hat{A}$ and $\hat{C}$ are, respectively, real antisymmetric $p \times p$ and $q \times q$ matrices, and $\hat{B}$ is an arbitrary real $p \times q$ matrix. So the number of parameters is

for $\hat{A}$: $p(p-1)/2$

for $\hat{C}$: $q(q-1)/2$

for $\hat{B}$: pq .

Together,

$$\begin{aligned} \frac{1}{2}[p^2 + 2pq + q^2 - p - q] &= \frac{1}{2}[(p+q)^2 - (p+q)] \\ &= \frac{1}{2} n(n-1) \quad (n = p+q) \quad . \end{aligned}$$

Thus SO(p,q) with $p+q=n$ has as many parameters as SO(n) (cf. Exercise 3.17). Now we determine the generators. Let

Exercise 15.4

$$\hat{A} = \begin{pmatrix} 0 & -a_{12} & \dots & & -a_{1p} \\ a_{12} & & & & \vdots \\ \vdots & & \ddots & & \vdots \\ \vdots & & & & -a_{p-1,p} \\ a_{1p} & & \dots & a_{p-1,p} & 0 \end{pmatrix}$$

be an antisymmetric matrix. Then the corresponding generator is

$$\begin{aligned}\hat{S}^A_{rt} &= \mathrm{i}\frac{\partial}{\partial a_{rt}}\hat{A} \\ &= \mathrm{i}\begin{pmatrix} 0 & \dots & -1 \\ \vdots & \ddots & \vdots \\ 1 & \dots & 0 \end{pmatrix} \quad \text{for} \quad r,t=1,\dots,p;\ r<t \quad .\end{aligned}$$

Analytically,

$$(\hat{S}^A_{rt})_{ij} = \mathrm{i}(-\delta_{ri}\delta_{tj} + \delta_{rj}\delta_{ti}) \quad (r,t=1,\dots,p;\ r<t;\ \text{and}\ i,j=1,\dots,n) \quad .$$

In a completely analogous way one gets for the generator of the matrix $\hat{C}$

$$(\hat{S}^C_{rt})_{ij} = \mathrm{i}(-\delta_{ri}\delta_{tj} + \delta_{rj}\delta_{ti}) \quad (r,t=p+1,\dots,n;\ r<t;\ \text{and}\ i,j=1,\dots,n) \quad .$$

The remaining part of $S\hat{U}$ can be parametrized as follows:

$$\begin{pmatrix} 0 & \hat{B} \\ \hat{B}^T & 0 \end{pmatrix} = \begin{pmatrix} & & & b_{1,p+1} & \dots & b_{1,p+q=n} \\ & & & \vdots & & \vdots \\ & & & b_{p,p+1} & \dots & b_{p,p+q} \\ b_{1,p+1} & \dots & b_{p,p+1} & & & \\ \vdots & & \vdots & & & \\ b_{1,p+q=n} & \dots & b_{p,p+q} & & & \end{pmatrix}$$

Correspondingly, we have

$$(\hat{S}^B_{rt})_{ij} = \mathrm{i}(\delta_{ri}\delta_{tj} + \delta_{rj}\delta_{ti}) \quad (r=1,\dots,p;\ t=p+1,\dots,n;\ \text{and}\ i,j=1,\dots,n).$$

Now the infinitesimal coordinate transformation can be written as

$$\begin{aligned}x'_k &= \hat{U}_{kj}x_j = (\mathbb{1}_{(n)} + \delta\hat{U})_{kj}x_j \\ &= x_k - \mathrm{i}\sum_{j=1}^{n} x_j\left(\sum_{\substack{r,t=1\\(r<t)}}^{p}(a_{rt}\hat{S}^A_{rt})_{kj} + \sum_{\substack{r,t=p+1\\(r<t)}}^{n}(c_{rt}\hat{S}^C_{rt})_{kj}\right. \\ &\qquad \left. + \sum_{r=1}^{p}\sum_{t=p+1}^{n}(b_{rt}\hat{S}^B_{rt})_{kj}\right) \quad .\end{aligned}$$

Exercise 15.4

Hence, we get the following infinitesimal operators $\hat{L}_{rt}$ (3.71). For $r, t = 1, \dots, p$ and $r < t$,

$$\hat{L}_{rt} = \mathrm{i} \sum_{k+1}^{n} \frac{\partial}{\partial a_{rt}} x'_k \bigg|_{\delta\hat{U}=0} \frac{\partial}{\partial x_k} = \mathrm{i} \sum_{k+1}^{n} (-\mathrm{i}) \sum_{j=1}^{n} (\hat{S}^A_{rt})_{kj} \, x_k \frac{\partial}{\partial x_k}$$

$$= \sum_{k,j=1}^{n} \mathrm{i}(-\delta_{ri}\delta_{tj} + \delta_{rj}\delta_{ti}) x_j \frac{\partial}{\partial x_k} = \mathrm{i}\left(x_r \frac{\partial}{\partial x_t} - x_t \frac{\partial}{\partial x_r} \right) .$$

In an analogous way for $r, t = (p+1), \dots, (p+q=n)$ and $r < t$

$$\hat{L}_{rt} = \mathrm{i} \sum_{k+1}^{n} \frac{\partial}{\partial c_{rt}} x'_k \bigg|_{\delta\hat{U}=0} \frac{\partial}{\partial x_k} = \mathrm{i}\left(x_r \frac{\partial}{\partial x_t} - x_t \frac{\partial}{\partial x_r} \right) .$$

For $r = 1, \dots, p$ and $t = p+1, \dots, n$ the minus sign is just replaced by a plus sign:

$$\hat{L}_{rt} = \mathrm{i} \sum_{k+1}^{n} \frac{\partial}{\partial b_{rt}} x'_k \bigg|_{\delta\hat{U}=0} \frac{\partial}{\partial x_k} = \mathrm{i}\left(x_r \frac{\partial}{\partial x_t} + x_t \frac{\partial}{\partial x_r} \right) .$$

The commutators of these infinitesimal operators can be calculated easily knowing that $[\partial/\partial x_i, x_j] = \delta_{ij}$.

15.2 The Lie Group SO(2, 1)

In order to have a simple and clear example in mind, we concentrate in the following on SO(2, 1), i.e. $p = 2$ and $q = 1$. Here we have three generators ($\hat{L}_{12}, \hat{L}_{13}$ and $\hat{L}_{23}$) and their commutation relations

$$[\hat{L}_{12}, \hat{L}_{13}] = -\mathrm{i}\hat{L}_{23} \quad , \quad [\hat{L}_{12}, \hat{L}_{23}] = \mathrm{i}\hat{L}_{13} \quad , \quad [\hat{L}_{13}, \hat{L}_{23}] = \mathrm{i}\hat{L}_{12} \quad . \tag{15.4}$$

However, usually we do not use these three generators, but the following:

$$\hat{J}_x = -\hat{L}_{23} = -\mathrm{i}\left(y \frac{\partial}{\partial z} + z \frac{\partial}{\partial y} \right)$$

$$\hat{J}_y = \hat{L}_{13} = \mathrm{i}\left(x \frac{\partial}{\partial z} + z \frac{\partial}{\partial x} \right)$$

$$\hat{J}_z = -\hat{L}_{12} = -\mathrm{i}\left(x \frac{\partial}{\partial y} + y \frac{\partial}{\partial x} \right) . \tag{15.5}$$

(this choice is allowed, because the Lie algebras $\{\hat{L}_{23}, \hat{L}_{13}, \hat{L}_{12}\}$ and $\{\hat{L}_{23}, \hat{L}_{13}, \hat{L}_{12}\}$ are isomorphic; see Example 3.20).

Now, the commutation relations adopt a very suggestive form:

$$[\hat{J}_x, \hat{J}_y] = -\mathrm{i}\hat{J}_z \quad , \quad [\hat{J}_y, \hat{J}_z] = \mathrm{i}\hat{J}_x \quad , \quad [\hat{J}_z, \hat{J}_x] = \mathrm{i}\hat{J}_y \quad . \tag{15.6}$$

But for the minus sign in the first commutator, these are just the well-known commutation relations of SO(3) ($[\hat{\hat{J}}_i, \hat{\hat{J}}_j] = \mathrm{i}\sum_k \varepsilon_{ijk}\hat{\hat{J}}_k$). Formally one gets the commutation relations of SO(2, 1) from those of SO(3) by making the replacement

$$\hat{\hat{J}}_x \to \hat{J}_x = \mathrm{i}\hat{\hat{J}}_x \quad , \quad \hat{\hat{J}}_y \to \hat{J}_y = \mathrm{i}\hat{\hat{J}}_y \quad , \quad \hat{\hat{J}}_z \to \hat{J}_z = \mathrm{i}\hat{\hat{J}}_z \quad .$$

(However, this projection of the *real* Lie algebra of SO(3) on the *real* Lie algebra of SO(2, 1) is no isomorphism, because imaginary coefficients are used.)

This relation between SO(2, 1) and SO(3) makes it possible to recognize immediately that SO(2, 1) is not compact, because the group elements of the SO(3) transform as follows:

$$\exp[-\mathrm{i}(\alpha_x\hat{\hat{J}}_x + \alpha_y\hat{\hat{J}}_y + \alpha_z\hat{\hat{J}}_z)] \to \exp(\alpha_x\hat{\hat{J}}_x + \alpha_y\hat{\hat{J}}_y - \mathrm{i}\alpha_z\hat{\hat{J}}_z) \quad .$$

The first two terms in the exponent of the SO(2, 1) group element are responsible for the fact that the absolute value of the group element is no longer bounded. Hence, it follows immediately that SO(2, 1) is not compact.

EXERCISE

15.5 Casimir Operator of SO(2, 1)

Problem. a) Determine the Casimir operator of SO(2, 1).
b) Express this Casimir operator in terms of

$$\hat{J}_\pm = \hat{J}_x \pm \mathrm{i}\hat{J}_y \text{ and } \hat{J}_z \quad .$$

Solution. a) SO(2, 1) has rank 1 and therefore only one Casimir operator. This Casimir operator is given by (cf. Sect. 3.9)

$$\hat{C} = \sum_{ij} g^{ij}\hat{J}_i\hat{J}_j \quad ,$$

where g^{ij} is the inverse Cartan metric tensor: $g^{ij} = (g_{ij})^{-1}$. In its turn, g_{ij} depends on the structure constants C_{ijk} (recall that $[\hat{J}_i, \hat{J}_j] = \mathrm{i}\sum_k C_{ijk}\hat{J}_k$) as follows (cf. Example 3.9):

$$g_{ij} = \sum_{kl} C_{ikl}C_{jlk} \quad .$$

For SO(2, 1), because $C_{ijk} = -C_{jik}$, one has only three linearly independent non-vanishing structure constants:

$$C_{123} = -\mathrm{i} \quad , \quad C_{231} = \mathrm{i} \quad , \quad C_{132} = -\mathrm{i} \quad .$$

Exercise 15.5

All structure constants with two or three equal indices vanish, and so g_{ij} can easily be determined. For example, we have

$$g_{11} = \sum_{kl} C_{1kl} C_{1lk} = C_{123} C_{132} + C_{132} C_{123} = 2(-\mathrm{i})(-\mathrm{i}) = -2 \quad .$$

$$g_{ij} = \begin{pmatrix} -2 & \dots & 0 \\ \vdots & -2 & \vdots \\ 0 & \dots & 2 \end{pmatrix} \Rightarrow g_{ij} = \begin{pmatrix} -1/2 & \dots & 0 \\ \vdots & -1/2 & \vdots \\ 0 & \dots & 1/2 \end{pmatrix} \quad .$$

Hence, $\hat{C} = \frac{1}{2}(-\hat{J}_x^2 - \hat{J}_y^2 + \hat{J}_z^2)$, or, if one suppresses the factor 1/2 for convenience.

$$\hat{C} = -\hat{J}_x^2 - \hat{J}_y^2 + \hat{J}_z^2 \quad .$$

b) We get

$$\hat{J}_+\hat{J}_- = \hat{J}_x^2 + \hat{J}_y^2 - \mathrm{i}[\hat{J}_x, \hat{J}_y] = \hat{J}_x^2 + \hat{J}_y^2 - \hat{J}_z$$

and thus

$$\hat{C} = -(\hat{J}_+\hat{J}_- + \hat{J}_z) + \hat{J}_z^2 = -\hat{J}_+\hat{J}_- + \hat{J}_z^2 - \hat{J}_z \quad .$$

Alternatively,

$$\hat{J}_-\hat{J}_+ = \hat{J}_x^2 + \hat{J}_y^2 + \mathrm{i}[\hat{J}_x, \hat{J}_y] = \hat{J}_x^2 + \hat{J}_y^2 + \hat{J}_z$$
$$\Rightarrow \hat{C} = -\hat{J}_-\hat{J}_+ + \hat{J}_z^2 + \hat{J}_z \quad .$$

From these two representations of the Casimir operator in terms of $\hat{J}_+, \hat{J}_-$ and $\hat{J}_z$ one can deduce important consequences for *unitary representations* (i.e. $\hat{J}_i^\dagger = \hat{J}_i$ for $i = x, y, z$) of SO(2, 1). As in the case of the unitary representations of SO(3) one can choose the corresponding states to be eigenstates of $\hat{C}$ and $\hat{J}_z$:

$$\hat{C}|Xa\rangle = X|Xa\rangle \quad , \quad \hat{J}_z|Xa\rangle = a|Xa\rangle \tag{15.7}$$

with real numbers X and a, because $\hat{C}$ and $\hat{J}_z$ are Hermitian for unitary representations. In the case of SO(3) we had the relations $X = j(j+1)$ and $a = m$, and by this we got $(j \le 0)$

$$j(j+1) - m(m \pm 1) \ge 0, \quad \text{thus}$$
$$X - a(a \pm 1) \ge 0 \quad \text{with} \quad X \ge 0 \quad . \tag{15.8}$$

This yielded $-j \le m \le j$, and so the existence of a finite number $(= 2j + 1)$ of different states $|X_0, a = m\rangle$ for a fixed $X_0 = j(j+1)$ was demonstrated. Here one speaks of *finite-dimensional irreducible unitary representations.*

The relation for unitary representations $|Xa\rangle$ of SO(2, 1), which is analogous to (15.8), can be found as follows.

We know from Exercise 15.5 that

$$\hat{C} = -\hat{J}_+\hat{J}_- + \hat{J}_z^2 - \hat{J}_z \Rightarrow \hat{J}_+\hat{J}_- = -\hat{C} + \hat{J}_z^2 - \hat{J}_z \quad .$$

Now, $\hat{J}_- = \hat{J}_+^\dagger$, which leads to $\hat{J}_+\hat{J}_- = \hat{J}_+(\hat{J}_+)^\dagger$. But this means that $\hat{J}_+\hat{J}_-$ is a positive definite operator, because applying it to an arbitrary state $|\psi\rangle$ yields

$$\langle\psi|\hat{J}_+\hat{J}_+^\dagger|\psi\rangle = \langle\varphi|\varphi\rangle \geq 0$$

with $|\varphi\rangle = \hat{J}_+^\dagger|\psi\rangle$. From this it follows for eigenstates $|\chi\rangle$ with $(\hat{J}_+\hat{J}_+^\dagger)|\chi\rangle = b|\chi\rangle$

$$\langle\chi|\hat{J}_+\hat{J}_+^\dagger|\chi\rangle = b\langle\chi|\chi\rangle \geq 0 \Rightarrow b > 0$$

(b is real, as $\hat{J}_+\hat{J}_+^\dagger$ is Hermitian). In particular, when applied to $|Xa\rangle$ we get:

$$\hat{J}_+\hat{J}_-|Xa\rangle = (-\hat{C} + \hat{J}_z^2 - \hat{J}_z)|Xa\rangle = (-X + a(a+1))|Xa\rangle \quad ,$$

i.e. $|Xa\rangle =$ is an eigenstate of $\hat{J}_+\hat{J}_-$ with the eigenvalue $-X + a(a+1) \geq 0$. In an analogous way it follows from Exercise 15.5 that

$$\hat{J}_-\hat{J}_+ = \hat{J}_-\hat{J}_-^\dagger = -\hat{C} + \hat{J}_z^2 + \hat{J}_z$$

and thus

$$-X + a(a+1) \geq 0 \quad .$$

Altogether we have

$$-X + a(a+1) \geq 0 \quad . \tag{15.9}$$

As $\hat{C}$ is Hermitian, X is real. However, contrary to the case of SO(3), one *cannot* say anything about the sign of X, because here $\hat{C}$ is not a sum of positive definite Hermitian operators.

The relation (15.9) has an important consequence for the dimension of irreducible unitary representations of SO(2, 1), because for both positive and negative X the value $|a|$ has no upper bound. Hence, it follows that for every X there is an arbitrary number of a's and so we have an *infinite-dimensional irreducible unitary representation.* The fact that the compact SO(3) possesses finite-dimensional, but the non-compact SO(2, 1) infinite-dimensional, irreducible unitary representations is not pure chance. On the contrary, one can show that the following theorem is valid:

i) Every irreducible unitary representation of a *compact* Lie group is *finite* dimensional.
ii) A coherent, simple, *non-compact* Lie-group permits *no* finite-dimensional unitary representation except the trivial one (where all generators are represented by the unit matrix).

We notice that SO(2, 1) is coherent (i.e. possesses a coherent, connected parameter range) and also simple (see Sect. 3.3).

But let us return to the irreducible unitary representations of SO(2, 1). The relation (15.9) furthermore states that the j in $X = j(j+1)$ need no longer be an integer or half-integer number. Again, the reason for this is that there is not always a breaking condition for a which gives a relation between a and X or j. Therefore one distinguishes between *discrete and continuous* series for X. In the latter, X can adopt arbitrary real values within a certain range. For details, please refer to the literature[1]; we mention here only a special continuous series, namely that with $X < -1/4$. With $X = j(j+1)$, this leads to

$$j^2 + j + \tfrac{1}{4} = (j + \tfrac{1}{2})^2 < 0 \quad . \tag{15.10}$$

Therefore one chooses here

$$j = -1/2 + \mathrm{i}k \quad , \quad k \in \mathbb{R} \quad . \tag{15.11}$$

Then we have

$$X = j(j+1) = -k^2 - 1/4 \quad .$$

Because of $-X > 1/4$ and (15.9), for every fixed k, a has to fulfil the condition $a(a+1) \geq -1/4$. But as we have

$$a(a \pm 1) + \tfrac{1}{4} = (a \pm \tfrac{1}{2})^2 \geq 0$$

(for $a \in \mathbb{R}$), arbitrary real a are permitted. However, as shown in the literature, adjacent values of a differ by 1; this can be derived from the properties of $\hat{J}_+$ *and* $\hat{J}_-$.

15.3 Application to Scattering Problems

Continuous series such as $j = -\frac{1}{2} + \mathrm{i}k$ are predestined for the group-theoretical treatment of problems in quantum mechanics, where *continuous (energy) eigenvalues* appear: for example *scattering problems*. The procedure will be outlined in the final exercise.

EXERCISE

15.6 Coordinate Representation of SO(2, 1) Operators

Problem. Determine the explicit coordinate representation of the SO(2, 1) operators $\hat{C}$ and $\hat{J}_z$ in *polar hyperbolic coordinates*:

$$x = r\cosh\varrho\cos\phi \quad y = r\cosh\varrho\sin\phi \quad z = r\sinh\varrho$$

$$r \geq 0; -\infty < \varrho < \infty; 0 \leq \phi < 2\pi \quad .$$

(Why do $\hat{J}_\pm$ and $\hat{J}_z$ not depend on r?) Then perform a similarity transformation $\hat{O} \to \hat{U}\hat{O}\hat{U}^{-1}$ with $\hat{U} = \cosh^{1/2}\varrho$ and find the coordinate representation of the

---

[1] L.C. Biedenharn in *Noncompact Groups in Particle Physics*, ed. by Y. Chow (Benjamin, New York 1966) p. 23.

eigenvalue equation

Exercise 15.6

$$\hat{C}|jm\rangle = j(j+1)|jm\rangle \quad (\text{with } \hat{J}_z|jm\rangle = m|jm\rangle) \quad .$$

Solution. The following differentials are to be calculated:

$$\frac{\partial}{\partial x} = \frac{\partial r}{\partial x}\frac{\partial}{\partial r} + \frac{\partial \varrho}{\partial x}\frac{\partial}{\partial \varrho} + \frac{\partial \phi}{\partial x}\frac{\partial}{\partial \phi} \quad ,$$

$$\frac{\partial}{\partial y} \text{ and } \frac{\partial}{\partial y} \quad .$$

Using the inverse transformations

$$r = \sqrt{x^2+y^2-z^2} \quad , \quad \varrho = \sinh^{-1}(z/\sqrt{x^2+y^2-z^2}) \quad ,$$

$$\phi = \tan^{-1}(y/x) \quad ,$$

one gets for these differentials

$$\frac{\partial}{\partial x} = \cosh\varrho\cos\phi\frac{\partial}{\partial r} - \frac{\sinh\varrho\cos\phi}{r}\frac{\partial}{\partial \varrho} - \frac{\sin\phi}{r\cosh\varrho}\frac{\partial}{\partial \phi}$$

$$\frac{\partial}{\partial y} = \cosh\varrho\sin\phi\frac{\partial}{\partial r} - \frac{\sinh\varrho\sin\phi}{r}\frac{\partial}{\partial \varrho} + \frac{\cos\phi}{r\cosh\varrho}\frac{\partial}{\partial \phi}$$

$$\frac{\partial}{\partial z} = -\sinh\varrho\cos\phi\frac{\partial}{\partial r} + \frac{\cosh\varrho}{r}\frac{\partial}{\partial \varrho}$$

and, hence,

$$\hat{J}_z = -\mathrm{i}\left(x\frac{\partial}{\partial y} - y\frac{\partial}{\partial x}\right) = -\mathrm{i}\frac{\partial}{\partial \phi}$$

$$\begin{aligned}\hat{J}_\pm &= \hat{J}_x \pm \mathrm{i}\hat{J}_y \\ &= -\mathrm{i}\left(y\frac{\partial}{\partial z} + z\frac{\partial}{\partial y}\right) \mp \left(x\frac{\partial}{\partial z} + z\frac{\partial}{\partial x}\right) \\ &= \mathrm{e}^{\pm\mathrm{i}\phi}\left(\mp\frac{\partial}{\partial \varrho} - \mathrm{i}\tanh\varrho\frac{\partial}{\partial \phi}\right) \quad .\end{aligned}$$

Thus these SO(2, 1) generators are r-independent. This is clear insofar as in the given coordinates we have for $r = $ const. (i.e. no r-dependence)

$$x^2+y^2-z^2 = r^2(\cosh^2\varrho - \sinh^2\varrho) = r^2 = \text{const.}$$

Thus the two coordinates ϱ and ϕ alone completely parametrize a domain of $\mathbb{R}^3$ (a hyperboloid), which is *invariant* under SO(2, 1) transformations. Therefore the generators $\hat{J}_\pm$ and $\hat{J}_z$, with whose help these transformations can be executed, have to be r-independent, too. (The analogue to this is spherical coordinates (r,θ,ϕ) with $r = $ const. in the case of SO(3), whose transformations leave the surface of a sphere invariant.)

Exercise 15.6

The similarity transformation with $\hat{U} = \cosh^{1/2} \varrho$ only concerns the operator $(\partial/\partial\varrho)$, because $(\partial\varrho/\partial\phi) = 0$. The transformation is

$$\frac{\partial}{\partial\varrho} \to \left(\frac{\partial}{\partial\varrho}\right)' = \cosh^{1/2} \varrho \frac{\partial}{\partial\varrho} \cosh^{-1/2} \varrho$$

$$= -\tfrac{1}{2} \tanh \varrho + \frac{\partial}{\partial\varrho} \quad .$$

(This similarity transformation is convenient, because the volume element for the hyperboloid $x^2 + y^2 - z^2 = \text{const.}$ (calculated with the norm $\|\boldsymbol{r}\|^2 = x^2 + y^2 - z^2$) is proportional to $\cosh \varrho$. So the essential part of the volume element is included in the wave functions $\psi' = \cosh^{1/2} \varrho\psi$ so that one can use a volume element proportional to $d\varrho d\phi$ for integrations.) Thus we get

$$\hat{J}_z = \hat{J}'_z = -\mathrm{i}\frac{\partial}{\partial\phi} \quad \hat{J}'_\pm = \mathrm{e}^{\pm \mathrm{i}\phi}\left[\mp\frac{\partial}{\partial\varrho} + \tanh \varrho\left(\pm\tfrac{1}{2} - \mathrm{i}\frac{\partial}{\partial\phi}\right)\right]$$

and so

$$\hat{C}' = -\hat{J}_+\hat{J}_- + \hat{J}'^2_z - \hat{J}'_z = \frac{\partial^2}{\partial\varrho^2} - \frac{\partial^2/\partial\phi^2 + 1/4}{\cosh^2 \varrho} - \tfrac{1}{4} \quad .$$

To solve the eigenvalue problem

$$\hat{C}'|jm\rangle = j(j+1)|jm\rangle \quad , \quad \hat{J}'_z|jm\rangle = m|jm\rangle$$

we make the following ansatz:

$$\langle\varrho\phi|jm\rangle = U_{jm}(\varrho)\mathrm{e}^{\mathrm{i}m\phi} \quad .$$

This immediately solves the second eigenvalue equation. Substituting for $\hat{C}$ in the first equation yields

$$\left(-\frac{\partial^2}{\partial\varrho^2} - \frac{m^2 - 1/4}{\cosh^2\varrho}\right)u_{jm}(\varrho) = -[j(j+1) + \tfrac{1}{4}]\,u_{jm} = -(j+\tfrac{1}{2})^2\,u_{jm} \quad .$$

But this is just a one-dimensional Schrödinger equation (with $\hbar = m = 1$) for the so-called Pöschl-Teller potential

$$V(\varrho) = -V_0/\cosh^2 \varrho$$

with strength $V_0 = m^2 - 1/4$! The eigenenergy of this potential-problem is $E_j = -(j+1/2)^2$. Using the continuous series (15.11), $j = 1/2 + \mathrm{i}k (k \in \mathbb{R})$, one gets

$$E_j = E_{k(j)} = k^2 > 0 \quad ,$$

and can thus treat the *scattering problem* ($E > 0$) for the Pöschl-Teller potential

Exercise 15.6

in a group-theoretical way. The infinite-dimensional irreducible unitary representations of SO(2, 1) (i.e. k fixed, m variable) describe a scattering problem with fixed energy k^2 and variable potential $m^2 - 1/4$.

Generalizations for scattering problems with up to three space dimensions and other potentials are given in the literature[2].

[2] Y. Alhassid, F. Gürsey, F. Iachello: Ann. Phys. (N.Y.) **167**, 181 (1986); J. Wu, F. Iachello, Y. Alhassid: Ann. Phys. (N.Y.) **173**, 68 (1987).

Subject Index

Names given here in *italic* will also be found in the biographical notes